Quantum Signatures in 0D Systems

Series on the Foundations of Natural Science and Technology

ISSN: 2010-1961

Published:

Vol. 20 *Quantum Signatures in 0D Systems*
by K. P. Ghatak and J. Pal

Vol. 19 *A New Approach to Quantum Theory: The Central Role of Probability*
by S. B. Sontz

Vol. 18 *Density-Of-States Function and Related Applications in Quantized Structures*
by K. P. Ghatak and A. Biswas

Vol. 17 *Quantum Capacitance in Quantized Transistors*
by K. P. Ghatak and J. Pal

Vol. 16 *Electron Statistics in Quantum Confined Superlattices*
by K. P. Ghatak and A. Biswas

Vol. 15 *Topics in Nanoscience*
In 2 Parts
Part I: Basic Views, Complex Nanosystems: Typical Results and Future
Part II: Quantized Structures, Nanoelectronics, Thin Films
edited by W. Schommers

Vol. 14 *Elastic Constants in Heavily Doped Low Dimensional Materials*
by K. P. Ghatak and M. Mitra

Vol. 13 *Scalar Field Cosmology*
by S. Chervon, I. Fomin, V. Yurov and A. Yurov

*For further details, please visit: http://www.worldscientific.com/series/sfnst

(Continued at end of book)

Series on the Foundations of Natural Science and Technology — Vol. 20

Quantum Signatures in 0D Systems

Kamakhya Prasad Ghatak

University of Engineering and Management, India
and
Institute of Engineering and Management, India

NanoTechnology Center of Excellence

Jayita Pal

Meghnad Saha Institute of Technology, India

World Scientific

NEW JERSEY • LONDON • SINGAPORE • BEIJING • SHANGHAI • HONG KONG • TAIPEI • CHENNAI • TOKYO

Published by

World Scientific Publishing Co. Pte. Ltd.
5 Toh Tuck Link, Singapore 596224
USA office: 27 Warren Street, Suite 401-402, Hackensack, NJ 07601
UK office: 57 Shelton Street, Covent Garden, London WC2H 9HE

Library of Congress Cataloging-in-Publication Data
Names: Ghatak, Kamakhya Prasad author | Pal, Jayita author
Title: Quantum signatures in 0D systems / authors, Kamakhya Prasad Ghatak,
Institute of Engineering and Management, India,
Jayita Pal, Meghnad Saha Institute of Technology, India.
Description: Singapore ; Hackensack, NJ : World Scientific, [2026] |
Series: Series on the foundations of natural science and technology, 2010-1961 ; volume 20 |
Includes bibliographical references and index.
Identifiers: LCCN 2025041969 | ISBN 9789819823772 hardcover |
ISBN 9789819823789 ebook for institutions | ISBN 9789819823796 ebook for individuals
Subjects: LCSH: Quantum dots | Superlattices as materials | Semiconductors
Classification: LCC QC611.6.Q35 G43 2026
LC record available at https://lccn.loc.gov/2025041969

British Library Cataloguing-in-Publication Data
A catalogue record for this book is available from the British Library.

For any available supplementary material, please visit
https://www.worldscientific.com/worldscibooks/10.1142/14603#t=suppl

Desk Editors: Soundararajan Raghuraman/Muhammad Ihsan

Typeset by Stallion Press
Email: enquiries@stallionpress.com

The Gurudev is everything

Gratitude, they say, is a measure of wiseness. Although I, far from being wise, wish to record my speechless gratitude to Baba and Dadabhai for allowing me to touch their Lotus feet

— K. P. Ghatak

The author dedicates this book to her loving parents and sweet family

— J. Pal

After 50 years of continuous effort we have finally realized that we have learned enough to understand how little we know and to see dimly how much more there is yet to discover in the wonderland entitled Quantum Signatures in 0D Systems

Preface

Reading of good books is like conversation with the finest persons of the world.

It is well known that the symmetry of the wave-vector space of the electrons in electronic materials having various band structures is being reduced from a constant energy 3D closed volume in wave-vector space, to a quantized 2D closed surface, quantized non-parabolas and fully quantized wave-vector space, leading to the formation of 0D systems such as quantum dots (QDs), magneto-quantum MOSFETs and QWFETs, magneto-size quantized systems, magneto-quantum well superlattices (SLs), magneto-NIPIs, QDSLs and other field aided 0D systems which are being experimentally realized with the advent of modern experimental methods, namely, fine line lithography (FLL), metallo-organic chemical vapor deposition (MOCVD), molecular beam epitaxy (MBE), etc. for fabricating such structures.

As the concept of quantization increases from 1D to 3D, the degree of freedom of the free carriers decreases drastically and the total density-of-states (DOS) function changes from Heaviside step function to the Dirac's delta function forming QDs which, in turn, depend on the carrier energy spectra in different materials. An enormous range of important applications of such low-dimensional structures for modern low-dimensional materials science in the quantum regime together with a rapid increase in computing power have generated considerable interest in the study of the optical properties of quantum effect devices based on various new materials of reduced dimensionality. Examples of such new applications include quantum switches, quantum registers, quantum sensors, quantum logic gates, quantum well and quantum wire transistors, quantum cascade lasers and other nano-electronic devices [1–14].

With the advent of Quantum Hall Effect (QHE) for which Professor Klaus von Klitzing was awarded the Nobel prize in physics in 1985 [15,16], there has been considerable interest in studying the magneto-thermoelectric power (MTP) in various types of nanostructured materials having quantum confinement of their charge carriers in one, two and three dimensions of the respective wave-vector space, leading to different carrier energy spectra [17–49]. It is well known that the MTP is a very important quantity [22], since the change in entropy (a vital concept in thermodynamics) can be known from this relation by determining the experimental values of the change of electron concentration. The analysis of MTP generates information regarding the effective mass of the carriers in materials which occupies a central position in the whole field of materials science in general [23]. The classical MTP (G_0) equation is valid only under the non-degenerate carrier concentration and the magnitude of the MTP is given by $G_0 = (\pi^2 k_B/3e)$ (k_B and e are Boltzmann's constant and the magnitude of the carrier charge, respectively [24]). Thus, it is readily inferred that this conventional form is a function of three fundamental constants only, being independent of the signature of the charge carriers in materials.

Zawadzki [22] demonstrated that the MTP for electronic materials having degenerate electron concentration is essentially determined by their respective energy band structures. It has, therefore, different values in different materials and changes with the doping, magnitude of the reciprocal quantizing magnetic field under magnetic quantization, quantizing electric field as in inversion layers, nanothickness as in quantum wells, wires and dots, with superlattice period as in quantum confined semiconductor superlattices with graded interfaces having various carrier energy spectra and also in other types of field-assisted nanostructured materials.

In Chapter 1, we have investigated the MTP in quantum dots of III–V, II–VI, n-gallium phosphide, n-germanium, tellurium, graphite, platinum antimonide, zero-gap, lead germanium telluride, gallium antimonide, stressed materials, bismuth, IV–VI, II–V, zinc and cadmium diphosphides, bismuth telluride and antimony on the basis of respective carrier energy spectra, respectively. The technical applications of the said technologically important materials have been written in detail in Chapter 1 where enough elementary analyses have been introduced. This process is being followed for all the remaining chapters of this book. The materials for the advanced readers have been shown by $*$ sign in the various sections of the content from Chapter 2 up to the rest. Besides, in this book, we have given 250 open research problems which is by far the most important part for the readers to initiate their own contributions in this pin-pointed deep topic of research of 0D systems. These problems are useful for both PhD aspirants and researchers, and it is written for post-graduate students, engineers and professionals in the fields of nanoscience, nanotechnology, quantum materials, solid state electronics and semiconductor physics, respectively. We feel that this approach will enhance the wider acceptability of this book by the readers.

We suggest in Chapter 1, the methods of experimental determinations of the second- and third-order carrier contribution to the elastic constants, Debye Screening Length, the Einstein's Relation and we further explore 24 different applications: Debye screening length, carrier contribution to the elastic constants, Einstein relation, diffusion coefficient of the minority carriers, nonlinear optical response, third-order nonlinear optical susceptibility, generalized Raman gain, plasma frequency, activity coefficient, magneto-thermal effect in quantized structures, normalized hall coefficient, reflection coefficient, heat capacity, magnetic susceptibilities, Faraday rotation, Fowler–Nordheim field emission, optical effective mass, righi–leduc coefficient, electric susceptibility, electric susceptibility mass, electron diffusion thermo-power, hydrostatic piezo-resistance coefficient, relaxation time for acoustic mode scattering and gate capacitance, respectively.

Although many new effects in quantized structures have already been reported, the interest for further research of other aspects of such quantum-confined materials is becoming increasingly important. One such significant property is Einstein's photoemission (EP) which is a physical phenomenon and occupies a singular position in the whole arena of 0D systems and related disciplines in general and whose importance has already been established since the inception of Einstein's photoelectric effect (for which Einstein won Nobel Prize in 1921), which in recent years finds extensive applications in modern optoelectronics, characterization and investigation of condensed matter systems, photoemission spectroscopy and related aspects in connection with the investigations of the optical properties of nanostructures [50–54]. It is worth remarking that Albert Einstein's description of the cause of photoelectric effect by absorption of quanta of light (photons) appears in his classic research paper entitled Von. A. Einstein, "On a Heuristic Viewpoint Concerning the Production and Transformation of Light" (Über einen die Erzeugung und Verwandlung des Lichtes betreffenden heuristischen Gesichtspunkt), Annalen der Physik Volume 322 Issue 6, Pages 132–148, (1905). Incidentally, 1905 has been recognized as the world year of physics and the said contribution is the first of the four revolutionary papers published by the then 26-years-old Einstein in the same year. His explanation of the features of the photoelectric effect and the characteristic frequency won him the Nobel Prize in 1921. Interest in low-dimensional silicon nanostructures also grew up and gained momentum after the discovery of room temperature photoluminescence and electroluminescence of silicon nanowires in porous silicon [50]. Work on ultrathin layers of $SiSiO_2$ superlattices resulting into visible light emission at room temperature clearly exhibited low-dimensional quantum confinement effect [51] and one of the most popular techniques for analyzing the low-dimensional structures is to employ photoemission techniques. Recent observation of room temperature photoluminescence and electroluminescence in porous silicon has stimulated vigorous research activities in silicon nanostructures [52]. In Chapter 2, we study the EP from quantum dots of all the materials of Chapter 1.

Incidentally, even after 30 years of continuous effort, we see that the complete investigation of the quantum signatures in 0D systems comprising the whole set of the materials and allied sciences is really a sea and permanently enjoys the domain of impossibility theorems. The quantum signatures have different forms for different materials which change under one-, two- and three-dimensional quantum confinements of the charge carriers. In this context, it may be noted that the available reports on the said areas cannot afford even a book containing the detailed investigations regarding the quantum signatures in 0D systems. It is important to note that the effects of quantizing magnetic field (B) on the band structures of compound materials are most striking than that of the parabolic one and are easily observed in experiments. A number of interesting physical features originate from the significant changes in the basic energy wave-vector relation of the carriers caused by the magnetic field. The valuable information could also be obtained from experiments under magnetic quantization regarding the important physical properties such as Fermi energy and effective masses of the carriers, which affect almost all the transport properties of the electron devices [55] of various materials having different carrier dispersion relations [56, 57].

It is worth remarking that the effects of crossed electric and quantizing magnetic fields on the transport properties of materials having various band structures have relatively been less investigated compared to the corresponding magnetic quantization, although the study of the cross-fields are of fundamental importance with respect to the addition of new physics and the related experimental findings in modern quantum effect devices. It is well known that in the presence of electric field (E_0) along x-axis, B along z-axis, the dispersion laws of the carriers in materials become modified and for which the carrier moves in both the z- and y-directions respectively. The motion along y-direction is purely due to the presence of E_0 along x-axis and in the absence of electric field, the effective electron mass along y-axis tends to infinity, indicating the fact that the electron motion along y-axis is forbidden. The effective electron mass of the isotropic, bulk materials having parabolic energy bands exhibits mass anisotropy in the presence of cross-fields and this anisotropy depends on the electron energy, the magnetic quantum number, the electric and the magnetic fields, respectively, although the effective electron mass along z-axis is a constant quantity. Incidentally, Zawadzki and Lax [58] derived the expression for the dispersion relation of the conduction electrons for III–V materials in accordance with the two-band model of Kane under cross-field configuration which generates the interest to study this particular topic in the present context in general [59].

With the advent of modern experimental techniques of fabricating nanomaterials as already noted, it is also possible to grow semiconductor superlattices (SLs) composed of alternative layers of two different degenerate layers with controlled thickness [60]. These structures have found wide applications in many new devices, such as photodiodes [61], photoresistors [61], transistors [62], light emitters [63], and tunneling devices [64–68]. The investigations of the physical properties of narrow gap SLs have increased extensively; since they are important for optoelectronic devices and because of the quality of hetero-structures, involving narrow gap materials have been improved. It may be written in this context that the doping superlattices are crystals with a periodic sequence of ultrathin film layers [69–79] of the same semiconductor with the intrinsic layer in between together with the opposite sign of doping. All the donors will be positively charged and all the acceptors negative. This periodic space charge causes a periodic space charge potential which quantizes the motions of the carriers in the z-direction together with the formation of the sub-band energies. In this book, the activity coefficient (AC) and carrier contribution to the elastic constants (CEC) in doping superlattices under magnetic quantization made of nonlinear optical and tetragonal, III–V, II–VI, IV–VI, and stressed Kane-type materials have been investigated in Chapters 3 and 4, respectively.

It is well known that the electrons in bulk materials, in general, have three-dimensional freedom of motion. When these electrons are confined in a one-dimensional potential well whose width is of the order of the carrier wavelength, the motion in that particular direction gets quantized while that along the other two directions remains free. Thus, the energy spectrum appears in the shape of discrete levels for the one-dimensional quantization, each of which has a continuum for the two-dimensional free motion. The transport phenomena of such one-dimensional confined carriers have been recently studied [71] with great interest. For the metal–oxide–semiconductor (MOS) structures, the work functions of the metal and the semiconductor substrate are different and the application of an external voltage at the metal gate causes the change in the charge density at the oxide semiconductor interface, leading to a bending of the energy bands of the semiconductor near the surface. As a result, a one-dimensional potential well is formed at the semiconductor interface. The spatial variation of the potential profile is so sharp that for considerable large values of the electric field, the width of the potential well becomes of the order of the de Broglie wavelength of the carriers. The Fermi energy, which is near the edge of the conduction band in the bulk, becomes nearer to the edge of the valance band at the surface creating inversion layers. The energy levels of the carriers bound within the potential well get quantized and form electric sub bands. Each of the sub-bands corresponds to a quantized level in a plane perpendicular to the surface leading to a quasi-two-dimensional electron gas. Thus, the extreme band bending at low temperatures allows us to observe the quantum effects at the surface [71]. Although considerable work has already been done regarding the various physical properties of different types of inversion layers having various band structures, it appears from the literature that there lies scope in the investigations made while the interest for studying different other features of accumulation layers is becoming increasingly important. In Chapter 5, we have studied the quantum capacitance (QC) under magnetic quantization in quantized transistors (QTs) of nonlinear optical and tetragonal compounds, II–VI, IV–VI, stressed Kane-type materials, bismuth telluride, germanium, gallium aresenide, gallium aresenide, indium antimonide and tellurium materials, respectively. Besides, in the same chapter, we study the QC in magneto-quantum well FETs of nonlinear optical and tetragonal compounds, III–V, ternary and quaternary, II–VI, IV–VI, stressed Kane-type materials, bismuth telluride, germanium, gallium aresenide, gallium antimonide, indium antimonide, tellurium, platinum antimonide and gallium phosphide, respectively. In Chapter 6, we have studied the $I_D - V_D$ equation under magnetic quantization in quantum MOSFETs and QTs, respectively, for all the materials, as discussed in Chapter 5.

It is well known that Keldysh [80] first suggested the fundamental concept of an SL, although it was successfully experimentally realized by Esaki and Tsu [81]. The importance of SLs in the field of nano-electronics has already been described in Refs. [82–86]. The most extensively studied III–V SL is the one consisting of alternate layers of GaAs and $Ga_{1-x}Al_xAs$ owing to the relative ease of fabrication. The GaAs layer forms quantum wells and $Ga_{1-x}Al_xAs$ form potential barriers. The III–V SLs are attractive for the realization of high speed electronic and optoelectronic devices [87]. In addition to SLs with the usual structure, SLs with more complex structures such as II–VI [88], IV–VI [89] and HgTe/CdTe [90] SLs have also been proposed. The IV–VI SLs exhibit quite different properties compared to the III–V SL due to the peculiar band structure of the constituent materials [91]. The epitaxial growth of II–VI SL is a relatively recent development and the primary motivation for studying the mentioned SLs made of materials with large band gap is in their potential for optoelectronic operation in the blue [71]. HgTe/CdTe SLs have raised a great deal of attention since 1979 as promising new materials for long wavelength infrared detectors and other electro-optical applications [92]. Interest in Hg-based SLs has been further increased as new properties with potential device applications were revealed [93]. These features arise from the unique zero band gap material HgTe [93] and the direct band gap semiconductor CdTe which can be described by the three-band mode of Kane [93]. The combination of the aforementioned materials with specified dispersion relation makes HgTe/CdTe SL very attractive, especially because of the possibility to tailor the material properties for various applications by varying the energy band constants of the SLs. In addition to it, for effective mass SLs, the electronic sub-bands appear continually in real space [94].

We note that all the aforementioned SLs have been proposed with the assumption that the interfaces between the layers are sharply defined, of zero thickness, i.e., devoid of any interface effects. The SL potential distribution may then be considered as a one-dimensional array of rectangular potential wells. The aforementioned advanced experimental techniques may produce SLs with physical interfaces between the two materials crystallographically abrupt; adjoining their interface will change at least on an atomic scale. As the potential form changes from a well (barrier) to a barrier (well), an intermediate potential region exists for the electrons. The influence of finite thickness of the interfaces on the electron dispersion law is very important, since the electron energy spectrum governs the electron transport in SLs. In Chapter 7, we have studied the MTP in quantum dot superlattices. In Chapters 8–10, we have further investigated EP, HC and CEC in QWSL under magnetic quantization, respectively.

It is well known that heavy doping and carrier degeneracy are the keys to unlock the important properties of materials and they are especially instrumental in dictating the characteristics of Ohomic contacts and Schottky contacts, respectively [50]. It is an amazing fact that although the heavily doped materials (HDS) have been investigated in the literature, the study of the corresponding dispersion relations (DRs) of HDS is still one of the open research problems. We have obtained the exact E–k dispersion relations in HD nonlinear optical, III–V, II–VI, gallium phosphide, germanium, platinum antimonide, stressed, IV–VI, lead germanium telluride, tellurium, II–V, zinc and cadmium di-phosphides, bismuth telluride, III–V, II–VI, IV–VI and HgTe/CdTe quantum well HD superlattices with graded interfaces under magnetic quantization, III–V, II–VI, IV–VI and HgTe/CdTe HD effective mass quantum well superlattices under magnetic quantization, quantum-confined effective mass superlattices and superlattices of HD optoelectronic materials with graded interfaces, respectively. Our method is not at all related with the DOS technique as used in the aforementioned works. From the electron energy spectrum, one can obtain the DOS, but the DOS technique as used in the literature cannot generate the dispersion laws. Therefore, our study is more fundamental than those in the existing literature because the Boltzmann transport equation, which controls the study of the charge transport properties of the semiconductor devices, can be solved if and only if the carrier energy spectra is known. From Chapters 11–20, we have studied the CEC in heavily doped QDs, the EP from HDQD of various non-parabolic materials, the AC and the HC in HD doping superlattices under magnetic quantization, the QC in HD quantized transistors (HDQTs) under magnetic quantization, the $I_D - V_D$ equation for ballistic HDQTFETs under magnetic quantization, the MTP in HDQDSLs, the EP in HDQWSLs under magnetic quantization, the Righi–Leduc coefficient (RLC) in HD non-parabolic materials under magneto-size quantization and Landau's Dia (χ_d) and Pauli's para (χ_p) magnetic susceptibilities for quantized HD 0D structures, respectively.

With the advent of nanophotonics, there has been considerable interest in studying the optical processes in 0D systems in the presence of intense light waves [95]. It appears from the literature that the investigations in the presence of external intense photoexcitation have been carried out on the assumption that the carrier energy spectra are invariant quantities under strong external light waves, which is not fundamentally true. The physical properties of materials in the presence of strong light waves which alter the basic dispersion relations have relatively been much less investigated in Refs. [96, 97] compared to the cases of other external fields, and in optoelectronics, the influence of strong light waves is needed for the characterization of the low-dimensional opto-electronic devices. Chapter 21 investigates the magneto-CEC in 0D systems. With the advent of nanodevices, the built-in electric field becomes so large that the electron energy spectrum changes fundamentally [90–92] instead of being invariant, and Chapter 22 studies the relaxation time under acoustic mode of scattering (RAS) in the presence of intense electric field. In Chapter 23, we propose a complex distribution function for HD materials, leading to three new concepts. In Chapter 24, we have studied χ_d and χ_p in cylindrical QDs of III–V materials in the presence of crossed electric and magnetic fields. The book ends with Chapter 25, which contains the conclusion and scope for future research in this exciting topic of quantum science and technology in general.

It is needless to say that this monograph is based on the *iceberg principle* [98] and the rest of which will be explored by the researchers of different appropriate fields. Since, there is no existing report devoted solely to the study of the quantum signatures in 0D systems to the best of our knowledge, we earnestly hope that the present book will provide a useful reference source for the present and next generations of the readers, the researchers of materials and allied sciences in general. We have discussed enough regarding the quantum signatures in different 0D systems although lot of new computer-oriented numerical analysis are being left for the purpose of being computed by the readers to generate the new graphs and the inferences from them, which altogether is a sea in itself. The production of error-free first edition of any book from every point of view is a permanent member of impossibility theorems. In spite of our joint concentrated efforts for couple of years together with the seasoned team of World-Scientific, the same stands very true for this monograph also. Various expressions and few chapters of this book are appearing for the first time in printed form. The suggestions from the readers for the development of the book will be highly appreciated for the purpose of inclusion in the future edition if any. In this book, as noted already, from Chapter 1 till the end, we have presented circa 250 open research problems for the graduate students, PhD aspirants, researchers and engineers in this pin-pointed research topic. We strongly hope that alert readers of this monograph will not only solve the said problems by removing all the mathematical approximations and establishing the appropriate uniqueness conditions but also will generate new research problems, both theoretical and experimental, thereby transforming this monograph into a solid book. Incidentally, our readers after reading this book will easily understand how little is presented and how much more is yet to be investigated in this exciting topic, which is the signature of coexistence of new physics, advanced mathematics combined with the inner fire for performing creative researches in this context from the young scientists. Like Kikoin [99], we feel that "A young scientist is no good if his teacher learns nothing new from him and gives his teacher nothing to be proud of". We emphatically write that the problems presented here form the integral part of this book and will be useful for the readers to initiate their own contributions on the quantum signatures of different 0D systems since like Sakurai [100], we firmly believe "The reader who has read the book but cannot do the exercise has learned nothing. It is nice to note that if we assign the alphabets A–Z, the positive integers from 1 to 26, alphabetically, then the word ATTITUDE" receives the perfect score of 100 and is the vital quality needed from the readers since attitude is the ladder on which all the other virtues mount.

In this monograph, we have investigated the expressions of the electron statistics and the sub-band energy have been formulated throughout this monograph as further study for the purpose of in-depth investigations of the said important pinpointed research topic. Thus, in this first-of-its-kind book, the readers will get much information regarding the quantum signatures on the different electronic properties of the different 0D systems having various band structures. Although the name of the book is an example of extremely high Q-factor, from the content, one can easily infer that it should be useful in graduate and post-graduate courses on materials science, condensed matter physics, solid states electronics, nanoscience and technology and solid-state sciences and devices in many universities and the institutions in addition to both PhD students and researchers in the aforementioned fields. It is needless to mention, had this book been available 50 years ago when the senior author started his research career to see the sea of semiconductor science in the famous Institute of Radio Physics and Electronics of the University of Calcutta, India, then he would have been a far better scientist than he is today. Finally, the authors jointly hope that their humble effort will kindle the desire to delve deeper into this fascinating and deep topic by anyone engaged in materials research and device development either in academics or in industries.

References

[1] S. Bhattacharya and K. P. Ghatak, *Fowler–Nordheim Field Emission: Effects in Semiconductor Nanostructures*, Springer Series in Solid State Sciences, Vol. 170, (Springer, **2012**); K. P. Ghatak and A. Biswas, *Density-of-States Function and Related Applications in Quantized Structures*, Series on the Foundations of Natural Science and Technology, (World Scientific, **2024**); K. P. Ghatak and J. Pal, *Quantum Capacitance in Quantized Transistors*, Series on the Foundations of Natural Science and Technology, Vol. 17, (World Scientific, **2024**); K. P. Ghatak and A. Biswas, *Electron Statistics in Quantum Confined Superlattices*, Series on the Foundations of Natural Science and Technology, Vol. 16, (World Scientific, **2023**); K. P. Ghatak and M. Mitra, *Electronic Properties*, Series in Nanomaterials, Vol. 1, (De Guyter, Germany, **2020**); K. P. Ghatak and M. Mitra, *Quantization and Entropy*, Series in Nanomaterials, Vol. 2, (De Guyter, Germany, **2020**).

[2] T. N. Sen, K. P. Ghatak, *J. Nanosci. Nanotech.*, 16, 1229 **(2016)**.

[3] S. Debbarma, K. P. Ghatak, *J. Nanosci. Nanotech.*, 16, 1095 **(2016)**; S. Chakrabarti, B. Chatterjee, S. Debbarma, K. P. Ghatak, *J. Nanosci. Nanotech.*, 15, 6460 **(2015)**.

[4] M. Mitra, B. Chatterjee, K. P. Ghatak, *J. Comput. Theor. Nanosci.*, 12, 1898 **(2015)**; K. P. Ghatak, L. S. Singh, K. Sarkar, N. Debbarma, M. Debbarma, *Mater. Focus*, 4, 85 **(2015)**.

[5] K. P. Ghatak and M. Mitra, *Elastic Constants in Heavily Doped Low Dimensional Materials*, Series on the Foundations of Natural Science and Technology, Vol. 14, (World Scientific, **2021**); K. P. Ghatak (Ed.), *Quantum Wires: An Overview*, Series on the NanoTechnology Science and Technology, (NOVA Science Publishers, USA, **2020**); K. P. Ghatak, M. Mitra and A. Biswas, *Heisenberg's Uncertainty Principle and the Electron Statistics in Quantized Structures*, (Springer, Germany, **2022**).

[6] K. P. Ghatak, *Quantum Effects, Heavy Doping, and the Effective Mass*, Series on the Foundations of Natural Science and Technology, Vol. 8, (World Scientific, **2017**); K. P. Ghatak, *Magneto Thermoelectric Power in Heavily Doped Quantized Structures*, Series on the Foundations of Natural Science and Technology, Vol. 7, (World Scientific, **2016**); K. P. Ghatak, *Dispersion Relations in Heavily-Doped Nanostructures*, Springer Tracts in Modern Physics, Vol. 265, (Springer, **2016**); K. P. Ghatak,

Einstein's Photo-emission: Emission from Heavily-Doped Quantized Structures, Springer Tracts in Modern Physics, Vol. 262, (Springer, **2015**).

[7] K. P. Ghatak and S. Bhattacharya, *Heavily Doped 2D Quantized Structures and the Einstein Relation*, Springer Tracts in Modern Physics, Vol. 260, (Springer, **2015**); K. P. Ghatak and S. Bhattacharya, *Debye Screening Length: Effects of Nanostructured Materials*, Springer Tracts in Modern Physics, Vol. 255, (Springer, **2014**); S. Bhattacharya and K. P. Ghatak, *Effective Electron Mass in Low-Dimensional Semiconductors*, Springer Series in Materials Sciences, Vol. 167, (Springer, **2013**); D. De, S. Bhattacharya and K. P. Ghatak (Eds.), *Quantum Dots and Quantum Cellular Automata*, Series in Physics Research and Technology, (NOVA Science Publishers, USA, **2013**).

[8] K. P. Ghatak and S. Bhattacharaya (Eds.), *Bismuth: Characteristics, Production and Applications*, Nova Series in Materials Science and Technologies, (NOVA Science Publishers, USA, **2012**); K. P. Ghatak and S. Bhattacharya, *Thermoelectric Power in Nanostructured Materials Strong Magnetic Fields*, Springer Series in Materials Science, Vol. 137, (Springer, **2010**); K. P. Ghatak, D. De and S. Bhattacharya, *Photoemission from Optoelectronic Materials and their Nanostructures*, Springer Series in Nanostructure Science and Technology, (Springer, **2009**); K. P. Ghatak, S. Bhattacharya and D. De, *Einstein Relation in Compound Semiconductors and their Heterostructures*, Springer Series in Materials Science, Vol. 116, (Springer, **2009**).

[9] N. Paitya, K. P. Ghatak, *J. Nanosci. Nanotech.*, 12, 8985 **(2012)**; N. Paitya, K. P. Ghatak, *J. Nanoeng. Nanomanuf.*, 2, 347 **(2012)**; P. K. Bose, N. Paitya, S. Bhattacharya, D. De, S. Saha, K. M. Chatterjee, S. Pahari, K. P. Ghatak, *Quant. Matt.*, 1, 89 **(2012)**; S. Bhattacharya, D. De, N. Paitya, S. M. Adhikari, K. P. Ghatak, *Adv. Sci. Lett.*, 16, 348 **(2012)**.

[10] S. Bhattacharya, D. De, S. M. Adhikari, K. P. Ghatak, *Superlatt. Microstruct.*, 51, 203 **(2012)**; S. Bhattacharya, D. De, S. M. Adhikari, S. Saha, K. M. Chatterjee, S. Choudhury, K. P. Ghatak, *Superlatt. Microstruct.*, 50, 609 **(2011)**; D. De, S. Bhattacharya, S. M. Adhikari, A. Kumar, P. K. Bose, K. P. Ghatak, *Beilstein J. Nanotech.*, 2, 339 **(2011)**; S. Bhattacharya, S. Choudhury, K. P. Ghatak, *Superlatt. Microstruct.*, 48, 257 **(2010)**; S. Pahari, S. Bhattacharya, D. De, S. M. Adhikari, A. Niyogi, A. Dey, N. Paitya, S. C. Saha, K. P. Ghatak, P. K. Bose, *Phys. B: Conden. Matter*, 405, 4064 **(2010)**; K. P. Ghatak, S. Bhattacharya, S. S. Roy, L. J. Singh, *Nonlinear Opt.*, 32, 307 **(2010)**; K. P. Ghatak, S. Bhattacharya, S. K. Biswas, A. Dey, A. K. Dasgupta, *Phys. Scr.*, 75, 820 **(2007)**; S. Mukherjee, S. N. Mitra, P. K. Bose, A. R. Ghatak, A. Neogi, J. P. Banerjee, A. Sinha, M. Pal, S. Bhattacharya, K. P. Ghatak, *J. Comput. Theor. Nanosci.*, 4, 550 **(2007)**; S. Mukherjee, D. De, D. J. Mukherjee, S. Bhattacharya, A. Sinha, K. P. Ghatak, *Phys. B: Conden. Matter*, 393, 347 **(2007)**.

[11] S. Bhattarchya, D. De, S. Ghosh, P. Banerjee, S. Saha, M. Mitra, B. Nag, M. Pal, S. K. Biswas, K. P. Ghatak, *J. Comput. Theor. Nanosci.*, 7, 1066 **(2010)**; D. De, A. Kumar, S. M. Adhikari, S. Pahari, N. Islam, P. Banerjee, S. K. Biswas, S. Bhattacharya, K. P. Ghatak, *Superlatt. Microstruct.*, 47, 377 **(2010)**; A. Kumar, S. Chowdhury, S. M. Adhikari, S. Ghosh, M. Mitra, D. De, A. Sharma, S. Bhattacharya, A. Dey, K. P. Ghatak, *J. Comput. Theor. Nanosci.*, 7, 115 **(2010)**; A. Kumar, S. Choudhury, S. Saha, S. Pahari, D. De, S. Bhattacharya, K. P. Ghatak, *Phys. B: Conden. Matter*, 405, 472 **(2010)**.

[12] S. Pahari, S. Bhattacharya, S. Roy, A. Saha, D. De, K. P. Ghatak, *Superlatt. Microstruct.*, 46, 760 **(2009)**; S. Pahari, S. Bhattacharya, K. P. Ghatak, *J. Comput. Theor. Nanosci.*, 6, 2088 **(2009)**; K. P. Ghatak, S. Bhattacharya, S. Pahari, D. De, R. Benedictus, *Superlatt. Microstruct.*, 46, 387 **(2009)**; K. P. Ghatak, S. Bhattacharya, S. Pahari, S. N. Mitra, P. K. Bose, D. De, *J. Phys. Chem. Solids*, 70, 122 **(2009)**.

[13] D. De, K. P. Ghatak, *J. Mater. Sci. Mater. Electron.*, 20, 185 **(2009)**; S. Bhattacharya, R. Sarkar, D. De, S. Mukherjee, S. Pahari, A. Saha, S. Roy, N. C. Paul, S. Ghosh, K. P. Ghatak, *J. Comput. Theor. Nanosci.*, 6, 112 **(2009)**; S. Bhattacharya, N. C. Paul, D. De, K. P. Ghatak, *Phys. B: Conden. Matter*, 403, 4139 **(2008)**; S. Bhattacharya, S. Pahari, R. Sarkar, S. Ghosh, K. P. Ghatak, *Phys. B: Conden. Matter*, 403, 3635 **(2008)**; K. P. Ghatak, S. Bhattacharya, D. De, P. K. Bose, S. N. Mitra, S. Pahari, *Phys. B: Conden. Matter*, 403, 2930 **(2008)**.

[14] K. P. Ghatak, S. Bhattacharya, D. De, R. Sarkar, S. Pahari, A. Dey, A. K. Dasgupta, S. N. Biswas, *J. Comput. Theor. Nanosci.*, 5, 1345 **(2008)**; K. P. Ghatak, S. Bhattacharya, K. M. Singh, S. Choudhury, S. Pahari, *Phys. B: Conden. Matter*, 403, 2116 **(2008)**; K. P. Ghatak, S. Bhattacharya, S. Bhowmik, R. Benedictus, S. Choudhury, *J. Appl. Phys.*, 103, 94314 **(2008)**; S. Choudhury, D. De, S. Mukherjee, A. Neogi, A. Sinha, M. Pal, S. K. Biswas, S. Pahari, S. Bhattacharya, K. P. Ghatak, *J. Comput. Theor. Nanosci.*, 5, 375 **(2008)**; K. P. Ghatak, S. Bhattacharya, S. Bhowmik, R. Benedictus, S. Choudhury, *J. Appl. Phys.*, 103, 34303 **(2008)**; K. P. Ghatak, S. Bhattacharya, *J. Appl. Phys.*, 102, 73704 **(2007)**.

[15] K. v. Klitzing, G. Dorda, M. Pepper, *Phys. Rev. Lett.*, 45, 494 **(1980)**.

[16] K. v. Klitzing, *Rev. Mod. Phys.*, 58, 519 **(1986)**.

[17] J. Hajdu, G. Landwehr, in *Strong and Ultrastrong Magnetic Fields and their Applications*, F. Herlach (Ed.), (Springer, **1985**), p. 17.

[18] I. M. Tsidilkovskii, *Band Structure of Semiconductors*, (Pergamon Press, Oxford, **1982**).

[19] E. A. Arushanov, A. F. Knyazev, A. N. Natepov, S. I. Radautsan, *Sov. Phys. Semicond.*, 15, 828 **(1981)**.

[20] S. P. Zelenim, A. S. Kondratev, A. E. Kuchma, *Sov. Phys. Semicond.*, 16, 355 **(1982)**.

[21] F. M. Peeters, P. Vasilopoulos, *Phys. Rev. B*, 46, 4667 **(1992)**.

[22] W. Zawadzki, in *Two-Dimensional Systems, Heterostructures and Superlattices*, Springer Series in Solid State Sciences, Vol. 53, Ed. by G. Bauer, F. Kuchar, H. Heinrich, (Springer, Berlin, Heidelberg, **1984**).

[23] B. M. Askerov, N. F. Gashimzcede, M. M. Panakhov, *Sov. Phys. Sol. Stat.*, 29, 465 **(1987)**.

[24] G. P. Chuiko, *Sov. Phys. Semicond.*, 19, 1279 **(1985)**.

[25] S. Pahari, S. Bhattacharya, K. P. Ghatak, *J. Comput. Theor. Nanosci.*, 6, 2088 **(2009)**.

[26] K. P. Ghatak, S. Bhattacharya, S. Bhowmik, R. Benedictus, S. Choudhury, *J. Appl. Phys.*, 103, 034303 **(2008)**.

[27] K. P. Ghatak, S. Bhattacharya, S. Pahari, D. De, S. Ghosh, M. Mitra, *Ann. der Physik*, 17, 195 **(2008)**.

[28] K. P. Ghatak, S. N. Biswas, *J. Appl. Phys.*, 70, 299 **(1991)**.

[29] K. P. Ghatak, S. N. Biswas, *J. Low Temp. Phys.*, 78, 219 **(1990)**.

[30] K. P. Ghatak, M. Mondal, *J. Appl. Phys.*, 65, 3480 **(1989)**.

[31] K. P. Ghatak, B. Nag, *Nanostruct. Mater.*, 5, 769 **(1995)**.

[32] K. P. Ghatak, M. Mondal, *Phys. Stat. Sol. (b)*, 185, K5 **(1994)**.

[33] K. P. Ghatak, B. Mitra, *Il Nuovo Cimento D*, 15, 97 **(1993)**.

[34] K. P. Ghatak, S. N. Biswas, *Phys. Stat. Sol. (b)*, 140, K107 **(1987)**.

[35] K. P. Ghatak, *Il Nuovo Cimento D*, 13, 1321 **(1991)**.

[36] K. P. Ghatak, A. Ghoshal, *Phys. Stat. Sol. (b)*, 170, K27 **(1992)**.

[37] K. P. Ghatak, B. De, B. Nag, P. K. Chakraborty, *Molecular Crystals and Liquid Crystals Science and Technology Section B: Nonlinear Optics*, 16, 221 **(1996)**.

[38] K. P. Ghatak, M. Mitra, B. Goswami, B. Nag, *Molecular Crystals and Liquid Crystals Science and Technology Section B: Nonlinear Optics*, 16, 167 **(1996)**.

[39] K. P. Ghatak, D. K. Basu, D. Basu, B. Nag, *Nuovo Cimento della Societa Italiana di Fisica D – Condensed Matter, Atomic, Molecular and Chemical Physics, Biophysics*, 18, 947 **(1996)**.

[40] B. Mitra, K. P. Ghatak, *Phys. Lett. A*, 141, 81 **(1989)**; S. K. Biswas, A. R. Ghatak, A. Neogi, A. Sharma, S. Bhattacharya, K. P. Ghatak, *Phys. E: Low-dimen. Syst. Nanostruct.*, 36, 163 **(2007)**.

[41] M. Mondal, A. Ghoshal, K. P. Ghatak, *Il Nuovo Cimento D*, 14, 63 **(1992)**; K. P. Ghatak, M. Mondal, *Phys. Stat. Sol. (b)*, 135, 819 **(1986)**.

[42] L. J. Singh, S. Choudhury, D. Baruah, S. K. Biswas, S. Pahari, K. P. Ghatak, *Phys. B*, 368, 188 **(2005)**.

[43] K. P. Ghatak, B. De, M. Mondal, S. N. Biswas, *Mat. Res. Soc. Symp. Proc.*, 198, 327 **(1990)**.

[44] K. P. Ghatak, *Proc. SPIE*, 1584, 435 **(1992)**.

[45] K. P. Ghatak, B. De, *Mat. Res. Soc. Symp. Proc.*, 234, 55 and 59 **(1991)**.

[46] K. P. Ghatak, B. De, M. Mondal, S. N. Biswas, *Mat. Res. Soc. Symp. Proc.*, 184, 261 **(1990)**.

[47] K. P. Ghatak, *Influence of Band Structure on Some Quantum Processes in Tetragonal Semiconductors*, D. Eng. Thesis, (Jadavpur University, Kolkata, India, **1991**).

[48] K. P. Ghatak, S. N. Biswas, *Mat. Res. Soc. Symp. Proc.*, 216, 465 **(1990)**.
[49] T. M. Tritt, M. Kanatzidis, G. Mahan, H. B. Lyon Jr. (Eds.), *New Materials for Small Scale Thermoelectric Refrigeration and Power Generation Applications*, Vol. 545, Proceedings of the 1998 Materials Research Society.
[50] L. T. Canham, *Appl. Phys. Lett.*, 57, 1046 **(1990)**.
[51] Z. H. Lu, D. J. Lockwood and J. M. Baribeam, *Nature*, 378, 258 **(1995)**.
[52] A. G. Cullis, L. T. Canham, P. D. O. Calocott, *J. Appl. Phys.*, 82, 909 **(1997)**.
[53] M. Cardona, L. Ley, *Photoemission in Solids 1 and 2, Topics in Applied Physics*, Vols. 26, 27, (Springer-Verlag, Germany, **1978**); S. Hüfner, *Photoelectron Spectroscopy*, (Springer, Germany, **2003**); S. Hüfner, (Ed.), *Very High Resolution Photoelectron Spectroscopy*, Lecture Notes in Physics, Vol. 715, (Springer-Verlag, Germany, **2007**); D. W. Lynch, C. G. Olson, *Photoemission Studies of High-Temperature Superconductors*, (Cambridge University Press, UK, **1999**); *Photoemission and the Electronic Properties of Surfaces*, (Eds.), B. Feuerbacher, B. Fitton, R. F. Willis, (Wiley, New York, **1978**); W. Schattke, M. A. V. Hove, *Solid-State Photoemission and Related Methods: Theory and Experiment*, (Wiley, USA, **2003**); V. V. Afanas'ev, *Internal Photoemission Spectroscopy: Principles and Applications*, (Elsevier, North Holland, **2010**); D. J. Lockwood, *Light Emission in Silicon in Silicon based Materials and Devices*, Vol. 2, Ed. H. S. Nalwa, (Academic Press, San Diego, USA, **2001**).
[54] K. P. Ghatak, D. De, S. Bhattacharya, *Photoemission from Optoelectronic Materials and their Nanostructures*, Springer Series in Nanostructure Science and Technology, (Springer, **2009**).
[55] N. Miura, *Physics of Materials in High Magnetic Fields*, Series on Semiconductor Science and Technology, (Oxford University Press, USA, **2007**); K. H. J. Buschow, F. R. de Boer, *Physics of Magnetism and Magnetic Materials*, (Springer, New York, **2003**); D. Sellmyer, R. Skomski (Eds.), *Advanced Magnetic Nanostructures*, (Springer, New York, **2005**); J. A. C. Bland, B. Heinrich (Eds.), *Ultrathin Magnetic Structures III: Fundamentals of Nanomagnetism (Pt. 3)*, (Springer-Verlag, Germany, **2005**); B. K. Ridley, *Quantum Processes in Materials*, Fourth Edition, (Oxford publications, Oxford, **1999**); J. H. Davies, *Physics of Low Dimensional Materials*, (Cambridge University Press, UK, **1998**); S. Blundell, *Magnetism in Condensed Matter*, Oxford Master Series in Condensed Matter Physics, (Oxford University Press, USA, **2001**); C. Weisbuch, B. Vinter, *Quantum Semiconductor Structures: Fundamentals and Applications*, (Academic Publishers, USA, **1991**); D. Ferry, *Semiconductor Transport*, (CRC, USA, **2000**); M. Reed (Ed.), *Materials and Semimetals: Nanostructured Systems*, (Academic Press, USA, **1992**); T. Dittrich, *Quantum Transport and Dissipation*, (Wiley-VCH Verlag GmbH, Germany, **1998**); A. Y. Shik, *Quantum Wells: Physics & Electronics of Two Dimensional Systems*, (World Scientific, USA, **1997**).

[56] K. P. Ghatak, M. Mondal, *Zietschrift fur Naturforschung A*, 41a, 881 **(1986)**; K. P. Ghatak, M. Mondal, *J. Appl. Phys.*, 62, 922 **(1987)**; K. P. Ghatak, S. N. Biswas, *Phys. Stat. Sol. (b)*, 140, K107 **(1987)**; K. P. Ghatak, M. Mondal, *Phys. Stat. Sol. (b)*, 139, 195 **(1987)**; K. P. Ghatak, M. Mondal, *Phys. Stat. Sol. (b)*, 148, 645 **(1988)**; K. P. Ghatak, S. N. Biswas, *J. Low Temp. Phys.*, 78, 219 **(1990)**; K. P. Ghatak, M. Mondal, *Phys. Stat. Sol. (b)*, 160, 673 **(1990)**; K. P. Ghatak, A. Ghoshal, B. Mitra, *Nuovo Cimento D*, 13, 867 **(1991)**; K. P. Ghatak, *Nuovo Cimento D*, 13, 1321 **(1992)**; K. P. Ghatak, S. N. Biswas, *Nonlinear Opt.*, 4, 347 **(1993)**.

[57] A. N. Chakravarti, K. P. Ghatak, A. Dhar and S. Ghosh, *Phys. Stat. Sol. (b)*, 105, K55 **(1981)**; A. N. Chakravarti, A. K. Choudhury, K. P. Ghatak, *Phys. Stat. Sol. (a)*, 63, K97 **(1981)**; A. N. Chakravarti, A. K. Choudhury, K. P. Ghatak, S. Ghosh, A. Dhar, *Appl. Phys.*, 25, 105 **(1981)**; A. N. Chakravarti, K. P. Ghatak, G. B. Rao, K. K. Ghosh, *Phys. Stat. Sol. (b)*, 112, 75 **(1982)**; A. N. Chakravarti, K. P. Ghatak, K. K. Ghosh, H. M. Mukherjee, *Phys. Stat. Sol. (b)*, 116, 17 **(1983)**; M. Mondal, K. P. Ghatak, *Phys. Stat. Sol. (b)*, 133, K143 **(1984)**; M. Mondal, K. P. Ghatak, *Phys. Stat. Sol. (b)*, 126, K47 **(1984)**; M. Mondal, K. P. Ghatak, *Phys. Stat. Sol. (b)*, 129, K745 **(1985)**; M. Mondal, K. P. Ghatak, *Phys. Scr.*, 31, 615 **(1985)**; M. Mondal, K. P. Ghatak, *Phys. Stat. Sol. (b)*, 135, 239 **(1986)**; M. Mondal, K. P. Ghatak, *Phys. Stat. Sol. (b)*, 135, K21 **(1986)**; M. Mondal, S. Bhattacharyya, K. P. Ghatak, *Appl. Phys. A*, 42, 331 **(1987)**; B. Mitra, A. Ghoshal, K. P. Ghatak, *Phys. Stat. Sol. (b)*, 150, K67 **(1988)**; M. Mondal, K. P. Ghatak, *Phys. Stat. Sol. (b)*, 146, K97 **(1988)**; B. Mitra, K. P. Ghatak, *Sol. Stat. Electron.*, 32, 515 **(1989)**; B. Mitra, K. P. Ghatak, *Phys. Lett.*, 135, 397 **(1989)**.

[58] W. Zawadzki, B. Lax, *Phys. Rev. Lett.*, 16, 1001 **(1966)**.

[59] B. Mitra, A. Ghoshal, K. P. Ghatak, *Phys. Stat. Sol. (b)*, 154, K147 **(1989)**; K. P. Ghatak, B. Goswami, M. Mitra, B. Nag, *Non. Opt. Quant. Opt.*, 16, 9 **(1996)**; M. Mondal, K. P. Ghatak, *Ann. der Physik*, 46, 502 **(1989)**; M. J. Harrison, *Phys. Rev. A*, 29, 2272 **(1984)**; J. Zak and W. Zawadzki, *Phys. Rev.*, 146, 536 **(1966)**; W. Zawadzki, Q. H. Vrehen, and B. Lax, *Phys. Rev.*, 148, 849 **(1966)**; Q. H. Vrehen, W. Zawadzki and M. Reine, *Phys. Rev.*, 158, 702 **(1967)**; M. H. Weiler, W. Zawadzki and B. Lax, *Phys. Rev.*, 163, 733 **(1967)**; W. Zawadzki and J. Kowalski, *Phys. Rev. Lett.*, 27, 1713 **(1971)**; C. Chu, M. -S. Chu, and T. Ohkawa, *Phys. Rev. Lett.*, 41, 653 **(1978)**; P. Hu and C. S. Ting, *Phys. Rev. B*, 36, 9671 **(1987)**; E. I. Butikov, A. S. Kondratev and A. E. Kuchma, *Sov. Phys. Sol. Stat.*, 13, 2594 **(1972)**; M. Mondal and K. P. Ghatak, *Phys. Stat. Sol. (b)*, 133, K67 **(1986)**; M. Mondal, N. Chattopadhyay and K. P. Ghatak, *J. Low Temp. Phys.*, 66, 131 **(1987)**; K. P. Ghatak and M. Mondal, *Zeitschrift fur Physik B*, 69, 471 **(1988)**; M. Mondal and K. P. Ghatak, *Phys. Lett. A*, 131, 529 **(1988)**; M. Mondal and K. P. Ghatak, *Phys. Stat. Sol. (b)*, Germany, 147, K179 **(1988)**; B. Mitra and K. P. Ghatak, *Phys. Lett.*, 137, 413 **(1989)**; B. Mitra and K. P. Ghatak, *Phys. Stat. Sol. (b)*, 164, K13

(1991); K. P. Ghatak and B. Mitra, *Int. J. Electron.*, 70, 345 **(1991)**; K. P. Ghatak, N. Chattopadhyay and S. N. Biswas, *SPIE, Optoelectr. Mater, Devi, Packa, Interconn.*, 836, 203 **(1988)**; K. P. Ghatak, M. Mondal and S. Bhattacharyya, *SPIE*, 1284, 113 **(1990)**; K. P. Ghatak, *SPIE*, 1280, 53 **(1990)**.

[60] N. G. Anderson, W. D. Laidig, R. M. Kolbas, Y. C. Lo, *J. Appl. Phys.*, 60, 2361 **(1986)**; D. De, S. Bhattacharya, S. M. Adhikari, A. Kumar, P. K. Bose, K. P. Ghatak, *Beilstein J. Nanotech.*, 2, 339 **(2012)**; D. De, A. Kumar, S. M. Adhikari, S. Pahari, N. Islam, P. Banerjee, S. K. Biswas, S. Bhattacharya, K. P. Ghatak, *Superlatt. Microstruct.*, 47, 377 **(2010)**; S. Pahari, S. Bhattacharya, K. P. Ghatak, *J. Comput. Theor. Nanosci.*, 6, 2088 **(2009)**; K. P. Ghatak, J. Mukhopadhyay, J. P. Banerjee, *SPIE Proc. Ser.*, 4746, 1292 **(2002)**; K. P. Ghatak, S. Dutta, D. K. Basu, B. Nag, *Il Nuovo Cimento D*, 20, 227 **(1998)**; K. P. Ghatak, B. De, *Mat. Res. Soc. Proc.*, 300, 513 **(1993)**; K. P. Ghatak, B. Mitra, *Il Nuovo Cimento D*, 15, 97 **(1993)**; K. P. Ghatak, *Inter. Soc. Opt. Photon. Proc. Soc. Photo Opt. Instru. Engg.*, 1626, 115 **(1992)**; S. N. Biswas, K. P. Ghatak, *Int. J. Electron. Theor. Exp.*, 70, 125 **(1991)**; B. Mitra, K. P. Ghatak, *Phys. Lett. A*, 142, 401 **(1989)**; K. P. Ghatak, B Mitra, A. Ghoshal, *Phys. Stat. Sol. (b)*, 154, K121 **(1989)**; B. Mitra, K. P. Ghatak, *Phys. Stat. Sol. (b)*, 149, K117 **(1988)**.

[61] F. Capasso, *Semicond. Semimetals*, 22, 2 **(1985)**.

[62] F. Capasso, K. Mohammed, A. Y. Cho, R. Hull, A. L. Hutchinson, *Appl. Phys. Lett.*, 47, 420 **(1985)**.

[63] F. Capasso, R. A. Kiehl, *J. Appl. Phys.*, 58, 1366 **(1985)**.

[64] K. Ploog, G. H. Doheler, *Adv. Phys.*, 32, 285 **(1983)**.

[65] F. Capasso, K. Mohammed, A. Y. Cho, *Appl. Phys. Lett.*, 48, 478 **(1986)**.

[66] R. Grill, C. Metzner, G. H. Döhler, *Phys. Rev. B*, 63, 235316 **(2001)**; R. Grill, C. Metzner, G. H. Döhler, *Phys. Rev. B*, 61, 15614 **(2000)**.

[67] A. R. Kost, M. H. Jupina, T. C. Hasenberg, E. M. Garmire, *J. Appl. Phys.*, 99, 023501 **(2006)**.

[68] A. G. Smirnov, D. V. Ushakov, V. K. Kononenko, *Proc. SPIE*, 4706, 70 **(2002)**.

[69] D. V. Ushakov, V. K. Kononenko, I. S. Manak, *Proc. SPIE*, 4358, 171 **(2001)**.

[70] J. Z. Wang, Z. G. Wang, Z. M. Wang, S. L. Feng, Z. Yang, *Phys. Rev. B*, 62, 6956 **(2000)**.

[71] A. R. Kost, L. West, T. C. Hasenberg, J. O. White, M. Matloubian, G. C. Valley, *Appl. Phys. Lett.*, 63, 3494 **(1993)**.

[72] A. S. Bastola, S. J. Chua, S. J. Xu, *J. Appl. Phys.*, 83, 1476 **(1998)**.

[73] Z. J. Yang, E. M. Garmire, D. Doctor, *J. Appl. Phys.*, 82, 3874 **(1997)**.

[74] G. H. Avetisyan, V. B. Kulikov, I. D. Zalevsky, P. V. Bulaev, *Proc. SPIE*, 2694, 216 **(1996)**.

[75] U. Pfeiffer, M. Kneissl, B. Knüpfer, N. Müller, P. Kiesel, G. H. Döhler, J. S. Smith, *Appl. Phys. Lett.*, 68, 1838 **(1996)**.

[76] H. L. Vaghjiani, E. A. Johnson, M. J. Kane, R. Grey, C. C. Phillips, *J. Appl. Phys.*, 76, 4407 (**1994**).
[77] P. Kiesel, K. H. Gulden, A. Hoefler, M. Kneissl, B. Knuepfer, S. U. Dankowski, P. Riel, X. X. Wu, J. S. Smith, G. H. Doehler, *Proc. SPIE*, 1985, 278 (**1993**).
[78] G. H. Doheler, *Phys. Script.*, 24, 430 (**1981**).
[79] S. Mukherjee, S. N. Mitra, P. K. Bose, A. R. Ghatak, A. Neoigi, J. P. Banerjee, A. Sinha, M. Pal, S. Bhattacharya, K. P. Ghatak, *J. Comput. Theor. Nanosci.*, 4, 550 (**2007**).
[80] L. V. Keldysh, *Sov. Phys. Sol. Stat.*, 4, 1658 (**1962**).
[81] L. Esaki and R. Tsu, *IBM J. Res. Dev.*, 14, 61 (**1970**).
[82] R. Tsu, *Superlattices to Nanoelectronics*, (Elsevier, **2005**); E. L. Ivchenko and G. Pikus, *Superlattices and Other Heterostructures*, (Springer-Berlin, **1995**).
[83] G. Bastard, *Wave Mechanics Applied to Heterostructures*, (Editions de Physique, Les Ulis, France, **1990**).
[84] K. P. Ghatak, B. De, *MRS Proceedings*, Cambridge University Press, 299, 65 (**1994**); K. P Ghatak, B. De, *MRS Proceedings*, Cambridge University Press, 242, 377 (**1992**); K. P. Ghatak, San Diego,'91, San Diego, CA, *Inter. Soc. Opt. Photon.*, 282 (**1991**); K. P. Ghatak, *Properties of II–VI Semiconductors: Bulk Crystals, Epitaxial Films, Quantum Well Structures, and Dilute Magnetic Systems, Materials Research Society*, 161, 371 (**1990**); K. P. Ghatak, S. N. Biswas, *MRS Proceedings*, Cambridge University Press, 161 (**1989**); S. Bhattacharyya, K. P. Ghatak, S. Biswas, *OE/Fibers*' 87, 73 (**1987**).
[85] K. V. Vaidyanathan, R. A. Jullens, C. L. Anderson and H. L. Dunlap, *Sol. Stat. Electron.*, 26, 717 (**1983**).
[86] B. A. Wilson, *IEEE J. Quant. Electron.*, 24, 1763 (**1988**).
[87] M. Krichbaum, P. Kocevar, H. Pascher and G. Bauer, *IEEE J. Quant. Electron.*, 24, 717 (**1988**).
[88] J. N. Schulman and T. C. McGill, *Appl. Phys. Lett.*, 34, 663 (**1979**).
[89] H. Kinoshita, T. Sakashita and H. Fajiyasu, *J. Appl. Phys.*, 52, 2869 (**1981**).
[90] L. Ghenin, R. G. Mani, J. R. Anderson and J. T. Cheung, *Phys. Rev. B*, 39, 1419 (**1989**).
[91] C. A. Hoffman, J. R. Mayer, F. J. Bartoli, J. W. Han, J. W. Cook, J. F. Schetzina and J. M. Schubman, *Phys. Rev. B*, 39, 5208 (**1989**).
[92] V. A. Yakovlev, *Sov. Phys. Semicond.*, 13, 692 (**1979**).
[93] E. O. Kane, *J. Phys. Chem. Solids*, 1, 249 (**1957**).
[94] H. Sasaki, *Phys. Rev. B*, 30, 7016 (**1984**).
[95] P. K. Bose, N. Paitya, S. Bhattacharya, D. De, S. Saha, K. M. Chatterjee, S. Pahari, K. P. Ghatak, *Quant. Matt.*, 1, 89 (**2012**).
[96] S. Bhattarchya, D. De, S. Ghosh, P. Banerjee, S. Saha, M. Mitra, B. Nag, M. Pal, S. K. Biswas, K. P. Ghatak, *J. Comput. Theor. Nanosci.*, 7, 1066 (**2010**); K. P. Ghatak, S. Bhattacharya, S. Pahari, S. N. Mitra, P. K. Bose, D. De, *J. Phys. Chem. Solids*, 70, 122 (**2009**).

[97] S. Bhattacharya, N . C. Paul, D. De, K. P. Ghatak, *Phys. B: Conden. Matter*, 403, 4139 **(2008)**.

[98] A. Pais, *J. Robert Oppenheimer*, (Oxford University Press, UK, **2006**), p. xviii.

[99] I. K. Kikoin, *Science for Everyone: Encounters with Physicists and Physics*, (Mir Publishers, Russia, **1989**), p. 154.

[100] J. J. Sakurai, *Advanced Quantum Mechanics*, (Addison–Wesley Publishing Company, USA, **1967**).

About the Authors

Education is learning, what you didn't even know you didn't know.

Kamakhya Prasad Ghatak (h-index: 38; i-10 index: 194; maximum citations of a research paper: 329; total citations: 7285) obtained his D. Phil. (Tech) degree (based on 27 published research papers in eminent SCI journals, which is still a record in the Institute of Radio Physics and Electronics) from the University of Calcutta, India, in 1988. He is the first recipient of the Degree of Doctor of Engineering of Jadavpur University, Kolkata, in 1991 since the university's inception in 1955, and in the same year, he received the INSA visiting fellowship to IIT Kharagpur. Prof. Ghatak was listed by Stanford University under the field of applied physics in the World Ranking of Top 2% Scientists list. He is the principal co-author of more than 200 research papers in SCI Journals and 385 research publications in SCOPUS. He was Assistant Professor in the Department of Radio Physics and Electronics of the University of Calcutta from 1983 to 1987, Associate Professor in the Department of ETCE of Jadavpur University from 1987 to 1994, Professor and HOD of Electronic Science of the University of Calcutta from 1994 to 2011, HOD and Dean of the National Institute of Technology, Agartala, from 2012 to 2014 and from January 2015, he joined in the UEM as Dean, Engineering, where he is currently an Emeritus Professor since 1st July, 2024. He was at the top of the merit lists in all the above assignments. Prof. Ghatak has also co-authored 10 research monographs on various topics of nanotechnology for Springer-Verlag Germany (Vol. 170 in *Solid State Sciences*, Vols. 116, 137, and 167 in *Materials Science*, Vols. 255 and 260 in *Tracts in Modern Physics (TMP)*, 1 in *Nanostructured Science and Technology series*) and 1 book in Springer Nature, Vols. 1 and 2 in *Nanomaterials Series* of De Gruyter, Germany. He is the sole author of Vols. 262 and 265 in *TMP* of Springer-

Verlag, Germany, and five research monographs (Vols. 7, 8, 14, 16, and 17) in the *World Scientific Series on the Foundations of Natural Science and Technology*. The selling cross-sections of the said 19 research monographs are very high. He is also the Chief Editor of the three edited monographs in the *Nanotechnology Series* of NOVA, USA. This first-time achievement of Prof. Ghatak is still a record in West Bengal. The Vidwan portal unit of the Central Government of India gives Prof. K. P. Ghatak a score of 8.9 out of 10, the highest score among the nanotechnologists of private engineering colleges and universities in West Bengal.

Prof. Ghatak has successfully supervised more than 100 PhD students in different topics of nanotechnology, and the list includes directors, vice-chancellors, principals, professors, and CEOs of different reputed organizations. The AICTE has selected the first R&D project of his life for the Best Project Award in Electronics and the second Best Research Project Award, considering all the branches of engineering for the year 2006. In 2011, the UGC awarded a research project to him and placed him at the top of the list of awardees. He is the first recipient of the academic excellence award of the University of Engineering & Management, Kolkata, from the then Governor of West Bengal at its first convocation in 2017. He is the Referee, Editor, and Guest Editor of various reputed international journals. He is the referee of international and intercontinental experimental research projects of very high value between Austria and France. His present research interest is quantum materials. From 2015 to 2017, in accordance with Google Scholar, his number of research publications in three consecutive years exceeded 100. His brief CV has been enlisted in many biographical references of USA and UK.

Jayita Pal is an Assistant Professor in the Department of Physics of Meghnad Saha Institute of Technology, Kolkata. She obtained her PhD degree in 2010 and is the author of 20 research papers/chapters in eminent journals and edited monographs published by Springer. She has 20 years of teaching experience, and her present teaching and research interests are quantum mechanics and quantized structures. She is a prominent member of the research group of Prof. Ghatak.

Acknowledgments

Optimism is the output of wiseness.

Acknowledgments by Kamakhya Prasad Ghatak

I am grateful to late Professor A. N. Chakravarti, formerly Dean and Head of the Department of the Institute of Radio Physics and Electronics of the University of Calcutta, my mentor and an ardent advocator of the concept of theoretical minimum of Landau, who convinced a 21-year-old circuit theorist in 1974 that the wonderland of *quantum phenomena*, in general, is the hidden dual dance of quantum mechanics, statistical mechanics together with advanced mathematical techniques, and even to appreciate the incredible beauty, he not only placed a stiff note for me to derive all the equations in the monumental course of theoretical physics, the classics of Landau–Lifshitz together with the two-volume classics of Morse–Feshbach 50 years ago but also forced me to stay in the creative critical zone of research till date. I express my gratitude to late Professor B. R. Nag, formerly Head of the Departments of Radio Physics and Electronics and Electronic Science of the University of Calcutta, to whom I am ever grateful as a student and research worker, the Semiconductor Grandmaster of India, for his three books on semiconductor science in general and more than 200 research papers (many of them are absolutely in honor's class), which still fire my imagination. I consider myself to be rather fortunate to study mathematics under the direct influence of late Professor S. C. Dasgupta, formerly Head of the Department of Mathematics of the then Bengal Engineering College, Shibpur (presently the Indian Institute of Engineering Science and Technology) and late Professor M. Mitra (both of whom were formidable applied mathematicians with deep physical insight and could solve any problem from the two-volume classics of Morse–Feshbach and the famous Whittaker/Watson instantly on the blackboard) also of the Department of Mathematics for teaching me deeply the various methods of different branches of applied mathematics with special emphasis on analytic number theory when, in my late teens, I was pursuing the bachelor's degree in the branch of electronics and telecommunication engineering (ETCE) 55 years ago. I offer my thanks to late Professor S. S. Boral, formerly Founding Head of the ETCE Department of the then Bengal Engineering College, Shibpur for teaching me the theoretical acoustics from the classic of Morse/Ingard by urging me to solve the problems. I am grateful to late Professor S. K. Sen, formerly Head of the Department of Electrical Engineering of Bengal Engineering College, Shibpur, Ex-Vice-Chancellor of Jadavpur University and Ex-Power Minister of West Bengal State for teaching me profoundly nonlinear network analysis and synthesis and nonlinear control systems in general. I am indebted to late Professor C. K. Majumdar, the first Director of S. N. Bose Institute of Kolkata to light the fire for theoretical physics.

I am grateful to all my friends and colleagues for the last 55 years from my college days till date for forming my inner fire to solve independently the problems from three volumes of Sakurai, 10 volumes of Addison–Wesley Modular Series on Solid State Devices, two volumes of Cohen–Tannoudji *et al.*, three volumes of Carslaw, six volumes of Guillemin, three volumes of Tuttle, two volumes of Scott, two volumes of Budak, five volumes of Barkley physics course, three volumes of Feynman's lectures in physics, etc. It is interesting to note that they are insisting me in the real sense of the term to verify all the entries of Gradshteyn–Ryzhik and three volumes of Bateman manuscript project for the last 50 years. It may be noted that academic output of a nice post-graduate teacher is the response of a good quality RC-coupled amplifier in the mid-band zone, whereas the same for a fine research worker is the Dirac delta function. Incidentally, I can safely say that I belong to neither as a consequence of the academic surroundings which this life presents to me. I express my warm gratitude to Professors J. S. Blakemore, H. L. Hartnagel, D. Bimberg, W. L. Freeman and W. Schommers for various academic interactions spanning over the last four decades. I remember the sweet memories of late Professors P. N. Robson, I. M. Stephenson, V. S. Letokhov, J. Bodnar and K. F. Brennan with true reverence for inspiring me with the ebullient idea that the publications of the research monographs and advanced books from internationally reputed publishers containing innovative concepts is of prime importance to excel in creative research activity.

From the flashback of memory, I wish to note that the late Professor P. T. Landsberg (popularly known as PTL to his scientific friends) visited the Institute of Radio Physics and Electronics of the University of Calcutta in late seventies where I was a PhD student of the said institute. I was very impressed by his crystal-clear lecture on semiconductor physics and I formed permanent covalent bonds with him. Whatever small I have learned with respect to the fundamentals of advanced quantum science is mainly due to him. Incidentally, his wife Mrs. Sophia Landsberg in her letter of desperate sadness dated January 12, 2009, informed me that from the last phase of 2008 onward due to Alzheimer's disease her husband had to give up scientific works. The disappearance of PTL from my research scenario is the biggest academic loss for me which I have to bear till my own physical disappearance. I still remember his power-packed words that "The definition of Kamakhya Prasad Ghatak (KPG) = PhD + D. Engg. + more than 300 research chapters in nanoscience and technology in SCOPUS enlisted journals, + PhD guide of more than 50 PhD students + many research books + HOD + Dean ...." is a big zero to PTL if KPG cannot live in a world filled with new creative concepts and not merely repeating the past successes, since past success is a dead concept, at least to PTL.

In 21st December, 1974, late Professor A. N. Chakravarti, my mentor in my first interaction with him emphatically energized me by making me acquainted with the famous Fermi–Dirac Integrals and telling me that I must master *Semiconductor Statistics* (Pergamon Press, London, 1962) by J. S. Blakemore for my initiation in semiconductor physics. Later on, I came in close touch with late Professor K. Seeger, the well known author of the book *Semiconductor Physics* (Springer Series in Solid State Sciences, Vol. 40, Springer-Verlag, Germany, 1982). The solid mathematical physicist, late Professor P. N. Butcher has been a steady hidden force since 1983 before his sad demise with respect to our scripting the series in band structure-dependent properties of nanostructured materials. Both PTL and P. N. Butcher visited the Institute of Radio Physics and Electronics of the University of Calcutta, my alma mater where I started my research as a M. Tech. student and later on as a faculty member. I formed permanent covalent bonds with late Professors J. S. Blakemore, K. Seeger, P. N. Butcher and PTL through letters (pre-email era) and these four first-class semiconductor physicists, in turn, shared with pleasure their vast creative knowledge on semiconductors and related sciences with a novice like me.

I offer special thanks to late Mr. N. Guhachoudhury of Jadavpur University (popularly known as Niren babu) for instilling in me the thought that the academic output = ((desire × determination × dedication) − (false enhanced self-ego pretending like a true friend while being a real unrecognizable foe)) although a thank you falls in the band gap regime for my beloved better half See who really formed the backbone of my long unperturbed research career, since in accordance with "Sanatan" Hindu Dharma, the fusion of marriage has transformed us to form a single entity, where the individuality is being lost. I am grateful to all the members of my research group (from 1983 till date) for not only quantum confining me in the infinitely deep quantum wells of Ramanujan and Rabindranath but also inspiring me to teach quantum mechanics and related topics from the eight volumes classics of Greiner *et al.*

In this context, I wish to offer my indebtedness to M. Mondal, the first member of my research team who in 1983 joined with me to complete his PhD work under my guidance when late Professor R. K. Poddar, the then Vice-Chancellor of the University of Calcutta selected me as a Lecturer (presently Assistant Professor) of the famous Department of Radio Physics and Electronics. In 1987, late Professor S. K. Sen, the then Vice-Chancellor of Jadavpur University accepted me as the Reader (presently Associate Professor) in the Department of Electronics and Telecommunication Engineering and since then a flood of young researchers (more than 12 in number consisting of B. Mitra, A. Ghoshal, D. Bhattacharya, A. R. Ghatak, etc. around me created my research team and insisted me to work with them at 16 hours per day including holidays in different directions of research for the purpose of my creative conversion from an ordinary engineer to a 360° research scientist and consequently I enjoyed the high rate of research publications in different reputed international journals in various aspects of band structure-dependent properties of quantized structures. It is nice to note that the said young talented researchers obtained their respective PhD degree under my direct supervision. Incidentally in 1994, late Professor R. N. Basu, the then Vice-Chancellor of the University of Calcutta selected me as a Professor in the Department of Electronic Science and another flood of research helmed me in a new direction of materials science. The persons responsible for this change include S. Datta, S. Sengupta, A. Ali, etc. The 11th and 12th names of this golden series are, respectively, S. Bhattacharya and D. De, who in turn formed permanent covalent bonds with me not only with respect to research (S. Bhattacharya and D. De are, respectively, my respected co-authors of seven and two research monographs in different series of Springer) and also in all aspects of life in general.

It is interesting to note that after serving 18 years as a Professor in the Department of Electronic Science, in 2012, late Professor P. K. Bose, the first Director of the National Institute of Technology, Agartala requested me to join as a Professor and Departmental Head in Electronics and Communication Engineering. Being my life-long friend, I accepted his offer (and later on as a DEAN) and more than 10 young scholars around me again directed my research in an altogether new direction. In 2015, the Honorable Pro-Vice Chancellor Professor S. Chakrabarti of University of Engineering and Management (Institute of Engineering and Management), Kolkata, had kindly offered me the position of Research Director, Senior Professor and Dean Engineering in his famous Institute in the academic fag end of my life to complete my last run toward the creative knowledge temple with my new young research workers and I am grateful to him for his creative gesture. In my 50+ years of academic life (I have the wide experience of teaching quantum mechanics (from non-relativistic quantum mechanics up to string theory), engineering mathematics, applied mechanics (from engineering statics up to nonlinear mechanics including indeterminate structures) and 50% of the courses in Electrical Engineering) and 50+ years of research life (mostly in Materials Science and Quantized Structures), I have finally realized that the teaching without research is body without brain and research without teaching is body without blood although my all-time hero, creatively prolific number theorist Godfrey Harold Hardy in his classic book entitled *A Mathematician's Apology* (Cambridge University Press, 1990) tells us "I hate teaching".

Incidentally, one young theoretician friend of mine often tells me that many works in theoretical semiconductor science are based on the following seven principles:

- principles of placing the necessary and sufficient conditions of a proof in the band gap regime;
- principles of interchange of the summation and the integration processes and unconditioned convergences of the series and the integrals;
- principles of random applications of one electron theory and superposition theorem in studying the properties of materials, although the many-body effects are very important together with the fact that the nature is fundamentally nonlinear in nature;

- principles of using the invariant band structure concept of materials even in the presence of strong external fields (magnetic, electric, heavy doping etc.) and the random applications of perturbation theory, which is in a sense quantum-mechanical Taylor series without considering the related consequences;
- principle of random applications of the binomial theorem without considering the important concept of branch cut;
- principle of little discussion regarding the whole set of materials science comprising different compounds having various band structures under different physical conditions as compared with the simplified two-band model of Kane for III–V materials;
- principle of using the Fermi golden rule, the band structure, and the related features, which are valid for non-degenerate compounds to materials having degenerate carrier concentrations directly.

Although my young friend is a purist in his conjecture, there is no doubt certain elements of truth inside his beautiful comments. I can only hope that our creative readers will present their intricate and advanced theories after paying due weightage of the aforementioned seven principles of my friend.

I am grateful to late Professor Dr. S. Chakrabarti the Honorable Chancellor of the University of Engineering and Management for his encouragement and enthusiastic support, his keen academic insight and the creation of congenial atmosphere to promote research and higher learning among the faculty members. I am extremely grateful to Professor Dr. R. Paul, Director IQAC of UEM, Kolkata for really helping me in times of academic turbulence. I offer special thanks to Professor Dr. P. K. Das, HOD, Basic Science and other members of my present research team for placing their combined effort toward the development of this book in the DO-WHILE-LOOP of a computer program and critically reading the manuscript in its last phase before sending it to the Desk Editor, M. I. Putra World Scientific Publishing. I am grateful to M. Ghosh, my personal assistant, for typing the manuscript of this book for three times very diligently. I am also grateful to D. Bhattacharjee for overall supervision and word by word checking of this book. At this point, we offer our best wishes of our readers as they begin this trip into the new world of low-dimensional physics, with us as their guides, where the senior author has spent partially a lifetime and still he is thinking about new ideas and planning to write new books with the prominent members of his research group. Last but not the least, I am grateful for ever to my life-long time-tested friend Dr. S. Sanyal,

Ex-Principal, Lake School for Girls, Kolkata for not only motivating me at rather turbulent moments of my academic carrier but also instilling in me the concept that the ratio of total accumulated knowledge in my chosen field of research to my own contribution in my topic of research tends to infinity at any time and is the definition of non-removable essential pole in the transfer function of my life.

Acknowledgments by Jayita Pal

It is a great pleasure to express my deep gratitude to Professor K. P. Ghatak for inspiring me to carry out extensive research in the field of nanoscience and nanotechnology. My family members, especially my husband and son, deserve a very special mention for constant support, cooperation and also for forming the backbone of my long research carrier. Lastly, I wish to offer special thanks to my parents for their motivation and guidance in the midst of all difficulties.

Joint Acknowledgments by Both the Authors

We are grateful to Mr. Rhaimie B. Wahap, Editor of World Scientific Publishing, who single handedly carried out this project starting from the very inception of the duly filled in book proposal form for this book. He arranged for referring of the book proposal by the learned referees for critical analysis. We are also grateful to Mr. M. I. Putra who kindly agreed to act as a Desk Editor for this book in spite of his 360° heavy commitment and other academic duties. Besides, he helped us from every point of view.

We offer special thanks to Professor K. K. Phua, Chairman, World Scientific Publishing for further necessary action in this context. We have given enough fundamentals together with various graphs right from the very first chapter so that it can be accessible to graduate students in addition to scientists and engineers. The authors are responsible for non-imaginative shortcomings. We firmly believe that our Mother Nature has propelled this project in Her own unseen ways in spite of several insurmountable obstacles.

K. P. Ghatak
J. Pal

Contents

What lies before us and what lies behind us are tiny matters as compared to what lies WITHIN US.

Chapter 1

The Magneto Thermoelectric Power in Quantum Dots

Try to get what you like or you would be forced to like what you get.

1.1 Introduction

Zawadzki [1] demonstrated that the Magneto Thermoelectric Power (MTP) for electronic materials having degenerate electron concentration is essentially determined by their respective energy band structures. It has, therefore, different values in different materials and changes with the doping, magnitude of the reciprocal quantizing magnetic field under magnetic quantization, quantizing electric field as in inversion layers, nanothickness as in quantum wells, wires and dots, with superlattice period as in quantum confined semiconductor superlattices with graded interfaces having various carrier energy spectra and also in other types of field assisted nanostructured materials. Some of the significant features, which have emerged from these studies are listed as follows:

(a) The MTP decreases with the increase in carrier concentration;
(b) The MTP decreases with increasing doping in heavily doped semiconductors forming band tails;
(c) The nature of variations is significantly influenced by the spectrum constants of various materials having different band structures;
(d) The MTP exhibits oscillatory dependence with inverse quantizing magnetic field because of the Shubnikov de Haas effect;
(e) The MTP decreases with the magnitude of the quantizing electric field in inversion layers;
(f) The MTP exhibits composite oscillations with significantly different values in superlattices and various other quantized field aided structures.

In Section 1.2.1, we shall present a generalized formulation of MTP in this context. We shall use the same to investigate the MTP in QDs of III–V, II–VI, GaP, Ge, Te, Graphite, $PtSb_2$, zero-gap, II–V, GaSb, stressed materials, Bismuth, IV–VI, Lead Germanium Telluride, Zinc and Cadmium diphosphides, Bi_2Te_3 and Antimony from Sections 1.2.2 to 1.2.18 respectively.

The Quantum Dots (QDs) can be used for visualizing and tracking molecular processes in cells using standard fluorescence microscopy [2–4]. They display minimal photo-bleaching [5], thus allowing molecular tracking over prolonged periods and consequently, single molecule can be tracked by using optical fluorescence microscopy [6]. The salient features of quantum dot lasers [7] include low threshold currents, higher power, and great stability as compared with the conventional one and the QDs find extensive applications in nano-robotics [8], neural networks [9] and high density memory or storage media [10]. The QDs are also used in nano-photonics [11] because of their theoretically high quantum yield and have been suggested as implementations of qubits for quantum information processing [12]. The QDs also find applications in diode lasers [13, 14], amplifiers [15], and optical sensors [16]. High-quality QDs are well suited for optical encoding [16] because of their broad excitation profiles and narrow emission spectra. The new generations of QDs have far-reaching potential for the accurate investigations of intracellular processes at the single-molecule level, high-resolution cellular imaging, long-term in vivo observation of cell trafficking, tumor targeting, and diagnostics [17]. The QD nanotechnology is one of the most promising candidates for use in solid-state quantum computation [18]. It may also be noted that the QDs are being used in single electron transistors [19], photovoltaic devices [20], photoelectrics [21], ultrafast all-optical switches and logic gates [22], organic dyes [23] and in other types of nano devices.

In Section 1.2.1, we shall present the generalized formulation of MTP in this context. It may be noted that the III–V compounds find potential applications in infrared detectors, quantum dot light emitting diodes [24], quantum cascade lasers [25], quantum well wires [26], optoelectronic sensors [27], high electron mobility transistors [28] etc. The III–V, ternary and quaternary materials are called the Kane type compounds since their electron energy spectra are being defined by the three-band model of Kane [29]. In Section 1.2.2, the MTP from QDs of III–V materials has been studied on the basis of simplified parabolic band model of Kane with enough details. Then we study the same for two-band model of Kane special cases. Again the generalized three band model of Kane is being employed to investigate MTP. Besides, Kane, the conduction electrons of III–V materials also obey another nine different types of electron dispersion laws as given in the literature. The MTP has also been investigated for all the cases for the purpose of complete presentation and relative assessment among the energy band models of III–V compounds.

The II–VI compounds are being extensively used in nanoribbons, blue green diode lasers, photosensitive thin films, infrared detectors, ultrahigh-speed bipolar transistors, fiber optic communications, microwave devices, photovoltaic and solar cells, semiconductor gamma-ray detector arrays, semiconductor detector gamma camera and allow for a greater density of data storage on optically addressed compact discs [30–32]. The carrier energy spectra in II–VI materials are defined by the Hopfield model [33], where the splitting of the two-spin states by the spin-orbit coupling and the crystalline field has been taken into account. Section 1.2.3 contains the investigation of the MTP in QDs of II–VI compounds, taking p-CdS as an example. The n-Gallium Phosphide (n-GaP) is being used in QD light emitting diode [34], high efficiency yellow solid state lamps, light sources and high peak current pulse for high gain tubes. The green and yellow light emitting diodes made of nitrogen-doped n-GaP possess a longer device life at high drive currents [35]. In Section 1.2.4, the MTP in QDs of n-GaP is studied. The importance of Germanium is already well-known since the inception of transistor technology and in recent years, memory circuits, single photon detectors, single photon avalanche diode, ultrafast all-optical switch, THz lasers and THz spectrometers [36, 37] are made of Ge. In Section 1.2.5, the MTP has been studied in QDs of Ge on the basis of two different types of band models.

Tellurium (Te) is also an elemental semiconductor which has been used as the semiconductor layer in thin-film transistors (TFT) [38]. Te also finds extensive applications in CO_2 laser detectors [39], electronic imaging, strain sensitive devices [40] and multichannel Bragg cell [41]. Section 1.2.6 contains the investigation of MTP in QDs of Tellurium on the basis of two different types of band models. The importance of graphite is already well-known in the whole spectrum of materials science and the low dimensional graphite is used instead of carbon wire in many practical applications. Graphite intercalation compounds are often used as suitable model for investigation of low dimensional systems and in particular for investigation of phase transition in such systems [42]. In Section 1.2.7, the MTP in QDs of graphite has been explored. Platinum Antimonide ($PtSb_2$) finds applications in device miniaturization, colloidal nanoparticle synthesis, sensors and detector materials and thermo-photovoltaic devices [43]. The MTP in QDs of p-$PtSb_2$ has been investigated in Section 1.2.8. Zero-gap compounds are used in optical wave-guide switch or modulators that can be fabricated by using the electro-optic and thermo-optic effects for facilitating optical communications and signal processing. The gapless materials also find extensive applications in infra-red detectors and night vision cameras [44]. Section 1.2.9 contains the study of MTP in QDs of the same taking p-HgTe as an example.

The II–V materials are used in photovoltaic cells constructed of single crystal materials in contact with electrolyte solutions. Cadmium selenide shows an open-circuit voltage of 0.8 V and power conservation coefficients near 6 percent for 720-nm light [45]. They are also used in ultrasonic amplification [46]. The thin film transistor using cadmium selenide as the semiconductor has been developed [47, 48]. In Section 1.2.10, the MTP in QDs of II–V materials has been studied taking CdSb as an example. Gallium antimonide (GaSb) finds applications in the fiber optic transmission window, heterojunctions and quantum wells. A complementary heterojunction field effect transistor (CHFET) in which the channels for the p-FET device and the n-FET device forming the complementary FET are formed from GaSb. The band gap energy of GaSb makes it suitable for low power operation [49]. In Section 1.2.11, the MTP in QDs of GaSb has been studied.

It may be noted that the stressed materials are being widely investigated for strained silicon transistors, quantum cascade lasers, semiconductor strain gauges, thermal detectors and strained-layer structures [50]. The MTP in QDs of stressed materials (taking stressed n-InSb as an example) has been investigated in Section 1.2.12. In recent years, Bismuth (Bi) nanolines are fabricated and Bi also finds use in array of antennas which leads to the interaction of electromagnetic waves with such Bi-nano-wires [51]. Several dispersion relations of the carriers have been proposed for Bi. Shoenberg [52] experimentally verified that the de Haas-Van Alphen and cyclotron resonance experiments supported the ellipsoidal parabolic model of Bi, although, the magnetic field dependence of many physical properties of Bi supports the two-band model [53]. The experimental investigations on the magneto-optical [54] and the ultrasonic quantum oscillations [55] support the Lax ellipsoidal non-parabolic model [53]. Kao [56], Dinger and Lawson [57] and Koch and Jensen [58] demonstrated that the Cohen model [59] is in conformity with the experimental results in a better way.

Besides, the Hybrid model of bismuth, as developed by Takoka *et al.* also finds use in the literature [60]. McClure and Choi [61] devised a new model of Bi and they showed that it can explain the data for a large number of magneto-oscillatory and resonance experiments. In Section 1.2.13, we have formulated the MTP in QDs of Bi in accordance with the aforementioned energy band models for the purpose of relative assessment.

Lead chalcogenides (PbTe, PbSe, and PbS) are IV–VI compounds whose studies over several decades have been motivated by their importance in infrared IR detectors, lasers, light-emitting devices, photovoltaics and high temperature thermoelectric [62]. PbTe, in particular, is the end compound of several ternary and quaternary high performance high temperature thermoelectric materials [63]. It has been used not only as bulk but also as films [64], quantum wells [65], superlattices [66], nanowires [67], and colloidal and embedded nanocrystals [68]. PbTe films doped with various impurities have also been investigated [69]. These studies revealed some of the interesting features that has been observed in bulk PbTe, such as Fermi level pinning and, in the case of superconductivity [70]. In Sections 1.2.14 and 1.2.15, the MTP in QDs of IV–VI materials and $Pb_{1-x}Ge_xTe$ has been studied.

The diphosphides find prominent role in biochemistry where the folding and structural stabilization of many important extra-cellular peptide and protein molecules, including hormones, enzymes, growth factors, toxins, and immunoglobulin are in concerned [71]. Besides, artificial introduction of extra diphosphides into peptides or proteins can improve biological activity [72] or confer thermo-stability [73]. The asymmetric diphosphide bond formation in peptides containing a free thiol group takes place over a wide pH range in aqueous buffers and can be crucially monitored by spectro-photometric titration of the released 3-nitro-2-pyridinethiol [74,75]. In Section 1.2.16, the MTP in QDs of zinc and cadmium diphosphides has been investigated.

Bismuth telluride (Bi_2Te_3) was first identified as a material for thermoelectric refrigeration in 1954 [76] and its physical properties were later improved by the addition of bismuth selenide and antimony telluride to form solid solutions [77]. The alloys of Bi_2Te_3 are very important compounds for the thermoelectric industry and have extensively been investigated in the literature [77]. In Section 1.2.17, the MTP in QDs of Bi_2Te_3 has been considered. In recent years, antimony has emerged to be very promising since glasses made from antimony are being extensively used in near infrared spectral range for third or second order nonlinear processes. The chalcogenide glasses are in general associated with high nonlinear properties for their Infrared transmission from 0.5–1μm to 12–18 μm [78]. Alloys of Sb are used as ultrahigh frequency indicators and in thin film thermocouple [78]. In Section 1.2.18, the MTP in QDs of Sb has been studied. Section 1.3 contains the detailed discussions of the twenty four different applications where we have also suggested the experimental methods of determinations of the Einstein's Relation, the Debye Screening Length and the carrier contribution to the elastic constants respectively. For the purpose of specific illustration, we have studied the different quantum signatures on the MTP in general in this context. Section 1.4 contains the results and discussion in this regard. Section 1.5 contains 15 open research problems in this context which are by far the most important part for the readers to initiate their own contribution in this pin pointed deep topic of 0D systems.

1.2 Theoretical Background

1.2.1 The generalized formulation of MTP

The MTP (G_0) can, in general, be expressed as [1]

$$G_0 = \frac{1}{e}\left(\frac{\partial \overline{S_0}}{\partial n_0}\right)_{E_F,T} \tag{1.1}$$

where E_F is the Fermi energy corresponding to the electron concentration n_0 and S_0 is the entropy per unit volume which can be written as

$$\overline{S_0} = -\left.\frac{\partial \Omega}{\partial T}\right|_{E=E_F} \tag{1.2}$$

in which T is temperature and Ω is the thermodynamic potential which, in turn, can be expressed in accordance with the Fermi–Dirac statistics as

$$\Omega = -k_B T \sum \ln\left|1 + \exp\left[\frac{E_F - E_{\delta_0}}{k_B T}\right]\right| \tag{1.3}$$

where the summation is carried out over all the possible δ_0 states and k_B is Boltzmann constant. Using (1.1)–(1.3), we get

$$G_0 = \frac{1}{eTn_0}\int_{-\infty}^{\infty}(E - E_F)R(E)\left[-\frac{\partial f_0}{\partial E}\right]dE \tag{1.4}$$

where $R(E)$ is the total number of states.

Equation (1.4) can be written under the condition of carrier degeneracy [1] as

$$G_0 = \left(\frac{\pi^2 k_B^2 T}{3en_0}\right)\left(\frac{\partial n_0}{\partial E_F}\right) \tag{1.5}$$

For inversion layers and NIPI structures, under the condition of electric quantum limit, (1.5) assumes the form

$$G_0 = \frac{\pi^2 k_B^2 T}{3en_{02\mathrm{D}}}\left[\frac{\partial n_{02\mathrm{D}}}{\partial(E_{F2\mathrm{D}} - E_{02\mathrm{D}})}\right] \tag{1.6}$$

where $n_{02\mathrm{D}}$, $E_{F2\mathrm{D}}$ and $E_{02\mathrm{D}}$ are the surface electron concentration, the Fermi energy and the sub-band energy for the said 2D systems at the electric quantum limit.

For heavily doped semiconductors (1.5) assumes the form

$$G_0 = \frac{\pi^2 k_B^2 T}{3en_{0\mathrm{HD}}}\left[\frac{\partial n_{0\mathrm{HD}}}{\partial(E_{\mathrm{FHD}} - E_{0\mathrm{HD}})}\right] \tag{1.7}$$

where $n_{0\mathrm{HD}}$, E_{FHD} and $E_{0\mathrm{HD}}$ are the electron concentration, the Fermi energy and the band tail energy for the heavily doped semiconductors where $E_{0\mathrm{HD}}$ should be obtained from the heavily doped dispersion relation of the semiconductor under the conditions $E = E_{0\mathrm{HD}}$ and $k = 0$.

It should be noted that being a thermodynamic relation and temperature induced phenomena, the MTP as expressed above in general, is valid for electronic materials having arbitrary dispersion relations and their nanostructures. In addition to bulk materials in the presence of classical large magnetic field, the expression of MTP is valid under one, two and three dimensional quantum confinement of the charge carriers (such as quantum wells in ultrathin films, NIPI structures, inversion and accumulation layers, quantum well superlattices, carbon nanotubes, quantum wires, quantum wire superlattices, quantum dots, magneto inversion and accumulation layers, quantum dot superlattices, magneto NIPIs, quantum well superlattices under magnetic quantization, ultrathin films under magnetic quantization, etc.). The formulation of G_0 requires the relation between electron statistics and the corresponding Fermi energy which is basically the band structure dependent quantity and changes under different physical conditions. It is worth remarking to note that the number $(\pi^2/3)$ has occurred as a consequence of mathematical analysis and is not connected with the well-known Lorenz number. For quantum wells in ultrathin films, NIPI structures, inversion and accumulation layers, quantum well superlattices, magneto inversion and accumulation layers, magneto NIPIs, quantum well superlattices under magnetic quantization and magneto size quantization, the carrier concentration is measured per unit area whereas, for quantum wires, quantum wires under magnetic field, quantum wire superlattices and such allied systems, the same can be measured per unit length. Besides, for bulk materials under strong magnetic field, quantum dots, quantum dots under magnetic field, quantum dot superlattices and quantum dot superlattices under magnetic field, the carrier concentration is expressed per unit volume.

1.2.2 The MTP in quantum dots of III–V materials

(a) The model of the simplified parabolic energy bands

In this case the electron dispersion relation is given by

$$E = \frac{\hbar^2 k^2}{2m_c} \tag{1.8}$$

where E is the electron energy as measured from at the edge of the conduction band in the vertically upward direction in the absence of any quantization, $\hbar = (h/2\pi)$, h is Planck's constant and m_c is the effective electron mass at the edge of the conduction band.

In the presence of 3D quantization as in quantum dots (1.8), the totally quantized energy (E_1) assumes the form

$$E_1 = \frac{\hbar^2 \pi^2}{2m_c}[(n_x/d_x)^2 + (n_y/d_y)^2 + (n_z/d_z)^2] \tag{1.9}$$

where n_i and $d_i (i = x, y, z)$ are size quantum number and nano thickness in the ith direction respectively.

The density-of-states function assumes the form

$$N(E) = \frac{2g_v}{d_x d_y d_z} \sum_{n_x=1}^{n_{x\max}} \sum_{n_y=1}^{n_{y\max}} \sum_{n_z=1}^{n_{z\max}} \delta'(E - E_1) \tag{1.10}$$

where δ' is the Dirac delta function and g_v is valley degeneracy.

The electron concentration is given by

$$n_{0D} = \frac{2g_v}{d_x d_y d_z} \sum_{n_x=1}^{n_{x\max}} \sum_{n_y=1}^{n_{y\max}} \sum_{n_z=1}^{n_{z\max}} \int_{E_{1,1}}^{\infty} \delta'(E - E_1) f_0 dE \tag{1.11}$$

where

$$f_0 = \left[1 + \exp\left(\frac{E - E_F}{k_B T}\right)\right]^{-1} \quad \text{and} \quad E_F$$

is the electron energy as measured from at the edge of the conduction band in the vertically upward direction in the absence of any quantization.

From (1.11), we can write

$$n_{0\text{D}} = \frac{2g_v}{d_x d_y d_z} \sum_{n_x=1}^{n_{x\max}} \sum_{n_y=1}^{n_{y\max}} \sum_{n_z=1}^{n_{z\max}} F_{-1}(\eta_{1,1}) \tag{1.12}$$

where $\eta_{1,1} = (k_B T)^{-1}[E_F - E_1]$ and $F_{-1}(\eta_{1,1})$ is the special case of the one parameter Fermi–Dirac integral of order j which can be written as [79]

$$F_j(\eta) = \left(\frac{1}{\Gamma(j+1)}\right) \int_0^\infty \frac{x^j\,dx}{1 + \exp(x - \eta)} \tag{1.13}$$

or for all j, analytically continued as a complex contour integral around the negative x-axis

$$F_j(\eta) = \left(\frac{\Gamma(-j)}{2\pi\sqrt{-1}}\right) \int_{-\infty}^{+0} \frac{x^j\,dx}{1 + \exp(-x - \eta)} \tag{1.14}$$

where

$$\Gamma(j+1) = j\Gamma(j), \quad \Gamma\left(\frac{1}{2}\right) = \sqrt{\pi} \quad \text{and} \quad \Gamma(0) = 1$$

Incidentally, few important properties of the Fermi–Dirac integral are listed in this context without proof [79] since these are very useful in obtaining direct results, without cumbersome mathematics and needed for further progress.

$$1.\ \frac{d}{d\eta}[F_j(\eta)] = F_{j-1}(\eta) \tag{1.15a}$$

$$2.\ \int F_j(\eta)d\eta = F_{j+1}(\eta) \tag{1.15b}$$

$$3.\ F_{1/2}(\eta) \approx \frac{4}{3\sqrt{\pi}}\eta^{3/2}\left(1 + \frac{\pi^2}{8\eta^2}\right), \eta > 0 \tag{1.15c}$$

$$4.\ F_0(\eta) = \ln|1 + \exp(\eta)| \tag{1.15d}$$

$$5.\ F_j(\eta) \approx e^n, n < 0 \quad \text{for all } j \tag{1.15e}$$

Using (1.5) and (1.12), we get

$$G_0 = \frac{\pi^2 k_B}{3e} \left[\sum_{n_x=1}^{n_{x_{\max}}} \sum_{n_y=1}^{n_{y_{\max}}} \sum_{n_z=1}^{n_{z_{\max}}} F_{-1}(\eta_{1,1}) \right]^{-1} \left[\sum_{n_x=1}^{n_{x_{\max}}} \sum_{n_y=1}^{n_{y_{\max}}} \sum_{n_z=1}^{n_{z_{\max}}} F_{-2}(\eta_{1,1}) \right] \tag{1.16}$$

The dispersion relation of the conduction electrons of III–V compounds are described by the models of Kane (both two and three bands) [79], Stillman *et al.* [80], Newson and Kurobe [81], Rossler [82], Palik *et al.* [83], Johnson and Dickey [84], Agafonov *et al.* [85], Kolodziejczak *et al.* [J. Kolodziejczak, S. Zukotynski and H. Stramska, *Phys. Stat. Sol.*, 14, 471 (1966)], Kyrachkov *et al.* [S. V. Kryachkov and V. I. Porkhun, *Sov. Phys. Semicond.*, 12, 1206 (1978)] and Chaikovskii [I. A. Chaikovskii, *Sov. Phys. Semicond.*, 6, 1 (1972)] respectively. For the purpose of complete and coherent presentation, the MTP in QDs of III–V compounds have also been investigated in accordance with the aforementioned different dispersion relations for the purpose of relative comparison as follows:

(b) The two band model of Kane

Under the inequalities $\Delta >> E_{g_0}$ or $\Delta << E_{g_0}$ (E_{g_0} is the band gap in the absence of any field and Δ is the spin orbit splitting of the valance band), the Dispersion Relation (DR) assumes the form

$$E(1 + \alpha E) = (\hbar^2 k^2 / 2m_c), \quad \alpha \equiv 1/E_{g_0} \tag{1.17}$$

Equation (1.17) is known as the two-band model of Kane and should be as such for studying the electronic properties of the materials whose band-structures obey the above inequalities [33].

The total quantized energy E_2 in this case is given by

$$E_2(1 + \alpha E_2) = \frac{\hbar^2 \pi^2}{2m_c} \left[\left(\frac{n_x}{d_x} \right)^2 + \left(\frac{n_y}{d_y} \right)^2 + \left(\frac{n_z}{d_z} \right)^2 \right] \tag{1.18}$$

The electron concentration can be written as

$$n_0 = \frac{2g_v}{d_x d_y d_z} \sum_{n_x=1}^{n_{x_{\max}}} \sum_{n_y=1}^{n_{y_{\max}}} \sum_{n_z=1}^{n_{z_{\max}}} F_{-1}(\eta_{1,2}) \tag{1.19}$$

where

$$\eta_{1,2} = (k_B T)^{-1} (E_F - E_2)$$

Using (1.5) and (1.19), we get

$$G_0 = \frac{\pi^2 k_B}{3e}\left[\sum_{n_x=1}^{n_{x_{\max}}}\sum_{n_y=1}^{n_{y_{\max}}}\sum_{n_z=1}^{n_{z_{\max}}} F_{-1}(\eta_{1,2})\right]^{-1}\left[\sum_{n_x=1}^{n_{x_{\max}}}\sum_{n_y=1}^{n_{y_{\max}}}\sum_{n_z=1}^{n_{z_{\max}}} F_{-2}(\eta_{1,2})\right] \tag{1.20}$$

(c) The three band model of Kane

In accordance with three band model of Kane the DR in III–V materials can be expressed [79] as

$$\frac{\hbar^2 k^2}{2m_c} = I(E), I(E) \equiv \frac{E(E+E_{g_0})(E+E_{g_0}+\Delta)(E_{g_0}+\frac{1}{3}\Delta)}{E_{g_0}(E_{g_0}+\Delta)(E+E_{g_0}+\frac{2}{3}\Delta)} \tag{1.21}$$

which is known as the three band model of Kane [79] and is often used to study the electronic properties of III–V materials.

The total quantized energy (E_3) in this case assumes the form

$$I(E_3) = \frac{\hbar^2\pi^2}{2m_c}\left[\left(\frac{n_x}{d_x}\right)^2 + \left(\frac{n_y}{d_y}\right)^2 + \left(\frac{n_z}{d_z}\right)^2\right] \tag{1.22}$$

The electron concentration is given by

$$n_0 = \frac{2g_v}{d_x d_y d_z}\sum_{n_x=1}^{n_{x_{\max}}}\sum_{n_y=1}^{n_{y_{\max}}}\sum_{n_z=1}^{n_{z_{\max}}} F_{-1}(\eta_{1,3}) \tag{1.23}$$

where

$$\eta_{1,3} = (k_B T)^{-1}(E_F - E_3)$$

Using (1.5) and (1.23), we get

$$G_0 = \frac{\pi^2 k_B}{3e}\left[\sum_{n_x=1}^{n_{x_{\max}}}\sum_{n_y=1}^{n_{y_{\max}}}\sum_{n_z=1}^{n_{z_{\max}}} F_{-1}(\eta_{1,3})\right]^{-1}\left[\sum_{n_x=1}^{n_{x_{\max}}}\sum_{n_y=1}^{n_{y_{\max}}}\sum_{n_z=1}^{n_{z_{\max}}} F_{-2}(\eta_{1,3})\right] \tag{1.24}$$

(d) The model of Stillman *et al.*

In accordance with the model of Stillman *et al.* [80], the electron dispersion law of III–V materials assumes the form

$$E = t_{11}k^2 - t_{12}k^4 \tag{1.25}$$

where

$$t_{11} \equiv \frac{\hbar^2}{2m_c} \quad \text{and} \quad t_{12} \equiv \left(1 - \frac{m_c}{m_0}\right)^2 \left(\frac{\hbar^2}{2m_c}\right)^2 \left[\left(3E_{g_0} + 4\Delta + \frac{2\Delta^2}{E_{g_0}}\right) \times \{(E_{g_0} + \Delta)(2\Delta + 3E_{g_0})\}^{-1}\right]$$

Equation (1.25) can be expressed as

$$\frac{\hbar^2 k^2}{2m_c} = I_{11}(E) \tag{1.26}$$

where

$$I_{11}(E) \equiv a_{11}[1 - (1 - a_{12}E)^{1/2}], \quad a_{11} \equiv \left(\frac{\hbar^2 t_{11}}{4m_c t_{12}}\right) \quad \text{and} \quad a_{12} \equiv \frac{4t_{12}}{t_{11}^2}$$

The E_4 in this case can be defined as

$$I_{11}(E_4) = \frac{\hbar^2\pi^2}{2m_c}\left[\left(\frac{n_x}{d_x}\right)^2 + \left(\frac{n_y}{d_y}\right)^2 + \left(\frac{n_z}{d_z}\right)^2\right] \tag{1.27}$$

The electron concentration is given by

$$n_0 = \frac{2g_v}{d_x d_y d_z} \sum_{n_x=1}^{n_{x_{\max}}} \sum_{n_y=1}^{n_{y_{\max}}} \sum_{n_z=1}^{n_{z_{\max}}} F_{-1}(\eta_{1,4}) \tag{1.28}$$

where

$$\eta_{1,4} = (k_B T)^{-1}(E_F - E_4)$$

Using (1.5) and (1.28), we get

$$G_0 = \frac{\pi^2 k_B}{3e}\left[\sum_{n_x=1}^{n_{x_{\max}}} \sum_{n_y=1}^{n_{y_{\max}}} \sum_{n_z=1}^{n_{z_{\max}}} F_{-1}(\eta_{1,4})\right]^{-1} \left[\sum_{n_x=1}^{n_{x_{\max}}} \sum_{n_y=1}^{n_{y_{\max}}} \sum_{n_z=1}^{n_{z_{\max}}} F_{-2}(\eta_{1,4})\right] \tag{1.29}$$

(e) The model of Newson and Kurobe

In accordance with the model of Newson and Kurobe, the electron dispersion law in this case assumes the form as [81]

$$E = a_{13}k_z^4 + \left[\frac{\hbar^2}{2m_c} + a_{14}k_s^2\right]k_z^2 + \frac{\hbar^2}{2m_c}k_s^2 + a_{14}k_x^2 k_y^2 + a_{13}\left(k_x^4 + k_y^4\right) \tag{1.30}$$

where a_{13} is the non-parabolicity constant, $a_{14}(\equiv 2a_{13} + a_{15})$ and a_{15} is known as the warping parameter.

The total quantized energy E_5 in this case can be written as

$$E_5 = a_{13}\left(\frac{\pi n_z}{d_z}\right)^4 + \left[\frac{\hbar^2}{2m_c} + a_{14}\left(\left(\frac{\pi n_x}{d_x}\right)^2 + \left(\frac{\pi n_y}{d_y}\right)^2\right)\right] \cdot \left(\frac{\pi n_z}{d_z}\right)^2 + \frac{\hbar^2}{2m_c}\left[\left(\left(\frac{\pi n_x}{d_x}\right)^2 + \left(\frac{\pi n_y}{d_y}\right)^2\right)\right] + a_{14}\pi^4\left(\frac{n_x n_y}{d_x d_y}\right)^2 + a_{13}\pi^4\left[\left(\frac{n_x}{d_x}\right)^4 + \left(\frac{n_y}{d_y}\right)^4\right] \tag{1.31}$$

The electron concentration is given by

$$n_0 = \frac{2g_v}{d_x d_y d_z}\sum_{n_x=1}^{n_{x_{\max}}}\sum_{n_y=1}^{n_{y_{\max}}}\sum_{n_z=1}^{n_{z_{\max}}} F_{-1}(\eta_{1,5}) \tag{1.32}$$

where

$$\eta_{1,5} = (k_B T)^{-1}(E_F - E_5)$$

Using (1.5) and (1.32), we get

$$G_0 = \frac{\pi^2 k_B}{3e}\left[\sum_{n_x=1}^{n_{x_{\max}}}\sum_{n_y=1}^{n_{y_{\max}}}\sum_{n_z=1}^{n_{z_{\max}}} F_{-1}(\eta_{1,5})\right]^{-1}\left[\sum_{n_x=1}^{n_{x_{\max}}}\sum_{n_y=1}^{n_{y_{\max}}}\sum_{n_z=1}^{n_{z_{\max}}} F_{-2}(\eta_{1,5})\right] \tag{1.33}$$

(f) The model of Rossler

The dispersion relation of the conduction electrons in accordance with the model of Rossler can be written as [82]

$$E = \frac{\hbar^2 k^2}{2m_c} + [\alpha_{11} + \alpha_{12}k]k^4 + (\beta_{11} + \beta_{12}k)[k_x^2 k_y^2 + k_y^2 k_z^2 + k_z^2 k_x^2] \pm [\gamma_{11} + \gamma_{12}k][k^2(k_x^2 k_y^2 + k_y^2 k_z^2 + k_z^2 k_x^2) - 9k_x^2 k_y^2 k_z^2]^{1/2} \tag{1.34}$$

where $\alpha_{11}, \alpha_{12}, \beta_{11}, \beta_{12}, \gamma_{11}$ and γ_{12} are energy–band constants.

The total quantized energy $E_{6,\pm}$ in this case can be written as

$$E_{6,\pm} \equiv \frac{\hbar^2\pi^2}{2m_c}\left[\left(\frac{n_x}{d_x}\right)^2+\left(\frac{n_y}{d_y}\right)^2+\left(\frac{n_z}{d_z}\right)^2\right]$$
$$+\left[\alpha_{11}+\alpha_{12}\left[\left(\frac{\pi n_x}{d_x}\right)^2+\left(\frac{\pi n_y}{d_y}\right)^2+\left(\frac{\pi n_z}{d_z}\right)^2\right]^{1/2}\right]$$
$$\left[\left(\frac{\pi n_x}{d_x}\right)^2+\left(\frac{\pi n_y}{d_y}\right)^2+\left(\frac{\pi n_z}{d_z}\right)^2\right]^2$$
$$+\left[\beta_{11}+\beta_{12}\left[\left(\frac{\pi n_x}{d_x}\right)^2+\left(\frac{\pi n_y}{d_y}\right)^2+\left(\frac{\pi n_z}{d_z}\right)^2\right]^{1/2}\right]$$
$$\left[\pi^4\left(\frac{n_x n_y}{d_x d_y}\right)^2+\pi^4\left(\frac{n_y n_z}{d_y d_z}\right)^2+\pi^4\left(\frac{n_z n_x}{d_z d_x}\right)^2\right]$$
$$\pm\left[\gamma_{11}+\gamma_{12}\left[\left(\frac{\pi n_x}{d_x}\right)^2+\left(\frac{\pi n_y}{d_y}\right)^2+\left(\frac{\pi n_z}{d_z}\right)^2\right]^{1/2}\right]$$
$$\left[\left[\left(\frac{\pi n_x}{d_x}\right)^2+\left(\frac{\pi n_y}{d_y}\right)^2+\left(\frac{\pi n_z}{d_z}\right)^2\right]\left[\pi^4\left(\frac{n_x n_y}{d_x d_y}\right)^2\right.\right.$$
$$\left.\left.+\pi^4\left(\frac{n_y n_z}{d_y d_z}\right)^2+\pi^4\left(\frac{n_z n_x}{d_z d_x}\right)^2\right]-9\pi^6(n_x n_y n_z/d_x d_y d_z)^6\right] \tag{1.35}$$

The electron concentration is given by

$$n_0=\frac{g_v}{d_x d_y d_z}\sum_{n_x=1}^{n_{x_{\max}}}\sum_{n_y=1}^{n_{y_{\max}}}\sum_{n_z=1}^{n_{z_{\max}}}F_{-1}(\eta_{1,6\pm}) \tag{1.36}$$

where

$$\eta_{1,6\pm}=(k_BT)^{-1}(E_F-E_{6\pm})$$

Using (1.5) and (1.36), we get

$$G_0=\frac{\pi^2 k_B}{3e}\left[\sum_{n_x=1}^{n_{x_{\max}}}\sum_{n_y=1}^{n_{y_{\max}}}\sum_{n_z=1}^{n_{z_{\max}}}F_{-1}(\eta_{1,6\pm})\right]^{-1}\left[\sum_{n_x=1}^{n_{x_{\max}}}\sum_{n_y=1}^{n_{y_{\max}}}\sum_{n_z=1}^{n_{z_{\max}}}F_{-2}(\eta_{1,6\pm})\right] \tag{1.37}$$

(g) The model of Palik *et al.*

In accordance with the model of Palik *et al.* [83], the energy spectrum of the conduction electrons in III–V materials, up to the fourth order in effective mass theory, taking into account the interactions the heavy hole, the light hole and the split-off bands can be written as [83]

$$E = \frac{\hbar^2 k^2}{2m_c} - b_{11}k^4 \tag{1.38}$$

where

$$b_{11} \equiv \left[\frac{\hbar^4}{4E_{g_0}(m_c)^2}\right]\left[\frac{1+\frac{x_{11}^2}{2}}{1+\frac{x_{11}}{2}}\right](1-y_1)^2, \quad x_{11} \equiv \left[1+\left(\frac{\Delta}{E_{g_0}}\right)\right]^{-1}$$

and

$$y_1 \equiv m_c/m_0$$

From (1.38), we get

$$\frac{\hbar^2 k^2}{2m_c} = I_{12}(E) \tag{1.39}$$

where

$$I_{12}(E) \equiv b_{12}[a_{12} - \sqrt{a_{12}^2 - 4Eb_{11}}], \quad b_{12} \equiv \frac{a_{12}}{2b_{11}} \quad \text{and} \quad a_{12} \equiv \frac{\hbar^2}{2m_c}$$

The total quantized energy (E_7) in this case can be defined as

$$I_{12}(E_7) = \frac{\hbar^2}{2m_c}\left[\left(\frac{\pi n_x}{d_x}\right)^2 + \left(\frac{\pi n_y}{d_y}\right)^2 + \left(\frac{\pi n_z}{d_z}\right)^2\right] \tag{1.40}$$

The electron concentration is given by

$$n_0 = \frac{2g_v}{d_x d_y d_z}\sum_{n_x=1}^{n_{x_{\max}}}\sum_{n_y=1}^{n_{y_{\max}}}\sum_{n_z=1}^{n_{z_{\max}}} F_{-1}(\eta_{1,7}) \tag{1.41}$$

where

$$\eta_{1,7} = (k_B T)^{-1}(E_F - E_7)$$

Using (1.5) and (1.41), we get

$$G_0 = \frac{\pi^2 k_B}{3e}\left[\sum_{n_x=1}^{n_{x_{\max}}}\sum_{n_y=1}^{n_{y_{\max}}}\sum_{n_z=1}^{n_{z_{\max}}} F_{-1}(\eta_{1,7})\right]^{-1}\left[\sum_{n_x=1}^{n_{x_{\max}}}\sum_{n_y=1}^{n_{y_{\max}}}\sum_{n_z=1}^{n_{z_{\max}}} F_{-2}(\eta_{1,7})\right] \tag{1.42}$$

(h) The model of Johnson and Dickey

In accordance with the model of Johnson and Dickey [84] the electron dispersion law in III–V materials assumes the form

$$E = -\frac{E_{g_0}}{2} + \frac{\hbar^2 k^2}{2}\left[\frac{1}{m_c} - \frac{1}{m_0}\right] + \frac{E_{g_0}}{2}[1 + a_{15}(E)k^2]^{1/2} \tag{1.43}$$

where

$$a_{15}(E) \equiv \frac{4\hbar^2}{2m_c E_{g_0}} a_{14}(E), \quad a_{14}(E) \equiv \frac{(E_{g_0} + \Delta)\left(E + E_{g_0} + \frac{2}{3}\Delta\right)}{\left(E_{g_0} + \frac{2}{3}\Delta\right)(E + E_{g_0} + \Delta)},$$

$$\frac{m_0}{m_c} = (\overline{P})^2 \left[\frac{\left(E_{g_0} + \frac{2}{3}\Delta\right)}{(E_{g_0})(E_{g_0} + \Delta)}\right]$$

and $(\overline{P})$ is the energy band constant in this case.

The E_8 in this case is given by

$$E_8 = -\frac{E_{g_0}}{2} + \frac{\hbar^2}{2}\left[\frac{1}{m_c} - \frac{1}{m_0}\right]\left[\left(\frac{\pi n_x}{d_x}\right)^2 + \left(\frac{\pi n_y}{d_y}\right)^2 + \left(\frac{\pi n_z}{d_z}\right)^2\right]$$
$$+ \frac{E_{g_0}}{2}\left[1 + a_{15}(E_8)\left[\left(\frac{\pi n_x}{d_x}\right)^2 + \left(\frac{\pi n_y}{d_y}\right)^2 + \left(\frac{\pi n_z}{d_z}\right)^2\right]\right]^{1/2} \tag{1.44}$$

The electron concentration can be written as

$$n_0 = \frac{2g_v}{d_x d_y d_z} \sum_{n_x=1}^{n_{x_{\max}}} \sum_{n_y=1}^{n_{y_{\max}}} \sum_{n_z=1}^{n_{z_{\max}}} F_{-1}(\eta_{1,8}) \tag{1.45}$$

where

$$\eta_{1,8} = (k_B T)^{-1}(E_F - E_8)$$

Using (1.5) and (1.45), we get

$$G_0 = \frac{\pi^2 k_B}{3e}\left[\sum_{n_x=1}^{n_{x_{\max}}} \sum_{n_y=1}^{n_{y_{\max}}} \sum_{n_z=1}^{n_{z_{\max}}} F_{-1}(\eta_{1,8})\right]^{-1}\left[\sum_{n_x=1}^{n_{x_{\max}}} \sum_{n_y=1}^{n_{y_{\max}}} \sum_{n_z=1}^{n_{z_{\max}}} F_{-2}(\eta_{1,8})\right] \tag{1.46}$$

(i) The model of Agafonov *et al.*

In accordance with the model of Agafonov *et al.* [85], the electron dispersion law in III–V materials can be written as

$$E = \frac{\overline{y} - E_{g_0}}{2}\left[1 - T_5\left(\frac{k_x^4 + k_y^4 + k_z^4}{\overline{y}k^2}\right)\right] \tag{1.47}$$

where

$$(\overline{y})^2 = \left[E_{g_0}^2 + \frac{8}{3}P_0^2k^2\right], \quad T_5 = \frac{\overline{D}\sqrt{3} - 3\overline{B}}{2}, \quad \overline{D} = -40\left(\frac{\hbar^2}{2m_0}\right),$$

$$\overline{B} = -21\left(\frac{\hbar^2}{2m_0}\right)$$

and P_0 is the momentum matrix element.

The total quantized energy (E_9) in this case can be written as

$$E_9 = \left(\frac{\psi_{30} - E_{g_0}}{2}\right)\left[1 - \left[T_5\psi_{30}\left[\left(\frac{\pi n_x}{d_x}\right)^4 + \left(\frac{\pi n_y}{d_y}\right)^4 + \left(\frac{\pi n_z}{d_z}\right)^4\right]\right] \cdot \left[\psi_{30}\left[\left(\frac{\pi n_x}{d_x}\right)^2 + \left(\frac{\pi n_y}{d_y}\right)^2 + \left(\frac{\pi n_z}{d_z}\right)^2\right]\right]^{-1}\right] \tag{1.48}$$

where

$$\psi_{30} = \left[E_{g_0}^2 + \frac{8}{3}P_0^2\left[\left(\frac{\pi n_x}{d_x}\right)^2 + \left(\frac{\pi n_y}{d_y}\right)^2 + \left(\frac{\pi n_z}{d_z}\right)^2\right]\right]^{1/2}$$

The electron concentration is given by

$$n_0 = \frac{2g_v}{d_x d_y d_z}\sum_{n_x=1}^{n_{x_{\max}}}\sum_{n_y=1}^{n_{y_{\max}}}\sum_{n_z=1}^{n_{z_{\max}}} F_{-1}(\eta_{1,9}) \tag{1.49}$$

where

$$\eta_{1,9} = (k_B T)^{-1}(E_F - E_9)$$

Using (1.5) and (1.49), we get

$$G_0 = \frac{\pi^2 k_B}{3e}\left[\sum_{n_x=1}^{n_{x_{\max}}}\sum_{n_y=1}^{n_{y_{\max}}}\sum_{n_z=1}^{n_{z_{\max}}} F_{-1}(\eta_{1,9})\right]^{-1}\left[\sum_{n_x=1}^{n_{x_{\max}}}\sum_{n_y=1}^{n_{y_{\max}}}\sum_{n_z=1}^{n_{z_{\max}}} F_{-2}(\eta_{1,9})\right] \tag{1.50a}$$

(j) The model of Kolodziejczak *et al.*

In accordance with the model of Kolodziejczak *et al.* [242], the electron energy spectrum of III–V compounds can be expressed, taking into account the interaction of the higher bands as

$$E = \frac{\hbar^2 k^2}{2m_0} + \frac{\chi_0 - E_g}{2} + a\left[\frac{\chi_0 + E_g}{2\chi_0}\right]\frac{\hbar^2 k^2}{2m_0} + \left[\frac{\chi_0 - E_g}{2}\right]\frac{\hbar^2 k^2}{2m_0}\left[b + \frac{c}{5}\right] \tag{1.50b}$$

where

$$\chi_0 \equiv \left[E_g^2 - 4FE_g\left(\frac{E_g + \frac{2}{3}\Delta}{E_g + \Delta}\right)\left(\frac{\hbar^2 k^2}{2m_c}\right)\right]^{1/2}$$

F represents the interaction between the $\Gamma_{25'}$ and Γ_{15} states,

$$a \equiv \left[\frac{-p^2[E(\Gamma_{15c}) - E(\Gamma_{15v})] - [E(\Gamma_{15c}) - E(\Gamma_{25v})]}{2[E(\Gamma_{15c}) - E(\Gamma_{15v})][E(\Gamma_{15c}) - E(\Gamma_{15v}) - E_g]}\right], \quad b \equiv \frac{1}{3}[M + 4G]$$

M represents the interaction between $\Gamma_{25'}$, and Γ_{15} states, G represents the interaction between $\Gamma_{25'}$ and $\Gamma_{12'}$ states,

$$c \equiv \frac{1}{2}\left[\frac{(F - G + M)^2 - (F + 2G - M)^2}{F + 2G - M}\right]$$

and the other notations are the same as in the above reference.

For $n - AlSb$, $E(\Gamma_{1c}) - E(\Gamma_{15v}) = 1.80\,\text{eV}$, $\delta = 0.075\,\text{eV}$, $E(\Gamma_{15c}) - E(\Gamma_{15v}) = 4.30\,\text{eV}$, $E(\Gamma_{15}) - E(\Gamma_{25'}) = 3.21\,\text{eV}$, $F = -12.610$, $M = -2.235$, $G = -1.031$.

The QDDR in this case assumes the form

$$\begin{aligned} E_{150} = {} & \frac{\hbar^2 k_1^2}{2m_0} + \frac{\chi_{0,1} - E_g}{2} + a\left[\frac{\chi_{0,1} + E_g}{2\chi_{0,1}}\right]\frac{\hbar^2 k_1^2}{2m_0} \\ & + \left[\frac{\chi_{0,1} - E_g}{2}\right]\frac{\hbar^2 k_1^2}{2m_0}\left[b + \frac{c}{5}\right] \end{aligned} \tag{1.50c}$$

where

$$\chi_{0,1} \equiv \left[E_g^2 - 4FE_g\left(\frac{E_g + \frac{2}{3}\Delta}{E_g + \Delta}\right)\left(\frac{\hbar^2 k_1^2}{2m_c}\right)\right]^{1/2}$$

E_{150} is the totally quantized energy in this case and

$$k_1^2 = \left(\frac{\pi n_x}{d_x}\right)^2 + \left(\frac{\pi n_y}{d_y}\right)^2 + \left(\frac{\pi n_z}{d_z}\right)^2$$

The electron concentration is given by

$$n_0 = \frac{2g_v}{d_x d_y d_z} \sum_{n_x=1}^{n_{x_{\max}}} \sum_{n_y=1}^{n_{y_{\max}}} \sum_{n_z=1}^{n_{z_{\max}}} F_{-1}(\eta_{1,9,150}) \tag{1.50d}$$

where

$$\eta_{1,9,150} = (k_B T)^{-1}(E_F - E_{150})$$

Using (1.5) and (1.50d), we get

$$G_0 = \frac{\pi^2 k_B}{3e} \left[\sum_{n_x=1}^{n_{x_{\max}}} \sum_{n_y=1}^{n_{y_{\max}}} \sum_{n_z=1}^{n_{z_{\max}}} F_{-1}(\eta_{1,9,150}) \right]^{-1} \times \left[\sum_{n_x=1}^{n_{x_{\max}}} \sum_{n_y=1}^{n_{y_{\max}}} \sum_{n_z=1}^{n_{z_{\max}}} F_{-2}(\eta_{1,9,150}) \right] \tag{1.50e}$$

(k) The model of Kryachkov *et al.*

The E–k relation of the conduction electrons in semiconductors in the presence of electron-phonon interaction assumes the form [243]

$$E = \frac{\hbar^2 k^2}{2m_c} - \alpha_c \hbar\omega_0 \frac{p_0}{\hbar k} \tan^{-1}\left[\frac{\hbar^2 k^2}{2m_c(\hbar\omega_0 - E)} \right]^{1/2} \tag{1.50f}$$

where α_c is the dimensionless coupling constant, $\rho_0 = (2m_c\hbar\omega_0)^{1/2}$ and ω_0 is the angular frequency of the optical phonon.

The QDDR assumes the form

$$E_{151} = \frac{\hbar^2 k_1^2}{2m_c} - \alpha_c \hbar\omega_0 \frac{p_0}{\hbar k_1} \tan^{-1}\left[\frac{\hbar^2 k_1^2}{2m_c(\hbar\omega_0 - E_{151})} \right]^{1/2} \tag{1.50g}$$

where E_{151} is the totally quantized energy in this case.

The electron concentration is given by

$$n_0 = \frac{2g_v}{d_x d_y d_z} \sum_{n_x=1}^{n_{x_{\max}}} \sum_{n_y=1}^{n_{y_{\max}}} \sum_{n_z=1}^{n_{z_{\max}}} F_{-1}(\eta_{1,9,151}) \tag{1.50h}$$

where

$$\eta_{1,9,151} = (k_B T)^{-1}(E_F - E_{151})$$

Using (1.5) and (1.50h), we get

$$G_0 = \frac{\pi^2 k_B}{3e} \left[\sum_{n_x=1}^{n_{x_{\max}}} \sum_{n_y=1}^{n_{y_{\max}}} \sum_{n_z=1}^{n_{z_{\max}}} F_{-1}(\eta_{1,9,151}) \right]^{-1} \times \left[\sum_{n_x=1}^{n_{x_{\max}}} \sum_{n_y=1}^{n_{y_{\max}}} \sum_{n_z=1}^{n_{z_{\max}}} F_{-2}(\eta_{1,9,151}) \right] \tag{1.50i}$$

(l) The model of Chaikovskii

The DR in the presence of a strong oscillatory electric field along z-axis for a semiconductor having parabolic energy band is given by [244]

$$E = \frac{\hbar^2 k^2}{2m_c} + \frac{e^2 E_0^2}{4(\omega_0)^2 m_c} \tag{1.50j}$$

where E_0 and ω_0 are the amplitude and the angular frequency of the electric field.

The QDDR can be expressed as

$$E_{152} = \frac{\hbar^2 k_1^2}{2m_c} + \frac{e^2 E_0^2}{4(\omega_0)^2 m_c} \tag{1.50k}$$

where E_{152} is the totally quantized energy in this case.

The electron concentration is given by

$$n_0 = \frac{2g_v}{d_x d_y d_z} \sum_{n_x=1}^{n_{x_{\max}}} \sum_{n_y=1}^{n_{y_{\max}}} \sum_{n_z=1}^{n_{z_{\max}}} F_{-1}(\eta_{1,9,152}) \tag{1.50l}$$

where

$$\eta_{1,9,152} = (k_B T)^{-1}(E_F - E_{152})$$

Using (1.5) and (1.50l), we get

$$G_0 = \frac{\pi^2 k_B}{3e} \left[\sum_{n_x=1}^{n_{x_{\max}}} \sum_{n_y=1}^{n_{y_{\max}}} \sum_{n_z=1}^{n_{z_{\max}}} F_{-1}(\eta_{1,9,152}) \right]^{-1} \times \left[\sum_{n_x=1}^{n_{x_{\max}}} \sum_{n_y=1}^{n_{y_{\max}}} \sum_{n_z=1}^{n_{z_{\max}}} F_{-2}(\eta_{1,9,152}) \right] \tag{1.50m}$$

1.2.3 The MTP in quantum dots of II–VI materials

(a) The Hopfield model

The carrier energy spectra in bulk specimens of II–VI compounds in accordance with Hopfield model can be written as [33]

$$E = a_o' k_s^2 + b_o' k_z^2 \pm \bar{\lambda}_o k_s \tag{1.51}$$

where

$$a_0' \equiv \hbar^2/2m_\perp^*, \quad b_0' \equiv \hbar^2/2m_\parallel^* \quad \text{and} \quad \bar{\lambda}_o$$

represents the splitting of the two – spin states by the spin orbit coupling and the crystalline field.

Using (1.51), the total quantized energy ($E_{QD10,\pm}$) in this case can be expressed as

$$E_{10,\pm} = a_0' \left[\left(\frac{\pi n_x}{d_x} \right)^2 + \left(\frac{\pi n_y}{d_y} \right)^2 \right] + b_0' \left(\frac{\pi n_z}{d_z} \right)^2$$
$$\pm \bar{\lambda}_o \left[\left(\frac{\pi n_x}{d_x} \right)^2 + \left(\frac{\pi n_y}{d_y} \right)^2 \right]^{1/2} \quad (1.52)$$

The carrier concentration is given by

$$n_0 = \frac{g_v}{d_x d_y d_z} \sum_{n_x=1}^{n_{x_{\max}}} \sum_{n_y=1}^{n_{y_{\max}}} \sum_{n_z=1}^{n_{z_{\max}}} F_{-1}(\eta_{1,10}) \quad (1.53)$$

where

$$\eta_{1,10} = (k_B T)^{-1} (E_F - E_{10,\pm})$$

Using (1.5) and (1.53), we get

$$G_0 = \frac{\pi^2 k_B}{3e} \left[\sum_{n_x=1}^{n_{x_{\max}}} \sum_{n_y=1}^{n_{y_{\max}}} \sum_{n_z=1}^{n_{z_{\max}}} F_{-1}(\eta_{1,10}) \right]^{-1} \left[\sum_{n_x=1}^{n_{x_{\max}}} \sum_{n_y=1}^{n_{y_{\max}}} \sum_{n_z=1}^{n_{z_{\max}}} F_{-2}(\eta_{1,10}) \right] \quad (1.54)$$

1.2.4 The MTP in quantum dots of n-gallium phosphide

(a) The model of Rees

The dispersion relation of the conduction electrons in bulk specimens of n-GaP can be written following Rees as [86]

$$E = \frac{\hbar^2 k_s^2}{2m_\perp^*} + \frac{\hbar^2}{2m_\parallel^*} [k_s^2 + k_z^2] - \left[\frac{\hbar^4 k_0^2}{m_\parallel^{*2}} (k_s^2 + k_z^2) + |V_G|^2 \right]^{1/2} + |V_G| \quad (1.55)$$

where k_0 and $|V_G|$ are constants of the energy spectrum.

Using (1.55), the total quantized energy (E_{11}) in this case is given by

$$E_{11} = \frac{\hbar^2}{2m_\perp^*} \left[\left(\frac{\pi n_x}{d_x} \right)^2 + \left(\frac{\pi n_x}{d_x} \right)^2 \right] + \frac{\hbar^2}{2m_\parallel^*} \left[\left(\frac{\pi n_x}{d_x} \right)^2 + \left(\frac{\pi n_y}{d_y} \right)^2 \right.$$
$$\left. + \left(\frac{\pi n_z}{d_z} \right)^2 \right] - \left[\frac{\hbar^4 k_0^2}{m_\parallel^{*2}} \left[\left(\frac{\pi n_x}{d_x} \right)^2 + \left(\frac{\pi n_y}{d_y} \right)^2 + \left(\frac{\pi n_z}{d_z} \right)^2 \right] \right]$$
$$+ |V_G|^{2^{1/2}} + |V_G| \quad (1.56)$$

The electron concentration can be written as

$$n_0 = \frac{2g_v}{d_x d_y d_z} \sum_{n_x=1}^{n_{x_{\max}}} \sum_{n_y=1}^{n_{y_{\max}}} \sum_{n_z=1}^{n_{z_{\max}}} F_{-1}(\eta_{1,11}) \tag{1.57}$$

where

$$\eta_{1,11} = (k_B T)^{-1}(E_F - E_{11})$$

Using (1.5) and (1.57), we get

$$G_0 = \frac{\pi^2 k_B}{3e} \left[\sum_{n_x=1}^{n_{x_{\max}}} \sum_{n_y=1}^{n_{y_{\max}}} \sum_{n_z=1}^{n_{z_{\max}}} F_{-1}(\eta_{1,11}) \right]^{-1} \left[\sum_{n_x=1}^{n_{x_{\max}}} \sum_{n_y=1}^{n_{y_{\max}}} \sum_{n_z=1}^{n_{z_{\max}}} F_{-2}(\eta_{1,11}) \right] \tag{1.58a}$$

(b) The model of Ivchenko and Pikus

The dispersion relation of the conduction electrons in bulk specimens of n-GaP can be written by following Ivechenko and Pikus as [86]

$$E = \frac{\hbar^2 k_z^2}{2m_\parallel^*} + \frac{\hbar^2 k_s^2}{2m_\perp^*} \mp \frac{\overline{\Delta}_0}{2} \pm \left[\left(\frac{\overline{\Delta}_0}{2} \right)^2 + P_1 k_z^2 + D_1 k_x^2 k_y^2 \right]^{1/2} \tag{1.58b}$$

where $\overline{\Delta}_0 = 335\,\text{meV}$, $P_1 = 2 \times 10^{-10}\,\text{eVm}$, $D_1 = P_1 a_1$ and $a_1 = 5.4 \times 10^{-10}\,m$.

The total quantized energy $(\overline{E}_{11})$ in this case assumes the form

$$\overline{E}_{11} = \frac{\hbar^2}{2m_\parallel^*} \left(\frac{n_z \pi}{d_z} \right)^2 + \frac{\hbar^2}{2m_\perp^*} \left[\left(\frac{n_x \pi}{d_x} \right)^2 + \left(\frac{n_y \pi}{d_y} \right)^2 \right] \mp \frac{\overline{\Delta}_0}{2}$$
$$\pm \left[\left(\frac{\overline{\Delta}_0}{2} \right)^2 + P_1 \left(\frac{n_z \pi}{d_z} \right)^2 + D_1 \left(\frac{n_x \pi}{d_x} \right)^2 \left(\frac{n_y \pi}{d_y} \right)^2 \right]^{1/2} \tag{1.58c}$$

The electron concentration can be written as

$$n_0 = \frac{g_v}{d_x d_y d_z} \sum_{n_x=1}^{n_{x_{\max}}} \sum_{n_y=1}^{n_{y_{\max}}} \sum_{n_z=1}^{n_{z_{\max}}} F_{-1}(\overline{\eta}_{1,11}) \tag{1.58d}$$

where

$$\overline{\eta}_{1,11} = (k_B T)^{-1}(E_F - \overline{E}_{11})$$

Using (1.5) and (1.58d), we get

$$G_0 = \frac{\pi^2 k_B}{3e}\left[\sum_{n_x=1}^{n_{x_{\max}}}\sum_{n_y=1}^{n_{y_{\max}}}\sum_{n_z=1}^{n_{z_{\max}}} F_{-1}(\overline{\eta}_{1,11})\right]^{-1}\left[\sum_{n_x=1}^{n_{x_{\max}}}\sum_{n_y=1}^{n_{y_{\max}}}\sum_{n_z=1}^{n_{z_{\max}}} F_{-2}(\overline{\eta}_{1,11})\right] \tag{1.58e}$$

1.2.5 The MTP in quantum dots of n-germanium

It is well known that the conduction electrons of n-Ge obey two different types of dispersion laws since band non-parabolicity has been included in two different ways as given in the literature [87–89].

(a) The model of Cardona *et al.*

The energy spectrum of the conduction electrons in bulk specimens of n-Ge can be expressed in accordance with Cardona *et al.* [87, 88] as

$$E = -\frac{E_{g_0}}{2} + \frac{\hbar^2 k_z^2}{2m_{\|}^*} + \left[\frac{E_{g_0}^2}{4} + E_{g_0} k_s^2\left(\frac{\hbar^2}{2m_{\perp}^*}\right)\right]^{1/2} \tag{1.59}$$

where in this case $m_{\|}^*$ and $m_{\perp}^*$ are the longitudinal and transverse effective masses along $\langle 111\rangle$ direction at the edge of the conduction band respectively.

The totally quantized energy E_{12} in this case is given by

$$E_{12} = -\frac{E_{g_0}}{2} + \frac{\hbar^2}{2m_{\|}^*}\left(\frac{\pi n_z}{d_z}\right)^2 + \left[\frac{E_{g_0}^2}{4} + E_{g_0}\left(\frac{\hbar^2}{2m_{\perp}^*}\right)\left\{\left(\frac{\pi n_x}{d_x}\right)^2 + \left(\frac{\pi n_y}{d_y}\right)^2\right\}\right]^{1/2} \tag{1.60}$$

The electron concentration can be written as

$$n_0 = \frac{2g_v}{d_x d_y d_z}\sum_{n_x=1}^{n_{x_{\max}}}\sum_{n_y=1}^{n_{y_{\max}}}\sum_{n_z=1}^{n_{z_{\max}}} F_{-1}(\eta_{1,12}) \tag{1.61}$$

where

$$\eta_{1,12} = (k_B T)^{-1}(E_F - E_{12})$$

Using (1.5) and (1.61), we get

$$G_0 = \frac{\pi^2 k_B}{3e}\left[\sum_{n_x=1}^{n_{x_{\max}}}\sum_{n_y=1}^{n_{y_{\max}}}\sum_{n_z=1}^{n_{z_{\max}}} F_{-1}(\eta_{1,12})\right]^{-1}\left[\sum_{n_x=1}^{n_{x_{\max}}}\sum_{n_y=1}^{n_{y_{\max}}}\sum_{n_z=1}^{n_{z_{\max}}} F_{-2}(\eta_{1,12})\right] \tag{1.62}$$

(b) The model of Wang and Ressler

The dispersion relation of the conduction electron in bulk specimens of n-Ge can be expressed in accordance with the model of Wang and Ressler [89] can be written as

$$E = \frac{\hbar^2 k_z^2}{2m_{||}^*} + \frac{\hbar^2 k_s^2}{2m_\perp^*} - c_1\left(\frac{\hbar^2 k_s^2}{2m_\perp^*}\right)^2 - d_1\left(\frac{\hbar^2 k_s^2}{2m_\perp^*}\right)\left(\frac{\hbar^2 k_z^2}{2m_{||}^*}\right) - e_1\left(\frac{\hbar^2 k_z^2}{2m_{||}^*}\right)^2 \tag{1.63}$$

where

$$c_1 = C(2m_\perp^*/\hbar^2)^2, \quad C = 1.4A, A = \frac{1}{4}(\hbar^4/E_{g_0}m_\perp^{*^2})\left(1 - \frac{m_\perp^*}{m_0}\right)^2,$$

$$d_1 = d\left(\frac{4m_\perp^* m_{||}^*}{\hbar^4}\right), \quad d = 0.8A, e_1 = e_0(2m_{||}^*/\hbar^2)^2 \quad \text{and} \quad e_0 = 0.005A$$

Using (1.63), the total quantized energy E_{13} in this case is given by

$$\begin{aligned} E_{13} = {} & \frac{\hbar^2}{2m_{||}^*}\left(\frac{\pi n_z}{d_z}\right)^2 + \frac{\hbar^2}{2m_\perp^*}\left\{\left(\frac{\pi n_x}{d_x}\right)^2 + \left(\frac{\pi n_y}{d_y}\right)^2\right\} \\ & - c_1\left(\frac{\hbar^2}{2m_\perp^*}\right)^2\left\{\left(\frac{\pi n_x}{d_x}\right)^2 + \left(\frac{\pi n_y}{d_y}\right)^2\right\}^2 \\ & - d_1\left(\frac{\hbar^2}{2m_\perp^*}\right)\left\{\left(\frac{\pi n_x}{d_x}\right)^2 + \left(\frac{\pi n_y}{d_y}\right)^2\right\}\left(\frac{\hbar^2}{2m_{||}^*}\right)\left(\frac{\pi n_z}{d_z}\right)^2 \\ & - e_1\left(\frac{\hbar^2}{2m_{||}^*}\right)^2\left(\frac{\pi n_z}{d_z}\right)^4 \end{aligned} \tag{1.64}$$

The electron concentration can be written as

$$n_0 = \frac{2g_v}{d_x d_y d_z}\sum_{n_x=1}^{n_{x_{\max}}}\sum_{n_y=1}^{n_{y_{\max}}}\sum_{n_z=1}^{n_{z_{\max}}} F_{-1}(\eta_{1,13}) \tag{1.65}$$

where

$$\eta_{1,13} = (k_B T)^{-1}(E_F - E_{13})$$

Using (1.5) and (1.65), we get

$$G_0 = \frac{\pi^2 k_B}{3e}\left[\sum_{n_x=1}^{n_{x_{\max}}}\sum_{n_y=1}^{n_{y_{\max}}}\sum_{n_z=1}^{n_{z_{\max}}} F_{-1}(\eta_{1,13})\right]^{-1}\left[\sum_{n_x=1}^{n_{x_{\max}}}\sum_{n_y=1}^{n_{y_{\max}}}\sum_{n_z=1}^{n_{z_{\max}}} F_{-2}(\eta_{1,13})\right] \tag{1.66}$$

1.2.6 The MTP in quantum dots of tellurium

The carriers of Tellurium find various descriptions for the energy-wave vector dispersion relations in the literature. Among them we shall use the *E–k* dispersion relations as given by Bouat *et al.* [90] and Ortenberg and Button [91] respectively.

(a) The model of Bouat *et al.*

The dispersion relation of the conduction electrons of Tellurium can be written in accordance with the Bouat *et al.* as [90]

$$E = A_6 k_z^2 + A_7 k_s^2 \pm [A_8 k_z^2 + A_9 k_s^2]^{1/2} \tag{1.67}$$

where A_6, A_7, A_8, A_9 are the energy band constants.

The total quantized energy can be written as

$$\begin{aligned} E_{14,\pm} &= A_6\left(\frac{\pi n_z}{d_z}\right)^2 + A_7\left[\left(\frac{\pi n_x}{d_x}\right)^2 + \left(\frac{\pi n_y}{d_y}\right)^2\right] \\ &\pm \left[A_8\left(\frac{\pi n_z}{d_z}\right)^2 + A_9\left[\left(\frac{\pi n_x}{d_x}\right)^2 + \left(\frac{\pi n_y}{d_y}\right)^2\right]\right]^{1/2} \end{aligned} \tag{1.68}$$

The electron concentration is given by

$$n_0 = \frac{g_v}{d_x d_y d_z} \sum_{n_x=1}^{n_{x_{\max}}} \sum_{n_y=1}^{n_{y_{\max}}} \sum_{n_z=1}^{n_{z_{\max}}} F_{-1}(\eta_{1,14}) \tag{1.69}$$

where

$$\eta_{1,14} = (k_B T)^{-1}(E_F - E_{14,\pm})$$

Using (1.5) and (1.69), we get

$$G_0 = \frac{\pi^2 k_B}{3e}\left[\sum_{n_x=1}^{n_{x_{\max}}} \sum_{n_y=1}^{n_{y_{\max}}} \sum_{n_z=1}^{n_{z_{\max}}} F_{-1}(\eta_{1,14})\right]^{-1}\left[\sum_{n_x=1}^{n_{x_{\max}}} \sum_{n_y=1}^{n_{y_{\max}}} \sum_{n_z=1}^{n_{z_{\max}}} F_{-2}(\eta_{1,14})\right] \tag{1.70}$$

(b) The model of Ortenberg and Button

The dispersion relation of the conduction electrons of Tellurium can be written in accordance with the model of Ortenberg and Button as [91]

$$E = \mp t_1 + t_2 k_z^2 + t_3 k_s^2 + t_4 k_s^4 + t_5 k_s^2 k_z^2 \pm [(t_1 + t_6 k_s^2)^2 + t_7 k_z^2]^{1/2} \tag{1.71}$$

where $t_1, t_2, t_3, t_4, t_5, t_6$ and t_7 are the energy band constants.

The QDDR can be written as

$$E_{15,\pm} = \mp t_1 + t_2\left(\frac{\pi n_z}{d_z}\right)^2 + t_3\left[\left(\frac{\pi n_x}{d_x}\right)^2 + \left(\frac{\pi n_y}{d_y}\right)^2\right] + t_4\left[\left(\frac{\pi n_x}{d_x}\right)^2 + \left(\frac{\pi n_y}{d_y}\right)^2\right]^2 + t_5\left[\left(\frac{\pi n_x}{d_x}\right)^2 + \left(\frac{\pi n_y}{d_y}\right)^2\right]\left(\frac{\pi n_z}{d_z}\right)^2 \pm \left[\left[t_1 + t_6\left\{\left(\frac{\pi n_x}{d_x}\right)^2 + \left(\frac{\pi n_y}{d_y}\right)^2\right\}\right]^2 + t_7\left(\frac{\pi n_z}{d_z}\right)^2\right]^{1/2} \quad (1.72)$$

where $E_{15,\pm}$ is the total quantized energy in this case.

The electron concentration is given by

$$n_0 = \frac{g_v}{d_x d_y d_z}\sum_{n_x=1}^{n_{x_{\max}}}\sum_{n_y=1}^{n_{y_{\max}}}\sum_{n_z=1}^{n_{z_{\max}}} F_{-1}(\eta_{1,15}) \quad (1.73)$$

where

$$\eta_{1,15} = (k_B T)^{-1}(E_F - E_{15,\pm})$$

Using (1.5) and (1.73), we get

$$G_0 = \frac{\pi^2 k_B}{3e}\left[\sum_{n_x=1}^{n_{x_{\max}}}\sum_{n_y=1}^{n_{y_{\max}}}\sum_{n_z=1}^{n_{z_{\max}}} F_{-1}(\eta_{1,15})\right]^{-1}\left[\sum_{n_x=1}^{n_{x_{\max}}}\sum_{n_y=1}^{n_{y_{\max}}}\sum_{n_z=1}^{n_{z_{\max}}} F_{-2}(\eta_{1,15})\right] \quad (1.74)$$

1.2.7 The MTP in quantum dots of graphite

(a) The model of Ushio *et al.*

The carrier dispersion law in graphite can be written as [92]

$$E = \frac{1}{2}(E_2 + E_3) \pm \left[\frac{1}{4}(E_2 - E_3)^2 + \eta_{46}^2 k^2\right]^{1/2} \quad (1.75)$$

where $E_2 = (\Delta_1 - 2\gamma_1\cos\phi_0 + 2\gamma_5\cos^2\phi_0)$, $\phi_0 = (\bar{c}k_z/2)$, $E_3 = (2\gamma_2\cos^2\phi_0)$, $\eta_{46} = \left(\frac{\sqrt{3}}{2}\right)\bar{a}(\gamma_0 + 2\gamma_4\cos\phi_0)$, $\Delta_1, \gamma_1, \gamma_5, \bar{c}$ and $\bar{a}$ are energy band constants.

The QDDR can be expressed as

$$E_{16,\pm} = \frac{1}{2}\left[\Delta_1 - 2\gamma_1 \cos\left(\frac{\overline{c}(\pi n_z)}{2d_z}\right) + 2\gamma_5 \cos^2\left(\frac{\overline{c}(\pi n_z)}{2d_z}\right) + 2\gamma_2 \cos^2\left(\frac{\overline{c}(\pi n_z)}{2d_z}\right)\right] \pm \left[\frac{1}{4}\left(\Delta_1 - 2\gamma_1 \cos^2\left(\frac{\overline{c}(\pi n_z)}{2d_z}\right) + 2\gamma_5 \cos^2\left(\frac{\overline{c}(\pi n_z)}{2d_z}\right) - 2\gamma_2 \cos^2\left(\frac{\overline{c}(\pi n_z)}{2d_z}\right)\right)^2 + \frac{3}{4}(\overline{a})^2\left(\gamma_0 + 2\gamma_4 \cos\left(\frac{n_z \pi \overline{c}}{2d_z}\right)\right)^2 \times \left[\left(\frac{\pi n_x}{d_x}\right)^2 + \left(\frac{\pi n_y}{d_y}\right)^2 + \left(\frac{\pi n_z}{d_z}\right)^2\right]\right]^{1/2} \tag{1.76}$$

where $E_{16,\pm}$ is the total quantized energy in this case.

The electron concentration is given by

$$n_0 = \frac{g_v}{d_x d_y d_z} \sum_{n_x=1}^{n_{x_{\max}}} \sum_{n_y=1}^{n_{y_{\max}}} \sum_{n_z=1}^{n_{z_{\max}}} F_{-1}(\eta_{1,16}) \tag{1.77}$$

where

$$\eta_{1,16} = (k_B T)^{-1}(E_F - E_{16,\pm})$$

Using (1.5) and (1.77), we get

$$G_0 = \frac{\pi^2 k_B}{3e}\left[\sum_{n_x=1}^{n_{x_{\max}}} \sum_{n_y=1}^{n_{y_{\max}}} \sum_{n_z=1}^{n_{z_{\max}}} F_{-1}(\eta_{1,16})\right]^{-1}\left[\sum_{n_x=1}^{n_{x_{\max}}} \sum_{n_y=1}^{n_{y_{\max}}} \sum_{n_z=1}^{n_{z_{\max}}} F_{-2}(\eta_{1,16})\right] \tag{1.78}$$

*1.2.8 The MTP in quantum dots of platinum antimonide

(a) The model of Emtage

The dispersion relation of the carriers in $PtSb_2$ can be written as [93]

$$\left(E + \lambda_0 \frac{(\overline{a})^2}{4}k^2 - l k_S^2 \frac{(\overline{a})^2}{4}\right)\left(E + \delta_0 - \upsilon \frac{(\overline{a})^2}{4}k^2 - \overline{\overline{n}} \frac{(\overline{a})^2}{4}k_S^2\right) = I \frac{(\overline{a})^4}{16}k^4 \tag{1.79}$$

where $\lambda_0, l, \delta_0, \upsilon$ and $\overline{\overline{n}}$ are the band constants and $\overline{\overline{a}}$ is the lattice constant.

Equation (1.79) can be expressed as

$$[E + \omega_1 k_s^2 + \omega_2 k_z^2][E + \delta_0 - \omega_3 k_s^2 - \omega_4 k_z^2] = I_1 (k_z^2 + k_s^2)^2 \tag{1.80}$$

where

$$\omega_1 \equiv \left(\lambda_0 \frac{(\overline{\overline{a}})^2}{4} - l \frac{(\overline{\overline{a}})^2}{4}\right), \quad \omega_2 \equiv \lambda_0 \frac{(\overline{\overline{a}})^2}{4}, \quad \omega_3 \equiv \left(\overline{\overline{n}} \frac{(\overline{\overline{a}})^2}{4} + \upsilon \frac{(\overline{\overline{a}})^2}{4}\right),$$

$$\omega_4 \equiv \upsilon \frac{(\overline{\overline{a}})^2}{4} \quad \text{and} \quad I_1 \equiv I \left(\frac{(\overline{\overline{a}})^2}{4}\right)^2$$

The total quantized energy E_{17} is given by

$$\left[E_{17} + \omega_1 \left[\left(\frac{\pi n_x}{d_x}\right)^2 + \left(\frac{\pi n_y}{d_y}\right)^2\right] + \omega_2 \left(\frac{\pi n_z}{d_z}\right)^2\right]$$
$$\cdot \left[E_{17} + \delta_0 + \omega_3 \left[\left(\frac{\pi n_x}{d_x}\right)^2 + \left(\frac{\pi n_y}{d_y}\right)^2\right] - \omega_4 \left(\frac{\pi n_z}{d_z}\right)^2\right]$$
$$= I_1 \left[\left(\frac{\pi n_x}{d_x}\right)^2 + \left(\frac{\pi n_y}{d_y}\right)^2 + \left(\frac{\pi n_z}{d_z}\right)^2\right]^2 \tag{1.81}$$

The hole concentration is given by

$$p_0 = \frac{2g_v}{d_x d_y d_z} \sum_{n_x=1}^{n_{x_{\max}}} \sum_{n_y=1}^{n_{y_{\max}}} \sum_{n_z=1}^{n_{z_{\max}}} F_{-1}(\eta_{1,17}) \tag{1.82}$$

where

$$\eta_{1,17} = (k_B T)^{-1}(E_F - E_{17})$$

Using (1.5) and (1.82), we get

$$G_0 = \frac{\pi^2 k_B}{3e} \left[\sum_{n_x=1}^{n_{x_{\max}}} \sum_{n_y=1}^{n_{y_{\max}}} \sum_{n_z=1}^{n_{z_{\max}}} F_{-1}(\eta_{1,17})\right]^{-1} \left[\sum_{n_x=1}^{n_{x_{\max}}} \sum_{n_y=1}^{n_{y_{\max}}} \sum_{n_z=1}^{n_{z_{\max}}} F_{-2}(\eta_{1,17})\right] \tag{1.83}$$

1.2.9 The MTP in quantum dots of zero-gap materials

(a) The model of Ivanov-Omskii *et al.*

The dispersion relation of the holes in gapless materials [94] assumes the form

$$\overline{\overline{E}} = \frac{\hbar^2 k^2}{2m_v^*} + \frac{3e^2 k}{128\varepsilon_{sc}} - \frac{2}{\pi} E_B \ln |k \cdot (\overline{k_0})^{-1}| \tag{1.84}$$

where $\overline{\overline{E}}$ is the energy of the hole as measured from the top of the valence band in the vertically downward direction, m_v^* is the effective mass of the holes at the top of the valence band, ε_{sc} is the semiconductor permittivity, E_B is the Bohr electron energy and $\overline{k_0}$ is the inverse Bohr radius.

The total quantized energy E_{18} is given by

$$E_{18} = \frac{\hbar^2\pi^2}{2m^*}\left[\left(\frac{n_x}{d_x}\right)^2 + \left(\frac{n_y}{d_y}\right)^2 + \left(\frac{n_z}{d_z}\right)^2\right] - \frac{3e^2}{128\varepsilon_{sc}}\left[\left(\frac{n_x}{d_x}\right)^2 + \left(\frac{n_y}{d_y}\right)^2 + \left(\frac{n_z}{d_z}\right)^2\right]^{1/2} - \frac{E_B}{\pi}\ln\left|\frac{(\pi n_x/d_x)^2 + (\pi n_y/d_y)^2 + (\pi n_z/d_z)^2}{(\overline{k_0})^2}\right| \tag{1.85}$$

The hole concentration can be written as

$$p_0 = \frac{2g_v}{d_x d_y d_z}\sum_{n_x=1}^{n_{x_{\max}}}\sum_{n_y=1}^{n_{y_{\max}}}\sum_{n_z=1}^{n_{z_{\max}}} F_{-1}(\eta_{1,18}) \tag{1.86}$$

where

$$\eta_{1,18} = (k_B T)^{-1}(E_F - E_{18})$$

Using (1.5) and (1.86), we get

$$G_0 = \frac{\pi^2 k_B}{3e}\left[\sum_{n_x=1}^{n_{x_{\max}}}\sum_{n_y=1}^{n_{y_{\max}}}\sum_{n_z=1}^{n_{z_{\max}}} F_{-1}(\eta_{1,18})\right]^{-1}\left[\sum_{n_x=1}^{n_{x_{\max}}}\sum_{n_y=1}^{n_{y_{\max}}}\sum_{n_z=1}^{n_{z_{\max}}} F_{-2}(\eta_{1,18})\right] \tag{1.87}$$

*1.2.10 The MTP in quantum dots of II–V materials

(a) The model of Yamada

The dispersion relation of the holes in II–V compounds in accordance with Yamada [95, 96] can be expressed as

$$E = A_{10}k_x^2 + A_{11}k_y^2 + A_{12}k_z^2 + A_{13}k_x \pm \left[(A_{14}k_x^2 + A_{15}k_y^2 + A_{16}k_z^2 + A_{17}k_x)^2 + A_{18}k_y^2 + A_{19}^2\right]^{1/2} \mp A_{19} \tag{1.88}$$

where $A_{10}, A_{11}, A_{12}, A_{13}, A_{14}, A_{15}, A_{16}, A_{17}, A_{18}$ and A_{19} are energy band constants.

The total quantized energy $E_{19,\pm}$ is given by

$$E_{19,\pm} = A_{10}\left(\frac{\pi n_x}{d_x}\right)^2 + A_{11}\left(\frac{\pi n_y}{d_y}\right)^2 + A_{12}\left(\frac{\pi n_z}{d_z}\right)^2 + A_{13}\left(\frac{\pi n_x}{d_x}\right)$$
$$\pm\left[\left(A_{14}\left(\frac{\pi n_x}{d_x}\right)^2 + A_{15}\left(\frac{\pi n_y}{d_y}\right)^2 + A_{16}\left(\frac{\pi n_z}{d_z}\right)^2\right.\right.$$
$$\left.\left. + A_{17}\left(\frac{\pi n_x}{d_x}\right)\right)^2 + A_{18}\left(\frac{\pi n_y}{d_y}\right)^2 + A_{19}^2\right]^{1/2} \mp A_{19} \tag{1.89}$$

The hole concentration can be written as

$$p_0 = \frac{g_v}{d_x d_y d_z}\sum_{n_x=1}^{n_{x_{\max}}}\sum_{n_y=1}^{n_{y_{\max}}}\sum_{n_z=1}^{n_{z_{\max}}} F_{-1}(\eta_{1,19}) \tag{1.90}$$

where

$$\eta_{1,19} = (k_BT)^{-1}(E_F - E_{19,\pm})$$

Using (1.5) and (1.90), we get

$$G_0 = \frac{\pi^2 k_B}{3e}\left[\sum_{n_x=1}^{n_{x_{\max}}}\sum_{n_y=1}^{n_{y_{\max}}}\sum_{n_z=1}^{n_{z_{\max}}} F_{-1}(\eta_{1,19})\right]^{-1}\left[\sum_{n_x=1}^{n_{x_{\max}}}\sum_{n_y=1}^{n_{y_{\max}}}\sum_{n_z=1}^{n_{z_{\max}}} F_{-2}(\eta_{1,19})\right] \tag{1.91}$$

*1.2.11 The MTP in quantum dots of gallium antimonide

The conduction electrons of n-GaSb obey the three dispersion relations as provided by Seiler *et al.* [97], Mathur *et al.* [98] and Zhang [99] respectively. The MTP in QDs of GaSb is being presented in accordance with the aforementioned models for the joint purpose of coherent presentation and relative assessment.

(a) The model of Seiler *et al.*

In accordance with the model of Seiler *et al.* [97] the dispersion relation of the conduction electrons in n-GaSb assume the form

$$E = \left[-\frac{E_g}{2} + \frac{E_g}{2}\left[1+\alpha_4 k^2\right]^{1/2} + \frac{\bar{\varsigma}_0\hbar^2k^2}{2m_0} + \frac{\bar{\nu}_0 f_1(k)\hbar^2}{2m_0} \pm \frac{\bar{\varpi}_0 f_2(k)\hbar^2}{2m_0}\right] \tag{1.92}$$

where $\alpha_4 \equiv 4P_0^2\left(E_g + \frac{2}{3}\Delta\right)\left[E_g^2\left(E_g+\Delta\right)\right]^{-1}$, P_0 is the isotropic momentum matrix element, $f_1(k) \equiv k^{-2}[k_x^2k_y^2 + k_y^2k_z^2 + k_z^2k_x^2]$ represents the warping of Fermi surface, $f_2(k) \equiv [\{k^2(k_x^2k_y^2 + k_y^2k_z^2 + k_z^2k_x^2) - 9k_x^2k_y^2k_z^2\}^{1/2}k^{-1}]$ represents the inversion asymmetry splitting of the conduction band, $\overline{\varsigma}_0$, $\overline{\nu}_0$ and ϖ_0 represent the constants of the electron spectrum in this case.

The total quantized energy $E_{20,\pm}$ is given by

$$E_{20,\pm} = \left[-\frac{E_{g0}}{2} + \frac{E_{g0}}{2}\left[1 + \alpha_4\left[\left(\frac{\pi n_x}{d_x}\right)^2 + \left(\frac{\pi n_y}{d_y}\right)^2 + \left(\frac{\pi n_z}{d_z}\right)^2\right]\right]\right]^{1/2} + \left(\frac{\overline{\zeta_0}\hbar^2}{2m_0}\right)\left[\left(\frac{\pi n_x}{d_x}\right)^2 + \left(\frac{\pi n_y}{d_y}\right)^2 + \left(\frac{\pi n_z}{d_z}\right)^2\right] + \left(\frac{\overline{v_0}\hbar^2}{2m_0}\right)S_6 \pm \left(\frac{\overline{w_0}S_7\hbar^2}{2m_0}\right)\Bigg] \tag{1.93}$$

where

$$S_6 = \left[\left(\frac{\pi n_x}{d_x}\right)^2 + \left(\frac{\pi n_y}{d_y}\right)^2 + \left(\frac{\pi n_z}{d_z}\right)^2\right]^{-2}\left[\left(\frac{\pi^2 n_x n_y}{d_x d_y}\right)^2 + \left(\frac{\pi^2 n_y n_z}{d_y d_z}\right)^2 + \left(\frac{\pi^2 n_z n_x}{d_x d_z}\right)^2\right]$$

$$S_7 = \left[\left\{\left(\left(\frac{\pi n_x}{d_x}\right)^2 + \left(\frac{\pi n_y}{d_y}\right)^2 + \left(\frac{\pi n_z}{d_z}\right)^2\right)\left(\frac{n_x^2n_y^2\pi^4}{d_x^2d_y^2} + \frac{n_y^2n_z^2\pi^4}{d_y^2d_z^2} + \frac{n_z^2n_x^2\pi^4}{d_z^2d_x^2}\right) - \frac{9\pi^6 n_x^2n_y^2n_z^2}{d_x^2d_y^2d_z^2}\right\}^{1/2}\left[\left(\frac{\pi n_x}{d_x}\right)^2 + \left(\frac{\pi n_y}{d_y}\right)^2 + \left(\frac{\pi n_z}{d_z}\right)^2\right]^{-1/2}\right]$$

The electron concentration is given by

$$p_0 = \frac{g_v}{d_x d_y d_z}\sum_{n_x=1}^{n_{x_{\max}}}\sum_{n_y=1}^{n_{y_{\max}}}\sum_{n_z=1}^{n_{z_{\max}}} F_{-1}(\eta_{1,20}) \tag{1.94}$$

where

$$\eta_{1,20} = (k_BT)^{-1}(E_F - E_{20,\pm})$$

Using (1.5) and (1.94), we get

$$G_0 = \frac{\pi^2 k_B}{3e}\left[\sum_{n_x=1}^{n_{x_{\max}}}\sum_{n_y=1}^{n_{y_{\max}}}\sum_{n_z=1}^{n_{z_{\max}}} F_{-1}(\eta_{1,20})\right]^{-1}\left[\sum_{n_x=1}^{n_{x_{\max}}}\sum_{n_y=1}^{n_{y_{\max}}}\sum_{n_z=1}^{n_{z_{\max}}} F_{-2}(\eta_{1,20})\right] \tag{1.95}$$

(b) The model of Mathur *et al.*

In accordance with the model of Mathur *et al.* [98], the electron dispersion law in n-GaSb can be written as

$$E = \alpha_9 k^2 + (E_{g1}/2)\left[\sqrt{1+\alpha_{10}k^2} - 1\right] \tag{1.96}$$

where $\alpha_9 = (\hbar^2/2m_0)$, $E_{g1} = E_{g_0} + [(5\times 10^{-S}T^2)/2(112+T)]$ and $\alpha_{10} = ((2\hbar^2)/(E_{g1}))[(1/m^*) - (1/m_0)]$.

From (1.96), we get

$$k^2 = \frac{E}{\alpha_9} + \alpha_{11} - [\alpha_{12}E + \alpha_{13}]^{1/2} \equiv \gamma_7(\overline{E}) \tag{1.97}$$

where $\alpha_{11} = (E_{g1}^2/8\alpha_9^2)[\alpha_{10} + (4\alpha_9/E_{g1})]$, $\alpha_{12} = (E_{g1}^2/\alpha_9^3)$, $\alpha_{13} = (E_{g1}^4/64\alpha_9^4)[\alpha_{10}^2 + (16\alpha_9^2/E_{g1}^2) + (8\alpha_9\alpha_{10}/E_{g1})]$.

The total quantized energy E_{21} is given by

$$\begin{aligned} E_{21} &= \alpha_9[(\pi n_x/d_x)^2 + (\pi n_y/d_y)^2 + (\pi n_z/d_z)^2] \\ &\quad + (E_{g1}/2)\left[\sqrt{1+\alpha_{10}\left[(\pi n_x/d_x)^2 + (\pi n_y/d_y)^2 + (\pi n_z/d_z)^2\right]} - 1\right] \end{aligned} \tag{1.98}$$

The electron concentration is given by

$$n_0 = (2g_v/d_x d_y d_z)\sum_{n_x=1}^{n_{x_{\max}}}\sum_{n_y=1}^{n_{y_{\max}}}\sum_{n_z=1}^{n_{z_{\max}}} F_{-1}(\eta_{1,21}) \tag{1.99}$$

where

$$\eta_{1,21} = (k_B T)^{-1}(E_F - E_{21})$$

Using (1.5) and (1.99), we get

$$G_0 = \frac{\pi^2 k_B}{3e}\left[\sum_{n_x=1}^{n_{x_{\max}}}\sum_{n_y=1}^{n_{y_{\max}}}\sum_{n_z=1}^{n_{z_{\max}}} F_{-1}(\eta_{1,21})\right]^{-1}\left[\sum_{n_x=1}^{n_{x_{\max}}}\sum_{n_y=1}^{n_{y_{\max}}}\sum_{n_z=1}^{n_{z_{\max}}} F_{-2}(\eta_{1,21})\right] \tag{1.100}$$

(c) The model of Zhang

The dispersion relation of the conduction electrons in n-GaSb can be expressed in accordance with Zhang [99] a

$$\begin{aligned} E &= [\overline{E_1} + \overline{E_2}Z_1(k)]k^2 + [\overline{E_3} + \overline{E_4}Z_1(k)]k^4 \\ &\quad + k^6[\overline{E_5} + \overline{E_6}Z_1(k) + \overline{E_7}Z_2(k)] \end{aligned} \tag{1.101}$$

where

$$Z_1(k) = \frac{5\sqrt{21}}{4}\left[\frac{k_x^4+k_y^4+k_z^4}{k^4}-\frac{3}{5}\right]$$

$$Z_2(k) = \left[\frac{639639}{32}\right]^{1/2}\left[\frac{k_x^2k_y^2k_z^2}{k^6}+\frac{1}{12}\left(\frac{k_x^4+k_y^4+k_z^4}{k^4}-\frac{3}{5}\right)-\frac{1}{105}\right]$$

the coefficients are in eV, the values of k are in $10\left(\frac{a}{2\pi}\right)$ times of k in atomic units (a is lattice constant),

$$\overline{E_1},\quad \overline{E_2},\quad \overline{E_3},\quad \overline{E_4},\quad \overline{E_5},\quad \overline{E_6}\quad \text{and}\quad \overline{E_7}$$

are energy–band constants.

The total quantized energy E_{22} is given by

$$E_{22} = [\overline{E_1}+\overline{E_2}Z_3][k_1^2]+[k_1^4]\cdot[\overline{E_3}+\overline{E_4}Z_3]+[k_1^6]\cdot[\overline{E_5}+\overline{E_6}Z_3+\overline{E_7}Z_4] \tag{1.102}$$

where

$$Z_3 = \frac{5\sqrt{21}}{4}\left[\left[\left(\frac{\pi n_x}{d_x}\right)^4+\left(\frac{\pi n_y}{d_y}\right)^4+\left(\frac{\pi n_z}{d_z}\right)^4\right]\cdot[k_1^{-4}]-\frac{3}{5}\right]$$

and

$$Z_4 = \left[\frac{639639}{32}\right]^{1/2}\left[\left(\frac{n_x}{d_x}\frac{n_y}{d_y}\frac{n_z}{d_z}\right)^2[k_1^{-6}]+\frac{1}{(15)\sqrt{21}}Z_3-\frac{1}{105}\right]$$

The electron concentration is given by

$$n_0 = \frac{2g_v}{d_xd_yd_z}\sum_{n_x=1}^{n_{x_{\max}}}\sum_{n_y=1}^{n_{y_{\max}}}\sum_{n_z=1}^{n_{z_{\max}}}F_{-1}(\eta_{1,22}) \tag{1.103}$$

where

$$\eta_{1,22} = (k_BT)^{-1}(E_F-E_{22})$$

Using (1.5) and (1.103), we get

$$G_0 = \frac{\pi^2k_B}{3e}\left[\sum_{n_x=1}^{n_{x_{\max}}}\sum_{n_y=1}^{n_{y_{\max}}}\sum_{n_z=1}^{n_{z_{\max}}}F_{-1}(\eta_{1,22})\right]^{-1}\left[\sum_{n_x=1}^{n_{x_{\max}}}\sum_{n_y=1}^{n_{y_{\max}}}\sum_{n_z=1}^{n_{z_{\max}}}F_{-2}(\eta_{1,22})\right] \tag{1.104}$$

1.2.12 MTP in quantum dots of stressed materials

(a) The model of Seiler *et al.* [97, 99]

The dispersion relation of the conduction electrons in bulk specimens of stressed materials can be written as [99]

$$\frac{k_x^2}{[a^*(E)]^2} + \frac{k_y^2}{[b^*(E)]^2} + \frac{k_z^2}{[c^*(E)]^2} = 1 \tag{1.105}$$

where

$$[a^*(E)]^2 = \frac{\overline{K}_0(E)}{\overline{A}_0(E) + \frac{1}{2}\overline{D}_0(E)}, \quad \overline{K}_0(E) = \left[E - C_1\varepsilon - \frac{2C_2^2\varepsilon_{xy}^2}{3E'_{g_0}}\right]\left(\frac{3E'_{g_0}}{2B_2^2}\right),$$

C_1 is the conduction band deformation potential, ε is the trace of the strain tensor $\hat{\varepsilon}$ which can be written as $\hat{\varepsilon} = \begin{bmatrix} \varepsilon_{xx} & \varepsilon_{xy} & 0 \\ \varepsilon_{xy} & \varepsilon_{yy} & 0 \\ 0 & 0 & \varepsilon_{zz} \end{bmatrix}$, C_2 is a constant which describes the strain interaction between the conduction and valance bands, $E'_{g_0} = E_{g_0} + E - C_1\varepsilon$, E_{g_0} is the band gap in the absence of stress, B_2 is the momentum-matrix element,

$$\overline{A}_0(E) = \left[1 - \frac{(\bar{a}_0 + C_1)}{E'_{g_0}} + \frac{3\bar{b}_0\varepsilon_{xx}}{2E'_{g_0}} - \frac{\bar{b}_0\varepsilon}{2E'_{g_0}}\right],$$

$$\bar{a}_0 = -\frac{1}{3}(\bar{b}_0 + 2\overline{m}), \quad \bar{b}_0 = \frac{1}{3}(\bar{l} - \overline{m}), \quad \bar{d}_0 = \frac{2\overline{n}}{\sqrt{3}}, \quad \bar{l}, \quad \overline{m}, \quad \overline{n}$$

are the matrix elements of the strain perturbation operator,

$$\overline{D}_0(E) = (\bar{d}_0\sqrt{3})\frac{\varepsilon_{xy}}{E'_{g_0}}, \quad [b^*(E)]^2 = \frac{\overline{K}_0(E)}{\overline{A}_0(E) - \frac{1}{2}\overline{D}_0(E)},$$

$$[c^*(E)]^2 = \frac{\overline{K}_0(E)}{\overline{L}_0(E)} \quad \text{and} \quad \overline{L}_0(E) = \left[1 - \frac{(\bar{a}_0 + C_1)}{E'_{g_0}} + \frac{3\bar{b}_0\varepsilon_{zz}}{E'_{g_0}} - \frac{\bar{b}_0\varepsilon}{2E'_{g_0}}\right]$$

The total quantized energy E_{23} in this case assumes the form

$$\left(\frac{\pi n_x}{d_x}\right)^2 [a^*(E_{23})]^{-2} + \left(\frac{\pi n_y}{d_y}\right)^2 [b^*(E_{23})]^{-2} + \left(\frac{\pi n_z}{d_z}\right)^2 [c^*(E_{23})]^{-2} = 1 \tag{1.106}$$

The electron concentration is given by

$$n_0 = \frac{2g_v}{d_x d_y d_z} \sum_{n_x=1}^{n_{x_{\max}}} \sum_{n_y=1}^{n_{y_{\max}}} \sum_{n_z=1}^{n_{z_{\max}}} F_{-1}(\eta_{1,23}) \tag{1.107}$$

where

$$\eta_{1,23} = (k_B T)^{-1}(E_F - E_{23})$$

Using (1.5) and (1.107), we get

$$G_0 = \frac{\pi^2 k_B}{3e}\left[\sum_{n_x=1}^{n_{x_{\max}}}\sum_{n_y=1}^{n_{y_{\max}}}\sum_{n_z=1}^{n_{z_{\max}}} F_{-1}(\eta_{1,23})\right]^{-1}\left[\sum_{n_x=1}^{n_{x_{\max}}}\sum_{n_y=1}^{n_{y_{\max}}}\sum_{n_z=1}^{n_{z_{\max}}} F_{-2}(\eta_{1,23})\right] \tag{1.108}$$

1.2.13 The MTP in quantum dots of bismuth

The carrier energy spectra in Bismuth are distributed by the model of McClure and Choi [61], Hybrid [60], Cohen [59], Lax [53] and parabolic ellipsoidal respectively and for the purpose of coherent presentation the MTP in QD of Bi in accordance with the said models have been formulated.

(a) The McClure and Choi model

The dispersion relation of the carriers in Bi can be written, following the McClure and Choi [61], as

$$E(1+\alpha E) = \frac{p_x^2}{2m_1} + \frac{p_y^2}{2m_2} + \frac{p_z^2}{2m_3} + \frac{p_y^2}{2m_2}\alpha E\left\{1-\left(\frac{m_2}{m'_2}\right)\right\} + \frac{p_y^4\alpha}{4m_2m'_2} - \frac{\alpha p_x^2 p_y^2}{4m_1m_2} - \frac{\alpha p_y^2 p_z^2}{4m_2m_3} \tag{1.109}$$

where $p_i \equiv \hbar k_i, i = x, y, z, m_1, m_2$ and m_3 are the effective carrier masses at the band-edge along x-, y- and z-directions respectively and m'_2 is the effective-mass tensor component at the top of the valence band (for electrons) or at the bottom of the conduction band (for holes).

The total quantized energy E_{24} is given by

$$E_{24}(1+\alpha E_{24}) = \frac{\hbar^2}{2m_1}\left(\frac{\pi n_x}{d_x}\right)^2 + \frac{\hbar^2}{2m_2}\left(\frac{\pi n_y}{d_y}\right)^2 + \frac{\hbar^2}{2m_3}\left(\frac{\pi n_z}{d_z}\right)^2 + \frac{\hbar^2\alpha E_{24}}{2m_2}\left(\frac{\pi n_y}{d_y}\right)^2\left(1-\frac{m_2}{m'_2}\right) + \frac{\alpha\hbar^4}{4m_2m'_2}\left(\frac{\pi n_y}{d_y}\right)^4 - \frac{\alpha\hbar^4}{4m_1m_2}\left(\frac{\pi^2 n_x^2}{d_x^2}\cdot\frac{\pi^2 n_y^2}{d_y^2}\right) - \frac{\alpha\hbar^4\pi^4}{4m_2m_3}\left(\frac{n_y n_z}{d_y d_z}\right)^2 \tag{1.110}$$

The electron concentration can be written as

$$n_0 = \frac{2g_v}{d_x d_y d_z}\sum_{n_x=1}^{n_{x_{\max}}}\sum_{n_y=1}^{n_{y_{\max}}}\sum_{n_z=1}^{n_{z_{\max}}} F_{-1}(\eta_{1,24}) \tag{1.111}$$

where

$$\eta_{1,24} = (k_B T)^{-1}(E_F - E_{24})$$

Using (1.5) and (1.111), we get

$$G_0 = \frac{\pi^2 k_B}{3e}\left[\sum_{n_x=1}^{n_{x_{\max}}}\sum_{n_y=1}^{n_{y_{\max}}}\sum_{n_z=1}^{n_{z_{\max}}} F_{-1}(\eta_{1,24})\right]^{-1}\left[\sum_{n_x=1}^{n_{x_{\max}}}\sum_{n_y=1}^{n_{y_{\max}}}\sum_{n_z=1}^{n_{z_{\max}}} F_{-2}(\eta_{1,24})\right] \tag{1.112}$$

(b) The Hybrid model

The dispersion relation of the carriers in bulk specimens of Bi in accordance with the Hybrid model can be represented as [60]

$$E(1+\alpha E) = \frac{\theta_0(E)(\hbar k_y)^2}{2M_2} + \frac{\alpha\gamma_0\hbar^4 k_y^4}{4M_2^2} + \frac{\hbar^2 k_x^2}{2m_1} + \frac{\hbar^2 k_z^2}{2m_3}, \tag{1.113}$$

in which $\theta_0(E) \equiv [1 + \alpha E(1-\gamma_0) + \overline{\delta}_0]$, $\gamma_0 \equiv \frac{M_2}{M'_2}$, $\overline{\delta}_0 \equiv \frac{M_2}{m_2}$ and the other notations are defined in [60].

The total quantized energy E_{25} is given by

$$E_{25}(1+\alpha E_{25}) = \frac{\hbar^2}{2m_1}\left(\frac{\pi n_z}{d_z}\right)^2 + \frac{\hbar^2}{2m_3}\left(\frac{\pi n_z}{d_z}\right)^2 + \frac{\hbar^2}{2M_2}\left(\frac{\pi n_y}{d_y}\right)^2$$

$$[1 + \overline{\delta}_0 + \alpha E_{25}(1-\overline{\gamma}_0)] + \frac{\overline{\gamma}_0\hbar^4}{4M_2^2 E_{g_0}}\left(\frac{\pi n_y}{d_y}\right)^4 \tag{1.114}$$

The electron concentration can be written as

$$n_0 = \frac{2g_v}{d_x d_y d_z}\sum_{n_x=1}^{n_{x_{\max}}}\sum_{n_y=1}^{n_{y_{\max}}}\sum_{n_z=1}^{n_{z_{\max}}} F_{-1}(\eta_{1,25}) \tag{1.115}$$

where

$$\eta_{1,25} = (k_B T)^{-1}(E_F - E_{25})$$

Using (1.5) and (1.115), we get

$$G_0 = \frac{\pi^2 k_B}{3e}\left[\sum_{n_x=1}^{n_{x_{\max}}}\sum_{n_y=1}^{n_{y_{\max}}}\sum_{n_z=1}^{n_{z_{\max}}} F_{-1}(\eta_{1,25})\right]^{-1}\left[\sum_{n_x=1}^{n_{x_{\max}}}\sum_{n_y=1}^{n_{y_{\max}}}\sum_{n_z=1}^{n_{z_{\max}}} F_{-2}(\eta_{1,25})\right] \tag{1.116}$$

(c) The Cohen model

In accordance with the Cohen model [59], the dispersion law of the carriers in Bi is given by

$$E(1+\alpha E) = \frac{p_x^2}{2m_1} + \frac{p_z^2}{2m_3} - \frac{\alpha E p_y^2}{2m'_2} + \frac{p_y^2(1+\alpha E)}{2m_2} + \frac{\alpha p_y^4}{4m_2 m'_2} \tag{1.117}$$

The total quantized energy (E_{26}) assumes the form

$$E_{26}(1+\alpha E_{26}) = \frac{\hbar^2}{2m_1}\left(\frac{\pi n_x}{d_x}\right)^2 + \frac{\hbar^2}{2m_3}\left(\frac{\pi n_z}{d_z}\right)^2 + \frac{\alpha\hbar^4}{4m_2m_2'}\left(\frac{\pi n_y}{d_y}\right)^4 + \frac{\hbar^2}{2m_2}\left(\frac{\pi n_y}{d_y}\right)^2\left[1+\alpha E_{26}\left(1-\frac{m_2}{m_2'}\right)\right] \quad (1.118)$$

The electron concentration is given by

$$n_0 = \frac{2g_v}{d_xd_yd_z}\sum_{n_x=1}^{n_{x_{\max}}}\sum_{n_y=1}^{n_{y_{\max}}}\sum_{n_z=1}^{n_{z_{\max}}} F_{-1}(\eta_{1,26}) \quad (1.119)$$

where

$$\eta_{1,26} = (k_BT)^{-1}(E_F - E_{25})$$

Using (1.5) and (1.119), we get

$$G_0 = \frac{\pi^2 k_B}{3e}\left[\sum_{n_x=1}^{n_{x_{\max}}}\sum_{n_y=1}^{n_{y_{\max}}}\sum_{n_z=1}^{n_{z_{\max}}} F_{-1}(\eta_{1,25})\right]^{-1}\left[\sum_{n_x=1}^{n_{x_{\max}}}\sum_{n_y=1}^{n_{y_{\max}}}\sum_{n_z=1}^{n_{z_{\max}}} F_{-2}(\eta_{1,26})\right] \quad (1.120)$$

(d) The Lax model

The electron energy spectra in bulk specimens of Bi in accordance with the Lax model can be written as [53]

$$E(1+\alpha E) = \frac{p_x^2}{2m_1} + \frac{p_y^2}{2m_2} + \frac{p_z^2}{2m_3} \quad (1.121)$$

The total quantized energy E_{27} is given by

$$E_{27}(1+\alpha E_{27}) = \frac{\hbar^2}{2m_1}\left(\frac{\pi n_x}{d_x}\right)^2 + \frac{\hbar^2}{2m_2}\left(\frac{\pi n_y}{d_y}\right)^2 + \frac{\hbar^2}{2m_3}\left(\frac{\pi n_z}{d_z}\right)^2 \quad (1.122)$$

The electron concentration can be written as

$$n_0 = \frac{2g_v}{d_xd_yd_z}\sum_{n_x=1}^{n_{x_{\max}}}\sum_{n_y=1}^{n_{y_{\max}}}\sum_{n_z=1}^{n_{z_{\max}}} F_{-1}(\eta_{1,27}) \quad (1.123)$$

where

$$\eta_{1,27} = (k_BT)^{-1}(E_F - E_{27})$$

Using (1.5) and (1.123), we get

$$G_0 = \frac{\pi^2 k_B}{3e}\left[\sum_{n_x=1}^{n_{x_{\max}}}\sum_{n_y=1}^{n_{y_{\max}}}\sum_{n_z=1}^{n_{z_{\max}}} F_{-1}(\eta_{1,27})\right]^{-1}\left[\sum_{n_x=1}^{n_{x_{\max}}}\sum_{n_y=1}^{n_{y_{\max}}}\sum_{n_z=1}^{n_{z_{\max}}} F_{-2}(\eta_{1,27})\right] \quad (1.124a)$$

(e) Ellipsoidal Parabolic model

The carrier energy spectra for this type of band model of Bi can be written as [52]

$$E = \frac{p_x^2}{2m_1} + \frac{p_y^2}{2m_2} + \frac{p_z^2}{2m_3} \tag{1.124b}$$

The total quantized energy $\overline{E}_{27}$ for this model is given by

$$\overline{E}_{27} = \frac{\hbar^2}{2m_1}\left(\frac{\pi n_x}{d_x}\right)^2 + \frac{\hbar^2}{2m_2}\left(\frac{\pi n_y}{d_y}\right)^2 + \frac{\hbar^2}{2m_3}\left(\frac{\pi n_z}{d_z}\right)^2 \tag{1.125}$$

The electron concentration can be written as

$$n_0 = \frac{2g_v}{d_x d_y d_z} \sum_{n_x=1}^{n_{x_{\max}}} \sum_{n_y=1}^{n_{y_{\max}}} \sum_{n_z=1}^{n_{z_{\max}}} F_{-1}(\overline{\eta}_{1,27}) \tag{1.126}$$

where

$$\overline{\eta}_{1,27} = (k_B T)^{-1}(E_F - \overline{E}_{27})$$

Using (1.5) and (1.126), we get

$$G_0 = \frac{\pi^2 k_B}{3e}\left[\sum_{n_x=1}^{n_{x_{\max}}} \sum_{n_y=1}^{n_{y_{\max}}} \sum_{n_z=1}^{n_{z_{\max}}} F_{-1}(\overline{\eta}_{1,27})\right]^{-1}\left[\sum_{n_x=1}^{n_{x_{\max}}} \sum_{n_y=1}^{n_{y_{\max}}} \sum_{n_z=1}^{n_{z_{\max}}} F_{-2}(\overline{\eta}_{1,27})\right] \tag{1.127}$$

1.2.14 The MTP in quantum dots of IV–VI materials

In addition to Cohen model, the carriers of IV–VI materials are also described by the three more types of dispersion relations as provided by Dimmock [101], Bangert and Kastner [102] and Foley *et al.* [103,104] respectively. The MTP in QDs of IV–VI materials in accordance with the aforementioned models have further been discussed for the purpose of complete presentation:

(a) The model of Dimmock

The dispersion relation of the conduction electrons in IV–VI materials can be expressed in accordance with Dimmock [101] as

$$\left[\overline{\varepsilon} - \frac{E_{g_0}}{2} - \frac{\hbar^2 k_s^2}{2m_t^-} - \frac{\hbar^2 k_z^2}{2m_l^-}\right]\left[\overline{\varepsilon} + \frac{E_{g_0}}{2} + \frac{\hbar^2 k_s^2}{2m_t^-} + \frac{\hbar^2 k_z^2}{2m_l^-}\right] = P_\perp^2 k_s^2 + P_{||}^2 k_z^2 \tag{1.128}$$

where ε is the energy as measured from the center of the band gap E_{g_0}, $m_t^\pm$ and $m_l^\pm$ represent the contributions to the transverse and longitudinal effective masses of the external L_6^+ and L_6^- bands arising from the $\overrightarrow{k} \cdot \overrightarrow{p}$ perturbations with the other bands taken to the second order.

Using

$$\bar{\varepsilon} = E + (E_{g_0}/2), \quad P_\perp^2 = \frac{\hbar^2 E_{g_0}}{2m_t^*}, \quad P_{\|}^2 = \frac{\hbar^2 E_{g_0}}{2m_l^*}$$

(m_t^* and m_l^* are the transverse and longitudinal effective masses at $k = 0$) in (1.128).

The total quantized energy E_{28} in this case can be written as

$$\left[E_{28} - \frac{\hbar^2}{2m_t^-}\left\{\left(\frac{\pi n_x}{d_x}\right)^2 + \left(\frac{\pi n_y}{d_y}\right)^2\right\} - \frac{\hbar^2}{2m_l^-}\left(\frac{\pi n_z}{d_z}\right)^2\right]$$
$$\times\left[1 + \alpha E_{28} + \frac{\alpha\hbar^2}{2m_t^+}\left\{\left(\frac{\pi n_x}{d_x}\right)^2 + \left(\frac{\pi n_y}{d_y}\right)^2\right\} + \alpha\frac{\hbar^2}{2m_l^+}\left(\frac{\pi n_z}{d_z}\right)^2\right]$$
$$= \frac{\hbar^2}{2m_t^*}\left[\left(\frac{\pi n_x}{d_x}\right)^2 + \left(\frac{\pi n_y}{d_y}\right)^2\right] + \frac{\hbar^2}{2m_l^*}\left(\frac{\pi n_z}{d_z}\right)^2 \tag{1.129}$$

The electron concentration is given by

$$n_0 = \frac{2g_v}{d_x d_y d_z}\sum_{n_x=1}^{n_{x_{\max}}}\sum_{n_y=1}^{n_{y_{\max}}}\sum_{n_z=1}^{n_{z_{\max}}} F_{-1}(\eta_{1,28}) \tag{1.130}$$

where

$$\eta_{1,28} = (k_B T)^{-1}(E_F - E_{28})$$

Using (1.5) and (1.130), we get

$$G_0 = \frac{\pi^2 k_B}{3e}\left[\sum_{n_x=1}^{n_{x_{\max}}}\sum_{n_y=1}^{n_{y_{\max}}}\sum_{n_z=1}^{n_{z_{\max}}} F_{-1}(\eta_{1,28})\right]^{-1}\left[\sum_{n_x=1}^{n_{x_{\max}}}\sum_{n_y=1}^{n_{y_{\max}}}\sum_{n_z=1}^{n_{z_{\max}}} F_{-2}(\eta_{1,28})\right] \tag{1.131}$$

(b) The model of Bangert and Kastner [102]

The electron dispersion law in IV–VI compounds can be written in accordance with Bangert and Kastner [102] as

$$\Gamma(E) = \overline{F}_1(E)k_s^2 + \overline{F}_2(E)k_z^2 \tag{1.132}$$

where

$$\Gamma(E) \equiv 2E, \overline{F}_1(E) \equiv \left[\frac{(\overline{R})^2}{E+E_{g_0}} + \frac{(\overline{s})^2}{E+\Delta_c'} + \frac{(\overline{Q})^2}{E+\Delta_c''}\right],$$

$$\overline{F}_2(E) \equiv \left[\frac{2(\overline{A})^2}{E+E_{g_0}} + \frac{\left(\overline{s}+\overline{Q}\right)^2}{E+\Delta_c''}\right],$$

$\overline{R}, \overline{s}, \Delta_c', \overline{Q}, \Delta_c''$ and $\overline{A}$ are the energy spectrum constants.

The total quantized energy E_{29} can be expressed as

$$\Gamma(E_{29}) = F_1(E_{29})\left[\left(\frac{\pi n_x}{d_x}\right)^2 + \left(\frac{\pi n_y}{d_y}\right)^2\right] + F_2(E_{29})\left(\frac{\pi n_z}{d_z}\right)^2 \quad (1.133)$$

The electron concentration is given by

$$n_0 = \frac{2g_v}{d_x d_y d_z}\sum_{n_x=1}^{n_{x_{\max}}}\sum_{n_y=1}^{n_{y_{\max}}}\sum_{n_z=1}^{n_{z_{\max}}} F_{-1}(\eta_{1,29}) \quad (1.134)$$

where

$$\eta_{1,29} = (k_B T)^{-1}(E_F - E_{29})$$

Using (1.5) and (1.134), we get

$$G_0 = \frac{\pi^2 k_B}{3e}\left[\sum_{n_x=1}^{n_{x_{\max}}}\sum_{n_y=1}^{n_{y_{\max}}}\sum_{n_z=1}^{n_{z_{\max}}} F_{-1}(\eta_{1,29})\right]^{-1}\left[\sum_{n_x=1}^{n_{x_{\max}}}\sum_{n_y=1}^{n_{y_{\max}}}\sum_{n_z=1}^{n_{z_{\max}}} F_{-2}(\eta_{1,29})\right] \quad (1.135)$$

(c) The model of Foley *et al.*

In accordance with Foley *et al.* [103, 104], the electron dispersion relation assumes the form

$$E + \frac{E_{g_0}}{2} = \frac{\hbar^2 k_s^2}{2m_\perp^-} + \frac{\hbar^2 k_z^2}{2m_{||}^-} + \left[\left[\frac{\hbar^2 k_s^2}{2m_\perp^+} + \frac{\hbar^2 k_z^2}{2m_{||}^+} + \frac{E_{g_0}}{2}\right]^2 + P_{||}^2 k_z^2 + P_\perp^2 k_s^2\right]^{1/2} \quad (1.136)$$

where

$$\frac{1}{m_\perp^\pm} = \frac{1}{2}\left[\frac{1}{m_{tc}} \pm \frac{1}{m_{tv}}\right], \quad \frac{1}{m_{||}^\pm} = \frac{1}{2}\left[\frac{1}{m_{lc}} \pm \frac{1}{m_{lv}}\right],$$

m_{tc} and m_{lc} are the transverse and longitudinal effective electron masses of the conduction electrons at the edge of the conduction band, m_{tv} and

m_{lv} are the transverse and longitudinal effective hole masses of the holes at the edge of the valence band.

The total quantized energy E_{30} for this model is given by

$$E_{30} = -\frac{E_{go}}{2} + \frac{\hbar^2}{2m_{\perp}^{-}}[(\pi n_x/d_x)^2 + (\pi n_y/d_y)^2] + (\hbar^2/2m_{||}^{-})(n_z\pi/d_z)^2 + \left[\left[\frac{E_{go}}{2} + \frac{\hbar}{2m_{||}^{+}}(n_z\pi/d_z) + \frac{\hbar^2}{2m_{\perp}^{+}}\{(\pi n_x/d_x)^2 + (\pi n_y/d_y)^2\}\right]^2 + P_{||}^2(n_z\pi/d_z)^2 + P_{\perp}^2\{(\pi n_x/d_x)^2 + (\pi n_y/d_y)^2\}\right]^{1/2} \quad (1.137)$$

The electron concentration can be expressed as

$$n_0 = \frac{2g_v}{d_x d_y d_z} \sum_{n_x=1}^{n_{x_{\max}}} \sum_{n_y=1}^{n_{y_{\max}}} \sum_{n_z=1}^{n_{z_{\max}}} F_{-1}(\eta_{1,30}) \quad (1.138)$$

where

$$\eta_{1,30} = (k_B T)^{-1}(E_F - E_{30})$$

Using (1.5) and (1.138), we get

$$G_0 = \frac{\pi^2 k_B}{3e} \left[\sum_{n_x=1}^{n_{x_{\max}}} \sum_{n_y=1}^{n_{y_{\max}}} \sum_{n_z=1}^{n_{z_{\max}}} F_{-1}(\eta_{1,30})\right]^{-1} \left[\sum_{n_x=1}^{n_{x_{\max}}} \sum_{n_y=1}^{n_{y_{\max}}} \sum_{n_z=1}^{n_{z_{\max}}} F_{-2}(\eta_{1,30})\right] \quad (1.139)$$

*1.2.15 The MTP in quantum dots of lead germanium telluride

(a) The model of Vassilev

The dispersion law of n-type $Pb_{1-x}Ge_xTe$ with $x = 0.01$ can be expressed following Vassilev [105] as

$$[E - 0.606k_s^2 - 0.722k_z^2][E + \overline{E}_{go} + 0.411k_s^2 + 0.377k_z^2] = 0.23k_s^2 + 0.02k_z^2 \pm [0.06\overline{E}_{go} + 0.061k_s^2 + 0.0066k_z^2]k_s \quad (1.140)$$

where $\overline{E}_{go} = 0.21\,\text{eV}$, k_x, k_y and k_z are in the units of are in the units of $10^9\,m^{-1}$.

The total quantized energy ($E_{31,\pm}$) is given by

$$\left[E_{31,\pm} - 0.606\left(\left(\frac{\pi n_x}{d_x}\right)^2 + \left(\frac{\pi n_y}{d_y}\right)^2\right) - 0.0722\left(\frac{\pi n_z}{d_z}\right)^2\right]$$

$$\times\left[E_{31,\pm} + 0.471\left(\left(\frac{\pi n_x}{d_x}\right)^2 + \left(\frac{\pi n_y}{d_y}\right)^2\right) + \overline{E_{g_0}} + 0.0377\left(\frac{\pi n_z}{d_z}\right)^2\right]$$

$$= 0.23\left(\left(\frac{\pi n_x}{d_x}\right)^2 + \left(\frac{\pi n_y}{d_y}\right)^2\right) + 0.02\left(\frac{\pi n_z}{d_z}\right)^2$$

$$\pm\left[\left(\left(\frac{\pi n_x}{d_x}\right)^2 + \left(\frac{\pi n_y}{d_y}\right)^2\right)\right]^{1/2}$$

$$\left[0.06\overline{E_{g_0}} + 0.061\left(\left(\frac{\pi n_x}{d_x}\right)^2 + \left(\frac{\pi n_y}{d_y}\right)^2\right) + 0.0066\left(\frac{\pi n_z}{d_z}\right)^2\right] \quad (1.141)$$

The electron concentration is given by

$$n_0 = \frac{g_v}{d_x d_y d_z}\sum_{n_x=1}^{n_{x_{\max}}}\sum_{n_y=1}^{n_{y_{\max}}}\sum_{n_z=1}^{n_{z_{\max}}} F_{-1}(\eta_{1,31}) \quad (1.142)$$

where

$$\eta_{1,31} = (k_B T)^{-1}(E_F - E_{31,\pm})$$

Using (1.5) and (1.142), we get

$$G_0 = \frac{\pi^2 k_B}{3e}\left[\sum_{n_x=1}^{n_{x_{\max}}}\sum_{n_y=1}^{n_{y_{\max}}}\sum_{n_z=1}^{n_{z_{\max}}} F_{-1}(\eta_{1,31})\right]^{-1}\left[\sum_{n_x=1}^{n_{x_{\max}}}\sum_{n_y=1}^{n_{y_{\max}}}\sum_{n_z=1}^{n_{z_{\max}}} F_{-2}(\eta_{1,31})\right] \quad (1.143)$$

*1.2.16 The MTP in quantum dots of zinc and cadmium diphosphides

(a) The model of Chuiko

The dispersion relation of the holes of Cadmium and Zinc diphosphides can approximately be written following Chuiko [106] as

$$E = \left[\beta_1 + \frac{\beta_2\beta_3(k)}{8\beta_4}\right]k^2 \pm \left\{\left[\beta_4\beta_3(k)\left(\beta_5 - \frac{\beta_2\beta_3(k)}{8\beta_4}\right)k^2\right]\right.$$

$$\left. + 8\beta_4^2\left(1 - \frac{\beta_3^2(k)}{4}\right) - \beta_2\left(1 - \frac{\beta_3^2(k)}{4}\right)k^2\right\}^{1/2} \quad (1.144)$$

where $\beta_1, \beta_2, \beta_4$ and β_5 are system constants and

$$\beta_3(k) = \frac{k_x^2 + k_y^2 - 2k_z^2}{k^2}$$

The total quantized energy $E_{32,\pm}$ is given by

$$E_{32,\pm} = \left[\beta_1 + \frac{\beta_2\beta_6}{8\beta_4}\right]\beta_7^2 \pm \left\{\left[\beta_4\beta_6\left(\beta_5 - \frac{\beta_2\beta_6}{8\beta_4}\right)\beta_7^2\right] + 8\beta_4^2\left(1 - \frac{\beta_6^2}{4}\right) - \beta_2\left(1 - \frac{\beta_6^2}{4}\right)\beta_7^2\right\}^{1/2} \tag{1.145}$$

where

$$\beta_6 = \frac{1}{\beta_7^2}\left[\left(\frac{\pi n_x}{d_x}\right)^2 + \left(\frac{\pi n_y}{d_y}\right)^2 + \left(\frac{\pi n_z}{d_z}\right)^2\right] \quad \text{and}$$

$$\beta_7^2 = \left[\left(\frac{\pi n_x}{d_x}\right)^2 + \left(\frac{\pi n_y}{d_y}\right)^2 + \left(\frac{\pi n_z}{d_z}\right)^2\right]$$

The hole concentration can be written as

$$p_0 = \frac{g_v}{d_x d_y d_z} \sum_{n_x=1}^{n_{x_{\max}}} \sum_{n_y=1}^{n_{y_{\max}}} \sum_{n_z=1}^{n_{z_{\max}}} F_{-1}(\eta_{1,32}) \tag{1.146}$$

where

$$\eta_{1,32} = (k_B T)^{-1}(E_F - E_{32,\pm})$$

Using (1.5) and (1.146), we get

$$G_0 = \frac{\pi^2 k_B}{3e}\left[\sum_{n_x=1}^{n_{x_{\max}}} \sum_{n_y=1}^{n_{y_{\max}}} \sum_{n_z=1}^{n_{z_{\max}}} F_{-1}(\eta_{1,32})\right]^{-1}\left[\sum_{n_x=1}^{n_{x_{\max}}} \sum_{n_y=1}^{n_{y_{\max}}} \sum_{n_z=1}^{n_{z_{\max}}} F_{-2}(\eta_{1,32})\right] \tag{1.147}$$

1.2.17 The MTP in quantum dots of bismuth telluride

(a) The model of Kohler *et al.*

The dispersion relation of the holes in Bi_2Te_3 can be expressed as [107,108]

$$E(1 + \alpha E) = \frac{\hbar^2}{2m_0}((\overline{\alpha}_{11})k_x^2 + \alpha_{22}k_y^2 + \alpha_{33}k_z^2 + 2\alpha_{23}k_y k_z) \tag{1.148}$$

where $\overline{\alpha}_{11}, \alpha_{22}, \alpha_{33}$ and α_{23} are spectrum constants.

The total quantized energy E_{33} can be written as

$$E_{33}(1+\alpha E_{33}) = \frac{\hbar^2}{2m_0}\left[(\overline{\alpha}_{11})\left(\frac{n_x\pi}{d_x}\right)^2 + \alpha_{22}\left(\frac{n_y\pi}{d_y}\right)^2 + \alpha_{33}\left(\frac{n_z\pi}{d_z}\right)^2 + 2\alpha_{23}\left(\frac{\pi^2 n_y n_z}{d_y d_z}\right)\right] \quad (1.149)$$

The hole concentration is given by

$$p_0 = \frac{2g_v}{d_x d_y d_z}\sum_{n_x=1}^{n_{x_{\max}}}\sum_{n_y=1}^{n_{y_{\max}}}\sum_{n_z=1}^{n_{z_{\max}}} F_{-1}(\eta_{1,33}) \quad (1.150)$$

where

$$\eta_{1,33} = (k_B T)^{-1}(E_F - E_{33})$$

Using (1.5) and (1.150), we get

$$G_0 = \frac{\pi^2 k_B}{3e}\left[\sum_{n_x=1}^{n_{x_{\max}}}\sum_{n_y=1}^{n_{y_{\max}}}\sum_{n_z=1}^{n_{z_{\max}}} F_{-1}(\eta_{1,33})\right]^{-1}\left[\sum_{n_x=1}^{n_{x_{\max}}}\sum_{n_y=1}^{n_{y_{\max}}}\sum_{n_z=1}^{n_{z_{\max}}} F_{-2}(\eta_{1,33})\right] \quad (1.151)$$

1.2.18 The MTP in quantum dots of antimony

(a) The model of Ketterson

The dispersion relation of the conduction electrons in Antimony (Sb) can be written following Ketterson [109] as

$$\frac{2m_0 E}{\hbar^2} = \overline{\alpha_{11}}k_x^2 + \overline{\alpha_{22}}k_y^2 + \overline{\alpha_{33}}k_z^2 + 2\overline{\alpha_{23}}k_y k_z \quad (1.152)$$

$$\frac{2m_0 E}{\hbar^2} = a_1 k_x^2 + a_2 k_y^2 + a_3 k_z^2 + a_4 k_y k_z + a_5 k_z k_x + a_6 k_x k_y \quad (1.153)$$

$$\frac{2m_0 E}{\hbar^2} = a_1 k_x^2 + a_2 k_y^2 + a_3 k_z^2 + a_4 k_y k_z - a_5 k_z k_x - a_6 k_x k_y \quad (1.154)$$

where

$$a_1 = \frac{1}{4}(\overline{\alpha_{11}} + 3\overline{\alpha_{22}}), \quad a_2 = \frac{1}{4}(\overline{\alpha_{22}} + 3\overline{\alpha_{11}}), \quad a_3 = \overline{\alpha_{33}}, \quad a_4 = \overline{\alpha_{33}},$$

$$a_5 = \sqrt{3}, \quad a_6 = \sqrt{3}(\overline{\alpha_{22}} - \overline{\alpha_{11}}),$$

$\overline{\alpha_{11}}, \overline{\alpha_{22}}, \overline{\alpha_{33}}$ and $\overline{\alpha_{23}}$ and are the system constants.

The total quantized energy E_{34}, whose bulk carrier dispersion law is described by (1.152) can be written as

$$\frac{2m_0 E_{34}}{\hbar^2} = \overline{\alpha}_{11}\left(\frac{\pi n_{x1}}{d_x}\right)^2 + \overline{\alpha}_{22}\left(\frac{\pi n_{y1}}{d_y}\right)^2 + \overline{\alpha}_{33}\left(\frac{\pi n_{z1}}{d_y}\right)^2 + 2\overline{\alpha}_{23}\left(\frac{\pi^2 n_{y1}\cdot n_{z1}}{d_y d_z}\right) \tag{1.155}$$

Therefore, the electron concentration for the dispersion relation as described by (1.152) in this case assumes the form

$$n_{01} = \frac{2g_v}{d_x d_y d_z}\sum_{n_{x1}=1}^{n_{x1_{\max}}}\sum_{n_{y1}=1}^{n_{y1_{\max}}}\sum_{n_{z1}=1}^{n_{z1_{\max}}} F_{-1}(\eta_{1,34}) \tag{1.156}$$

where

$$\eta_{1,34} = (k_B T)^{-1}(E_F - E_{34})$$

The total quantized energy E_{35}, whose bulk carrier dispersion law is described by (1.153) can be written as

$$E_{35} = \frac{\hbar^2}{2m_0}\left[a_1\left(\frac{n_{x2}\pi}{d_x}\right)^2 + a_2\left(\frac{n_{y2}\pi}{d_y}\right)^2 + a_3\left(\frac{n_{z2}\pi}{d_z}\right)^2 + a_4\frac{n_{y2}n_{z2}\pi^2}{d_y d_z} + a_5\frac{n_{x2}n_{z2}\pi^2}{d_x d_z} + a_6\frac{\pi^2 n_{x2}n_{y2}}{d_x d_y}\right] \tag{1.157}$$

Similarly for QDs, whose bulk carrier dispersion law is given by (1.153), the corresponding electron concentration can be expressed as

$$n_{02} = \frac{2g_v}{d_x d_y d_z}\sum_{n_{x2}=1}^{n_{x2_{\max}}}\sum_{n_{y2}=1}^{n_{y2_{\max}}}\sum_{n_{z2}=1}^{n_{z2_{\max}}} F_{-1}(\eta_{1,35}) \tag{1.158}$$

where

$$\eta_{1,35} = (k_B T)^{-1}(E_F - E_{35})$$

The totally quantized energy E_{36}, whose bulk carrier dispersion law is described by (1.154) can be written as

$$E_{36} = \frac{\hbar^2}{2m_0}\left[a_1\left(\frac{n_{x3}\pi}{d_x}\right)^2 + a_2\left(\frac{n_{y3}\pi}{d_y}\right)^2 + a_3\left(\frac{n_{z3}\pi}{d_z}\right)^2 + a_4\frac{n_{y3}n_{z3}\pi^2}{d_y d_z} - a_5\frac{n_{x3}n_{z3}\pi^2}{d_x d_z} - a_6\frac{\pi^2 n_{x3}n_{y3}}{d_x d_y}\right] \tag{1.159}$$

Similarly for QDs, whose bulk carrier dispersion law is given by (1.154), the corresponding electron concentration can be written as

$$n_{03} = \frac{2g_v}{d_x d_y d_z} \sum_{n_{x3}=1}^{n_{x3_{\max}}} \sum_{n_{y3}=1}^{n_{y3_{\max}}} \sum_{n_{z3}=1}^{n_{z3_{\max}}} F_{-1}(\eta_{1,36}) \tag{1.160}$$

where

$$\eta_{1,36} = (k_B T)^{-1}(E_F - E_{36})$$

Therefore, the total electron concentration in this case is given by

$$n_0 = \frac{2g_v}{d_x d_y d_z} \left[\sum_{n_{x1}=1}^{n_{x1_{\max}}} \sum_{n_{y1}=1}^{n_{y1_{\max}}} \sum_{n_{z1}=1}^{n_{z1_{\max}}} [F_{-1}(\eta_{1,34})] + \sum_{n_{x2}=1}^{n_{x2_{\max}}} \sum_{n_{y2}=1}^{n_{y2_{\max}}} \sum_{n_{z2}=1}^{n_{z2_{\max}}} [F_{-1}(\eta_{1,35})] + \sum_{n_{x3}=1}^{n_{x3_{\max}}} \sum_{n_{y3}=1}^{n_{y3_{\max}}} \sum_{n_{z3}=1}^{n_{z3_{\max}}} [F_{-1}(\eta_{1,36})] \right] \tag{1.161}$$

Using (1.5) and (1.161), we get

$$G_0 = \frac{\pi^2 k_B}{3e} \left[\sum_{n_{x1}=1}^{n_{x1_{\max}}} \sum_{n_{y1}=1}^{n_{y1_{\max}}} \sum_{n_{z1}=1}^{n_{z1_{\max}}} [F_{-1}(\eta_{1,34})] + \sum_{n_{x2}=1}^{n_{x2_{\max}}} \sum_{n_{y2}=1}^{n_{y2_{\max}}} \sum_{n_{z2}=1}^{n_{z2_{\max}}} [F_{-1}(\eta_{1,35})] + \sum_{n_{x3}=1}^{n_{x3_{\max}}} \sum_{n_{y3}=1}^{n_{y3_{\max}}} \sum_{n_{z3}=1}^{n_{z3_{\max}}} [F_{-1}(\eta_{1,36})] \right]^{-1} \times \left[\sum_{n_{x1}=1}^{n_{x1_{\max}}} \sum_{n_{y1}=1}^{n_{y1_{\max}}} \sum_{n_{z1}=1}^{n_{z1_{\max}}} [F_{-2}(\eta_{1,34})] + \sum_{n_{x2}=1}^{n_{x2_{\max}}} \sum_{n_{y2}=1}^{n_{y2_{\max}}} \sum_{n_{z2}=1}^{n_{z2_{\max}}} [F_{-2}(\eta_{1,35})] + \sum_{n_{x3}=1}^{n_{x3_{\max}}} \sum_{n_{y3}=1}^{n_{y3_{\max}}} \sum_{n_{z3}=1}^{n_{z3_{\max}}} [F_{-2}(\eta_{1,36})] \right] \tag{1.162a}$$

1.3 Few Related Applications

1.3.1 Introduction

In this book, we shall investigate different quantum signatures of different quantized structures made of various technologically important materials. The concept of quantum signatures is one of the main keys for investigating the carrier transport phenomena of quantum effect devices as already written in the preface. In this section, we shall discuss **twenty four different applications** in this context.

1.3.2 Debye screening length

It is well known that the Debye screening Length (DSL) of the carriers in semiconductors is a very important quantity characterizing the screening of the Coulomb field of the ionized impurity centers by the free carriers [110]. It affects many of the special features of modern nano-devices, the carrier motilities under different mechanisms of scattering, and the carrier plasmas in semiconductors [111]. The DSL is a very good approximation to the accurate self-consistent screening in presence of band tails and is also used to illustrate the interaction between the colliding carriers in Auger effect in solids [111]. The classical value of the DSL is equal to $[\varepsilon_{sc} k_B T/(e^2 n_0)]^{1/2}$ which is valid for both the carriers. In this conventional form, the DSL decreases with increasing carrier concentration at a constant temperature and this relation holds only under the condition of carrier non-degeneracy. It is interesting to note that the under the condition of extreme degeneracy, the expression of DSL for materials having parabolic energy bands can be written as $L_D = (\pi^{2/3}\hbar\sqrt{\varepsilon_{sc}})(e g_v^{1/3} 3^{1/6} n_0^{1/6}\sqrt{m_c})^{-1}$. Thus, we observed that in this case the result is independent of temperature, but depends on n_0, g_v, and m_c. Besides, the indices of inverse electron variation changes from half in the former case to one-sixth in the latter case. Since, the performance of the electron devices at the device terminals and the speed of operation of modern switching transistors are significantly influence by the degree of carrier degeneracy present in these devices, the simplest way of analyzing such devices taking into account of the degeneracy of the band is to use the appropriate DSL to express the performance at the device terminal and switching speed in terms of the carrier concentration [112].

The DSL depends on the DOS function which, in turn, is significantly affected by the different carrier energy spectra of different semiconductors having various band structures. In recent years, various energy wave vector dispersion relations of the carriers of different materials have been proposed [113, 114] which have created the interest in studying the DSL in such quantized structures under external conditions. It is well known from the fundamental study of Landsberg [111], that the DSL for electronic materials having degenerate electron concentration is essentially determined by their respective energy band structures. It has, therefore, different values in different materials and varies with the electron concentration, with the magnitude of the reciprocal quantizing magnetic field under magnetic quantization, with the quantizing electric field as in inversion layers, with the nano-thickness as in quantum wells, with superlattice period as in the quantum confined superlattices of small gap compounds with graded interfaces having various carrier energy spectra. The nature of these variations has been investigated by in the literature [115–123].

The 3D and 2D DSL can, in general, be expressed as [113]

$$L_{3\mathrm{D}} = \left[\frac{e^2}{\varepsilon_{sc}}\frac{\partial n_0}{\partial E_F}\right]^{-1/2} \tag{1.162b}$$

$$L_{2\mathrm{D}} = \left[\frac{e^2}{2\varepsilon_{sc}}\frac{\partial n_0}{\partial E_F}\right]^{-1} \tag{1.162c}$$

Using (1.5) and (1.162b), we can write

$$L_{3\mathrm{D}} = \left(\frac{\pi^2 k_B^2 T \varepsilon_{sc}}{3e^3 n_0 G_0}\right)^{-1/2} \tag{1.163}$$

The 3D DSL for the heavily doped systems is given by

$$L_{3\mathrm{D}} = \left[\frac{e^2}{\varepsilon_{sc}}\frac{\partial n_{0\mathrm{HD}}}{\partial (E_{\mathrm{FHD}} - E_{0\mathrm{HD}})}\right]^{-1/2} \tag{1.164}$$

Using (1.7) and (1.164), we obtain the same equation as given by (1.163). Thus, the 3D DSL for degenerate materials can be determined by knowing the experimental values of G.

For inversion layers and NIPI structures, the 2D DSL is given by

$$L_{2\mathrm{D}} = \left[\frac{e^2}{2\varepsilon_{sc}}\frac{\partial n_{02\mathrm{D}}}{\partial (E_{F2\mathrm{D}} - E_{02\mathrm{D}})}\right]^{-1} \tag{1.165}$$

Using (1.6) and (1.165), we get

$$L_{2\mathrm{D}} = \frac{2\pi^2 k_B^2 T \varepsilon_{sc}}{3e^3 n_{02\mathrm{D}} G_0} \tag{1.166}$$

For QWs, the G_0 assumes the form

$$G_0 = \frac{\pi^2 k_B^2 T}{3en_{2D}} \cdot \frac{\partial n_{2D}}{\partial E_{Fs}} \tag{1.167}$$

where n_{2D} and E_{F_s} are 2D electron concentration and Fermi energy in this case. Using (1.162c) and (1.167), we get

$$L_{2D} = \frac{2\pi^2 k_B^2 T \varepsilon_{sc}}{3e^3 n_{2D} G_0} \tag{1.168}$$

From the suggestion for the experimental determination of the 3D DSL for degenerate materials having arbitrary dispersion laws as given by (1.163), we observe that for a constant T, the DSL varies inversely with the square root of $G_0 n_0$. Only the experimental values of G_0 for any material as a function of electron concentration will generate the experimental values of the 3D DSL for that range of n_0 for that material. Since $(G_0 n_0)^{-1/2}$ decreases with increasing n_0 for constant T, from (1.163), we can conclude that the 3D DSL will decrease with increasing n_0. For 2D DSL L_D at a constant temperature varies inversely with $n_{02D}G_0$ or $n_{2D}G_0$. Since $n_{02D}G_0$ and $n_{2D}G_0$ increase with decreasing surface concentration, we can infer that 2D DSL will increase with decreasing n_{02D} or n_{2D} for the appropriate 2D systems. This statement provides a compatibility test of our theoretical analysis. Thus, (1.163), (1.166) and (1.168) provide experimental checks of both 3D and 2D DSLs and also a technique for probing the band structures of the different materials having arbitrary band structures.

1.3.3 Carrier contribution to the elastic constants

The knowledge of the carrier contribution to the elastic constants is very important in studying the mechanical properties of the materials and has been investigated in the literature [124–127].

The electronic contribution to the second- and third-order elastic constants can be written as [124–127]

$$\Delta C_{44} = \left[\frac{-(\overline{G}_0)^2}{9}\frac{\partial n_0}{\partial (E_F)}\right] \tag{1.169a}$$

and

$$\Delta C_{456} = \left[\frac{(\overline{G}_0)^3}{27}\frac{\partial^2 n_0}{\partial (E_F)^2}\right] \tag{1.169b}$$

where $\overline{G}_0$ is the deformation potential constant.

The electronic contribution to the second- and third-order elastic constants for HD materials can be written as [124–127]

$$\Delta C_{44} = \left[\frac{-(\overline{G}_0)^2}{9}\frac{\partial n_{0\mathrm{HD}}}{\partial(E_{\mathrm{FHD}} - E_{0\mathrm{HD}})}\right] \tag{1.170a}$$

and

$$\Delta C_{456} = \left[\frac{(\overline{G}_0)^3}{27}\frac{\partial^2 n_{0\mathrm{HD}}}{\partial(E_{\mathrm{FHD}} - E_{0\mathrm{HD}})^2}\right] \tag{1.170b}$$

Thus, using (1.7), (1.169) and (1.170), we can write

$$\Delta C_{44} = \left[-n_0(\overline{G}_0)^2|e|\frac{G}{3\pi^2 k_B^2 T}\right] \tag{1.171}$$

and

$$\Delta C_{456} = \left[n_0|e|(\overline{G}_0)^3\frac{G^2}{3\pi^4 k_B^3 T}\right]\left(1 + \frac{n_0}{G}\frac{\partial G}{\partial n_0}\right) \tag{1.172}$$

Therefore, by using (1.171) and (1.172), we can investigate ΔC_{44} and ΔC_{456} for all the cases of this monograph. Besides, the experimental graph of (G_0) versus n_0 allows us to determine the electronic contribution to the elastic constants for materials having arbitrary spectra.

1.3.4 Einstein relation

It is well known that the Einstein relation (ER) for the diffusivity-mobility ratio (DMR) occupies a central position in the whole field of solid state device electronics and the related sciences since the diffusion constant (a quantity very useful for device analysis where exact experimental determination is rather difficult) can be obtained from this ratio by knowing the experimental values of the mobility. The classical value of the DMR is equal to $(k_B T|e|)$. This relation in this form was first introduced by Einstein to study of the diffusion of gas particles and is known as the Einstein relation [128, 129]. It appears that the DMR increases linearly with increasing T and is independent of electron concentration. This relation is applicable for both types of charge carriers only under non-degenerate carrier concentration although its validity has been suggested erroneously for degenerate materials [130]. Landsberg first pointed out that the DMR for degenerate semiconductors is essentially determined by their energy band structures [131,132]. This relation is useful for semiconductor homo-structures [131,132], semiconductor-semiconductor hetero-structures [135, 136], metals-semiconductor hetero-structures and insulator-semiconductor hetero-structures [146–151]. The nature of the variations of the DMR under different physical conditions has been studied in the literature [128–152]. Incidentally, A. N. Chakravarti (a recognized leading expert of DMR in general) and has passed away in 2023 although his extremely productive research group (where the first author of this book is still a powerful member) are still contributing significantly regarding this pin pointed research topic on DMR from 1972 [127, 150–173] and some of the significant features, which have emerged from these studies, are:

The DMR increases monotonically with increasing carrier concentration in bulk semiconductors and the nature of these variations are significantly influenced by the band structures of different materials.

The DMR increases with the increasing quantizing electric field as in inversion layers.

The DMR oscillates with the inverse quantizing magnetic field under magnetic quantization due to the Shubnikov-de Haas effect.

The DMR shows composite oscillations with the various controlled quantities of semiconductor superlattices.

In ultrathin films, quantum wires and other field assisted low-dimensional systems, the value of the DMR changes appreciably with the external variables depending on the nature of quantum confinements of different materials.

The DMR depends on the DOS function, which, in turn, is significantly affected by the different carrier energy spectra of different semiconductors having various band structures. It can, in general, be proved that for bulk specimens the DMR is given by [149]

$$\frac{D}{\mu} = \frac{n_{0\mathrm{HD}}}{e}\left[\frac{\partial n_{0\mathrm{HD}}}{\partial(E_{\mathrm{FHD}} - E_{0\mathrm{HD}})}\right]^{-1} \tag{1.173}$$

The electric quantum limit as in inversion layers and NIPI structures refers to the lowest electric sub-band and the DMR assumes the form

$$\frac{D}{\mu} = \frac{n_{02\mathrm{D}}}{e}\left[\frac{\partial n_{02\mathrm{D}}}{\partial(E_{\mathrm{F2D}} - E_{02\mathrm{D}})}\right]^{-1} \tag{1.174}$$

Using (1.7), (1.173) and (1.174), one obtains

$$\frac{D}{\mu} = \left(\frac{\pi^2 k_B^2 T}{3|e|^2 G_0}\right) \tag{1.175}$$

Thus, the DMR for degenerate materials can be determined by knowing the experimental values of G_0.

The suggestion for the experimental determination of the DMR for degenerate semiconductors having arbitrary dispersion laws as given by (1.175) does not contain any energy band constants. For a fixed temperature, the DSL varies inversely as G_0. Only the experimental values of G_0 for any material as a function of electron concentration will generate the experimental values of the DMR for that range of n_0 for that system. Since G_0 decreases with increasing n_0, from (1.175) one can infer that the DMR will increase with increase in n_0. This statement is the compatibility test so far as the suggestion for the experimental determination of DMR for degenerate materials is concerned.

1.3.5 Diffusion coefficient of the minority carriers

This particular coefficient in quantum confined lasers can be expressed as [174]

$$\frac{D_i}{D_0} = \frac{dE_{Fi}}{dE_F} \tag{1.176}$$

where D_i and D_0 are the diffusion coefficients of the minority carriers both in the presence and absence of quantum confinements and E_{Fi} and E_F are the Fermi energies in the respective cases. It appears then that, the formulation of the above ratio requires a relation between E_{Fi} and E_F, which, in turn, can be formed through the respective electron statistics as discussed in various cases in this monographs.

1.3.6 Nonlinear optical response

The nonlinear response from the optical excitation of the free carriers is given by [175]

$$Z_0 = \frac{-e^2}{\omega^2\hbar^2}\int_0^\infty \left(k_x\frac{\partial k_x}{\partial E}\right)^{-1} f_0 N(E)dE \tag{1.177}$$

where ω is the optical angular frequency. From the various E–k relations of different materials under different physical conditions and from band structure, we can derive the term $\left(k_x\frac{\partial k_x}{\partial E}\right)$ and thus by using the DOS function as formulated, we can study the Z_0 for all types of materials as considered in this book.

1.3.7 Third-order nonlinear optical susceptibility

This particular susceptibility can be written as [176]

$$\chi_{\mathrm{NP}}(\omega_1, \omega_2, \omega_3) = \frac{n_0 e^4 \langle \varepsilon^4 \rangle}{24\omega_1\omega_2\omega_3(\omega_1 + \omega_2 + \omega_3)\hbar^4} \tag{1.178}$$

where $n_0\langle\varepsilon^4\rangle = \int_0^\infty \frac{\partial^4 E}{\partial k_z^4} N(E) f_0 dE$ and the other notations are defined in [176]. The term $\left(\frac{\partial^4 E}{\partial k_z^4}\right)$ can be formulated by using the DRs of different materials as given in appropriate sections of this book. Thus, one can investigate the $\chi_{\mathrm{NP}}(\omega_1\omega_2\omega_3)$ by using the DRs for all materials as considered in this book.

1.3.8 Generalized Raman gain

The generalized Raman gain in optoelectronic materials can be expressed as [177]

$$R_G = \overline{I}\left(\frac{16\pi^2 c^2}{\hbar\omega\rho g\omega_s^2 n_s n_p}\right)\left(\frac{\Gamma_\rho}{\Gamma}\right)\left(\left(\frac{e^2}{mc^2}\right)^2 m^2 R^2\right) \tag{1.179}$$

where $\overline{I} = \sum_{n,t_z}[f_0(n, k_z, \uparrow) - f_0(n, k_z, \downarrow)]$, $f_0(n, k_z, \uparrow)$ is the Fermi factor for spin-up Landau levels, $f_0(n, k_z, \uparrow)$ is the Fermi factor for spin down Landau levels, n is the Landau quantum number and the other notations are defined in [177]. It appears then the formulation of R_G is determined by the appropriate derivation requires the magneto-dispersion relations. By using the different appropriate formulas as formulated in various chapters of this monograph R_G can, in general, be investigated.

1.3.9 The plasma frequency

The plasma frequency ω_p can be expressed as [178]

$$\omega_P^2 = \frac{e^2}{\varepsilon_{sc}} \int_E^\infty N(E) v_x^2 \frac{\partial f_0(E)}{\partial E} dE \tag{1.180}$$

where E' can be obtained from $N(E') = 0$.

Thus, by using the expressions of DOS functions from different chapters, we can study the Plasma Frequency for various quantized structures.

1.3.10 The activity coefficient

Dilute solution transport equations introduced by Landsberg [179] have been very effective in describing the fluxes of electrons and holes in semiconductors. These expressions, however, are inaccurate at high carrier concentrations due to non-idealities associated with carrier degeneracy and band gap narrowing. Within the frame work of dilute solution theory, these non-idealities can be included by introduction of concentration-dependent activity coefficients. Expressions for activity coefficients should also be useful in the area of chemical kinetics where mass-action principles are applied. Reviews of the literature on activity coefficients are available. Of special interest is a paper by Hwangand Brews [179], where majority carrier activity coefficients were presented that included the effects of Fermi statistics and band gap narrowing due to exchange interactions and impurity effects. Explicit expressions and plots of electron activity coefficients as a function of concentration for GaAs were included. Activity coefficients were derived under the assumption of complete ionization. No mention was made, however, of minority carrier screening by majority carriers, which are shown here to have a strong influence upon the majority carrier activity coefficients.

The activity coefficient in semiconductors can, in general be expressed as [179]

$$f_e^* = \frac{k_B T}{n_0} \int_0^\infty \frac{N(E)}{(1-f_0)^2} \left[-\frac{\partial f_0}{\partial E} \right] dE \tag{1.181}$$

Thus, by using the expressions of DOS functions from different chapters, we can study the activity coefficient for various quantized structures.

1.3.11 Magneto-thermal effect in quantized structures

The magneto-thermal effect can be expressed by the dimensionless quantity as [180]

$$\phi = \frac{B}{T}\frac{\partial T}{\partial B} = -\frac{B}{T}\left[\left(\frac{\partial S}{\partial \beta}\right)_{(T,n)}\right]\left[\left(\frac{\partial S}{\partial T}\right)_{(\beta,n)}\right]^{-1} \tag{1.182}$$

where S is the entropy which can be written as [180]

$$S = k_B \int_0^\infty -\left(\frac{\partial f_0}{\partial E}\right)\left(\frac{E-E_F}{k_B T}\right)\left[\int_0^E N(E)dE\right] dE \tag{1.183}$$

Under magnetic quantization we know, the DOS function assumes the form

$$N(E) = \left(\frac{eBg_v}{2\pi^2\hbar}\right) \sum_{n=0}^{n_{\max}} \frac{\partial k_z'(E)}{\partial E} H(E - E') \tag{1.184}$$

where E' is the Landau sub-band energy or Landau level.

Therefore,

$$S = k_B \sum_{n=0}^{n_{\max}} \int_{E'}^{\alpha} -\left(\frac{\partial f_0}{\partial E}\right)\left(\frac{E - E_F}{k_B T}\right) \frac{eBg_v}{2\pi^2\hbar} k_z dE \tag{1.185}$$

Since

$$\int N(E)dE = \frac{eBg_v}{2\pi^2\hbar} \sum_{n=0}^{n_{\max}} k_z(E) \tag{1.186}$$

$$S = c_0 \sum_{n=0}^{n_{\max}} \left[\int_{E'}^{\alpha} \left\{k_z E\left(\frac{\partial f_0}{\partial E}\right) dE\right\} - E_F \int_{E'}^{\alpha} \left\{k_z \left(\frac{-\partial f_0}{\partial E}\right) dE\right\}\right] \tag{1.187}$$

where $c_0 = (eBg_v/2\pi^2\hbar T)$.

Using the concept of 3D Quantization (magneto-size quantization, Magneto NIPI, Magneto Inversion layers, Magneto accumulation Layers, quantum dot, Quantum dot superlattices, magneto-size quantized superlattices and the respective heavily doped cases together with influence of light waves and electric fields on the 3D quantized structures), the DOS functions can respectively be written as

$$N(E) = \frac{g_v eB}{2\pi\hbar} \sum_{n_z=0}^{n_{z\max}} \sum_{n=0}^{n_{\max}} \delta'(E - E'') \tag{1.188}$$

$$N(E) = \frac{g_v eB}{2\pi\hbar} \sum_{i=0}^{i_{\max}} \sum_{n=0}^{n_{\max}} \delta'(E - E'') \tag{1.189}$$

and

$$N(E) = \frac{2g_v}{(d_x d_y d_z)} \sum_{n_x=1}^{n_{x\max}} \sum_{n_y=1}^{n_{y\max}} \sum_{n_z=1}^{n_{z\max}} \delta'(E - E''') \tag{1.190}$$

where E'' is the totally quantized energy in 2D systems in the presence of applied quantizing magnetic field along z-direction and E''' is the totally quantized energy in 0D systems in the absence of quantizing magnetic field along z-direction, (1.188) represents all the cases corresponding to E'',

(1.189) represents all the cases corresponding to Magneto Inversion layers and Magneto accumulation Layers and (1.190) represents the rest cases.

In the three cases, we can show

$$S = c_0'' \sum\sum E''[F_0(\eta'') - (\eta'')F_{-1}(\eta'')] \tag{1.191}$$

$$S = c_0''' \sum\sum E'''[F_0(\eta''') - (\eta''')F_{-1}(\eta''')] \tag{1.192}$$

where

$$c_0'' = \frac{g_v eBk_B}{2\pi\hbar}, \quad c_0''' = \frac{2g_v}{d_x d_y d_z}, \quad \eta'' = \frac{E_F - E''}{k_B T}, \eta''' = \frac{E_F - E'''}{k_B T}$$

and E_F in this case is the Fermi energy for all the respective cases of 3D quantization of wave vector space.

Thus, by using the appropriate equations, we can study the magneto-thermal effect for different quantized structures.

1.3.12 Normalized Hall coefficient

The normalized Hall coefficient in a semiconductor can be expressed as [181]

$$\frac{R(T)}{R(0)} = 1 + \frac{(\pi k_B T)^2}{3}\left[\frac{1}{N(E)}\frac{\partial N(E)}{\partial E}\right]^2\Bigg|_{E=E_{FT}} \tag{1.193}$$

where $R(T)$ is the Hall coefficient at $T, R(0)$ is the same at $T \to 0$ and E_{FT} is the Fermi energy at T.

Thus, by using the expressions of DOS functions from different chapters, we can study the Normalized Hall coefficient for various quantized structures.

1.3.13 Reflection coefficient

In the presence of light waves, the reflection coefficient (R) is given by [182]

$$R = \frac{(x-1)^2}{(x+1)^2} \tag{1.194}$$

where $x = \sqrt{\varepsilon'_{sc} - \frac{n_0 e^2 \lambda^2}{m\pi c^2}}$, m is the effective electron mass at the Fermi level and c is the velocity of light.

Thus, by using the appropriate expressions from different chapters, we can study the reflection coefficient for various quantized structures.

1.3.14 Heat capacity

The heat capacity of the carriers in semiconductors is a very important quantity.

The total heat capacity c_t is given by

$$c_t = c_e + c_l \tag{1.195}$$

where c_e is the electronic heat capacity and is given by [183]

$$c_e = \int_0^{\infty} -\frac{\partial f_0}{\partial E}\left[\frac{E-E_F}{T}+\frac{\partial E_F}{\partial T}\right](E-E_F)N(E)dE \tag{1.196}$$

Thus, (1.196) can be written as

$$c_t = \frac{\pi^2 k_B^2 T N(E_F)}{3}\left[1+T\left(\frac{\partial E_F}{\partial T}\right)\left(\frac{N'(E_F)}{N(E_F)}\right)\right]+c_l \tag{1.197}$$

Thus, by using the expressions of DOS functions from different chapters, we can study the Heat Capacity for various quantized structures.

1.3.15 Magnetic susceptibilities

The Landau dia-magnetic susceptibility (χ_d) can be written as

$$\chi_d = -\mu_0 \frac{\partial^2 F(B)}{\partial B^2} \tag{1.198}$$

where μ_0 is the free space permeability and $F(B)$ is the free energy which can be written as

$$F(B) = n_0 E_{FB} - \int_{k_0}^{\infty} D_2(E) f_0 dE \tag{1.199}$$

where k_0 is the lower limit of integration,

$$D_2(E) = \int N(E)dE \tag{1.200}$$

and $N(E)$ is the corresponding DOS function.

The Pauli's paramagnetic susceptibility (χ_p) can be expressed [184] through equations (1.199) and (1.200) where the electron spin has to be incorporated in $N(E)$ by using the respective DR.

Thus, by using the expressions of DOS functions from different chapters, we can study the Magnetic susceptibilities for various quantized structures.

1.3.16 Faraday rotation

The Faraday rotation of the electrons in semiconductors can in general, be expressed under the condition $\omega\tau \gg 1$, as [185]

$$\theta = AI_F \tag{1.201}$$

where ω is the angular frequency of the incident light, τ is the momentum relaxation time,

$$A \equiv \frac{e^3 g_0}{8\pi^3\hbar^4\omega^2}\left(\frac{1}{2n_r\varepsilon_0 c}\right),$$

g_0 is the thickness of the semiconductor specimen, n_r is the real part of the refractive index at that frequency and

$$I_F = \frac{(2\pi)^3}{2g_v}\int_0^\infty \frac{\partial E}{\partial k_x}\left[\frac{\partial E}{\partial k_y}\cdot\frac{\partial^2 E}{\partial k_x \partial k_y} - \frac{\partial E}{\partial k_x}\cdot\frac{\partial^2 E}{\partial k_y^2}\right]\frac{\partial f_0}{\partial E}N(E)dE \tag{1.202}$$

Thus, from different DRs as formulated in this book, we can formulate the partial differentials and from the expressions of $N(E)$ as derived in different chapters, we can study the Faraday rotations for various quantized structures.

1.3.17 Fowler–Nordheim Field emission

The Fowler–Nordheim field emission (FNFE) is a well-known quantum-mechanical phenomenon that involves tunneling of electrons through a surface barrier due to the application of an intense external electric field. Normally, at field strengths of the order of 10^8 V/m (below the electrical breakdown), the potential barriers at the surfaces of metals and semiconductors usually become very thin and result in field emission of electrons due to the tunnel effect [186]. This has been well-investigated with reference to three dimensional electron gases in metals and semiconductors and the FNFE from quantum confined structures has also been studied in this context [187–193]

The current (I) due to Fowler–Nordheim $\{FN\}$ field emission can be written as

$$I = \frac{1}{2}\sum_{n_x=1}^{n_{x\max}}\sum_{n_y=1}^{n_{y\max}}\left[\int_{E_{11}}^{i}[e.vn_1t_{11}]\right] \tag{1.203}$$

where e is the magnitude of the electron charge, v is the velocity of the electron and given by

$$v = \frac{1}{\hbar}\frac{\partial E}{\partial k_z}n_1 = N_{1D}(E)\cdot f_0 dE, \quad N_{1D}(E)$$

is the 1D DOS function per sub-bands and can be written as

$$N_{1D}(E) = \frac{2g_v}{\pi}\frac{\partial k_z}{\partial E},$$

t_{11} is the transmission coefficient which can be expressed as $t_{11} = \exp[-2\int_i |k_i|di]$ in which $i = x, i, z$.

Thus, by using the expressions of DOS functions from different chapters, we can study the Fowler–Nordheim Field Emission for various quantized structures.

1.3.18 Optical effective mass

The investigation of the energy band structure of the noble metals is one of the standard problems in solid state physics. The topology and geometry of the Fermi surfaces of the noble metals are very well known in detail from the de Hass–van Alphen measurements of Shoenberg. Very accurate values for the cyclotron masses of the noble metals were experimentally obtained from de Hass–van Alphen investigations [194]. The optical effective mass being used to study the different optical properties of semiconductors can be expressed following Madelung [194–226] as

$$m_{opt} = 3\hbar^2 n_0 \left[\int_0^\infty N(E)\frac{\partial E}{\partial k} f_0 dE\right]^{=1} \tag{1.204}$$

Thus, by using the expressions of DOS functions from different chapters, we can study the Optical Effective Mass for various quantized structures.

1.3.19 Righi–Leduc coefficient

It is well known that some transport effects under strong magnetic field are independent of relaxation mechanism and are determined by the DRs only. Among them one is thermoelectric power and the other one is Righi–Leduc Coefficient, which can in turn, be written as [227]

$$R = \frac{cn_0}{eB^2\chi_{ph}}\left[\langle(E-E_F)^2\rangle - \langle E-E_F\rangle^2\right] \tag{1.205}$$

where χ_{ph} is the phonon thermal conductivity,

$$\langle(E-E_F)^2\rangle = \frac{1}{n}\int (E-E_F)^2 G(E)\left[-\frac{\partial f_0}{\partial E}\right] dE$$

$G(E)$ is the total number of states and is given by

$$G(E) = \int_{E_{30}}^{E} N(\zeta)d\zeta$$

in which ζ is the variable of integration, E_{30} is the lower limit of integration and

$$\langle E - E_F \rangle = \frac{1}{n} \int (E - E_F) G(E) \left[-\frac{\partial f_0}{\partial E} \right] dE$$

Thus, by using the expressions of DOS functions from different chapters, we can study the Righi–Leduc coefficient for various quantized structures.

1.3.20 Electric susceptibility

In semiconductors the effect of free carriers on the optical properties becomes important at wavelengths longer than the intrinsic absorption edge. Free carriers produce absorption and affect also the dispersion at sufficiently long wave lengths electric susceptibility can be written as [228]

$$\chi_c = \frac{-e^2}{3\omega^2 \varepsilon_0 \hbar^2} \int_0^\infty N(E) \frac{\partial E}{\partial k} f_0 dE \tag{1.206}$$

Thus, by using the expressions of DOS functions from different chapters, we can study the Electric Susceptibility for various quantized structures.

1.3.21 Electric susceptibility mass

In recent years there has been considerable interest in studying the electric susceptibility mass of the carriers in narrow gap materials because of its direct influence in the study of carrier scattering mechanisms in semiconductor. Electric susceptibility mass can be determined from measurements in the infrared region of the frequency dependence of the spectral reflectivity at normal incidence and provides useful information regarding band structures. Electric Susceptibility Mass can be written as [229]

$$m_s^* = 3A_{20} \left[\int_{E_{20}}^\infty N(E) \frac{\partial E}{\partial k} f_0 dE \right]^{-1} \tag{1.207}$$

where $A_{20} = n_0 \varepsilon_0 \hbar^2$ and E_{20} is the lower limit of integration.

Thus, by using the expressions of DOS functions from different chapters, we can study the Electric Susceptibility Mass for various quantized structures.

1.3.22 Electron diffusion thermo-power

The electron diffusion thermo power (S_e) is an important property of AlGaN/GaN QLD structures since it yields useful information about the carrier transport mechanism. S_e can be written as [230]

$$S_e = -\frac{1}{n_0 e}\frac{df}{dT} \tag{1.208}$$

where

$$f = n_0 E_F - \sum \int_{E_{21}}^{\infty} G(E) f_0 dE$$

in which

$$G(E) = \int_{E_{30}}^{E} N(\zeta) d\zeta, \quad E_{21} \quad \text{and} \quad E_{30}$$

are the lower limits of integrations.

Thus, by using the expressions of DOS functions from different chapters, we can study the Electron Diffusion Thermo-power for various quantized structures.

1.3.23 Hydrostatic Piezo-resistance coefficient

The importance of Hydrostatic Piezo-resistance Coefficient (H_c) is well known in the sensor technology and H_c can be expressed as [231]

$$H_c = \frac{-\partial n_0}{n_0 \partial P} \tag{1.209}$$

where P is the pressure in kilobar.

Thus, by using the expressions of integrated DOS functions from different chapters, we can study the Hydrostatic Piezo-resistance Coefficient for various quantized structures.

1.3.24 Relaxation time for acoustic mode scattering

The importance of momentum relaxation time for acoustic mode scattering in semiconductors is well known for the evaluation of the electron mobility and in the simplest scattering theory, the relaxation time for acoustic mode scattering ($\tau_{\mathrm{AMS}}(E)$) [231] is directly proportional to the inverse of $N(E)$ and can, in general be expressed as

$$\tau_{\mathrm{AMS}}(E) = \frac{\tau_0}{N(E)} \tag{1.210}$$

where τ_0 is a constant.

Thus, using various expressions of $N(E)$ in different cases, we can study the $\tau_{am}(E)$ for various quantized structures.

1.3.25 Gate capacitance

In recent years there has been considerable interest in studying the surface capacitance of inversion layers in Si MOS structures. The dependence of the MOS capacitance on gate voltage and on a quantizing magnetic field has been studied both theoretically and experimentally. The fact that the MOS capacitance can be very easily controlled by varying the gate voltage is of much importance from the point of view of technical applications. It may also be stated that, as far as the determination of the effective mass under the degenerate electron distribution at the surface is concerned, measurement of magneto-capacitance as compared to that of conductivity or cyclotron resonance would not be more advantageous regarding the experimental facilities required or accuracies achieved. Nevertheless, it is felt that the theoretical investigation presented here would be of much significance as the interest on gate capacitance has been growing very much in recent years from the point of view of exploration of other fundamental aspects of semiconductor surfaces of MOS structures [232, 233].

The gate capacitance (C_g) can in general be expressed as

$$C_g = \frac{e\partial n_s}{\partial V_g} \tag{1.211}$$

where n_s is the surface electron concentration per unit area and V_g is the gate voltage. It may be noted that although C_g for inversion layers have been studied in literature but investigations of C_g in accumulation layers are rather small.

Thus, by using the expressions of the electron statistics of inversion and accumulation layers as given in this monograph, we can study the gate capacitance for different important cases.

1.4 Result and Discussions

Using the appropriate values of the energy band constants as given in Table 1.1, in Figs. 1.1, 1.3 and 1.5, the normalized MTP in QDs of InAs, GaAs and InSb has been plotted as a function of film thickness in accordance with the three (using (1.23) and (1.24)) and two (using (1.19) and (1.20)) band models of Kane [79] together with parabolic energy bands (using (1.12) and (1.16)) as shown by curves (a), (b) and (c) in the

Table 1.1. The numerical values of the energy band constants of the dispersion relations of few materials.

S. no	Materials	Numerical Values of the Energy Band Constants
1	The conduction electrons of n-Cadmium Germanium Arsenide can be described by three types of band models	The values of the energy band constants in accordance with the generalized electron dispersion relation of nonlinear optical materials are as follows: $E_{g0} = 0.57\,\text{eV}$, $\Delta_{\parallel} = 0.30\,\text{eV}$, $\Delta_{\perp} = 0.36\,\text{eV}$, $m_{\parallel}^* = 0.034m_0$, $m_{\perp}^* = 0.039m_0$, $T = 4K$, $\delta = -0.21\,\text{eV}$ and $g_v = 1$ [100]. In accordance with the three band model of Kane, the spectrum constants are given by $\Delta = (\Delta_{\parallel} + \Delta_{\perp})/2 = 0.33\,\text{eV}$, $E_{g0} = 0.57eV$, $m_c = (m_{\parallel}^* + m_{\perp}^*)/2 = 0.0365m_0$ and $\delta = 0\,\text{eV}$. In accordance with two band model of Kane, $E_{g0} = 0.57\,\text{eV}$ and $m_c = 0.0365m_0$.
2	n-Indium Arsenide	The values $E_{g0} = 0.36\,\text{eV}$, $\Delta = 0.43\,\text{eV}$, $m_c = 0.026m_0$ and $g_v = 1$ [79] are valid for three band model of Kane.
3	n-Gallium Arsenide	The values $E_{g0} = 1.55\,\text{eV}$, $\Delta = 0.35\,\text{eV}$, $m_c = 0.07m_0$ and $g_v = 1$ are valid for three band model of Kane [79]. The values $\alpha_{13} = -1.97 \times 10^{-37}\text{eVm}^4$ and $a_{15} = -2.3 \times 10^{-34}\text{eVm}^4$ are valid for the model of Newson and Kurobe [81]. The values $\alpha_{11} = -2132 \times 10^{-40}\text{eVm}^4$, $\alpha_{12} = 9030 \times 10^{-50}\text{eVm}^5$, $\beta_{11} = -2493 \times 10^{-40}\text{eVm}^4$, $\beta_{12} = 12594 \times 10^{-50}\text{eVm}^5$, $\gamma_{11} = 30 \times 10^{-30}\text{eVm}^3$, $\gamma_{12} = -154 \times 10^{-42}\text{eVm}^4$ are valid for the model of Rossler [82].
4	n-Gallium Aluminium Arsenide	$E_{g0} = (1.424 + 1.266x + 0.26x^2)\text{eV}$, $\Delta = (0.34 - 0.5x)\text{eV}$, $m_c = [0.066 + 0.088x]m_0$ and $g_v = 1$ [234].
5	n-Mercury Cadmium Telluride	$E_{g0} = (-0.302 + 1.93x + 5.35 \times 10^{-4}(1 - 2x)T - 0.810x^2 + 0.832x^3)\text{eV}, \Delta = (0.63 + 0.24x - 0.27x^2)\text{eV}$, $m_c = 0.1m_0 E_{g0}(\text{eV})^{-1}$ and $g_v = 1$ [235].
6	n-Indium Gallium Arsenide Phosphide lattice matched to Indium Phosphide	$E_{g0} = (1.337 - 0.73y + 0.13y^2)\text{eV}$, $\Delta = (0.114 + 0.26y - 0.22y^2)\text{eV}, m_c = (0.08 - 0.039y)m_0$, $y = (0.1896 - 0.4052x)/(0.1896 - 0.0123x)$ and $g_v = 1$ [236].
7	n-Indium Antimonide	$E_{g0} = 0.2352\text{eV}$, $\Delta = 0.81\,\text{eV}$, $m_c = 0.01359m_0$ and $g_v = 1$ [99].
8	n-Gallium Antimonide	The values of $E_{g0} = 0.81\,\text{eV}$, $\Delta = 0.80\,\text{eV}$, $P = 9.48 \times 10^{-10}\text{eVm}$, $\bar{\varsigma}_0 = -2.1$, $\bar{v}_0 = -1.49$, $\bar{\omega}_0 = 0.42$ and $g_v = 1$ are valid for the model of Seiler *et al.* [97]. The values $\overline{E_1} = 1.024\,\text{eV}$, $\overline{E_2} = 0\,\text{eV}$, $\overline{E_3} = -1.132\,\text{eV}$, $\overline{E_4} = 0.05\,\text{eV}$, $\overline{E_5} = 1.107\,\text{eV}$, $\overline{E_6} = -0.113\,\text{eV}$ and $\overline{E_7} = -0.0072\,\text{eV}$ are valid for the model of Zhang [99].
9	p-Cadmium Sulphide	$m_{\parallel}^* = 0.7m_0, m_{\perp}^* = 1.5m_0, \bar{\lambda}_0 = 1.4 \times 10^{-10}\text{eVm}$ and $g_v = 1$ [33].

(Continued)

Table 1.1. (*Continued*)

S. no	Materials	Numerical Values of the Energy Band Constants
10	n-Lead Telluride	The values $m_t^- = 0.070m_0$, $m_l^- = 0.54m_0$, $m_t^+ = 0.010m_0$, $m_l^+ = 1.4m_0$, $P_{\|} = 141\,\text{meVnm}$, $P_{\perp} = 486\,\text{meVnm}$, $E_{g_0} = 190\,\text{meV}$ and $g_v = 4$ [79] are valid for the Dimmock model [101].
		The values $(\overline{R})^2 = 2.3 \times 10^{-19}(\text{eVm})^2$, $E_{g_0} = 0.16\,\text{eV}$, $(\overline{s})^2 = 4.6(\overline{R})^2$, $\Delta'_c = 3.07\,\text{eV}$, $(\overline{Q})^2 = 1.3(\overline{R})^2$, $\Delta''_c = 3.28\,\text{eV}$, $(\overline{A})^2 = 0.83 \times 10^{-19}(\text{eVm})^2$ are valid for the model of Bangert and Kastner [102].
		The values $m_{tv} = 0.0965m_0$, $m_{lv} = 1.33m_0$, $m_{tc} = 0.088m_0$, $m_{lc} = 0.83m_0$ are valid for the model of Foley *et al.* [103, 104].
		The values $m_1 = 0.0239m_0$, $m_2 = 0.024m_0$, $m'_2 = 0.31m_0$, $m_3 = 0.24m_0^{184}$ are valid for the Cohen model [59].
11	n-PbSnTe	$m_t^- = 0.070m_0, m_l^- = 0.54m_0$, $m_t^+ = 0.010m_0$, $m_l^+ = 1.4m_0$, $P_{\|} = 141\,\text{meVnm}$, $P_{\perp} = 486\,\text{meVnm}$, $E_g = 190\,\text{meV}$, $g_v = 4$ [163] and $\varepsilon_{sc} = 33\varepsilon_0$ [100, 162, 237].
12	n-$Pb_{1-x}Sn_xSe$	$m_t^- = 0.063m_0$, $m_l^- = 0.41m_0$, $m_t^+ = 0.089m_0$, $m_l^+ = 1.6m_0$, $P_{\|} = 137\,\text{meVnm}$, $P_{\perp} = 464\,\text{meVnm}$, $E_g = 90\,\text{meV}$, $g_v = 4$ [163] and $\varepsilon_{sc} = 60\varepsilon_0$ [100, 162, 237].
13	Stressed n-Indium Antimonide	The values $m_c = 0.048m_o$, $E_{g_0} = 0.081\,\text{eV}$, $B_2 = 9 \times 10^{-10}\text{eVm}$, $C_1^c = 3\,\text{eV}$, $C_2^c = 2\,\text{eV}$, $a_0 = -10\,\text{eV}$, $b_0 = -1.7\,\text{eV}$, $d = -4.4\,\text{eV}$, $S_{xx} = 0.6 \times 10^{-3}(kbar)^{-1}$, $S_{yy} = 0.42 \times 10^{-3}(kbar)^{-1}$, $S_{zz} = 0.39 \times 10^{-3}(kbar)^{-1}$, $S_{xy} = 0.5 \times 10^{-3}(kbar)^{-1}$, $\varepsilon_{xx} = \sigma S_{xx}$, $\varepsilon_{yy} = \sigma S_{yy}$, $\varepsilon_{zz} = \sigma S_{zz}$, $\varepsilon_{xy} = \sigma S_{xy}$, σ is the stress in kilobar, and $g_v = 1$ are valid for the model of Seiler *et al.* [97].
14	Bismuth	$E_{g_0} = 0.0153\,\text{eV}$, $m_1 = 0.00194m_0, m_2 = 0.313m_0$, $m_3 = 0.00246m_0$, $m'_2 = 0.36m_0$, $g_v = 3, g_s = 2$ [60], $M_2 = 0.128m_0$ and $M'_2 = 0.80m_0$ [60].
15	Mercury Telluride	$m_v^* = 0.028m_0, g_v = 1$ and $\varepsilon_\infty = 15.2\varepsilon_0$ [94].
16	Platinum Antimonide	For valence bands, along <111> direction, $\lambda_1 = 0.33\,\text{eV}$, $l_1 = 1.09\,\text{eV}$, $\nu_1 = 0.17\,\text{eV}$, $\overline{\overline{n}} = 0.22\,\text{eV}$, $\overline{\overline{a}} = 0.643nm$, $I_0 = 0.30(\text{eV})^2$, $\delta'_0 = 0.33\,\text{eV}$ and $g_v = 8$ [93].
17	n-Gallium Phosphide	$m_{\|}^* = 0.92m_0$, $m_{\perp}^* = 0.25m_0$, $k_0 = 1.7 \times 10^{19}m^{-1}$, $\|V_G\| = 0.21\,\text{eV}$, $g_v = 6$ and $g_s = 2$ [86].
18	Germanium	$E_{g_0} = 0.785\,\text{eV}$, $m_{\|}^* = 1.57m_0$ and $m_{\perp}^* = 0.0807m_0$ [237–240].

Table 1.1. (*Continued*)

S. no	Materials	Numerical Values of the Energy Band Constants
19	Tellurium	The values $A_6 = 6.7 \times 10^{-16}$meVm^2, $A_7 = 4.2 \times 10^{-16}$meVm^2, $A_8 = (6 \times 10^{-8}\text{meVm})^2$ and $A_9 = (3.6 \times 10^{-8}\text{meVm})^2$ are valid for the model of Bouat *et al.* [90]. The values $t_1 = 0.06315$ eV, $t_2 = -10.0\hbar^2/2m_0$, $t_3 = -5.55\hbar^2/2m_0$, $t_4 = 0.3 \times 10^{-36}$ eVm^4, $t_5 = 0.3 \times 10^{-36}$ eVm^4, $t_6 = -5.55\hbar^2/2m_0$ and $t_7 = 6.18 \times 10^{-20}(\text{eVm})^2$ are valid for the model of Ortenberg and Button [91].
20	Graphite	The values $\Delta_1 = -0.0002$ eV, $\gamma_1 = 0.392$ eV, $\gamma_5 = 0.194$ eV, $\bar{c} = 0.674nm$, $\gamma_2 = -0.019$ eV, $\bar{a} = 0.246nm$, $\gamma_0 = 3$ eV, $\gamma_4 = 0.193$ eV are valid for the model of Brandt *et al.* [239].
21	Lead Germanium Telluride	The values $\overline{E_{g_0}} = 0.21$ eV and $g_v = 4$ [105] are valid for the model of Vassilev.
22	Cadmium Antimonide	The values $A_{10} = -4.65 \times 10^{-19}$ eVm^2, $A_{11} = -2.035 \times 10^{-19}$ eVm^2, $A_{12} = -5.12 \times 10^{-19}$ eVm^2, $A_{13} = -0.25 \times 10^{-10}$eVm, $A_{14} = 1.42 \times 10^{-19}$ eVm^2, $A_{15} = 0.405 \times 10^{-19}$ eVm^2, $A_{16} = -4.07 \times 10^{-19}$ eVm^2, $A_{17} = 3.22 \times 10^{-10}$eVm, $A_{18} = 1.69 \times 10^{-20}(\text{eVm})^2$ and $A_{19} = 0.070$ eV [95, 96] are valid for the model of Yamada.
23	Cadmium Diphosphide	The values $\beta_1 = 8.6 \times 10^{-21}$ eVm^2, $\beta_2 = 1.8 \times 10^{-21}(\text{eVm})^2$, $\beta_4 = 0.0825$ eV and $\beta_5 = -1.9 \times 10^{-19}$ eVm^2 are valid for the model of Chuiko [106].
24	Zinc Diphosphide	The values $\beta_1 = 8.7 \times 10^{-21}$ eVm^2, $\beta_2 = 1.9 \times 10^{-21}(\text{eVm})^2$, $\beta_4 = 0.0875$ eV and $\beta_5 = -1.9 \times 10^{-19}$ eVm^2 are valid for the model of Chuiko [106].
25	Bismuth Telluride	The values $E_{g_0} = 0.145$ eV, $\overline{\overline{\alpha_{11}}} = 3.25$, $\alpha_{22} = 4.81$, $\alpha_{33} = 9.02$, $\alpha_{23} = 4.15$, $g_s = 2$ and $g_v = 6$ are valid for the model of Stordeur *et al.* [108, 109].
26	Antimony	The values $\overline{\alpha}_{11} = 16.7$, $\overline{\alpha}_{22} = 5.98$, $\overline{\alpha}_{33} = 11.61$ and $\overline{\alpha}_{23} = 7.54$ are valid for the model of Ketterson [110].
27	Zinc Selenide	$m_2^* = 0.16m_0, \Delta_2 = 0.42$ eV and $E_{g02} = 2.82$ eV [237].
28	Lead Selenide	$m_t^- = 0.23m_0$, $m_l^- = 0.32m_0$, $m_t^+ = 0.115m_0$, $m_l^+ = 0.303m_0$, $P_{\|\|} \approx 138$ meVnm, $P_\perp = 471$ meVnm and $E_{g_0} = 0.28$ eV [239, 240].

respective figures. The Figs. 1.2, 1.4 and 1.6 demonstrate the concentration dependence of the normalized MTP in QDs of InAs, GaAs and InSb for all the cases of Figs. 1.1, 1.3 and 1.5 respectively.

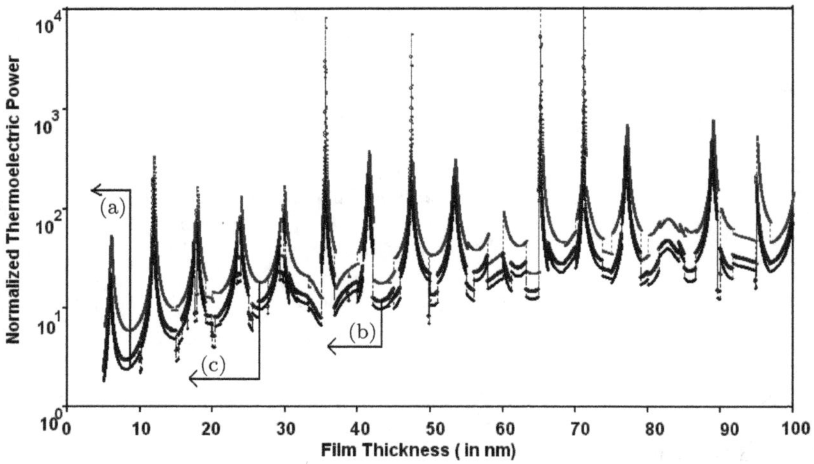

Fig. 1.1. Plot of the normalized MTP in QDs of InAs as a function of film thickness in accordance with the (a) three and (b) two band models of Kane together with (c) parabolic energy bands.

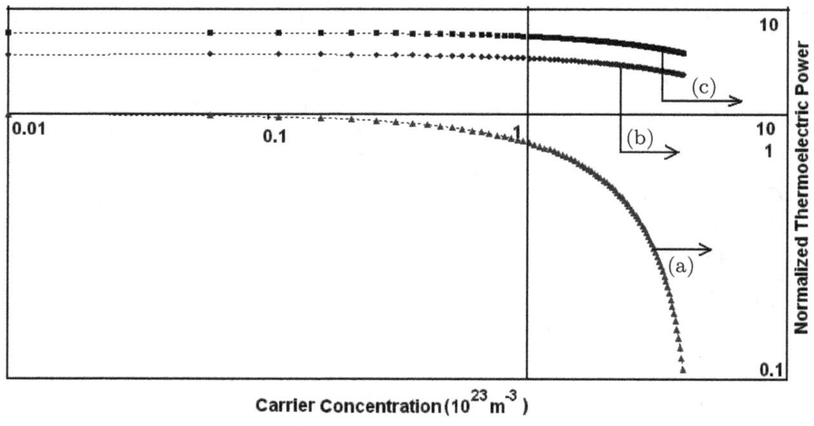

Fig. 1.2. Plot of the normalized MTP in QDs of InAs as a function of carrier concentration in accordance with the (a) three and (b) two band models of Kane together with (c) parabolic energy bands.

The Figs. 1.7, 1.8 and 1.10 illustrate the film thickness dependence of the normalized MTP in QDs of InAs, GaAs and InSb in accordance with the models of Stillman *et al.* [80] (using (1.28) and (1.29)), Newson *et al.* [81] (using (1.32) and (1.33)) and Rossler *et al.* [82] (using (1.36) and (1.37))

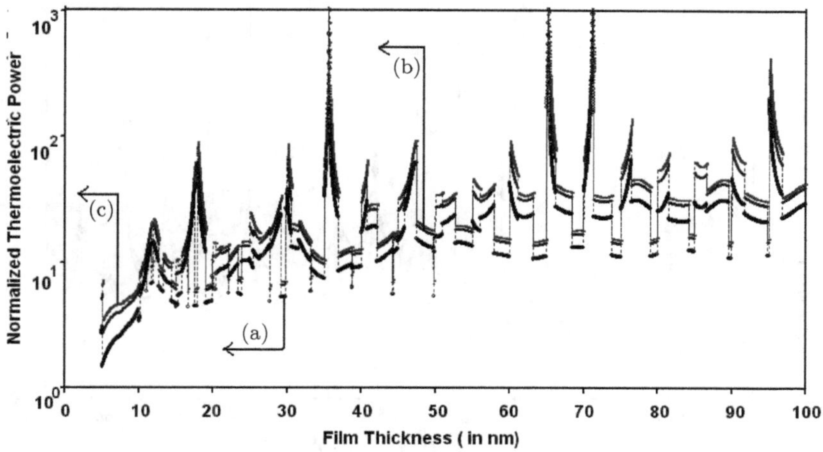

Fig. 1.3. Plot of the normalized MTP from QDs of GaAs as a function of film thickness in accordance with the (a) three and (b) two band models of Kane together with (c) the parabolic energy bands.

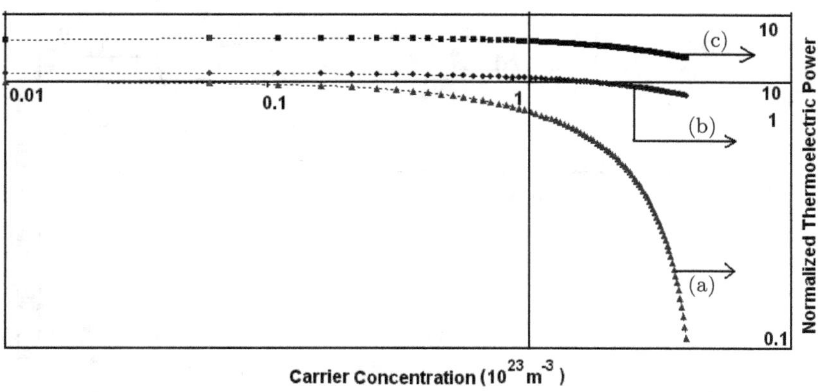

Fig. 1.4. Plot of the normalized MTP in QDs of GaAs as a function of carrier concentration in accordance with the (a) three and (b) two band models of Kane together with (c) parabolic energy bands.

as represented by curves (a)–(c) respectively. The Figs. 1.9, 1.11 and 1.13 demonstrate the influence of carrier concentration on the normalized MTP in QDs of GaAs, InSb and InAs for all the cases of Fig. 1.7. Figure 1.12 exhibits the film thickness dependence of the normalized MTP for QDs of InSb and InAs in accordance with the model of Agafonov *et al.* [85] (using (1.49) and (1.50a)) as represented by the curves (a) and (b) respectively.

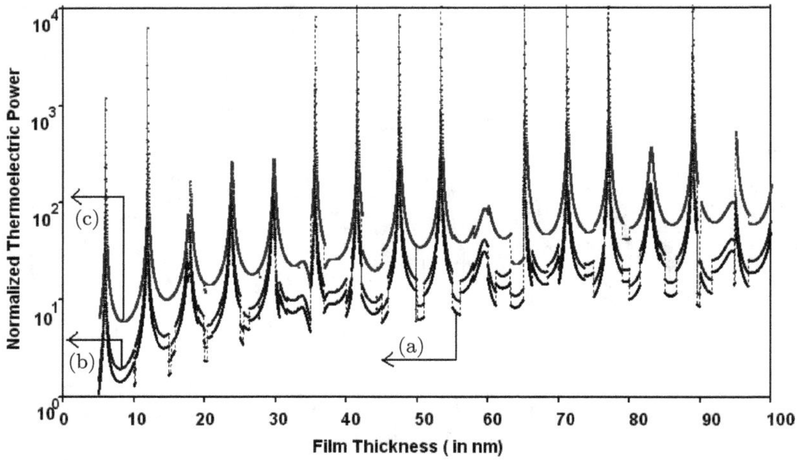

Fig. 1.5. Plot of the normalized MTP in QDs of InSb as a function of film thickness in accordance with the (a) three and (b) two band models of Kane together with (c) parabolic energy bands.

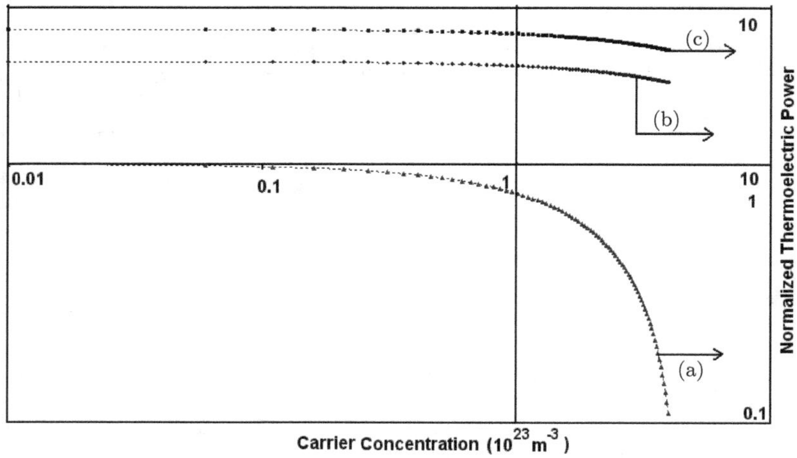

Fig. 1.6. Plot of the normalized MTP in QDs of InSb as a function of carrier concentration in accordance with the (a) three and (b) two band models of Kane together with (c) parabolic energy bands.

Figure 1.13 illustrates the concentration dependence of normalized MTP for all the cases of Fig. 1.12. Figure 1.14 demonstrates the film thickness dependence of the normalized MTP for QDs of InSb and InAs in accordance with the model of Johnson *et al.* [84] (using (1.45) and (1.46)) as represented

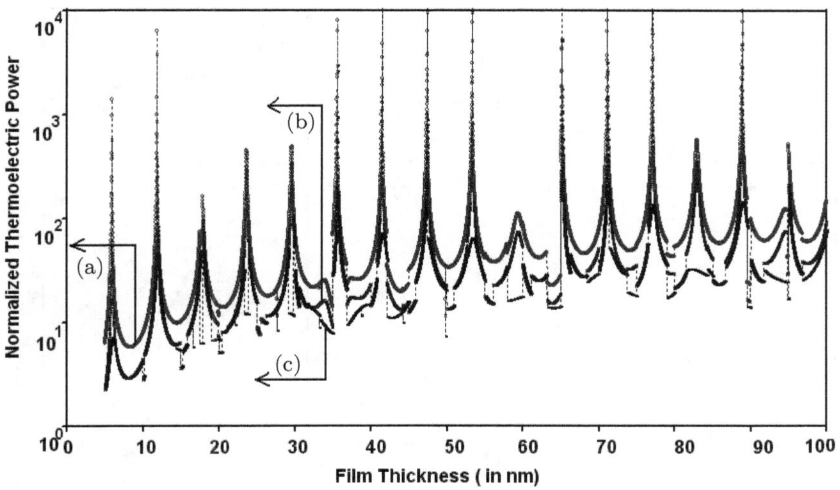

Fig. 1.7. Plot of the normalized MTP in QDs of InAs as a function of film thickness in accordance with the models of (a) Stillman *et al.* (b) Newson *et al.* and (c) Rossler *et al.* respectively.

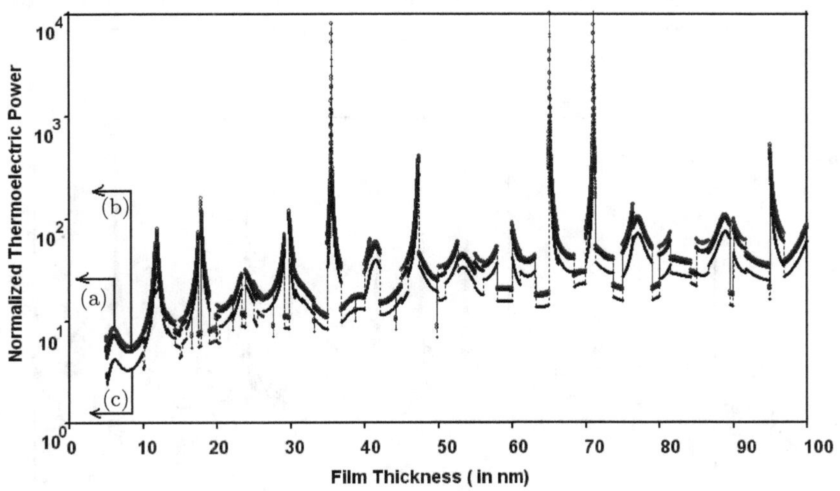

Fig. 1.8. Plot of the normalized MTP in QDs of GaAs as a function of film thickness in accordance with the models of (a) Stillman *et al.*, (b) Newson *et al.* and (c) Rossler *et al.* respectively.

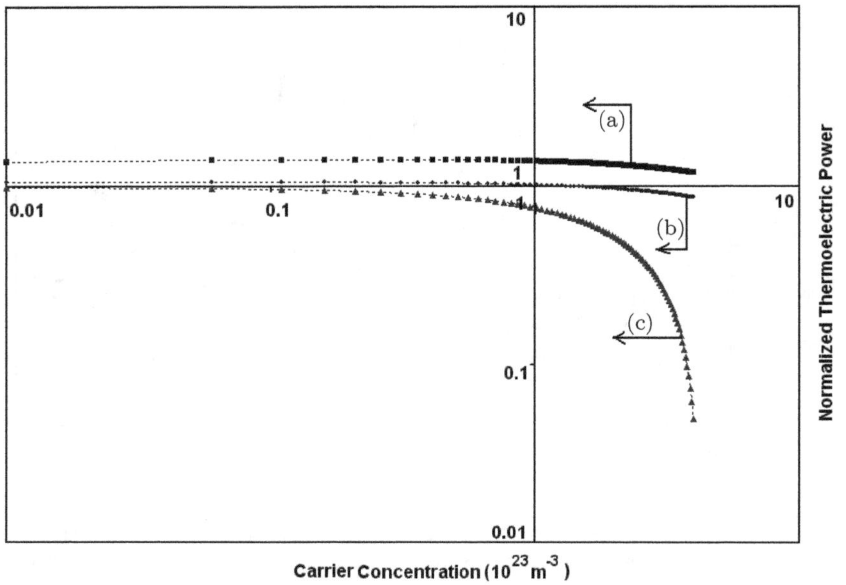

Fig. 1.9. Plot of the normalized MTP in QDs of GaAs as a function of carrier concentration in accordance with the models of (a) Stillman *et al.*, (b) Newson *et al.* and (c) Rossler *et al.* respectively.

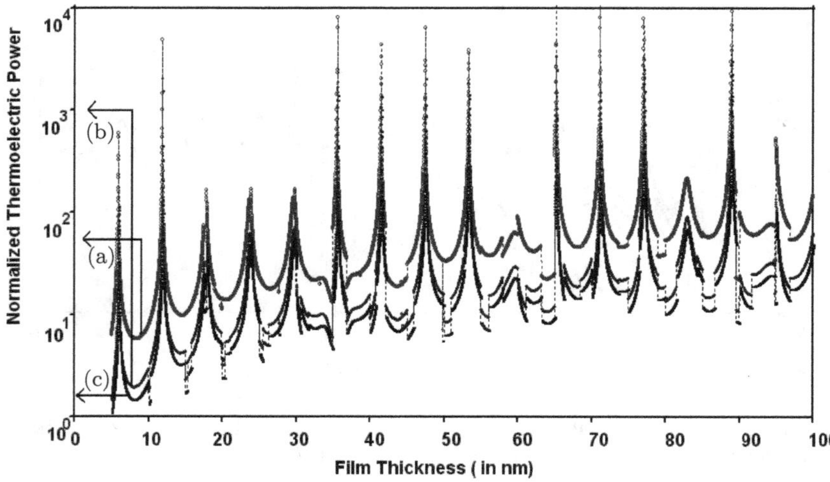

Fig. 1.10. Plot of the normalized MTP in QDs of InSb as a function of film thickness in accordance with the models of (a) Stillman *et al.*, (b) Newson *et al.* and (c) Rossler *et al.* respectively.

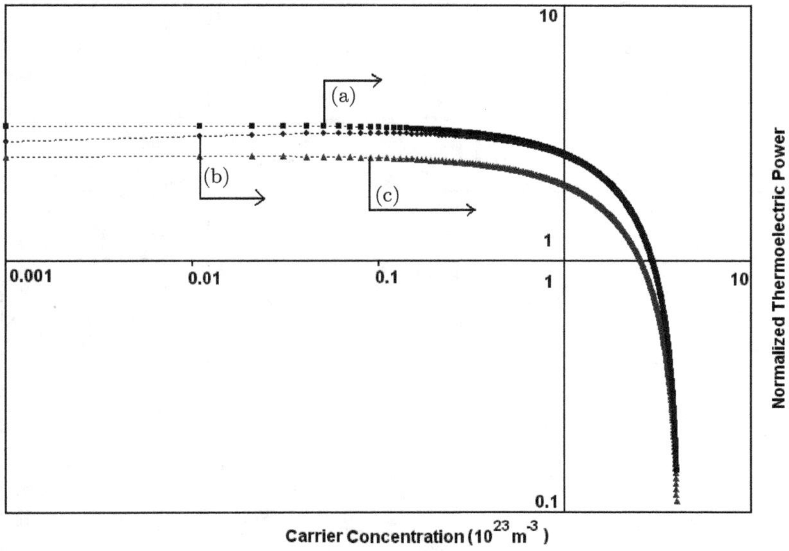

Fig. 1.11. Plot of the normalized MTP in QDs of InSb as a function of carrier concentration in accordance with the (a) three and (b) two band models of Kane together with (c) parabolic energy bands.

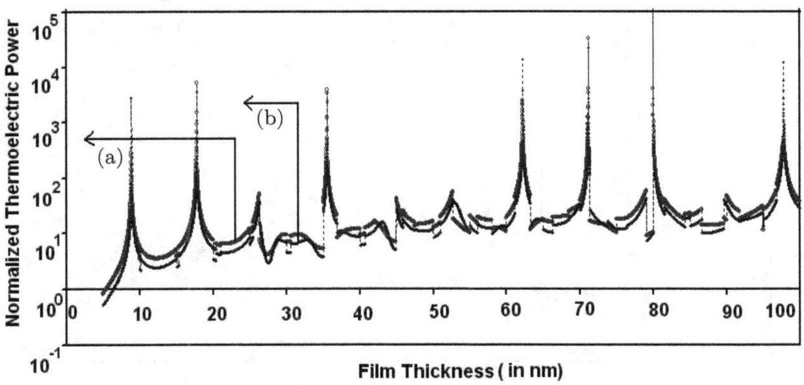

Fig. 1.12. Plot of the normalized MTP in QDs of (a) InSb and (b) InAs as a function of film thickness in accordance with model of Agafonov *et al.*

by the curves (a) and (b) respectively. Figure 1.15 shows the concentration dependence of normalized MTP for all the cases of Fig. 1.14.

Figure 1.16 depicts the film thickness dependence of the normalized MTP for QDs of InSb and InAs in accordance with the model of Palik

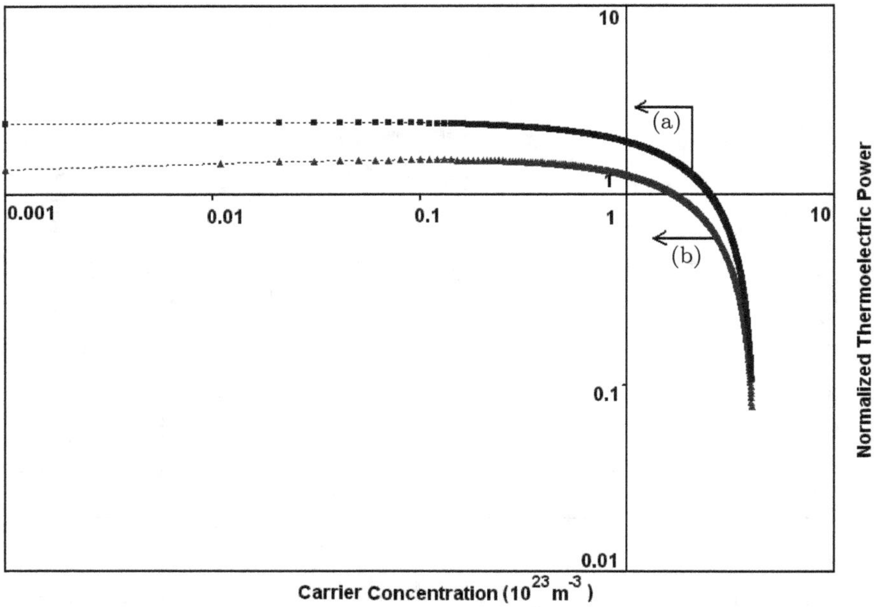

Fig. 1.13. Plot of the normalized MTP in QDs of (a) InSb and (b) InAs as a function of carrier concentration in accordance with model of Agafonov *et al.*

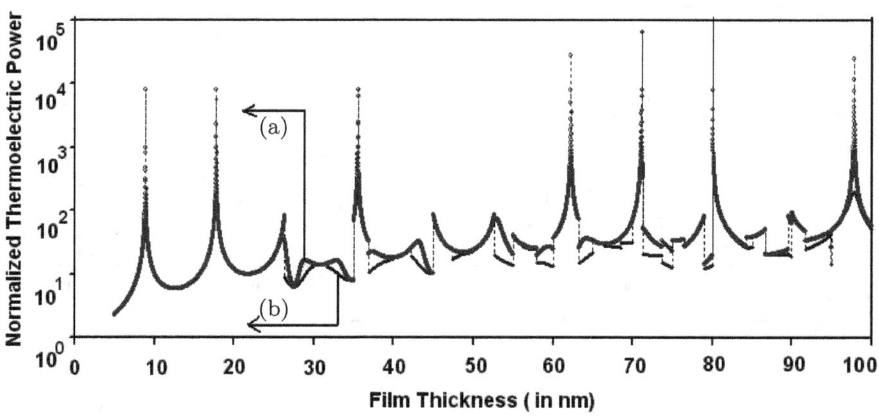

Fig. 1.14. Plot of the normalized MTP in QDs of (a) InSb and (b) InAs as a function of film thickness in accordance with the model of Johnson *et al.*

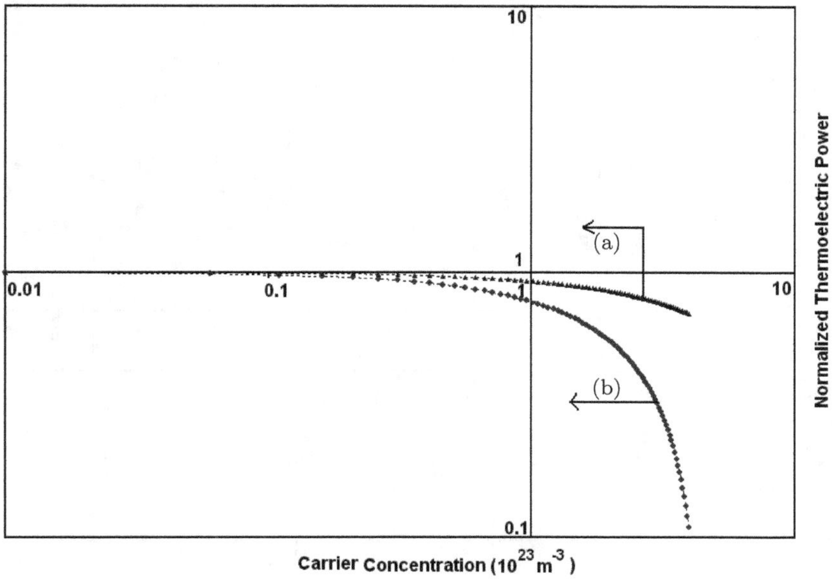

Fig. 1.15. Plot of the normalized MTP in QDs of (a) InSb and (b) InAs as a function of carrier concentration in accordance with the model of Johnson *et al.*

et al. [83] (using (1.41) and (1.42)) as represented by the curves (a) and (b) in this context. Figure 1.17 models the concentration dependence of normalized MTP for all the cases of Fig. 1.16. Figure 1.18 exhibits the variation of the normalized MTP with the film thickness in QDs of II–VI materials in accordance with Hopfield model (using (1.53) and (1.54), taking *p*-CdS as an example and considering both the cases $\overline{\lambda}_0 \neq 0$ and $\overline{\lambda}_0 = 0$) and GaP (using (1.57) and (1.58a) in accordance with the model of Rees) as shown by curves (a), (b) and (c) respectively. Figure 1.19 demonstrates the concentration variation of the normalized MTP for all cases of Fig. 1.18. In Fig. 1.20, the normalized MTP has been plotted as a function of film thickness for QDs of Germanium for both the models of Wang *et al.* [134] (using (1.65) and (1.66)) and Cardona *et al.* [87] (using (1.61) and (1.62)) as shown by curves (a) and (b) respectively.

Figure 1.21 exhibits the normalized MTP as a function of film thickness in QDs of Tellurium (by using the models of Bouat *et al.* [90] (using (1.69) and (1.70)) and Ortenberg *et al.* [91] (using (1.73) and (1.74)) and stressed

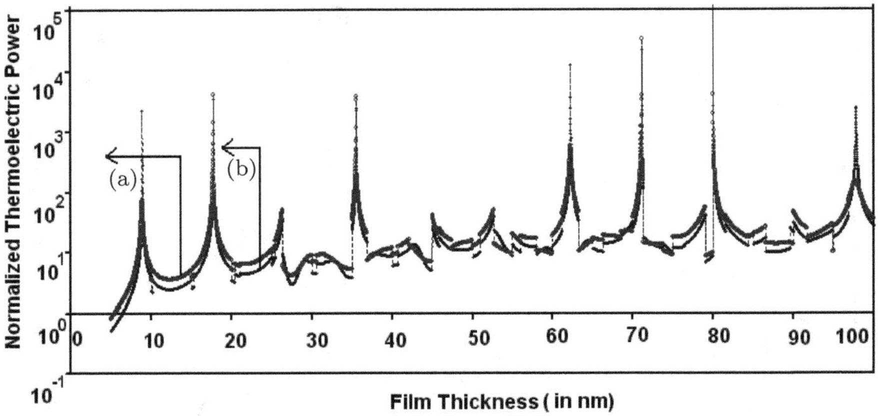

Fig. 1.16. Plot of the normalized MTP in QDs of (a) InSb and (b) InAs as a function of film thickness in accordance with the model of Palik *et al.*

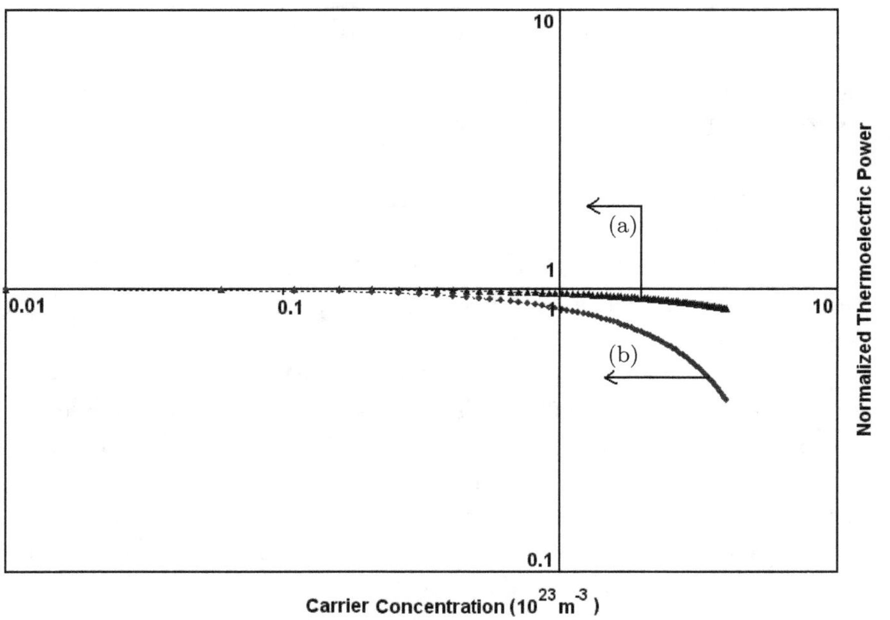

Fig. 1.17. Plot of the normalized MTP in QDs of (a) InSb and (b) InAs as a function of carrier concentration in accordance with the model of Palik *et al.*

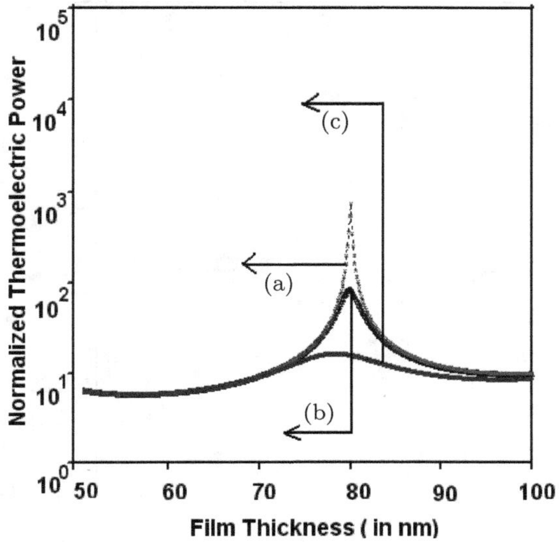

Fig. 1.18. Plot of the normalized MTP in QDs of CdS with (a) $\overline{\lambda}_0 \neq 0$, (b) $\overline{\lambda}_0 = 0$ and (c) GaP as a function of film thickness in accordance with the models of Hopfield and Rees respectively.

Kane type materials (taking n-InSb as an example) in accordance with the models of Seiler *et al.* [97] (using (1.107) and (1.108)) as shown by curves (a)–(c) respectively. Figure 1.22 shows the dependence of the normalized MTP on the carrier concentration for all the cases of Fig. 1.21. In Fig. 1.23, the normalized MTP has been plotted as a function of carrier concentration for QDs of Graphite (using (1.77) and (1.78)), Platinum Antimonide (using (1.82) and (1.83)), zero-gap (using (1.86) and (1.87)) and Lead Germanium Telluride (using (1.142) and (1.143)) in accordance with the models of Ushio *et al.* [92], Emtage [93], Ivanov-Omskii *et al.* [94] and Vassilev [105] as shown by curves (a)–(d) respectively.

Figure 1.24 depicts the plot of normalized MTP as a function of film thickness of GaSb in accordance with the models of Seiler *et al.* [97] (using (1.94) and (1.95)), Mathur *et al.* [98] (using (1.99) and (1.100)) and Zhang [99] (using (1.103) and (1.104)) as shown by curves (a)–(c) respectively. Figure 1.25 exhibits the carrier concentration dependence of the normalized MTP for all the cases of Fig. 1.24. Figure 1.26 illustrates the variation of the normalized MTP with film thickness in QDs of Bismuth in accordance with the models of McClure *et al.* [61] (using (1.111) and 1.112)), Hybrid

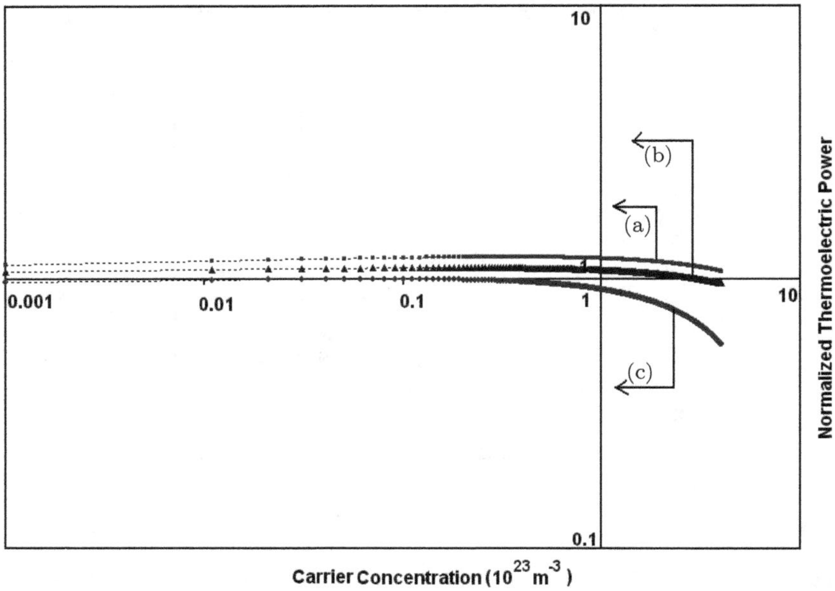

Fig. 1.19. Plot of the normalized MTP in QDs of CdS with (a) $\bar{\lambda}_0 \neq 0$, (b) $\bar{\lambda}_0 = 0$ and (c) GaP as a function of carrier concentration in accordance with the model of Hopfield and Rees.

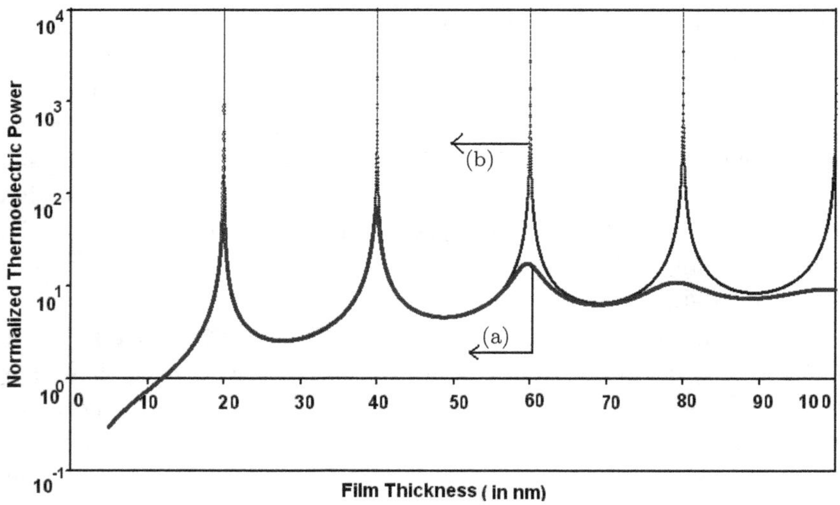

Fig. 1.20. Plot of the normalized MTP in QDs of Germanium as a function of film thickness in accordance with the models of (a) Wang *et al.* and (b) Cardona *et al.* respectively.

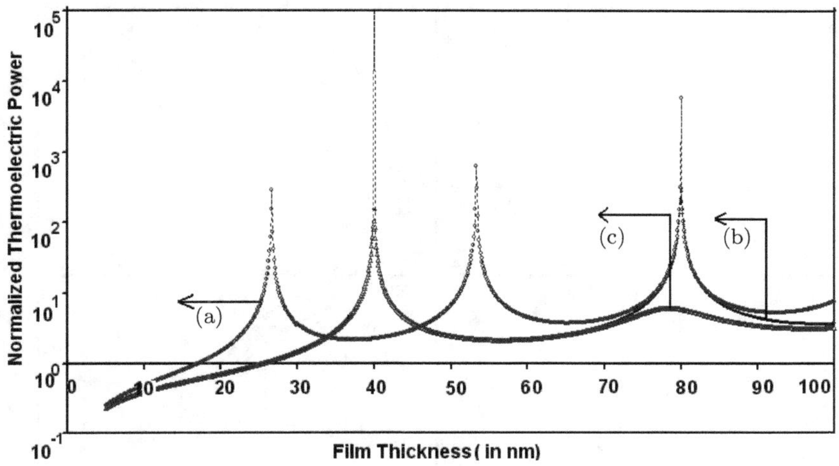

Fig. 1.21. Plot of the normalized MTP in QDs of Tellurium and stressed Kane type material (n-InSb) as a function of film thickness in accordance with the models of (a) Bouat *et al.* (b) Ortenberg *et al.* and (c) Seiler *et al.* respectively.

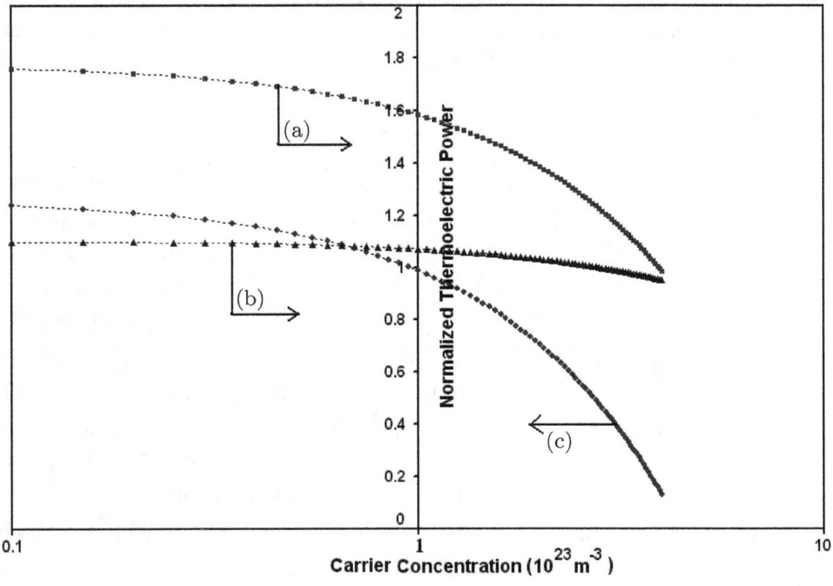

Fig. 1.22. Plot of the normalized MTP in QDs of Tellurium and stressed Kane type material (n-InSb) as a function of carrier concentration in accordance with the model of (a) Bouat *et al.*, (b) Ortenberg *et al.* and (c) Seiler *et al.* respectively.

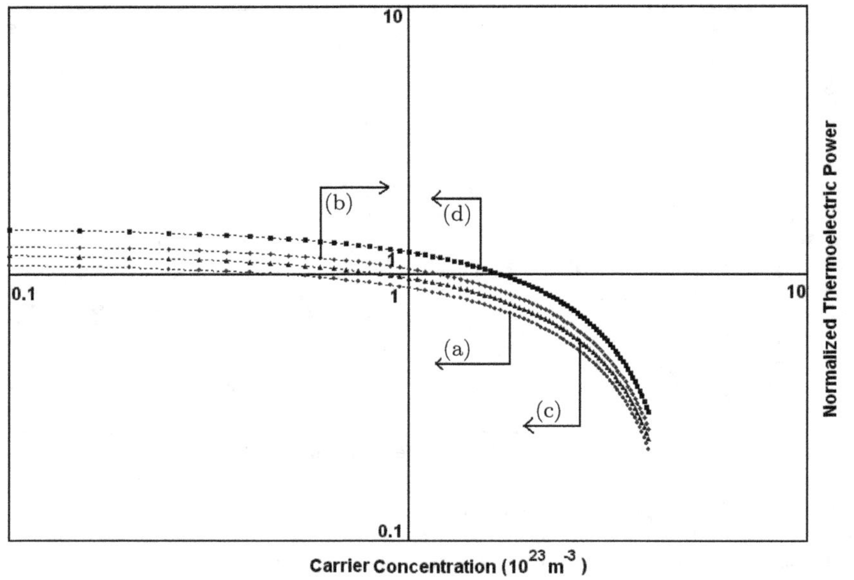

Fig. 1.23. Plot of the normalized MTP in QDs of (a) Graphite, (b) Platinum antimonide, (c) zero-gap (HgTe) and (d) $Pb_{1-x}Ge_xTe$ as a function of carrier concentration in accordance with the models of Ushio *et al.*, Emtage, Ivanov-Omskii *et al.* and Vassilev respectively.

[60] (using (1.115) and (1.116)), Cohen [59] (using (1.119) and (1.120)) and Lax *et al.* [53] (using (1.123) and (1.124a) as shown by curves (a)–(d) respectively. Figure 1.27 shows the dependence of normalized MTP on the carrier concentration in QDs of Bismuth for all types of band models as stated in Fig. 1.26. Figure 1.28 shows the variation of MTP on film thickness for QDs of IV–VI materials (taking PbTe as an example) in accordance with the models of Dimmock [101] (using (1.130) and (1.131)), Bangert *et al.* [102] (using (1.134) and (1.135)) and Foley *et al.* [103] (using (1.138) and (1.139)) together with Bismuth Telluride by using the model of Stordeur *et al.* [107] (using (1.150) and (1.151)) as shown by the curves (a)–(d) respectively. Figure 1.29 models the variation of the MTP with the carrier concentration for QDs of IV–VI materials and Bismuth Telluride in accordance with the band models of Fig. 1.28. Figure 1.30 shows the plot of the normalized MTP as a function of film thickness in QDs of II–V materials (taking CdSb as an example) using the model of Yamada [95]

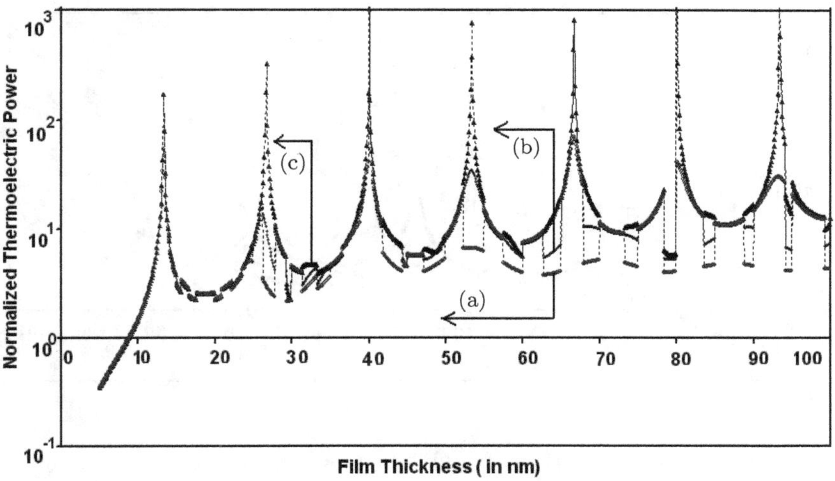

Fig. 1.24. Plot of the normalized MTP in QDs of Gallium antimonide as a function of film thickness in accordance with the models of (a) Seiler *et al.*, (b) Mathur *et al.* and (c) Zhang respectively.

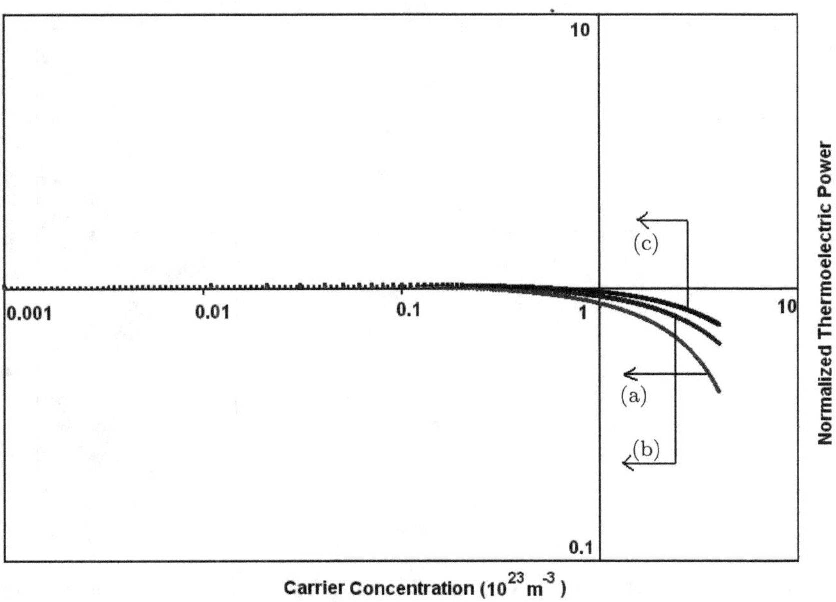

Fig. 1.25. Plot of the normalized MTP in QDs of Gallium antimonide as a of normalized carrier concentration in accordance with the model of (a) Seiler *et al.*, (b) Mathur *et al.* and (c) Zhang respectively.

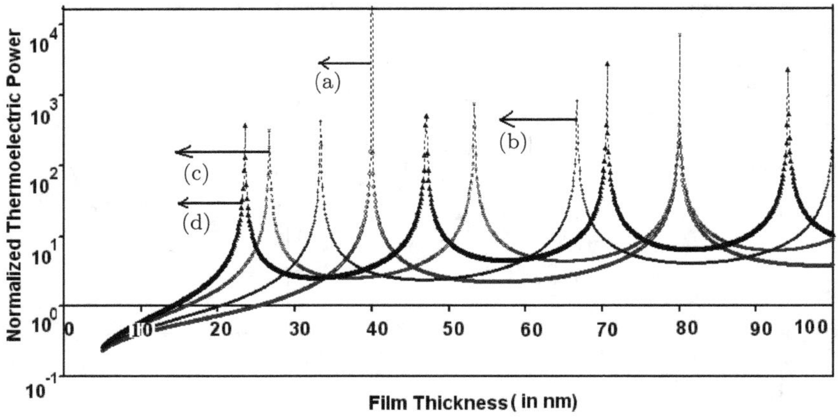

Fig. 1.26. Plot of the normalized MTP in QDs of Bismuth as a function of film thickness in accordance with the models of (a) McClure *et al.*, (b) Takaoka *et al.* (Hybrid model), (c) Cohen and (d) Lax *et al.* respectively.

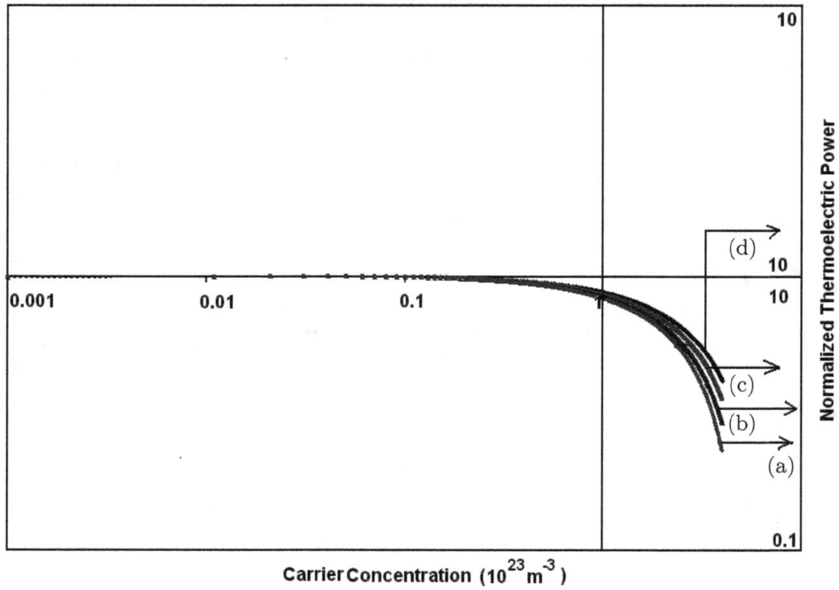

Fig. 1.27. Plot of the normalized MTP in QDs of Bismuth as a function of carrier concentration in accordance with the model of (a) McClure *et al.*, (b) Takaoka *et al.* (Hybrid model), (c) Cohen and (d) Lax *et al.* respectively.

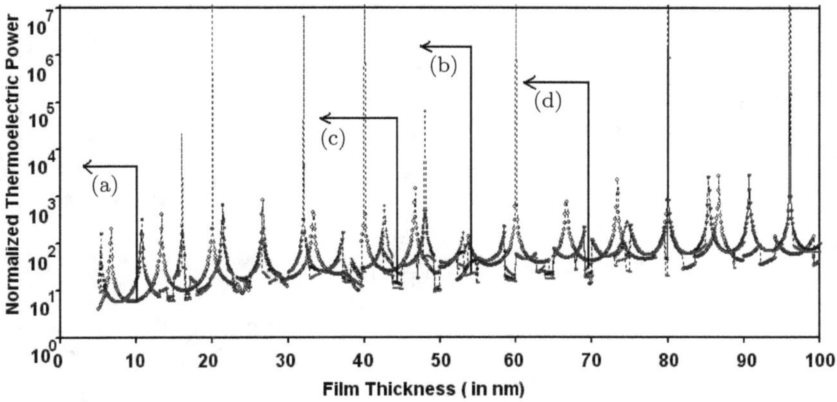

Fig. 1.28. Plot of the normalized MTP in QDs of PbTe as a function of film thickness in accordance with the models of (a) Dimmock, (b) Bangert *et al.* and (c) Foley *et al.* The plot (d) refers to Bi_2Te_3 in accordance with the model of Stordeur *et al.*

(using (1.90) and (1.91)), Zinc and Cadmium diphosphides using the model of Chuiko [106] (using (1.146) and (1.147)) and Antimony in accordance with the model of Ketterson [109] (using (1.161) and (1.162a)) as shown by the curves (a)–(d) respectively. For the purpose of simplified numerical computation, broadening has been neglected for obtaining all the plots. The inclusion of broadening will change the numerical magnitudes without altering the physics inside.

The signature of 3D quantization is forthwith evident from the Figs. 1.1, 1.3, 1.5, 1.7, 1.8, 1.10, 1.12, 1.14, 1.16, 1.18, 1.20, 1.21, 1.24, 1.26, 1.28 and 1.30 for all materials as discussed having different band structures. It can be facilely discerned from the same that the normalized MTP oscillates with film thickness exhibiting spikes for various values of film thickness which are totally band structure dependent. The occurrence of peaks in the said figures originates from the totally quantized energy levels of the carriers of the concerned dots. The MTP spectra are found bearing composite oscillations as function of the nanothickness. These are generally due to selection rules in the quantum numbers along the three confined directions. The dependence of the normalized MTP on the carrier concentration is manifested by the Figs. 1.2, 1.4, 1.6, 1.9, 1.11, 1.13, 1.15, 1.17, 1.19, 1.22, 1.23, 1.25, 1.27 and 1.29 for the different materials as considered here.

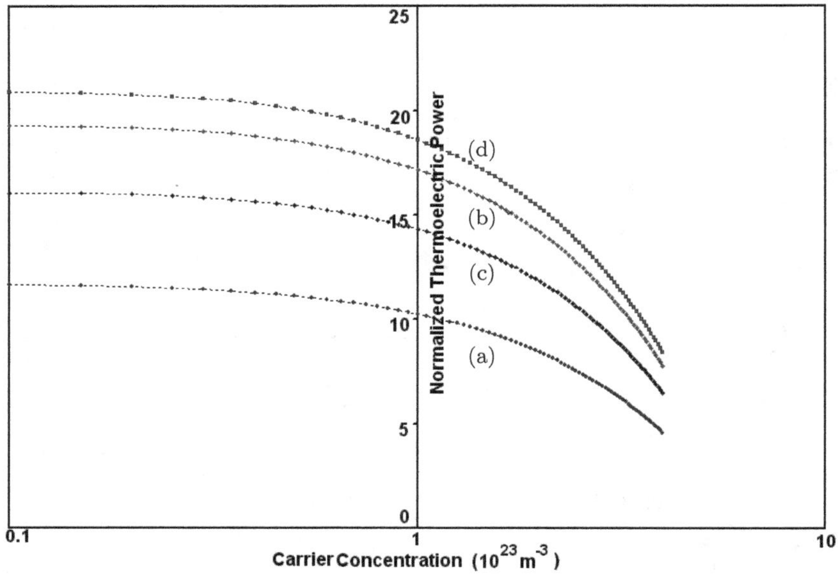

Fig. 1.29. Plot of the normalized MTP in QDs of PbTe and Bi_2Te_3 as a function of carrier concentration in accordance with the model of (a) Dimmock, (b) Bangert *et al.* and (c) Foley *et al.* The plot (d) refers to Bi2Te3 in accordance with the model of Stordeur *et al.*

It can be ascertained from the same figures that the MTP of all the corresponding materials decreases with increasing carrier concentration for relatively higher values of the carrier degeneracy. Although the MTP varies in various manners with all the variables in all the limiting cases as evident from all the figures, the rate of variations in each case are totally band-structure dependent.

The quantum oscillations of the MTP in QDs exhibit different numerical magnitudes as compared to the same in QWs and NWs. It may be comprehended that the QDs lead to the discrete energy levels, somewhat like atomic energy levels, which produce very large changes. This is in accordance to the inherent nature of the quantum confinement of the carrier gas as dealt with here. In QDs, there remain no free carrier states in between any two allowed sets of size-quantized levels unlike that found for QWs and NWs where the quantum confinements are 1D and 2D respectively. Consequently, the crossing of the Fermi level by the size-quantized levels in QDs would have much greater impact on the redistribution of the carriers

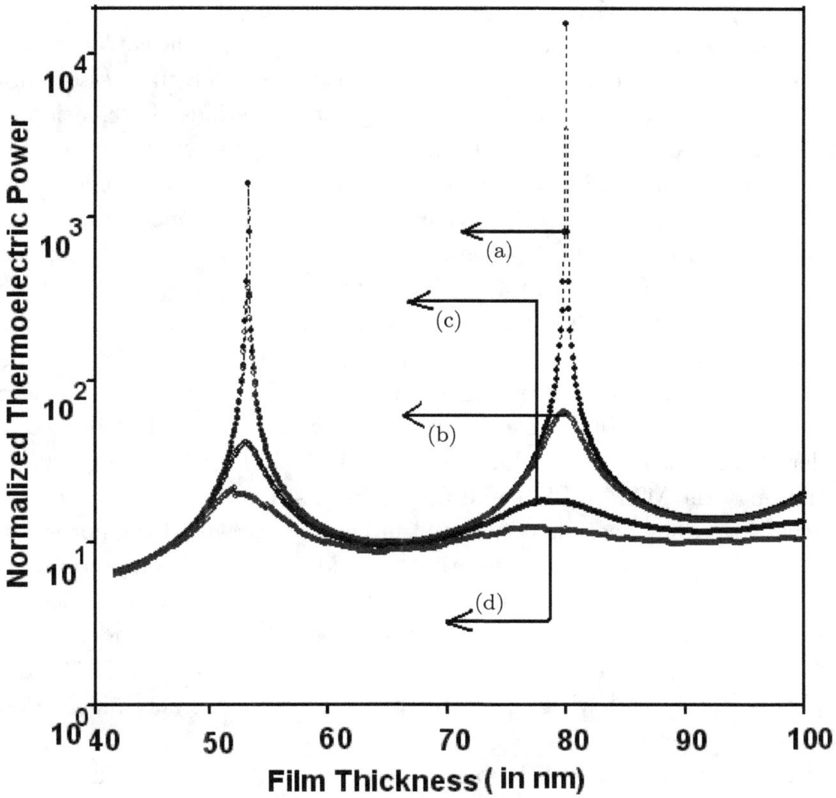

Fig. 1.30. Plot of the normalized MTP in QDs of (a) II–V compound (CdSb), (b) zinc diphosphide, (c) cadmium diphosphide and (d) antimony as a function of film thickness in accordance with the models of Yamada, Chuiko and Ketterson respectively.

among the allowed levels, as compared to that found for QWs and NWs respectively.

The effect of spin splitting of the carriers in QDs of *p*-CdS on MTP can be numerically investigated from Figs. 1.18 and 1.19. It appears that the absence of the spin splitting constant decreases the numerical value of the MTP in CdS for a particular range of film thickness. It appears from Fig. 1.20 that the MTP in Germanium in accordance with the model of Wang *et al.* is comparatively low with that of the Cardona *et al.* model.

The MTPs in QDs of Te and stressed Kane type materials are depicted in Figs. 1.21 and 1.22 with respect to film thickness and carrier concentration respectively. It appears that at extremely low and high film thicknesses, the MTP in QDs of Te dominates over that of the corresponding stressed InSb, although, at mid zone thickness, the MTP in stressed InSb exhibits a high peak together with the fact that the periods of oscillations of the MTP are comparatively higher in Te than that of stressed compounds. In Fig. 1.23, the MTP in QDs of graphite, Platinum antimonide, zero gap and $Pb_{1-x}Ge_xTe$ materials have been plotted as a function of carrier concentration. It appears that the MTP in QDs of $Pb_{1-x}Ge_xTe$ are higher than that of the QDs of graphite, $PtSb_2$ and zero-gap respectively. From Fig. 1.26, it appears that the MTP in Bi exhibits the high sharp peak in the mid thickness zone in accordance with the model of McClure and Choi. From Fig. 1.24, it is evident that the MTP in GaSb oscillates with film thickness exhibiting spikes in accordance with the models proposed by Seiler *et al.*, Mathur *et al.* and Zhang respectively. In Figs. 1.28 and 1.29, the variation of the MTP in PbTe and Bi_2Te_3 against film thickness and carrier concentration has been demonstrated in accordance with the appropriate band models. In Fig. 1.30, the variation of MTP in II–V compound, zinc and cadmium diphosphides and antimony as a function of film thickness has been shown in accordance with the appropriate band models. From Fig. 1.28, it appears that the numerical values of MTP in QDs of PbTe and Bi_2Te_3 in accordance with all the band models are extremely higher than all other materials.

In this chapter, investigations have been made of MTP in QDs of III–V, II–VI, GaP, Germanium, Te, Graphite, $PtSb_2$, zero-gap, II–V, GaSb, stressed, IV–VI, Lead Germanium Telluride, Zinc and Cadmium diphosphides, Bi_2Te_3 and Antimony together with Bismuth in accordance with the appropriate carrier dispersion laws as available in the literature by formulating the respective carrier statistics from Sections 1.2.2 to 1.2.18. It is imperative to state that our investigations excludes the many-body, hot electron, spin and the allied quantum dot effects in this simplified theoretical formalism due to the absence of proper analytical techniques for including them for the generalized systems as considered here. Our simplified approach will be appropriate for the purpose of comparison when the methods of tackling the formidable problems after inclusion of the said effects for the generalized systems emerge. It is vital to elucidate that the results of our simple theory get transformed to the well-known formulation of MTP for wide-gap materials having parabolic energy bands. This indirect test not only exhibits the mathematical compatibility of our formulation but also shows the fact that our simple analysis is a more generalized one, since one can obtain the corresponding results for materials having parabolic energy bands under certain limiting conditions from our present derivation. The experimental results for the verification of the theoretical analyses of this chapter are still not available in the literature. It is worth noting that our generalized formulation would be useful to analyze the experimental results when they materialize. The inclusion of the said effects would certainly increase the accuracy of the results although the qualitative features of the MTP would not change in the presence of the aforementioned effects. It is worth remarking that the influence of the broadening parameter has not been included in numerical computations for the purpose of simplified approach, although readers will enjoy the computer analysis in this context by including the said feature.

It can be noted that on the basis of the dispersion relations of the various quantized structures as discussed above the heat capacity, the dia and para-magnetic susceptibilities and the various important DC/AC transport coefficients can be probed for all types of quantized structures as considered here. Thus, our theoretical formulation comprises the dispersion relation dependent properties of various technologically important quantum-confined materials having different band-structures. We have not considered other types of compounds in order to keep the presentation concise and succinct. With different sets of energy band parameters, we shall get different numerical values of the MTP. The nature of variations of the MTP as shown here would be similar for the other types of materials and the simplified analysis of this chapter exhibits the basic qualitative features of the MTP in such quantized structures. It may be noted that the basic aim of this chapter is not solely to demonstrate the influence of quantum confinement on the MTP for a wide class of quantized materials but also to formulate the appropriate carrier statistics in the most generalized form, since the transport and other phenomena in modern nanostructured devices having different band structures and the derivation of the expressions of many important carrier properties are based on the temperature-dependent carrier statistics of such materials.

1.5 Open Research Problems

The problems under these sections which end with the end of this monograph form the integral part of the same and are intended for researchers together with advanced readers from a variety of disciplines as indicated in the Preface for rendering their own contribution in this pin-pointed research topic of thermo power in nanostructured materials. The numerical values of the energy band constants of the materials needed for appropriate numerical computations are given in Table 1.1.

(R.1.1) Investigate the diffusion thermoelectric power (DTP), phonon-drag thermoelectric power (PTP) and thermoelectric figure-of-merit (Z) in the absence of magnetic field by considering all types of scattering mechanisms for QDs in the presence of three finite potential wells and three parabolic potential wells applied separately in the three orthogonal directions for all the materials whose unperturbed carrier energy spectra are defined in this chapter.

(R.1.2) Investigate the DTP, PTP and Z in the absence of magnetic field by considering all types of scattering mechanisms for QDs in the presence of three finite potential wells and three parabolic potential wells applied separately in the three orthogonal directions under an arbitrarily oriented (a) non-uniform electric field and (b) alternating electric field respectively for all the materials whose unperturbed carrier energy spectra are defined in this chapter.

(R.1.3) Investigate the DTP, PTP and Z for QDs by considering all types of scattering mechanisms in the presence of three finite potential wells and three parabolic potential wells applied separately in the three orthogonal directions under an arbitrarily oriented alternating magnetic field by including broadening and the electron spin for all the materials whose unperturbed carrier energy spectra are defined in this chapter.

(R.1.4) Investigate the DTP, PTP and Z for QDs by considering all types of scattering mechanisms by considering the presence of three finite potential wells and three parabolic potential wells applied separately in the three orthogonal directions under an arbitrarily oriented alternating magnetic field and crossed alternating electric field by including broadening and the electron spin for all the materials whose unperturbed carrier energy spectra are defined in this chapter.

(R.1.5) Investigate the DTP, PTP and Z for QDs by considering all types of scattering mechanisms in the presence of three finite potential wells and three parabolic potential wells applied separately in the three orthogonal directions under an arbitrarily oriented alternating magnetic field and crossed alternating non-uniform electric field by including broadening and the electron spin whose for all the materials unperturbed carrier energy spectra are defined in this chapter.

(R.1.6) Investigate the DTP, PTP and Z in the absence of magnetic field for QDs by considering all types of scattering mechanisms in the presence of three finite potential wells and three parabolic potential wells applied separately in the three orthogonal directions under exponential, Kane, Halperin, Lax and Bonch-Bruevich band tails [125] for all the materials whose unperturbed carrier energy spectra are defined in this chapter.

(R.1.7) Investigate the DTP, PTP and Z in the absence of magnetic field for QDs by considering all types of scattering mechanisms in the presence of three finite potential wells and three parabolic potential wells applied separately in the three orthogonal directions for all the materials as defined in (R.1.6) under an arbitrarily oriented (a) non-uniform electric field and (b) alternating electric field respectively whose unperturbed carrier energy spectra are defined in this chapter.

(R.1.8) Investigate the DTP, PTP and Z for QDs by considering all types of scattering mechanisms in the presence of three finite potential wells and three parabolic potential wells applied separately in the three orthogonal directions for all the materials as described in (R.1.6) under an arbitrarily oriented alternating magnetic field by including broadening and the electron spin whose unperturbed carrier energy spectra are defined in this chapter.

(R.1.9) Investigate the DTP, PTP and Z for QDs by considering all types of scattering mechanisms in the presence of three finite potential wells and three parabolic potential wells applied separately in the three orthogonal directions as discussed in (R.1.6) under an arbitrarily oriented alternating magnetic field and crossed alternating electric field by including broadening and the electron spin for all the materials whose unperturbed carrier energy spectra are defined in this chapter.

(R.1.10) Investigate all the appropriate problems of this chapter for QDs of the following materials:

(a) The energy spectrum of the valance bands of CuCl in accordance with Yekimov *et al.* [166] can be written as

$$E_h = (\gamma_6 - 2\gamma_7)\frac{\hbar^2 k^2}{2m_0} \tag{R.1.1}$$

and

$$E_{l,s} = (\gamma_6 + \gamma_7)\frac{\hbar^2 k^2}{2m_0} - \frac{\Delta_1}{2} \pm \left[\frac{\Delta_1^2}{4} + \gamma_7\Delta_1\frac{\hbar^2 k^2}{2m_0} + 9\left(\frac{\gamma_7\hbar^2 k^2}{2m_0}\right)^2\right]^{1/2} \tag{R.1.2}$$

where $\gamma_6 = 0.53$, $\gamma_7 = 0.07$, $\Delta_1 = 70\,\text{meV}$.

(b) In the presence of stress, χ_6 along $\langle 001\rangle$ and $\langle 111\rangle$ directions, the energy spectra of the holes in semiconductors having diamond structure valance bands can be respectively expressed following Roman [167] *et al.* as

$$E = A_6 k^2 \pm [\overline{B}_7^2 k^4 + \delta_6^2 + B_7\delta_6(2k_z^2 - k_s^2)]^{1/2} \tag{R.1.3}$$

and

$$E = A_6 k^2 \pm \left[\overline{B}_7^2 k^4 + \delta_7^2 + \frac{D_6}{\sqrt{3}}\delta_7(2k_z^2 - k_s^2)\right]^{1/2} \tag{R.1.4}$$

where A_6, B_7, D_6 and C_6 are inverse mass band parameters in which $\delta_6 \equiv l_7(\overline{S}_{11} - \overline{S}_{12})\chi_6$, $\overline{S}_{ij}$ are the usual elastic compliance constants,

$$\overline{B}_7^2 \equiv \left(B_7^2 + \frac{c_6^2}{5}\right) \quad \text{and} \quad \delta_7 \equiv \left(\frac{d_8 S_{44}}{2\sqrt{3}}\right)\chi_6.$$

For gray tin,

$$d_8 = -4.1\,\text{eV}, \quad l_7 = -2.3\,\text{eV}, \quad A_6 = 19.2\frac{\hbar^2}{2m_0}, \quad B_7 = 26.3\frac{\hbar^2}{2m_0},$$

$$D_6 = 31\frac{\hbar^2}{2m_0} \quad \text{and} \quad c_6^2 = -1112\frac{\hbar^2}{2m_0}.$$

(c) The dispersion relation of the holes in p-InSb is given by [168]

$$\overline{E} = c_4(1 + \gamma_4 f_4)k^2 \pm \frac{1}{3}\left[2\sqrt{2}\sqrt{c_4}\sqrt{16 + 5\gamma_4}\sqrt{E_4 g_4 k}\right] \quad \text{(R.1.5)}$$

where $\overline{E}$ is the energy of the hole as measured from the top of the valance and within it,

$$c_4 \equiv \frac{\hbar^2}{2m_0} + \theta_4, \quad \theta_4 \equiv 4.7\frac{\hbar^2}{2m_0}, \quad \gamma_4 \equiv \frac{b_4}{c_4}, \quad b_4 \equiv \frac{3}{2}b_5 + 2\theta_4,$$

$$b_5 \equiv 2.4\frac{\hbar^2}{2m_0}, \quad f_4 \equiv \frac{1}{4}[\sin^2 2\theta + \sin^4\theta \sin^2 2\phi],$$

θ is measured from the positive z-axis, ϕ is measured from positive x-axis,

$$g_4 \equiv \sin\theta\left[\cos^2\theta + \frac{1}{4}\sin^4\theta\sin^2 2\phi\right]$$

and

$$E_4 = 5 \times 10^{-4}\,\text{eV}.$$

(R.1.11) Investigate all the appropriate problems of this chapter after proper modifications for wedge shaped, cylindrical, ellipsoidal, conical, triangular, colloidal, pyramidal, circular, lateral parabolic rotational, parabolic cylindrical and position-controlled QDs respectively.

(R.1.12) Investigate all the problems of (R.1.11) in the presence of strain after proper modifications.

(R.1.13) Investigate all the appropriate problems of this chapter for quantum rods.

(R.1.14) Investigate the influence of deep traps and surface states separately for all the appropriate problems of this chapter after proper modifications.

(R.1.15) Investigate the influence of the localization of carriers for all the appropriate problems of this chapter after proper modifications.

References

[1] W. Zawadzki, in *Two-Dimensional Systems, Heterostructures and Superlattices*, Springer Series in Solid State Sciences, Vol. 53, Ed. by G. Bauer, F. Kuchar, H. Heinrich, (Springer, Berlin, Heidelberg, **1984**).

[2] J. K. Jaiswal, H. Mattoussi, J. M. Mauro, S. M. Simon, *Nat. Biotechnol.*, 21, 47 **(2003)**.

[3] A. Watson, X. Wu, M. Bruchez, *Biotechniques*, 34, 296 **(2003)**; J. Nakanishi, Y. Kikuchi, T. Takarada, H. Nakayama, K. Yamaguchi, M. Maeda, *J. Am. Chem. Soc.*, 126, 16314 **(2004)**.
[4] X. Michalet, F. F. Pinaud, L. A. Bentolila, J. M. Tsay, S. Doose, J. J. Li, G. Sundaresan, A. M. Wu, S. S. Gambhir, S. Weiss, *Science*, 307, 538 **(2005)**.
[5] W. G. J. H. M. van Sark, P. L. T. M. Frederix, D. J. Van den Heuvel, H. C. G. A. Bol, J. N. J. van Lingen, C. de Mello Donegá, A. Meijerink, *J. Phys. Chem. B*, 105, 8281 **(2001)**.
[6] E. J. Sánchez, L. Novotny, X. S. Xie, *Phys. Rev. Lett.*, 82, 4014 **(1999)**; B. Bailey, D. L. Farkas, D. L. Taylor, F. Lanni, *Nature*, 366, 44 **(1993)**.
[7] L. V. Asryan, R. A. Suris, in *Selected Topics in Electronics and Systems*, 25, Ed. by E. Borovitskaya, M. S. Shur, (World Scientific, Singapore, **2002**); L. V. Asryan, R. A. Suris, *Int. J. High Speed Electron. Syst.*, Special Issue on "Quantum Dot Heterostructures — Fabrication, Application, Theory," 12 (1), 111 **(2002)**; L. V. Asryan, S. Luryi, *Future Trends in Microelectronics: The Nano Millennium*, Ed. by S. Luryi, J. M. Xu, A. Zaslavsky, (Wiley Interscience, New York, 219, **2002**).
[8] R. A. Freitas, Jr., *J. Comput. Theor. Nanosci.*, 2, 1 **(2005)**; A. Ferreira, C. Mavroidis, *IEEE Robot. Autom. Magaz.*, 13, 78 **(2006)**; A. Dubey, G. Sharma, C. Mavroidis, S. M. Tomassone, K. Nikitczuk, M. L. Yarmush, *J. Comput. Theor. Nanosci.*, 1, 18 **(2004)**; C. Mavroidis, A. Dubey, M. L. Yarmush, *Annu. Rev. Biomed. Engg.*, 6, 363 **(2004)**.
[9] Y. Liu, J. A. Starzyk, Z. Zhu, *IEEE Trans. Neural Netw.*, 19, 983–995, **(2008)** [In the press]; J. A. Starzyk, H. He, *IEEE Trans. Neural Netw.*, 18, 344 **(2007)**; J. A. Starzyk, H. He, *IEEE Trans. Circuits Syst. II*, 54, 176 **(2007)**.
[10] E. -S. Hasaneen, E. Heller, R. Bansal, W. Huang, F. Jain, *Sol. Stat. Electron.*, 48, 2055 **(2004)**.
[11] T. Kawazoe, S. Tanaka, M. Ohtsu, *J. Nanophoton.*, 2, 029502 **(2008)**.
[12] H. J. Krenner, S. Stufler, M. Sabathil, E. C. Clark, P. Ester, M. Bichler, G. Abstreiter, J. J Finley, A. Zrenner, *New J. Phys.*, 7, 184 **(2005)**.
[13] A. E. Zhukov, A. R. Kovsh, *Quant. Electron.*, 38, 409 **(2008)**.
[14] M. Sugawara, T. Akiyama, N. Hatori, Y. Nakata, H. Ebe, H. Ishikawa, *Meas. Sci. Technol.*, 13, 1683 **(2002)**; M. van der Poel, D. Birkedal, J. Hvam, M. Laemmlin, D. Bimberg, *Conf. Lasers Electro-Opt. (CLEO)*, 1, 16 **(2004)**.
[15] J. M. Costa-Fernandez, *Anal. Bioanal. Chem.*, 384, 37 **(2006)**; H. S. Djie, C. E. Dimas, D.-N. Wang, B.-S. Ooi, J. C. M. Hwang, G. T. Dang, W. H. Chang, *IEEE Sensors J.*, 7, 251 **(2007)**.
[16] X.-X. Zhu, Y.-C. Cao, X. Jin, J. Yang, X.-F. Hua, H.-Q. Wang, B. Liu, Z. Wang, J.-H. Wang, L. Yang, Y.-D. Zhao, *Nanotechnology*, 19, 025708 **(2008)**; X. Gao, W. C. W. Chan, S. Nie, *J. Biomed. Opt.*, 7, 532 **(2002)**.
[17] J. K. Jaiswal, E. R. Goldman, H. Mattoussi, S. M. Simon, *Nat. Methods*, 1, 73 **(2004)**.

[18] H. Matsueda, *Int. J. Circuit Theor. Appl.*, 31, 23 **(2003)**; X. Hu, S. Das Sarma, *Phys. Stat. Sol. (b)*, 238, 360 **(2003)**.

[19] G.-L. Chen, D. M. T. Kuo, W.-T. Lai, P.-W. Li, *Nanotechnology*, 18, 475402 **(2007)**; A. G. Pogosov, M. V. Budantsev, A. A. Shevyrin, A. E. Plotnikov, A. K. Bakarov, A. I. Toropov, *JETP Lett.*, 87, 150 **(2008)**.

[20] K. W. Johnston, A. G. Pattantyus-Abraham, J. P. Clifford, S. H. Myrskog, D. D. MacNeil, L. Levina, E. H. Sargent, *Appl. Phys. Lett.*, 92, 151115 **(2008)**; K. S. Leschkies, R. Divakar, J. Basu, E. E.-Pommer, J. E. Boercker, C. B. Carter, U. R. Kortshagen, D. J. Norris, E. S. Aydil, *Nano Lett.*, 7, 1793 **(2007)**.

[21] I.-S. Liu, H.-H. Lo, C.-T. Chien, Y.-Y. Lin, C.-W. Chen, Y.-F. Chen, W.-F. Su, S.-C. Liou, *J. Mater. Chem.*, 18, 675 **(2008)**.

[22] N. Hitoshi, Y. Sugimoto, K. Nanamoto, N. Ikeda, Y. Tanaka, Y. Nakamura, S. Ohkouchi, Y. Watanabe, K. Inoue, H. Ishikawa, K. Asakawa, *Opt. Express*, 12, 6606 **(2004)**; N. Yamamoto, T. Matsuno, H. Takai, N. Ohtani, *Jpn. J. Appl. Phys.*, 44, 4749 **(2005)**; T. Yamada, Y. Kinoshita, S. Kasai, H. Hasegawa, Y. Amemiya, *Jpn. J. Appl. Phys.*, 40, 4485 **(2001)**; K. Asakawa, Y. Sugimoto, Y. Watanabe, N. Ozaki, A. Mizutani, Y. Takata, Y. Kitagawa, H. Ishikawa, N. Ikeda, K. Awazu, X. Wang, A. Watanabe, S. Nakamura, S. Ohkouchi, K. Inoue, M. Kristensen, O. Sigmund, P. I. Borel, R. Baets, *New J. Phys.*, 8, 208 **(2006)**.

[23] A. R. Clapp, I. L. Medintz, B. R. Fisher, G. P. Anderson, H. Mattoussi, *J. Am. Chem. Soc.*, 127, 1242 **(2005)**; L. Shi, B. Hernandez, M. Selke, *J. Am. Chem. Soc.*, 128, 6278 **(2006)**; C. Wu, J. Zheng, C. Huang, J. Lai, S. Li, C. Chen, Y. Zhao, *Angew. Chem. Int. Ed.*, 46, 5393 **(2007)**.

[24] J. L. Shay, J. W. Wernick, *Ternary Chalcopyrite Semiconductors-Growth, Electronic Properties and Applications*, (Pergamon Press, London, **1975**).

[25] J. E. Rowe, J. L. Shay, *Phys. Rev. B*, 3, 451 **(1973)**.

[26] H. Kildal, *Phys. Rev. B*, 10, 5082 **(1974)**.

[27] J. Bodnar, in *Proceedings of the International Conference on Physics of Narrow-gap Semiconductors*, (Polish Science Publishers, Warsaw, **1978**).

[28] G. P. Chuiko, N. N. Chuiko, *Sov. Phys. Semicond.*, 15, 739 **(1981)**.

[29] K. P. Ghatak and S. N. Biswas, *Proc. SPIE*, 1484, 149 **(1991)**.

[30] J. A. Zapien, Y. K. Liu, Y. Y. Shan, H. Tang, C. S. Lee, S. T. Lee, *Appl. Phys. Lett.*, 90, 213114 **(2007)**.

[31] R. M. Park, *Proc. SPIE*, 2524, 142 **(1995)**.

[32] S.-G. Hur, E.-T. Kim, J.-H. Lee, G.-H. Kim, S.-G. Yoon, *Electrochem. Solid-State Lett.*, 11, H176 **(2008)**; H. Kroemer, *Rev. Mod. Phys.*, 73, 783 **(2001)**; T. Nguyen Duy, J. Meslage, G. Pichard, *J. Crys. Growth*, 72, 490 **(1985)**; T. Aramoto, F. Adurodija, Y. Nishiyama, T. Arita, A. Hanafusa, K. Omura, A. Morita, *Solar Energy Mat. Solar Cells*, 75, 211 **(2003)**; H. B. Barber, *J. Electron. Mater.*, 25, 1232 **(1996)**; S. Taniguchi, T. Hino, S. Itoh, K. Nakano, N. Nakayama, A. Ishibashi, M. Ikeda, *Electron. Lett.*, 32, 552 **(1996)**.

[33] J. J. Hopfield, *J. Appl. Phys.*, 32, 2277 **(1961)**.

[34] F. Hatami, V. Lordi, J. S. Harris, H. Kostial, W. T. Masselink, *J. Appl. Phys.*, 97, 096106 **(2005)**.
[35] B. W. Wessels, *J. Electrochem. Soc.*, 122, 402 **(1975)**; D. W. L. Tolfree, *J. Sci. Instrum.*, 41, 788 **(1964)**; P. B. Hart, *Proc. IEEE*, 61, 880 **(1973)**.
[36] H. Choi, M. Chang, M. Jo, S. J. Jung, H. Hwang, *Electrochem. Solid-State Lett.*, 11, H154 **(2008)**.
[37] S. Cova, M. Ghioni, A. Lacaita, C. Samori, F. Zappa, *Appl. Opt.*, 35, 1956 **(1996)**; H. W. H. Lee, B. R. Taylor, S. M. Kauzlarich, *Nonlinear Optics: Materials, Fundamentals, and Applications*, (Technical Digest, 12, **2000**); E. Brundermann, U. Heugen, A. Bergner, R. Schiwon, G. W. Schwaab, S. Ebbinghaus, D. R. Chamberlin, E. E. Haller, M. Havenith, *29th International Conference on Infrared and Millimeter Waves and 12th International Conference on Terahertz Electronics*, 283 **(2004)**.
[38] P. K. Weimer, *Proc. IEEE*, 52, 608 **(1964)**.
[39] G. Ribakovs, A. A. Gundjian, *IEEE J. Quant. Electron.*, QE-14, 42 **(1978)**.
[40] S. K. Dey, *J. Vac. Sci. Technol.*, 10, 227 **(1973)**; S. J. Lynch, *Thin Solid Films*, 102, 47 **(1983)**.
[41] V. V. Kudzin, V. S. Kulakov, D. R. Pape', S. V. Kulakov, V. V. Molotok, *IEEE Ultrasonics Symposium*, 1, 749 **(1997)**.
[42] M. S. Dresselhaus, J. G. Mavroides, *IBM J. Res. Dev.*, 8, 262 **(1964)**; P. R. Schroeder, M. S. Dresselhaus, A. Javan, *Phys. Rev. Lett.*, 20, 1292 **(1968)**; M. S. Dresselhaus, in *Proceedings of the Conference on the Physics of Semimetals and Narrow-gap Semiconductors*, Ed. by D. L. Carter, R. T. Bate, (Pergamon Press, New York, **1970**); M. S. Dresselhaus, G. Dresselhaus, *Adv. Phys.*, 51, 1 **(2002)**.
[43] M. A. Hines, G. D. Scholes, *Adv. Mater.*, 15, 1844 **(2003)**; C. A. Wang, R. K. Huang, D. A. Shiau, M. K. Connors, P. G. Murphy, P. W. O'Brien, A. C. Anderson, D. M. DePoy, G. Nichols, M. N. Palmisiano, *Appl. Phys. Lett.*, 83, 1286 **(2003)**; C. W. Hitchcock, R. J. Gutmann, J. M. Borrego, I. B. Bhat, G. W. Charache, *IEEE Trans. Electron. Dev.*, 46, 2154 **(1999)**.
[44] S. Matsushita, K. Kawai, H. Uchida, *IEEE J. Lightwave Technol.*, LT-3, 533 **(1985)**; N. K. Ailawadi, R. C. Alferness, G. D. Bergland, R. A. Thompson, *IEEE Mag. Lightwave Telecommun. Syst.*, 2, 38 **(1991)**; K. Kubota, J. Noda, O. Mikami, *IEEE J. Quant. Electron.*, QE-16, 754 **(1980)**; R. C. Alferness, C. H. Joyner, L. L. Buhl, S. K. Korotky, *IEEE J. Quant. Electron.*, QE-19, 1339 **(1983)**; R. G. Walker, *IEEE J. Quant. Electron.*, 27, 654 **(1991)**.
[45] G. C. Young, W. W. Anderson, L. B. Anderson, *IEEE Trans. Electron. Dev.*, 24, 492 **(1977)**.
[46] R. L. Gordon, V. I. Neeley, H. R. Curtin, *Proc. IEEE*, 54, 2014 **(1966)**.
[47] P. K. Weimer, *Proc. IRE*, 50, 1462 **(1962)**.
[48] M. J. Lee, S. W. Wright, C. P. Judge, P. Y. Cheung, *Display Res. Conf., Int. Conf. Rec.*, 211 **(1991)**.
[49] A. N. Baranov, T. I. Voronina, N. S. Zimogorova, L. M. Kauskaya, Y. P. Yakoviev, *Sov. Phys. Semicond.*, 19, 1676 **(1985)**; M. Yano,

Y. Suzuki, T. Ishii, Y. Matsushima, M. Kimata, *Jpn. J. Appl. Phys.*, 17, 2091 **(1978)**; F. S. Yuang, Y. K. Su, N. Y. Li, *Jpn. J. Appl. Phys.*, 30, 207 **(1991)**; F. S. Yuang, Y. K. Su, N. Y. Li, K. J. Gan, *J. Appl. Phys.*, 68, 6383 **(1990)**; Y. K. Su, S. M. Chen, *J. Appl. Phys.*, 73, 8349 **(1993)**; S. K. Haywood, A. B. Henriques, N. J. Mason, R. J. Nicholas, P. J. Walker, *Semicond. Sci. Technol.*, 3, 315 **(1988)**.

[50] F. Hüe, M. Hÿtch, H. Bender, F. Houdellier, A. Claverie, *Phys. Rev. Lett.*, 100, 156602 **(2008)**; S. Banerjee, K. A. Shore, C. J. Mitchell, J. L. Sly, M. Missous, *IEE Proc. Circuits Devices Syst.*, 152, 497 **(2005)**; M. Razeghi, A. Evans, S. Slivken, J. S. Yu, J. G. Zheng, V. P. Dravid, *Proc. SPIE*, 5840, 54 **(2005)**; R. A. Stradling, *Semicond. Sci. Technol.*, 6, C52 **(1991)**.

[51] R. V. Belosludov, A. A. Farajian, H. Mizuseki, K. Miki, Y. Kawazoe, *Phys. Rev. B*, 75, 113411 **(2007)**; J. Heremans, C. M. Thrush, Y. -M. Lin, S. Cronin, Z. Zhang, M. S. Dresselhaus, J. F. Mansfield, *Phys. Rev. B*, 61, 2921 **(2000)**.

[52] D. Shoenberg, *Proc. Roy. Soc.* (London), 170, 341 **(1939)**; B. Abeles, S. Meiboom, *Phys. Rev.*, 101, 544 **(1956)**.

[53] B. Lax, J. G. Mavroides, H. J. Zieger, R. J. Keyes, *Phys. Rev. Lett.*, 5, 241 **(1960)**.

[54] M. Maltz, M. S. Dresselhaus, *Phys. Rev. B*, 2, 2877 **(1970)**.

[55] M. Cankurtaran, H. Celik, T. Alper, *J. Phys. F: Metal Phys.*, 16, 853 **(1986)**.

[56] Y. -H. Kao, *Phys. Rev.*, 129, 1122 **(1963)**.

[57] R. J. Dinger, A. W. Lawson, *Phys. Rev. B*, 3, 253 **(1971)**.

[58] J. F. Koch, J. D. Jensen, *Phys. Rev.*, 184, 643 **(1969)**.

[59] M. H. Cohen, *Phys. Rev.*, 121, 387 **(1961)**.

[60] S. Takaoka, H. Kawamura, K. Murase, S. Takano, *Phys. Rev. B*, 13, 1428 **(1976)**; C. C. Wu, C. J. Lin, *J. Low. Temp. Phys.*, 57, 469 **(1984)**.

[61] J. W. McClure, K. H. Choi, *Sol. Stat. Commun.*, 21, 1015 **(1977)**.

[62] G. P. Agrawal, N. K. Dutta, *Semiconductor Lasers*, (Van Nostrand Reinhold, New York, **1993**); S. Chatterjee, U. Pal, *Opt. Engg.*, 32, 2923 **(1993)**; T. K. Chaudhuri, *Int. J. Energy Res.*, 16, 481 **(1992)**; J. H. Dughaish, *Phys. B*, 322, 205 **(2002)**; C. Wood, *Rep. Prog. Phys.*, 51, 459 **(1988)**.

[63] K. -F. Hsu, S. Loo, F. Guo, W. Chen, J. S. Dyck, C. Uher, T. Hogan, E. K. Polychroniadis, M. G. Kanatzidis, *Science*, 303, 818 **(2004)**; J. Androulakis, K. F. Hsu, R. Pcionek, H. Kong, C. Uher, J. J. D'Angelo, A. Downey, T. Hogan, M. G. Kanatzidis, *Adv. Mater.*, 18, 1170 **(2006)**; P. F. P. Poudeu, J. D'Angelo, A. D. Downey, J. L. Short, T. P. Hogan, M. G. Kanatzidis, *Angew. Chem. Int. Ed.*, 45, 3835 **(2006)**; P. F. Poudeu, J. D'Angelo, H. Kong, A. Downey, J. L. Short, R. Pcionek, T. P. Hogan, C. Uher, M. G. Kanatzidis, *J. Am. Chem. Soc.*, 128, 14347 **(2006)**; J. R. Sootsman, R. J. Pcionek, H. Kong, C. Uher, M. G. Kanatzidis, *Chem. Mater.*, 18, 4993 **(2006)**.

[64] A. J. Mountvala, G. Abowitz, *J. Am. Ceram. Chem. Soc.*, 48, 651 **(1965)**; E. I. Rogacheva, I. M. Krivulkin, O. N. Nashchekina, A. Yu. Sipatov, V. A. Volobuev, M. S. Dresselhaus, *Appl. Phys. Lett.*, 78, 3238 **(2001)**; H. S. Lee, B. Cheong, T. S. Lee, K. S. Lee, W. M. Kim, J. W. Lee, S. H. Cho, J. Y. Huh, *Appl. Phys. Lett.*, 85, 2782 **(2004)**; K. Kishimoto, M. Tsukamoto, T. Koyanagi, *J. Appl. Phys.*, 92, 5331 **(2002)**.

[65] E. I. Rogacheva, O. N. Nashchekina, S. N. Grigorov, M. A. Us, M. S. Dresselhaus, S. B. Cronin, *Nanotechnology*, 14, 53 **(2003)**.

[66] E. I. Rogacheva, O. N. Nashchekina, A. V. Meriuts, S. G. Lyubchenko, M. S. Dresselhaus, G. Dresselhaus, *Appl. Phys. Lett.*, 86, 063103 **(2005)**; E. I. Rogacheva, S. N. Grigorov, O. N. Nashchekina, T. V. Tavrina, S. G. Lyubchenko, A. Yu. Sipatov, V. V. Volobuev, A. G. Fedorov, M. S. Dresselhaus, *Thin Solid Films*, 493, 41 **(2005)**.

[67] X. Qiu, Y. Lou, A. C. S. Samia, A. Devadoss, J. D. Burgess, S. Dayal, C. Burda, *Angew. Chem. Int. Ed.*, 44, 5855 **(2005)**.

[68] C. Wang, G. Zhang, S. Fan, Y. Li, *J. Phys. Chem. Solids*, 62, 1957 **(2001)**; B. Poudel, W. Z. Wang, D. Z. Wang, J. Y. Huang, Z. F. Ren, *J. Nanosci. Nanotech.*, 6, 1050 **(2006)**; B. Zhang, J. He, T. M. Tritt, *Appl. Phys. Lett.*, 88, 043119 **(2006)**; W. Heiss, H. Groiss, E. Kaufmann, G. Hesser, M. Böberl, G. Springholz, F. Schäffler, K. Koike, H. Harada, M. Yano, *Appl. Phys. Lett.*, 88, 192109 **(2006)**.

[69] F. Smektala, C. Quemard, V. Couderc, A. Barthélémy, *J. Non-Crys. Sol.*, 274, 232 **(2000)**.

[70] B. A. Volkov, L. I. Ryabova, D. R. Khokhlov, *Phys. Usp.*, 45, 819 **(2002)**, and references therein.

[71] J. M. Thornton, *J. Mol. Biol.*, 151, 261 **(1981)**; T. E. Creighton, *Methods Enzymol.*, 131, 83 **(1986)**; T. E. Creighton, *Bio Essays*, 8, 57 **(1988)**.

[72] V. J. Hruby, *Life Sci.*, 31, 189 **(1982)**; V. J. Hruby, F. Al-Obeidi, W. Kazmierski, *Biochem. J.*, 268, 249 **(1990)**.

[73] R. Wetzel, *Trends Biochem. Sci.*, 12, 478 **(1987)**.

[74] T. Kimura, R. Matsueda, Y. Nakagawa, E. T. Kaiser, *Anal. Biochem.*, 122, 274 **(1982)**.

[75] D. Andreu, F. Albericio, N. A. Sole, M. C. Munson, M. Ferrer, G. Barany, *Methods in Molecular Biology, 35 Peptide Synthesis Protocols*, Ed. M. W. Pennington, B. M. Dunn, (Humana Press Inc., New Jersey, **1994**).

[76] H. J. Goldsmid, R. W. Douglas, *Brit. J. Appl. Phys.*, 5, 386 **(1954)**.

[77] F. D. Rosi, B. Abeles, R. V. Jensen, *J. Phys. Chem. Solids*, 10, 191 **(1959)**; T. M. Tritt (Ed.), *Semiconductors and Semimetals, Vols. 69, 70 and 71: Recent Trends in Thermoelectric Materials Research I, II and III*, (Academic Press, USA, **2000**); D. M. Rowe (Ed.), *CRC Handbook of Thermoelectrics*, (CRC Press, USA, **1995**); D. M. Rowe, C. M. Bhandari, *Modern Thermoelectrics*, (Reston Publishing Company, Virginia, **1983**); D. M. Rowe (Ed.), *Thermoelectrics Handbook: Macro to Nano*, (CRC Press, USA, **2006**).

[78] G. Boudebs, F. Sanchez, J. Troles, F. Smektala, *Opt. Commun.*, 199, 425 **(2001)**; J. M. Harbold, F. O. Ilday, F. W. Wise, J. S. Sanghera, V. Q. Nguyen, L. B. Shaw, I. D. Aggarwal, *Opt. Lett.*, 27, 119 **(2002)**; J. M.

Harbold, F. O. Ilday, F. W. Wise, B. Aitken, *IEEE Photon. Technol. Lett.*, 14, 822 **(2002)**; V. T. Plaksiy, V. M. Svetlichniy, *Symposium on Physics and Engineering of Millimeter and Sub-Millimeter Waves, 2001. The Fourth International Kharkov*, 1, 331 **(2001)**; Z. Helin, D. M. Rowe, S. G. K. Williams, *2001. Proc. XX International Conference on Thermoelectrics*, 314 **(2001)**.

[79] B. R. Nag, *Electron Transport in Compound Semiconductors*, (Springer-Verlag, Germany, **1980**); E. O. Kane, in *Semiconductors and Semimetals*, Vol. 1, Ed. by R. K. Willardson, A. C. Beer, (Academic Press, New York, **1966**).

[80] G. E. Stillman, C. M. Wolfe, J. O. Dimmock, in S*emiconductors and Semi metals*, Ed. R. K. Willardon, A. C. Beer, 12, (Academic Press, IV, V, USA, **1977**).

[81] D. J. Newson, A. Kurobe, *Semicond. Sci. Technol.*, 3, 786 **(1988)**.

[82] U. Rossler, *Sol. Stat. Commun.*, 49, 943 **(1984)**.

[83] E. D. Palik, G. S. Picus, S. Teither, R. E. Wallis, *Phys. Rev.*, 122, 475 **(1961)**.

[84] E. J. Johnson, D. H. Dickey, *Phys. Rev.*, 1, 2676 **(1970)**.

[85] V. G. Agafonov, P. M. Valov, B. S. Ryvkin, I. D. Yarashetskin, *Sov. Phys. Semicond.*, 12, 1182 **(1978)**.

[86] G. J. Rees, *Physics of Compounds*, in Proc. of the 13th Inter. Nat. Conf. Ed. by F. G. Fumi, 1166, (North Holland Company, The Netherlands, **1976**); E. L. Ivechenko, G. E. Pikus, *Sov. Phys. Semicond.*, 13, 579 **(1979)**.

[87] M. Cardona, W. Paul, H. Brooks, *Helv. Acta Phys.*, 33, 329 **(1960)**.

[88] A. F. Gibson, in *Proceeding of International School of Physics "ENRICO FERMI"* course XIII, Ed. R. A. Smith, (Academic Press, 171, **1963**).

[89] C. C. Wang, N. W. Ressler, *Phys. Rev.*, 2, 1827 **(1970)**.

[90] J. Bouat, J. C. Thuillier, *Surface Sci.*, 73, 528 **(1978)**.

[91] M. V. Ortenberg, K. J. Button, *Phys. Rev. B*, 16, 2618 **(1977)**.

[92] H. Ushio, T. Dau, Y. Uemura, *J. Phys. Soc. Jap.*, 33, 1551 **(1972)**.

[93] P. R. Emtage, *Phys. Rev.*, 138, A246 **(1965)**.

[94] V. I. Ivanov-Omskii, A. Sh. Mekhtisev, S. A. Rustambekova, E. N. Ukraintsev, *Phys. Stat. Sol. (b)*, 119, 159 **(1983)**.

[95] Y. Yamada, *J. Phys. Soc. Jap.*, 35, 1600 **(1973)**.

[96] M. Singh, P. R. Wallace, S. D. Jog and E. Arushanov, *J. Phys. Chem. Solids*, 45, 409 **(1984)**.

[97] D. G. Seiler, B. D. Bajaj and A. E. Stephens, *Phys. Rev. B*, 16, 2822 **(1977)**; A. V. Germaneko and G. M. Minkov, *Phys. Stat. Sol. (b)*, 184, 9 **(1994)**; G. L. Bir and G. E. Pikus, *Symmetry and Strain–Induced Effects in Semiconductors Nauka, Russia* **(1972)**. (in Russian); M. Mondal and K. P. Ghatak, *Phys. Stat. Sol. (b)*, 135, K21 **(1986)**.

[98] P. C. Mathur and S. Jain, *Phys. Rev.*, 19, 3159 **(1979)**.

[99] H. I. Zhang, *Phys. Rev. B*, 1, 3450 **(1970)**.

[100] K. P. Ghatak, S. Bhattacharya, D. De, *Einstein Relation in Compound Semiconductors and Their Nanostructures*, Springer Series in Materials Science, Vol. 116, (Springer-Verlag, Germany, **2008**).

[101] J. O. Dimmock, in *The Physics of Semimetals and Narrowgap Semiconductors*, Ed. by D. L. Carter, R. T. Bates, (Pergamon Press, Oxford, **1971**).

[102] E. Bangert, P. Kastner, *Phys. Stat. Sol. (b)*, 61, 503 (**1974**).

[103] G. M. T. Foley, P. N. Langenberg, *Phys. Rev. B*, 15, 4850 (**1977**).

[104] W. E. Spicer, G. J. Lapeyre, *Phys. Rev.*, 139, A565 (**1965**).

[105] L. A. Vassilev, *Phys. Stat. Sol. (b)*, 121, 203 (**1984**); S. Takaoka, K. Murase, *Phys. Rev. B*, 20, 2823 (**1979**).

[106] G. P. Chuiko, *Sov. Phys. Semicond.*, 19, 1381 (**1985**).

[107] M. Stordeur, W. Kuhnberger, *Phys. Stat. Sol. (b)*, 69, 377 (**1975**).

[108] H. Köhler, *Phys. Stat. Sol. (b)*, 74, 591 (**1976**).

[109] J. B. Ketterson, *Phys. Rev.*, 129, 18 (**1963**).

[110] K. P. Ghatak, A. Biswas, *Electron Statistics in Quantum Confined Superlattices*, Vol. 16, (World Scientific, 1–726, **2023**).

[111] P. T. Landsberg, *Eur. J. Phys.*, 2, 213 (**1981**).

[112] R. B. Dingle, *Philos. Mag.*, 46, 813 (**1955**); D. Redfield, M. A. Afromowitz, *ibid.*, 19, 831 (**1969**); H. C. Casey, F. Stern, *J. Appl. Phys.*, 47, 631 (**1976**).

[113] S. N. Mohammad, *J. Phys. C*, 13, 2685 (**1980**).

[114] K. P. Ghatak, S. Bhattacharya, S. Pahari, S. N. Mitra, P. K. Bose, D. De, *J. Phys. Chem. Solids*, 70, 122 (**2009**); K. P. Ghatak, B. Mitra, *Phys. Scr.*, 46, 182 (**1992**).

[115] S. Bhattacharya, N. C. Paul, K. P. Ghatak, *Phys. B*, 403, 4139 (**2008**); K. P. Ghatak, S. Bhattacharya, H. Saikia, D. Baruah, A. Saikia, K. M. Singh, A. Ali, S. N. Mitra, P. K. Bose, A. Sinha, *J. Comput. Theor. Nanosci.*, 3, 727 (**2006**).

[116] A. N. Chakravarti, D. Mukherjee, *Phys. Lett. A*, 53, 403 (**1975**).

[117] B. Mitra, D. K. Basu, B. Nag, K. P. Ghatak, *Nonlinear Opt.*, 17, 171 (**1997**).

[118] M. Mondal, K. P. Ghatak, *Phys. Stat. Sol. (b)*, 135, 239 (**1986**).

[119] A. N. Chakravarti, S. Swaminathan, *Phys. Stat. Sol. (a)*, 23, K191 (**1974**); A. N. Chakravarti, *ibid.*, 25, K105 (**1974**).

[120] A. N. Chakravarti, K. P. Ghatak, K. K. Ghosh, A. Dhar, *Phys. Stat. Sol. (b)*, 103, K55 (**1981**).

[121] T. Ando, A. H. Fowler and F. Stern, *Rev. Mod. Phys.*, 54, 437 (**1982**); P. K. Basu, *Optical Processes in Semiconductors*, (Oxford University Press, New York, **2001**).

[122] N. Kampf, D. Ben-Yaakov, D. Andelman, S. A. Safran, J. Klein, *Phys. Rev. Lett.*, 103, 118304 (**2009**); P. Arnold, L. G. Yaffe, *Phys. Rev. D*, 52, 7208 (**1995**); G. S. Kulkarni and Z. Zhong, *Nano Lett.*, 12, 719 (**2012**); E. Stern, R. Wagner, F. J. Sigworth, R. Breaker, T. M. Fahmy, M. A. Reed, *Nano Lett.*, 7, 3405 (**2007**).

[123] A. N. Chakravarti, D. Mukherji, *Phys. Lett. A*, 53, 57 (**1975**); W. Zawadzki, *Adv. Phys.*, 23, 435 (**1974**).

[124] A. K. Sreedhar and S. C. Gupta, *Phys. Rev. B*, 5, 3160 (**1972**); R. W. Keyes, *IBM J. Res. Dev.*, 5, 266 (**1961**).

[125] S. Bhattacharya, S. Chowdhury and K. P. Ghatak, *J. Comput. Theor. Nanosci.*, 3, 423 **(2006)**; B. Nag and K. P. Ghatak, *J. Phys. Chem. Solids*, 58, 427 **(1997)**.

[126] K. P. Ghatak, J. P. Banerjee, D. Bhattacharyya and B. Nag, *Nanotechnology*, 7, 110 **(1996)**; K. P. Ghatak, J. P. Banerjee, M. Mitra and B. Nag, *Nonlinear Opt.*, 17, 193 **(1996)**; K. P. Ghatak, *Int. J. Electron.*, 71, 239 **(1991)**.

[127] K. P. Ghatak, B. De, S. N. Biswas and M. Mondal, *Mechanical Behavior of Materials and Structures in Microelectronics, MRS Symposium Proceedings*, Spring Meeting, 2216, 191 **(1991)**; K. P. Ghatak and B. De, *MRS Symposium Proceedings*, 226, 191 **(1991)**; K. P. Ghatak, B. Nag and G. Majumdar, *MRS*, 379, 109 **(1995)**; D. Baruah, S. Choudhury, K. M. Singh and K. P. Ghatak, *J. Phys., Conf. Series*, 61, 80 **(2007)**.

[128] K. P. Ghatak, M. Mitra, A. Biswas, *Heisenberg's Uncertainty Principle and the Electron Statistics in Quantized Structures*, (Springer, **2022**).

[129] A. Einstein, *Ann. der Physik*, 17, 549 **(1905)**; H. Kroemer, *IEEE Trans. Electron. Dev.*, 25, 850 **(1978)**; W. Nernst, *Z. Phys. Chem.*, 2, 613 **(1888)**; J. S. Townsend, *Trans. Roy. Soc. A*, 193, 129 **(1900)**; C. Wagner, *Z. Physik, Chem. B*, 21, 24 **(1933)**; C. Herring and M. H. Nichols, *Rev. Mod. Phys.*, 21, 185 **(1949)**; P. T. Landsberg, *Thermodynamics and Statistical Mechanics*, (Oxford University Press, Oxford **1978**); P. T. Landesberg, *In Recombination in Semiconductors*, (Cambridge University Press, U.K., **1991**).

[130] R. W. Lade, *Proc. IEEE*, 52, 743 **(1965)**.

[131] P. T. Landsberg, *Proc. Roy. Soc. A*, 213, 226 **(1952)**.

[132] C. H. Wang, A. Neugroschel, *IEEE Electron. Dev. Lett. ED*, 11, 576 **(1990)**.

[133] I. Y. Leu, A. Neugroschel, *IEEE Trans. Electron. Dev. ED*, 40, 1872 **(1993)**.

[134] F. Stengel, S. N. Mohammad, H. Morkoç, *J. Appl. Phys.*, 80, 3031 **(1996)**.

[135] H. J. Pan, W. C. Wang, K. B. Thai, C. C. Cheng, K. H. Yu, K. W. Lin, C. Z. Wu, W. C. Liu, *Semicond. Sci. Technol.*, 15, 1101 **(2000)**.

[136] S. N. Mohammad, *J. Appl. Phys.*, 95, 4856 **(2004)**.

[137] V. K. Arora, *Appl. Phys. Lett.*, 80, 3763 **(2002)**.

[138] S. N. Mohammad, *J. Appl. Phys.*, 95, 7940 **(2004)**.

[139] S. N. Mohammad, *Philos. Mag.*, 84, 2559 **(2004)**.

[140] S. N. Mohammad, *J. Appl. Phys.*, 97, 063703 **(2005)**.

[141] K. Suzue, S. N. Mohammad, Z. F. Fan, W. Kim, O. Aktas, A. E. Botchkarev, H. Morkoç, *J. Appl. Phys.*, 80, 4467 **(1996)**.

[142] S. N. Mohammad, Z. F. Fan, W. Kim, O. Aktas, A. E. Botchkarev, A. Salvador, H. Morkoç, *Electron. Lett.*, 32, 598 **(1996)**.

[143] Z. Fan, S. N. Mohammad, W. Kim, O. Aktas, A. E. Botchkarev, K. Suzue, H. Morkoç, *J. Electron. Mater.*, 25, 1703 **(1996)**.

[144] C. Lu, H. Chen, X. Lv, X. Xia, S. N. Mohammad, *J. Appl. Phys.*, 91, 9216 **(2002)**.

[145] S. G. Dmitriev, Yu V. Markin, *Semiconductors*, 34, 931 **(2000)**.

[146] M. Tao, D. Park, S. N. Mohammad, D. Li, A. E. Botchkerav, H. Morkoç, *Philos. Mag. B*, 73, 723 **(1996)**.

[147] D. G. Park, M. Tao, D. Li, A. E. Botchkarev, Z. Fan, S. N. Mohammad, H. Morkoç, *J. Vac. Sci. Technol. B*, 14, 2674 **(1996)**.

[148] Z. Chen, D. G. Park, S. N. Mohammad, H. Morkoç, *Appl. Phys. Lett.*, 69, 230 **(1996)**.

[149] A. N. Chakravarti, D. P. Parui, *Phys. Lett. A*, 40, 113 **(1972)**.

[150] A. N. Chakravarti, D. P. Parui, *Phys. Lett. A*, 43, 60 **(1973)**; B. R. Nag, A. N. Chakravarti, *Sol. Stat. Electron.*, 18, 109 **(1975)**.

[151] B. R. Nag, A. N. Chakravarti, P. K. Basu, *Phys. Stat. Sol. (a)*, 68, K75 **(1981)**.

[152] B. R. Nag, A. N. Chakravarti, *Phys. Stat. Sol. (a)*, 67, K113 **(1981)**.

[153] A. N. Chakravarti, D. P. Parui, *Canad. J. Phys.*, 51, 451 **(1973)**; S. Ghosh, A. N. Chakravarti, *Phys. Stat. Sol. (b)*, 147, 355 **(1988)**; A. N. Chakravarti, K. K. Ghosh, K. P. Ghatak, H. M. Mukherjee, *Phys. Stat. Sol. (b)*, 118, 843 **(1983)**; A. N. Chakravarti, K. P. Ghatak, K. K. Ghosh, G. B. Rao, *Phys. Stat. Sol. (b)*, 111, K61 **(1982)**; A. N. Chakravarti, K. P. Ghatak, K. K. Ghosh, H. M. Mukherjee, S. Ghosh, *Phys. Stat. Sol. (b)*, 108, 609 **(1981)**; A. N. Chakravarti, K. P. Ghatak, S. Ghosh, A. Dhar, *Phys. Stat. Sol. (b)*, 105, K55 **(1981)**; A. N. Chakravarti, A. K. Choudhury, K. P. Ghatak, D. Roy Choudhury, *Phys. Stat. Sol. (b)*, 59, K211 **(1980)**; A. N. Chakravarti, A. K. Chowdhury, K. P. Ghatak, D. R. Choudhury, *Acta Phys. Polonica A*, 58, 251 **(1980)**; K. P. Ghatak, A. K. Chowdhury, S. Ghosh, A. N. Chakravarti, *Phys. Stat. Sol. (b)*, 99, K55 **(1980)**; A. N. Chakravarti, A. K. Chowdhury, K. P. Ghatak, D. R. Choudhury, *Czech. J. Phys. B*, 30, 1161 **(1980)**; A. N. Chakravarti, K. P. Ghatak, A. Dhar, K. K. Ghosh, S. Ghosh, *Acta Phys. Polonica A*, 60, 151 **(1981)**; A. N. Chakravarti, A. K. Chowdhury, K. P. Ghatak, A. Dhar, D. R. Choudhury, *Czech. J. Phys. B*, 31, 905 **(1981)**; A. N. Chakravarti, K. P. Ghatak, K. K. Ghosh, S. Ghosh, H. M. Mukherjee, *Czech. J. Phys. B*, 31, 1138 **(1981)**.

[154] P. T. Landsberg, *J. Appl. Phys.*, 56, 1696 **(1984)**; P. T. Landsberg, *Phys. Rev. B*, 33, 8321 **(1986)**; Y. Roichman, N. Tessler, *Appl. Phys. Lett.*, 80, 1948 **(2002)**; J. M. H. Peters, *Eur. J. Phys.*, 3, 19 **(1982)**; H. Van Cong, S. Brunet, S. Charar, *Phys. Stat. Sol. (b)*, 109, K1 **(1982)**; H. Van Cong, *Phys. Stat. Sol. (a)*, 56, 395 **(1979)**; H. Van Cong, *Sol. Stat. Electron.*, 24, 495 **(1981)**.

[155] S. N. Mohammad, S. T. H. Abidi, *J. Appl. Phys.*, 61, 4909 **(1987)**; S. N. Mohammad,S. T. H. Abidi *Sol. Stat. Electron.*, 27, 1153 **(1985)**; *J. Appl. Phys.*, 56, 3341 **(1984)**; M. A. Sobhan, S. N. Mohammad, *J. Appl. Phys.*, 58, 2634 **(1985)**; S. N. Mohammad, A. V. Bemis, *IEEE Trans. Electron. Dev. ED*, 39, 282 **(1992)**; S. N. Mohammad, R. L. Carter, *Philos. Mag. B*, 72, 13 **(1995)**; S. N. Mohammad, *Sol. Stat. Electron.*, 46, 203 **(2002)**; S. N. Mohammad, J. Chen, J. I. Chyi, H. Morkoç, *Appl. Phys. Lett.*, 56, 937 **(1990)**.

[156] P. T. Landsberg, S. A. Hope, *Sol. Stat. Electron.*, 20, 421 **(1977)**; S. A. Hope, G. Feat, P. T. Landsberg, *J. Phys. A. Math. Gen.*, 14, 2377 **(1981)**.

[157] W. Elsäber, E. O. Göbel, *Electron. Lett.*, 19, 335 **(1983)**; R. Hilfer, A. Blumen, *Phys. Rev. A*, 37, 578 **(1988)**; T. G. Castner, *Phys. Rev. B*, 55, 4003 **(1997)**; E. Barkai, V. N. Fleurov, *Phys. Rev. E*, 58, 1296 **(1998)**; T. H. Nguyen, S. K. O'Leary, *Appl. Phys. Lett.*, 83, 1998 **(2003)**; T. H. Nguyen, S. K. O'Leary, *J. Appl. Phys.*, 98, 076102 **(2005)**; C. G. Rodrigues, Á. R. Vasconcellos, R. Luzzi, *J. Appl. Phys.*, 99, 073701 **(2006)**.

[158] R. K. Jain, *Phys. Stat. Sol. (a)*, 42, K221 **(1977)**; B. A. Aronzon, E. Z. Meilikhov, *Phys. Stat. Sol. (a)*, 19, 313 **(1973)**.

[159] S. Choudhury, D. De, S. Mukherjee, A. Neogi, A. Sinha, M. Pal, S. K. Biswas, S. Pahari, S. Bhattacharya, K. P. Ghatak, *J. Comput. Theor. Nanosci.*, 5, 375 **(2008)**; S. M. Adhikari, K. P. Ghatak, *Quant. Matt.*, 2, 296 **(2013)**; S. Pahari, S. Bhattacharya, D. De, S. M. Adhikari, A. Niyogi, A. Dey, N. Paitya, K. P. Ghatak, *Phys. B: Conden. Matter*, 405, 4064 **(2010)**; K. P. Ghatak, S. Bhattacharya, S. Pahari, D. De, R. Benedictus, *Superlatt. Microstruct.*, 46, 387 **(2009)**.

[160] J. P. Bouchaud, A. Georges, *Phys. Rep.*, 195, 127 **(1996)**; V. Blickle, T. Speck, C. Lutz, U. Seifert, C. Bechinger, *Phys. Rev. Lett.*, 98, 210601 **(2007)**; Y. Kang, E. Jean, C. M. Fortmonn, *Appl. Phys. Lett.*, 88, 112110 **(2006)**; F. Neumann, Y. A. Genenko, H. V. Seggern, *J. Appl. Phys.*, 99, 013704 **(2006)**; J. van de Lagemaat, *Phys. Rev. B*, 73, 235319 **(2005)**; Q. Gu, E. A. Schiff, S. Grneber, F. Wang, R. Schwarz, *Phys. Rev. Lett.*, 76, 3196 **(1996)**; M. Y. Azbel, *Phys. Rev. B*, 46, 15004 **(1992)**.

[161] A. H. Marshak, *Sol. Stat. Electron.*, 30, 1089 **(1987)**; A. H. Marshak, C. M. V. Vliet, *Proc. IEEE*, 72, 148 **(1984)**; C. M. V. Vliet, A. V. D. Zeil, *Sol. Stat. Electron.*, 20, 931 **(1977)**.

[162] A. Khan, A. Das, *Appl. Phys. A*, 89, 695 **(2007)**.

[163] M. Krieehbaum, P. Kocevar, H. Pascher, G. Bauer, *IEEE QE*, 24, 1727 **(1988)**; J. R. Lowney, S. D. Senturia, *J. Appl. Phys.*, 47, 1771 **(1976)**.

[164] K. P. Ghatak, S. N. Biswas, *J. Appl. Phys.*, 70, 4309 **(1991)**; K. P. Ghatak, B. Mitra, M. Mondal, *Ann. der Physik*, 48, 283 **(1991)**; B. Mitra, K. P. Ghatak, *Phys. Lett. A*, 135, 397 **(1989)**; K. P. Ghatak, N. Chattopadhyay, M. Mondal, *J. Appl. Phys.*, 63, 4536 **(1988)**; *J. Low Temp. Phys.*, 73, 321 **(1988)**.

[165] K. P. Ghatak, D. Bhattacharyya, *Phys. Lett. A*, 184, 366 **(1994)**.

[166] A. I. Yekimov, A. A. Onushchenko, A. G. Plyukhim, A. L. Efors, *J. Exp. Theor. Phys.*, 88, 1490 **(1985)**.

[167] B. J. Roman, A. W. Ewald, *Phys. Rev. B*, 5, 3914 **(1972)**.

[168] K. P. Ghatak, M. Mondal, *Thin Solid Films*, 148, 219 **(1987)**.

[169] K. P. Ghatak, S. N. Biswas, *Nanostruct. Mater.*, 2, 91 **(1993)**.

[170] K. P. Ghatak, *Influence of Band Structure on Some Quantum Processes in Nonlinear Optical Semiconductors*, D. Eng. Thesis, (Jadavpur University, Kolkata, India, **1991**).

[171] K. P. Ghatak, N. Chattropadhyay, M. Mondal, *Appl. Phys. A*, 44, 305 **(1987)**.

[172] S. N. Biswas, K. P. Ghatak, *Proceedings of the Society of Photo-Optical and Instrumentation Engineers (SPIE), Quantum Well and Superlattice Physics*, 792, 239 **(1987)**.

[173] M. Mondal, K. P. Ghatak, *J. Phys. C, (Sol. Stat.)*, 20, 1671 **(1987)**; M. Mondal. S. N. Banik, K. P. Ghatak, *Canad. J. Phys.*, 67, 72 **(1989)**.

[174] P. K. Chakraborty, S. Bhattacharya, K. P. Ghatak, *J. Appl. Phys.*, 98, 053517 **(2005)**.

[175] K. P. Ghatak, S. Bhattacharyya, M. Mondal, *SPIE*, 1307, 588 **(1990)**.

[176] W. S. Koh, L. K. Ang, *Nanotechnology*, 19, 235402 **(2008)**.

[177] K. P. Ghatak, M. Mondal, *J. Appl. Phys.*, 70, 1277 **(1992)**.

[178] S. Bhattacharya, K. P. Ghatak, S. N. Biswas, *Optoelectr. Mater, Devi, Packa, Interconn.*, *SPIE*, 836, 72 **(1988)**.

[179] W. Shockley, Electrons and wholes in Semiconductors. Semiconductors, (D. van Nostrand Co., Inc., New York, **1950**); H. Reiss, C. S. Fuller and F. J. Morin, *Bell Syst. Tech. J.*, 35, 535 **(1956)**; P. T. Landsberg and A. G. Guy, *Phys. Rev. B*, 28, 1187 **(1983)**; R. Zallen, Handbook on Semiconductors, W. Paul, Ed., (North-Holland Publishing Co, **1982**), pp. 359–449; C. J. Hwang and J. R. Brews, *J. Phys. Chem. Solids*, 32, 837 **(1971)**.

[180] A. Isihara, Y. Shiwa, *J. Phys. C*, 18, 4703 **(1985)**.

[181] K. P. Ghatak, B. De, *Defect Engg. Semi. Growth, Process. Device Tech. Symp., Mat. Res. Soc.*, MRS Spring Meeting, 262, 911 **(1992)**.

[182] F. J. Blatt, *J. Phys. Chem. Solids*, 17, 177 **(1961)**.

[183] D. K. Roy, *Quantum Mechanical Tunnelling and its Applications*, (World Scientific Publication Co. Pte. Ltd., **1986**), p. 309.

[184] J. M. Beebe, B. Kim, J. W. Gadzuk, C. D. Frisbie, J. G. Kushmerick, *Phys. Rev. Lett.*, 97, 026801 **(1999)**.

[185] E. V. Rozenfeld, Yu. P. Irkhin and P. G. Guletskii, *Sov. Phys. Sol. Stat.*, 29, 1134 **(1987)**.

[186] S. I. Baranchuk, N. V. Mileshkina, *Sov. Phys. Sol. Stat.*, 23, 1715 **(1981)**.

[187] R. H. Fowler, L. Nordheim, *Proc. Roy. Soc. London, Series-A*, 119, 173 **(1928)**; A. V. D. Ziel, *Solid State Physical Electronics*, (Prentice-Hall, Inc., Englewood Cliffs, USA, **1957**), p. 176.

[188] B. Mitra, K. P. Ghatak, *Phys. Lett. A*, 357, 146 **(1990)**; K. P. Ghatak, M. Mondal, *J. Mag. Mag. Mat.*, 74, 203 **(1988)**; K. P. Ghatak, B. Mitra, *Phys. Lett. A*, 156, 233 **(1991)**.

[189] V. T. Binh, Ch. Adessi, *Phys. Rev. Lett.*, 85, 864 **(2000)**; R. G. Forbes, *Ultramicroscopy*, 79, 11 **(1999)**; J. W. Gadzuk, E. W. Plummer, *Rev. Mod. Phys.*, 45, 487 **(1973)**; Y. Feng, J. P. Verboncoeur, *Phys. Plasmas*, 12, 103301 **(2005)**; M. Razavy, *Quantum Theory of Tunneling*, (World Scientific Publishing Co. Pte. Ltd., Singapore, **2003**).

[190] P. G. Borzyak, A. A. dadykin, *Sov. Phys. Docklady*, 27, 335 **(1982)**; S. Bono, R. H. Good, Jr., *Surface Sci.*, 134, 272 **(1983)**; S. M. Lyth, S. R. P. Silva, *Appl. Phys. Lett.*, 90, 173124 **(2007)**; C. Xu, X. Sun, *Int. J. Nanotech.*, 1, 452 **(2004)**; S. D. Liang, L. Chen, *Phys. Rev. Lett.*, 101, 027602 **(2008)**.

[191] E. C. Heeres, E. P. A. M. Bakkers, A. L. Roest, M. Kaiser, T. H. Oosterkamp, N. de Jonge, *Nano Lett.*, 7, 536 **(2007)**; L. Dong, J. Jiao, D. W. Tuggle, J. M. Petty, S. A. Elliff, M. Coulter, *Appl. Phys. Lett.*, 82, 1096 **(2003)**; S. Y. Li, P. Lin, C. Y. Lee, T. Y. Tseng, *J. Appl. Phys.*, 95, 3711 **(2004)**; N. N. Kulkarni, J. Bae, C. K. Shih, S. K. Stanley, S. S. Coffee, J. G. Ekerdt, *Appl. Phys. Lett.*, 87, 213115 **(2005)**.

[192] K. Senthil, K. Yong, *Mat. Chem. Phys.*, 112, 88 **(2008)**; R. Zhou, H. C. Chang, V. Protasenko, M. Kuno, A. K. Singh, D. Jena, H. Xing, *J. Appl. Phys.*, 101, 073704 **(2007)**; K. S. Yeong, J. T. L. Thong, *J. Appl. Phys.*, 100, 114325 **(2006)**; C. H. Oon, S. H. Khong, C. B. Boothroyd, J. T. L. Thong, *J. Appl. Phys.*, 99, 064309 **(2006)**.

[193] B. H. Kim, M. S. Kim, K. T. Park, J. K. Lee, D. H. Park, J. Joo, S. G. Yu, S. H. Lee, *Appl. Phys. Lett.*, 83, 539 **(2003)**; Z. S. Wu, S. Z. Deng, N. S. Xu, J. Chen, J. Zhou, J. Chen, *Appl. Phys. Lett.*, 80, 3829 **(2002)**; Y. W. Zhu, T. Yu, F. C. Cheong, X. J. Xu, C. T. Lim, V. B. C. Tan, J. T. L. Thong, C. H. Sow, *Nanotechnology*, 16, 88 **(2005)**; Y. W. Zhu, H. Z. Zhang, X. C. Sun, S. Q. Feng, J. Xu, Q. Zhao, B. Xiang, R. M. Wang, D. P. Yu, *Appl. Phys. Lett.*, 83, 144 **(2003)**; S. Bhattacharjee, T. Chowdhury, *Appl. Phys. Lett.*, 95, 061501 **(2009)**; S. Kher, A. Dixit, D. N. Rawat, M. S. Sodha, *Appl. Phys. Lett.*, 96, 044101 **(2010)**.

[194] O. Madelung, *Introduction to Solid-State Theory*, (Springer Berlin Heidelberg, Germany, **2012**), p-1–491; I. Shigeo, W. Teruo, O. Kazuyoshi, T. Masateru, U. Satoshi, N. Norio, *J. Vac. Sci. Technol. B*, 13, 487 **(2009)**; C. A. Spindt, I. Brodie, L. Humphrey, E. R. Westerberg, *J. Appl. Phys.*, 47, 5248 **(2009)**; Q. Fu, A. V. Nurmikko, L. A. Kolodziejski, R. L. Gunshor, J. W. Wu, *Appl. Phys. Lett.*, 51, 578 **(2009)**.

[195] I. Mertig, E. Mrosan, V. N. Antonov, V. I. Antonov, P. Ziesche, *Phys. Stat. Sol. (b)*, 135, K13 **(1986)**.

[196] L. T. Canham, *Appl. Phys. Lett.*, 57, 1046 **(1990)**.

[197] Z. H. Lu, D. J. Lockwood and J. M. Baribeam, *Nature*, 378, 258 **(1995)**.

[198] A. G. Cullis, L. T. Canham, P. D. O. Calocott, *J. Appl. Phys.*, 82, 909 **(1997)**.

[199] M. Cardona, L. Ley, *Photoemission in Solids 1 and 2, Topics in Applied Physics*, Vols. 26, 27, (Springer-Verlag, Germany, **1978**); S. Hüfner, (Ed.), *Very High Resolution Photoelectron Spectroscopy,* Lecture Notes in Physics, Vol. 715, (Springer-Verlag, Germany, **2007**); D. W. Lynch, C. G. Olson, *Photoemission Studies of High-Temperature Superconductors*, (Cambridge University Press, UK, **1999**); *Photoemission and the Electronic Properties of Surfaces*, (Eds.), B. Feuerbacher, B. Fitton, R. F. Willis, (Wiley, New York, **1978**); W. Schattke, M. A. V. Hove, *Solid-State Photoemission and Related Methods: Theory and Experiment*, (Wiley, USA, **2003**); V. V. Afanas'ev, *Internal Photoemission Spectroscopy: Principles and Applications*, (Elsevier, North Holland, **2010**); D. J. Lockwood, *Light Emission in Silicon in Silicon based Materials and Devices*, Vol. 2, Ed. H. S. Nalwa, (Academic Press, San Diego, USA, **2001**); C. C. Wang, N. W. Ressler, *Phys. Rev.*, 2, 1827 **(1970)**.

[200] K. P. Ghatak, D. De, S. Bhattacharya, *Photoemission from Optoelectronic Materials and Their Nanostructures*, Springer Series in Nanostructure Science and Technology, (Springer, Germany, **2009**).

[201] B. Mitra, K. P. Ghatak, *Sol. Stat. Electron.*, 32, 810 **(1989)**; S. Choudhury, L. J. Singh, K. P. Ghatak, *Nanotechnology*, 15, 180 **(2004)**; K. P. Ghatak, A. K. Chowdhury, S. Ghosh, A. N. Chakravarti, *Appl. Phys.*, 23, 241 **(1980)**; K. P. Ghatak, S. Bhattacharya, S. Bhowmik, R. Benedictus, S. Choudhury, *J. Appl. Phys.*, 103, 094314 **(2008)**; A. N. Chakravarti, K. P. Ghatak, A. Dhar, K. K. Ghosh, S. Ghosh, *Appl. Phys. A*, 26, 165 **(1981)**.

[202] R. K. Pathria, *Statistical Mechanics*, 2nd ed. (Butterworth-Heinemann, Oxford, **1996**).

[203] D. De, S. Bhattacharya, S. M. Adhikari, A. Kumar, P. K. Bose, K. P. Ghatak, *Beilstein J. Nanotech.*, 2, 339 **(2012)**; D. De, Kumar, S. M. Adhikari, S. Pahari, N. Islam, P. Banerjee, S. K. Biswas, S. Bhattacharya, K. P. Ghatak, *Superlatt. Microstruct.*, 47, 377 **(2010)**; K. P. Ghatak, M. Mondal, S. N. Biswas, *J. Appl. Phys.*, 68, 3032 **(1990)**; K. P. Ghatak, D. Bhattacharyya, B. Nag, S. N. Biswas, *J. Nonlin. Opt. Quant. Opt.*, 13, 267 **(1995)**; L. Torres, L. Lopez-Diaz, J. Iniguez, *Appl. Phys. Lett.*, 73, 3766 **(1996)**; A. Y. Toporov, R. M. Langford, A. K. Petford-Long, *Appl. Phys. Lett.*, 77, 3063 **(2000)**.

[204] S. M. Adhikari, K. P. Ghatak, *J. Adv. Phys.*, 2, 130 **(2013)**; D. De, S. Bhattacharaya, S. Ghosh, K. P. Ghatak, *Adv. Sci. Engg. Medi.*, 4, 211 **(2012)**; A. Kumar, S. Chowdhury, S. M. Adhikari, S. Ghosh, M. Mitra, D. De, A. Sharma, S. Bhattacharya, A. Dey, K. P. Ghatak, *J. Comput. Theor. Nanosci.*, 7, 115 **(2010)**; R. Houdré, C. Hermann, G. Lampel, P. M. Frijlink, *Surface Sci.*, 168, 538 **(1986)**; T. C. Chiang, R. Ludeke, D. E. Eastman, *Phys. Rev. B*, 25, 6518 **(1982)**; S. P. Svensson, J. Kanski, T. G. Andersson, P. O. Nilsson, *J. Vac. Sci. Technol. B*, 2, 235 **(1984)**; S. F. Alvarado, F. Ciccacci, M. Campagna, *Appl. Phys. Lett.*, 39, 615 **(1981)**; L. Fleming, M. D. Ulrich, K. Efimenko, J. Genzer, A. S. Y. Chan, T. E. Madey, S. J. Oh, O. Zhou, J. E. Rowe, *Vac. Sci. Technol. B*, 22, 2000 **(2004)**.

[205] B. Mitra, K. P. Ghatak, *Phys. Scr.*, 40, 776 **(1989)**; D. De, S. Bhattacharya, K. P. Ghatak, *International Workshop in Physics of Semiconductor Devices, IEEE*, 897 **(2007)**; S. Bhattacharya, D. De, S. Chowdhury, S. Karmakar, D. K. Basu, S. Pahari, K. P. Ghatak, *J. Comput. Theor. Nanosci.*, 3, 280 **(2006)**.

[206] I. Matsuda, S. Hasegawa, A. Konchenko, Y. Nakayama, Y. Nakamura, M. Ichikawa, *Phys. Rev. B*, 73, 113311 **(2006)**; V. L. Colvin, A. P. Alivisatos, J. G. Tobin, *Phys. Rev. Lett.*, 66, 2786 **(1991)**; B. Schroeter, K. Komlev, W. Richter, *Mat. Sci. and Engg. B*, 88, 259 **(2002)**.

[207] K. P. Ghatak, S. N. Biswas, *Nonlinear Opt.*, 4, 39 **(1993)**.

[208] K. P. Ghatak, S. N. Biswas, *SPIE, Growth and Characterization of Materials for Infrared Detectors and Nonlinear Optical Switches,* USA, Vol. 1484, p. 136 (**1991**).

[209] K. P. Ghatak, B. De, *Polymeric Materials for Integrated Optics and Information Storage, Materials Research Society (MRS) Symposium Proceedings*, MRS Spring Meeting, Vol. 228, p. 237 **(1991)**; R. P. Cowburn, A. O. Adeyeye, J. A. C. Bland, *Appl. Phys. Lett.*, 70, 2309 **(1997)**.

[210] K. P. Ghatak, B. Nag, G. Majumdar, *Strained Layer Expitaxy – Materials, Processing, and Device Applications, MRS Symposium Proceedings*, MRS Spring Meeting, Vol. 379, p. 85 **(1995)**.

[211] K. P. Ghatak, *SPIE, High Speed Phenomena in Photonic Materials and Optical Bistability*, USA, Vol. 1280, p. 53 **(1990)**.

[212] K. P. Ghatak, *Long Wave Length Semiconductor Devices, Materials and Processes Symposium Proceedings, MRS Symposium Proceedings*, MRS Spring Meeting, Vol. 216, p. 469 **(1990)**.

[213] K. P. Ghatak, A. Ghoshal, S. Bhattacharyya, *SPIE, Nonlinear Optical Materials and Devices for Photonic Switching*, USA, Vol. 1216, p. 282 **(1990)**.

[214] K. P. Ghatak, *SPIE, Nonlinear Optics III*, USA, Vol. 1626, p. 115 **(1992)**.

[215] K. P. Ghatak, A. Ghoshal, B. De, *SPIE, Optoelectronic Devices and Applications*, USA, Vol. 1338, p. 111 **(1990)**.

[216] K. P. Ghatak, S. N. Biswas, in *Proceedings of the Society of Photo-Optical and Instrumentation Engineers (SPIE)*, Nonlinear Optics II, USA, Vol. 1409, p. 28 **(1991)**; K. P. Ghatak, *SPIE, Process Module Metrology*, USA, Vol. 1594, p. 110 **(1992)**; K. P. Ghatak, *SPIE, International Conference on the Application and Theory of Periodic Structures*, Vol. 1545, p. 282 **(1991)**.

[217] K. P. Ghatak, M. Mondal, *Sol. Stat. Electron.*, 31, 1561 **(1988)**.

[218] K. P. Ghatak, M. Mondal, *J. Appl. Phys.*, 69, 1666 **(1991)**.

[219] C. Majumdar, A. B. Maity, A. N. Chakravarti, *Phys. Stat. Sol. (b)*, 140, K7 **(1987)**.

[220] C. Majumdar, A. B. Maity, A. N. Chakravarti, *Phys. Stat. Sol. (b)*, 141, K35 **(1987)**.

[221] N. R. Das, K. K. Ghosh, D. Ghoshal, *Phys. Stat. Sol. (b)*, 197, 97 **(1996)**.

[222] C. Majumdar, A. B. Maity, A. N. Chakravarti, *Phys. Stat. Sol. (b)*, 144, K13 **(1987)**.

[223] N. R. Das, A. N. Chakravarti, *Phys. Stat. Sol. (b)*, 176, 335 **(1993)**.

[224] S. Sen, N. R. Das and A. N. Chakravarti, *J. Phys. Conden. Mat.*, 19, 186205 **(2007)**; N. R. Das, S. Ghosh, A. N. Chakravarti, *Phys. Stat. Sol. (b)*, 174, 45 **(1992)**.

[225] A. B. Maity, C. Majumdar, A. N. Chakravarti, *Phys. Stat. Sol. (b)*, 144, K93 **(1987)**.

[226] A. B. Maity, C. Majumdar, A. N. Chakravarti, *Phys. Stat. Sol. (b)*, 149, 565 **(1988)**.

[227] A. V. D. Ziel, *Solid State Physical Electronics*, (Prentice-Hall, Inc., Englewood Cliffs, **1957**); A. Modinos, *Field, Thermionic and Secondary Electron Emission Spectroscopy*, (Plenum Press, New York, **1984**).

[228] L. J. Singh, S. Choudhary, A. Mallik, K. P. Ghatak, *J. Comput. Theor. Nanosci.*, 2, 287 **(2005)**.

[229] K. P. Ghatak and M. Mitra, *Nanomaterials, Volume 2: Quantization and Entropy*, (De Gruyter, Berlin, Boston, **2020**), p-1–450.

[230] B. Mitra, A. Ghoshal and K. P. Ghatak, *Phys. Stat. Sol. (b)*, 155, K23 **(1989)**.

[231] J. A. Woollam, *Phys. Rev.*, 3, 1148 **(1971)**.

[232] V. V. Kaminoskii, N. N. Stepanav, I. M. Smirnov, *Sov. Phys. Sol. Stat.*, 27, 1295 **(1985)**.

[233] P. I. Baranskii, V. V. Kolomoets and S. S. Korolyuk, *Phys. Stat Sol. (b)*, 116, K109 **(1983)**.

[234] D. R. Choudhury, A. K. Chowdhury, A. N. Chakravarti, *Phys. Scr.*, 22, 656 **(1981)**; M. Mondal, K. P. Ghatak, *Phys. Stat. Sol. (a)*, 93, 377 **(1986)**.

[235] S. Adachi, *J. Appl. Phys.*, 58, R1 **(1985)**; S. Adachi, *GaAs and Related Materials: Bulk Semiconductors and Superlattice Properties*, (World Scientific, USA, **1994**).

[236] G. L. Hansen, J. L. Schmit, T. N. Casselman, *J. Appl. Phys.*, 63, 7079 **(1982)**.

[237] S. Adachi, *J. Appl. Phys.*, 53, 8775 **(1982)**.

[238] O. Madelung, *Semiconductors: Data Handbook*, 3rd ed. (Springer-Verlag, Germany, **2003**); S. Adachi, *Properties of Group-IV, III–V and II–VI Semiconductors*, (John Wiley and Sons, USA, **2005**).

[239] N. B. Brandt, V. N. Davydov, V. A. Kulbachinskii, O. M. Nikitina, *Sov. Phys. Sol. Stat.*, 29, 1014 **(1987)**.

[240] I. Kang, F. W. Wise, *J. Opt. Soc. Am. B*, 14, 1632 **(1997)**.

[241] M. S. Lundstrom, J. Guo, *Nanoscale Transistors, Device Physics, Modeling and Simulation*, (Springer, USA, **2006**).

[242] J. Kolodziejczak, S. Zukotynski and H. Stramska, *Phys. Stat. Sol.*, 14, 471 **(1966)**.

[243] S. V. Kryachkov and V. I. Porkhun, *Sov. Phys. Semicond.*, 12, 1206 **(1978)**.

[244] I. A. Chaikovskii, *Sov. Phys. Semicond.*, 6, 1 **(1972)**.

Chapter 2

Einstein's Photoemission from Quantum Dots of Various Non-Parabolic Materials

Education's purpose is to replace our closed empty mind with an open one.

2.1 Introduction

It is well known that the Einstein's photoelectric effect occupies a singular position in the whole arena of materials science and related disciplines in general together with the fact that the photoemission from the electronic materials is also a vital physical phenomenon from the viewpoint of modern optoelectronics and photoemission spectroscopy [1]. The classical equation of the photo-emitted current density is [1, 2] $J = [4\pi e m_c g_v (k_B T)^2/h^3]\exp[(h\nu - \phi)/(k_B T)]$, where hv and ϕ are the incident photon energy along z-axis and the work function respectively. The aforementioned equation is valid for both the charge carriers and in this conventional form it appears that, the photoemission changes with the effective mass, temperature, work function and the incident photon energy respectively. This relation holds true only under the condition of carrier non-degeneracy.

The EP has different values for different materials and varies with doping and with external fields which creates quantization of the wave-vector space of the carriers leading to various types of quantized structures. The nature of these variations has been studied in [2–35] and some of the significant features are as follows:

1. The photoemission from bulk materials increases with the increase in doping.

2. The photoemission exhibits oscillatory dependence with inverse quantizing magnetic field because of the Shubnikov de Haas (SdH) effect.
3. The photoemission changes significantly with the magnitude of the externally applied quantizing electric field in electronic materials.
4. The photoemission from quantum confined Bismuth, nonlinear optical, III–V, II–VI and IV–VI materials oscillate with nanothickness in various manners which are totally band structure dependent.
5. The nature of variations is significantly influenced by the energy band constants of various materials having different band structures.
6. The photoemission has significantly different values in quantum confined semiconductor superlattices and various other quantized structures.

The Section 2.2.1 of this chapter investigates the EP in QDs of nonlinear optical and tetragonal materials (taking n-$CdGeAs_2$ as an example) which find applications in nonlinear optics and light emitting diodes [36]. The quasi-cubic model can be used to investigate the symmetric properties of both the bands at the zone center of wave vector space of the same compound [37]. Including the anisotropic crystal potential in the Hamiltonian, and special features of the nonlinear optical compounds, Kildal [38] formulated the electron dispersion law under the assumptions of isotropic momentum matrix and the isotropic spin orbit splitting constant, respectively, although the anisotropies in the two aforementioned band constants are the significant physical features of the said materials [39–41]. In Sections 2.2.2–2.2.18, we study the EP from QDs of III–V, II–VI, GaP, Ge, Te, Graphite, $PtSb_2$, Zero-gap, II–V, GaSb, Stressed materials, Bi, IV–VI, Lead Germanium Telluride, Zinc and Cadmium diphosphides, Bismuth Telluride and Antimony respectively. In Section 2.3, contains the result and discussion pertinent to this chapter. Section 2.4 contains 15 open research problems.

2.2 Theoretical Background

2.2.1 The EP from QDs of nonlinear optical and tetragonal materials

The form of k · p matrix for nonlinear optical compounds can be expressed extending Bodnar [39] as

$$H = \begin{bmatrix} H_1 & H_2 \\ H_2^+ & H_1 \end{bmatrix} \tag{2.1}$$

where

$$H_1 \equiv \begin{bmatrix} E_{g_0} & 0 & P_{\|}k_z & 0 \\ 0 & (-2\Delta_{\|}/3) & (\sqrt{2}\Delta_{\perp}/3) & 0 \\ P_{\|}k_z & (\sqrt{2}\Delta_{\perp}/3) & -(\delta + \frac{1}{3}\Delta_{\|}) & 0 \\ 0 & 0 & 0 & 0 \end{bmatrix} \quad \text{and}$$

$$H_2 \equiv \begin{bmatrix} 0 & -f_{,+} & 0 & f_{,-} \\ f_{,+} & 0 & 0 & 0 \\ 0 & 0 & 0 & 0 \\ f_{,+} & 0 & 0 & 0 \end{bmatrix}$$

in which E_{g_0} is the band gap in the absence of any field, $P_{\|}$ and $P_{\perp}$ are the momentum matrix elements parallel and perpendicular to the direction of crystal axis respectively, δ is the crystal field splitting constant, $\Delta_{\|}$ and $\Delta_{\perp}$ are the spin-orbit splitting constants parallel and perpendicular to the c-axis respectively, $f_{,\pm} \equiv (P_{\perp}/\sqrt{2})(k_x \pm ik_y)$ and $i = \sqrt{-1}$. Thus, neglecting the contribution of the higher bands and the free electron term, the diagonalization of the above matrix leads to the dispersion relation of the conduction electrons in bulk specimens of nonlinear optical compounds [41] as

$$\gamma(E) = f_1(E)k_s^2 + f_2(E)k_z^2 \tag{2.2}$$

where

$$\gamma(E) \equiv E(E + E_{g_0})\left[(E + E_{g_0})(E + E_{g_0} + \Delta_{\|}) + \delta\left(E + E_{g_0} + \frac{2}{3}\Delta_{\|}\right)\right.$$
$$\left. + \frac{2}{9}(\Delta_{\|}^2 - \Delta_{\perp}^2)\right], \quad k_s^2 = k_x^2 + k_y^2,$$

$$f_1(E) \equiv \frac{\hbar^2 E_{g_0}(E_{g_0} + \Delta_{\perp})}{\left[2m_{\perp}^*\left(E_{g_0} + \frac{2}{3}\Delta_{\perp}\right)\right]}\left[\delta\left(E + E_{g_0} + \frac{1}{3}\Delta_{\|}\right)\right.$$
$$\left. + (E + E_{g_0})\left(E + E_{g_0} + \frac{2}{3}\Delta_{\|}\right) + \frac{1}{9}\left(\Delta_{\|}^2 - \Delta_{\|}^2\right)\right]$$

$$f_2(E) \equiv \frac{\hbar^2 E_{g_0}(E_{g_0} + \Delta_{\|})}{\left[2m_{\|}^*\left(E_{g_0} + \frac{2}{3}\Delta_{\|}\right)\right]}\left[(E + E_{g_0})\left(E + E_{g_0} + \frac{2}{3}\Delta_{\|}\right)\right]$$

and $m_{\parallel}^*$ and $m_{\perp}^*$ are the longitudinal and transverse effective electron masses at the edge of the conduction band respectively.

Let E_{n_i} ($i = x, y$ and z) be the quantized energy levels due to infinitely deep potential well along ith-axis with n_i (=1, 2, 3, ...) as the size quantum numbers.

Therefore, from (2.2), one can write

$$\gamma(E_{n_x}) = f_1(E_{n_x})\left(\frac{\pi n_x}{d_x}\right)^2 \tag{2.3}$$

$$\gamma(E_{n_y}) = f_1(E_{n_y})\left(\frac{\pi n_y}{d_y}\right)^2 \tag{2.4}$$

$$\gamma(E_{n_z}) = f_2(E_{n_z})\left(\frac{\pi n_z}{d_z}\right)^2 \tag{2.5}$$

From (2.2), the total quantized energy (E_{QD1}) can be expressed as

$$\gamma(E_{\text{QD1}}) = f_1(E_{\text{QD1}})\left[\left(\frac{\pi n_x}{d_x}\right)^2 + \left(\frac{\pi n_y}{d_y}\right)^2\right] + f_2(E_{\text{QD1}})\left[\left(\frac{\pi n_z}{d_z}\right)^2\right] \tag{2.6}$$

The total density-of-states function in this case is given by

$$N_{\text{0DT}}(E) = \frac{2g_v}{d_x d_y d_z}\sum_{n_x=1}^{n_{x\max}}\sum_{n_y=1}^{n_{y\max}}\sum_{n_z=1}^{n_{z\max}} \delta'(E - E_{\text{QD1}}) \tag{2.7}$$

Using (2.5) and Fermi–Dirac occupation probability factor, the total electron concentration can be written as

$$n_{\text{0D}} = \frac{2g_v}{d_x d_y d_z}\sum_{n_x=1}^{n_{x\max}}\sum_{n_y=1}^{n_{y\max}}\sum_{n_z=1}^{n_{z\max}} F_{-1}(\eta_{31}) \tag{2.8}$$

where $\eta_{31} \equiv (k_B T)^{-1}(E_{\text{F0D}} - E_{\text{QD1}})$.

Therefore, the electron concentration per sub-band is given by

$$\Delta n_{\text{0D}} = \left(\frac{2g_v}{d_x d_y d_z}\right) F_{-1}(\eta_{31}) \tag{2.9}$$

The expression of the total photo-emitted current density in this case is

$$J_{\text{0D}} = \frac{e\alpha_0 g_v}{2}\sum_{n_x=1}^{n_{x\max}}\sum_{n_y=1}^{n_{y\max}}\sum_{n_{z\min}}^{n_{z\max}} (\Delta n_{\text{0D}}) v_z(E_{n_z}) \tag{2.10}$$

where the $n_{z_{\min}}$ should be the nearest integer of the following inequality

$$n_{z_{\min}} \geq \frac{d_z}{\pi} \frac{\sqrt{\gamma(W - h\nu)}}{\sqrt{f_2(W - h\nu)}} \tag{2.11}$$

and the velocity of the photo-emitted electrons in the n_z^{th} sub-band can be written as

$$v_z(E_{n_z}) = \frac{1}{\hbar} Q_1(E_{n_z}) \tag{2.12}$$

in which

$$Q_1(E_{n_z}) \equiv \frac{2 f_2^{3/2}(E_{n_z})[\gamma(E_{n_z})]^{1/2}}{[f_2(E_{n_z})\gamma'(E_{n_z}) - \gamma(E_{n_z}) f_2'(E_{n_z})]}$$

Using (2.10), (2.11) and (2.12), we get

$$J_{\mathrm{OD}} = \left(\frac{e\alpha_0 g_v}{\hbar d_x d_y d_z}\right) \sum_{n_x=1}^{n_{x\max}} \sum_{n_y=1}^{n_{y\max}} \sum_{n_{z\min}}^{n_{z\max}} F_{-1}(\eta_{31}) Q_1(E_{n_z}) \tag{2.13}$$

2.2.2 The EP from QDs of III–V materials

(a) The three band model of Kane

Under the conditions, $\delta = 0$, $\Delta_{\|} = \Delta_{\perp} = \Delta$ and $m_{\|}^* = m_{\perp}^* = m_c$, (2.2) gets simplified to three band Kane model as given by (1.21). The quantized energy levels in QDs of III–V semiconductors in accordance with the three band model of Kane can be expressed as

$$I(E_{n_x}) = (\hbar^2/2m_c)(\pi n_x/d_x)^2 \tag{2.14}$$

$$I(E_{n_y}) = \frac{\hbar^2}{2m_c}\left(\frac{\pi n_y}{d_y}\right)^2 \tag{2.15}$$

and

$$I(E_{n_z}) = \frac{\hbar^2}{2m_c}\left(\frac{\pi n_z}{d_z}\right)^2 \tag{2.16}$$

The velocity of the photoelectron in the n_z^{th} quantized level is given by

$$v_z(E_{n_z}) = \left(\sqrt{\frac{2}{m_c}}\right) Q_2(E_{n_z}) \tag{2.17}$$

where

$$Q_2(E_{n_z}) \equiv [(I'(E_{n_z}))']^{-1}(\sqrt{I(E_{n_z})})$$

Thus, the photo-emitted current density can be expressed as

$$J_{0D} = \left[\frac{\alpha_0 e g_v}{(d_x d_y d_z)}\sqrt{\frac{2}{m_c}}\right] \sum_{n_x=1}^{n_{x\max}} \sum_{n_y=1}^{n_{y\max}} \sum_{n_{z\min}}^{n_{z\max}} Q_2(E_{n_z}) F_{-1}(\eta_{1,3}) \tag{2.18}$$

in which $n_{z_{\min}}$ is the nearest integer of the following inequality,

$$n_{z_{\min}} \geq \left(\frac{d_z}{\pi}\frac{\sqrt{2m_c}}{\hbar}[I(W - h\nu)]^{1/2}\right) \tag{2.19}$$

In this case, the electron statistics is given by (1.23).

(b) The two band model of Kane

Under the constraints $\Delta \gg E_{g_0}$ or $\Delta \ll E_{g_0}$ the three band model of Kane gets simplified to the two band model of Kane as given by (1.17) and the photo-emitted current density assumes the form

$$J_{0D} = \frac{\alpha_0 g_v e}{d_x d_y d_z}\left(\sqrt{\frac{2}{m_c}}\right) \sum_{n_x=1}^{n_{x\max}} \sum_{n_y=1}^{n_{y\max}} \sum_{n_{z\min}}^{n_{z\max}} Q_3(E_{n_z}) F_{-1}(\eta_{1,2}) \tag{2.20}$$

where

$$n_{z_{\min}} \geq \left(\frac{d_z}{\pi}\frac{\sqrt{2m_c}}{\hbar}[(W - h\nu)[1 + \alpha(W - h\nu)]]^{1/2}\right),$$

$$Q_3(E_{n_z}) \equiv [\sqrt{E_{n_z}(1 + \alpha E_{n_z})}/(1 + 2\alpha E_{n_z})]$$

The E_{n_z} obeys the equation

$$E_{n_z}(1 + \alpha E_{n_z}) = \frac{\hbar^2\pi^2}{2m_c}\left(\frac{n_z}{d_z}\right)^2 \tag{2.21}$$

The form of n_0 as given by (1.19). For $\alpha \to 0$, the expressions of the 0D electron concentration per sub-band (Δn_{0D}), E_{n_z}, $n_{z_{\min}}$ and J_{0D} for QDs of wide-gap materials can, respectively, be written as

$$\Delta n_{0D} = \frac{2g_v}{d_x d_y d_z} F_{-1}(\eta_{1,1}) \tag{2.22}$$

$$E_{n_z} = \frac{\hbar^2}{2m_c}\left(\frac{n_z\pi}{d_z}\right)^2 \tag{2.23}$$

$$n_{z_{\min}} \geq \left(\frac{d_z}{\pi}\right)\frac{\sqrt{2m_c}}{\hbar}\sqrt{(W - hv)} \tag{2.24}$$

$$J_{\mathrm{OD}} = \frac{\alpha_o e\hbar\pi g_v}{m_c d_x d_y d_z^2}\sum_{n_x=1}^{n_{x\max}}\sum_{n_y=1}^{n_{y\max}}\sum_{n_{z_{\min}}}^{n_{z\max}}[n_z F_{-1}(\eta_{1,1})] \tag{2.25}$$

(c) The model of Stillman *et al.*

The photoelectric current density in this case can be written as

$$J_{\mathrm{OD}} = \frac{\alpha_0 g_v e}{d_x d_y d_z}\left(\frac{2}{m_c}\right)^{1/2}\sum_{n_x=1}^{n_{x\max}}\sum_{n_y=1}^{n_{y\max}}\sum_{n_{z_{\min}}}^{n_{z\max}} Q_5(E_{n_z})F_{-1}(\eta_{1,4}) \tag{2.26}$$

where

$$n_{z_{\min}} \geq \frac{d_z}{\pi}\frac{\sqrt{2m_c}}{\hbar}[I_{11}(W - h\nu)]^{1/2} Q_5(E_{n_z}) \equiv [[I_{11}(E_{n_z})]^{1/2}[I'_{11}(E_{n_z})]^{-1}]$$

and

$$\eta_{34} \equiv (k_B T)^{-1}[E_{\mathrm{F0D}} - E_{\mathrm{QD4}}]$$

The E_{n_z} obeys the equation

$$(\hbar^2/2m_c)(\pi n_z/d_z)^2 \equiv I_{11}(E_{n_z}) \tag{2.27}$$

In this case, the electron statistics is given by (1.28).

(d) The model of Newson and Kurobe

The v_z and E_{n_z} for this model are given by

$$v_z = 4a_{13}^{1/4}(E_{n_z})^{3/4}/\hbar \tag{2.28}$$

$$E_{n_z} = a_{13}\left(\frac{\pi n_z}{d_z}\right)^4 \tag{2.29}$$

The photo-emitted current density in this case is given by

$$J_{\mathrm{OD}} = \frac{4\alpha_0 e g_v}{\hbar d_x d_y d_z}(a_{13}^{1/4})\sum_{n_x=1}^{n_{x\max}}\sum_{n_y=1}^{n_{y\max}}\sum_{n_{z_{\min}}}^{n_{z\max}}(E_{n_z})^{3/4}F_{-1}(\eta_{1,5}) \tag{2.30}$$

where

$$n_{z_{\min}} \geq \frac{d_z}{\pi}(a_{13})^{-1/4}(W - h\nu)^{1/4}$$

In this case, the electron statistics is given by (1.32)

(e) The model of Rossler

The ν_z and E_{n_z} in this case are given by

$$\nu_z = \frac{1}{\hbar}\left[\frac{\hbar^2}{m_c}\left(\frac{\pi n_z}{d_z}\right) + \alpha_{11}\left(\frac{\pi n_z}{d_z}\right)^4 + 4\left(\frac{\pi n_z}{d_z}\right)^3\left[\alpha_{11} + \alpha_{12}\frac{\pi n_z}{d_z}\right]\right]^{-1} \tag{2.31}$$

where E_{n_z} obeys the equation

$$E_{n_z} = \left[\frac{\hbar^2}{2m_c}\left(\frac{\pi n_z}{d_z}\right)^2 + \left[\alpha_{11} + \alpha_{12}\frac{\pi n_z}{d_z}\right]\left(\frac{\pi n_z}{d_z}\right)^4\right] \tag{2.32}$$

The photo-emitted current density can be expressed as

$$J_{0\mathrm{D}} = \frac{\alpha_0 e g_v}{2\hbar}\left(\frac{1}{d_x d_y d_z}\right)\sum_{n_x=1}^{n_{x\max}}\sum_{n_y=1}^{n_{y\max}}\sum_{n_{z_{\min}}}^{n_{z\max}} Q_6(E_{n_z})F_{-1}(\eta_{1,6,\pm}) \tag{2.33}$$

The $n_{z_{\min}}$ is the nearest integer of the equation

$$(W - h\nu) = \left(\frac{\hbar^2}{2m_c}\right)\left(\frac{\pi n_{z_{\min}}}{d_z}\right)^2 + \left[\alpha_{11} + \alpha_{12}\frac{\pi n_{z_{\min}}}{d_z}\right] \times \left(\frac{\pi n_{z_{\min}}}{d_z}\right)^4 \tag{2.34}$$

In this case, the electron statistics is given by (1.36).

(f) The model of Palik *et al.*

The photoemission current density in this case is given by

$$J_{0\mathrm{D}} = \frac{\alpha_0 e g_v}{\hbar d_x d_y d_z}\left(\frac{2}{m_c}\right)^{1/2}\sum_{n_x=1}^{n_{x\max}}\sum_{n_y=1}^{n_{y\max}}\sum_{n_{z_{\min}}}^{n_{z\max}} Q_7(E_{n_z})F_{-1}(\eta_{1,7}) \tag{2.35}$$

where

$$n_{z_{\min}} \geq \left(\frac{\sqrt{2m_c}}{\hbar}\right)\left(\frac{d_z}{\pi}\right)[I_{12}(W - h\nu)]^{1/2} \quad \text{and}$$

$$Q_7(E_{n_z}) \equiv [[I_{12}(E_{n_z})]^{1/2}/I'_{12}(E_{n_z})]$$

The E_{n_z} can be defined by the following equation:

$$I_{12}(E_{n_z}) = \frac{\hbar^2}{2m_c}\left(\frac{\pi n_z}{d_z}\right)^2 \tag{2.36}$$

In this case, the electron statistics is given by (1.41).

(g) The model of Johnson and Dickey

The $\nu_z(E_{n_z})$ in this case can be written as,

$$\nu_z(E_{n_z}) = \frac{1}{\hbar}Q_8(E_{n_z}) \tag{2.37}$$

where

$$Q_8(E_{n_z}) \equiv \left[\left[1 - \frac{E_{g0}a'_{15}(E_{n_z})}{4}\left(\frac{\pi n_z}{d_z}\right)^2\left[1 + a_{15}(E_{n_z})\left(\frac{\pi n_z}{d_z}\right)^2\right]^{-1/2}\right]^{-1}\right.$$
$$\times\left[\hbar^2\left(\frac{1}{m_c} - \frac{1}{m_0}\right)\left(\frac{\pi n_z}{d_z}\right) + \frac{E_{g0}a_{15}(E_{n_z})}{4}\left(\frac{\pi n_z}{d_z}\right)\right.$$
$$\left.\left.\times\left[1 + a_{15}(E_{n_z})\left(\frac{\pi n_z}{d_z}\right)^2\right]^{-1/2}\right]\right],$$

and

$$E_{n_z} = -\frac{E_{g0}}{2} + \frac{\hbar^2}{2}\left(\frac{\pi n_z}{d_z}\right)^2\left(\frac{1}{m_c} - \frac{1}{m_0}\right)$$
$$+ \frac{E_{g0}}{2}\left[1 + a_{15}(E_{n_z})\left(\frac{\pi n_z}{d_z}\right)^2\right]^{1/2} \tag{2.38}$$

The photo-emitted current density can be written as

$$J_{0D} = \frac{\alpha_0 g_v e}{\hbar d_x d_y d_z}\sum_{n_x=1}^{n_{x\max}}\sum_{n_y=1}^{n_{y\max}}\sum_{n_{z\min}}^{n_{z\max}} Q_8(E_{n_z})F_{-1}(\eta_{1,8}) \tag{2.39}$$

where $n_{z_{\min}}$ is the nearest integer of the following equation

$$(W - h\nu) = -\frac{E_{g0}}{2} + \frac{\hbar^2}{2}\left(\frac{\pi n_{z_{\min}}}{d_z}\right)^2\left(\frac{1}{m_c} - \frac{1}{m_0}\right)$$
$$+ \frac{E_{g0}}{2}\left[1 + a_{15}(W - h\nu)\left(\frac{\pi n_{z_{\min}}}{d_z}\right)^2\right]^{1/2} \tag{2.40}$$

In this case, the electron statistics is given by (1.45).

(h) The model of Agafonov *et al.*

The $\nu_z(E_{n_z})$ in this case can be written as

$$\nu_z(E_{n_z}) = \frac{1}{\hbar} Q_9(E_{n_z}) \tag{2.41}$$

where

$$\begin{aligned} Q_9(E_{n_z}) \equiv & \left[\left[\frac{1}{2}\left[\sqrt{E_{g_0}^2 + \frac{8}{3}P^2\left(\frac{\pi n_z}{d_z}\right)^2} - E_{g_0}\right]\left[\left[-2T_5\left(\frac{\pi n_z}{d_z}\right)\right]\right.\right.\right. \\ & \times \left[E_{g_0}^2 + \frac{8}{3}P_0^2\left(\frac{\pi n_z}{d_z}\right)^2\right]^{-1/2} + \left[\frac{8}{3}P^2 T_5\left(\frac{\pi n_z}{d_z}\right)^3\right] \\ & \left.\left.\times \left(E_{g_0}^2 + \frac{8}{3}P_0^2\left(\frac{\pi n_z}{d_z}\right)^2\right)^{-3/2}\right]\right] \\ & + \frac{8}{3}P_0^2\left(\frac{\pi n_z}{d_z}\right)\left[E_{g_0}^2 + \frac{8}{3}P_0^2\left(\frac{\pi n_z}{d_z}\right)^2\right]^{-1/2} \\ & \left.\times \left[1 - \left[T_5\left(\frac{\pi n_z}{d_z}\right)^2\right] \times \left[\left[E_{g_0}^2 + \frac{8}{3}P_0^2\left(\frac{\pi n_z}{d_z}\right)^2\right]^{1/2}\right]^{-1}\right]\right] \end{aligned}$$

and E_{n_z} is defined by the following equation

$$\begin{aligned} E_{n_z} = & \frac{1}{2}\left[\left(E_{g_0}^2 + \frac{8}{3}P_0^2\left(\frac{\pi n_z}{d_z}\right)^2\right)^{1/2} - E_{g_0}\right] \\ & \times \left[1 - T_5 \times \left(E_{g_0}^2 + \frac{8}{3}P_0^2\left(\frac{\pi n_z}{d_z}\right)^2\right)^{-1/2}\left(\frac{\pi n_z}{d_z}\right)^2\right] \end{aligned} \tag{2.42}$$

The photo-emitted current density is given by

$$J_{0\mathrm{D}} = \frac{\alpha_0 g_v e}{\hbar d_x d_y d_z} \sum_{n_x=1}^{n_{x\max}} \sum_{n_y=1}^{n_{y\max}} \sum_{n_{z\min}}^{n_{z\max}} Q_9(E_{n_z}) F_{-1}(\eta_{1,9}) \tag{2.43}$$

The $n_{z_{\min}}$ should be determined from the following equation

$$(W - h\nu) = \frac{1}{2}\left[\sqrt{E_{g_0}^2 + \frac{8}{3}P^2\left(\left(\frac{\pi n_{z_{\min}}}{d_z}\right)^2\right)} - E_{g_0}\right] \times \left[1 - T_5\left(\frac{\pi n_{z_{\min}}}{d_z}\right)^2\left[E_{g_0}^2 + \frac{8}{3}P^2\left(\left(\frac{\pi n_{z_{\min}}}{d_z}\right)^2\right)\right]^{-1/2}\right] \tag{2.44}$$

In this case, the electron statistics is given by (1.49).

2.2.3 The EP from QDs of II–VI materials

(a) Hopefield model

The $\nu_z(E_{n_z})$ and E_{n_z} for this case are given by

$$v_z(E_{n_z}) = \frac{\hbar}{m_{\parallel}^*}\left(\frac{\pi n_z}{d_z}\right) \tag{2.45}$$

$$E_{n_z} = \frac{\hbar^2}{2m_{\parallel}^*}\left(\frac{\pi n_z}{d_z}\right)^2 \tag{2.46}$$

The photo-emitted current density is given by

$$J_{\mathrm{OD}} = \frac{\alpha_0 g_v \hbar \pi e}{2d_x d_y d_z^2 m_{\parallel}^*}\sum_{n_x=1}^{n_{x\max}}\sum_{n_y=1}^{n_{y\max}}\sum_{n_{z_{\min}}}^{n_{z\max}} n_z F_{-1}(\eta_{1,10}) \tag{2.47}$$

where

$$n_{z_{\min}} \geq \frac{d_z}{\pi}\frac{\sqrt{2m_{\parallel}^*}}{\hbar}(W - h\nu)^{1/2} \tag{2.48}$$

In this case, the electron statistics is given by (1.53).

2.2.4 The EP from QDs of *n*-gallium phosphide

(a) The model of Rees *et al.*

The $\nu_z(E_{n_z})$ and E_{n_z} in this case can be respectively written as

$$\nu_z(E_{n_z}) = \frac{\hbar}{m_{\parallel}^*}\left(\frac{\pi n_z}{d_z}\right)Q_{10}(E_{n_z}) \tag{2.49}$$

where

$$Q_{10}(E_{n_z}) = \left[1 - \left(\frac{\hbar^2 k_0^2}{m_\parallel^*}\right)\left\{|V_G|^2 + \frac{\hbar^4 k_0^2}{(m_\parallel^*)^2}\left(\frac{\pi n_z}{d_z}\right)^2\right\}^{-1/2}\right]$$

and

$$E_{n_z} = \frac{\hbar^2}{2m_\parallel^*}\left(\frac{\pi n_z}{d_z}\right)^2 - \left[\frac{\hbar^4 k_0^2}{m_\parallel^{*2}}\left(\frac{\pi n_z}{d_z}\right)^2 + |V_G|^2\right]^{1/2} + |V_G| \quad (2.50)$$

The photo-emitted current density is given by

$$J_{0\mathrm{D}} = \frac{\alpha_0 \pi \hbar g_v e}{d_x d_y d_z^2 m_\parallel^*}\sum_{n_x=1}^{n_{x\max}}\sum_{n_y=1}^{n_{y\max}}\sum_{n_{z\min}}^{n_{z\max}} Q_{10}(E_{n_z})F_{-1}(\eta_{1,11}) \quad (2.51)$$

where $n_{z_{\min}}$ is the nearest integer of the following equation

$$(W - h\nu) = \frac{\hbar^2}{2m_\parallel^*}\left(\frac{\pi n_{z_{\min}}}{d_z}\right)^2 + |V_G| - \left[|V_G|^2 + \frac{\hbar^4 k_0^2}{(m_\parallel^*)^2}\left(\frac{\pi n_{z_{\min}}}{d_z}\right)^2\right]^{1/2} \quad (2.52)$$

In this case, the electron statistics is given by (1.57).

2.2.5 The EP from QDs of *n*-germanium

(a) The model of Cardona *et al.*

Using the model of Cardona *et al.*, as discussed in (1.59) of Chapter 1, the $\nu_z(E_{n_z})$ and E_{n_z} can be respectively written as

$$\nu_z(E_{n_z}) = \frac{\hbar}{m_\parallel^*}\left(\frac{\pi n_z}{d_z}\right) \quad (2.53)$$

$$E_{n_z} = \frac{\hbar^2}{2m_\parallel^*}\left(\frac{\pi n_z}{d_z}\right)^2 \quad (2.54)$$

The photo-emitted current density is given by

$$J_{0\mathrm{D}} = \frac{\alpha_0 \pi \hbar g_v e}{d_x d_y d_z^2 m_\parallel^*}\sum_{n_x=1}^{n_{x\max}}\sum_{n_y=1}^{n_{y\max}}\sum_{n_{z\min}}^{n_{z\max}} n_z F_{-1}(\eta_{1,12}) \quad (2.55)$$

where $n_{z_{\min}}$ is the nearest integer of the following inequality

$$n_{z_{\min}} \geq \frac{d_z}{\pi} \frac{\sqrt{2m_{\parallel}^*}}{\hbar} (W - h\nu)^{1/2} \tag{2.56}$$

In this case, the electron statistics is given by (1.69).

(b) The model of Wang and Ressler

Using the model of Wang and Ressler, as discussed in (1.63) of Chapter 1, the $\nu_z(E_{n_z})$ and E_{n_z} can be respectively written as

$$v_z(E_{n_z}) = \frac{\hbar}{m_{\parallel}^*} \left(\frac{\pi n_z}{d_z}\right) Q_{11}(E_{n_z}) \tag{2.57}$$

where

$$Q_{11}(E_{n_z}) = \left[1 - \bar{e}_1 \frac{\hbar^2}{(m_{\parallel}^*)^2} \left(\frac{\pi n_z}{d_z}\right)^2\right]$$

and

$$E_{n_z} = \frac{\hbar^2}{2m_{\parallel}^*} \left(\frac{\pi n_z}{d_z}\right)^2 - e_1 \left(\frac{\hbar^2}{2m_{\parallel}^*}\right)^2 \left(\frac{\pi n_z}{d_z}\right)^4 \tag{2.58}$$

The photo-emitted current density is given by

$$J_{\mathrm{OD}} = \frac{\alpha_0 \pi \hbar g_v e}{d_x d_y d_z^2 m_{\parallel}^*} \sum_{n_x=1}^{n_{x\max}} \sum_{n_y=1}^{n_{y\max}} \sum_{n_{z\min}}^{n_{z\max}} Q_{11}(E_{n_z}) F_{-1}(\eta_{1,13}) \tag{2.59}$$

where $n_{z_{\min}}$ is the nearest integer of the following inequality

$$(W - h\nu) = \frac{\hbar^2}{2m_{\parallel}^*} \left(\frac{\pi n_{z_{\min}}}{d_z}\right)^2 - e_1 \left(\frac{\hbar^2}{2m_{\parallel}^*}\right)^2 \left(\frac{\pi n_{z_{\min}}}{d_z}\right)^4 \tag{2.60}$$

In this case, the electron statistics is given by (1.65).

2.2.6 The EP from QDs of tellurium

(a) The model of Bouat *et al.*

In accordance with Bouat *et al.* as given in (1.67) the $v_z(E_{n_{z,\pm}})$ in this case is given by

$$v_z(E_{n_{z,\pm}}) = \frac{1}{\hbar} Q_{12}(E_{n_{z,\pm}}) \tag{2.61}$$

where

$$Q_{12}(E_{n_{z,\pm}}) = 2A_6\left(\frac{n_z\pi}{d_z}\right) \pm \sqrt{A_8} \quad \text{in which}$$

$$E_{n_{z,\pm}} = A_6\left(\frac{n_z\pi}{d_z}\right)^2 \pm \left(\frac{n_z\pi}{d_z}\right)\sqrt{A_8}$$

The photo-emitted current density is given by

$$J_{\text{0D}} = \frac{\alpha_0 e g_v}{2\hbar d_x d_y d_z} \sum_{n_x=1}^{n_{x\max}} \sum_{n_y=1}^{n_{y\max}} \sum_{n_z=1}^{n_{z\max}} Q_{12}(E_{n_{z,\pm}}) F_{-1}(\eta_{1,14}) \tag{2.62}$$

where $n_{z_{\min}}$ can be determined from

$$W - h\nu = A_6\left(\frac{\pi n_{z_{\min}}}{d_z}\right)^2 \pm \sqrt{A_8} \cdot \frac{\pi}{d_z} \cdot n_{z_{\min}} \tag{2.63}$$

In this case, the electron statistics is given by (1.61).

(b) The model of Ortenberg and Button

In accordance with the model of Ortenberg and Button as given by (1.71) the $v_z(E_{n_{z,\pm}})$ in this case is given by

$$v_z(E_{n_{z,\pm}}) = \frac{1}{\hbar} Q_{13}(E_{n_{z,\pm}}) \tag{2.64}$$

where

$$Q_{13}(E_{n_{z,\pm}}) = \left[2t_2\left(\frac{\pi n_z}{d_z}\right) \pm t_7\left(\frac{\pi n_z}{d_z}\right) \times \left[t_1^2 + t_7\left(\frac{\pi n_z}{d_z}\right)^2\right]^{-1/2}\right] \quad \text{and}$$

$$E_{n_{z,\pm}} = t_1 + t_2\left(\frac{\pi n_z}{d_z}\right)^2 \pm \left[t_1^2 + t_7\left(\frac{\pi n_z}{d_z}\right)^2\right]^{1/2}$$

The photo-emitted current density can be written as

$$J_{\text{0D}} = \frac{\alpha_0 e g_v}{2\hbar d_x d_y d_z} \sum_{n_x=1}^{n_{x\max}} \sum_{n_y=1}^{n_{y\max}} \sum_{n_z=1}^{n_{z\max}} Q_{13}(E_{n_{z,\pm}}) F_{-1}(\eta_{1,15}) \tag{2.65}$$

where $n_{z_{\min}}$ should be determined from the following equation

$$W - h\nu = t_1 + t_2\left(\frac{\pi n_{z_{\min}}}{d_z}\right)^2 \pm \left[t_1^2 + t_7\left(\frac{\pi n_{z_{\min}}}{d_z}\right)^2\right]^{1/2} \tag{2.66}$$

In this case, the electron statistics is given by (1.73).

2.2.7 The EP from QDs of graphite

(a) The model of Ushio *et al.*

The $v_z(E_{n_{z,\pm}})$ in this case is given by

$$v_z(E_{n_{z,\pm}}) = \frac{1}{\hbar} Q_{14}(E_{n_{z,\pm}}) \tag{2.67}$$

where

$$\begin{aligned}
Q_{14}(E_{n_{z,\pm}}) &= \frac{1}{2}\left[\gamma_1 \bar{c} \sin\left(\frac{\bar{c}(\pi n_z)}{2d_z}\right) - \gamma_5 \bar{c} \sin\left(\frac{\bar{c}(\pi n_z)}{2d_z}\right) - \gamma_2 \bar{c} \sin\left(\frac{\bar{c}(\pi n_z)}{2d_z}\right)\right] \\
&\pm \frac{1}{2}\left[\frac{1}{4}\left(\Delta_1 - 2\gamma_1 \cos\left(\frac{\bar{c}(\pi n_z)}{2d_z}\right) + 2\gamma_5 \cos^2\left(\frac{\bar{c}(\pi n_z)}{2d_z}\right)\right.\right. \\
&\left. - 2\gamma_2 \cos\left(\frac{\bar{c}(\pi n_z)}{2d_z}\right)\right)^2 + \frac{3(\bar{a})^2}{4}\left(\frac{n_z \pi}{d_z}\right)^2 \\
&\times \left.\left(\gamma_0 + 2\gamma_4 \cos\left(\frac{\bar{c}(\pi n_z)}{2d_z}\right)\right)^2\right]^{-1/2} \left[\frac{1}{2}\left[\Delta_1 - 2\gamma_1 \cos\left(\frac{\bar{c}(\pi n_z)}{2d_z}\right)\right.\right. \\
&\left. + 2\gamma_5 \cos^2\left(\frac{\bar{c}(\pi n_z)}{2d_z}\right) - 2\gamma_2 \cos^2\left(\frac{\bar{c}(\pi n_z)}{2d_z}\right)\right] \\
&\times \left[\bar{c}\gamma_1 \sin\left(\frac{\bar{c}(\pi n_z)}{2d_z}\right) - \bar{c}\gamma_5 \sin\left(\frac{\bar{c}(\pi n_z)}{2d_z}\right) + \gamma_2 \bar{c} \sin\left(\frac{\bar{c}(\pi n_z)}{2d_z}\right)\right] \\
&+ (\bar{a})^2 \frac{3}{4}\left(\frac{\pi n_z}{d_z}\right)\left[\gamma_0 + 2\gamma_4 \cos\left(\frac{\bar{c}(\pi n_z)}{2d_z}\right)\right]^2 \\
&\left. - (\bar{a})^2 \frac{3}{2} \bar{c}\gamma_4 \left(\frac{\pi n_z}{d_z}\right)^2 \left[\gamma_0 + 2\gamma_4 \cos\left(\frac{\bar{c}(\pi n_z)}{2d_z}\right)\right]\right]
\end{aligned}$$

in which

$$\begin{aligned}
E_{n_{z,\pm}} &= \frac{1}{2}\left[\Delta_1 - 2\gamma_1 \cos\left(\frac{\bar{c}(\pi n_z)}{2d_z}\right) + 2\gamma_5 \cos^2\left(\frac{\bar{c}(\pi n_z)}{2d_z}\right)\right. \\
&\left. + 2\gamma_2 \cos^2\left(\frac{\bar{c}(\pi n_z)}{2d_z}\right)\right] \pm \left[\frac{1}{4}\left(\Delta_1 - 2\gamma_1 \cos\left(\frac{\bar{c}(\pi n_z)}{2d_z}\right)\right.\right. \\
&\left. + 2\gamma_5 \cos^2\left(\frac{\bar{c}(\pi n_z)}{2d_z}\right) - 2\gamma_2 \cos^2\left(\frac{\bar{c}(\pi n_z)}{2d_z}\right)\right)^2 \\
&\left. + \frac{3(\bar{a})^2}{4}\left(\gamma_0 + 2\gamma_4 \cos\left(\frac{\bar{c}(\pi n_z)}{2d_z}\right)\right)^2 \left(\frac{\pi n_z}{d_z}\right)^2\right]^{1/2}
\end{aligned}$$

The photo-emitted current density can be written as

$$J_{0\mathrm{D}} = \frac{\alpha_0 g_v e}{2d_x d_y d_z \hbar} \sum_{n_x=1}^{n_{x\max}} \sum_{n_y=1}^{n_{y\max}} \sum_{n_{z\min}}^{n_{z\max}} F_{-1}(\eta_{1,16}) Q_{14}(E_{n_z,\pm}) \tag{2.68}$$

where $n_{z_{\min}}$ should be determined from the following equation

$$(W - h\nu) = \frac{1}{2}\left[\Delta_1 - 2\gamma_1 \cos\left(\frac{\bar{c}(\pi n_{z_{\min}})}{2d_z}\right) + 2\gamma_5 \cos^2\left(\frac{\bar{c}(\pi n_{z_{\min}})}{2d_z}\right) + 2\gamma_2 \cos^2\left(\frac{\bar{c}(\pi n_{z_{\min}})}{2d_z}\right)\right] \pm \left[\frac{1}{4}\left(\Delta_1 - 2\gamma_1 \cos\left(\frac{\bar{c}(\pi n_{z_{\min}})}{2d_z}\right) + 2\gamma_5 \cos^2\left(\frac{\bar{c}(\pi n_{z_{\min}})}{2d_z}\right) - 2\gamma_2 \cos^2\left(\frac{\bar{c}(\pi n_{z_{\min}})}{2d_z}\right)\right)^2 + \frac{3(\bar{a})^2}{4}\left(\gamma_0 + 2\gamma_4 \cos\left(\frac{\bar{c}(\pi n_{z_{\min}})}{2d_z}\right)\right)^2 \left(\frac{\pi n_{z_{\min}}}{d_z}\right)^2\right]^{1/2} \tag{2.69}$$

In this case, the electron statistics is given by (1.77).

*2.2.8 The EP from QDs of platinum antimonide

(a) The model of Emtage

The $v_z(E_{n_z})$ in this case is given by

$$v_z(E_{n_z}) = \frac{1}{\hbar} Q_{15}(E_{n_z}) \tag{2.70}$$

where

$$Q_{15}(E_{n_z}) = \left[2E_{n_z} + \delta_0 - \omega_4 \left(\frac{\pi n_z}{d_2}\right)^2 + \omega_2 \left(\frac{\pi n_z}{d_2}\right)^2\right]^{-1} \times \left[4I_1 \left(\frac{\pi n_z}{d_2}\right)^3 + 2\omega_4 \left(\frac{\pi n_z}{d_2}\right)\left(E_{n_z} + \omega_2 \left(\frac{\pi n_z}{d_2}\right)^2\right) - 2\omega_2 \left(\frac{\pi n_z}{d_2}\right)\left(E + \delta_0 - \omega_4 \left(\frac{\pi n_z}{d_2}\right)^2\right)\right]$$

in which E_{n_z} should be determined from the following equation

$$\left[E_{n_z} + \omega_2 \left(\frac{\pi n_z}{d_z}\right)^2\right]\left[E_{n_z} + \delta_0 - \omega_4 \left(\frac{\pi n_z}{d_z}\right)^2\right] = I_1 \left(\frac{\pi n_z}{d_z}\right)^4 \tag{2.71}$$

The photo-emitted current density assumes the form

$$J_{\mathrm{OD}} = \frac{\alpha_0 e g_v}{d_x d_y d_z \hbar} \sum_{n_x=1}^{n_{x\max}} \sum_{n_y=1}^{n_{y\max}} \sum_{n_{z\min}}^{n_{z\max}} Q_{15}(E_{n_z}) F_{-1}(\eta_{1,17}) \tag{2.72}$$

where $n_{z_{\min}}$ should be determined from the following equation

$$\left[(W - h\nu) + \omega_2 \left(\frac{\pi n_{z_{\min}}}{d_z}\right)^2\right] \left[(W - h\nu) + \delta_0 - \omega_4 \left(\frac{\pi n_{z_{\min}}}{d_z}\right)^2\right]$$

$$= I_1 \left[\frac{\pi n_{z_{\min}}}{d_z}\right]^4 \tag{2.73}$$

In this case, the hole statistics is given by (1.82).

2.2.9 The EP from QDs of zero-gap materials

(a) The model of Ivanov-Omskii *et al.*

The $v_z(E_{n_z})$ in this case is given by

$$v_z(E_{n_z}) = \frac{1}{\hbar} Q_{16}(E_{n_z}) \tag{2.74}$$

where

$$Q_{16}(E_{n_z}) = \left[\frac{\hbar}{m} \cdot \left(\frac{\pi n_z}{d_z}\right) + \frac{1}{\hbar}\left[\frac{3e^2}{128 t_s} - \frac{2 E_B d_z}{\pi^2 n_z}\right]\right]$$

in which E_{n_z} should be determined from the following equation

$$E_{n_z} = \left[\frac{\hbar}{2m^*}\left(\frac{\pi n_z}{d_z}\right)^2 + \frac{3e^2}{128\varepsilon_s} \times \left(\frac{\pi n_z}{d_z}\right) - \frac{2}{\pi} E_B \ln\left|\frac{\pi n_z}{d_z k_0}\right|\right] \tag{2.75}$$

The photo-emitted current density can be written as

$$J_{\mathrm{OD}} = \frac{\alpha_0 g_v e}{\hbar d_x d_y d_z} \sum_{n_x=1}^{n_{x\max}} \sum_{n_y=1}^{n_{y\max}} \sum_{n_z=1}^{n_{z\max}} Q_{16}(E_{n_z}) F_{-1}(\eta_{1,18}) \tag{2.76}$$

where $n_{z_{\min}}$ should be determined from the following equation

$$(W - h\nu) = \frac{\hbar}{2m^*}\left[\frac{\pi n_{z_{\min}}}{d_z}\right]^2 + \frac{3e^2}{128\varepsilon_s}\left(\frac{\pi n_{z_{\min}}}{d_z}\right) - \frac{2}{\pi} E_B \ln\left|\frac{\pi n_{z_{\min}}}{d_z}\right| \tag{2.77}$$

In this case, the hole statistics is given by (1.86).

*2.2.10 The EP from QDs of II–V materials

(a) The model of Yamada

The $v_z(E_{n_z})$ in this case is given by

$$v_z(E_{n_z}) = \frac{1}{\hbar} Q_{24,\pm}(E_{n_z}) \tag{2.78}$$

where

$$Q_{24,\pm}(E_{n_z}) = \left[2A_{12}\left(\frac{\pi n_z}{d_z}\right) \pm 2A_{16}^2\left(\frac{\pi n_z}{d_z}\right)^3 \left[A_{16}^2\left(\frac{\pi n_z}{d_z}\right)^4 + A_{19}^2\right]^{-1/2}\right]$$

in which E_{n_z} should be determined from the following equation

$$E_{n_z} = A_{12}\left(\frac{\pi n_z}{d_z}\right)^2 \pm \left[A_{16}^2\left(\frac{\pi n_z}{d_z}\right)^4 + A_{19}^2\right]^{1/2} \mp A_{19} \tag{2.79}$$

The photo-emitted current density can be written as

$$J_{0\mathrm{D}} = \frac{\alpha_0 e g_v}{2\hbar d_x d_y d_z} \sum_{n_x=1}^{n_{x\max}} \sum_{n_y=1}^{n_{y\max}} \sum_{n_{z\min}}^{n_{z\max}} Q_{24,\pm}(E_{n_z})[1+\exp(\eta_{1,19})]^{-1} \tag{2.80}$$

where $n_{z_{\min}}$ should be the nearest integer of the following equation

$$W - h\nu = A_{12}\left(\frac{\pi n_{z_{\min}}}{d_z}\right)^2 \pm \left[A_{16}^2\left(\frac{\pi n_{z_{\min}}}{d_z}\right)^4 + A_{19}^2\right]^{1/2} \mp A_{19} \tag{2.81}$$

In this case, the hole statistics is given by (1.90).

*2.2.11 The EP from QDs of gallium antimonide

(a) The model of Seiler *et al.*

In accordance with the model of Seiler *et al.* (1.92) the $v_z(E_{n_z})$ in this case is given by

$$v_z(E_{n_{z,\pm}}) = \frac{1}{\hbar} Q_{18}(E_{n_{z,\pm}}) \tag{2.82}$$

where

$$Q_{18}(E_{n_{z,\pm}}) = \left[\alpha_4 E_{g_0}\left(\frac{\pi n_z}{d_z}\right)\left[1+\alpha_4\left(\frac{\pi n_z}{d_z}\right)^2\right]^{-1/2} + \left(\frac{\overline{\zeta}_0\hbar^2}{m_0}\right)\left(\frac{\pi n_z}{d_z}\right)\right]$$

in which $E_{n_{z,\pm}}$ should be determined from the following equation

$$E_{n_{z,\pm}} = -\frac{E_{g0}}{2} + \frac{E_{g0}}{2}\left[1+\alpha_4\left(\frac{\pi n_z}{d_z}\right)^2\right]^{1/2} + \frac{\overline{\zeta_0}\hbar^2}{2m_0}\left(\frac{\pi n_z}{d_z}\right)^2 \tag{2.83}$$

The photo-emitted current density can be written as

$$J_{0D} = \frac{\alpha_0 e g_v}{2\hbar d_x d_y d_z}\sum_{n_x=1}^{n_{x\max}}\sum_{n_y=1}^{n_{y\max}}\sum_{n_{z\min}}^{n_{z\max}} Q_{18}(E_{n_{z,\pm}})F_{-1}(\eta_{1,20}) \tag{2.84}$$

where $n_{z_{\min}}$ should be determined from the following equation

$$(W-h\nu) = -\frac{E_{g0}}{2} + \frac{E_{g0}}{2}\left[1+\alpha_4\left(\frac{\pi n_{z_{\min}}}{d_z}\right)^2\right]^{1/2} + \overline{\zeta_0}\left(\frac{\hbar^2}{2m_0}\right)\left(\frac{\pi n_{z_{\min}}}{d_z}\right)^2 \tag{2.85}$$

In this case, the electron statistics is given by (1.94).

(b) The model of Mathur *et al.*

In accordance with the model of Mathur *et al.* (1.96), the $v_z(E_{n_z})$ in this case is given by

$$v_z(E_{n_z}) = \frac{1}{\hbar}Q_{19}(E_{n_z}) \tag{2.86}$$

where

$$Q_{19}(E_{n_z}) = \left[2\alpha_9\left(\frac{\pi n_z}{d_z}\right) + \frac{E_{g1}}{2}\left[\alpha_{10}\left(\frac{\pi n_z}{d_z}\right)\left(1+\alpha_{10}\left(\frac{\pi n_z}{d_z}\right)^2\right)^{-1/2}\right]\right]$$

in which E_{n_z} should be determined from the following equation

$$E_{n_z} = \left(\frac{2\pi n_z}{d_z}\right)\left[\alpha_9 + \frac{\alpha_{10}E_{g1}}{4}\left[1+\alpha_{10}\left(\frac{\pi n_z}{d_z}\right)^2\right]^{-1/2}\right] \tag{2.87}$$

The photo-emitted current density can be written as

$$J_{0D} = \frac{\alpha_0 e g_v}{\hbar d_x d_y d_z}\sum_{n_x=1}^{n_{x\max}}\sum_{n_y=1}^{n_{y\max}}\sum_{n_{z\min}}^{n_{z\max}} Q_{19}(E_{n_z})F_{-1}(\eta_{1,21}) \tag{2.88}$$

where $n_{z_{\min}}$ should be determined from the following equation

$$W - h\nu = \alpha_9 \left(\frac{\pi n_{z_{\min}}}{d_z}\right)^2 + \frac{E_{g1}}{2}\left[-1 + \sqrt{1 + \alpha_{10}\left(\frac{\pi n_{z_{\min}}}{d_z}\right)^2}\right] \tag{2.89}$$

In this case, the electron statistics is given by (1.99).

(c) The model of Zhang

The dispersion relation of the conduction electrons in n-GaSb can be expressed in accordance with Zhang (1.101) and the $v_z(E_{n_z})$ in this case is given by

$$v_z(E_{n_z}) = \frac{1}{\hbar} Q_{20}(E_{n_z}) \tag{2.90}$$

where

$$Q_{20}(E_{n_z}) = 2\left(\frac{\pi n_z}{d_z}\right)\left[\overline{E_1} + \frac{\sqrt{21}}{2}\cdot\overline{E_2}\right] + 4\left(\frac{\pi n_z}{d_z}\right)^3\left[\overline{E_3} + \frac{\sqrt{21}}{2}\cdot\overline{E_4}\right]$$
$$+6\left(\frac{\pi n_z}{d_z}\right)^5 \times \left[\overline{E_5} + \frac{\sqrt{21}}{2}\cdot\overline{E_6} + \left(\frac{\overline{E_7}}{42}\right)\left(\frac{639639}{32}\right)^{1/2}\right]$$

in which E_{n_z} should be determined from the following equation

$$E_{n_z} = \left[\overline{E_1} + \overline{E_2}\left(\frac{\sqrt{21}}{2}\right)\right]\left(\frac{\pi n_z}{d_z}\right)^2 + \left(\overline{E_3} + \overline{E_4}\left(\frac{\sqrt{21}}{2}\right)\right)\left(\frac{\pi n_z}{d_z}\right)^4$$
$$+\left(\frac{\pi n_z}{d_z}\right)^6\left[\overline{E_5} + \overline{E_6}\left(\frac{\sqrt{21}}{2}\right) + \overline{E_7}\left(\frac{639639}{32}\right)^{1/2}\left[\frac{1}{42}\right]\right] \tag{2.91}$$

The photo-emitted current density can be written as

$$J_{0\mathrm{D}} = \frac{\alpha_0 e g_v}{2\hbar d_x d_y d_z} \sum_{n_x=1}^{n_{x_{\max}}} \sum_{n_y=1}^{n_{y_{\max}}} \sum_{n_{z_{\min}}}^{n_{z_{\max}}} Q_{20} F_{-1}(E_{n_z}) F_{-1}(\eta_{1,22}) \tag{2.92}$$

where $n_{z_{\min}}$ should be determined from the following equation

$$W - h\nu = \left[\overline{E_1} + \overline{E_2}\left(\frac{\sqrt{21}}{2}\right)\right]\left(\frac{\pi n_{z_{\min}}}{d_z}\right)^2 + \left(\frac{\pi n_{z_{\min}}}{d_z}\right)^4\left[\overline{E_3} + \overline{E_4}\left(\frac{\sqrt{21}}{2}\right)\right]$$
$$+\left(\frac{\pi n_{z_{\min}}}{d_z}\right)^6\left[\overline{E_5} + \overline{E_6}\left[\frac{\sqrt{21}}{2}\right] + \overline{E_7}\cdot\left(\frac{1}{42}\right)\left(\frac{639639}{32}\right)^{1/2}\right] \tag{2.93}$$

In this case, the electron statistics is given by (1.103).

2.2.12 The EP from QDs of stressed materials

(a) Model of Seiler *et al.*

The $v_z(E_{n_z})$ for QDs of stressed materials in this case is given by

$$v_z(E_{n_z}) = \frac{Q_{21}(E_{n_z})}{\hbar} \tag{2.94}$$

where

$$Q_{21}(E_{n_z}) = \frac{1}{[c^*(E_{n_z})]'}$$

The E_{n_z} can be expressed through the equation

$$c^*(E_{n_z}) = \frac{n_z \pi}{d_z} \tag{2.95}$$

The photo-emitted current density can be written as

$$J_{\text{OD}} = \frac{\alpha_0 e g_v}{d_x d_y d_z \hbar} \sum_{n_x=1}^{n_{x\max}} \sum_{n_y=1}^{n_{y\max}} \sum_{n_z=1}^{n_{z\max}} Q_{21}(E_{n_z}) F_{-1}(\overline{\eta}_{1,23}) \tag{2.96}$$

where $n_{z_{\min}}$ is given by the following inequality

$$n_{z_{\min}} \geq \left(\frac{d_z}{\pi}\right) [c^*(W - h\nu)] \tag{2.97}$$

In this case, the electron statistics is given by (1.107).

2.2.13 The EP from QDs of bismuth

(a) The McClure and Choi model

The $v_z(E_{n_z})$ of the conduction electrons in QDs of Bi in accordance with McClure and Choi (1.109) can be expressed as

$$v_z(E_{n_z}) = \frac{\pi \hbar n_z}{d_z m_3 (1 + 2\alpha E_{n_z})} \tag{2.98}$$

E_{n_z} should be determined from the following equation

$$E_{n_z}(1 + \alpha E_{n_z}) = \frac{\hbar^2}{2m_3}\left(\frac{\pi n_z}{d_z}\right)^2 \tag{2.99}$$

The photo-emitted current density can be written as

$$J_{\text{OD}} = \frac{a e g_v \hbar \pi}{m_3 \cdot d_x d_y d_z^2} \sum_{n_x=1}^{n_{x\max}} \sum_{n_y=1}^{n_{y\max}} \sum_{n_{z_{\min}}}^{n_{z\max}} \frac{n_z F_{-1}(\eta_{1,24})}{(1 + 2\alpha E_{n_z})} \tag{2.100}$$

where $n_{z_{\min}}$ should be determined from the following equation

$$(W - hv)[1 + \alpha(W - hv)] = \frac{\hbar^2\pi^2}{2m_3}\left(\frac{n_{z_{\min}}}{d_z}\right)^2 \tag{2.101}$$

In this case, the electron statistics is given by (1.111).

(b) The Hybrid model

The $v_z(E_{n_z})$ for QDs of Bi in accordance with Hybrid model (1.113) can be written as

$$v_z(E_{n_z}) = \frac{\pi\hbar n_z}{d_z m_3(1 + 2\alpha E_{n_z})} \tag{2.102}$$

in which E_{n_z} should be determined from the following equation

$$E_{n_z}(1 + \alpha E_{n_z}) = \frac{\hbar^2}{2m_3}(\pi n_z/d_z)^2 \tag{2.103a}$$

The photo-emitted current density can be written as

$$J_{0\mathrm{D}} = \frac{aeg_v\hbar\pi}{m_3 \cdot d_x d_y d_z^2}\sum_{n_x=1}^{n_{x\max}}\sum_{n_y=1}^{n_{y\max}}\sum_{n_{z_{\min}}}^{n_{z\max}}\frac{n_z F_{-1}(\eta_{1,25})}{(1 + 2\alpha E_{n_z})} \tag{2.103b}$$

where $n_{z_{\min}}$ should be determined from the following equation

$$(W - hv)[1 + \alpha(W - hv)] = \frac{\hbar^2}{2m_3}[\pi n_{z_{\min}}/d_z]^2 \tag{2.104}$$

In this case, the electron statistics is given by (1.115).

(c) The Cohen model

The $v_z(E_{n_z})$ for QDs of Bi in accordance with the model of Cohen (1.117) can be written as

$$v_z(E_{n_z}) = \frac{\pi\hbar n_z}{(1 + 2\alpha E_{n_z})d_z m_3} \tag{2.105}$$

in which E_{n_z} should be determined from the following equation

$$E_{n_z}(1 + \alpha E_{n_z}) = \frac{\hbar^2}{2m_3}\left(\frac{\pi n_z}{d_z}\right)^2 \tag{2.106}$$

The photo-emitted current density can be written as

$$J_{0\mathrm{D}} = \frac{\hbar\pi aeg_v}{m_3 \cdot d_x d_y d_z^2}\sum_{n_x=1}^{n_{x\max}}\sum_{n_y=1}^{n_{y\max}}\sum_{n_{z_{\min}}}^{n_{z\max}}\frac{n_z F_{-1}(\eta_{1,26})}{(1 + 2\alpha E_{n_z})} \tag{2.107}$$

where $n_{z_{\min}}$ should be determined from the following equation

$$(W - hv)[1 + \alpha(W - hv)] = \frac{\hbar^2}{2m_3}\left(\frac{\pi n_{z_{\min}}}{d_z}\right)^2 \tag{2.108}$$

In this case, the electron statistics is given by (1.119).

(d) The Lax model

The $v_z(E_{n_z})$ for QDs of Bi in accordance with the model of Lax (1.121) is given by

$$v_z(E_{n_z}) = \frac{\pi\hbar n_z}{d_z \cdot m_3(1 + 2\alpha E_{n_z})} \tag{2.109}$$

in which E_{n_z} should be determined from the following equation

$$E_{n_z}(1 + \alpha E_{n_z}) = \frac{\hbar^2}{2m_3}(\pi n_z/d_z)^2 \tag{2.110}$$

The photo-emitted current density can be written as

$$J_{\mathrm{0D}} = \frac{\hbar\pi a e g_v}{m_3 \cdot d_x d_y d_z^2} \sum_{n_x=1}^{n_{x\max}} \sum_{n_y=1}^{n_{y\max}} \sum_{n_{z_{\min}}}^{n_{z\max}} \frac{n_z F_{-1}(\eta_{1,27})}{(1 + 2\alpha E_{n_z})} \tag{2.111}$$

where $n_{z_{\min}}$ should be determined from the following equation

$$(W - hv)[1 + \alpha(W - hv)] = \frac{\hbar^2}{2m_3}[\pi n_{z_{\min}}/d_z]^2 \tag{2.112}$$

In this case, the electron statistics is given by (1.123).

(e) Ellipsoidal Parabolic model

The $v_z(E_{n_z})$ for QDs of Bi in accordance with the Ellipsoidal Parabolic model of (1.125) is given by

$$v_z(E_{n_z}) = \frac{\pi\hbar n_z}{d_z \cdot m_3} \tag{2.113}$$

in which E_{n_z} should be determined from the following equation

$$E_{n_z} = \frac{\hbar^2}{2m_3}(\pi n_z/d_z)^2 \tag{2.114}$$

The photo-emitted current density can be written as

$$J_{\mathrm{0D}} = \frac{\hbar\pi a e g_v}{m_3 \cdot d_x d_y d_z^2} \sum_{n_x=1}^{n_{x\max}} \sum_{n_y=1}^{n_{y\max}} \sum_{n_{z_{\min}}}^{n_{z\max}} n_z F_{-1}(\bar{\eta}_{1,27}) \tag{2.115}$$

where $n_{z_{\min}}$ should be determined from the following equation

$$(W - hv) = \frac{\hbar^2}{2m_3}[\pi n_{z_{\min}}/d_z]^2 \tag{2.116}$$

In this case, the electron statistics is given by (1.126).

2.2.14 The EP from QDs of IV–VI materials

In this section, the photoemission from QDs of IV–VI materials has been presented in accordance with the models of Dimmock (1.128), Bangert and Kastner (1.132) and Foley *et al.* (1.136) respectively, although the carrier dispersion laws in this compound also follows Cohen and Lax models.

(a) Dimmock model

In the present case, the $v_z(E_{n_z})$ in given by

$$v_z(E_{n_z}) = \frac{\pi\hbar n_z}{d_z m_l^*} Q_{21}(E_{n_z}) \tag{2.117}$$

where

$$Q_{21}(E_{n_z}) = \left[1 + \frac{\alpha\hbar^2 m_l^*}{m_l^+ m_l^-}\left(\frac{\pi n_z}{d_z}\right)^2 + \frac{m_l^*}{m_l^-}(1 + \alpha E_{n_z})\right] \times \left[1 + 2\alpha E_{n_z} + \frac{\alpha\hbar^2}{2m_l^+}\left(\frac{\pi n_z}{d_z}\right)^2 (1 - \frac{m_l^+}{m_l^-})\right]^{-1}$$

in which E_{n_z} should be determined from the following equation

$$E_{n_z}(1 + \alpha E_{n_z}) + \alpha E_{n_z}\left(\frac{\hbar^2}{2m_l^+}\left(\frac{\pi n_z}{d_z}\right)^2\right) - \frac{\hbar^2}{2m_l^-}\left(\frac{\pi n_z}{d_z}\right)^2 \cdot (1 + \alpha E_{n_z}) - \frac{\alpha\hbar^4}{4m_l^+ m_l^-}\left(\frac{\pi n_z}{d_z}\right)^4 = \frac{\hbar^2}{2m_l^*}\left(\frac{\pi n_z}{d_z}\right)^2 \tag{2.118}$$

The photo-emitted current density can be written as

$$J_{0\text{D}} = \frac{\pi\hbar\alpha_0 e g_v}{d_x d_y d_z^2 m_l^*} \sum_{n_x=1}^{n_{x\max}} \sum_{n_y=1}^{n_{y\max}} \sum_{n_{z_{\min}}}^{n_{z\max}} Q_{21}(E_{n_z}) F_{-1}(\eta_{1,28}) \tag{2.119}$$

where $n_{z_{\min}}$ should be the nearest integer of the following equation

$$(W - hv)[1 + \alpha(W - hv)] + \alpha(W - hv)\frac{\hbar^2}{2m_l^+}\left(\frac{\pi n_{z_{\min}}}{d_z}\right)^2 - \frac{\hbar^2}{2m_l^-}\left(\frac{\pi n_{z_{\min}}}{d_z}\right)^2 (1 + \alpha(W - hv)) - \frac{\alpha\hbar^4}{4m_l^+ m_l^-}\left(\frac{\pi n_{z_{\min}}}{d_z}\right)^4 = \left(\frac{\hbar^2}{2m_l^*}\right)\left[\frac{\pi n_{z_{\min}}}{d_z}\right]^2 \tag{2.120}$$

In this case, the electron statistics is given by (1.130).

(b) Model of Bangert and Kastner

In this case, the $v_z(E_{n_z})$ in given by

$$v_z(E_{n_z}) = \frac{Q_{22}(E_{n_z})}{\hbar} \tag{2.121}$$

where

$$Q_{22}(E_{n_z}) = 2\left(\frac{\pi n_z}{d_z}\right)\overline{F}_2(E_{n_z})\left[\Gamma'(E_{n_z}) - \left(\frac{\pi n_z}{d_z}\right)^2 (\overline{F}_2(E_{n_z}))'\right]^{-1}$$

in which E_{n_z} should be determined from the following equation

$$\Gamma(E_{n_z}) = \overline{F}_2(E_{n_z})\left(\frac{\pi n_z}{d_z}\right)^2 \tag{2.122}$$

The photo-emitted current density can be written as

$$J_{\mathrm{0D}} = \frac{\alpha_0 e g_v}{\hbar d_x d_y d_z}\sum_{n_x=1}^{n_{x\max}}\sum_{n_y=1}^{n_{y\max}}\sum_{n_{z_{\min}}}^{n_{z\max}} Q_{22}(E_{n_z})F_{-1}(\eta_{1,29}) \tag{2.123}$$

where $n_{z_{\min}}$ should be the nearest integer of the following equation

$$\Gamma(W - hv) = \overline{F}_2(W - hv)\left(\frac{\pi n_{z_{\min}}}{d_z}\right)^2 \tag{2.124}$$

In this case, the electron statistics is given by (1.134).

(c) Model of Foley *et al.*

In this case, the $v_z(E_{n_z})$ in given by

$$v_z(E_{n_z}) = \frac{Q_{23}(E_{n_z})}{\hbar} \tag{2.125}$$

where

$$Q_{23}(E_{n_z}) = \frac{\hbar}{m_{\parallel}^{-}}\left(\frac{\pi n_z}{d_z}\right) + \frac{1}{\hbar}\left[\frac{E_{g0}}{2} + \frac{\hbar}{2m_{\parallel}^{+}}\left(\frac{\pi n_z}{d_z}\right)^2 + P_{\parallel}^2\left(\frac{\pi n_z}{d_z}\right)^2\right]^{-1/2}$$
$$\times \left[\left(\frac{E_{g0}}{2} + \frac{\hbar^2}{2m_{\parallel}^{+}}\left(\frac{\pi n_z}{d_z}\right)\right)\left(\frac{\hbar^2 \pi n_z}{m_{\parallel}^{+} d_z}\right) + P_{\parallel}^2\left(\frac{\pi n_z}{d_z}\right)\right]$$

in which E_{n_z} should be determined from the following equation

$$E_{n_z} = -\frac{E_{g0}}{2} + \frac{\hbar^2}{2m_{\parallel}^{-}}\left(\frac{\pi n_z}{d_z}\right)^2$$
$$+ \left[\left(\frac{E_{g0}}{2} + \frac{\hbar^2}{2m_{\parallel}^{+}}\left(\frac{\pi n_z}{d_z}\right)^2\right)^2 + P_{\parallel}^2\left(\frac{\pi n_z}{d_z}\right)^2\right]^{1/2} \tag{2.126}$$

The photo-emitted current density can be written as

$$J_{0\mathrm{D}} = \frac{\alpha_0 e g_v}{d_x d_y d_z \hbar} \sum_{n_x=1}^{n_{x\max}} \sum_{n_y=1}^{n_{y\max}} \sum_{n_{z\min}}^{n_{z\max}} Q_{23}(E_{n_z}) F_{-1}(\eta_{1,30}) \tag{2.127}$$

where $n_{z\min}$ should be the nearest integer of the following equation

$$(W - hv) = -\frac{E_{g0}}{2} + \frac{\hbar^2}{2m_{\parallel}^{-}}\left(\frac{\pi n_{z\min}}{d_z}\right)^2$$
$$+ \left[\left(\frac{E_{g0}}{2} + \frac{\hbar^2}{2m_{\parallel}^{+}}\left(\frac{\pi n_{z\min}}{d_z}\right)^2\right)^2 + P_{\parallel}^2\left(\frac{\pi n_{z\min}}{d_z}\right)^2\right]^{1/2} \tag{2.128}$$

In this case, the electron statistics is given by (1.138).

*2.2.15 The EP from QDs of lead germanium telluride

In this case, the $v_z(E_{n_z})$ is given by

$$v_z(E_{n_z}) = \frac{1}{\hbar} Q_{17}(E_{n_z}) \tag{2.129}$$

where

$$Q_{17}(E_{n_z}) = \left[2E_{n_z} - 0.0345\left(\frac{\pi n_z}{d_z}\right)^2\right]^{-1}$$
$$\times\left[0.04\left(\frac{\pi n_z}{d_z}\right) + 0.1444\left(\frac{\pi n_z}{d_z}\right)\left[E_{n_z} + 0.0377\left(\frac{\pi n_z}{d_z}\right)^2\right]\right.$$
$$\left.-0.0754\left(\frac{\pi n_z}{d_z}\right)\left[E_{n_z} - 0.0722\left(\frac{\pi n_z}{d_z}\right)^2\right]\right]$$

in which E_{n_z} should be determined from the following equation

$$\left[0.0722\left(\frac{\pi n_z}{d_z}\right)^2 - E_{n_z}\right]\left[-E_{n_z} - \overline{E_{g0}} - 0.0377\left(\frac{\pi n_z}{d_z}\right)^2\right]$$
$$= 0.02\left(\frac{\pi n_z}{d_z}\right)^2 \pm \left[\left(\frac{\pi n_x}{d_x}\right)^2 + \left(\frac{\pi n_y}{d_y}\right)^2\right]^{1/2}$$
$$\times\left[0.06\overline{E_{g0}} + 0.0066\left(\frac{\pi n_z}{d_z}\right)^2\right] \tag{2.130}$$

The photo-emitted current density can be written as

$$J_{0\mathrm{D}} = \frac{\alpha_0 g_v e}{2\hbar d_x d_y d_z}\sum_{n_x=1}^{n_{x\max}}\sum_{n_y=1}^{n_{y\max}}\sum_{n_z=1}^{n_{z\max}} Q_{17}(E_{n_z})F_{-1}(\eta_{1,31}) \tag{2.131}$$

where $n_{z_{\min}}$ should be determined from the following equation

$$\left[(W-h\nu) - 0.0722\left(\frac{\pi n_{z_{\min}}}{d_z}\right)^2\right]\left[(W-h\nu) + \overline{E_{g0}} + 0.0377\left(\frac{\pi n_{z_{\min}}}{d_z}\right)^2\right]$$
$$= 0.0\left(\frac{\pi n_{z_{\min}}}{d_z}\right)^2 \tag{2.132}$$

In this case, the electron statistics is given by (1.142).

*2.2.16 The EP from QDs of zinc and cadmium diphosphides

The $v_z(E_{n_z})$ in this case is given by

$$v_z(E_{n_z}) = \frac{1}{\hbar} Q_{25}(E_{n_z}) \tag{2.133}$$

where

$$\begin{aligned} Q_{25}(E_{n_z}) &= \left[2\left(\frac{\pi n_z}{d_z}\right)\left(\beta_1 - \frac{\beta_2}{4\beta_4}\right) \mp \left(\frac{\pi n_z}{d_z}\right)\left[\left(-2\beta_4\left(\beta_5 + \frac{\beta_2}{4\beta_4}\right) - \frac{3}{2}\beta_2\right)\right]\right. \\ &\quad \times \left[12\beta_4^2 + \left[\left(+2\beta_4\left(\beta_5 + \frac{\beta_2}{4\beta_4}\right) + \frac{3}{2}\beta_2\right)\right]\left(\frac{\pi n_z}{d_z}\right)^2\right]^{-1/2}\Bigg] \end{aligned}$$

in which E_{n_z} should be determined from the following equation

$$\begin{aligned} E_{n_z} &= \left(\beta_1 - \frac{\beta_2}{4\beta_4}\right)\left(\frac{\pi n_z}{d_z}\right)^2 \\ &\quad \pm \left\{\left(-2\beta_4\left(\beta_5 + \frac{\beta_2}{4\beta_4}\right) - \frac{3}{2}\beta_2\right)\left(\frac{\pi n_z}{d_z}\right)^2 + 12\beta_4^2\right\}^{1/2} \end{aligned} \tag{2.134}$$

The photo-emitted current density can be written as

$$J_{0\mathrm{D}} = \frac{\alpha_0 e g_v}{2\hbar d_x d_y d_z} \sum_{n_x=1}^{n_{x\max}} \sum_{n_y=1}^{n_{y\max}} \sum_{n_{z_{\min}}}^{n_{z\max}} Q_{25}(E_{n_z})[1 + \exp(\eta_{1,32})]^{-1} \tag{2.135}$$

where $n_{z_{\min}}$ should be the nearest integer of the following equation

$$\begin{aligned} W - h\nu &= \left(\beta_1 - \frac{\beta_2}{4\beta_4}\right)\left(\frac{\pi n_{z_{\min}}}{d_z}\right)^2 \\ &\quad \pm \left\{\left(-2\beta_4\left(\beta_5 + \frac{\beta_2}{4\beta_4}\right) - \frac{3}{2}\beta_2\right)\left(\frac{\pi n_{z_{\min}}}{d_z}\right)^2 + 12\beta_4^2\right\}^{1/2} \end{aligned} \tag{2.136}$$

In this case, the hole statistics is given by (1.146).

2.2.17 The EP from QDs of bismuth telluride

In this case, the $v_z(E_{n_z})$ is given by

$$v_z(E_{n_z}) = \frac{\pi\hbar\alpha_{33}n_z}{d_z m_0(1+2\alpha E_{n_z})} \tag{2.137}$$

in which E_{n_z} should be determined from the following equation

$$E_{n_z}(1+\alpha E_{n_z}) = \frac{\hbar^2}{2m_0}\left(\alpha_{33}\left(\frac{\pi n_z}{d_z}\right)^2\right) \tag{2.138}$$

The photo-emitted current density assumes the form

$$J_{\mathrm{OD}} = \frac{\pi\alpha_0 e g_v \hbar\alpha_{33}}{m_0 d_x d_y d_z^2}\sum_{n_x=1}^{n_{x\max}}\sum_{n_y=1}^{n_{y\max}}\sum_{n_{z_{\min}}}^{n_{z\max}}\frac{n_z[1+\exp(\eta_{1,33})]^{-1}}{(1+2\alpha E_{n_z})} \tag{2.139}$$

where $n_{z_{\min}}$ should be the nearest integer of the following equation

$$(W-hv)[1+\alpha(W-hv)] = \frac{\hbar^2\alpha_{33}}{2m_0}[\pi n_{z_{\min}}/d_z]^2 \tag{2.140}$$

In this case, the hole statistics is given by (1.150).

2.2.18 The EP from QDs of antimony

For the QDs, whose bulk carrier dispersion law is described by (1.152), the $v_{z1}(E_{n_{z1}})$ can be expressed as

$$v_{z1}(E_{n_{z1}}) = \overline{\alpha_{33}}\frac{\hbar}{m_0}\times\frac{\pi n_{z1}}{d_z} \tag{2.141}$$

where

$$E_{n_{z1}} = \bar{\alpha}_{33}\frac{\hbar^2}{2m_0}\times\left(\frac{\pi n_{z1}}{d_z}\right)^2 \tag{2.142}$$

For the QDs, whose bulk carrier dispersion law is described by (1.152), the photo-emitted current density can be expressed as

$$J_{\mathrm{OD1}} = \frac{\alpha_0 e\hbar\pi\bar{\alpha}_{33}}{m_0\cdot d_x d_y d_z^2}\sum_{n_{x1}=1}^{n_{x1\max}}\sum_{n_{y1}=1}^{n_{y1\max}}\sum_{n_{z1_{\min}}}^{n_{z1\max}} n_{z1}F_{-1}(\eta_{1,34}) \tag{2.143}$$

where $n_{z1_{\min}}$ should be the nearest integer of the following equation

$$W - hv = \bar{\alpha}_{33}\frac{\hbar^2}{2m_0} \times \left(\frac{\pi n_{z1_{\min}}}{d_z}\right)^2 \tag{2.144}$$

Similarly for QDs, whose bulk carrier dispersion law is given by (1.153), the corresponding photocurrent density can, respectively, be expressed as

$$J_{0D2} = \frac{\pi\hbar\alpha_0 e a_3}{m_0 \cdot d_x d_y d_z^2} \sum_{n_{x2}=1}^{n_{x2_{\max}}} \sum_{n_{y2}=1}^{n_{y2_{\max}}} \sum_{n_{z2_{\min}}}^{n_{z2_{\max}}} n_{z2} F_{-1}(\eta_{64}) \tag{2.145}$$

The $n_{z2_{\min}}$ should be the nearest integer of the following equation

$$W - h\nu = a_3\left(\frac{\hbar^2}{2m_0}\right)\left(\frac{\pi n_{z2_{\min}}}{d_z}\right)^2 \tag{2.146}$$

Similarly for QDs, whose bulk carrier dispersion law is given by (1.154), the corresponding photocurrent density can, respectively, be written as

$$J_{0D3} = \frac{\pi\hbar\alpha_0 e a_3}{m_0 \cdot d_x d_y d_z^2} \sum_{n_{x3}=1}^{n_{x3_{\max}}} \sum_{n_{y3}=1}^{n_{y3_{\max}}} \sum_{n_{z3_{\min}}}^{n_{z3_{\max}}} n_{z3} F_{-1}(\eta_{65}) \tag{2.147}$$

The $n_{z2_{\min}}$ should be the nearest integer of the following equation

$$W - hv = a_3\left(\frac{\hbar^2}{2m_0}\right)\left(\frac{\pi n_{z3_{\min}}}{d_z}\right)^2 \tag{2.148}$$

Therefore, the respective simplified expression of the current density from QDs of antimony can be written as

$$J_{0D} = J_{0D1} + J_{0D2} + J_{0D3} \tag{2.149}$$

In this case, the electron statistics is given by (1.161).

2.3 Result and Discussions

Using (2.8) and (2.13) and the band constants from the Table 1.1, we have plotted the normalized photo-emitted current density from QDs of $CdGeAs_2$ as a function of film thickness as shown in plot (a) of Fig. 2.1, where the plot (b) indicates the case for $\delta = 0$. The plots (c) and (d) represent the photoemission in this case in accordance with three (using (2.18)) and two (using (2.20)) band models of Kane. In Figs. 2.2 and 2.3,

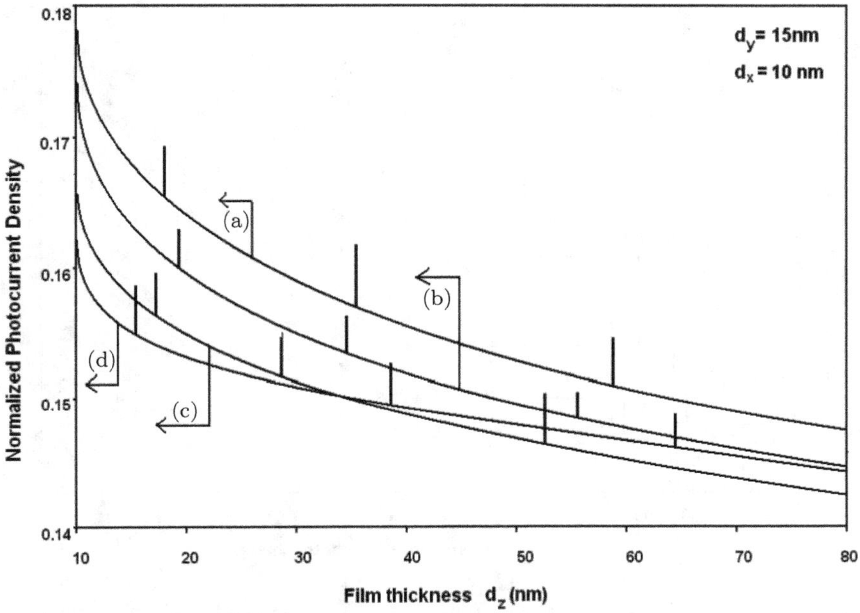

Fig. 2.1. Plot of the normalized J_{0D} from QDs of $CdGeAs_2$ as a function of d_z in accordance with (a) generalized band model (b) $\delta = 0$, (c) the three-band model of Kane and (d) the two band model of Kane.

the aforementioned variable has been plotted as functions of normalized incident photon energy and normalized electron degeneracy respectively for all cases of Fig. 2.1. In Fig. 2.4, the normalized photo-emitted current density from QDs of n-InSb has been plotted as a function of d_z in accordance with the (a) three band model of Kane, (b) two band model of Kane, the models of (c) Stillman *et al.* (using (2.26 and (1.28)), (d) Palik *et al.* (using (2.35 and (1.41)), (e) Johnson *et al.* (using (2.39) and (1.45)) and (f) Agafonov *et al.* (using (2.43) and (1.49)) respectively. The plots (g) and (h) refer to QDs of GaAs in accordance with the models of Newson *et al.* (using (2.30) and (1.32)) and Rossler (using (2.33) and (1.36)) respectively in the same figure. In Figs. 2.5 and 2.6, the normalized photoemission for all cases of Fig. 2.4 has been plotted as functions of normalized incident photon energy and the normalized electron degeneracy respectively. Figure 2.7 exhibits the normalized photo-emitted current density from QDs of CdS

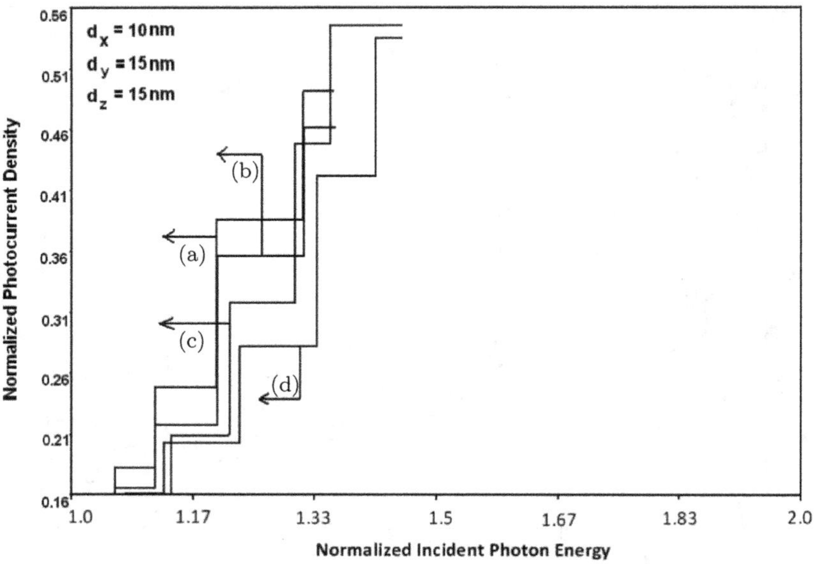

Fig. 2.2. Plot of the normalized J_{0D} from QDs of $CdGeAs_2$ as a function of normalized incident photon energy in accordance with the (a) generalized band model (b), (c) the three-band model of Kane and (d) the two band model of Kane.

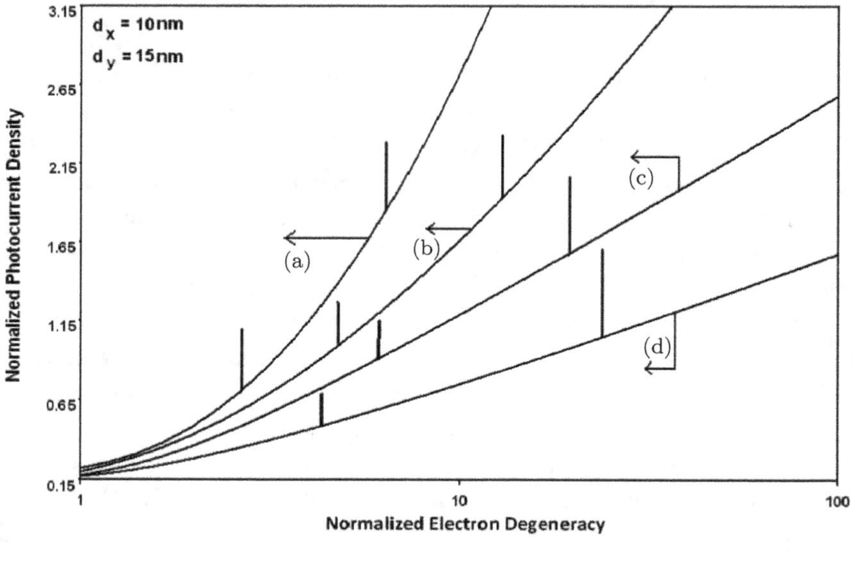

Fig. 2.3. Plot of the normalized J_{0D} from QDs of $CdGeAs_2$ as a function of normalized electron degeneracy in accordance with the (a) generalized band model (b), (c) the three-band model of Kane and (d) the two band model of Kane.

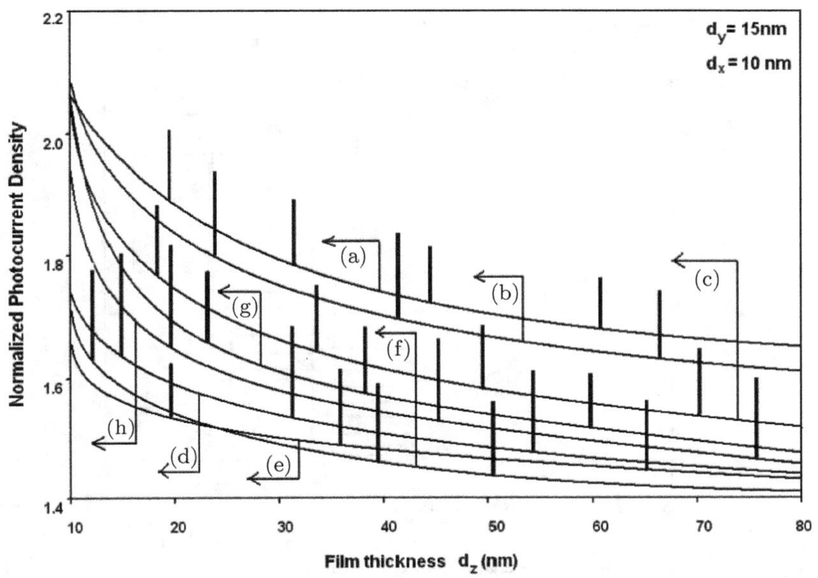

Fig. 2.4. Plot of the normalized J_{0D} as a function of d_z from QDs of n-InSb in accordance with the (a) three band model of Kane, (b) two band model of Kane, (c) model of Stillman *et al.* (d) model of Palik *et al.* (e) model of Johnson *et al.* (f) model of Agafonov *et al.* The plots (g) and (h) refer to QDs of GaAs in accordance with the models of Newson *et al.* and Rossler respectively.

(using (2.47) and (1.53)) as a function of d_z as shown in plot (a), in which the plot (b) is valid for $\bar{\lambda}_0 = 0$. The plot (c) in the same figure is valid for QDs of GaP and has been drawn by using (2.51) and (1.57) respectively. In Figs. 2.8 and 2.9, the aforementioned variable has been plotted as functions of normalized incident photon energy and normalized electron degeneracy respectively for all cases of Fig. 2.7.

Figure 2.10 exhibits the plot of the normalized photo-emitted current density as a function of film thickness from QDs of n-Ge in accordance with the models of (a) Cardona *et al.* (using (2.55) and (1.61) and (b) Wang *et al.* (using (2.59) and (1.65)) respectively. The curves (c) and (d) in the same figure refer to the photoemission from QDs of Tellurium in accordance with the models of Bouat *et al.* (using (2.62) and (1.69)) and Ortenberg *et al.* (using (2.65) and (1.73)) respectively. Besides, the curve (e) of the same figure has been plotted for QDs of Graphite (using (2.68)

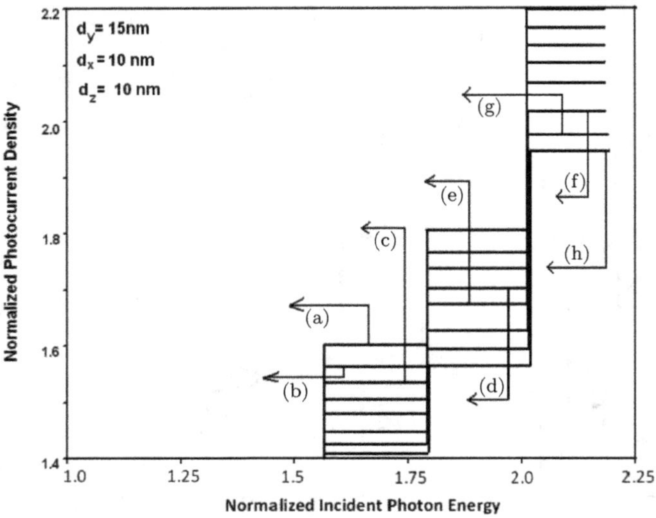

Fig. 2.5. Plot of the normalized J_{0D} from QDs of n-InSb and n-GaAs as a function of normalized incident photon energy in accordance with the (a) three band model of Kane, (b) two band model of Kane, (c) model of Stillman *et al.* (d) model of Palik *et al.* (e) model of Johnson *et al.* (f) model of Agafonov *et al.* The plots (g) and (h) refer to QDs of GaAs in accordance with the models of Newson *et al.* and Rossler respectively.

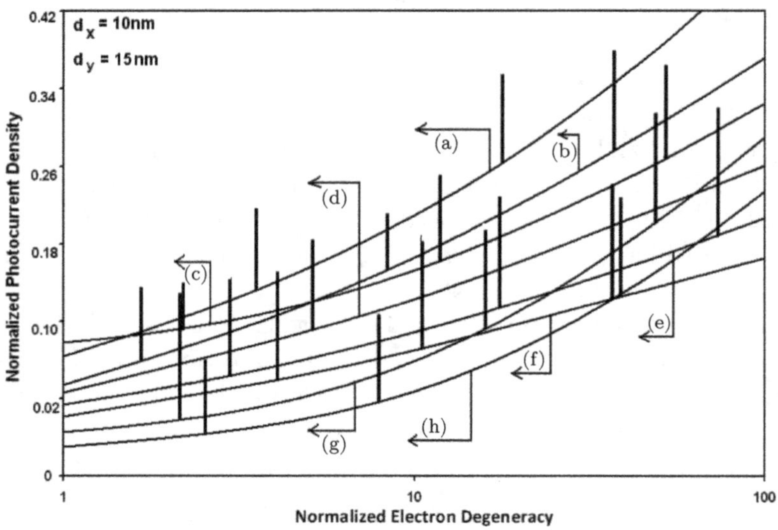

Fig. 2.6. Plot of the normalized J_{0D} from QDs of n-InSb and n-GaAs as a function of normalized electron degeneracy in accordance with the (a) three band model of Kane, (b) two band model of Kane, (c) model of Stillman *et al.* (d) model of Palik *et al.* (e) model of Johnson *et al.* (f) model of Agafonov *et al.* The plots (g) and (h) refer to QDs of GaAs in accordance with the models of Newson *et al.* and Rossler respectively.

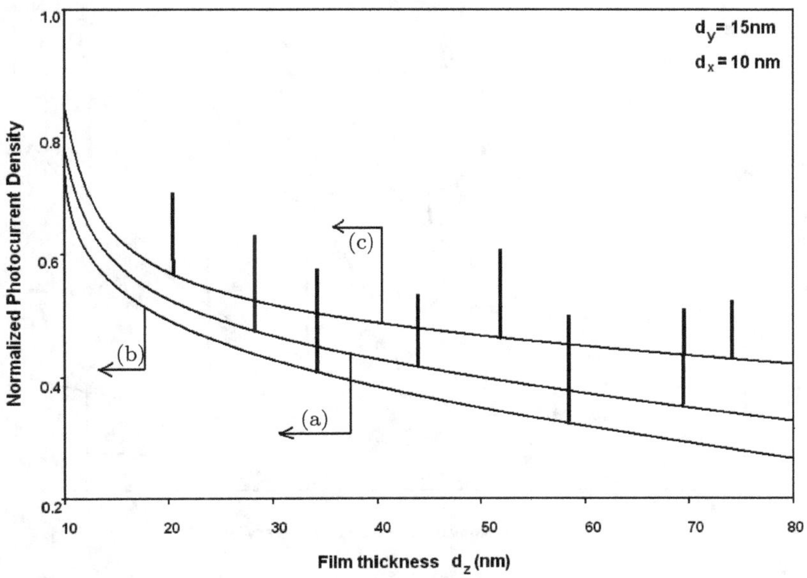

Fig. 2.7. Plot of the normalized J_{0D} from QDs of (a) CdS with $\bar{\lambda}_0 \neq 0$, (b) $\bar{\lambda}_0 = 0$ respectively and (c) GaP as a function of d_z.

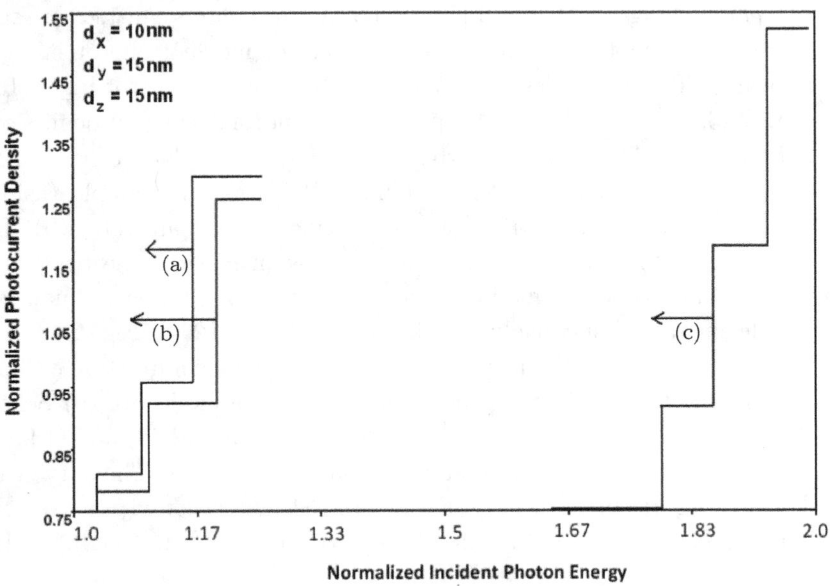

Fig. 2.8. Plot of the normalized J_{0D} from QDs of (a) CdS with $\bar{\lambda}_0 \neq 0$, (b) $\bar{\lambda}_0 = 0$ respectively and (c) GaP as a function of normalized incident photon energy.

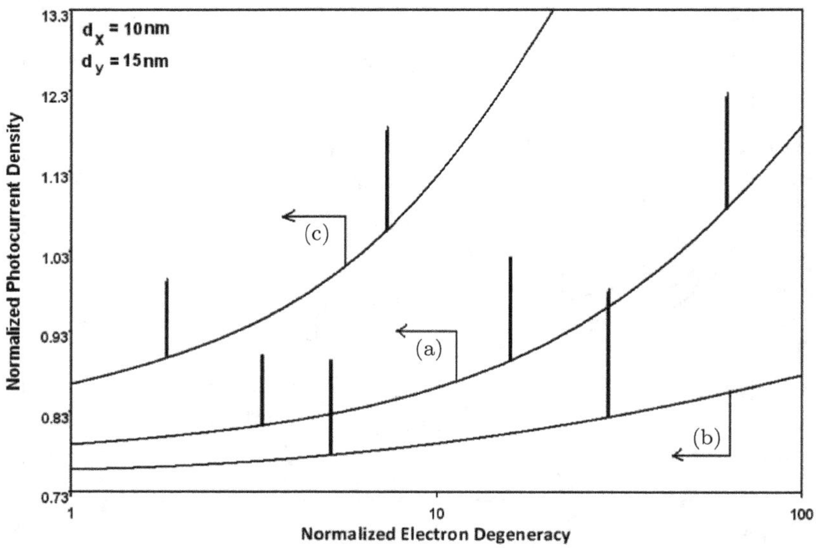

Fig. 2.9. Plot of the normalized J_{0D} from QDs of (a) CdS with $\bar{\lambda}_0 \neq 0$, (b) $\bar{\lambda}_0 = 0$ respectively and (c) GaP as a function of normalized electron degeneracy.

and (1.77)). In Figs. 2.11 and 2.12, the normalized photo-emitted current density has been plotted as functions of normalized incident photon energy and normalized electron degeneracy respectively for all cases of Fig. 2.10. In Fig. 2.13, we have drawn the plots of the normalized photoemission as a function of film thickness from QDs of (a) n-$PtSb_2$ (using (2.72) and (1.82)), (b) zero-gap materials, taking HgTe as an example (using (2.76) and (1.86)) and (c) $Pb_{1-x}Ge_xTe$ (using (2.131) and (1.142)). In Figs. 2.14 and 2.15, the normalized photoemission in this case has been plotted as functions of normalized incident photon energy and normalized carrier degeneracy respectively for all cases of Fig. 2.12. In Fig. 2.16, we have drawn the plots of the normalized photoemission as a function of film thickness from QDs of GaSb in accordance with the models of (a) Seiler *et al.* (using (2.84) and (1.94)), (b) Mathur *et al.* (using (2.88) and (1.99)) and (c) Zhang *et al.* (using (2.92) and (1.103)) respectively. The curve (d) in the same figure refers to the QDs of stressed materials, where stressed n-InSb has been considered as an example (using (2.96) and (1.107)). In

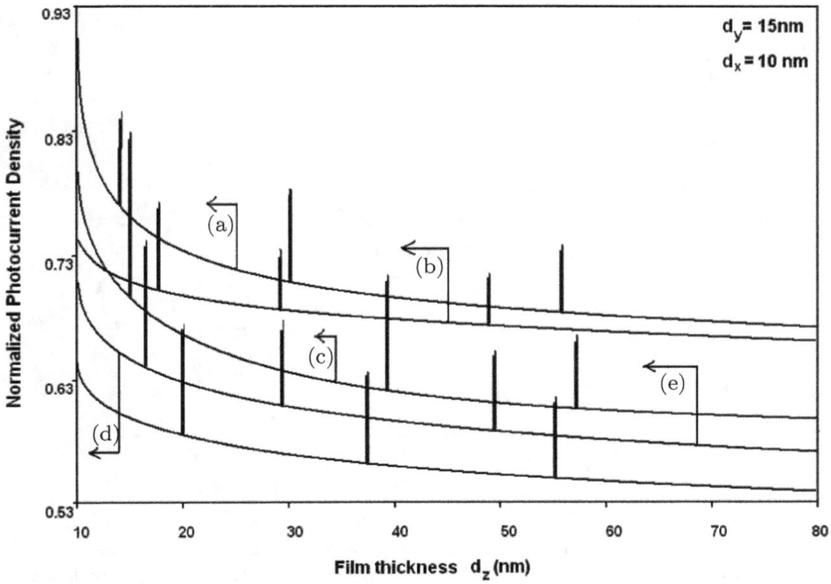

Fig. 2.10. Plot of the normalized J_{0D} from QDs of (a) n-Ge in accordance with the model of Cardona *et al.*, (b) n-Ge in accordance with the model of Wang *et al.* as a function of film thickness The curves (c) and (d) refer to the same plot for QDs of Tellurium in accordance with the model of Bouat *et al.* and Ortenberg *et al.* respectively. The curve (e) stands for the QDs of Graphite.

Figs. 2.17 and 2.18, the normalized photocurrent density in this case has been plotted as functions of normalized incident photon energy and the normalized electron degeneracy respectively for all cases of Fig. 2.16.

Figure 2.19 exhibits the plot of the normalized photocurrent density as a function of film thickness from QDs of Bismuth in accordance with the models of (a) McClure and Choi (using (2.100) and (1.111)), (b) the Cohen (using (2.107) and (1.119)), (c) the Hybrid (using (2.103b) and (1.115)) and (d) the Lax (using (2.111) and (1.123)) respectively. In Figs. 2.20 and 2.21, the same variable has been plotted as functions of normalized incident photon energy and normalized electron degeneracy respectively for all cases of Fig. 2.19. Figure 2.22 depicts the plot of the normalized photoemission current density as a function of film thickness from QDs of IV–VI materials taking n-PbTe as an example in accordance with the

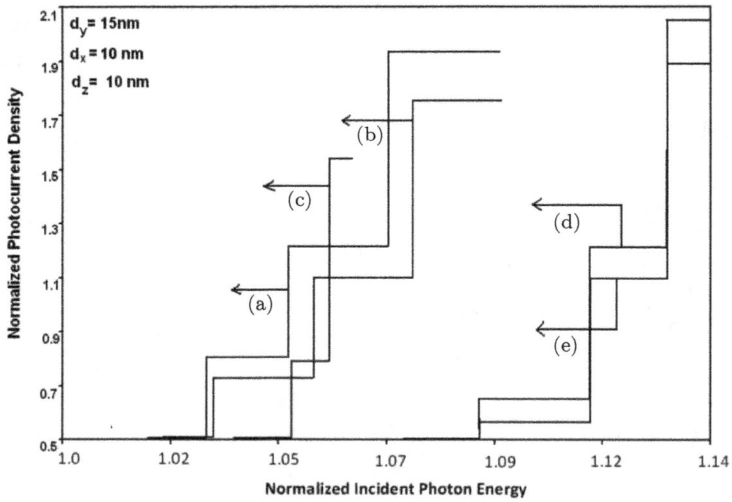

Fig. 2.11. Plot of the normalized J_{0D} as a function of normalized incident photon energy in accordance with the model of Wang *et al.* as a function of film thickness The curves (c) and (d) refer to the same plot for QDs of Tellurium in accordance with the model of Bouat *et al.* and Ortenberg *et al.* respectively. The curve (e) stands for the QDs of Graphite.

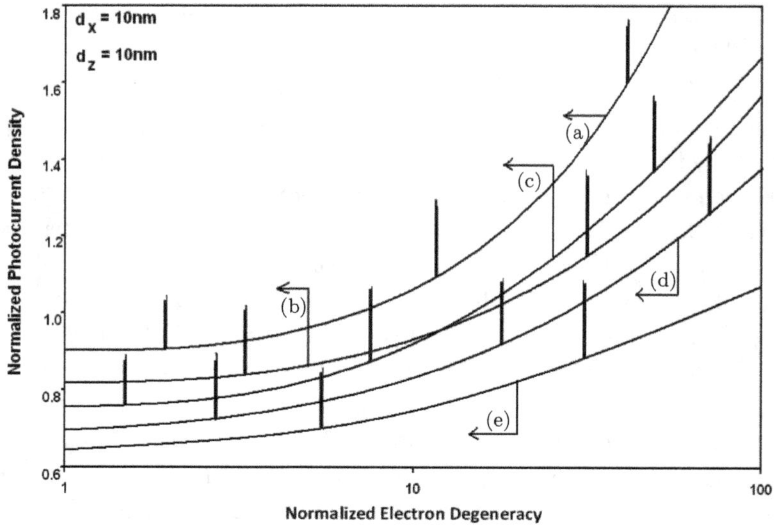

Fig. 2.12. Plot of the normalized J_{0D} as a function of normalized electron degeneracy in accordance with the model of Wang *et al.* as a function of film thickness The curves (c) and (d) refer to the same plot for QDs of Tellurium in accordance with the model of Bouat *et al.* and Ortenberg *et al.* respectively. The curve (e) stands for the QDs of Graphite" respectively.

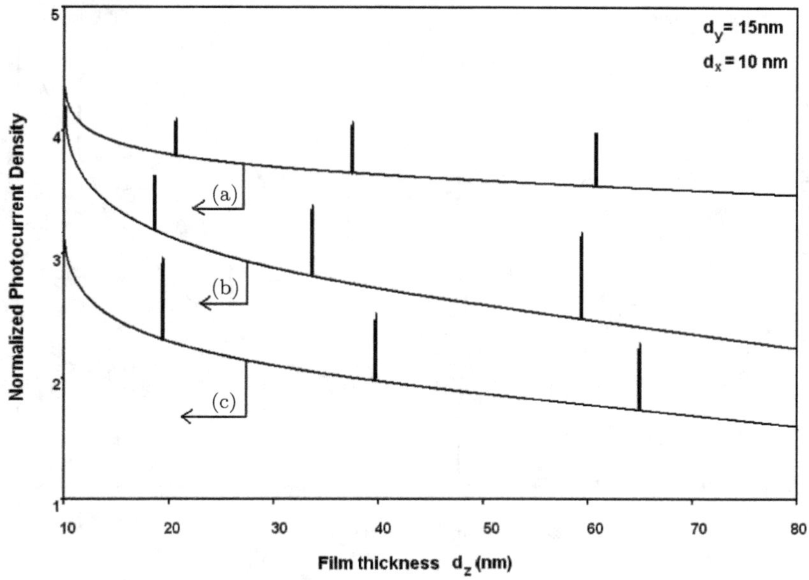

Fig. 2.13. Plot of the normalized J_{0D} from QDs of (a) $PtSb_2$, (b) HgTe and (c) $Pb_{1-x}Ge_xTe$ as a function of film thickness.

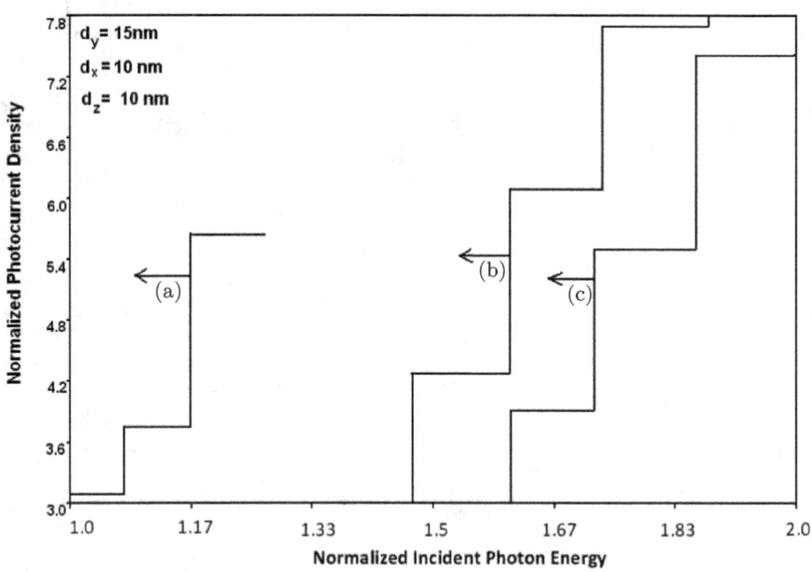

Fig. 2.14. Plot of the normalized J_{0D} from QDs of (a) $PtSb_2$, (b) HgTe and (c) $Pb_{1-x}Ge_xTe$ as a function of normalized incident photon energy.

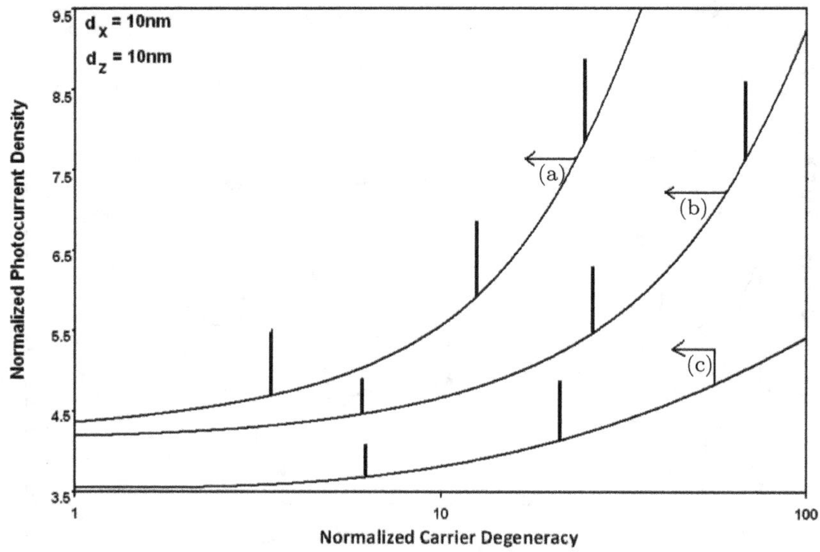

Fig. 2.15. Plot of the normalized J_{0D} from QDs of (a) $PtSb_2$, (b) HgTe and (c) $Pb_{1-x}Ge_xTe$ as a function of normalized carrier degeneracy.

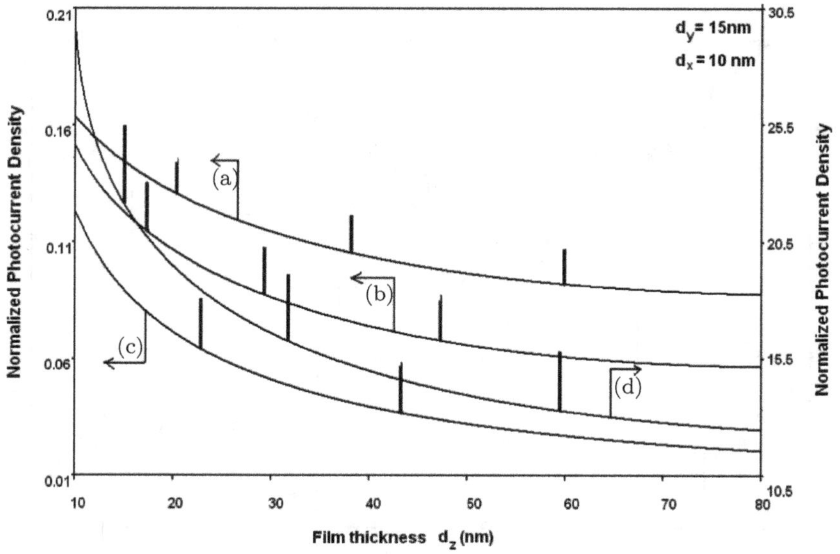

Fig. 2.16. Plot of the normalized J_{0D} from QDs of GaSb in accordance with the models of (a) Seiler *et al.* (b) Mathur *et al.* and (c) Zhang *et al.* respectively as a function of film thickness. The plot (d) refers to QDs of stressed n-InSb.

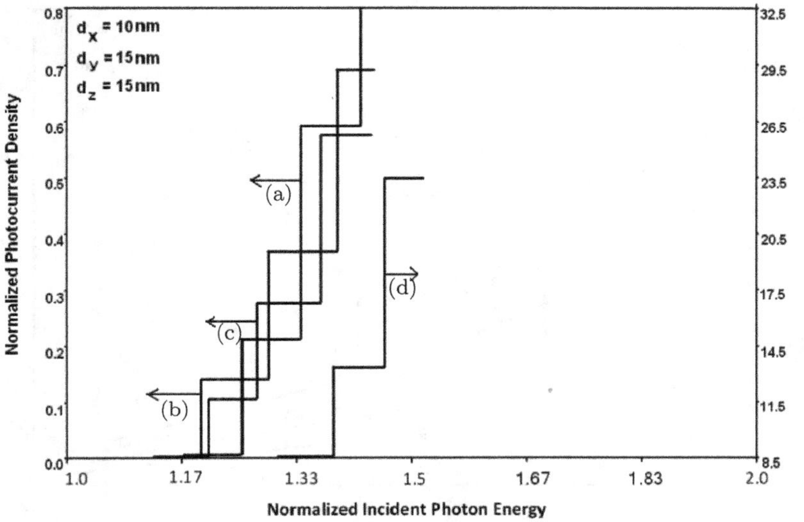

Fig. 2.17. Plot of the normalized J_{0D} as a function of normalized incident photon energy in accordance with the models of (a) Seiler *et al.*, (b) Mathur *et al.* and (c) Zhang *et al.* respectively as a function of film thickness. The plot (d) refers to QDs of stressed n-InSb.

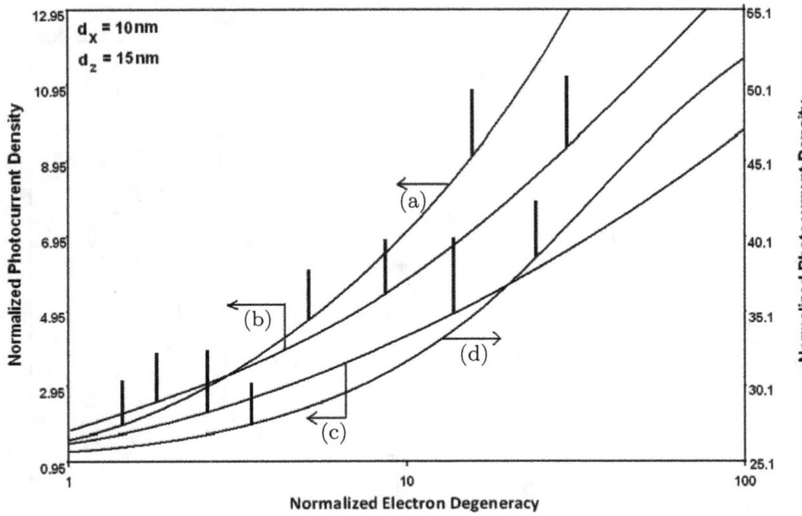

Fig. 2.18. Plot of the normalized J_{0D} as a function of normalized electron degeneracy in accordance with the models of (a) Seiler *et al.*, (b) Mathur *et al.* and (c) Zhang *et al.* respectively as a function of film thickness. The plot (d) refers to QDs of stressed n-InSb.

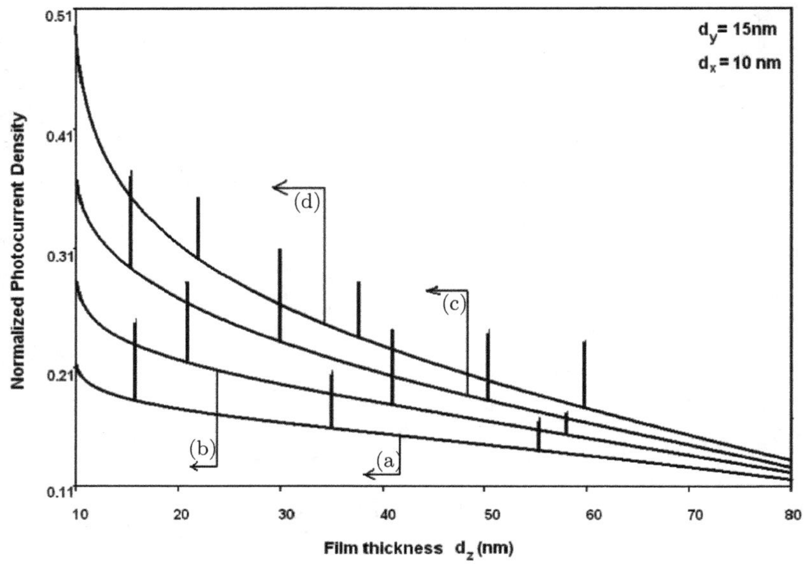

Fig. 2.19. Plot of the normalized $J_{0\mathrm{D}}$ as a function of film thickness from QDs of Bismuth in accordance with the models of (a) McClure and Choi, (b) the Cohen, (c) the Hybrid and (d) the Lax respectively.

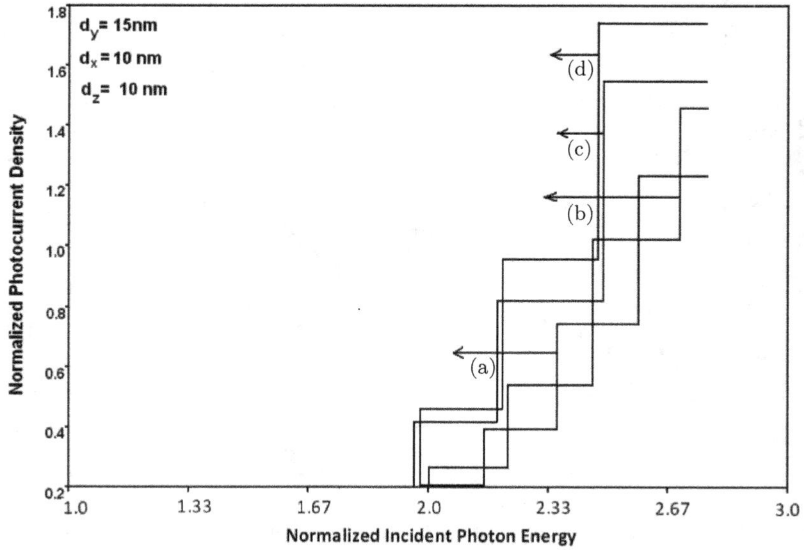

Fig. 2.20. Plot of the normalized $J_{0\mathrm{D}}$ as a function of normalized incident photon energy in accordance with models of (a) McClure and Choi, (b) the Cohen, (c) the Hybrid and (d) the Lax respectively.

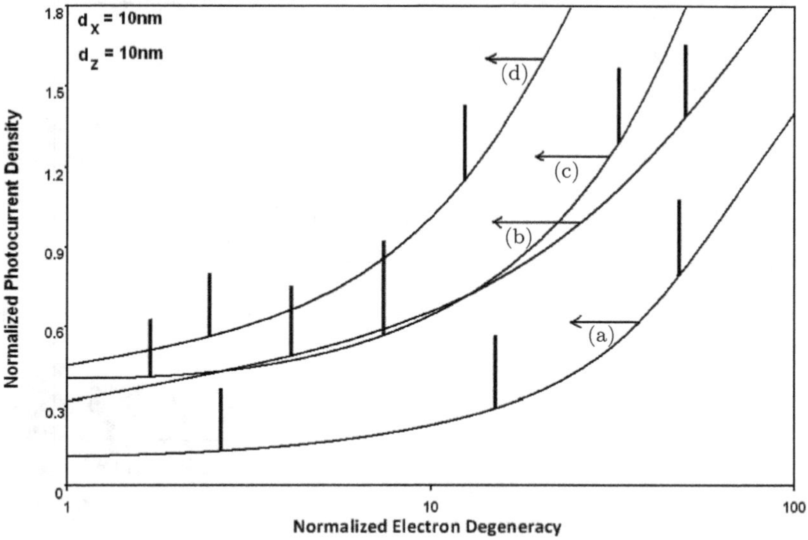

Fig. 2.21. Plot of the normalized J_{0D} as a function of normalized electron degeneracy in accordance with the model of Johnson *et al.*

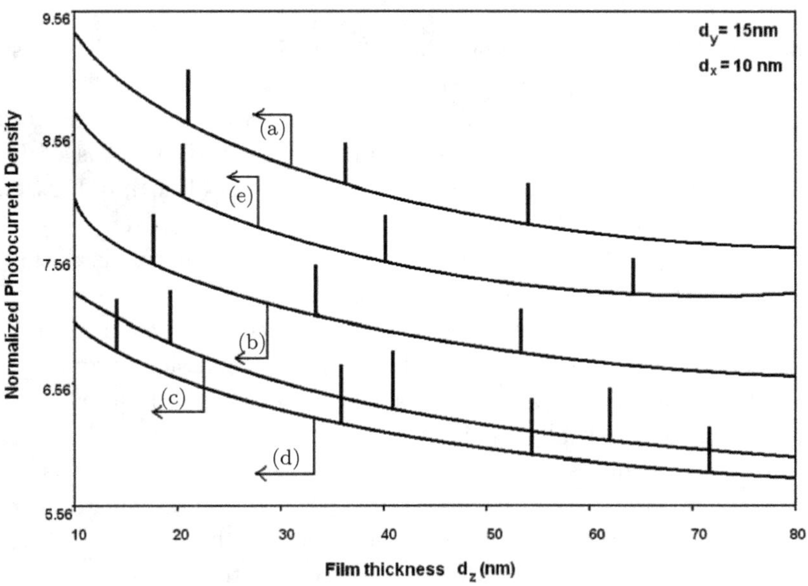

Fig. 2.22. Plot of the normalized J_{0D} from QDs of PbTe in accordance with the models of (a) Dimmock (b) Cohen, (c) Bangert *et al.* and (d) Foley *et al.* respectively as a function of film thickness. The plot (e) refers to QDs of CdSb.

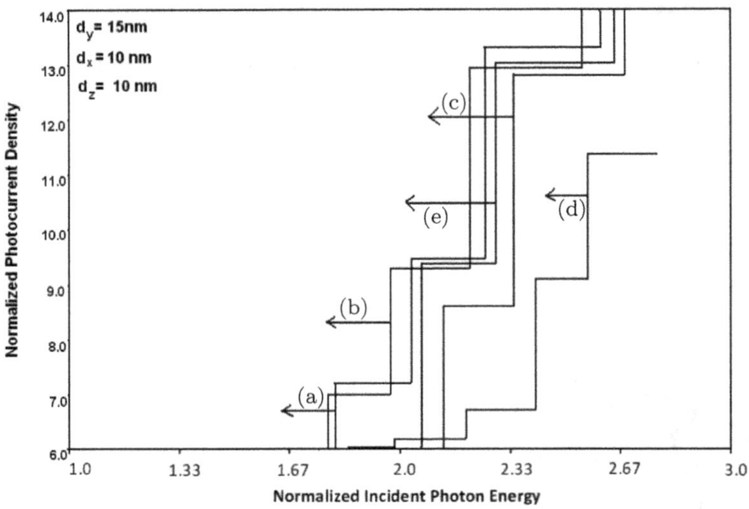

Fig. 2.23. Plot of the normalized J_{0D} as a function of normalized incident photon energy in accordance with the models of (a) Dimmock (b) Cohen, (c) Bangert *et al.* and (d) Foley *et al.* respectively as a function of film thickness. The plot (e) refers to QDs of CdSb.

models of (a) Dimmock (using (2.119) and (1.130)), (b) Cohen (using (2.107) and (1.119)), (c) Bangert *et al.* (using (2.123) and (1.134)), (d) Foley *et al.* (using (2.127) and (1.138)) respectively. Besides, the plot (e) of Fig. 2.22 exhibits the photoemission current density from QDs of II–V materials taking CdSb as an example (using (2.80) and (1.90)). In Figs. 2.23 and 2.24, the normalized photoemission in this case has been plotted as functions of normalized incident photon energy and normalized carrier degeneracy respectively for all cases of Fig. 2.22.

Finally, we can write that it is the band structure which changes in a fundamental way and consequently alters the photoemission together with the fact that in general all the physical properties of all the electronic materials changes radically leading to new physical concepts.

Figure 2.25 exhibits the plot of the normalized photoemission current density as a function of film thickness from QDs of (a) cadmium and (b) zinc diphosphides respectively which has been obtained by using (2.135) and (1.146). In Figs. 2.26 and 2.27, the normalized photoemission in this case has been plotted as functions of normalized photon energy and normalized

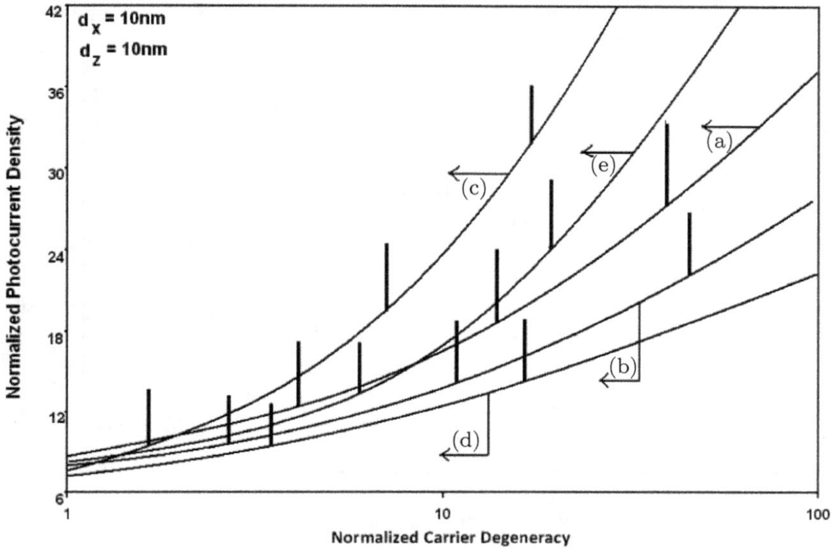

Fig. 2.24. Plot of the normalized J_{0D} as a function of normalized carrier degeneracy in accordance with the models of (a) Dimmock (b) Cohen, (c) Bangert *et al.* and (d) Foley *et al.* respectively as a function of film thickness. The plot (e) refers to QDs of CdSb.

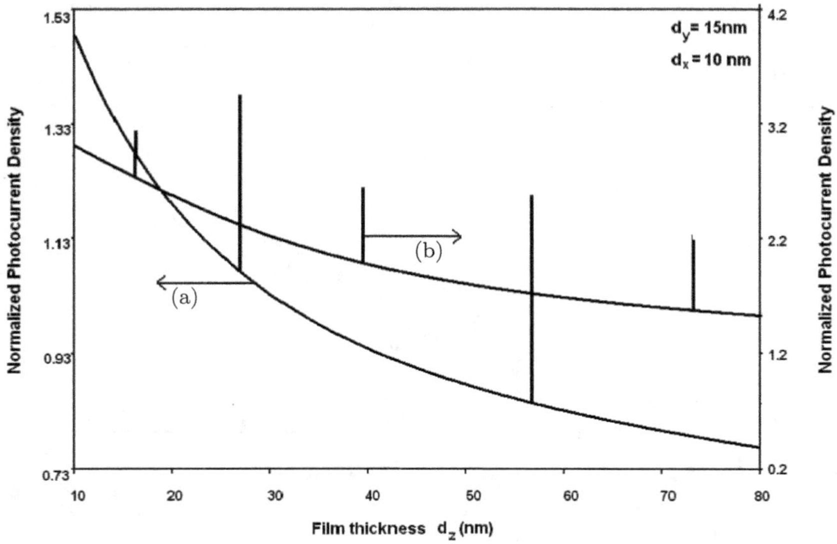

Fig. 2.25. Plot of the normalized J_{0D} from QDs of (a) cadmium and (b) zinc diphosphides as a function of film thickness.

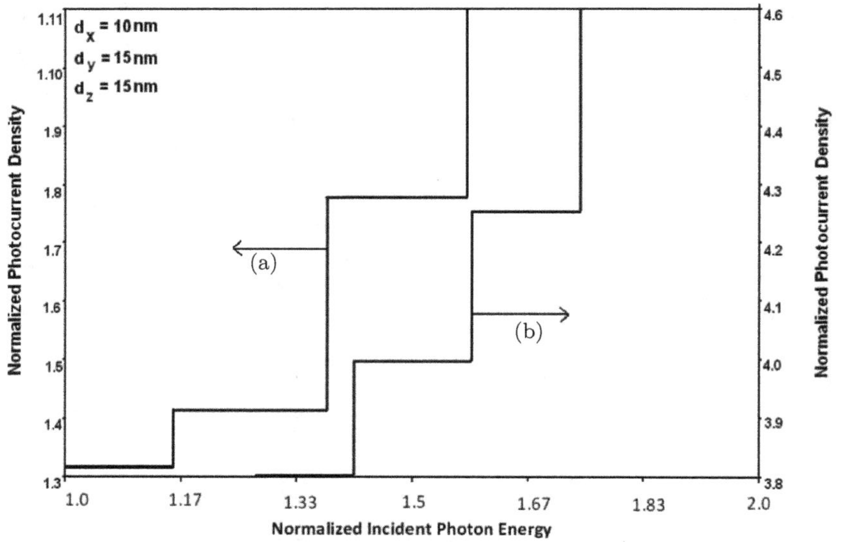

Fig. 2.26. Plot of the normalized $J_{0\text{D}}$ from QDs of (a) cadmium and (b) zinc diphosphides as a function of normalized incident photon energy.

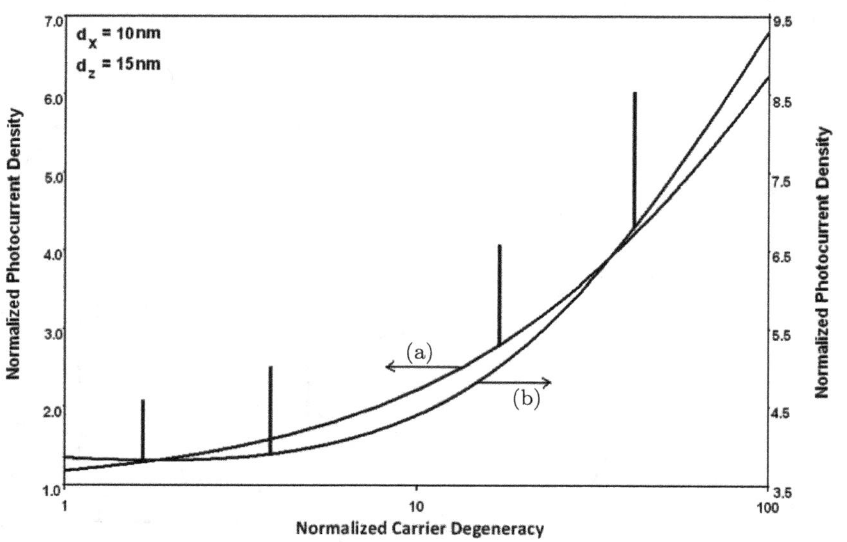

Fig. 2.27. Plot of the normalized $J_{0\text{D}}$ from QDs of (a) Cadmium and (b) zinc diphosphides as a function of normalized carrier degeneracy.

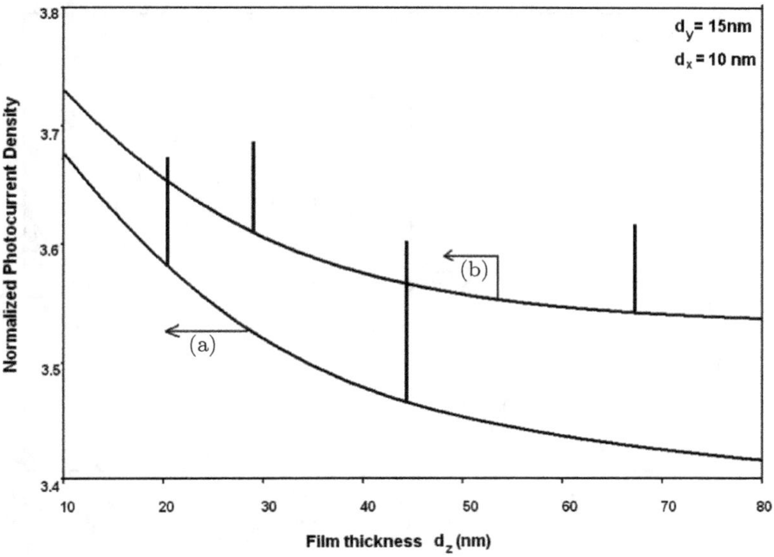

Fig. 2.28. Plot of the normalized J_{0D} from QDs of (a) Bi_2Te_3 and (b) Sb as a function of film thickness.

carrier degeneracy respectively for all cases of Fig. 2.25. Figure 2.28 shows the plot of the normalized photoemission as a function of film thickness from QDs of (a) Bi_2Te_3 (using (2.139) and (1.150)) and (b) Sb (using (2.149) and (1.161)) respectively. In Figs. 2.29 and 2.30, the normalized photoemission in this case has been plotted as functions of normalized incident photon energy and normalized carrier degeneracy respectively for all cases of Fig. 2.28.

It appears from Figs. 2.1, 2.4, 2.7, 2.10, 2.13, 2.16, 2.19, 2.22, 2.25 and 2.28 that the normalized photo-emitted current density increases with decreasing film thickness and exhibits spikes for various values of d_z which are totally band structure dependent. Figures 2.2, 2.5, 2.8, 2.11, 2.14, 2.17, 2.20, 2.23, 2.26 and 2.29 exhibit the step-functional dependence of the normalized photo-emitted current density from the QDs of different materials with the incident photon energy. It is apparent from Figs. 2.3, 2.6, 2.9, 2.12, 2.15, 2.18, 2.21, 2.24, 2.27 and 2.30 that the normalized photo-emitted current density from the QDs of various materials increases with increasing normalized carrier degeneracy and exhibits spikes for different

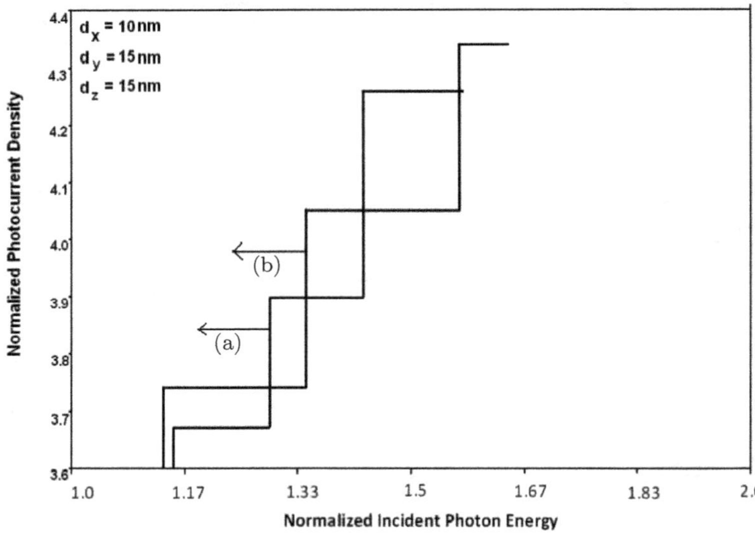

Fig. 2.29. Plot of the normalized J_{0D} from QDs of (a) Bi_2Te_3 and (b) Sb as a function of normalized incident photon energy.

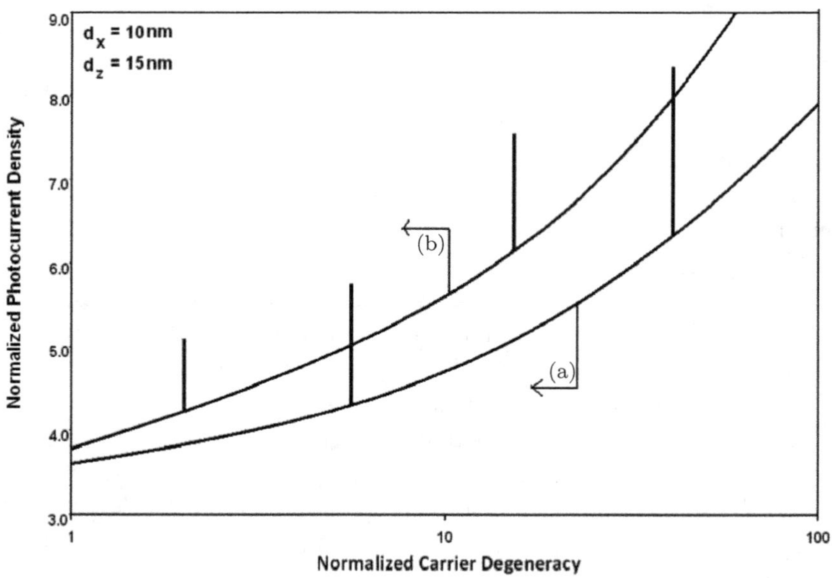

Fig. 2.30. Plot of the normalized J_{0D} from QDs of (a) Bi_2Te_3 and (b) Sb as a function of normalized carrier degeneracy.

values of carrier concentration which are again band structure dependent. It may be noted that the QDs lead to the discrete energy levels, somewhat like atomic energy levels, which produce very large changes. This follows from the inherent nature of the quantum confinement of the carrier gas dealt with here. In QDs, there remain no free carrier states in between any two allowed sets of size-quantized levels unlike that found for QWs in UFs and QWWs where the quantum confinements are 1D and 2D, respectively. Consequently, the crossing of the Fermi level by the size-quantized levels in QDs would have much greater impact on the redistribution of the carriers among the allowed levels, as compared to that found for QWs in UFs and QWWs respectively. Although the photoemission varies in various manners with all the variables in all the limiting cases as evident from all the figures, the rates of variations are totally band-structure dependent. The quantum signature of QDs for Einstein's photo-electric effect is rather prominent as compared to the same from QWs in UFs and QWWs. Finally, we can write that it is the band structure which changes in a fundamental way and consequently alters the photoemission together with the fact that in general all the physical properties of all the electronic materials changes radically leading to new physical concepts.

2.4 Open Research Problems

(R.2.1) Investigate the EP for QDs in the presence of three finite potential wells and three parabolic potential wells applied separately in the three orthogonal directions for all the materials whose unperturbed carrier energy spectra are defined in this chapter.

(R.2.2) Investigate the EP for QDs in the presence of three finite potential wells and three parabolic potential wells applied separately in the three orthogonal directions under an arbitrarily oriented (a) non-uniform electric field and (b) alternating electric field respectively for all the materials whose unperturbed carrier energy spectra are defined in this chapter.

(R.2.3) Investigate the EP in the presence of three finite potential wells and three parabolic potential wells applied separately in the three orthogonal directions under an arbitrarily oriented alternating magnetic field by including broadening and the electron spin for all the materials whose unperturbed carrier energy spectra are defined in this chapter.

(R.2.4) Investigate the EP by considering the presence of three finite potential wells and three parabolic potential wells applied separately in the three orthogonal directions under an arbitrarily oriented alternating magnetic field and crossed alternating electric field by including broadening and the electron spin for all the materials whose unperturbed carrier energy spectra are defined in this chapter.

(R.2.5) Investigate the EP in the presence of three finite potential wells and three parabolic potential wells applied separately in the three orthogonal directions under an arbitrarily oriented alternating magnetic field and crossed alternating non-uniform electric field by including broadening and the electron spin whose for all the materials unperturbed carrier energy spectra are defined in this chapter.

(R.2.6) Investigate the EP for QDs by considering all types of scattering mechanisms in the presence of three finite potential wells and three parabolic potential wells applied separately in the three orthogonal directions under exponential, Kane, Halperin, Lax and Bonch-Bruevich band tails [125] for all the materials whose unperturbed carrier energy spectra are defined in this chapter.

(R.2.7) Investigate the EP in the presence of three finite potential wells and three parabolic potential wells applied separately in the three orthogonal directions for all the materials as defined in Chapter 1 under an arbitrarily oriented (a) non-uniform electric field and (b) alternating electric field respectively whose unperturbed carrier energy spectra are defined in this chapter.

(R.2.8) Investigate the EP in the presence of three finite potential wells and three parabolic potential wells applied separately in the three orthogonal directions for all the materials as described in Chapter 1 under an arbitrarily oriented alternating magnetic field by including broadening and the electron spin whose unperturbed carrier energy spectra are defined in this chapter.

(R.2.9) Investigate the EP in the presence of three finite potential wells and three parabolic potential wells applied separately in the three orthogonal directions as discussed in Chapter 1 under an arbitrarily oriented alternating magnetic field and crossed alternating electric field by including broadening and the electron spin for all the materials whose unperturbed carrier energy spectra are defined in this chapter.

(R.2.10) Investigate all the appropriate problems of this chapter for QDs of the following materials:

(a) The energy spectrum of the valance bands of CuCl as given in Chapter 1.
(b) In the presence of stress, χ_6 along $\langle 001 \rangle$ and $\langle 111 \rangle$ directions, the energy spectra of the holes in semiconductors having diamond structure valance bands as given in Chapter 1.

The dispersion relation of the holes in *p*-InSb is given Chapter 1.

(R.2.11) Investigate all the appropriate problems of this chapter after proper modifications for wedge shaped, cylindrical, ellipsoidal, conical, triangular, colloidal, pyramidal, circular, lateral parabolic rotational, parabolic cylindrical and position-controlled QDs respectively.

(R.2.12) Investigate all the problems of (R.2.11) in the presence of strain after proper modifications.

(R.2.13) Investigate all the appropriate problems of this chapter for quantum rods.

(R.2.14) Investigate the influence of deep traps and surface states separately for all the appropriate problems of this chapter after proper modifications.

(R.2.15) Investigate the influence of the localization of carriers for all the appropriate problems of this chapter after proper modifications.

References

[1] M. Cardona, L. Ley, *Photoemission in Solids 1 and 2, Topics in Applied Physics*, Vols. 26, 27, (Springer-Verlag, Germany, **1978**); D. J. Lockwood, *Light Emission in Silicon in Silicon based Materials and Devices*, Vol. 2, Ed. H. S. Nalwa, (Academic Press, San Diego, USA, **2001**).

[2] R. K. Pathria, *Statistical Mechanics*, 2nd ed. (Butterworth-Heinemann, Oxford, **1996**).

[3] K. P. Ghatak, S. Bhattacharya, K. M. Singh, S. Choudhury, S. Pahari, *Phys. B*, 403, 2116 **(2008)**.

[4] K. P. Ghatak, S. N. Biswas, in *Proceedings of the Society of Photo-Optical and Instrumentation Engineers (SPIE)*, Nonlinear Optics II, USA, Vol. 1409, p. 28 **(1991)**.

[5] K. P. Ghatak, *SPIE, Process Module Metrology*, USA, Vol. 1594, p. 110 (**1992**); K. P. Ghatak, *SPIE, International Conference on the Application and Theory of Periodic Structures*, Vol. 1545, p. 282 (**1991**).
[6] K. P. Ghatak, M. Mondal, *Sol. Stat. Electron.*, 31, 1561 (**1988**).
[7] K. P. Ghatak, M. Mondal, *J. Appl. Phys.*, 69, 1666 (**1991**).
[8] K. P. Ghatak, D. Bhattacharyya, B. Nag, S. N. Biswas, *J. Nonlin. Opt. Quant. Opt.*, 13, 267 (**1995**).
[9] M. Mondal, S. Banik, K. P. Ghatak, *J. Low Temp. Phys.*, 74, 423 (**1989**).
[10] B. Mitra, A. Ghoshal, K. P. Ghatak, *Phys. Stat. Sol. (b)*, 150, K67 (**1988**).
[11] B. Mitra, K. P. Ghatak, *Phys. Scr.*, 40, 776 (**1989**).
[12] K. P. Ghatak, M. Mondal, S. N. Biswas, *J. Appl. Phys.*, 68, 3032 (**1990**).
[13] K. P. Ghatak, S. N. Biswas, *Nonlinear Opt.*, 4, 39 (**1993**).
[14] K. P. Ghatak, S. N. Biswas, *SPIE, Growth and Characterization of Materials for Infrared Detectors and Nonlinear Optical Switches*, USA, Vol. 1484, p. 136 (**1991**).
[15] K. P. Ghatak, B. De, *Polymeric Materials for Integrated Optics and Information Storage, Materials Research Society (MRS) Symposium Proceedings*, MRS Spring Meeting, Vol. 228, p. 237 (**1991**).
[16] K. P. Ghatak, B. Nag, G. Majumdar, *Strained Layer Expitaxy – Materials, Processing, and Device Applications, MRS Symposium Proceedings*, MRS Spring Meeting, Vol. 379, p. 85 (**1995**).
[17] K. P. Ghatak, *SPIE, High Speed Phenomena in Photonic Materials and Optical Bistability*, USA, Vol. 1280, p. 53 (**1990**).
[18] K. P. Ghatak, *Long Wave Length Semiconductor Devices, Materials and Processes Symposium Proceedings, MRS Symposium Proceedings*, MRS Spring Meeting, Vol. 216, p. 469 (**1990**).
[19] K. P. Ghatak, A. Ghoshal, S. Bhattacharyya, *SPIE, Nonlinear Optical Materials and Devices for Photonic Switching*, USA, Vol. 1216, p. 282 (**1990**).
[20] K. P. Ghatak, *SPIE, Nonlinear Optics III*, USA, Vol. 1626, p. 115 (**1992**).
[21] K. P. Ghatak, A. Ghoshal, B. De, *SPIE, Optoelectronic Devices and Applications*, USA, Vol. 1338, p. 111 (**1990**).
[22] R. Houdré, C. Hermann, G. Lampel, P. M. Frijlink, *Surface Sci.*, 168, 538 (**1986**).
[23] T. C. Chiang, R. Ludeke, D. E. Eastman, *Phys. Rev. B*, 25, 6518 (**1982**).
[24] S. P. Svensson, J. Kanski, T. G. Andersson, P. O. Nilsson, *J. Vac. Sci. Technol. B*, 2, 235 (**1984**); S. F. Alvarado, F. Ciccacci, M. Campagna, *Appl. Phys. Lett.*, 39, 615 (**1981**).
[25] C. Majumdar, A. B. Maity, A. N. Chakravarti, *Phys. Stat. Sol. (b)*, 140, K7 (**1987**).
[26] C. Majumdar, A. B. Maity, A. N. Chakravarti, *Phys. Stat. Sol. (b)*, 141, K35 (**1987**).
[27] N. R. Das, K. K. Ghosh, D. Ghoshal, *Phys. Stat. Sol. (b)*, 197, 97 (**1996**).

[28] C. Majumdar, A. B. Maity, A. N. Chakravarti, *Phys. Stat. Sol. (b)*, 144, K13 **(1987)**.
[29] N. R. Das, A. N. Chakravarti, *Phys. Stat. Sol. (b)*, 176, 335 **(1993)**.
[30] S. Sen, N. R. Das and A. N. Chakravarti, *J. Phys. Conden. Mat.*, 19, 186205 **(2007)**; N. R. Das, S. Ghosh, A. N. Chakravarti, *Phys. Stat. Sol. (b)*, 174, 45 **(1992)**.
[31] A. B. Maity, C. Majumdar, A. N. Chakravarti, *Phys. Stat. Sol. (b)*, 144, K93 **(1987)**.
[32] A. B. Maity, C. Majumdar, A. N. Chakravarti, *Phys. Stat. Sol. (b)*, 149, 565 **(1988)**.
[33] N. R. Das, A. N. Chakravarti, *Phys. Stat. Sol. (b)*, 169, 97 **(1992)**.
[34] A. Modinos, *Field, Thermionic and Secondary Electron Emission Spectroscopy*, (Plenum Press, New York, **1984**).
[35] A. V. D. Ziel, *Solid State Physical Electronics*, (Prentice-Hall, Inc., Englewood Cliffs, **1957**).
[36] J. L. Shay, J. W. Wernick, *Ternary Chalcopyrite Semiconductors-Growth, Electronic Properties and Applications*, (Pergamon Press, London, **1975**).
[37] J. E. Rowe, J. L. Shay, *Phys. Rev. B*, 3, 451 **(1973)**.
[38] H. Kildal, *Phys. Rev. B*, 10, 5082 **(1974)**.
[39] J. Bodnar, in *Proceedings of the International Conference on Physics of Narrow-gap Semiconductors*, (Polish Science Publishers, Warsaw, **1978**).
[40] G. P. Chuiko, N. N. Chuiko, *Sov. Phys. Semicond.*, 15, 739 **(1981)**.
[41] K. P. Ghatak and S. N. Biswas, *Proc. SPIE*, 1484, 149 **(1991)**.

Chapter 3

The Activity Coefficient in Doping Superlattices of Non-Parabolic Materials Under Magnetic Quantization

Every day may not be good, but there is something good in everyday.

3.1 Introduction

With the advent of modern experimental techniques of fabricating nanomaterials, it is possible to grow semiconductor superlattices (SLs) composed of alternative layers of two different degenerate layers with controlled thickness [1]. These structures have found wide applications in many new devices such as photodiodes [2], photoresistors [3], transistors [4], light emitters [5], tunneling devices [6], etc [7–18]. The investigations of the physical properties of narrow gap SLs have increased extensively, since they are important for optoelectronic devices and also since the quality of heterostuctures involving narrow gap materials has been greatly improved. It may be noted that the doping superlattices, also called NIPI structures, are crystals with a periodic sequence of ultrathin film layers [19–29] of the same semiconductor with the intrinsic layer in between together with the opposite sign of doping. All the donors will be positively charged and all the acceptors negatively. This periodic space charge causes a periodic space charge potential which quantizes the motions of the carriers in the z-direction together with the formation of the sub-band energies. The electronic structures of the doping superlattices differ radically from the corresponding bulk semiconductors as stated as follows:

(a) each band is split into mini-bands
(b) the magnitude and the spacing of these minibands may be designed by the choice of the superlattice's parameters and

(c) the electron energy spectrum of the doping superlattices becomes two-dimensional leading to the step functional dependence of the density-of-states function.

The importance of studying the activity coefficient (AC) has already been written in Section 1.3.10 and the applications of the quantizing magnetic field for investigating the physical properties of quantized materials have also been stated in the preface. In Section 3.2.1, of the theoretical background, the AC under magnetic quantization in doping superlattices of tetragonal materials has been investigated. Section 3.2.2 contains the results for doping superlattices of III–V, ternary and quaternary compounds in accordance with the three and the two band models of Kane together with parabolic energy bands and they form the special cases of Section 3.2.1. Sections 3.2.3–3.2.5 contain the study of the magneto AC for doping superlattices of II–VI, IV–VI and stressed Kane type semiconductors respectively. Section 3.3 contains the discussion of this chapter. Besides, the last Section 3.4 contains few open research problems in this context.

3.2 Theoretical Background

3.2.1 The AC in doping superlattices of nonlinear optical and tetragonal materials under magnetic quantization

The Dispersion Relation (DR) of the conduction electrons in doping superlattices of nonlinear optical materials can be expressed by using (2.2) as

$$\gamma(E) = f_1(E)k_s^2 + f_2(E)\left(n_i + \frac{1}{2}\right)\frac{2m_{\|}^*}{\hbar}\omega_8(E) \tag{3.1}$$

where

$$\omega_8(E) \equiv \left(\frac{n_0|e|^2}{d_0\varepsilon_{sc}[\theta_1(E)]}\right)^{1/2} \quad \text{and}$$

$$\theta_1(E) \equiv \frac{\hbar^2}{2}\left\{\frac{f_2(E)[\psi_1(E)]' - \gamma(E)[f_2(E)]'}{[f_2(E)]^2}\right\}$$

$n_i(=0, 1, 2\ldots)$ is the miniband index for doping superlattices and d_0 is the superlattice width.

In the presence of a quantizing magnetic field (B) the magneto DR in this case assumes the form

$$\gamma(E_{3,1}) = f_1(E_{3,1})\left(\frac{2eB}{\hbar}\right)\left(n+\frac{1}{2}\right) + f_2(E_{3,1})\left(n_i+\frac{1}{2}\right)\frac{2m_{\parallel}^*}{\hbar}\omega_8(E_{3,1}) \tag{3.2}$$

where $E_{3,1}$ is the total quantized energy in this case and n is the Landau quantum number.

The DOS function is given by

$$N_{Bn_z} = \frac{g_v eB}{\pi\hbar}\sum_{n=0}^{n_{\max}}\sum_{n_z=1}^{n_{z\max}}\delta'(E-E_{3,1}) \tag{3.3}$$

The electron concentration can be written as

$$n_0 = \frac{g_v eB}{\pi\hbar}\sum_{n=0}^{n_{\max}}\sum_{n_z=1}^{n_{z\max}}F_{-1}(\eta_{3,1}) \tag{3.4}$$

where $\eta_{3,1} = (k_B T)^{-1}[E_{F3} - E_{3,1}]$ and E_{F3} is the Fermi energy in this case.

Using (1.181) and (3.4), we can write the magneto AC as

$$f_e^* = \frac{4}{n_0}\frac{g_v eB}{\pi\hbar}\sum_{n=0}^{n_{\max}}\sum_{n_z=1}^{n_{z\max}}F_{-2}(\eta_{3,1}) \tag{3.5}$$

3.2.2 The AC in doping superlattices of III–V, ternary and quaternary materials under magnetic quantization

(a) Three band model of Kane

The electron energy spectrum in doping superlattices of III–V, ternary and quaternary materials can be expressed from (3.1) under the conditions $\Delta_{\parallel} = \Delta_{\perp} = \Delta$, $\delta = 0$ and $m_{\parallel}^* = m_{\perp}^* = m_c$ assumes the form

$$I(E) = \left(n_i+\frac{1}{2}\right)\hbar\omega_9(E) + \frac{\hbar^2 k_s^2}{2m_c} \tag{3.6}$$

where

$$\omega_9(E) \equiv \left(\frac{n_0|e|^2}{d_0\varepsilon_{sc}I'(E)m_c}\right)^{1/2}$$

The magneto DR in this case assumes the form

$$I(E_{3,2}) = \left(n_i + \frac{1}{2}\right)\hbar\omega_9(E_{3,2}) + \left(n + \frac{1}{2}\right)\hbar\omega_0 \tag{3.7}$$

where $E_{3,2}$ is the total quantized energy and $\omega_0 = (eB/m_c)$.

The DOS function is given by

$$N_{Bn_z} = \frac{g_v eB}{\pi\hbar}\sum_{n=0}^{n_{\max}}\sum_{n_z=1}^{n_{z\max}}\delta'(E - E_{3,2}) \tag{3.8}$$

The electron concentration can be written as

$$n_0 = \frac{g_v eB}{\pi\hbar}\sum_{n=0}^{n_{\max}}\sum_{n_z=1}^{n_{z\max}}F_{-1}(\eta_{3,2}) \tag{3.9}$$

where $\eta_{3,2} = (k_B T)^{-1}[E_{F3} - E_{3,2}]$.

Using (1.181) and (3.9), we can write the magneto AC as

$$f_e^* = \frac{4}{n_0}\frac{g_v eB}{\pi\hbar}\sum_{n=0}^{n_{\max}}\sum_{n_z=1}^{n_{z\max}}F_{-2}(\eta_{3,2}) \tag{3.10}$$

(b) Two band model of Kane

The electron energy spectrum in doping superlattices of III–V, ternary and quaternary materials can be expressed from (3.6) under the conditions $\Delta \gg E_g$ or $\Delta \ll E_g$ as

$$E(1+\alpha E) = \left(n_i + \frac{1}{2}\right)\hbar\omega_{10}(E) + \frac{\hbar^2 k_s^2}{2m_c} \tag{3.11}$$

where

$$\omega_{10}(E) \equiv \left(\frac{n_0|e|^2}{d_0\varepsilon_{sc}(1+2\alpha E)m_c}\right)^{1/2}$$

The magneto DR in this case assumes the form

$$I(E_{3,3}) = \left(n_i + \frac{1}{2}\right)\hbar\omega_{10}(E_{3,3}) + \left(n + \frac{1}{2}\right)\hbar\omega_0 \tag{3.12}$$

where $E_{3,3}$ is the total quantized energy.

The DOS function is given by

$$N_{Bn_z} = \frac{g_v eB}{\pi\hbar}\sum_{n=0}^{n_{\max}}\sum_{n_z=1}^{n_{z\max}}\delta'(E - E_{3,3}) \tag{3.13}$$

The electron concentration can be written as

$$n_0 = \frac{g_v eB}{\pi\hbar} \sum_{n=0}^{n_{\max}} \sum_{n_z=1}^{n_{z\max}} F_{-1}(\eta_{3,3}) \tag{3.14}$$

where $\eta_{3,3} = (k_B T)^{-1}[E_{F3} - E_{3,3}]$.

Using (1.181) and (3.14), we can write the magneto AC as

$$f_e^* = \frac{4}{n_0} \frac{g_v eB}{\pi\hbar} \sum_{n=0}^{n_{\max}} \sum_{n_z=1}^{n_{z\max}} F_{-2}(\eta_{3,3}) \tag{3.15}$$

(c) Parabolic band model

The electron energy spectrum in doping superlattices of III–V, ternary and quaternary materials under magnetic quantization whose energy band structures in the absence of band tails are described by the parabolic energy bands can be expressed as

$$E = \left(n_i + \frac{1}{2}\right)\hbar\omega_{11} + \frac{\hbar^2 k_s^2}{2m_c} \tag{3.16}$$

where

$$\omega_{11} \equiv \left(\frac{n_0|e|^2}{d_0 \varepsilon_{sc} m_c}\right)^{1/2}$$

The magneto DR in this case assumes the form

$$E_{3,4} = \left(n_i + \frac{1}{2}\right)\hbar\omega_{11} + \left(n + \frac{1}{2}\right)\hbar\omega_0 \tag{3.17}$$

where $E_{3,4}$ is the total quantized energy.

The DOS function is given by

$$N_{Bn_z} = \frac{g_v eB}{\pi\hbar} \sum_{n=0}^{n_{\max}} \sum_{n_z=1}^{n_{z\max}} \delta'(E - E_{3,4}) \tag{3.18}$$

The electron concentration can be written as

$$n_0 = \frac{g_v eB}{\pi\hbar} \sum_{n=0}^{n_{\max}} \sum_{n_z=1}^{n_{z\max}} F_{-1}(\eta_{3,4}) \tag{3.19}$$

where $\eta_{3,4} = (k_B T)^{-1}[E_{F3} - E_{3,4}]$.

Using (1.181) and (3.19), we can write the magneto AC as

$$f_e^* = \frac{4}{n_0} \frac{g_v eB}{\pi\hbar} \sum_{n=0}^{n_{\max}} \sum_{n_z=1}^{n_{z\max}} F_{-2}(\eta_{3,4}) \tag{3.20}$$

3.2.3 The AC in doping superlattices of II–VI materials under magnetic quantization

The carrier dispersion law in doping superlattices of II–VI compounds can be expressed following equation (1.51) as

$$E = a_0' k_s^2 + \left(n_i + \frac{1}{2}\right) \hbar\omega_{12} \pm \bar{\lambda}_0 k_s, \omega_{12} \equiv \left(\frac{n_0|e|^2}{d_0 \varepsilon_{sc} m_{\parallel}^*}\right)^{1/2} \tag{3.21}$$

The magneto DR in this case can be written as

$$E_{3,5} = a_0' \left[\frac{2eB}{\hbar}\left(n + \frac{1}{2}\right)\right] + \left(n_i + \frac{1}{2}\right) \hbar\omega_{12} \pm \bar{\lambda}_0 \left[\frac{2eB}{\hbar}\left(n + \frac{1}{2}\right)\right]^{1/2} \tag{3.22}$$

where $E_{3,5}$ is the total quantized energy.

The DOS function is given by

$$N_{Bn_z} = \frac{g_v eB}{2\pi\hbar} \sum_{n=0}^{n_{\max}} \sum_{n_z=1}^{n_{z\max}} \delta'(E - E_{3,5}) \tag{3.23}$$

The electron concentration can be written as

$$n_0 = \frac{g_v eB}{2\pi\hbar} \sum_{n=0}^{n_{\max}} \sum_{n_z=1}^{n_{z\max}} F_{-1}(\eta_{3,5}) \tag{3.24}$$

where $\eta_{3,5} = (k_B T)^{-1}[E_{F3} - E_{3,5}]$.

Using (1.181) and (3.24), we can write the AC as

$$f_e^* = \frac{4}{n_0} \frac{g_v eB}{2\pi\hbar} \sum_{n=0}^{n_{\max}} \sum_{n_z=1}^{n_{z\max}} F_{-2}(\eta_{3,5}) \tag{3.25}$$

3.2.4 The AC in doping superlattices of IV–VI materials under magnetic quantization

Equation (1.128) can be written as

$$\left[E - \frac{\hbar^2 k_s^2}{2m_t^-} - \frac{\hbar^2 k_z^2}{2m_l^-}\right]\left[1 + \alpha E + \alpha\frac{\hbar^2 k_s^2}{2m_t^+} + \alpha\frac{\hbar^2 k_z^2}{2m_l^+}\right] = \frac{\hbar^2 k_s^2}{2m_t^*} + \frac{\hbar^2 k_z^2}{2m_l^*} \tag{3.26}$$

The carrier energy spectrum in doping superlattices of IV–VI compounds can be written using equation (3.26) as

$$k_s^2 = (\hbar^2 S_{19})^{-1}[-S_{20}(E, n_i) + \sqrt{S_{20}^2(E, n_i) + 4S_{19}S_{21}(E, n_i)}] \tag{3.27}$$

in which

$$S_{19} \equiv \left(\frac{\alpha}{m_t^+ m_t^-}\right), \quad S_{20}(E, n_i) \equiv \left\{\frac{1}{m_t^*} - \left(\frac{\alpha E}{m_t^+}\right) + \frac{1+\alpha E}{m_t^-}\right.$$

$$\left. + \frac{\alpha\hbar^2}{2m_l^+ m_t^-}\left(n_i + \frac{1}{2}\right)T(E) + \frac{\alpha\hbar^2}{2m_l^- m_t^+}\left(n_i + \frac{1}{2}\right)T(E)\right\}$$

$$T(E) \equiv \frac{2m^*(0)}{\hbar}\omega_{11}(E), \quad m^*(0) \equiv \left(\frac{m_l^* m_l^-}{m_l^* + m_l^-}\right),$$

$$\omega_{11}(E) \equiv \left(\frac{n_0|e|^2}{d_0 \varepsilon_{sc} m^*(E)}\right)^{1/2},$$

$$m^*(E) \equiv \frac{1}{4t_1}\left[-(t_2(E))' + \frac{t_2(E)(t_2(E))' + 2t_1(1+2\alpha E)}{\sqrt{t_2^2(E) + 4Et_1(1+\alpha E)}}\right],$$

$$t_1 \equiv \left(\frac{\alpha}{4m_l^+ m_l^-}\right), \quad t_2(E) \equiv \frac{1}{2}\left[\left(\frac{1}{m_l^*}\right) - \left(\frac{\alpha E}{m_l^+}\right) + \left(\frac{1+\alpha E}{m_l^-}\right)\right],$$

$$(t_2(E))' \equiv \frac{\alpha}{2}\left(\frac{1}{m_l^-} - \left(\frac{1}{m_l^+}\right)\right) \quad \text{and} \quad S_{21}(E, n_i) \equiv [E(1+\alpha E)$$

$$+ \frac{\alpha E\hbar^2}{2m_l^+}\left(n_i + \frac{1}{2}\right)T(E) + \frac{\hbar^2}{2m_l^-}\left(n_i + \frac{1}{2}\right)T(E)(1+\alpha E)]$$

The magneto DR can be written as

$$\left(\frac{2eB}{\hbar}\right)\left(n + \frac{1}{2}\right) = (\hbar^2 S_{19})^{-1}[-S_{20}(E_{3,6}, n_i) + \sqrt{S_{20}^2(E_{3,6}, n_i) + 4S_{19}S_{21}(E_{3,6}, n_i)}] \quad (3.28)$$

where $E_{3,6}$ is the total quantized energy.

The DOS function is given by

$$N_{Bn_z} = \frac{g_v eB}{\pi\hbar}\sum_{n=0}^{n_{\max}}\sum_{n_z=1}^{n_{z\max}} \delta'(E - E_{3,6}) \quad (3.29)$$

The electron concentration can be written as

$$n_0 = \frac{g_v eB}{\pi\hbar}\sum_{n=0}^{n_{\max}}\sum_{n_z=1}^{n_{z\max}} F_{-1}(\eta_{3,6}) \quad (3.30)$$

where $\eta_{3,6} = (k_B T)^{-1}[E_{F3} - E_{3,6}]$.

Using (1.181) and (3.30), we can write the AC as

$$f_e^* = \frac{4}{n_0}\frac{g_v eB}{\pi\hbar}\sum_{n=0}^{n_{\max}}\sum_{n_z=1}^{n_{z\max}} F_{-2}(\eta_{3,6}) \tag{3.31}$$

*3.2.5 The AC in doping superlattices of stressed Kane type materials under magnetic quantization

The electron dispersion law in the doping superlattices of stressed Kane type semiconductors can be written using equation (1.105) as

$$\frac{k_x^2}{[a^*(E)]^2} + \frac{k_y^2}{[b^*(E)]^2} + \frac{1}{[c^*(E)]^2}\frac{2m_z^*(0)}{\hbar}\left(n_i + \frac{1}{2}\right)\omega_{12}(E) = 1 \tag{3.32}$$

where

$$\omega_{12}(E) \equiv \left(\frac{n_0|e|^2}{d_0\varepsilon_{sc}m_z^*(E)}\right)^{1/2} \quad \text{and} \quad m_z^*(E) \equiv \hbar^2 c^*(E)\frac{\partial}{\partial E}[c^*(E)]$$

The electron dispersion law in the doping superlattices of stressed Kane type semiconductors can be written as

$$\left(n + \frac{1}{2}\right)\hbar\omega_{100}(E_{3,7}) + \frac{1}{[c^*(E_{3,7})]^2}\frac{2m_z^*(0)}{\hbar}\left(n_i + \frac{1}{2}\right)\omega_{12}(E_{3,7}) = 1 \tag{3.33}$$

where $E_{3,7}$ is the total quantized energy in this case and

$$\omega_{100}(E_{3,7}) = \frac{eB}{\hbar a^*(E_{3,7})b^*(E_{3,7})}$$

The DOS function is given by

$$N_{Bn_z} = \frac{g_\nu eB}{\pi\hbar}\sum_{n=0}^{n_{\max}}\sum_{n_z=1}^{n_{z\max}} \delta'(E - E_{3,7}) \tag{3.34}$$

The electron concentration can be written as

$$n_0 = \frac{g_v eB}{\pi\hbar}\sum_{n=0}^{n_{\max}}\sum_{n_z=1}^{n_{z\max}} F_{-1}(\eta_{3,7}) \tag{3.35}$$

where $\eta_{3,7} = (k_B T)^{-1}[E_{F3} - E_{3,7}]$.

Using (1.181) and (3.35), we can write the AC

$$f_e^* = \frac{4}{n_0} \frac{g_v eB}{\pi\hbar} \sum_{n=0}^{n_{\max}} \sum_{n_z=1}^{n_{z\max}} F_{-2}(\eta_{3,7}) \tag{3.36}$$

3.3 Summary and Conclusion

From the magneto DR in doping superlattices of nonlinear optical and tetragonal materials as given by (3.2), we observe that the electron energy is totally quantized in 3D wave vector surfaces in the energy plane. Equations (3.3)–(3.5) represent the DOS function, the electron concentration and the corresponding magneto AC in this case respectively. Equations (3.7)–(3.10) represent the DR, DOS function, electron concentration and corresponding AC in this case respectively for doping superlattices of III–V materials where the bulk electrons follow the three band model of Kane under magnetic quantization.

The 0D electrons in doping superlattices of III–V materials under magnetic quantization are also described by two band model of Kane and parabolic energy bands with the DRs as given by (3.12) and (3.17) respectively. Equations (3.13)–(3.15) represent the DOS function, electron concentration and the corresponding AC in this case respectively for two band model of Kane. Equations (3.18)–(3.20) represent the DOS function, electron concentration and the corresponding AC in this case respectively for parabolic energy bands.

Equations (3.22)–(3.25) represent the DR, DOS function, electron concentration and the corresponding AC respectively for doping superlattices of II–VI materials under magnetic quantization. Equations (3.28)–(3.31) represent the DR, DOS function, electron concentration and the corresponding AC respectively for doping superlattices of IV–VI materials under magnetic quantization. Equations (3.33)–(3.36) represent the DR, DOS function, electron concentration and the corresponding AC respectively for doping superlattices of stressed Kane type materials under magnetic quantization. Besides in all the cases the DOS functions are non-uniformly placed Dirac delta functions in the energy axis.

It is the band structure which changes in a fundamental way and consequently all the physical properties of all the electronic materials change radically leading to new physical concepts.

We have not given any plot of AC in any case of this chapter since we believe our creative readers will plot the same and enjoy intricate physics, computational analysis, advanced mathematics and will create new problems both theoretical and experimental.

3.4 Open Research Problems

(R.3.1) Investigate the AC in the presence of an arbitrarily oriented quantizing magnetic field for doping superlattices of nonlinear optical semiconductors by including the electron spin. Study all the special cases for III–V, ternary and quaternary materials in this context.

(R.3.2) Investigate the AC in doping superlattices of IV–VI, II–VI and stressed Kane type compounds in the presence of an arbitrarily oriented quantizing magnetic field by including the electron spin.

(R.3.3) Investigate the AC for doping superlattices of all the materials as stated of Chapter 1.

(R.3.4) Investigate the AC for all the problems from (R.3.1) to (R.3.3) in the presence of an additional arbitrarily oriented electric field.

(R.3.5) Investigate the AC for all the problems from (R.3.1) to (R.3.3) in the presence of arbitrarily oriented crossed electric and magnetic fields.

References

[1] N. G. Anderson, W. D. Laidig, R. M. Kolbas, Y. C. Lo, *J. Appl. Phys.*, 60, 2361 **(1986)**.

[2] F. Capasso, *Semicond. Semimetals*, 22, 2 **(1985)**.

[3] F. Capasso, K. Mohammed, A. Y. Cho, R. Hull, A. L. Hutchinson, *Appl. Phys. Lett.*, 47, 420 **(1985)**.

[4] F. Capasso, R. A. Kiehl, *J. Appl. Phys.*, 58, 1366 **(1985)**.

[5] K. Ploog, G. H. Doheler, *Adv. Phys.*, 32, 285 **(1983)**.

[6] F. Capasso, K. Mohammed, A. Y. Cho, *Appl. Phys. Lett.*, 48, 478 **(1986)**.

[7] R. Grill, C. Metzner, G. H. Döhler, *Phys. Rev. B*, 63, 235316 **(2001)**.

[8] A. R. Kost, M. H. Jupina, T. C. Hasenberg, E. M. Garmire, *J. Appl. Phys.*, 99, 023501 **(2006)**.

[9] G. Smirnov, D. V. Ushakov, V. K. Kononenko, *Proc. SPIE*, 4706, 70 **(2002)**.

[10] D. V. Ushakov, V. K. Kononenko, I. S. Manak, *Proc. SPIE*, 4358, 171 **(2001)**.

[11] J. Z. Wang, Z. G. Wang, Z. M. Wang, S. L. Feng, Z. Yang, *Phys. Rev. B*, 62, 6956 **(2000)**.

[12] A. R. Kost, L. West, T. C. Hasenberg, J. O. White, M. Matloubian, G. C. Valley, *Appl. Phys. Lett.*, 63, 3494 **(1993)**.
[13] A. S. Bastola, S. J. Chua, S. J. Xu, *J. Appl. Phys.*, 83, 1476 **(1998)**.
[14] Z. J. Yang, E. M. Garmire, D. Doctor, *J. Appl. Phys.*, 82, 3874 **(1997)**.
[15] G. H. Avetisyan, V. B. Kulikov, I. D. Zalevsky, P. V. Bulaev, *Proc. SPIE*, 2694, 216 **(1996)**.
[16] U. Pfeiffer, M. Kneissl, B. Knüpfer, N. Müller, P. Kiesel, G. H. Döhler, J. S. Smith, *Appl. Phys. Lett.*, 68, 1838 **(1996)**.
[17] H. L. Vaghjiani, E. A. Johnson, M. J. Kane, R. Grey, C. C. Phillips, *J. Appl. Phys.*, 76, 4407 **(1994)**.
[18] P. Kiesel, K. H. Gulden, A. Hoefler, M. Kneissl, B. Knuepfer, S. U. Dankowski, P. Riel, X. X. Wu, J. S. Smith, G. H. Doehler, *Proc. SPIE*, 1985, 278 **(1993)**.
[19] G. H. Doheler, *Phys. Script.*, 24, 430 **(1981)**.
[20] S. Mukherjee, S. N. Mitra, P. K. Bose, A. R. Ghatak, A. Neoigi, J. P. Banerjee, A. Sinha, M. Pal, S. Bhattacharya, K. P. Ghatak, *J. Comput. Theor. Nanosci.*, 4, 550 **(2007)**; N. Paitya, K. P. Ghatak, *J. Adv. Phys.*, 1, 161 **(2012)**; N. Paitya, S. Bhattacharya, D. De, K. P. Ghatak, *Adv. Sci. Engg. Medi.*, 4, 96 **(2012)**.
[21] S. Bhattacharya, D. De, S. M. Adhikari, K. P. Ghatak, *Superlatt. Microstruct.*, 51, 203 **(2012)**; D. De, S. Bhattacharya, S. M. Adhikari, A. Kumar, P. K. Bose, K. P. Ghatak, *Beilstein J. Nanotech.*, 2, 339 **(2012)**; D. De, A. Kumar, S. M. Adhikari, S. Pahari, N. Islam, P. Banerjee, S. K. Biswas, S. Bhattacharya, K. P. Ghatak, *Superlatt. Microstruct.*, 47, 377 **(2010)**.
[22] S. Pahari, S. Bhattacharya, S. Roy, A. Saha, D. De, K. P. Ghatak, *Superlatt. Microstruct.*, 46, 760 **(2009)**; S. Pahari, S. Bhattacharya, K. P. Ghatak, *J. Comput. Theor. Nanosci.*, 6, 2088 **(2009)**; S. K. Biswas, A. R. Ghatak, A. Neogi, A. Sharma, S. Bhattacharya, K. P. Ghatak, *Phys. E: Low-dimen. Syst. Nanostruct.*, 36, 163 **(2007)**.
[23] L. J. Singh, S. Choudhury, D. Baruah, S. K. Biswas, S. Pahari, K. P. Ghatak, *Phys. B: Conden. Matter*, 368, 188 **(2005)**; S. Chowdhary, L. J. Singh, K. P. Ghatak, *Phys. B: Conden. Matter*, 365, 5 **(2005)**; L. J. Singh, S. Choudhary, A. Mallik, K. P. Ghatak, *J. Comput. Theor. Nanosci.*, 2, 287 **(2005)**.
[24] K. P. Ghatak, J. Mukhopadhyay, J. P. Banerjee, *SPIE Proc. Ser.*, 4746, 1292 **(2002)**; K. P. Ghatak, S. Dutta, D. K. Basu, B. Nag, *Il Nuovo Cimento D*, 20, 227 **(1998)**.
[25] K. P. Ghatak, B. Mitra, *Il Nuovo Cimento D*, 15, 97 **(1993)**; K. P. Ghatak, *Inter. Soc. Opt. Photon. Proc. Soc. Photo Opt. Instru. Engg.*, 1626, 115 **(1992)**; K. P. Ghatak, A. Ghoshal, *Phys. Stat. Sol. (b)*, 170, K27 **(1992)**.
[26] K. P. Ghatak, S. Bhattacharya, S. N. Biswas, *Proc. Soc. Photo Opt. Instru. Engg.*, 836, 72 **(1988)**; K. P. Ghatak, A. Ghoshal, S. N. Biswas, M. Mondal, *Proc. Soc. Photo Opt. Instru. Engg.*, 1308, 356 **(1990)**; K. P. Ghatak, G. Mazumder, *Proc. Mat. Res. Soc.*, 484, 679 **(1998)**.
[27] K. P. Ghatak, B. De, *Defect Engg. Semi. Growth, Process. Device Tech. Symp., Mat. Res. Soc.*, 262, 911 **(1992)**; S. N. Biswas, K. P. Ghatak, *Int.*

J. Electron. Theor. Exp., 70, 125 **(1991)**; B. Mitra, K. P. Ghatak, *Phys. Lett. A*, 146, 357 **(1990)**; B. Mitra, K. P. Ghatak, *Phys. Lett. A*, 142, 401 **(1989)**.

[28] K. P. Ghatak, B. Mitra, A. Ghoshal, *Phys. Stat. Sol. (b)*, 154, K121 **(1989)**; B. Mitra, K. P. Ghatak, *Phys. Stat. Sol. (b)*, 149, K117 **(1988)**; K. P. Ghatak, S. N. Biswas, *Proc. Soc. Photo Opt. Instru. Engg.*, 792, 239 **(1987)**.

[29] S. Bhattacharyya, K. P. Ghatak, S. Biswas, *OE/Fibers' 87, Inter. Soc. Opt. Photon.*, 836, 73 **(1988)**; M. Mondal, K. P. Ghatak, *Czech. J. Phys. B*, 36, 1389 **(1986)**; K. P. Ghatak, A. N. Chakravarti, *Phys. Stat. Sol. (b)*, 117, 707 **(1983)**.

Chapter 4

The Carrier Contribution to the Elastic Constants in Doping Superlattices of Non-Parabolic Materials Under Magnetic Quantization

One week means ten thousand and eighty minutes. If by applying the maximum power transfer theorem, I can really use five thousand and forty minutes, then instead of weak I will be rather strong mentally.

4.1 Introduction

The importance of studying the physical properties of doping superlattices has been given in [1–19]. Besides in recent years, there has been considerable interest in investigating the carrier contribution to the elastic constants (CECs) in different technologically important materials [20–30]. In Section 4.2.1, of the theoretical background, the CECS' in doping superlattices of nonlinear optical and tetragonal materials has been investigated under magnetic quantization. Section 4.2.2 contains the results for doping superlattices under magnetic quantization of III–V, ternary and quaternary semiconductors in accordance with the three and the two band models of Kane together with parabolic energy bands and they form the special cases of Section 4.2.1. Sections 4.2.3–4.2.5 contain the study of the CECS' under magnetic quantization for doping superlattices of II–VI, IV–VI and stressed Kane type semiconductors respectively. Sections 4.3 and 4.4 contain the summary and conclusion and the five open research problems for this chapter.

4.2 Theoretical Background

4.2.1 The CECs' in doping superlattices of nonlinear optical and tetragonal materials under magnetic quantization

Using (1.169a), (1.169b) and (3.4), we can study the 2nd order CEC (ΔC_{44}) and the third order (ΔC_{456}) respectively in this case.

4.2.2 The CECs' in doping superlattices of III–V, ternary and quaternary materials under magnetic quantization

(a) Using (1.169a), (1.169b) and (3.9), we can study the ΔC_{44} and ΔC_{456} respectively in this case where the bulk electrons follow the three band model of Kane.
(b) Using (1.169a), (1.169b) and (3.14), we can study the ΔC_{44} and ΔC_{456} respectively in this case where the bulk electrons follow the two band model of Kane.
(c) Using (1.169a), (1.169b) and (3.19), we can study the ΔC_{44} and ΔC_{456} respectively in this case where the bulk electrons follow the parabolic energy band model.

4.2.3 The CECs' in doping superlattices of II–VI materials under magnetic quantization

Using (1.169a), (1.169b) and (3.24), we can study the ΔC_{44} and ΔC_{456} respectively in this case.

4.2.4 The CECs' in doping superlattices of IV–VI materials under magnetic quantization

Using (1.169a), (1.169b) and (3.30), we can study the ΔC_{44} and ΔC_{456} respectively in this case.

*4.2.5 The CECs' in doping superlattices of HD stressed Kane type materials under magnetic quantization

Using (1.169a), (1.169b) and (3.35), we can study the ΔC_{44} and ΔC_{456} respectively in this case.

4.3 Summary and Conclusion

It is the band structure which changes in a fundamental way and consequently all the physical properties of all the electronic materials change radically leading to new physical concepts.

We have not given any plot of CECs' in doping superlattices under magnetic quantization since we believe our able readers will plot the same and enjoy intricate physics, computational analysis and advanced mathematics and will create new problems both theoretical and experimental.

4.4 Open Research Problems

(R.4.1) Investigate the CEC's in the presence of an arbitrarily oriented quantizing magnetic field for doping superlattices of nonlinear optical and tetragonal materials by including the electron spin. Study all the special cases for HD III–V, ternary and quaternary materials in this context.

(R.4.2) Investigate the CEC's in doping superlattices of IV–VI, II–VI and stressed Kane type compounds in the presence of an arbitrarily oriented quantizing magnetic field by including the electron spin.

(R.4.3) Investigate the CEC's for doping superlattices under strong magnetic quantization of all the materials as stated in (R.1.1) of Chapter 1.

(R.4.4) Investigate the CEC's under strong magnetic quantization for all the problems from (R.4.1) to (R.4.3) in the presence of an additional arbitrarily oriented electric field.

(R.4.5) Investigate the CEC's for all the problems from (R.4.1) to (R.4.3) in the presence of arbitrarily oriented crossed quantizing electric and magnetic fields.

References

[1] N. G. Anderson, W. D. Laidig, R. M. Kolbas, Y. C. Lo, *J. Appl. Phys.*, 60, 2361 **(1986)**.
[2] F. Capasso, *Semicond. Semimetals*, 22, 2 **(1985)**.
[3] F. Capasso, K. Mohammed, A. Y. Cho, R. Hull, A. L. Hutchinson, *Appl. Phys. Lett.*, 47, 420 **(1985)**.
[4] F. Capasso, R. A. Kiehl, *J. Appl. Phys.*, 58, 1366 **(1985)**.
[5] K. Ploog, G. H. Doheler, *Adv. Phys.*, 32, 285 **(1983)**.
[6] F. Capasso, K. Mohammed, A. Y. Cho, *Appl. Phys. Lett.*, 48, 478 **(1986)**.
[7] R. Grill, C. Metzner, G. H. Döhler, *Phys. Rev. B*, 63, 235316 **(2001)**.

[8] A. R. Kost, M. H. Jupina, T. C. Hasenberg, E. M. Garmire, *J. Appl. Phys.*, 99, 023501 (**2006**).
[9] G. Smirnov, D. V. Ushakov, V. K. Kononenko, *Proc. SPIE*, 4706, 70 (**2002**).
[10] D. V. Ushakov, V. K. Kononenko, I. S. Manak, *Proc. SPIE*, 4358, 171 (**2001**).
[11] J. Z. Wang, Z. G. Wang, Z. M. Wang, S. L. Feng, Z. Yang, *Phys. Rev. B*, 62, 6956 (**2000**).
[12] A. R. Kost, L. West, T. C. Hasenberg, J. O. White, M. Matloubian, G. C. Valley, *Appl. Phys. Lett.*, 63, 3494 (**1993**).
[13] A. S. Bastola, S. J. Chua, S. J. Xu, *J. Appl. Phys.*, 83, 1476 (**1998**).
[14] Z. J. Yang, E. M. Garmire, D. Doctor, *J. Appl. Phys.*, 82, 3874 (**1997**).
[15] G. H. Avetisyan, V. B. Kulikov, I. D. Zalevsky, P. V. Bulaev, *Proc. SPIE*, 2694, 216 (**1996**).
[16] U. Pfeiffer, M. Kneissl, B. Knüpfer, N. Müller, P. Kiesel, G. H. Döhler, J. S. Smith, *Appl. Phys. Lett.*, 68, 1838 (**1996**).
[17] H. L. Vaghjiani, E. A. Johnson, M. J. Kane, R. Grey, C. C. Phillips, *J. Appl. Phys.*, 76, 4407 (**1994**).
[18] P. Kiesel, K. H. Gulden, A. Hoefler, M. Kneissl, B. Knuepfer, S. U. Dankowski, P. Riel, X. X. Wu, J. S. Smith, G. H. Doehler, *Proc. SPIE*, 1985, 278 (**1993**).
[19] G. H. Doheler, *Phys. Script.*, 24, 430 (**1981**).
[20] S. Mukherjee, S. N. Mitra, P. K. Bose, A. R. Ghatak, A. Neoigi, J. P. Banerjee, A. Sinha, M. Pal, S. Bhattacharya, K. P. Ghatak, *J. Comput. Theor. Nanosci.*, 4, 550 (**2007**).
[21] N. Paitya, K. P. Ghatak, *J. Adv. Phys.*, 1, 161 (**2012**); N. Paitya, S. Bhattacharya, D. De, K. P. Ghatak, *Adv. Sci. Engg. Medi.*, 4, 96 (**2012**); S. Bhattacharya, D. De, S. M. Adhikari, K. P. Ghatak, *Superlatt. Microstruct.*, 51, 203 (**2012**).
[22] D. De, S. Bhattacharya, S. M. Adhikari, A. Kumar, P. K. Bose, K. P. Ghatak, *Beilstein J. Nanotech.*, 2, 339 (**2012**); D. De, A. Kumar, S. M. Adhikari, S. Pahari, N. Islam, P. Banerjee, S. K. Biswas, S. Bhattacharya, K. P. Ghatak, *Superlatt. Microstruct.*, 47, 377 (**2010**); S. Pahari, S. Bhattacharya, S. Roy, A. Saha, D. De, K. P. Ghatak, *Superlatt. Microstruct.*, 46, 760 (**2009**).
[23] S. Pahari, S. Bhattacharya, K. P. Ghatak, *J. Comput. Theor. Nanosci.*, 6, 2088 (**2009**); S. K. Biswas, A. R. Ghatak, A. Neogi, A. Sharma, S. Bhattacharya, K. P. Ghatak, *Phys. E: Low-dimen. Syst. Nanostruct.*, 36, 163 (**2007**); L. J. Singh, S. Choudhury, D. Baruah, S. K. Biswas, S. Pahari, K. P. Ghatak, *Phys. B: Conden. Matter*, 368, 188 (**2005**).
[24] S. Chowdhary, L. J. Singh, K. P. Ghatak, *Phys. B: Conden. Matter*, 365, 5 (**2005**); L. J. Singh, S. Choudhary, A. Mallik, K. P. Ghatak, *J. Comput. Theor. Nanosci.*, 2, 287 (**2005**); K. P. Ghatak, J. Mukhopadhyay, J. P. Banerjee, *SPIE Proc. Ser.*, 4746, 1292 (**2002**).
[25] K. P. Ghatak, S. Dutta, D. K. Basu, B. Nag, *Il Nuovo Cimento D*, 20, 227 (**1998**); K. P. Ghatak, D. K. Basu, B. Nag, *J. Phys. Chem. Solids*, 58, 133 (**1997**); K. P. Ghatak, B. De, *Mat. Res. Soc. Proc.*, 300, 513 (**1993**).

[26] K. P. Ghatak, B. Mitra, *Il Nuovo Cimento D*, 15, 97 **(1993)**; K. P. Ghatak, *Inter. Soc. Opt. Photon. Proc. Soc. Photo Opt. Instru. Engg.*, 1626, 115 **(1992)**; K. P. Ghatak, A. Ghoshal, *Phys. Stat. Sol. (b)*, 170, K27 **(1992)**.

[27] K. P. Ghatak, S. Bhattacharya, S. N. Biswas, *Proc. Soc. Photo Opt. Instru. Engg.*, 836, 72 **(1988)**; K. P. Ghatak, A. Ghoshal, S. N. Biswas, M. Mondal, *Proc. Soc. Photo Opt. Instru. Engg.*, 1308, 356 **(1990)**; K. P. Ghatak, G. Mazumder, *Proc. Mat. Res. Soc.*, 484, 679 **(1998)**.

[28] K. P. Ghatak, B. De, *Defect Engg. Semi. Growth, Process. Device Tech. Symp., Mat. Res. Soc.*, 262, 911 **(1992)**; S. N. Biswas, K. P. Ghatak, *Int. J. Electron. Theor. Exp.*, 70, 125 **(1991)**; B. Mitra, K. P. Ghatak, *Phys. Lett. A*, 146, 357 **(1990)**.

[29] B. Mitra, K. P. Ghatak, *Phys. Lett. A*, 142, 401 **(1989)**; K. P. Ghatak, B. Mitra, A. Ghoshal, *Phys. Stat. Sol. (b)*, 154, K121 **(1989)**; B. Mitra, K. P. Ghatak, *Phys. Stat. Sol. (b)*, 149, K117 **(1988)**.

[30] K. P. Ghatak, S. N. Biswas, *Proc. Soc. Photo Opt. Instru. Engg.*, 792, 239 **(1987)**; S. Bhattacharyya, K. P. Ghatak, S. Biswas, *OE/Fibers' 87*, *Inter. Soc. Opt. Photon.*, 836, 73 **(1988)**; M. Mondal, K. P. Ghatak, *Czech. J. Phys. B*, 36, 1389 **(1986)**; K. P. Ghatak, A. N. Chakravarti, *Phys. Stat. Sol. (b)*, 117, 707 **(1983)**.

Chapter 5

The Quantum Capacitance in Quantized Transistors Under Magnetic Quantization

I must try to convert the love for power to the power of love.

5.1 Introduction

One vital quantity which generates considerable interest is the quantum capacitance (QC) of metal oxide semiconductor quantum field effect transistors (QMOSFETs) under the presence of a large gate voltage which creates a large surface electric field [1–5]. The 2D electron statistics in the inversion layers of such devices can rather easily be varied by changing the gate voltage which, in turn, brings a change of the surface electric field, the QC depends on the gate-voltage. This variation has been investigated for n-channel inversion layers on p-type Si which shows few important oscillatory characteristics due to the two-dimensional Density-of-States function of the electrons in the quantized inversion layers. There has been considerable progress [6–13] in the investigations of the physical properties of the 2D QMOSFET devices operated under inversion mode made of different materials whose bulk dispersion relations (DR) are defined by the two band model of Kane. It appears then that there remains considerable scope of investigations on the QC for other types of technologically important FETs made of various vital compounds of non-standard energy bands under magnetic quantization.

In this chapter, from Sections 5.2.1–5.2.11, we shall study the QC in magneto MOSFETs of nonlinear optical and tetragonal materials, III–V, II–VI, IV–VI, stressed Kane type, Bi_2Te_3, Ge, GaAs, GaSb, InSb and Te respectively. Besides, from Sections 5.2.12–5.2.24, we shall further study the QC in magneto QWFETs of nonlinear optical, III–V, II–VI, IV–VI, stressed Kane type, Bi_2Te_3, Ge, GaAs, GaSb, InSb, Te, $PtSb_2$ sand GaP respectively by using various band models on the basis of various band models. Section 5.3 contains result and discussions. In Section 5.4, we present 11 research problems.

5.2 Theoretical Background

5.2.1 The QC in magneto MOSFETs of nonlinear optical and tetragonal materials

In MOS structure, the QC is formed by the series combination of a fixed capacitance C_{OX} due to the oxide layer and a variable capacitance C_s (surface capacitance) due to the space charge layer which can be controlled by the gate voltage.

The relation among these capacitances is given by

$$\frac{1}{C_G} = \frac{1}{C_s} + \frac{1}{C_{\mathrm{OX}}} \tag{5.1}$$

where $C_{\mathrm{OX}} = \varepsilon_{\mathrm{OX}}/d_{\mathrm{OX}}$, $\varepsilon_{\mathrm{OX}}$ and d_{OX} being the permittivity and the thickness of the oxide layer respectively. The C_s can also be expressed as

$$C_s = e\frac{dN_s}{dV_0} = e\frac{dN_s}{dV_g}\left[\left\{1 - e\frac{d_{\mathrm{OX}}}{\varepsilon_{\mathrm{OX}}}\frac{dN_s}{dV_g}\right\}\right]^{-1} \tag{5.2}$$

where N_S is defined as the surface concentration per unit area, V_g is the gate voltage and V_0 defined as the surface potential and can be written as

$$V_0\left(= V_g - \frac{eN_s}{C_{\mathrm{OX}}} - (\bar{E}_{fB}/e)\right) \tag{5.3}$$

where $\bar{E}_{fB}$ is the energy separation between the Fermi level and the conduction band edge in the bulk of the substrate material under the present condition.

Using (5.1)–(5.3), the QC per unit area can be written as

$$C_G = e\frac{dN_s}{dV_g} \tag{5.4}$$

It appears, therefore, that we should express the surface concentration as a function of the surface electric field in order to study the QC. This in turn would require DRs for two-dimensional electron motion as appropriate in this case.

The quantization rule for inversion layers is given by [14–25]

$$\int_0^{z_t} k_z dz = \frac{2}{3}(S_i)^{3/2} \tag{5.5}$$

where z_t is the classical turning point and S_i is the zeros of the Airy function $(Ai(-S_i) = 0)$.

(a) Weak electric field limit

In the presence of a surface electric field F_s along z-direction and perpendicular to the surface, (2.2) assumes the form

$$\psi_1(E-|e|F_s z)=\psi_2(E-|e|F_s z)k_s^2+\psi_3(E-|e|F_s z)k_z^2 \tag{5.6}$$

where

$$\psi_1(E)=\gamma(E),\psi_2(E)=f_1(E) \text{ and } \psi_3(E)=f_2(E)$$

Using (5.5) and (5.6), under the weak electric field limit $\left(\frac{eF_s z}{E}\ll 1\right)$, we can write

$$\int_0^{z_t}\left(\sqrt{A_7(E)-|e|F_s zD_7(E)}\right)dz=\frac{2}{3}\left(S_i\right)^{3/2} \tag{5.7}$$

in which

$$z_t\equiv\frac{A_7(E)}{[|e|F_s D_7(E)]},\quad A_7(E)\equiv\left[\frac{\psi_1(E)-\psi_2(E)k_s^2}{\psi_3(E)}\right],$$

$$D_7(E)\equiv\left[B_7(E)-A_7(E)C_7(E)\right],\quad B_7(E)\equiv\left[\frac{(\psi_1(E))'-(\psi_2(E))'\,k_s^2}{\psi_3(E)}\right],$$

$$C_7(E)\equiv\left[\frac{(\psi_3(E))'}{\psi_3(E)}\right]$$

and the primes indicate the differentiation of the differentiable functions with respect to E.

Thus, the 2D electron dispersion law in QMOSFETs operated under inversion mode and made of nonlinear optical and tetragonal materials under the weak electric field limit can approximately be written as

$$\psi_1(E)=P_7(E,i)k_s^2+Q_7(E,i) \tag{5.8}$$

where

$$P_7(E,i)\equiv\left[\psi_2(E)-\left(\frac{2t_2(E)}{3\left[t_2(E)\right]^{1/3}}\right)\psi_3(E)\bar{S}_i\left(|\bar{e}|\bar{F}_s\right)^{2/3}\right],$$

$$t_2(E)\equiv\left[\frac{[\psi_2(E)]'}{\psi_3(E)}-\left(\frac{\psi_2(E)\left[\psi_3(E)\right]'}{[\psi_3(E)]^2}\right)\right]$$

$$t_1(E)\equiv\left[\frac{[\psi_1(E)]'}{\psi_3(E)}-\left(\frac{\psi_1(E)\left[\psi_3(E)\right]'}{[\psi_3(E)]^2}\right)\right] \quad \text{and}$$

$$Q_7(E,i)\equiv S_i\psi_3(E)\left[|e|F_s t_1(E)\right]^{2/3}$$

The electron dispersion law in QMOSFETs operated under inversion mode and made of tetragonal and nonlinear optical materials under the weak electric field limit in the presence of a quantizing magnetic field B along z-direction can approximately be written following the area quantization rule of Landau [14] as

$$\psi_1(E_{41}) = P_7(E_{41}, i)\left(\frac{2eB}{\hbar}\right)\left(n + \frac{1}{2}\right) + Q_7(E_{41}, i) \tag{5.9}$$

where E_{41} is the total quantized energy in this case.

The DOS function assumes the form

$$N(E) = \left(\frac{g_v eB}{\pi\hbar}\right)\sum_{i=0}^{i_{\max}}\sum_{n=0}^{n_{\max}} \delta'(E - E_{41}) \tag{5.10}$$

where δ' is the Dirac's delta function.

Combining (5.10) with the Fermi–Dirac occupation probability factor, the surface electron concentration can be written as

$$N_s = \left(\frac{g_v eB}{\pi\hbar}\right)\sum_{i=0}^{i_{\max}}\sum_{n=0}^{n_{\max}} F_{-1}(\eta_1) \tag{5.11}$$

where $\eta_1 = (k_B T)^{-1}[E_{F1} - E_{41}]$, $E_{F1}\left(= eV_g - \frac{e^2 N_s}{C_{\text{OX}}} - \bar{E}_{fB}\right)$ and E_{F1} is the Fermi energy as measured from the edge of the conduction band at the surface in the presence of magnetic quantization.

Combining (5.4) and (5.11), we can study the magneto QC in this case.

(b) Strong electric field limit

The electron dispersion law in QMOSFETs operated under inversion mode made of tetragonal and nonlinear optical materials under the strong electric field limit can be written using (2.2) and (5.5) in this case as

$$k_s^2 = P_2(E, i) \tag{5.12}$$

where

$$P_2(E,i) \equiv [F_7(E)]^{-1}[F_6(E) - F_8(E,i)], F_6(E) \equiv \left[\frac{\psi_1(E)}{\psi_3(E)}\right]\left[1 + \frac{[\psi_1(E)]'[\psi_3(E)]'}{\psi_3(E)[\psi_1(E)]''} + \frac{\psi_1(E)[\psi_3(E)]''}{2[\psi_1(E)]''[\psi_3(E)]}\right],$$

$$F_7(E) \equiv \left[\left(\frac{\psi_2(E)}{\psi_3(E)}\right) - \left(\frac{\psi_1(E)[\psi_2(E)]''}{2[\psi_1(E)]''[\psi_3(E)]}\right)\right.$$
$$+ \left(\frac{\psi_2(E)}{\psi_3(E)}\right)\left[\frac{[\psi_1(E)]'\,[\psi_3(E)]'}{\psi_3(E)\,[\psi_1(E)]''} + \frac{\psi_1(E)[\psi_3(E)]''}{2[\psi_1(E)]''[\psi_3(E)]}\right]$$
$$\left.+ \left(\frac{\psi_1(E)}{\psi_3(E)}\right)\left[\frac{[\psi_2(E)]'\,[\psi_3(E)]'}{\psi_3(E)\,[\psi_1(E)]''} + \frac{\psi_2(E)\,[\psi_3(E)]''}{2[\psi_1(E)]''[\psi_3(E)]}\right]\right] \quad \text{and}$$

$$F_8(E,i) \equiv \left[\frac{2\sqrt{2}}{3}\,(S_i)^{3/2}\,(|e|F_s)\,\sqrt{[\psi_1(E)]''}\right]$$

Using the area quantization rule of Landau [14], from (5.12) we get

$$\left(\frac{2eB}{\hbar}\right)\left(n+\frac{1}{2}\right) = P_2(E,i)|_{E=E_{42}} \tag{5.13}$$

where E_{42} is the total quantized energy in this case.

The DOS function assumes the form

$$N(E) = \left(\frac{g_v eB}{\pi\hbar}\right)\sum_{i=0}^{i_{\max}}\sum_{n=0}^{n_{\max}} \delta'\left(E - E_{42}\right) \tag{5.14}$$

Combining (5.14) with the Fermi–Dirac occupation probability factor, the surface electron concentration can be written as

$$N_s = \left(\frac{g_v eB}{\pi\hbar}\right)\sum_{i=0}^{i_{\max}}\sum_{n=0}^{n_{\max}} F_{-1}\left(\eta_2\right) \tag{5.15}$$

where $\eta_2 = (k_B T)^{-1}[E_{F1} - E_{42}]$.

Combining (5.5) and (5.15), we can study the magneto QC in this case.

5.2.2 The QC in magneto MOSFETs of III–V, ternary and quaternary materials

(a) Weak electric field limit

Under the condition of weak electric field limit $\left(\frac{eF_s z}{E} \ll 1\right)$, the 2D electron dispersion law assumes the form

$$I(E) = \frac{\hbar^2 k_s^2}{2m_c} + S_i\left[\frac{\hbar\,|e|\,F_s[I(E)]'}{\sqrt{2m_c}}\right]^{2/3} \tag{5.16}$$

Equation (5.16) represents the DR of the 2D electrons in QMOSFETs of III–V, ternary and quaternary materials operated under inversion layer mode in the presence of the weak electric field limit whose bulk electrons obey the three band model of Kane.

The 2D magneto electron dispersion law in this case can be written following (5.16) as

$$I\left(E_{43}\right)=\left(n+\frac{1}{2}\right)\hbar\omega_0+S_i\left[\frac{\hbar|e|F_s\left[I\left(E_{43}\right)\right]'}{\sqrt{2m_c}}\right]^{2/3} \tag{5.17}$$

where $\omega_0=\frac{eB}{m_c}$, E_{43} is the total quantized energy in this case.

Equation (5.17) represents the magneto dispersion relation of the 2D electrons in QMOSFETs of III–V, ternary and quaternary materials under the weak electric field limit in the presence of magnetic quantization whose bulk electrons obey the **three band model of Kane**.

The DOS function assumes the form

$$N(E)=\left(\frac{g_veB}{\pi\hbar}\right)\sum_{i=0}^{i_{\max}}\sum_{n=0}^{n_{\max}}\delta'\left(E-E_{43}\right) \tag{5.18}$$

Combining (5.18) with the Fermi–Dirac occupation probability factor, the surface electron concentration can be written as

$$N_s=\left(\frac{g_veB}{\pi\hbar}\right)\sum_{i=0}^{i_{\max}}\sum_{n=0}^{n_{\max}}F_{-1}\left(\eta_3\right) \tag{5.19}$$

where $\eta_3=(k_BT)^{-1}[E_{F1}-E_{43}]$.

Combining (5.4) and (5.19), we can study the magneto QC in this case.

Using the constraints $\Delta\Box E_{g_0}$ or $\Delta\Box E_{g_0}$, from the (5.17) the 2D magneto dispersion law can be written as

$$E_{44}\left(1+\alpha E_{44}\right)=\left(n+\frac{1}{2}\right)\hbar\omega_0+S_i\left[\frac{\hbar|e|F_s\left(1+2\alpha E_{44}\right)}{\sqrt{2m_c}}\right]^{2/3} \tag{5.20}$$

where E_{44} is the total quantized energy in this case.

It may be noted that (5.20) is valid for the well-known **two band model of Kane.**

The DOS function assumes the form

$$N(E)=\left(\frac{g_veB}{\pi\hbar}\right)\sum_{i=0}^{i_{\max}}\sum_{n=0}^{n_{\max}}\delta'\left(E-E_{44}\right) \tag{5.21}$$

Combining (5.21) with the Fermi–Dirac occupation probability factor, the surface electron concentration can be written as

$$N_s = \left(\frac{g_v eB}{\pi\hbar}\right)\sum_{i=0}^{i_{\max}}\sum_{n=0}^{n_{\max}} F_{-1}(\bar{\eta}_4) \tag{5.22}$$

where $\bar{\eta}_4 = (k_B T)^{-1}[E_{F1} - E_{44}]$.

Combining (5.4) and (5.22), we can study the magneto QC in this case.

For $\alpha \to 0$, as for inversion layers, whose bulk electrons are defined by the parabolic energy bands, the magneto 2D dispersion relation can be expressed from (5.20) as

$$E_{45} = \left(n + \frac{1}{2}\right)\hbar\omega_0 + S_i\left[\frac{\hbar|e|F_s}{\sqrt{2m_c}}\right]^{2/3} \tag{5.23}$$

where E_{45} is the total quantized energy in this case.

Equation (5.23) is valid for all values of the surface electric field [1].

The density-of-states function assumes the form

$$N(E) = \left(\frac{g_v eB}{\pi\hbar}\right)\sum_{i=0}^{i_{\max}}\sum_{n=0}^{n_{\max}} \delta'(E - E_{45}) \tag{5.24}$$

Combining (5.24) with the Fermi–Dirac occupation probability factor, the surface electron concentration can be written as

$$N_s = \left(\frac{g_v eB}{\pi\hbar}\right)\sum_{i=0}^{i_{\max}}\sum_{n=0}^{n_{\max}} F_{-1}(\bar{\eta}_4) \tag{5.25}$$

where $\bar{\eta}_4 = (k_B T)^{-1}[E_{F1} - E_{45}]$.

Combining (5.4) and (5.25), we can study the magneto QC in this case.

(b) Strong electric field limit

(i) The three band Kane model

Using the substitutions $\delta = 0, \Delta_{\parallel} = \Delta_{\perp} = \Delta$ and $m_{\parallel}^* = m_{\perp}^* = m_c$, the 2D E–k_s relation under the condition of strong electric field limit, can be written from (5.12) as

$$\left[I(E) - \left\{\frac{|e|F_s\hbar}{\sqrt{2m_c}}\left(\frac{2\sqrt{2}\,(S_i)^{3/2}}{3}\right)\sqrt{[I(E)]''}\right\}\right] = \frac{\hbar^2 k_s^2}{2m_c} \tag{5.26}$$

Equation (5.26) represents the dispersion relation of the 2D electrons in QMOSFETs made of III–V, ternary and quaternary materials operated in the presence of inversion layer mode and under the strong electric field limit whose bulk conduction electrons are defined by the three band model of Kane.

The 2D magneto dispersion relation in this case can be written as

$$\left[I(E_{47}) - \left\{\frac{|e|F_s\hbar}{\sqrt{2m_c}}\left(\frac{2\sqrt{2}(S_i)^{3/2}}{3}\right)\sqrt{[I(E_{47})]''}\right\}\right] = \left(n+\frac{1}{2}\right)\hbar\omega_0 \tag{5.27}$$

where E_{47} is the total quantized energy in this case.

The DOS function assumes the form

$$N(E) = \left(\frac{g_v eB}{\pi\hbar}\right)\sum_{i=0}^{i_{\max}}\sum_{n=0}^{n_{\max}}\delta'(E - E_{47}) \tag{5.28}$$

Combining (5.28) with the Fermi–Dirac occupation probability factor, the surface electron concentration can be written as

$$N_s = \left(\frac{g_v eB}{\pi\hbar}\right)\sum_{i=0}^{i_{\max}}\sum_{n=0}^{n_{\max}}F_{-1}(\eta_6) \tag{5.29}$$

where $\eta_6 = (k_B T)^{-1}[E_{F1} - E_{47}]$.

Combining (5.4) and (5.29), we can study the magneto QC in this case.

(ii) Two band model of Kane

The 2D E–k_s relation under strong electric field limit can be written from (5.26) under the condition $\Delta \gg E_{g_0}$ or $\Delta \ll E_{g_0}$

$$E(1+\alpha E) = \frac{\hbar^2 k_s^2}{2m_c} + \frac{2\sqrt{2}}{3}(S_i)^{3/2}\left[\frac{\hbar|e|F_s}{\sqrt{m_c E_{g_0}}}\right] \tag{5.30}$$

Using (5.30), the 2D magneto dispersion relation in this case can be written as

$$E_{48}(1+\alpha E_{48}) = \left(n+\frac{1}{2}\right)\hbar\omega_0 + \frac{2\sqrt{2}}{3}(S_i)^{3/2}\left[\frac{\hbar|e|F_s}{\sqrt{2m_c E_{g_0}}}\right] \tag{5.31}$$

where E_{48} is the total quantized energy in this case.

The DOS function assumes the form

$$N(E) = \left(\frac{g_v eB}{\pi\hbar}\right)\sum_{i=0}^{i_{\max}}\sum_{n=0}^{n_{\max}}\delta'(E - E_{48}) \tag{5.32}$$

Combining (5.32) with the Fermi–Dirac occupation probability factor, the surface electron concentration can be written as

$$N_s = \left(\frac{g_v eB}{\pi\hbar}\right)\sum_{i=0}^{i_{\max}}\sum_{n=0}^{n_{\max}} F_{-1}(\eta_7) \tag{5.33}$$

where $\eta_7 = (k_B T)^{-1}[E_{F1} - E_{48}]$.

Combining (5.4) and (5.33), we can study the magneto QC in this case.

5.2.3 The QC in magneto MOSFETs of II–VI materials

In the presence of the surface electric field F_S along z-direction, (1.51) assumes the form

$$E - eF_S z = \frac{\hbar^2 k_s^2}{2m_\perp^*} + \frac{\hbar^2 k_z^2}{2m_\parallel^*} \pm \bar{\lambda}_0 k_s \tag{5.34}$$

Using (5.5) and (5.34), we get

$$\frac{\hbar^2 k_s^2}{2m_\perp^*} \pm \bar{\lambda}_0 k_s + S_i\left(\frac{\hbar e F_S}{\sqrt{2m_\parallel^*}}\right)^{2/3} - E = 0 \tag{5.35}$$

The magneto dispersion relation in this case can be written from (5.35) as

$$\left(n+\frac{1}{2}\right)\left(\frac{\hbar eB}{m_\perp^*}\right) \pm \bar{\lambda}_0\left(\left(\frac{2eB}{\hbar}\right)\left(n+\frac{1}{2}\right)\right)^{1/2} + S_i\left(\frac{\hbar e F_S}{\sqrt{2m_\parallel^*}}\right)^{2/3} - E_{49} = 0 \tag{5.36}$$

where E_{49} is the total quantized energy in this case.

The DOS function assumes the form

$$N(E) = \left(\frac{g_v eB}{2\pi\hbar}\right)\sum_{i=0}^{i_{\max}}\sum_{n=0}^{n_{\max}} \delta'(E - E_{49}) \tag{5.37}$$

Combining (5.37) with the Fermi–Dirac occupation probability factor, the surface electron concentration can be written as

$$N_s = \left(\frac{g_v eB}{2\pi\hbar}\right)\sum_{i=0}^{i_{\max}}\sum_{n=0}^{n_{\max}} F_{-1}(\eta_8) \tag{5.38}$$

where $\eta_8 = (k_B T)^{-1}[E_{F1} - E_{49}]$.

Combining (5.5) and (5.38), we can study the magneto QC in this case.

5.2.4 The QC in magneto MOSFETs of IV–VI materials

(a) Weak electric field limit

In the low electric field limit the (1.117) assumes the form

$$E(1+\alpha E) - |e|F_s z(1+2\alpha E) = \frac{p_x^2}{2M_1} + \frac{p_z^2}{2M_3} + \left(\frac{\alpha p_y^4}{4M_2M_2'}\right) + \frac{p_y^2}{2M_2} + \frac{p_y^2}{2M_2}\alpha\left(E - |e|F_s z\right)\left(1 - \frac{M_2}{M_2'}\right) \quad (5.39)$$

where $M_1 = m_{\perp c}, m_{\perp c}$ is the transverse effective electron mass at the edge of the conduction band at $k = 0$, $M_2 = \left(\frac{m_{\perp c}+2m_{\parallel c}}{3}\right)$, $m_{\parallel c}$ is the longitudinal effective electron mass at the edge of the conduction band at $k = 0$, $M_3 = \left(\frac{3m_{\perp c}m_{\parallel c}}{2m_{\parallel c}+m_{\perp c}}\right)$, $M_2' = \left(\frac{m_{\perp v}+2m_{\parallel v}}{3}\right)$, $m_{\perp v}$ and $m_{\parallel v}$ are the effective transverse and longitudinal hole masses at the edge of the valance band at $k = 0$.

The use of (5.5) and (5.39) leads to the simplified expression of the 2D electron dispersion law in n-channel inversion layers of IV–VI materials under the weak electric field limit as

$$\gamma_{71}(E,i) = p_{71}k_x^2 + q_{71}(E,i)k_y^2 + r_{71}k_y^4 \quad (5.40)$$

where

$$\gamma_{71}(E,i) \equiv \left[E(1+\alpha E) - S_i\left(\frac{\hbar|e|F_s}{\sqrt{2M_3}}\right)^{2/3}\left(1+\frac{4}{3}\alpha E\right)\right],$$

$$p_{71} \equiv \frac{\hbar^2}{2M_1}, \quad q_{71}(E,i) \equiv \left(\frac{\hbar^2}{2M_2}\right)\left[1+\alpha E\left(1-\frac{M_2}{M_2'}\right) - \frac{2\alpha S_i}{3}\left(\frac{\hbar|e|F_s}{\sqrt{2M_3}}\right)^{2/3}\left(1-\frac{M_2}{M_2'}\right)\right] \quad \text{and}$$

$$r_{71} \equiv \left(\frac{\alpha\hbar^4}{4M_2M_2'}\right)$$

The area enclosed by (5.41) is given by

$$A(E,i) = \frac{4}{3}\left(\frac{r_{71}}{p_{71}}\right)^{1/2}\left[\{a_{71w}(E,i)\}^2 + \{b_{71w}(E,i)\}^2\right]^{1/2} \times \left[\{a_{71w}(E,i)\}^2 F\left[\frac{\pi}{2}, \ell_{71w}(E,i)\right] - (\{a_{71w}(E,i)\}^2 + \{b_{71w}(E,i)\}^2) E\left[\frac{\pi}{2}, \ell_{71w}(E,i)\right]\right] \tag{5.41}$$

in which

$$\{a_{71w}(E,i)\}^2 \equiv \left[\frac{q_{71}(E,i)}{2r_{71}} + \frac{1}{2}\left[\frac{\{q_{71}(E,i)\}^2}{(r_{71})^2} + \frac{4\gamma_{71}(E,i)}{r_{71}}\right]^{1/2}\right]$$

$$\{b_{71w}(E,i)\}^2 \equiv \left[\frac{1}{2}\left[\frac{\{q_{71}(E,i)\}^2}{(r_{71})^2} + \frac{4\gamma_{71}(E,i)}{r_{71}}\right]^{1/2} - \left(\frac{q_1(E,i)}{2r_{71}}\right)\right]$$

$$\ell_{71w}(E,i) \equiv \frac{b_{71w}(E,i)}{\sqrt{\{a_{71w}(E,i)\}^2 + \{b_{71w}(E,i)\}^2}},$$

$$F\left[\frac{\pi}{2}, \ell_{71w}(E,i)\right] \quad \text{and} \quad E\left[\frac{\pi}{2}, \ell_{71w}(E,i)\right]$$

are the complete elliptic integral of the first and second kinds respectively.

The 2D magneto dispersion relation in this case can be expressed following (5.41) as

$$\left(\frac{2\pi eB}{\hbar}\right)\left(n + \frac{1}{2}\right) = \frac{4}{3}\left(\frac{r_{71}}{p_{71}}\right)^{1/2}\left[\{a_{71w}(E_{50},i)\}^2 + \{b_{71w}(E_{50},i)\}^2\right]^{1/2} \times \left[\{a_{71w}(E_{50},i)\}^2 F\left[\frac{\pi}{2}, \ell_{71w}(E_{50},i)\right] - \left(\{a_{71w}(E_{50},i)\}^2 + \{b_{71w}(E_{50},i)\}^2\right) E\left[\frac{\pi}{2}, \ell_{71w}(E_{50},i)\right]\right] \tag{5.42}$$

where E_{50} is the total quantized energy in this case.

The DOS function assumes the form

$$N(E) = \left(\frac{g_v eB}{\pi\hbar}\right)\sum_{i=0}^{i_{\max}}\sum_{n=0}^{n_{\max}} \delta'(E - E_{50}) \tag{5.43}$$

Combining (5.43) with the Fermi–Dirac occupation probability factor, the surface electron concentration can be written as

$$N_s = \left(\frac{g_v eB}{\pi\hbar}\right)\sum_{i=0}^{i_{\max}}\sum_{n=0}^{n_{\max}} F_{-1}(\eta_9) \tag{5.44}$$

where $\eta_9 = (k_B T)^{-1}[E_{F1} - E_{50}]$.

Combining (5.4) and (5.44), we can study the magneto QC in this case.

(b) Strong electric field limit

Under the strong electric field limit, the dispersion relation assumes the form

$$\gamma_{72}(E,i) = p_{72}p_x^2 + q_{72}(E)k_y^2 + r_{72}k_y^4 \tag{5.45}$$

where

$$\gamma_{72}(E,i) \equiv \left[E(1+\alpha E) - \frac{2\sqrt{2\alpha}}{3}\frac{\hbar|e|F_s}{\sqrt{2M_3}}(S_i)^{3/2}\right], \quad p_{72} \equiv \left(\frac{\hbar^2}{2M_1}\right)$$

$$q_{72}(E) \equiv \left(\frac{\hbar^2}{2M_2}\right)\left[1+\alpha E\left(1-\frac{M_2}{M_2'}\right)\right] \quad \text{and} \quad r_{72} \equiv \left[\frac{\alpha\hbar^4}{4M_2M_2'}\right]$$

The area enclosed by (5.45) is given by

$$\begin{aligned} A(E,i) = {} & \frac{4}{3}\left(\frac{r_{72}}{p_{72}}\right)^{1/2}[\{a_{72s}(E,i)\}^2 + \{b_{72s}(E,i)\}^2]^{1/2} \\ & \times \left[\{a_{72s}(E,i)\}^2 F\left[\frac{\pi}{2},\ell_{72s}(E,i)\right] - (\{a_{72s}(E,i)\}^2\right. \\ & \left. - \{b_{72s}(E,i)\}^2)E\left[\frac{\pi}{2},\ell_{72s}(E,i)\right]\right] \end{aligned} \tag{5.46}$$

in which

$$\{a_{72s}(E,i)\}^2 \equiv \left[\frac{q_{72}(E,i)}{2r_{72}} + \frac{1}{2}\left[\frac{\{q_{72}(E,i)\}^2}{(r_{72})^2} + \frac{4\gamma_{72}(E,i)}{r_{72}}\right]^{1/2}\right]$$

$$\{b_{72s}(E,i)\}^2 \equiv \left[\frac{1}{2}\left[\frac{\{q_{72}(E,i)\}^2}{(r_{72})^2} + \frac{4\gamma_{72}(E,i)}{r_{72}}\right]^{1/2} - \left(\frac{q_{72}(E,i)}{2r_{72}}\right)\right],$$

$$\ell_{72s}(E,i) \equiv \frac{b_{72s}(E,i)}{\sqrt{\{a_{72s}(E,i)\}^2 + \{b_{72s}(E,i)\}^2}}, F\left[\frac{\pi}{2},\ell_{72s}(E,i)\right]$$

and $E\left[\frac{\pi}{2}, \ell_{72s}(E,i)\right]$ are the complete elliptic integral of the first and second kinds respectively.

The 2D magneto dispersion relation in this case can be written as

$$\left(\frac{2\pi eB}{\hbar}\right)\left(n+\frac{1}{2}\right)=\frac{4}{3}\left(\frac{r_{72}}{p_{72}}\right)^{1/2}[\{a_{72s}(E_{51},i)\}^2+\{b_{72s}(E_{51}s,i)\}^2]^{1/2}$$
$$\times\left[\{a_{72s}(E_{51},i)\}^2F\left[\frac{\pi}{2},\ell_{72s}(E_{51},i)\right]-(\{a_{72s}(E_{51},i)\}^2\right.$$
$$\left.-\{b_{72s}(E_{51},i)\}^2)E\left[\frac{\pi}{2},\ell_{72s}(E_{51},i)\right]\right] \quad (5.47)$$

in which

$$\{a_{72s}(E_{51},i)\}^2\equiv\left[\frac{q_{72}(E_{51})}{2r_{72}}+\frac{1}{2}\left[\frac{\{q_{72}(E_{51})\}^2}{(r_{72})^2}+\frac{4\gamma_{72}(E_{51},i)}{r_{72}}\right]^{1/2}\right]$$

$$\{b_{72s}(E_{51},i)\}^2\equiv\left[\frac{1}{2}\left[\frac{\{q_{72}(E_{51})\}^2}{(r_{72})^2}+\frac{4\gamma_{72}(E_{51},i)}{r_{72}}\right]^{1/2}-\left(\frac{q_{72}(E_{51})}{2r_{72}}\right)\right],$$

$$\ell_{72s}(E_{51},i)\equiv\frac{b_{72s}(E_{51},i)}{\sqrt{\{a_{72s}(E_{51},i)\}^2+\{b_{72s}(E_{51},i)\}^2}}$$

$$\gamma_{72}(E_{51},i)\equiv\left[E_{51}(1+\alpha E_{51})-\frac{2\sqrt{2\alpha}}{3}\frac{\hbar|e|F_s}{\sqrt{2M_3}}(S_i)^{3/2}\right],$$

$$p_{72}\equiv\left(\frac{\hbar^2}{2M_1}\right),\quad q_{72}(E_{51})\equiv\left(\frac{\hbar^2}{2M_2}\right)\left[1+\alpha E_{51}\left(1-\frac{M_2}{M_2'}\right)\right],$$

$r_{72}\equiv\left[\frac{\alpha\hbar^4}{4M_2M_2'}\right]$ and E_{51} is the total quantized energy in this case.

The DOS function assumes the form

$$N(E)=\left(\frac{g_v eB}{\pi\hbar}\right)\sum_{i=0}^{i_{\max}}\sum_{n=0}^{n_{\max}}\delta'(E-E_{51}) \quad (5.48)$$

Combining (5.48) with the Fermi–Dirac occupation probability factor, the surface electron concentration can be written as

$$N_s=\left(\frac{g_v eB}{\pi\hbar}\right)\sum_{i=0}^{i_{\max}}\sum_{n=0}^{n_{\max}}F_{-1}(\eta_{10}) \quad (5.49)$$

where $\eta_{10}=(k_BT)^{-1}[E_{F1}-E_{51}]$.

Combining (5.4) and (5.49), we can study the magneto QC in this case.

*5.2.5 The QC in magneto MOSFETs of stressed Kane type materials

Equation (1.105) can be written as

$$\begin{aligned}\left(\bar{E}-\alpha_1\right)k_x^2+\left(\bar{E}-\alpha_2\right)k_y^2+\left(\bar{E}-\alpha_3\right)k_z^2&=t_1(\bar{E})^3\\&-t_2(\bar{E})^2+t_3\bar{E}+t_4\end{aligned}\tag{5.50}$$

where

$$\alpha_1\equiv\left[E_g-C_1\varepsilon-\left(\bar{a}_0+C_1\right)\varepsilon+\frac{3}{2}\bar{b}_0\varepsilon_{xx}-\frac{\bar{b}_0}{2}\varepsilon+\left(\frac{\sqrt{3}}{2}\right)\varepsilon_{xy}\bar{d}_0\right],$$

$$\alpha_2\equiv\left[E_g-C_1\varepsilon-\left(\bar{a}_0+C_1\right)\varepsilon+\frac{3}{2}\bar{b}_0\varepsilon_{xx}-\frac{\bar{b}_0}{2}\varepsilon-\left(\frac{\sqrt{3}}{2}\right)\varepsilon_{xy}\bar{d}_0\right],$$

$$\alpha_3\equiv\left[E_g-C_1\varepsilon-\left(\bar{a}_0+C_1\right)\varepsilon+\frac{3}{2}\bar{b}_0\varepsilon_{zz}-\frac{\bar{b}_0}{2}\varepsilon\right],$$

$$t_1\equiv\left(\frac{3}{2B_2^2}\right),\quad t_2\equiv\left(\frac{1}{2B_2^2}\right)\left[6\left(E_g-C_1\varepsilon\right)+3C_1\varepsilon\right],$$

$$t_3\equiv\left(\frac{1}{2B_2^2}\right)\left[3\left(E_g-C_1\varepsilon\right)^2+6C_1\varepsilon\left(E_g-C_1\varepsilon\right)-2{C_2}^2{\varepsilon_{xy}}^2\right]\quad\text{and}$$

$$t_4\equiv\left(\frac{1}{2B_2^2}\right)\left[-3C_1\varepsilon\left(E_g-C_1\varepsilon\right)^2+2{C_2}^2{\varepsilon_{xy}}^2\right]$$

We use E and $\bar{E}$ interchangeably throughout the book.

(a) Weak electric field limit

The use of (5.50) and (5.5) leads to the expression of the dispersion relation of the 2D electrons in n-channel inversion layers of stressed III–V materials under the low electric field limit as

$$[T_{57}(E,i)]k_x^2+[T_{67}(E,i)]k_y^2=T_{77}(E,i)\tag{5.51}$$

where

$$T_{57}(E,i)\equiv\left[E-\alpha_1+\frac{2}{3}S_i\left(\frac{|e|^2}{\varepsilon_{sc}}\right)^{2/3}\left(n_{2Dw}\right)^{2/3}L_{17}(E)\right],$$

$$L_{17}(E)\equiv\left[\frac{(E-\alpha_1)}{(E-\alpha_3)^{2/3}\left[\bar{T}_{47}(E)\right]^{1/3}}-(E-\alpha_3)^{1/3}\left[\bar{T}_{47}(E)\right]^{-1/3}\right],$$

$$[\bar{T}_{47}(E)] \equiv \left[\{\rho_5(E)\}' - \left(\frac{\rho_5(E)}{E-\alpha_3}\right)\right], \quad \rho_5(E) = \left[t_1E^3 - t_2E^2 + t_3E + t_4\right]$$

$$T_{67}(E,i) \equiv \left[E - T_2 + \frac{2}{3}S_i\left(\frac{|e|^2}{\varepsilon_{sc}}\right)^{2/3}(n_{2Dw})^{2/3}L_{27}(E)\right],$$

$$L_{27}(E) \equiv \left[\frac{(E-\alpha_2)}{(E-\alpha_3)^{2/3}\left[\bar{T}_{47}(E)\right]^{1/3}} - \left(\frac{(E-\alpha_3)^{1/3}}{\left[\bar{T}_{47}(E)\right]^{1/3}}\right)\right]$$

$$T_{77}(E,i) \equiv \left[\rho_5(E) - S_i\left(\frac{|e|^2}{\varepsilon_{sc}}\right)^{2/3}(n_{2Dw})^{2/3}L_{37}(E)\right] \quad \text{and}$$

$$L_{37}(E) \equiv (E-\alpha_3)^{1/3}\left[\bar{T}_{47}(E)\right]^{2/3}$$

The area of the 2D surface under the weak electric field limit can be written as

$$A(E,i) = \frac{\pi T_{77}(E,i)}{\sqrt{T_{57}(E,i)T_{67}(E,i)}} \tag{5.52}$$

The 2D magneto dispersion relation in this case can be written using (5.52) as

$$\left(\frac{2eB}{\hbar}\right)\left(n+\frac{1}{2}\right) = \frac{T_{77}(E_{52},i)}{\sqrt{T_{57}(E_{52},i)\,T_{67}(E_{52},i)}} \tag{5.53}$$

where E_{52} is the total quantized energy in this case.

The DOS function assumes the form

$$N(E) = \left(\frac{g_v eB}{\pi\hbar}\right)\sum_{i=0}^{i_{\max}}\sum_{n=0}^{n_{\max}}\delta'(E-E_{52}) \tag{5.54}$$

Combining (5.54) with the Fermi–Dirac occupation probability factor, the surface electron concentration can be written as

$$N_s = \left(\frac{g_v eB}{\pi\hbar}\right)\sum_{i=0}^{i_{\max}}\sum_{n=0}^{n_{\max}}F_{-1}(\eta_{11}) \tag{5.55}$$

where $\eta_{11} = (k_BT)^{-1}[E_{F1} - E_{52}]$.

Combining (5.5) and (5.55), we can study the magneto QC in this case.

(b) Strong electric field limit

The use of (5.50) and (5.5) leads to the simplified dispersion relation of the 2D electrons in n-channel inversion layers of stressed III–V materials under the high electric field limit can be expressed as

$$[T_{117}(E,i)]k_x^2 + [T_{127}(E,i)]k_y^2 = T_{137}(E,i) \tag{5.56}$$

where

$$T_{117}(E,i) \equiv \left[E - \alpha_1 + \frac{4}{3}|e|F_s\,(S_i)^{2/3}\,a_{77}(E)\right],$$

$$a_{77}(E) \equiv \frac{1}{2}\left[\frac{1}{\sqrt{(E-\alpha_3)\,[T_{97}(E)]}} - \left\{\frac{(E-\alpha_1)^2}{\sqrt{T_{97}(E)}\,(E-\alpha_3)^{3/2}}\right\}\right],$$

$$[T_{97}(E)] \equiv \left[\frac{\{\rho_5(E)\}''}{2} - \left(\frac{(\rho_5(E))'}{E-\alpha_3}\right) + \left(\frac{\rho_5(E)}{(E-\alpha_3)^2}\right)\right]$$

$$T_{127}(E,i) \equiv \left[E - \alpha_2 + \frac{4}{3}|e|F_s\,(S_i)^{2/3}\,a_{87}(E)\right],$$

$$a_{87}(E) \equiv \frac{1}{2}\left[\frac{1}{\sqrt{(E-a_3)\,T_{97}(E)}} - \left(\frac{(E-\alpha_2)^{1/3}}{\sqrt{T_{97}(E)}\,(E-T_3)^{3/2}}\right)\right]$$

$$T_{137}(E,i) \equiv \left[\rho_5(E) - (S_i)^{3/2}F_s a_{97}(E)\right] \quad \text{and}$$

$$a_{97}(E) \equiv \frac{4}{3}\sqrt{T_{97}(E)(E-\alpha_3)}$$

The area of the 2D surface in this case is given by

$$A(E,i) = \frac{\pi T_{137}(E,i)}{\sqrt{T_{117}(E,i)T_{127}(E,i)}} \tag{5.57}$$

At the high electric field limit the 2D dispersion relation and the magneto 2D dispersion law in this case can be expressed as

$$\left(\frac{2eB}{\hbar}\right)\left(n+\frac{1}{2}\right) = \left[\frac{T_{137}\,(E_{53},i)}{\sqrt{T_{117}\,(E_{53},i)\,T_{127}\,(E_{53},i)}}\right] \tag{5.58}$$

where E_{53} is the total quantized energy in this case.

The DOS function assumes the form

$$N(E) = \left(\frac{g_v eB}{\pi\hbar}\right)\sum_{i=0}^{i_{\max}}\sum_{n=0}^{n_{\max}}\delta'\,(E-E_{53}) \tag{5.59}$$

Combining (5.59) with the Fermi–Dirac occupation probability factor, the surface electron concentration can be written as

$$N_s = \left(\frac{g_v eB}{\pi\hbar}\right) \sum_{i=0}^{i_{\max}} \sum_{n=0}^{n_{\max}} F_{-1}\left(\eta_{12}\right) \tag{5.60}$$

where $\eta_{12} = (k_B T)^{-1}\left[E_{F1} - E_{53}\right]$.

Combining (5.5) and (5.60), we can study the magneto QC in this case.

5.2.6 The QC in magneto MOSFETs of bismuth telluride

The DR of the valance bands in bulk p-type Bismuth telluride can be re-written by using (1.148) as

$$\bar{E}(1+\alpha\bar{E}) = \left(I_1 k_x^2 + I_2 k_y^2 + I_3 k_z^2 + I_4 k_y k_z\right) \tag{5.61}$$

where $I_1 = \frac{\hbar^2}{2m_0}\bar{\alpha}_{11}, I_2 = \frac{\hbar^2}{2m_0}\alpha_{22}, I_3 = \frac{\hbar^2}{2m_0}\alpha_{33}$ and $I_4 = \left(\frac{\hbar^2}{m_0}\alpha_{23}\right)$.

(a) Weak electric field limit

Using (5.5) and (5.61), the 2D dispersion relation can be written as

$$E(1+\alpha E) = \left(I_2 k_y^2 + I_3 k_z^2 + I_4 k_y k_z\right) + S_i[(eF_S)^{2/3}(I_1)^{1/3}(1+2\alpha E)^{2/3}] \tag{5.62}$$

The 2D area can be expressed from (5.62) as

$$A(E) = \frac{g_v}{2\pi}\left(I_2 I_3 - I_4^2\right)^{-1/2} T_{46}(E,i) \tag{5.63a}$$

where $T_{46}(E,i) = [E(1+\alpha E) - S_i(eF_S(1+2\alpha E))^{2/3}(I_1)^{1/3}]$.

The 2D magneto dispersion relation in this case can be written as

$$\left(\frac{2eB}{\hbar}\right)\left(n+\frac{1}{2}\right) = \frac{g_v}{2\pi}\left(I_2 I_3 - I_4^2\right)^{-1/2} T_{46}\left(E_{54}, i\right) \tag{5.63b}$$

where E_{54} is the total quantized energy in this case.

The DOS function assumes the form

$$N(E) = \left(\frac{g_v eB}{\pi\hbar}\right) \sum_{i=0}^{i_{\max}} \sum_{n=0}^{n_{\max}} \delta'\left(E - E_{54}\right) \tag{5.64}$$

Combining (5.64) with the Fermi–Dirac occupation probability factor, the surface electron concentration can be written as

$$N_s = \left(\frac{g_v eB}{\pi\hbar}\right)\sum_{i=0}^{i_{\max}}\sum_{n=0}^{n_{\max}} F_{-1}(\eta_{13}) \tag{5.65}$$

where $\eta_{13} = (k_B T)^{-1}[E_{F1} - E_{54}]$.

Combining (5.5) and (5.65), we can study the magneto QC in this case.

(b) Strong electric field limit

Using (5.5) and (5.61), the 2D dispersion relation can be written as

$$E(1+\alpha E) = \left[(I_2 k_y^2 + I_3 k_z^2 + I_4 k_y k_z) + \frac{4}{3} S_i^{3/2} eF_S (I_1\alpha)^{1/2}\right] \tag{5.66}$$

The 2D area can be written as

$$A(E) = \frac{g_v}{2\pi}\left(I_2 I_3 - I_4^2\right)^{-1/2} T_{47}(E,i) \tag{5.67}$$

where $T_{47}(E,i) = [E(1+\alpha E) - \frac{4}{3} S_i^{3/2} eF_S (I_1\alpha)^{1/2}]$.

The 2D magneto dispersion relation in this case can be written as

$$\left(\frac{2eB}{\hbar}\right)\left(n+\frac{1}{2}\right) = \frac{g_v}{2\pi}\left(I_2 I_3 - I_4^2\right)^{-1/2} T_{47}(E_{55}, i) \tag{5.68}$$

where E_{55} is the total quantized energy in this case.

The DOS function assumes the form

$$N(E) = \left(\frac{g_v eB}{\pi\hbar}\right)\sum_{i=0}^{i_{\max}}\sum_{n=0}^{n_{\max}} \delta'(E - E_{55}) \tag{5.69}$$

Combining (5.69) with the Fermi–Dirac occupation probability factor, the surface electron concentration can be written as

$$N_s = \left(\frac{g_v eB}{\pi\hbar}\right)\sum_{i=0}^{i_{\max}}\sum_{n=0}^{n_{\max}} F_{-1}(\eta_{14}) \tag{5.70}$$

where $\eta_{14} = (k_B T)^{-1}[E_{F1} - E_{55}]$.

Combining (5.5) and (5.70), we can study the magneto QC in this case.

5.2.7 The QC in magneto MOSFETs of germanium

(a) Model of Cardona *et al.*

Equation (1.59) can be written as

$$\frac{\hbar^2 k_s^2}{2m_\perp^*} = \bar{E}(1+\alpha\bar{E}) + \alpha\left(\frac{\hbar^2 k_z^2}{2m_\parallel^*}\right)^2 - (1+2\alpha\bar{E})\left(\frac{\hbar^2 k_z^2}{2m_\parallel^*}\right) \tag{5.71}$$

Using (5.5) and (5.71), the 2D electron dispersion law in this case can be written as

$$\frac{\hbar^2 k_x{}^2}{2m_1} + \frac{\hbar^2 k_y{}^2}{2m_2} = \left[E(1+\alpha E) + \alpha E_{i2}{}^2 - E_{i2}(1+2\alpha E)\right] \tag{5.72}$$

where $m_1 = m_\perp^*, m_2 = \frac{m_\perp^*+2m_\parallel^*}{3}, m_3 = \frac{3m_\parallel^* m_\perp^*}{2m_\parallel^*+m_\perp^*}$ and $E_{i2} = S_i(\hbar e F_S/\sqrt{2m_3})^{2/3}$.

The area of 2D space is

$$A = \frac{2\pi\sqrt{m_1 m_2}}{\hbar^2}\left[E(1+\alpha E) + \alpha E_{i2}{}^2 - E_{i2}(1+2\alpha E)\right] \tag{5.73}$$

The 2D magneto dispersion relation in this case can be written following (5.73) as

$$\left(\frac{eB}{\hbar}\right)\left(n+\frac{1}{2}\right) = \frac{\sqrt{m_1^* m_2^*}}{\hbar^2}[E_{57}(1+\alpha E_{57}) + \alpha E_{i2D}^2 - E_{i2D}(1+2\alpha E_{57})] \tag{5.74}$$

where E_{57} is the total quantized energy in this case.

The DOS function assumes the form

$$N(E) = \left(\frac{g_v eB}{\pi\hbar}\right)\sum_{i=0}^{i_{\max}}\sum_{n=0}^{n_{\max}}\delta'(E-E_{57}) \tag{5.75}$$

Combining (5.75) with the Fermi–Dirac occupation probability factor, the surface electron concentration can be written as

$$N_s = \left(\frac{g_v eB}{\pi\hbar}\right)\sum_{i=0}^{i_{\max}}\sum_{n=0}^{n_{\max}} F_{-1}(\eta_{15}) \tag{5.76}$$

where $\eta_{15} = (k_B T)^{-1}[E_{F1} - E_{57}]$.

Combining (5.5) and (5.76), we can study the magneto QC in this case.

(b) Model of Wang *et al.*

Using (1.63) and (5.5), the 2D dispersion law in this case can be written as

$$\left[\left[1-d_1\left(\frac{\hbar^2k_x^2}{2m_1}+\frac{\hbar^2k_y^2}{2m_2}\right)\right]3^{-1}\right]\times\left[\left(1-d_1\left(\frac{\hbar^2k_x^2}{2m_1}+\frac{\hbar^2k_y^2}{2m_2}\right)\right)\right.$$
$$\times\left\{\left(1-d_1\left(\frac{\hbar^2k_x^2}{2m_1}+\frac{\hbar^2k_y^2}{2m_2}\right)\right)^2-4e_1\left(E-\left(\frac{\hbar^2k_x^2}{2m_1}+\frac{\hbar^2k_y^2}{2m_2}\right)\right)\right.$$
$$\left.\left.+4\left(\frac{\hbar^2k_x^2}{2m_1}+\frac{\hbar^2k_y^2}{2m_2}\right)+4\left(\frac{\hbar^2k_x^2}{2m_1}+\frac{\hbar^2k_y^2}{2m_2}\right)^2\right\}^{1/2}\right]^{3/2}$$
$$-\frac{1}{5}\left[\left(1-d_1\left(\frac{\hbar^2k_x^2}{2m_1}+\frac{\hbar^2k_y^2}{2m_2}\right)\right)-\left[\left(1-d_1\left(\frac{\hbar^2k_x^2}{2m_1}+\frac{\hbar^2k_y^2}{2m_2}\right)\right)^2\right.\right.$$
$$\left.\left.-4e_1\left(E-\left(\frac{\hbar^2k_x^2}{2m_1}+\frac{\hbar^2k_y^2}{2m_2}\right)\right)+C_1\left(\frac{\hbar^2k_x^2}{2m_1}+\frac{\hbar^2k_y^2}{2m_2}\right)^2\right]^{1/2}\right]^{5/2}$$
$$=\frac{2}{3}\hbar eF_S\left(e_1S_i\right)^{3/2}\left(m_3\right)^{-1/2} \tag{5.77}$$

where

$$C_1=1.4A_1\left(2m_1/\hbar^2\right)^2,\quad m_1=m_\perp^*,\quad A_1=\frac{\hbar^4}{4E_g}\left(\frac{1}{m_1}-\frac{1}{m_0}\right)^2,$$
$$d_1=0.8A\left(\frac{4m_1m_3}{\hbar^4}\right),\quad m_2=\frac{m_\perp^*+2m_\parallel^*}{3},\quad m_3=\frac{3m_\parallel^*m_\perp^*}{2m_\parallel^*+m_\perp^*},$$
$$e_1=0.005A_1\left(2m_3/\hbar^2\right)^2$$

The 2D magneto dispersion relation in this case can be written as

$$\left[\left[1-d_1\left(\left(n+\frac{1}{2}\right)\hbar\omega_{01}\right)\right]3^{-1}\right]\times\left[\left(1-d_1\left(\left(n+\frac{1}{2}\right)\hbar\omega_{01}\right)\right)\right.$$
$$\times\left\{\left(1-d_1\left(\left(n+\frac{1}{2}\right)\hbar\omega_{01}\right)\right)^2-4e_1\left(E_{59}-\left(\left(n+\frac{1}{2}\right)\hbar\omega_{01}\right)\right)\right.$$

$$+4\left(n+\frac{1}{2}\right)\hbar\omega_{01}+4\left(\left(n+\frac{1}{2}\right)\hbar\omega_{01}\right)^{2}\Bigg\}^{1/2}\Bigg]^{3/2}$$

$$-\frac{1}{5}\left[\left(1-d_1\left(\left(n+\frac{1}{2}\right)\hbar\omega_{01}\right)\right)-\left[\left(1-d_1\left(\left(n+\frac{1}{2}\right)\hbar\omega_{01}\right)\right)^{2}\right.\right.$$

$$\left.\left.-4e_1\left(E_{59}-\left(\left(n+\frac{1}{2}\right)\hbar\omega_{01}\right)\right)+C_1\left(\left(n+\frac{1}{2}\right)\hbar\omega_{01}\right)^{2}\right]^{1/2}\right]^{5/2}$$

$$=\frac{2}{3}\hbar eF_S\left(e_1S_i\right)^{3/2}\left(m_3^*\right)^{-1/2} \tag{5.78}$$

where E_{59} is the total quantized energy in this case and $\omega_{01}=\frac{eB}{\sqrt{m_1^*m_2^*}}$.

The DOS function assumes the form

$$N(E)=\left(\frac{g_veB}{\pi\hbar}\right)\sum_{i=0}^{i_{\max}}\sum_{n=0}^{n_{\max}}\delta'\left(E-E_{59}\right) \tag{5.79}$$

Combining (5.79) with the Fermi–Dirac occupation probability factor, the surface electron concentration can be written as

$$N_s=\left(\frac{g_veB}{\pi\hbar}\right)\sum_{i=0}^{i_{\max}}\sum_{n=0}^{n_{\max}}F_{-1}\left(\eta_{16}\right) \tag{5.80}$$

where $\eta_{16}=(k_BT)^{-1}[E_{F1}-E_{59}]$.

Combining (5.5) and (5.80), we can study the magneto QC in this case.

5.2.8 The QC in magneto MOSFETs of gallium aresenide

(a) Rossler's model

The dispersion relation in n-GaAs including band non-parabolacity, warping and spin-splitting can approximately be written as a simplified version of (1.34) as

$$\bar{E}=\frac{\hbar^2k^2}{2m_c}+\alpha_0k^4+\beta_0\left[k_x^2k_y^2+k_y^2k_z^2+k_z^2k_x^2\right]$$
$$\pm\bar{\gamma}_0\left[k^2\left(k_x^2k_y^2+k_y^2k_z^2+k_z^2k_x^2\right)-9k_x^2k_y^2k_z^2\right]^{1/2} \tag{5.81}$$

where $\alpha_0=-1984\times10^{-40}\text{eVm}^4,\beta_0=-1380\times10^{-40}\text{eVm}^4$ and $\bar{\gamma}_0=19\times10^{-30}\text{eVm}^3$. Using the method of spherical averaging [12], we can write

$$k^2 = \gamma_{4\pm}(\bar{E}) \tag{5.82}$$

where $\gamma_{4\pm}(\bar{E}) = (2e_2)^{-1}[+e_1 - \sqrt{e_1^2 - 4e_2[\bar{E} \pm \Gamma_4(\bar{E})]}]$,

$$e_2 = \alpha_0, e_1 = \left(\frac{\hbar^2}{2m_c} + \frac{\beta_0}{\sqrt{5}}\right), \Gamma_4(\bar{E}) = \left(e_3\left(\frac{1}{2e_2}\right)^{3/2}\right)$$
$$\times \left[+e_1 - \sqrt{e_1^2 - 4\bar{E}e_2}\right]^{3/2} \quad \text{and} \quad e_3 = 0.856\bar{\gamma}_0$$

(b) Weak electric field limit

Under this condition using (5.5) and (5.82), the 2D electron dispersion law assumes the form

$$\gamma_{4\pm}(E) = k_s^2 + S_i\left[eF_s\gamma'_{4,\pm}(E)\right]^{2/3} \tag{5.83}$$

The 2D magneto dispersion relation in this case can be written as

$$\gamma_{4\pm}(E_{60}) = \left(\frac{2eB}{\hbar}\right)\left(n + \frac{1}{2}\right) + S_i\left[eF_s\gamma'_{4,\pm}(E_{60})\right]^{2/3} \tag{5.84}$$

where E_{60} is the total quantized energy in this case.

The DOS function assumes the form

$$N(E) = \left(\frac{g_v eB}{2\pi\hbar}\right)\sum_{i=0}^{i_{\max}}\sum_{n=0}^{n_{\max}}\delta'(E - E_{60}) \tag{5.85}$$

Combining (5.85) with the Fermi–Dirac occupation probability factor, the surface electron concentration can be written as

$$N_s = \left(\frac{g_v eB}{2\pi\hbar}\right)\sum_{i=0}^{i_{\max}}\sum_{n=0}^{n_{\max}}F_{-1}(\eta_{17}) \tag{5.86}$$

where $\eta_{17} = (k_B T)^{-1}[E_{F1} - E_{60}]$.

Combining (5.5) and (5.86), we can study the magneto QC in this case.

(c) Strong electric field limit

The 2D DR can be written in this case by using (5.5) and (5.82) as

$$\gamma_{4\pm}(E) = k_s^2 + \frac{2\sqrt{2}eF_s}{3}S_i^{3/2}\sqrt{\gamma''_{4\pm}(E)} \tag{5.87}$$

The 2D magneto dispersion relation in this case can be written following (5.87) as

$$\gamma_{4\pm}(E_{61}) = \left(\frac{2eB}{\hbar}\right)\left(n+\frac{1}{2}\right) + \frac{2\sqrt{2}eF_s}{3} S_i^{3/2}\left[\gamma''_{4,\pm}(E_{61})\right]^{1/2} \tag{5.88}$$

where E_{61} is the total quantized energy in this case.

The DOS function assumes the form

$$N(E) = \left(\frac{g_v eB}{2\pi\hbar}\right)\sum_{i=0}^{i_{\max}}\sum_{n=0}^{n_{\max}} \delta'(E-E_{61}) \tag{5.89}$$

Combining (5.89) with the Fermi–Dirac occupation probability factor, the surface electron concentration can be written as

$$N_s = \left(\frac{g_v eB}{2\pi\hbar}\right)\sum_{i=0}^{i_{\max}}\sum_{n=0}^{n_{\max}} F_{-1}(\eta_{18}) \tag{5.90}$$

where $\eta_{18} = (k_B T)^{-1}[E_{F1} - E_{61}]$.

Combining (5.5) and (5.90), we can study the magneto QC in this case.

(d) Model of Stillman *et al.* [5]

Equation (1.26) can be written as

$$k^2 = \gamma_5(\bar{E}) \tag{5.91}$$

where $\gamma_5(\bar{E}) = \left(\frac{2m_c}{\hbar^2}\right) I_{11}(\bar{E})$.

(e) Weak electric field limit

The 2D DR can be written in this case by using (5.5) and (5.91) as

$$\gamma_5(E) = k_s^2 + S_i\left[eF_S\gamma'_5(E)\right]^{2/3} \tag{5.92}$$

The 2D magneto dispersion relation in this case can be written following (5.92) as

$$\gamma_5(E_{62}) = \left(\frac{2eB}{\hbar}\right)\left(n+\frac{1}{2}\right) + S_i\left[eF_s\gamma'_5(E_{62})\right]^{2/3} \tag{5.93}$$

where E_{62} is the total quantized energy in this case.

The DOS function assumes the form

$$N(E) = \left(\frac{g_v eB}{\pi\hbar}\right)\sum_{i=0}^{i_{\max}}\sum_{n=0}^{n_{\max}} \delta'(E-E_{62}) \tag{5.94}$$

Combining (5.94) with the Fermi–Dirac occupation probability factor, the surface electron concentration can be written as

$$N_s = \left(\frac{g_v eB}{\pi\hbar}\right) \sum_{i=0}^{i_{\max}} \sum_{n=0}^{n_{\max}} F_{-1}(\eta_{19}) \tag{5.95}$$

where $\eta_{19} = (k_B T)^{-1}[E_{F1} - E_{62}]$.

Combining (5.5) and (5.95), we can study the magneto QC in this case.

(f) Strong electric field limit

Using (5.5) and (5.91), the 2D DR in this case is given by

$$\gamma_5(E) = k_s^2 + \frac{(2\sqrt{2})eF_s}{3} S_i^{3/2} \sqrt{\gamma_5''(E)} \tag{5.96}$$

The 2D magneto dispersion relation in this case can be written following (5.96) as

$$\gamma_5(E_{63}) = \left(\frac{2eB}{\hbar}\right)\left(n + \frac{1}{2}\right) + \frac{(2\sqrt{2})eF_s}{3} S_i^{3/2} \left[\gamma_5'(E_{63})\right]^{1/2} \tag{5.97}$$

where E_{63} is the total quantized energy in this case.

The DOS function assumes the form

$$N(E) = \left(\frac{g_v eB}{\pi\hbar}\right) \sum_{i=0}^{i_{\max}} \sum_{n=0}^{n_{\max}} \delta'(E - E_{63}) \tag{5.98}$$

Combining (5.98) with the Fermi–Dirac occupation probability factor, the surface electron concentration can be written as

$$N_s = \left(\frac{g_v eB}{\pi\hbar}\right) \sum_{i=0}^{i_{\max}} \sum_{n=0}^{n_{\max}} F_{-1}(\eta_{20}) \tag{5.99}$$

where $\eta_{20} = (k_B T)^{-1}[E_{F1} - E_{63}]$.

Combining (5.5) and (5.99), we can study the magneto QC in this case.

*5.2.9 The QC in magneto MOSFETs of gallium antimonide

(i) The model of Seiler *et al.*

In accordance with this model (1.92) after spherical averaging assumes the form

$$k^2 = \gamma_{6,\pm}(\bar{E}) \tag{5.100}$$

where $\gamma_{6,\pm}(\bar{E}) = (2S_{1,\pm}^2)^{-1}[S_{1,\pm}(1+2\alpha\bar{E}) - \alpha_4 - [\alpha_4^2 - 4S_{1,\pm}\alpha_4(1+2\alpha\bar{E}) + 4S_{1,\pm}^2(1-20\alpha\bar{E}-28\alpha^2(\bar{E})^2)]^{1/2}]$.
in which

$$S_{1,\pm} = \left(\frac{\alpha\hbar^2}{m_0}\right)\left[\bar{\xi}_0 + 0.1\bar{v}_0 \pm 0.856\bar{w}_0\right], \quad \alpha_4 = \frac{4P_0^2\left(E_g + \frac{2}{3}\Delta\right)}{E_g^2\left(E_g+\Delta\right)} \tag{5.101}$$

(a) Weak electric field limit

The 2D DR in this case can be written following (5.5) and (5.101) as

$$\gamma_{6,\pm}(E) = k_s^2 + S_i\left[eF_S\gamma_6(E)\right]^{2/3} \tag{5.102}$$

The 2D magneto dispersion relation in this case can be written following (5.102) as

$$\gamma_{6,\pm}\left(E_{64}\right) = \left(\frac{2eB}{\hbar}\right)\left(n+\frac{1}{2}\right) + S_i\left[eF_s\gamma_{6,\pm}'\left(E_{64}\right)\right]^{2/3} \tag{5.103}$$

where E_{64} is the total quantized energy in this case.

The DOS function assumes the form

$$N(E) = \left(\frac{g_v eB}{2\pi\hbar}\right)\sum_{i=0}^{i_{\max}}\sum_{n=0}^{n_{\max}}\delta'\left(E - E_{64}\right) \tag{5.104}$$

Combining (5.104) with the Fermi–Dirac occupation probability factor, the surface electron concentration can be written as

$$N_s = \left(\frac{g_v eB}{2\pi\hbar}\right)\sum_{i=0}^{i_{\max}}\sum_{n=0}^{n_{\max}}F_{-1}\left(\eta_{21}\right) \tag{5.105}$$

where $\eta_{21} = (k_B T)^{-1}[E_{F1} - E_{64}]$.

Combining (5.5) and (5.105), we can study the magneto QC in this case.

(b) Strong electric field limit

The 2D DR in this case can be written following (5.5) and (5.100) as

$$\gamma_{6,\pm}(E) = k_s^2 + \frac{(2\sqrt{2})eF_s}{3}S_i^{3/2}\sqrt{\gamma_{6,\pm}''(E)} \tag{5.106}$$

The 2D magneto dispersion relation in this case can be written following (5.106) as

$$\gamma_{6,\pm}\left(E_{65}\right) = \left(\frac{2eB}{\hbar}\right)\left(n+\frac{1}{2}\right) + \frac{(2\sqrt{2})}{3}eF_sS_i^{3/2}\left[\gamma_{6,\pm}''\left(E_{65}\right)\right]^{1/2} \tag{5.107}$$

where E_{65} is the total quantized energy in this case.

The DOS function assumes the form

$$N(E) = \left(\frac{g_v eB}{2\pi\hbar}\right) \sum_{i=0}^{i_{\max}} \sum_{n=0}^{n_{\max}} \delta'(E - E_{65}) \tag{5.108}$$

Combining (5.108) with the Fermi–Dirac occupation probability factor, the surface electron concentration can be written as

$$N_s = \left(\frac{g_v eB}{2\pi\hbar}\right) \sum_{i=0}^{i_{\max}} \sum_{n=0}^{n_{\max}} F_{-1}(\eta_{22}) \tag{5.109}$$

where

$$\eta_{22} = (k_B T)^{-1} [E_{F1} - E_{65}]. \tag{5.110}$$

Combining (5.5) and (5.110), we can study the magneto QC in this case.

(ii) The model of Mathur *et al.*

(a) Weak electric field limit

The 2D DR in this case can be written following (1.97) and (5.5) as

$$\gamma_7(E) = k_s^2 + S_i \left[eF_s\gamma_7'(E)\right]^{2/3} \tag{5.111}$$

The 2D magneto dispersion relation in this case can be written following (5.111) as

$$\left(\frac{2eB}{\hbar}\right)\left(n + \frac{1}{2}\right) = \gamma_7(E_{66}) - S_i \left[eF_s\gamma_7'(E_{66})\right]^{2/3} \tag{5.112}$$

where E_{66} is the total quantized energy in this case.

The DOS function assumes the form

$$N(E) = \left(\frac{g_v eB}{\pi\hbar}\right) \sum_{i=0}^{i_{\max}} \sum_{n=0}^{n_{\max}} \delta'(E - E_{66}) \tag{5.113}$$

Combining (5.113) with the Fermi–Dirac occupation probability factor, the surface electron concentration can be written as

$$N_s = \left(\frac{g_v eB}{\pi\hbar}\right) \sum_{i=0}^{i_{\max}} \sum_{n=0}^{n_{\max}} F_{-1}(\eta_{23}) \tag{5.114}$$

where $\eta_{23} = (k_B T)^{-1}[E_{F1} - E_{66}]$.

Combining (5.5) and (5.114), we can study the magneto QC in this case.

(b) Strong electric field limit

The 2D DR in this case can be written following (5.5) and (5.114) as

$$\gamma_7(E) = k_s^2 + \frac{(2\sqrt{2})eF_s}{3} S_i^{3/2} \sqrt{\gamma_7''(E)} \tag{5.115}$$

The 2D magneto dispersion relation in this case can be written following (5.116) as

$$\left(\frac{2eB}{\hbar}\right)\left(n+\frac{1}{2}\right) = \gamma_7\left(E_{67}\right) - \frac{(2\sqrt{2})}{3} eF_s S_i^{3/2} \left[\gamma_7''\left(E_{67}\right)\right]^{1/2} \tag{5.116}$$

where E_{67} is the total quantized energy in this case.

The DOS function assumes the form

$$N(E) = \left(\frac{g_v eB}{\pi\hbar}\right) \sum_{i=0}^{i_{\max}} \sum_{n=0}^{n_{\max}} \delta'\left(E - E_{67}\right) \tag{5.117}$$

Combining (5.117) with the Fermi–Dirac occupation probability factor, the surface electron concentration can be written as

$$N_s = \left(\frac{g_v eB}{\pi\hbar}\right) \sum_{i=0}^{i_{\max}} \sum_{n=0}^{n_{\max}} F_{-1}\left(\eta_{24}\right) \tag{5.118}$$

where

$$\eta_{24} = (k_B T)^{-1}[E_{F1} - E_{67}]. \tag{5.119}$$

Combining (5.5) and (5.118), we can study the magneto QC in this case.

(iii) The model of Zhang *et al.*

Using the (1.101) and the method of spherical averaging together with the method of successive approximation, we get

$$k^2 = \gamma_8(\bar{E}) \tag{5.120}$$

where $\gamma_8(\bar{E}) = (2\delta_2)^{-1}[\delta_1 - \sqrt{\delta_1^2 - 4\delta_2[\bar{E} - \delta_4(\bar{E})]}]$, $\delta_2 = E_4^{(1)}$, $\delta_1 = E_2^{(1)}$, $\delta_4(\bar{E}) = \delta_3 \left[\frac{\delta_1 - \sqrt{\delta_1^2 - 4\bar{E}\delta_3}}{2\delta_3}\right]^3$ and $\delta_3 = [E_6^{(1)} + 0.094E_6^{(3)}]$.

(a) Weak electric field limit

The 2D DR in this case can be written following (5.5) and (5.120) as

$$\gamma_8(E) = k_s^2 + S_i \left[eF_s \gamma_8'(E)\right]^{2/3} \tag{5.121}$$

The 2D magneto dispersion relation in this case can be written following (5.121) as

$$\left(\frac{2eB}{\hbar}\right)\left(n+\frac{1}{2}\right)=\gamma_8\left(E_{68}\right)-S_i\left[eF_s\gamma_8'\left(E_{68}\right)\right]^{2/3} \tag{5.122}$$

where E_{68} is the total quantized energy in this case.

The DOS function assumes the form

$$N(E)=\left(\frac{g_veB}{\pi\hbar}\right)\sum_{i=0}^{i_{\max}}\sum_{n=0}^{n_{\max}}\delta'\left(E-E_{68}\right) \tag{5.123}$$

Combining (5.123) with the Fermi–Dirac occupation probability factor, the surface electron concentration can be written as

$$N_s=\left(\frac{g_veB}{\pi\hbar}\right)\sum_{i=0}^{i_{\max}}\sum_{n=0}^{n_{\max}}F_{-1}\left(\eta_{25}\right) \tag{5.124}$$

where $\eta_{25}=(k_BT)^{-1}[E_{F1}-E_{68}]$.

Combining (5.5) and (5.124), we can study the magneto QC in this case.

(b) Strong electric field limit

The 2D DR in this case can be written following (5.5) and (5.120) as

$$\gamma_8(E)=k_s^2+\frac{(2\sqrt{2})eF_s}{3}S_i^{3/2}\left[\sqrt{\gamma_8''(E)}\right]^{1/2} \tag{5.125}$$

The 2D magneto dispersion relation in this case can be written following (5.125) as

$$\left(\frac{2eB}{\hbar}\right)\left(n+\frac{1}{2}\right)=\gamma_8\left(E_{69}\right)-\frac{(2\sqrt{2})}{3}eF_sS_i^{3/2}\left[\gamma_8''\left(E_{69}\right)\right]^{1/2} \tag{5.126}$$

where E_{69} is the total quantized energy in this case.

The DOS function assumes the form

$$N(E)=\left(\frac{g_veB}{\pi\hbar}\right)\sum_{i=0}^{i_{\max}}\sum_{n=0}^{n_{\max}}\delta'\left(E-E_{69}\right) \tag{5.127}$$

Combining (5.127) with the Fermi–Dirac occupation probability factor, the surface electron concentration can be written as

$$N_s=\left(g_veB/\pi\hbar\right)\sum_{i=0}^{i_{\max}}\sum_{n=0}^{n_{\max}}F_{-1}\left(\eta_{26}\right) \tag{5.128}$$

where $\eta_{26}=(k_BT)^{-1}[E_{F1}-E_{69}]$.

Combining (5.5) and (5.128), we can study the magneto QC in this case.

5.2.10 The QC in magneto MOSFETs of indium antimonide

(i) The model of Palik *et al.* [9]

The model of Palik *et al.* is given by (1.39) which can be expressed as

$$k^2 = \gamma_9(\bar{E}) \tag{5.129}$$

where

$$\gamma_9(\bar{E}) = \left(\frac{2m_c}{\hbar^2}\right) I_{12}(\bar{E}) \tag{5.130}$$

(a) Weak electric field limit

The 2D DR in this case can be written following (5.5) and (5.130) as

$$\gamma_9(E) = k_s^2 + S_i\left[eF_s\gamma_9'(E)\right]^{2/3} \tag{5.131}$$

The 2D magneto dispersion relation in this case can be written following (5.130) as

$$\left(\frac{2eB}{\hbar}\right)\left(n+\frac{1}{2}\right) = \gamma_9\left(E_{70}\right) - S_i\left[eF_s\gamma_9'\left(E_{70}\right)\right]^{2/3} \tag{5.132}$$

where E_{70} is the total quantized energy in this case.

The DOS function assumes the form

$$N(E) = \left(\frac{g_v eB}{\pi\hbar}\right)\sum_{i=0}^{i_{\max}}\sum_{n=0}^{n_{\max}} \delta'\left(E - E_{70}\right) \tag{5.133}$$

Combining (5.133) with the Fermi–Dirac occupation probability factor, the surface electron concentration can be written as

$$N_s = \left(\frac{g_v eB}{\pi\hbar}\right)\sum_{i=0}^{i_{\max}}\sum_{n=0}^{n_{\max}} F_{-1}\left(\eta_{27}\right) \tag{5.134}$$

where $\eta_{27} = (k_BT)^{-1}[E_{F1} - E_{70}]$.

Combining (5.5) and (5.134), we can study the magneto QC in this case.

(b) Strong electric field limit

The 2D DR in this case can be written following (5.5) and (5.130) as

$$\gamma_9(E) = k_s^2 + \frac{(2\sqrt{2})eF_s}{3} S_i^{3/2}\sqrt{\gamma_9''(E)} \tag{5.135}$$

The 2D magneto dispersion relation in this case can be written following (5.135) as

$$\left(\frac{2eB}{\hbar}\right)\left(n+\frac{1}{2}\right)=\gamma_9\left(E_{71}\right)-\frac{(2\sqrt{2})}{3}eF_sS_i^{3/2}\left[\gamma_9''\left(E_{71}\right)\right]^{1/2} \tag{5.136}$$

where E_{71} is the total quantized energy in this case.

The DOS function assumes the form

$$N(E)=\left(\frac{g_v eB}{\pi\hbar}\right)\sum_{i=0}^{i_{\max}}\sum_{n=0}^{n_{\max}}\delta'\left(E-E_{71}\right) \tag{5.137}$$

Combining (5.137) with the Fermi–Dirac occupation probability factor, the surface electron concentration can be written as

$$N_s=\left(\frac{g_v eB}{\pi\hbar}\right)\sum_{i=0}^{i_{\max}}\sum_{n=0}^{n_{\max}}F_{-1}\left(\eta_{28}\right) \tag{5.138}$$

where

$$\eta_{28}=(k_BT)^{-1}[E_{F1}-E_{71}]. \tag{5.139}$$

Combining (5.5) and (5.138), we can study the magneto QC in this case.

(ii) The model of Johnson *et al.* [10]

From (1.43), we can write

$$k^2=\gamma_{10}(\bar{E}) \tag{5.140}$$

where

$$\gamma_{10}(\bar{E})=\left[\left[E_{g_0}\bar{e}_7+2\bar{E}\bar{e}_7+\frac{E_{g_0}^2}{4}\bar{e}_8(\bar{E})\right]-\left[E_{g_0}^2\left(\bar{e}_7\right)^2+\frac{E_{g_0}^4}{16}\left(\bar{e}_8(\bar{E})\right)^2\right.\right.$$
$$\left.\left.+\bar{E}\left(\bar{e}_7\right)E_{g_0}^2\left(\bar{e}_8(\bar{E})\right)+\frac{\bar{e}_7}{2}E_{g_0}^3\left(\bar{e}_8(\bar{E})\right]^{1/2}\right]$$

$$\bar{e}_7=\frac{\hbar^2}{2}\left[\frac{1}{m_c'}-\frac{1}{m_c}\right]\quad\text{and}\quad\bar{e}_8(\bar{E})=\left(\frac{2\hbar^2}{m_c'E_{g_0}}\right)$$
$$\times\left[\frac{\left(E_{g_0}+\Delta\right)\left(\bar{E}+E_{g_0}+\frac{2}{3}\Delta\right)}{\left(E_{g_0}+\frac{2}{3}\Delta\right)\left(\bar{E}+E_{g_0}+\Delta\right)}\right]$$

(a) Weak electric field limit

The 2D DR in this case can be written following (5.5) and (5.140) as

$$\gamma_{10}(E) = k_s^2 + S_i \left[eF_s\gamma_{10}'(E)\right]^{2/3} \tag{5.141}$$

The 2D magneto dispersion relation in this case can be written following (5.141) as

$$\left(\frac{2eB}{\hbar}\right)\left(n+\frac{1}{2}\right) = \gamma_{10}\left(E_{72}\right) - S_i\left[eF_s\gamma_{10}'\left(E_{72}\right)\right]^{2/3} \tag{5.142}$$

where E_{72} is the total quantized energy in this case.

The DOS function assumes the form

$$N(E) = \left(\frac{g_v eB}{\pi\hbar}\right)\sum_{i=0}^{i_{\max}}\sum_{n=0}^{n_{\max}}\delta'\left(E - E_{72}\right) \tag{5.143}$$

Combining (5.143) with the Fermi–Dirac occupation probability factor, the surface electron concentration can be written as

$$N_s = \left(\frac{g_v eB}{\pi\hbar}\right)\sum_{i=0}^{i_{\max}}\sum_{n=0}^{n_{\max}}F_{-1}\left(\eta_{29}\right) \tag{5.144}$$

where $\eta_{29} = (k_BT)^{-1}[E_{F1} - E_{72}]$.

Combining (5.5) and (5.144), we can study the magneto QC in this case.

(b) Strong electric field limit

The 2D DR in this case can be written following (5.5) and (5.140) as

$$\gamma_{10}(E) = k_s^2 + \frac{(2\sqrt{2})}{3}\left(eF_s\right)\left(S_i\right)^{3/2}\sqrt{\gamma_{10}''(E)} \tag{5.145}$$

The 2D magneto dispersion relation in this case can be written following (2.115) as

$$\left(\frac{2eB}{\hbar}\right)\left(n+\frac{1}{2}\right) = \gamma_{10}\left(E_{73}\right) - \frac{(2\sqrt{2})}{3}eF_sS_i^{3/2}\left[\gamma_{10}''\left(E_{73}\right)\right]^{1/2} \tag{5.146}$$

where E_{73} is the total quantized energy in this case.

The DOS function assumes the form

$$N(E) = \left(\frac{g_v eB}{\pi\hbar}\right) \sum_{i=0}^{i_{\max}} \sum_{n=0}^{n_{\max}} \delta'(E - E_{73}) \tag{5.147}$$

Combining (5.143) with the Fermi–Dirac occupation probability factor, the surface electron concentration can be written as

$$N_s = \left(\frac{g_v eB}{\pi\hbar}\right) \sum_{i=0}^{i_{\max}} \sum_{n=0}^{n_{\max}} F_{-1}(\eta_{30}) \tag{5.148}$$

where

$$\eta_{30} = (k_B T)^{-1}[E_{F1} - E_{73}]. \tag{5.149}$$

Combining (5.5) and (5.148), we can study the magneto QC in this case.

(iii) The model of Agafonov *et al.*

In accordance with Agafonov *et al.* the electron energy spectrum is given by (1.47) and using the method of averaging and the method of successive approximation, we can write

$$k^2 = \gamma_{11}(\bar{E}) \tag{5.150}$$

where

$$\begin{aligned}\gamma_{11}(\bar{E}) &= [8(\bar{E})^2 N_{01}^2/(2\bar{E} + E_{g_0})^2]^{-1}[-[4\bar{E}N_{01} - (8P_0^2/3)] \\ &\quad + [(4\bar{E}N_{01} - (8P_0^2/3))^2 - \{64(\bar{E})^2 N_{01}^2((\bar{E})^2 + \bar{E}E_{g_0})/ \\ &\quad \times (2\bar{E} + E_{g_0})^2\}]^{1/2}] \quad \text{and} \quad N_{01} = \frac{3}{10}(D\sqrt{3} - 3B)\end{aligned}$$

(a) Weak electric field limit

The 2D DR in this case can be written following (5.5) and (5.150) as

$$\gamma_{11}(E) = k_s^2 + S_i\left[eF_s\gamma_{11}'(E)\right]^{2/3} \tag{5.151}$$

The 2D magneto dispersion relation in this case can be written following (5.151) as

$$\left(\frac{2eB}{\hbar}\right)\left(n + \frac{1}{2}\right) = \gamma_{11}(E_{74}) - S_i\left[eF_s\gamma_{11}'(E_{74})\right]^{2/3} \tag{5.152}$$

where E_{74} is the total quantized energy in this case.

The DOS function assumes the form

$$N(E) = \left(\frac{g_v eB}{\pi\hbar}\right) \sum_{i=0}^{i_{\max}} \sum_{n=0}^{n_{\max}} \delta'(E - E_{74}) \tag{5.153}$$

Combining (5.153) with the Fermi–Dirac occupation probability factor, the surface electron concentration can be written as

$$N_s = \left(\frac{g_v eB}{\pi\hbar}\right) \sum_{i=0}^{i_{\max}} \sum_{n=0}^{n_{\max}} F_{-1}\left(\eta_{31}\right) \tag{5.154}$$

where $\eta_{31} = (k_B T)^{-1}[E_{F1} - E_{74}]$.

Combining (5.5) and (5.154), we can study the magneto QC in this case.

(b) Strong electric field limit

The 2D DR in this case can be written following (5.5) and (5.154) as

$$\gamma_{11}(E) = k_s^2 + \frac{(2\sqrt{2})}{3}\left(eF_s\right)\left(S_i\right)^{3/2}\sqrt{\gamma_{11}''(E)} \tag{5.155}$$

The 2D magneto dispersion relation in this case can be written following (5.155) as

$$\left(\frac{2eB}{\hbar}\right)\left(n + \frac{1}{2}\right) = \gamma_{11}\left(E_{75}\right) - \frac{(2\sqrt{2})}{3} eF_s S_i^{3/2}\left[\gamma_{11}''\left(E_{75}\right)\right]^{1/2} \tag{5.156}$$

where E_{75} is the total quantized energy in this case.

The DOS function assumes the form

$$N(E) = \left(\frac{g_v eB}{\pi\hbar}\right) \sum_{i=0}^{i_{\max}} \sum_{n=0}^{n_{\max}} \delta'\left(E - E_{75}\right) \tag{5.157}$$

Combining (5.157) with the Fermi–Dirac occupation probability factor, the surface electron concentration can be written as

$$N_s = \left(\frac{g_v eB}{\pi\hbar}\right) \sum_{i=0}^{i_{\max}} \sum_{n=0}^{n_{\max}} F_{-1}\left(\eta_{32}\right) \tag{5.158}$$

where

$$\eta_{32} = \left(k_B T\right)^{-1}\left[E_{F1} - E_{75}\right]. \tag{5.159}$$

Combining (5.5) and (5.158), we can study the magneto QC in this case.

5.2.11 The QC in magneto MOSFETs of tellurium

In the presence of a surface electric field (F_S) along z-direction, the (1.67) we can be written as

$$k_z^2 = a_1(\bar{E}) - b_1 z \pm \sqrt{C_1(\bar{E}) - D_1 z} \tag{5.160}$$

where

$$a_1(\bar{E}) = \left(2\psi_1^2\right)^{-1}\left[\bar{E}\psi_1 + \psi_3^2 - 2\psi_1\psi_2 k_s^2\right], \quad b_1 = (eF_s/2\psi_1),$$
$$C_1(\bar{E}) = \left(4\psi_1^4\right)^{-1}\left[\psi_3^4 + 4\psi_1\psi_3^2\bar{E} - 4\psi_1\psi_2\psi_3^2 k_s^2 + 4\psi_1^2\psi_4^2 k_s^2\right]^{1/2},$$
$$D_1 = \left[\psi_3^2 eF_s/\psi_1^3\right], \quad A_6 = \phi_1, A_7 = \phi_2, A_8 = \phi_3^2 \quad \text{and} \quad A_9 = \phi_4^2$$

Using (5.160) and (5.5), the 2D dispersion law assumes the form

$$\left[\frac{a_1(E)D_1}{b_1} \pm \frac{D_1\sqrt{C_1(E)}}{b_1}\right]^{2/3} - \left[\frac{a_1(E)D_1}{b_1} - D_1 z_{t,\mp}\right.$$
$$\left.\pm\left(\frac{D_1}{b_1}\right)\sqrt{C_1(E) - D_1 z_{t,\mp}}\right]^{2/3} \mp \frac{3}{4}\left(\frac{D_1}{b_1}\right)\left[\beta_1(E)\sqrt{\beta_1^2(E) - e_1^2(E)}\right.$$
$$\left.-\beta_2(E)\sqrt{\beta_2^2(E) - e_1^2(E)} + e_1^2(E)\ln\left|\frac{\beta_2(E) + \sqrt{\beta_2^2(E) - e_1^2(E)}}{\beta_1(E) + \sqrt{\beta_1^2(E) - e_1^2(E)}}\right|\right]$$
$$= \frac{(S_i D_1)^{3/2}}{\sqrt{b_1}} \tag{5.161}$$

where

$$e_1^2(E) = \left[\frac{D_1^2}{4b_1^2} + C_1(E) - \frac{a_1(E)D_1}{b_1}\right], \quad \beta_1(E) = \sqrt{C_1(E)} \pm \frac{D_1}{2b_1},$$
$$\beta_2(E) = \sqrt{C_1(E) - D_1 z_{t,\mp}} \pm \frac{D_1}{2b_1}, \quad Z_{t,\mp} = (eF_s)^{-1}\left[E - \psi_2 k_s^2 \mp \psi_4 k_s\right]$$

The magneto 2D dispersion relation in this case can be written following (5.161) as

$$\left[\frac{a_1\left(E_{76}\right)D_1}{b_1} \pm \frac{D_1\sqrt{C_1\left(E_{76}\right)}}{b_1}\right]^{2/3}$$
$$-\left[\frac{a_1(E_{76})D_1}{b_1} - D_1\bar{z}_{t,\mp} \pm \left(\frac{D_1}{b_1}\right)\sqrt{C_1(E_{76}) - D_1\bar{z}_{t,\mp}}\right]^{2/3}$$
$$\mp\frac{3}{4}\left(\frac{D_1}{b_1}\right)\left[\beta_1(E_{76})\sqrt{\beta_1^2(E_{76}) - e_1^2(E_{76})}\right.$$

$$- \beta_2(E_{76})\sqrt{\beta_2^2(E_{76}) - e_1^2(E_{76})}$$

$$+ e_1^2(E_{76})\lambda_n \left|\frac{\beta_2(E_{76}) + \sqrt{\beta_2^2(E_{76}) - e_1^2(E_{76})}}{\beta_1(E_{76}) + \sqrt{\beta_1^2(E_{76}) - e_1^2(E_{76})}}\right|\Bigg] = \frac{(S_i D_1)^{3/2}}{\sqrt{b_1}} \tag{5.162}$$

where

$$a_1(E_{76}) = \left(2\psi_1^2\right)^{-1}\left[E_{76}\psi_1 + \psi_3^2 - 2\psi_1\psi_2\left(\frac{2eB}{\hbar}\right)\left(n+\frac{1}{2}\right)\right],$$

$$C_1(E_{76}) = \left(4\psi_1^4\right)^{-1}$$

$$\times \left[\psi_3^4 + 4\psi_1\psi_3^2 E_{76} - 4\psi_1\psi_2\psi_3^2 k_s^2 + 4\psi_1^2\psi_4^2\left(\frac{2eB}{\hbar}\right)\left(n+\frac{1}{2}\right)\right]^{1/2}$$

$$e_1^2(E_{76}) = \left[\frac{D_1^2}{4b_1} + C_1(E_{76}) - \frac{a_1(E_{76})D_1}{b_1}\right],$$

$$\beta_1(E_{76}) = \sqrt{C_1(E_{76})} \pm \frac{D_1}{2b_1}, \quad \beta_2(E_{76}) = \sqrt{C_1(E_{76}) - D_1\bar{Z}_{t,\mp}} \pm \frac{D_1}{2b_1},$$

$$\bar{z}_{t,\mp} = (eF_s)^{-1}\left[E_{76} - \psi_2\left(\frac{2eB}{\hbar}\right)\left(n+\frac{1}{2}\right) \mp \psi_4\sqrt{\left(\frac{2eB}{\hbar}\right)\left(n+\frac{1}{2}\right)}\right]$$

and E_{76} is the total quantized energy in this case.

The DOS function assumes the form

$$N(E) = \left(\frac{g_v eB}{2\pi\hbar}\right)\sum_{i=0}^{i_{\max}}\sum_{n=0}^{n_{\max}}\delta'(E - E_{76}) \tag{5.163}$$

Combining (5.163) with the Fermi–Dirac occupation probability factor, the surface electron concentration can be written as

$$N_s = \left(\frac{g_v eB}{2\pi\hbar}\right)\sum_{i=0}^{i_{\max}}\sum_{n=0}^{n_{\max}}F_{-1}(\eta_{33}) \tag{5.164}$$

where $\eta_{33} = (k_BT)^{-1}[E_{F1} - E_{76}]$.

Combining (5.5) and (5.164), we can study the magneto QC in this case.

5.2.12 The QC in magneto QWFETs of nonlinear optical and tetragonal materials

The magneto 2D dispersion relation in this case can be written following (2.2) as

$$\psi_1(E_{77}) = \psi_2(E_{77})\left(\frac{2eB}{\hbar}\right)\left(n+\frac{1}{2}\right) + \psi_3(E_{77})(n_z\pi/d_z)^2 \tag{5.165}$$

where E_{77} is the total quantized energy in this case.

$$\gamma(E_{77}) = \psi_1(E_{77}),\, f_1(E_{77}) = \psi_2(E_{77}) \quad \text{and} \quad f_2(E_{77}) = \psi_3(E_{77})$$

The symbols of (5.165) are defined in Chapter 2.

The DOS function assumes the form

$$N(E) = \left(\frac{g_v eB}{\pi\hbar}\right)\sum_{n=0}^{n_{\max}}\sum_{n_z=1}^{n_{z\max}} \delta'(E - E_{77}) \tag{5.166}$$

Combining (5.166) with the Fermi–Dirac occupation probability factor, the surface electron concentration can be written as

$$N_s = \left(\frac{g_v eB}{\pi\hbar}\right)\sum_{n=0}^{n_{\max}}\sum_{n_z=1}^{n_{z\max}} F_{-1}(\eta_{34}) \tag{5.167}$$

where $\eta_{34} = (k_BT)^{-1}[E_{F1} - E_{77}]$.

Combining (5.5) and (5.167), we can study the magneto QC in this case.

5.2.13 The QC in magneto QWFETs of III–V, ternary and quaternary materials

(a) Three band model of Kane

The magneto 2D dispersion relation in this case can be written using (1.21) as

$$\left(n+\frac{1}{2}\right)\hbar\omega_0 + \left(\hbar^2/2m_c\right)(n_z\pi/d_z)^2 = I(E_{78}) \tag{5.168}$$

where E_{78} is the total quantized energy in this case.

The DOS function assumes the form

$$N(E) = \left(\frac{g_v eB}{\pi\hbar}\right)\sum_{n=0}^{n_{\max}}\sum_{n_z=1}^{n_{z\max}} \delta'(E - E_{78}) \tag{5.169}$$

Combining (5.169) with the Fermi–Dirac occupation probability factor, the surface electron concentration can be written as

$$N_s = \left(\frac{g_v eB}{\pi\hbar}\right) \sum_{n=0}^{n_{\max}} \sum_{n_z=1}^{n_{z\max}} F_{-1}(\eta_{35}) \tag{5.170}$$

where $\eta_{35} = (k_B T)^{-1}[E_{F1} - E_{78}]$.

Combining (5.5) and (5.170), we can study the magneto QC in this case.

(b) Two band model of Kane

The magneto 2D dispersion relation in this case can be written using (1.17) as

$$E_{79}(1 + \alpha E_{79}) = \left(n + \frac{1}{2}\right)\hbar\omega_0 + \frac{\hbar^2}{2m_c}\left(\frac{n_z \pi}{d_z}\right)^2 \tag{5.171}$$

where E_{79} is the total quantized energy in this case.

The DOS function assumes the form

$$N(E) = \left(\frac{g_v eB}{\pi\hbar}\right) \sum_{n=0}^{n_{\max}} \sum_{n_z=1}^{n_{\max}} \delta'(E - E_{79}) \tag{5.172}$$

Combining (5.172) with the Fermi–Dirac occupation probability factor, the surface electron concentration can be written as

$$N_s = \left(\frac{g_v eB}{\pi\hbar}\right) \sum_{n=0}^{n_{\max}} \sum_{n_z=1}^{n_{\max}} F_{-1}(\eta_{36}) \tag{5.173}$$

where $\eta_{36} = (k_B T)^{-1}[E_{F1} - E_{79}]$.

Combining (5.5) and (5.173), we can study the magneto QC in this case.

(c) Parabolic band model

The magneto 2D dispersion relation in this case can be written as

$$E_{80} = \left(n + \frac{1}{2}\right)\hbar\omega_0 + \frac{\hbar^2}{2m_c}\left(\frac{n_z \pi}{d_z}\right)^2 \tag{5.174}$$

where E_{80} is the total quantized energy in this case.

The DOS function assumes the form

$$N(E) = \left(\frac{g_v eB}{\pi\hbar}\right) \sum_{n=0}^{n_{\max}} \sum_{n_z=1}^{n_{z\max}} \delta'(E - E_{80}) \tag{5.175}$$

Combining (5.175) with the Fermi–Dirac occupation probability factor, the surface electron concentration can be written as

$$N_s = \left(\frac{g_v eB}{\pi\hbar}\right) \sum_{n=0}^{n_{\max}} \sum_{n_z=1}^{n_{z\max}} F_{-1}(\eta_{37}) \tag{5.176}$$

where $\eta_{37} = (k_B T)^{-1} [E_{F1} - E_{80}]$.

Combining (5.5) and (5.176), we can study the magneto QC in this case.

5.2.14 The QC in magneto QWFETs of II–VI materials

The magneto 2D dispersion relation in this case can be written using (1.51) as

$$E_{81} = a_0' \left(\frac{2eB}{\hbar}\right)\left(n + \frac{1}{2}\right) + b_0' \left(\frac{n_z \pi}{d_z}\right)^2 \pm \bar{\lambda}_0 \sqrt{\left(\frac{2eB}{\hbar}\right)\left(n + \frac{1}{2}\right)} \tag{5.177}$$

where E_{81} is the total quantized energy in this case.

The DOS function assumes the form

$$N(E) = \left(\frac{g_v eB}{2\pi\hbar}\right) \sum_{n=0}^{n_{\max}} \sum_{n_z=1}^{n_{z\max}} \delta'(E - E_{81}) \tag{5.178}$$

Combining (5.178) with the Fermi–Dirac occupation probability factor, the surface electron concentration can be written as

$$N_s = \left(\frac{g_v eB}{2\pi\hbar}\right) \sum_{n=0}^{n_{\max}} \sum_{n_z=1}^{n_{\max}} F_{-1}(\eta_{38}) \tag{5.179}$$

where $\eta_{38} = (k_B T)^{-1} [E_{F1} - E_{81}]$.

Combining (5.5) and (5.179), we can study the magneto QC in this case.

*5.2.15 The QC in magneto QWFETs of IV–VI materials

(a) The McClure and Choi model

Using this model for the DR of IV–VI compounds, for QWs following (1.109), we can write

$$E(1+\alpha E) = \frac{p_x^2}{2m_1} + \frac{p_y^2}{2m_2} + \frac{\hbar^2}{2m_3}\left(\frac{n_z \pi}{d_z}\right)^2 + \frac{p_y^2}{2m_2}\alpha E\left\{1 - \left(\frac{m_2}{m_2'}\right)\right\} + \frac{p_y^4 \alpha}{4m_2 m_2'} - \frac{\alpha p_x^2 p_y^2}{4m_1 m_2} - \frac{\alpha p_y^2 \hbar^2}{4m_2 m_3}\left(\frac{n_z \pi}{d_z}\right)^2 \tag{5.180}$$

Equation (5.180) can approximately be written as

$$\gamma_1(E,n_z) = p_1k_x^2 + q_1(E)k_y^2 + R_1(E,n_z)\,k_y^4 \tag{5.181}$$

where

$$\gamma_1(E,n_z) \equiv \left[E(1+\alpha E) - \frac{\hbar^2}{2m_3}\left(\frac{n_z\pi}{d_z}\right)^2\right],\ p_1 \equiv \frac{\hbar^2}{2m_1},$$

$$q_1(E) \equiv \frac{\hbar^2}{2m_2}\left[1+\alpha E\left(1-\frac{m_2}{m_2'}\right) - \alpha E(1+\alpha E)\right]$$

and

$$R_1(E,n_z) \equiv \left[\frac{\alpha\hbar^4}{4m_2m_2'} + \alpha\left(\frac{\hbar^2}{2m_2}\right)^2\left\{1+\alpha E\left(1-\frac{m_2}{m_2'}\right) - \frac{\alpha\hbar^2}{2m_3}\left(\frac{n_z\pi}{d_z}\right)^2\right\}\right]$$

The area enclosed by (5.181) is defined by the following integral

$$A(E,n_z) = 4\left[\frac{R_1(E,n_z)}{p_1}\right]^{1/2} J_1(E,n_z) \tag{5.182}$$

where $J_1(E,n_z) \equiv \int_0^{u_0(E,n_z)}\left[\frac{\gamma_1(E,n_z)}{R_1(E,n_z)} - \frac{q_1(E)k_y^2}{R_1(E,n_z)} - k_y^4\right]^{1/2} dk_y$
and

$$u_0(E,n_z) \equiv \left[\sqrt{\frac{q_1^2(E)}{4R_1^2(E,n_z)} + \frac{\gamma_1(E,n_z)}{R_1(E,n_z)}} - q_1(E)\right]^{1/2}$$

Thus, the area enclosed can be written as

$$\begin{aligned} A(E,n_z) = {} & \frac{4}{3}\left[\frac{R_1(E,n_z)}{p_1}\right]^{1/2}\left[a^2(E,n_z)+b^2(E,n_z)\right]^{1/2} \\ & \times\left[a^2(E,n_z)F\left[\frac{\pi}{2},\ell(E,n_z)\right] - \left[a^2(E,n_z)-b^2(E,n_z)\right]\right. \\ & \left.\times E\left[\frac{\pi}{2},\ell(E,n_z)\right]\right] \end{aligned} \tag{5.183}$$

where

$$a^2(E,n_z) \equiv \frac{q_1(E)}{2R_1(E,n_z)} + \frac{1}{2}\left[\frac{q_1^2(E)}{R_1^2(E,n_z)} + \frac{4\gamma_1(E,n_z)}{R_1(E,n_z)}\right]^{1/2},$$

$$b^2(E,n_z) \equiv \left[\frac{1}{2}\left[\frac{q_1^2(E)}{R_1^2(E,n_z)} + \frac{4\gamma_1(E,n_z)}{R_1(E,n_z)}\right]^{1/2} - \left(\frac{q_1(E)}{2R_1(E,n_z)}\right)\right] \quad \text{and}$$

$$\ell(E,n_z) \equiv \frac{b(E,n_z)}{\sqrt{a^2(E,n_z)+b^2(E,n_z)}}$$

The magneto 2D dispersion relation in this case can be written using (5.183) as

$$\left(\frac{2eB}{\hbar}\right)\left(n+\frac{1}{2}\right)=\frac{4}{3}\left[\left[\frac{R_1(E,n_z)}{p_1}\right]^{1/2}\left[a^2(E,n_z)+b^2(E,n_z)\right]^{1/2}\right.$$
$$\times\left[a^2(E,n_z)F\left[\frac{\pi}{2},\ell(E,n_z)\right]-\left[a^2(E,n_z)-b^2(E,n_z)\right]\right.$$
$$\left.\left.\left.\times E\left[\frac{\pi}{2},\ell(E,n_z)\right]\right]\right]\right|_{E=E_{82}} \tag{5.184}$$

where E_{82} is the total quantized energy in this case.

The DOS function assumes the form

$$N(E)=\left(\frac{g_v eB}{\pi\hbar}\right)\sum_{n=0}^{n_{\max}}\sum_{n_z=1}^{n_{z\max}}\delta'(E-E_{82}) \tag{5.185}$$

Combining (5.185) with the Fermi–Dirac occupation probability factor, the surface electron concentration can be written as

$$N_s=\left(\frac{g_v eB}{\pi\hbar}\right)\sum_{n=0}^{n_{\max}}\sum_{n_z=1}^{n_{z\max}}F_{-1}(\eta_{39}) \tag{5.186}$$

where $\eta_{39}=(k_BT)^{-1}[E_{F1}-E_{82}]$.

Combining (5.5) and (5.186), we can study the magneto QC in this case.

(b) The Hybrid model

The magneto 2D dispersion relation in this case can be written following (1.113) as

$$\left(\frac{2eB}{\hbar}\right)\left(n+\frac{1}{2}\right)=\frac{2\pi\sqrt{m_1m_3}}{\hbar^2}t_{29}(E_{83},n_y) \tag{5.187}$$

where E_{83} is the total quantized energy in this case and

$$t_{29}(E_{83},n_y)=\left[E_{83}(1+E_{83})-\frac{\theta_0(E_{83})}{2M_2}\hbar^2\left(\frac{n_y\pi}{d_y}\right)^2-\frac{d\bar{\gamma}_0\hbar^4}{4M_2^2}\left(\frac{n_y\pi}{d_y}\right)^4\right]$$

The DOS function assumes the form

$$N(E)=\left(\frac{g_v eB}{\pi\hbar}\right)\sum_{n=0}^{n_{\max}}\sum_{n_y=1}^{n_{y\max}}\delta'(E-E_{83}) \tag{5.188}$$

Combining (5.188) with the Fermi–Dirac occupation probability factor, the surface electron concentration can be written as

$$N_s = \left(\frac{g_v eB}{\pi\hbar}\right) \sum_{n=0}^{n_{\max}} \sum_{n_y=1}^{n_{y\max}} F_{-1}\left(\eta_{40}\right) \tag{5.189}$$

where $\eta_{40} = (k_B T)^{-1}[E_{F1} - E_{83}]$.

Combining (5.5) and (5.189), we can study the magneto QC in this case.

(c) The Cohen model

The magneto 2D dispersion relation in this case can be written using (??) as

$$\begin{aligned}\left(\frac{2eB}{\hbar}\right)\left(n+\frac{1}{2}\right) &= \frac{4}{3}\left[\frac{R_{1c}\left(E_{84}, n_z\right)}{p_1}\right]^{1/2} [a_c^2(E_{84}, n_z) + b_c^2(E_{84}, n_z)]^{1/2} \\ &\times \left[a_c^2\left(E_{84}, n_z\right) F\left[\frac{\pi}{2}, \ell\left(E_{84}, n_z\right)\right] - \left[a_c^2\left(E_{84}, n_z\right) - b_c^2\left(E_{84}, n_z\right)\right]\right. \\ &\left.\times\ E\left[\frac{\pi}{2}, \ell\left(E_{84}, n_z\right)\right]\right]\end{aligned} \tag{5.190}$$

where

$$a_c^2\left(E_{84}, n_z\right) \equiv \frac{q_2\left(E_{84}\right)}{2R_2} + \frac{1}{2}\left[\frac{q_2^2\left(E_{84}\right)}{R_2^2} + \frac{4\gamma_1\left(E_{84}, n_z\right)}{R_2}\right]^{1/2}$$

$$b_c^2\left(E_{84}, n_z\right) \equiv \frac{1}{2}\left[\frac{q_2^2\left(E_{84}\right)}{R_2^2} + \frac{4\gamma_1\left(E_{84}, n_z\right)}{R_2}\right]^{1/2} - \left(\frac{q_2\left(E_{84}\right)}{2R_2}\right),$$

$$\ell(E_{84}, n_z) \equiv \frac{b_c(E_{84}, n_z)}{\sqrt{a_c^2\left(E_{84}, n_z\right) + b_c^2(E_{84}, n_z)}},$$

$$q_2(E_{84}) = \left[\frac{\hbar^2}{2m_2}(1+\alpha E_{84}) - \frac{\alpha E_{84}\hbar^2}{2m_2'}\right], \quad R_2 = \frac{\alpha\hbar^4}{4m_2 m_2'} \quad \text{and}$$

$$\gamma_1(E_{84}, n_z) = \left[E_{84}\left(1+\alpha E_{84}\right) - \frac{\hbar^2}{2m_3}\left(\frac{n_z\pi}{d_z}\right)^2\right]$$

where E_{84} is the total quantized energy in this case.

The DOS function assumes the form

$$N(E) = \left(\frac{g_v eB}{\pi\hbar}\right) \sum_{n=0}^{n_{\max}} \sum_{n_z=1}^{n_{z\max}} \delta'\left(E - E_{84}\right) \tag{5.191}$$

Combining (5.191) with the Fermi–Dirac occupation probability factor, the surface electron concentration can be written as

$$N_s = \left(\frac{g_v eB}{\pi\hbar}\right) \sum_{n=0}^{n_{\max}} \sum_{n_z=1}^{n_{z\max}} F_{-1}\left(\eta_{41}\right) \tag{5.192}$$

where $\eta_{41} = (k_B T)^{-1}[E_{F1} - E_{84}]$.

Combining (5.5) and (5.192), we can study the magneto QC in this case.

(d) The Lax model

The magneto 2D dispersion relation in this case can be written following (1.121) as

$$E_{85}\left(1 + \alpha E_{85}\right) = \left(n + \frac{1}{2}\right)\hbar \frac{eB}{\sqrt{m_1 m_2}} + \frac{\hbar^2}{2m_3}\left(\frac{n_z \pi}{d_z}\right)^2 \tag{5.193}$$

where E_{85} is the total quantized energy in this case.

The DOS function assumes the form

$$N(E) = \left(\frac{g_v eB}{\pi\hbar}\right) \sum_{n=0}^{n_{\max}} \sum_{n_z=1}^{n_{z\max}} \delta'\left(E - E_{85}\right) \tag{5.194}$$

Combining (5.194) with the Fermi–Dirac occupation probability factor, the surface electron concentration can be written as

$$N_s = \left(\frac{g_v eB}{\pi\hbar}\right) \sum_{n=0}^{n_{\max}} \sum_{n_z=1}^{n_{\max}} F_{-1}\left(\eta_{42}\right) \tag{5.195}$$

where $\eta_{42} = (k_B T)^{-1}[E_{F1} - E_{85}]$.

Combining (5.5) and (5.195), we can study the magneto QC in this case.

(e) The Ellipsoidal Parabolic model

The magneto 2D dispersion relation in this case can be written following (1.124b) as

$$E_{87} = \left(n + \frac{1}{2}\right)\hbar \frac{eB}{\sqrt{m_1 m_2}} + \left(\frac{\hbar^2}{2m_3}\right)\left(\frac{n_z \pi}{d_z}\right)^2 \tag{5.196}$$

where E_{87} is the total quantized energy in this case.

The DOS function assumes the form

$$N(E) = \left(\frac{g_v eB}{\pi\hbar}\right) \sum_{n=0}^{n_{\max}} \sum_{n_z=1}^{n_{z\max}} \delta'\left(E - E_{87}\right) \tag{5.197}$$

Combining (5.197) with the Fermi–Dirac occupation probability factor, the surface electron concentration can be written as

$$N_s = \left(\frac{g_v eB}{\pi\hbar}\right) \sum_{n=0}^{n_{\max}} \sum_{n_z=1}^{n_{z\max}} F_{-1}\left(\eta_{44}\right) \tag{5.198}$$

where $\eta_{44} = (k_B T)^{-1}[E_{F1} - E_{87}]$.

Combining (5.5) and (5.198), we can study the magneto QC in this case.

(f) Dimmock model

Using (1.128) and the related substitutions, the same equation can be recasted as

$$\left[E - \frac{\hbar^2 k_s^2}{2m_t^-} - \frac{\hbar^2 k_z^2}{2m_l^-}\right]\left[1 + \alpha E + \alpha\frac{\hbar^2 k_s^2}{2m_t^+} + \alpha\frac{\hbar^2 k_z^2}{2m_l^+}\right] = \frac{\hbar^2 k_s^2}{2m_t^*} + \frac{\hbar^2 k_z^2}{2m_l^*} \tag{5.199}$$

The 2D DR of the conduction electrons in QWTs of IV–VI materials in accordance with Dimmock model for the dimensional quantization along z-direction can be expressed as

$$\begin{aligned}
&E(1+\alpha E) + \alpha E\left(\frac{\hbar^2 k_x^2}{2x_4} + \frac{\hbar^2 k_y^2}{2x_5}\right) + \alpha E\frac{\hbar^2}{2x_6}\left(\frac{n_z\pi}{d_z}\right)^2 \\
&\quad - (1+\alpha E)\left(\frac{\hbar^2 k_x^2}{2x_1} + \frac{\hbar^2 k_y^2}{2x_2}\right) - \alpha\left(\frac{\hbar^2 k_x^2}{2x_1} + \frac{\hbar^2 k_y^2}{2x_2}\right)\left(\frac{\hbar^2 k_x^2}{2x_4} + \frac{\hbar^2 k_y^2}{2x_5}\right) \\
&\quad - \alpha\left(\frac{\hbar^2 k_x^2}{2x_1} + \frac{\hbar^2 k_y^2}{2x_2}\right)\frac{\hbar^2}{2x_6}\left(\frac{n_z\pi}{d_z}\right)^2 - (1+\alpha E)\frac{\hbar^2}{2x_3}\left(\frac{n_z\pi}{d_z}\right)^2 \\
&\quad - \alpha\frac{\hbar^2}{2x_3}\left(\frac{n_z\pi}{d_z}\right)^2\left(\frac{\hbar^2 k_x^2}{2x_4} + \frac{\hbar^2 k_y^2}{2x_5}\right) - \alpha\frac{\hbar^2}{2x_3}\left(\frac{n_z\pi}{d_z}\right)^2\frac{\hbar^2}{2x_6}\left(\frac{n_z\pi}{d_z}\right)^2 \\
&= \frac{\hbar^2 k_x^2}{2m_1} + \frac{\hbar^2 k_y^2}{2m_2} + \frac{\hbar^2}{2m_3}\left(\frac{n_z\pi}{d_z}\right)^2
\end{aligned} \tag{5.200}$$

where

$$x_4 = m_t^+, x_5 = \frac{m_t^+ + 2m_l^+}{3}, \quad x_6 = \frac{3m_t^+ m_l^+}{2m_l^+ + m_t^+}, \quad x_1 = m_t^-,$$

$$x_2 = \frac{m_t^- + 2m_l^-}{3}, \quad x_3 = \frac{3m_t^- m_l^-}{2m_l^- + m_t^-},$$

$$m_1 = m_t^*, \quad m_2 = \frac{m_t^* + 2m_l^*}{3} \quad \text{and} \quad m_3 = \frac{3m_l^* m_t^*}{m_t^* + 2m_l^*}$$

Substituting $k_x = r\,\mathrm{Cos}\,\theta$ and $k_y = r\,\mathrm{Sin}\,\theta$ (where r and θ are 2D polar coordinates in 2D wave vector space) in (5.200), we can write

$$\begin{aligned}
&r^4\left[\alpha\frac{1}{4}\left(\frac{\hbar^2\mathrm{Cos}^2\theta}{x_1}+\frac{\hbar^2\mathrm{Sin}^2\theta}{x_2}\right)\left(\frac{\hbar^2\mathrm{Cos}^2\theta}{x_4}+\frac{\hbar^2\mathrm{Sin}^2\theta}{x_5}\right)\right]\\
&+r^2\frac{1}{2}\left[\left(\left(\frac{\hbar^2\mathrm{Cos}^2\theta}{m_1}+\frac{\hbar^2\mathrm{Sin}^2\theta}{m_2}\right)+\alpha\frac{\hbar^2}{2x_3}\left(\frac{n_z\pi}{d_z}\right)^2\right.\right.\\
&\times\left(\frac{\hbar^2\mathrm{Cos}^2\theta}{x_4}+\frac{\hbar^2\mathrm{Sin}^2\theta}{x_5}\right)+\alpha\left(\frac{\hbar^2\mathrm{Cos}^2\theta}{x_1}+\frac{\hbar^2\mathrm{Sin}^2\theta}{x_2}\right)\frac{\hbar^2}{2x_6}\left(\frac{n_z\pi}{d_z}\right)^2\\
&\left.+\hbar^2(1+\alpha E)\left(\frac{\mathrm{Cos}^2\theta}{x_1}+\frac{\mathrm{Sin}^2\theta}{x_2}\right)-\hbar^2\alpha E\left(\frac{\mathrm{Cos}^2\theta}{x_4}+\frac{\mathrm{Sin}^2\theta}{x_5}\right)\right]\\
&-\left[E(1+\alpha E)+\alpha E\frac{\hbar^2}{2x_6}\left(\frac{n_z\pi}{d_z}\right)^2-(1+\alpha E)\frac{\hbar^2}{2x_3}\left(\frac{n_z\pi}{d_z}\right)^2\right.\\
&\left.-\alpha\left(\frac{\hbar^4}{4x_3x_6}\left(\frac{n_z\pi}{d_z}\right)^4\right)\right]=0
\end{aligned} \tag{5.201}$$

The area $A(E, n_z)$ of the 2D wave vector space can be expressed as

$$A(E, n_z) = \bar{J}_1 - \bar{J}_2 \tag{5.202}$$

where

$$\bar{J}_1 \equiv 2\int_0^{\pi/2} \frac{c}{b}\,d\theta \tag{5.203}$$

and

$$\bar{J}_2 \equiv 2\int_0^{\pi/2} \frac{ac^2}{b^3}\,d\theta \tag{5.204}$$

in which

$$a \equiv \left[\alpha\left(\frac{\hbar^4}{4}\right)\left(\frac{\mathrm{Cos}^2\theta}{x_1}+\frac{\mathrm{Sin}^2\theta}{x_2}\right)\left(\frac{\mathrm{Cos}^2\theta}{x_4}+\frac{\mathrm{Sin}^2\theta}{x_5}\right)\right],$$

$$b \equiv \left(\frac{\hbar^2}{2}\right)\left[\left(\frac{\mathrm{Cos}^2\theta}{m_1}+\frac{\mathrm{Sin}^2\theta}{m_2}\right)+\alpha\left(\frac{\hbar^2}{2x_3}\right)\left(\frac{n_z\pi}{d_z}\right)^2\left(\frac{\mathrm{Cos}^2\theta}{x_4}+\frac{\mathrm{Sin}^2\theta}{x_5}\right)\right.$$

$$+\alpha\left(\frac{\hbar^2}{2x_6}\right)\left(\frac{n_z\pi}{d_z}\right)^2\left(\frac{\mathrm{Cos}^2\theta}{m_1}+\frac{\mathrm{Sin}^2\theta}{m_2}\right)$$

$$\left.+(1+\alpha E)\left(\frac{\mathrm{Cos}^2\theta}{x_1}+\frac{\mathrm{Sin}^2\theta}{x_2}\right)-\alpha E\left(\frac{\mathrm{Cos}^2\theta}{x_4}+\frac{\mathrm{Sin}^2\theta}{x_5}\right)\right] \quad \text{and}$$

$$c \equiv \left[E(1+\alpha E)+\alpha E\left(\frac{\hbar^2}{2x_6}\right)\left(\frac{n_z\pi}{d_z}\right)^2-(1+\alpha E)\left(\frac{\hbar^2}{2x_3}\right)\left(\frac{n_z\pi}{d_z}\right)^2\right.$$

$$\left.-\alpha\left(\frac{\hbar^4}{4x_3x_6}\right)\left(\frac{n_z\pi}{d_z}\right)^4\right]$$

Equation (5.203) can be expressed as

$$\bar{J}_1 = 2\int_0^{\pi/2}\frac{t_3(E,n_z)\,d\theta}{A_1(E,n_z)\,\mathrm{Cos}^2\theta+B_1(E,n_z)\,\mathrm{Sin}^2\theta}$$

where $t_3(E,n_z) \equiv c,\ A_1(E,n_z) \equiv \frac{\hbar^2}{2m_1}t_1(E,n_z)$,

$$t_1(E,n_z) \equiv \left[1+m_1\left[\frac{1}{x_4}\frac{\alpha\hbar^2}{2x_3}\left(\frac{n_z\pi}{d_z}\right)^2+\frac{\alpha\hbar^2}{2x_1x_6}\left(\frac{n_z\pi}{d_z}\right)^2\right.\right.$$

$$\left.\left.+\frac{1+\alpha E}{x_1}-\frac{\alpha E}{x_4}\right]\right],$$

$$B_1(E,n_z) \equiv \frac{\hbar^2}{2m_2}t_2(E,n_z) \quad \text{and}$$

$$t_2(E,n_z) \equiv \left[1+m_2\left[\frac{\alpha\hbar^2}{2x_3x_5}\left(\frac{n_z\pi}{d_z}\right)^2+\frac{\alpha\hbar^2}{2x_2x_6}\left(\frac{n_z\pi}{d_z}\right)^2\right.\right.$$

$$\left.\left.+\frac{1+\alpha E}{x_2}-\frac{\alpha E}{x_5}\right]\right]$$

Performing the integration, we get

$$\bar{J}_1 = \pi t_3(E, n_z) \left[A_1(E, n_z) B_1(E, n_z)\right]^{-1/2} \tag{5.205}$$

From (5.205), we can write

$$\bar{J}_2 = \frac{\alpha t_3^2 (E, n_z) \hbar^4}{2B_1^3 (E, n_z)} I \tag{5.206}$$

where

$$I \equiv \int_0^\infty \frac{\left(a_1 + a_2 z^2\right)\left(a_3 + a_4 z^2\right) dz}{\left[(\bar{a})^2 + z^2\right]^3} \tag{5.207}$$

in which $a_1 \equiv \frac{1}{x_1}, a_2 \equiv \frac{1}{x_2}, z = \tan\theta, \theta$ is a new variable, $a_3 \equiv \frac{1}{x_4}, a_4 \equiv \frac{1}{x_5}$ and $(\bar{a})^2 \equiv \left(\frac{A_1(E,n_z)}{B_1(E,n_z)}\right)$. The use of the Residue theorem of Complex Analysis leads to the evaluation of the integral in (3.64) as

$$I = \frac{\pi}{4\bar{a}} \left[a_1 a_4 + 3a_2 a_4\right] \tag{5.208}$$

Therefore, the 2D area of the 2D wave vector space can be written as

$$A(E, n_z) = \frac{\pi t_3 (E, n_z)}{\sqrt{A_1 (E, n_z) B_1 (E, n_z)}} \left[1 - \frac{1}{x_5}\left(\frac{1}{x_1} + \frac{3}{x_2}\right) \frac{\alpha t_3 (E, n_z) \hbar^4}{8B_1^2 (E, n_z)}\right] \tag{5.209}$$

The magneto 2D dispersion relation in this case can be written following (2.131) as

$$\left(\frac{2eB}{\hbar}\right)\left(n + \frac{1}{2}\right) = \frac{t_{31} (E_{86}, n_z)}{\sqrt{A_{11} (E_{86}, n_z) B_{11} (E_{86}, n_z)}} \times \left[1 - \frac{1}{x^5}\left(\frac{1}{x_1} + \frac{3}{x_2}\right) \frac{\alpha t_{31} (E_{86}, n_z) \hbar^4}{8B_{11}^2 (E_{86}, n_z)}\right] \tag{5.210}$$

where E_{86} is the total quantized energy in this case.

The DOS function assumes the form

$$N(E) = \left(\frac{g_v eB}{\pi\hbar}\right) \sum_{n=0}^{n_{\max}} \sum_{n_z=1}^{n_{z\max}} \delta' (E - E_{86}) \tag{5.211}$$

Combining (5.211) with the Fermi–Dirac occupation probability factor, the surface electron concentration can be written as

$$N_s = \left(\frac{g_v eB}{\pi\hbar}\right) \sum_{n=0}^{n_{\max}} \sum_{n_z=1}^{n_{z\max}} F_{-1}(\eta_{43}) \tag{5.212}$$

where $\eta_{43} = (k_B T)^{-1}[E_{F1} - E_{86}]$.

Combining (5.5) and (5.212), we can study the magneto QC in this case.

5.2.16 The QC in magneto QWFETs of stressed Kane type materials

The 2D electron dispersion law in this case assumes the form

$$\left[\frac{k_x}{a^*(E)}\right]^2 + \left[\frac{k_y}{b^*(E)}\right]^2 + \left[\frac{n_z\pi}{d_z c^*(E)}\right]^2 = 1 \tag{5.213}$$

The 2D area enclosed by (5.213) can be expressed as

$$A(E, n_z) = \pi P^2(E, n_z) \tag{5.214}$$

where $P^2(E, n_z) = [1 - [n_z\pi/d_z c^*(E)]^2]$.

The magneto 2D dispersion relation in this case can be written from (1.105) as

$$\left(\frac{2eB}{\hbar}\right)\left(n + \frac{1}{2}\right) = P^2(E_{88}, n_z)\, a^*(E_{88})\, b^*(E_{88}) \tag{5.215}$$

where E_{88} is the total quantized energy in this case.

The DOS function assumes the form

$$N(E) = \left(\frac{g_v eB}{\pi\hbar}\right) \sum_{n=0}^{n_{\max}} \sum_{n_z=1}^{n_{z\max}} \delta'(E - E_{88}) \tag{5.216}$$

Combining (5.216) with the Fermi–Dirac occupation probability factor, the surface electron concentration can be written as

$$N_s = \left(\frac{g_v eB}{\pi\hbar}\right) \sum_{n=0}^{n_{\max}} \sum_{n_z=1}^{n_{z\max}} F_{-1}(\eta_{45}) \tag{5.217}$$

where $\eta_{45} = (k_B T)^{-1}[E_{F1} - E_{88}]$.

Combining (5.5) and (5.217), we can study the magneto QC in this case.

5.2.17 The QC in magneto QWFETs of bismuth telluride

The 2D electron dispersion law in the present case can be expressed from (1.148) as

$$E(1+\alpha E) = \frac{\hbar^2}{2m_0}\left((\bar{\alpha}_{11})\left(\frac{n_x\pi}{d_x}\right)^2 + \alpha_{22}{k_y}^2 + \alpha_{33}{k_z}^2 + 2\alpha_{23}k_yk_z\right) \tag{5.218}$$

The area of the ellipse is given by

$$A_n(E, n_x) = \frac{\pi}{\sqrt{\alpha_{22}\alpha_{33} - 4{\alpha_{23}}^2}}\left[\frac{2m_0E(1+\alpha E)}{\hbar^2} - \bar{\alpha}_{11}\left(\frac{n_x\pi}{d_x}\right)^2\right] \tag{5.219}$$

The magneto 2D dispersion relation in this case can be written following (5.217) as

$$\left(\frac{2eB}{\hbar}\right)\left(n+\frac{1}{2}\right) = \frac{\pi}{\sqrt{\alpha_{22}\alpha_{33} - 4{\alpha_{23}}^2}}\left[\frac{2m_0E(1+\alpha E)}{\hbar^2} - \bar{\alpha}_{11}\left(\frac{n_x\pi}{d_x}\right)^2\right] \tag{5.220}$$

where E_{89} is the total quantized energy in this case.

The DOS function assumes the form

$$N(E) = \left(\frac{g_v eB}{\pi\hbar}\right)\sum_{n=0}^{n_{\max}}\sum_{n_z=1}^{n_{z\max}} \delta'(E - E_{89}) \tag{5.221}$$

Combining (5.221) with the Fermi–Dirac occupation probability factor, the surface electron concentration can be written as

$$N_s = \left(\frac{g_v eB}{\pi\hbar}\right)\sum_{n=0}^{n_{\max}}\sum_{n_z=1}^{n_{z\max}} F_{-1}(\eta_{46}) \tag{5.222}$$

where $\eta_{46} = (k_BT)^{-1}[E_{F1} - E_{89}]$.

Combining (6.4) and (5.222), we can study the magneto QC in this case.

5.2.18 The QC in magneto QWFETs of germanium

(a) Model of Cardona *et al.*

Equation (1.59) can be written as

$$\frac{\hbar^2 k_s^2}{2m_\perp^*} = \bar{E}(1+\alpha\bar{E}) + \alpha\left(\frac{\hbar^2 k_z^2}{2m_\parallel^*}\right)^2 - (1+2\alpha\bar{E})\left(\frac{\hbar^2 k_z^2}{2m_\parallel^*}\right) \tag{5.223}$$

The 2D electron dispersion law in this case can be written from (5.221) as

$$\frac{\hbar^2 k_x^2}{2m_1} + \frac{\hbar^2 k_y^2}{2m_2} = \left[E(1+\alpha E) + \alpha E_{i2}^2 - E_{i2}(1+2\alpha E)\right] \tag{5.224}$$

where $m_1 = m_\perp^*, m_2 = \frac{m_\perp^*+2m_\parallel^*}{3}, m_3 = \frac{3m_\perp^* m_\perp^*}{2m_\parallel^*+m_\perp^*}$ and $E_{i2} = S_i(\hbar eF_S/\sqrt{2m_3})^{2/3}$.

The area of 2D space is

$$A = \frac{2\pi\sqrt{m_1 m_2}}{\hbar^2}\left[E(1+\alpha E) + \alpha E_{i2}^2 - E_{i2}(1+2\alpha E)\right] \tag{5.225}$$

The 2D magneto dispersion relation in this case can be written following (5.223) as

$$\left(\frac{eB}{\hbar}\right)\left(n+\frac{1}{2}\right) = \frac{\sqrt{m_1 m_2}}{\hbar^2}\left[E_{57}\left(1+\alpha E_{57}\right) + \alpha E_{i2}^2 - E_{i2}\left(1+2\alpha E_{57}\right)\right] \tag{5.226}$$

where E_{57} is the total quantized energy in this case.

The DOS function assumes the form

$$N(E) = \left(\frac{g_v eB}{\pi\hbar}\right)\sum_{i=0}^{i_{\max}}\sum_{n=0}^{n_{z\max}} \delta'(E-E_{57}) \tag{5.227}$$

Combining (5.227) with the Fermi–Dirac occupation probability factor, the surface electron concentration can be written as

$$N_s = \left(\frac{g_v eB}{\pi\hbar}\right)\sum_{i=0}^{i_{\max}}\sum_{n=0}^{n_{z\max}} F_{-1}\left(\eta_{15}\right) \tag{5.228}$$

where $\eta_{15} = (k_B T)^{-1}[E_{F1} - E_{57}]$.

Combining (5.5) and (5.228), we can study the magneto QC in this case where E_{90} is the totally quantized energy in this case.

(b) The model of Wang *et al.*

The 2D dispersion law can be expressed for this model using (1.63) as

$$E = A_5\left(n_z\right) + A_6\left(n_z\right)\beta - c_1\beta^2 \tag{5.229}$$

where

$$A_5(n_z) \equiv \frac{\hbar^2}{2m_3^*}\left(\frac{\pi n_z}{d_z}\right)^2\left[1 - e_1\left(\frac{\hbar^2}{2m_3^*}\right)\left(\frac{\pi n_z}{d_z}\right)^2\right],$$

$$A_6(n_z) \equiv \left[1 - d_1\left(\frac{\hbar^2}{2m_3^*}\right)\left(\frac{\pi n_z}{d_z}\right)^2\right] \quad \text{and} \quad \beta \equiv \frac{\hbar^2 k_x^2}{2m_1^*} + \frac{\hbar^2 k_y^2}{2m_2^*}$$

Equation (5.227) can be written as

$$\frac{\hbar^2 k_x^2}{2m_1^*} + \frac{\hbar^2 k_y^2}{2m_2^*} = I_1\left(E, n_z\right) \tag{5.230}$$

where $I_1(E, n_z) \equiv (2c_1)^{-1}[A_6(n_z) - [A_6^2(n_z) - 4c_1 E + 4c_1 A_5(n_z)]^{1/2}]$.

From (5.230), the area of the 2D k_s-space is given by

$$A\left(E, n_z\right) = \frac{2\pi\sqrt{m_1^* m_2^*}}{\hbar^2} I_1\left(E, n_z\right) \tag{5.231}$$

The magneto 2D dispersion relation in this case can be written following (2.263) as

$$(eB)\left(n + \frac{1}{2}\right) = \frac{\sqrt{m_1^* m_2^*}}{\hbar^2} I_1\left(E_{91}, n_z\right) \tag{5.232}$$

where E_{91} is the total quantized energy in this case.

The DOS function assumes the form

$$N(E) = \left(\frac{g_v eB}{\pi\hbar}\right)\sum_{n=0}^{n_{\max}}\sum_{n_z=1}^{n_{z\max}} \delta'\left(E - E_{91}\right) \tag{5.233}$$

Combining (5.233) with the Fermi–Dirac occupation probability factor, the surface electron concentration can be written as

$$N_s = \left(\frac{g_v eB}{\pi\hbar}\right)\sum_{n=0}^{n_{\max}}\sum_{n_z=1}^{n_{z\max}} F_{-1}\left(\eta_{48}\right) \tag{5.234}$$

in which

$$\eta_{48} = (k_B T)^{-1}[E_{F1} - E_{91}] \tag{5.235}$$

Combining (5.5) and (5.234), we can study the magneto QC in this case.

5.2.19 The QC in magneto QWFETs of gallium aresenide

(a) Rossler model

The magneto 2D dispersion relation in this case can be written following (5.82) as

$$\left(\frac{2eB}{\hbar}\right)\left(n+\frac{1}{2}\right) = \gamma_{4\pm}(E_{92}) - \left(\frac{n_z \pi}{d_z}\right)^2 \tag{5.236}$$

where E_{92} is the total quantized energy in this case.

The DOS function assumes the form

$$N(E) = \left(\frac{g_v eB}{2\pi\hbar}\right)\sum_{n=0}^{n_{\max}}\sum_{n_z=1}^{n_{z\max}} \delta'(E - E_{92}) \tag{5.237}$$

Combining (5.237) with the Fermi–Dirac occupation probability factor, the surface electron concentration can be written as

$$N_s = \left(\frac{g_v eB}{2\pi\hbar}\right)\sum_{n=0}^{n_{\max}}\sum_{n_z=1}^{n_{z\max}} F_{-1}(\eta_{49}) \tag{5.238}$$

where $\eta_{49} = (k_B T)^{-1}[E_{F1} - E_{92}]$.

Combining (5.5) and (5.238), we can study the magneto QC in this case.

(b) Model of Stillman *et al.*

The 2D DR of the conduction electrons in QWTs of GaSb can be written following (5.91) as

$$k_s^2 = \gamma_5(E) - \left(\frac{n_z \pi}{d_z}\right)^2 \tag{5.239}$$

The magneto 2D dispersion relation in this case can be written following (5.239) as

$$\left(\frac{2eB}{\hbar}\right)\left(n+\frac{1}{2}\right) = \gamma_5(E_{93}) - \left(\frac{n_z \pi}{d_z}\right)^2 \tag{5.240}$$

where E_{93} is the total quantized energy in this case.

The DOS function assumes the form

$$N(E) = \left(\frac{g_v eB}{\pi\hbar}\right) \sum_{n=0}^{n_{\max}} \sum_{n_z=1}^{n_{z\max}} \delta'(E - E_{93}) \tag{5.241}$$

Combining (5.241) with the Fermi–Dirac occupation probability factor, the surface electron concentration can be written as

$$N_s = \left(\frac{g_v eB}{\pi\hbar}\right) \sum_{n=0}^{n_{\max}} \sum_{n_z=1}^{n_{\max}} F_{-1}(\eta_{50}) \tag{5.242}$$

where $\eta_{50} = (k_B T)^{-1}[E_{F1} - E_{93}]$.

Combining (5.5) and (5.242), we can study the magneto QC in this case.

5.2.20 The QC in magneto QWFETs of gallium antimonide

(a) Model of Seiler *et al.*

The 2D DR of the conduction electrons in GaSb in this case can be written following (5.100) as

$$k_s^2 = \gamma_{6\pm}(E) - \left(\frac{n_z \pi}{d_z}\right)^2 \tag{5.243}$$

The magneto 2D dispersion relation in this case can be written following (5.243) as

$$\left(\frac{2eB}{\hbar}\right)\left(n + \frac{1}{2}\right) = \gamma_{6,\pm}(E_{94}) - \left(\frac{n_z \pi}{d_z}\right)^2 \tag{5.244}$$

where E_{94} is the total quantized energy in this case.

The DOS function assumes the form

$$N(E) = \left(\frac{g_v eB}{2\pi\hbar}\right) \sum_{n=0}^{n_{\max}} \sum_{n_z=1}^{n_{z\max}} \delta'(E - E_{94}) \tag{5.245}$$

Combining (5.245) with the Fermi–Dirac occupation probability factor, the surface electron concentration can be written as

$$N_s = \left(\frac{g_v eB}{2\pi\hbar}\right) \sum_{n=0}^{n_{\max}} \sum_{n_z=1}^{n_{\max}} F_{-1}(\eta_{51}) \tag{5.246}$$

where $\eta_{51} = (k_B T)^{-1}[E_{F1} - E_{94}]$.

Combining (5.5) and (5.246), we can study the magneto QC in this case.

(b) Model of Mathur *et al.*

The 2D DR of the conduction electrons in QWTs of GaSb in this case can be written following (1.97) as

$$k_s^2 = \gamma_7(E) - \left(\frac{n_z\pi}{d_z}\right)^2 \tag{5.247}$$

The magneto 2D dispersion relation in this case can be written following (5.247) as

$$\left(\frac{2eB}{\hbar}\right)\left(n+\frac{1}{2}\right) = \gamma_7\left(E_{95}\right) - \left(\frac{n_z\pi}{d_z}\right)^2 \tag{5.248}$$

where E_{95} is the total quantized energy in this case.

The DOS function assumes the form

$$N(E) = \left(\frac{g_v eB}{\pi\hbar}\right)\sum_{n=0}^{n_{\max}}\sum_{n_z=1}^{n_{z\max}} \delta'\left(E - E_{95}\right) \tag{5.249}$$

Combining (5.249) with the Fermi–Dirac occupation probability factor, the surface electron concentration can be written as

$$N_s = \left(\frac{g_v eB}{\pi\hbar}\right)\sum_{n=0}^{n_{\max}}\sum_{n_z=1}^{n_{z\max}} F_{-1}\left(\eta_{52}\right) \tag{5.250}$$

where $\eta_{52} = (k_B T)^{-1}[E_{F1} - E_{95}]$.

Combining (5.5) and (5.250), we can study the magneto QC in this case.

(c) Model of Zhang

The 2D dispersion relation of the conduction electrons in GaSb in this case can be written following (5.120) as

$$k_s^2 = \gamma_8(E) - \left(\frac{n_z\pi}{d_z}\right)^2 \tag{5.251}$$

The magneto 2D dispersion relation in this case can be written following (5.251) as

$$\left(\frac{2eB}{\hbar}\right)\left(n+\frac{1}{2}\right) = \gamma_8\left(E_{96}\right) - \left(\frac{n_z\pi}{d_z}\right)^2 \tag{5.252}$$

where E_{96} is the total quantized energy in this case.

The DOS function assumes the form

$$N(E) = \left(\frac{g_v eB}{\pi\hbar}\right) \sum_{n=0}^{n_{\max}} \sum_{n_z=1}^{n_{z\max}} \delta'\left(E - E_{96}\right) \tag{5.253}$$

Combining (5.253) with the Fermi–Dirac occupation probability factor, the surface electron concentration can be written as

$$N_s = \left(\frac{g_v eB}{\pi\hbar}\right) \sum_{n=0}^{n_{\max}} \sum_{n_z=1}^{n_{z\max}} F_{-1}\left(\eta_{53}\right) \tag{5.254}$$

where $\eta_{53} = (k_B T)^{-1}[E_{F1} - E_{96}]$.

Combining (5.5) and (5.254), we can study the magneto QC in this case.

5.2.21 The QC in magneto QWFETs of indium antimonide

(a) Model of Palik *et al.*

The 2D DR of the conduction electrons in QWTs of InSb in this case can be written following (5.130) as

$$k_s^2 = \gamma_9(E) - \left(\frac{n_z \pi}{d_z}\right)^2 \tag{5.255}$$

The magneto 2D dispersion relation in this case can be written following (5.255) as

$$\left(\frac{2eB}{\hbar}\right)\left(n + \frac{1}{2}\right) = \gamma_9\left(E_{97}\right) - \left(\frac{n_z \pi}{d_z}\right)^2 \tag{5.256}$$

where E_{97} is the total quantized energy in this case.

The DOS function assumes the form

$$N(E) = \left(\frac{g_v eB}{\pi\hbar}\right) \sum_{n=0}^{n_{\max}} \sum_{n_z=1}^{n_{z\max}} \delta'\left(E - E_{97}\right) \tag{5.257}$$

Combining (5.257) with the Fermi–Dirac occupation probability factor, the surface electron concentration can be written as

$$N_s = \left(\frac{g_v eB}{\pi\hbar}\right) \sum_{n=0}^{n_{\max}} \sum_{n_z=1}^{n_{z\max}} F_{-1}\left(\eta_{54}\right) \tag{5.258}$$

where $\eta_{54} = (k_B T)^{-1}[E_{F1} - E_{97}]$.

Combining (5.5) and (5.258), we can study the magneto QC in this case.

(b) Model of Johnson *et al.*

The 2D DR of the conduction electrons in QWTs of InSb in this case can be written following (5.140) as

$$k_s^2 = \gamma_{10}(E) - \left(\frac{n_z \pi}{d_z}\right)^2 \tag{5.259}$$

The magneto 2D dispersion relation in this case can be written following (5.259) as

$$\left(\frac{2eB}{\hbar}\right)\left(n + \frac{1}{2}\right) = \gamma_{10}\left(E_{98}\right) - \left(\frac{n_z \pi}{d_z}\right)^2 \tag{5.260}$$

where E_{98} is the total quantized energy in this case.

The DOS function assumes the form

$$N(E) = \left(\frac{g_v eB}{\pi\hbar}\right) \sum_{n=0}^{n_{\max}} \sum_{n_z=1}^{n_{z\max}} \delta'\left(E - E_{98}\right) \tag{5.261}$$

Combining (5.261) with the Fermi–Dirac occupation probability factor, the surface electron concentration can be written as

$$N_s = \left(\frac{g_v eB}{\pi\hbar}\right) \sum_{n=0}^{n_{\max}} \sum_{n_z=1}^{n_{z\max}} F_{-1}\left(\eta_{55}\right) \tag{5.262}$$

where $\eta_{55} = (k_B T)^{-1}[E_{F1} - E_{98}]$.

Combining (5.5) and (5.262), we can study the magneto QC in this case.

(c) Model of Agafonov *et al.*

The 2D DR of the conduction electrons in InSb in this case can be written following (5.150) as

$$k_s^2 = \gamma_{11}(E) - \left(\frac{n_z \pi}{d_z}\right)^2 \tag{5.263}$$

The magneto 2D dispersion relation in this case can be written following (5.263) as

$$\left(\frac{2eB}{\hbar}\right)\left(n + \frac{1}{2}\right) = \gamma_{11}\left(E_{99}\right) - \left(\frac{n_z \pi}{d_z}\right)^2 \tag{5.264}$$

where E_{99} is the total quantized energy in this case.

The DOS function assumes the form

$$N(E) = \left(\frac{g_v eB}{\pi\hbar}\right) \sum_{n=0}^{n_{\max}} \sum_{n_z=1}^{n_{z\max}} \delta'\left(E - E_{99}\right) \tag{5.265}$$

Combining (5.265) with the Fermi–Dirac occupation probability factor, the surface electron concentration can be written as

$$N_s = \left(\frac{g_v eB}{\pi\hbar}\right) \sum_{n=0}^{n_{\max}} \sum_{n_z=1}^{n_{z\max}} F_{-1}(\eta_{56}) \tag{5.266a}$$

where $\eta_{56} = (k_BT)^{-1}[E_{F1} - E_{99}]$.

Combining (5.5) and (5.266a), we can study the magneto QC in this case.

5.2.22 The QC in magneto QWFETs of tellurium

Equation (1.67) can be recasted as

$$E = \psi_1 k_z^2 - \psi_2 k_s^2 \pm \left[\psi_3^2 k_z^2 + \psi_4^2 k_s^2\right]^{1/2} \tag{5.266b}$$

where $\psi_1 = A_6, \psi_2 = A_7, \psi_3^2 = A_8$ and $\psi_4^2 = A_9$.

The 2D DR in this case can be written from (5.266b) as

$$k_s^2 = \psi_5(E) - \psi_6\left(\frac{\pi n_z}{d_z}\right)^2 \pm \psi_7\left[\psi_8^2(E) - \left(\frac{\pi n_z}{d_z}\right)^2\right]^{1/2} \tag{5.267}$$

where

$$\psi_5(E) = \left[\frac{E}{\psi_2} + \frac{\psi_4^2}{2\psi_2^2}\right], \psi_5 = \left[\frac{\psi_1}{\psi_2}\right], \psi_7 = \left[\frac{\psi_4\sqrt{\psi_1}}{\psi_2^{3/2}}\right] \quad \text{and}$$

$$\psi_8^2(E) = \left[\frac{\psi_4^4 + 4E\psi_2\psi_4^2 + 4\psi_2^2\psi_3^2}{4\psi_1\psi_2\psi_4^2}\right]$$

The magneto 2D dispersion relation in this case can be written following (5.267) as

$$\left(\frac{2eB}{\hbar}\right)\left(n + \frac{1}{2}\right) = \psi_5(E_{100}) - \psi_6\left(\frac{\pi n_z}{d_z}\right)^2$$
$$\pm \psi_7\left[\psi_8^2(E_{100}) - \left(\frac{\pi n_z}{d_z}\right)^2\right]^{1/2} \tag{5.268}$$

where E_{100} is the total quantized energy in this case.

The DOS function assumes the form

$$N(E) = \left(\frac{g_v eB}{2\pi\hbar}\right) \sum_{n=0}^{n_{\max}} \sum_{n_z=1}^{n_{z\max}} \delta'(E - E_{100}) \tag{5.269}$$

Combining (5.269) with the Fermi–Dirac occupation probability factor, the surface electron concentration can be written as

$$N_s = \left(\frac{g_v eB}{2\pi\hbar}\right)\sum_{n=0}^{n_{\max}}\sum_{n_z=1}^{n_{z\max}} F_{-1}\left(\eta_{57}\right) \tag{5.270}$$

where $\eta_{57} = (k_B T)^{-1}[E_{F1} - E_{100}]$.

Combining (5.5) and (5.270), we can study the magneto QC in this case.

*5.2.23 The QC in magneto QWFETs of platinum antimonide

The use of (1.80) leads to the expression of the 2D dispersion law in QWTs of $PtSb_2$ as

$$k_s^2 = t_{44}\left(E, n_z\right) \tag{5.271}$$

$$t_{44}\left(E, n_z\right) = \left[2A_9\right]^{-1}\left[-A_{10}\left(E, n_z\right) + \sqrt{A_{10}^2\left(E, n_z\right) + 4A_9 A_{11}\left(E, n_z\right)}\right] \tag{5.272}$$

where

$$A_9 \equiv \left[I_1 + \omega_1\omega_3\right], A_{10}\left(E, n_z\right) \equiv \left[\omega_3 E + \omega_1\left\{E + \delta_0 - \omega_4\left(\frac{\pi n_z}{d_z}\right)^2\right\}\right.$$
$$\left. + \omega_2\omega_3\left(\frac{\pi n_z}{d_z}\right)^2 + 2I_1\left(\frac{\pi n_z}{d_z}\right)^2\right] \quad \text{and}$$

$$A_{11}\left(E, n_z\right) \equiv \left[E\left[E + \delta_0 - \omega_4\left(\frac{\pi n_z}{d_z}\right)^2\right]\right.$$
$$\left. + \omega_2\left(\frac{\pi n_z}{d_z}\right)^2\left[E + \delta_0 - \omega_4\left(\frac{\pi n_z}{d_z}\right)^2\right] - I_1\left(\frac{\pi n_z}{d_z}\right)^4\right]$$

The area of k_s-space can be expressed as

$$A\left(E, n_z\right) = \frac{\pi}{2A_9} t_{44}\left(E, n_z\right) \tag{5.273}$$

The magneto 2D dispersion relation in this case can be written following (5.273) as

$$\left(\frac{2eB}{\hbar}\right)\left(n+\frac{1}{2}\right) = t_{44}\left(E_{101}, n_z\right) \tag{5.274}$$

where E_{101} is the total quantized energy in this case.

The DOS function assumes the form

$$N(E) = \left(\frac{g_v eB}{2\pi\hbar}\right)\sum_{n=0}^{n_{\max}}\sum_{n_z=1}^{n_{z\max}} \delta'\left(E - E_{101}\right) \tag{5.275}$$

Combining (5.275) with the Fermi–Dirac occupation probability factor, the surface electron concentration can be written as

$$N_s = \left(\frac{g_v eB}{2\pi\hbar}\right)\sum_{n=0}^{n_{\max}}\sum_{n_z=1}^{n_{z\max}} F_{-1}\left(\eta_{58}\right) \tag{5.276}$$

where $\eta_{58} = (k_B T)^{-1}[E_{F1} - E_{101}]$.

Combining (5.5) and (5.276), we can study the magneto QC in this case.

5.2.24 The QC in magneto MOSFETs of gallium phosphide

The 2D electron DR in QWTs GaP can be expressed using (1.55) as

$$E = ax + by - \sqrt{cx + cy + D^2} + D \tag{5.277}$$

in which $a = \frac{\hbar^2}{2m_\perp^*} + b, b = \frac{\hbar^2}{2m_\parallel^*}, x = k_s^2, y = \left(n_z\pi/d_z\right)^2, c = [\hbar^4 k_0^2/\left(m_\parallel^*\right)^2]$ and $D = |V_G|$.

Equation (5.277) can be written as

$$k_s^2 = X_{11}(E, n_z) \tag{5.278}$$

in which

$$X_{11}(E, n_z) \equiv [E\bar{t}_1 + \bar{t}_2 - \bar{t}_3(n_z\pi/d_z)^2 - \bar{t}_4[(n_z\pi/d_z)^2 + (\bar{t}_5(E)^2)]^{1/2}],$$

$$\bar{t}_1 = \left[\frac{1}{a}\right], \quad \bar{t}_2 = \left[\frac{g_1}{2a^2}\right], \quad g_1 = (2aD - c), \quad \bar{t}_3 = \left[\frac{b}{a}\right],$$

$\bar{t}_4 = \left[\frac{\sqrt{g_3}}{2a^2}\right]$, $g_3 = (4abc + 4a^2c)$, $(\bar{t}_5(E)^2) = (g_3)^{-1}[g_2 - 4acE]$ and $g_2 = [4a^2b^2 + c^2 + 4acD]$

The magneto 2D dispersion relation in this case can be written following (5.278) as

$$\left(\frac{2eB}{\hbar}\right)\left(n + \frac{1}{2}\right) = X_{11}(E_{102}, n_z) \tag{5.279}$$

where E_{102} is the total quantized energy in this case.

The DOS function assumes the form

$$N(E) = \left(\frac{g_v eB}{2\pi\hbar}\right)\sum_{n=0}^{n_{\max}}\sum_{n_z=1}^{n_{z\max}} \delta'(E - E_{102}) \tag{5.280}$$

Combining (5.280) with the Fermi–Dirac occupation probability factor, the surface electron concentration can be written as

$$N_s = \left(\frac{g_v eB}{2\pi\hbar}\right)\sum_{n=0}^{n_{\max}}\sum_{n_z=1}^{n_{z\max}} F_{-1}(\eta_{59}) \tag{5.281}$$

where $\eta_{59} = (k_BT)^{-1}[E_{F1} - E_{102}]$.

Combining (5.5) and (5.281), we can study the magneto QC in this case.

5.3 Result and Discussions

Using the appropriate equations and the parameters as given in Table 1.1, we have determined the dependence of the QC taking the MOSFETs of $CdGeAs_2$, InSb and $Hg_{1-x}Cd_xTe$ as examples operated under inversion layers modes on a quantizing magnetic field at 5.2 K in the weak electric field limit, as shown in Fig. 5.1. Figure 5.2 exhibits the same dependence in magneto MOSFETs of $In_{1-x}Ga_xAs_yP_{1-y}$ lattice matched to InP operated under inversion layer mode. The surface electric field is chosen for the weak field limit. The period of oscillations can easily expressed as low temperatures by $(1/B) = g_s g_\nu e/(2\pi\hbar n_s)$ and is found graphically to be 0.9 and 0.897 T^{-1}, as calculated from the relation for the period corresponding to a surface concentration $5.4 \times 10^{14}\,m^{-2}$ induced at the surface by the surface field considered. The depths of the spikes increase with increasing thickness of the insulting layer. The oscillatory nature of the magnetic field dependence of the surface and QCs is found to be somewhat similar to that observed experimentally in Si-MOS structures having n-channel inversion

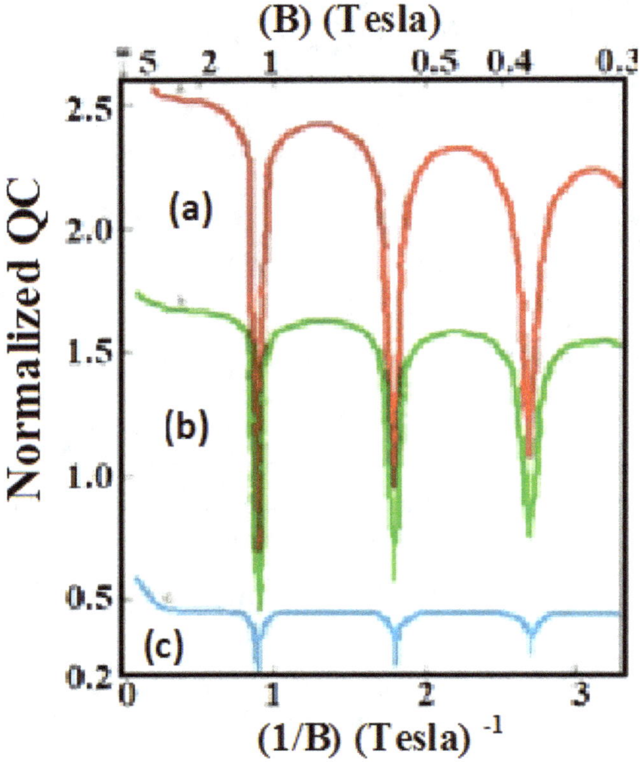

Fig. 5.1. Plot showing the oscillatory magnetic field dependence of the normalized QC in magneto MOSFETs of (a) $CdGeAs_2$ (b) InSb and (c) $Hg_{1-x}Cd_xTe$ operated under inversion layer mode.

layers, with decreasing magnetic field, i.e., with increasing $1/B$, the MOS capacitance should approach its value corresponding to zero magnetic field.

With varying magnetic field, each time a Landau level crosses the Fermi level, a change is reflected in the capacitance through redistribution of the carriers among the Landau levels. It may be noted that the three dimensional quantization list to discrete energy levels in the real sense of the term somewhat like atomic energy levels and, consequently, the crossing of the Fermi level by the Landau levels in steps produce very sharp changes. This is what is reflected in both the figures by the increasing sharpness of the spikes with increasing magnetic field.

Incidentally, however the collision broadening effects which have been neglected here would, of course, reduce the sharpness of the spikes to some

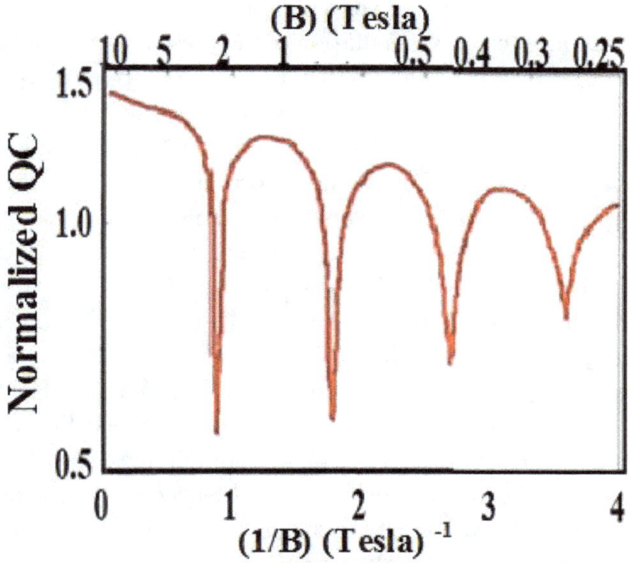

Fig. 5.2. Plot showing the oscillatory magnetic field dependence of the normalized QC in magneto MOSFETs of $In_{1-x}Ga_xAs_yP_{1-y}$ lattice matched to InP operated under inversion layer mode.

extent. Even then, the sharpness would be much greater than that observed for oscillatory phenomena based on the same physics in three dimensional electron gases of semiconductors. This follows from the inherent feature of the three dimensional quantization of the two dimensional electron gas dealt with here. Under such quantization three remains no free electron state between any two successive Landau levels unlike that found for three dimensional electron gases of semiconductors under two dimensional quantization in the wave vector space in the presence of a quantizing magnetic field. Consequently, the crossing of the Fermi level by the Landau levels under three dimensional quantization would have much greater impact on the redistribution of the electron amongst the available states, as compared to that found for two dimensional quantization. It is basically this impact which results in the increased sharpness of the oscillatory spikes. The neglect of collision broadening does it, therefore, not render meaningless to stress this point.

It may be noted that though the spin splitting of the Landau levels are usually large in small gap semiconductor these have not been taken into consideration in the present calculation. The spin effects would simple increasing the number of spike splitting each of the spikes into two and reduce their amplitudes. In a calculation of the present type, the amplitudes are really arbitrary and depend strictly on the resolution of computation. Thus, for simplicity, the spin effects have been neglected without losing much of the information on the oscillatory features of the magneto-quantization. In Section 5.2.3, we have formulated the N_s for magneto MOSFETs of II–VI materials as given by (5.38) and by combining the same with (5.5) we can study the QC in the present case. The expressions of 2D N_s for magneto MOSFET of IV–VI materials for both weak and strong electric field limits are given by (5.44) and (5.49) of Section 5.2.4 and combining each of them with (5.5) we can study the QC for both the limits respectively. The expressions of 2D N_s for magneto MOSFET of stressed materials for both weak and strong electric field limits are given by (5.55) and (5.60) of Section 5.2.5 and combining each of them with (5.5) we can study the QC for both the limits respectively. The expressions of 2D N_s for magneto MOSFET of Bismuth Telluride for both weak and strong electric field limits are given by (5.65) and (5.70) of Section 5.2.6 and combining each of them with (5.5) we can study the QC for both the limits respectively. The expressions of 2D N_s for magneto MOSFET of Germanium for the models of Cardona *et al.* and Wang *et al.* are given by (5.76) and (5.80) of Section 5.2.7 and combining each of them with (5.5) we can study the QC respectively in this case. In Section 5.2.8, we have formulated the expressions of 2D N_s for magneto MOSFET of Gallium Aresenide for the models of Rossler and Stillman *et al.* and they are given by (5.86) and (5.95) for weak electric field limit and (5.90) and (5.99) for strong electric field limit and combining each of them with (5.5) we can study the QC for both the limits respectively. In Section 5.2.9, we have formulated the expressions of 2D N_s for magneto MOSFET of Gallium Antimonide in accordance with the models of Seiler *et al.*, Mathur *et al.* and Zhang and they are given by (5.105), (5.114) and (5.124) for weak electric field limit and (5.109), (5.118) and (5.128) for strong electric field limit and combining each of them with (5.5) we can study the QC for both the limits respectively.

In Section 5.2.10, we have formulated the expressions of 2D N_s for magneto MOSFET of Indium Antimonide in accordance with the models of Palik *et al.*, Johnson *et al.*, and Agafonov *et al.* and they are given by (5.134), (5.144) and (5.154) for weak electric field limit and (5.138), (5.148) and (5.158) for strong electric field limit and combining each of them with (5.5), we can study the QC for both the limits respectively. In Section 5.2.11, we have formulated the N_s for magneto MOSFETs of Te as given by (5.164) and by combining the same with (5.5) we can study the QC in the present case.

In Section 5.2.12, we have formulated the N_s for magneto QWFETs of nonlinear optical materials as given by (5.167) and by combining the same with (5.5) we can study the QC in the present case. In Section 5.2.13, we have formulated the N_s for magneto QWFETs of III–V ternary and quaternary materials in accordance with the models of three and two band models of Kane together with parabolic energy bands as given by (5.170), (5.173) and (5.176) and by combining each of them with (5.5) we can study the QC in each case respectively. In Section 5.2.14, we have formulated the N_s for magneto QWFETs of II–VI materials as given by (5.179) and by combining the same with (5.5) we can study the QC in this case. The expressions of 2D N_s in magneto QWFETs of IV–VI materials in accordance with the models of the McClure and Choi, Hybrid, Cohen, Lax and the ellipsoidal parabolic, Dimmock are given by (5.186), (5.189), (5.192), (5.195), (5.198), and (5.212) in Section 5.2.15, respectively and by combining each of them with (5.5) we can study the QC of the magneto of IV–VI materials in accordance with the said types of band models.

In Section 5.2.16, we have formulated the N_s for QWFETs of stressed Kane type materials as given by (5.217) and by combining the same with (5.5) we can study the QC in the present case. In Section 5.2.17, we have formulated the N_s for QWFETs of Bismuth Telluride as given by (5.222) and by combining the same with (5.5) we can study the QC in the present case. The expressions of 2D N_s for magneto QWFETs of Germanium for the models of Cardona *et al.* and Wang *et al.* are given by (5.228) and (5.234) of Section 5.2.18 and combining each of them with (5.5) we can study the QC respectively in this case. In Section 5.2.19, we have formulated the expressions of 2D N_s for magneto QWFETs of Gallium Aresenide for the models of Rossler. and Stillman *et al.* and they are given by (5.238) and (5.242) and combining each of them with (5.5) we can study the QC respectively in this case. In Section 5.2.20, we have formulated the expressions of 2D N_s for magneto QWFETs of Gallium Antimonide in accordance with the models of Seiler *et al.*, Mathur *et al.* and Zhang and they are given by (5.246), (5.250) and (5.254) and combining each of them with (5.5) we can study the QC in this case respectively. In Section 5.2.21, we have formulated the expressions of 2D N_s for magneto QWFETs of Indium Antimonide in accordance with the models of Palik *et al.*, Johnson *et al.*, and Agafonov *et al.* and they are given by (5.258), (5.262) and

(5.266a) and combining each of them with (5.5) we can study the QC for all the said models respectively. In Section 5.2.22, we have formulated the N_s for magneto QWFETs of Te as given by (5.270) and by combining the same with (5.5) we can study the QC in the present case. In Section 5.2.23, we have formulated the N_s for magneto QWFETs of Platinum Antimonide as given by (5.276) and by combining the same with (5.5) we can study the QC in the present case. In Section 5.2.24, we have formulated the N_s for magneto QWFETs of GaP as given by (5.281) and by combining the same with (5.5) we can study the QC in the present case.

We have not given the graphs of the QCs' for MOSFETs and QWFETs made of many materials as discussed here since we feel that our readers will perform the intricate numerical computation to obtain the oscillatory plots for better assessment and joy of understanding. Incidentally, one can easily infer that how little is presented and how much more there is yet to be investigated in the research field of QCs' in this case for different 2D DRs having different band structures in general which is the signature of the co-existence of new physics and advanced mathematics. Finally we again realize that it is the band structure which changes in a fundamental way and consequently all the physical properties of all the electronic materials change radically leading to new physical concepts.

5.4 Open Research Problems

(R.5.1) Investigate the QCs' in the presence of an arbitrarily oriented electric quantization for QMOSFET and QWFET of nonlinear optical materials as studied in this chapter. Draw the corresponding graphs and explain the results.

(R.5.2) Investigate the QCs' in QMOSFET and QWFET of IV–VI, II–VI and stressed Kane type compounds in the presence of an arbitrarily oriented quantizing non-uniform electric field. Draw the corresponding graphs and explain the results.

(R.5.3) Investigate the QCs' in QMOSFET and QWFET of all the materials as stated in (R.5.1) of Chapter 1 in the presence of an arbitrarily oriented quantizing electric field. Draw the corresponding graphs and explain the results.

(R.5.4) Investigate the QCs' in the presence of an arbitrarily oriented non-quantizing magnetic field in QMOSFET and QWFET of all the materials as studied in this chapter by including the electron spin. Draw the corresponding graphs and explain the results.

(R.5.5) Investigate the QCs' in QMOSFET and QWFET of IV–VI, II–VI and stressed Kane type compounds in the presence of an arbitrarily oriented non-quantizing magnetic field by including the electron spin. Draw the corresponding graphs and explain the results.

(R.5.6) Investigate the QCs' in QMOSFET and QWFET of all the materials as stated in Chapter 1 in the presence of an arbitrarily oriented non-quantizing magnetic field by including electron spin. Draw the corresponding graphs and explain the results.

(R.5.7) Investigate the QCs' in QMOSFET and QWFET of all the problems from Chapter 2 in the presence of an additional arbitrarily oriented electric field. Draw the corresponding graphs and explain the results.

(R.5.8) Investigate the QCs' in QMOSFET and QWFET of all the problems from Chapter 2 in the presence of arbitrarily oriented crossed electric and magnetic fields. Draw the corresponding graphs and explain the results.

(R.5.9) Investigate the QCs' in QMOSFET and QWFET of all the problems from Chapter 2 in the presence of surface states. Draw the corresponding graphs and explain the results.

(R.5.10) Investigate the QCs' in QMOSFET and QWFET of all the problems from Chapter 2 in the presence of hot electron effects. Draw the corresponding graphs and explain the results.

(R.5.11) Investigate the QCs' in QMOSFET and QWFET of all the problems from Chapter 2 by including the occupancy of the electrons in various electric sub-bands. Draw the corresponding graphs and explain the results.

References

[1] A. S. Grove, D. J. Fitzgerald, *J. Sol. Stat. Electron.*, 9, 783 **(1966)**; K. P. Ghatak, S. N. Banik, A. Fizika, 5, 31 **(1996)**; B. Mitra, A. Ghoshal and K. P. Ghatak, *Phys. Stat. Sol. (b)*, 153, K209 **(1989)**.

[2] A. M. Voshchenkov, J. N. Zemel, *J. Phys. Rev. B*, 9, 4410 **(1974)**; K. P. Ghatak and M. Mondal, *J. Appl. Phys.*, 64, 918 **(1988)**; S. N. Biswas and K. P. Ghatak, *Adv. Proc. Semico. Dev.*, 797, 286 **(1987)**.

[3] Z. Djurid, Z. Spasojevid, D. Tjapkin, *J. Sol. Stat. Electron.*, 19, 931 **(1976)**; S. N. Biswas and K. P. Ghatak, *Megaga. Tech. Puls. Power. Appl.*, 8, 219 **(1987)**; K. P. Ghatak, N. Chattopadhyay and S. N. Biswas, *Optoelectr. Mater, Devi, Packa, Interconn.*, 836, 203 **(1987)**.

[4] M. Kaplit, Z. N. Zemel, *J. Phys. Rev. Lett.*, 21, 212 **(1963)**.
[5] A. Daerr, J. P. Kotthaus, J. F. Koch, *J. Sol. Stat. Commun.*, 17, 455 **(1975)**.
[6] A. Daerr, J. P. Kotthaus, *J. Surf. Sci.*, 73, 549 **(1978)**.
[7] H. A. Washburn and J. R. Sites, *J. Surf. Sci.*, 73, 537 **(1978)**.
[8] J. W. Rowe and J. L. Shay, *J. Phys. Rev. B*, 3, 451 **(1973)**.
[9] H. Kildal, *J. Phys. Rev. B*, 10, 5082 **(1974)**.
[10] J. Bodnar, in *Proceedings of the International Conference on Physics of Narrow-gap Semiconductors*, (Polish Science Publishers, Warsaw, **1978**).
[11] G. P. Chuiko, N. N. Chuiko, *J. Sov. Phys. Semicond.*, 15, 739 **(1981)**.
[12] K. P. Ghatak, S. N. Biswas, *J. Proc. SPIE*, 1484, 149 **(1991)**.
[13] G. Paasch, T. Fiedler, M. Kolar, I. Bartos, *J. Phys. Stat. Sol. (b)*, 118, 641 **(1983)**.
[14] L. D. Landau and E. M. Lifshitz, *Statistical Physics*, (Pergamon Press, London, **1970**).
[15] M. Cardona, W. Paul, H. Brooks, *Helv. Acta Phys.*, 33, 329 **(1960)**; A. F. Gibson, in *Proceeding of International School of Physics "ENRICO FERMI"* course XIII, Ed. R. A. Smith, (Academic Press, 171, **1963**).
[16] C. C. Wang, N. W. Ressler, *Phys. Rev.*, 2, 1827 **(1970)**.
[17] U. Rossler, *Sol. Stat. Commun.*, 49, 943 **(1984)**.
[18] G. E. Stillman, C. M. Wolfe and J. C. Dinmockin, in *Semiconductors and Semimetals*, Ed. R. K. Willardson and A. C. Beer, 12, 169 **(1977)**.
[19] D. G. Seiler, W. M. Beeker and L. M. Roth, *Phys. Rev.*, 1, 764 **(1970)**.
[20] P. C. Mathur and S. Jain, *Phys. Rev.*, 19, 3159 **(1979)**.
[21] P. R. Zhang, *Phys. Rev. B*, 1, 3450 **(1970)**.
[22] E. D. Palik, G. S. Picus, S. Teither and R. F. Wallis, *Phys. Rev.*, 122, 475 **(1961)**.
[23] E. J. Johnson and D. H. Dicley, *Phys. Rev.*, 1, 2676 **(1970)**.
[24] V. G. Agafonov, P. M. Valov, B. S. Ryvkin and I. D. Yaroshetskii, *Sov. Phys. Semicond.*, 12, 1182 **(1978)**.
[25] W. Szwozkiewicz, *Phys. Stat. Sol. (b)*, 79, 691 **(1977)**.

Chapter 6

The $I_D - V_D$ Equation in Magneto MOSFETs

Everything is predetermined in accordance with eastern philosophy.
Answer to all the problems is the 360 degree
mental attachment to the lotus feet of SADGURU.

6.1 Introduction

This is completely a new chapter where from Sections 6.2.1 to 6.2.11 we study the $I_D - V_D$ equation in magneto MOSFETs of nonlinear optical and tetragonal, III–V, ternary and quaternary, II–VI, IV–VI, stressed Kane type, Bi_2Te_3, Ge, GaAs, GaSb, InSb materials and Te respectively. From Sections 6.2.12 to 6.2.23, we study in the same chapter the $I_D - V_D$ equation in ballistic Magneto QWFETs of tetragonal and nonlinear optical, III–V ternary and quaternary, II–VI, IV–VI, stressed Kane type, Bi_2Te_3, Ge, GaAs, GaSb, InSb, Te, $PtSb_2$ and GaP materials respectively. In Section 6.3, we present summary and conclusion and last Section 6.4, we offer one multidimensional open research problems.

6.2 Theoretical Background

6.2.1 The $I_D - V_D$ equation in magneto MOSFETs of nonlinear optical and tetragonal materials

(a) Weak electric field limit

The current density can be expressed as

$$J_D = W[J_+(E_F) - J_-(E_F - eV_D)] \tag{6.1}$$

where W is the width,

$$J_+(E_F) = \frac{eg_v}{\pi} \sum_{i=0}^{i_{\max}} \int_{E_i}^{\infty} \frac{V_x N_{2D_i}(E)dE}{\left[1 + \exp\left(\frac{E-E_F}{k_B T}\right)\right]} \tag{6.2}$$

and

$$J_-(E_F - eV_D) = \frac{eg_v}{\pi} \sum_{i=0}^{i_{\max}} \int_{E_i}^{\infty} \frac{V_x N_{2D_i}(E) dE}{\left[1 + \exp\left(\frac{E - E_F + eV_D}{k_B T}\right)\right]} \tag{6.3}$$

where E_i is the sub-band energy of the ith sub-band, $N_{2D_i}(E)$ is the 2D Density-of-States function per sub-band and V_x is the velocity along z-direction.

From (2.2), we can write

$$k_x = g_{21,1}(E) \tag{6.4}$$

where $g_{21,1}(E) = \left[\frac{\gamma(E)}{f_1(E)}\right]^{1/2}$.

From (6.4), we can write

$$V_x = \frac{1}{\hbar} \frac{\partial E}{\partial k_x} = [\hbar g'_{21,1}(E)]^{-1} \tag{6.5}$$

Using (6.5), (6.1)–(6.3) and (5.10), we get

$$\frac{J_D}{W} = \frac{e^2 B g_v}{\pi^2 \hbar^2} \sum_{n=0}^{n_{\max}} \sum_{i=0}^{i_{\max}} [[g'_{21,1}(E_{41})]^{-1} [F_{-1}(\bar{\eta}_{4,1}) - F_{-1}(\bar{\theta}_{4,1})]] \tag{6.6}$$

where

$$\bar{\eta}_{4,1} = (k_B T)^{-1} [E_{F26,1} - E_{41}] \tag{6.7}$$

$\bar{\theta}_{4,1} = \left[\bar{\eta}_{4,1} - \frac{eV_D}{k_B T}\right]$, E_{41} is given by (5.9) and $E_{F26,1}$ is the Fermi energy in the present case which can be determined from

$$\frac{C_{ins}}{e}(V_T - V_{GS}) = \frac{g_v e B}{2\pi\hbar} \sum_{n=0}^{n_{\max}} \sum_{i=0}^{i_{\max}} [F_{-1}(\bar{\eta}_{4,1}) + F_{-1}(\bar{\theta}_{4,1})] \tag{6.8}$$

(b) Strong electric field limit

Using (6.5), (6.1)–(6.3) and (5.14), we get

$$\frac{J_D}{W} = \frac{e^2 B g_v}{\pi^2 \hbar^2} \sum_{n=0}^{n_{\max}} \sum_{i=0}^{i_{\max}} [[g'_{21,1}(E_{42})]^{-1} [F_{-1}(\bar{\eta}_{4,2}) - F_{-1}(\bar{\theta}_{4,2})]] \tag{6.9}$$

where $\bar{\eta}_{4,2} = (k_B T)^{-1} [E_{F26,2} - E_{42}]$, $\bar{\theta}_{4,2} = \left[\bar{\eta}_{4,2} - \frac{eV_D}{k_B T}\right]$, E_{42} is given by (5.13) and $E_{F26,2}$ is the Fermi energy in the present case which can be determined from

$$\frac{C_{ins}}{e}(V_T - V_{GS}) = \frac{g_v e B}{2\pi\hbar} \sum_{n=0}^{n_{\max}} \sum_{i=0}^{i_{\max}} [F_{-1}(\bar{\eta}_{4,2}) + F_{-1}(\bar{\theta}_{4,2})] \tag{6.10}$$

6.2.2 The $I_D - V_D$ equation in magneto MOSFETs of III–V, ternary and quaternary materials

(a) The three band model of Kane

(i) Weak electric field limit

Using (1.21), we can write

$$V_x = \left[\sqrt{\frac{2}{m_c}}\right][I'(E)]^{-1}[\sqrt{I(E)}] \tag{6.11}$$

Using (6.11), (6.1)–(6.3) and (5.18), we get

$$\frac{J_D}{W} = \frac{e^2 B g_v}{\pi^2 \hbar}\sqrt{\frac{2}{m_c}}\sum_{n=0}^{n_{\max}}\sum_{i=0}^{i_{\max}}[[\sqrt{I_{11}(E_{43})}I'_{11}(E_{43})]^{-1}] \times [F_{-1}(\bar{\eta}_{4,3}) - F_{-1}(\bar{\theta}_{4,3})] \tag{6.12}$$

where $\bar{\eta}_{4,3} = (k_B T)^{-1}[E_{F26,3} - E_{43}]$, $\bar{\theta}_{4,3} = \left[\bar{\eta}_{4,3} - \frac{eV_D}{k_B T}\right]$, E_{43} is given by (5.17) and $E_{F26,3}$ is the Fermi energy in the present case which can be determined from

$$\frac{C_{ins}}{e}(V_T - V_{GS}) = \frac{g_v e B}{2\pi\hbar}\sum_{n=0}^{n_{\max}}\sum_{i=0}^{i_{\max}}[F_{-1}(\bar{\eta}_{4,3}) + F_{-1}(\bar{\theta}_{4,3})] \tag{6.13}$$

(b) The two band model of Kane

(i) Weak electric field limit

Using (1.17), we can write

$$V_x = \left[\sqrt{\frac{2}{m_c}E(1+\alpha E)}\right][1 + 2\alpha E]^{-1} \tag{6.14}$$

Using (6.14), (6.1)–(6.3) and (5.21), we get

$$\frac{J_D}{W} = \frac{e^2 B g_v}{\pi^2 \hbar}\sqrt{\frac{2}{m_c}}\sum_{n=0}^{n_{\max}}\sum_{i=0}^{i_{\max}}[[\sqrt{E_{44}(1+\alpha E_{44})}][1 + 2\alpha E_{44}]^{-1}] \times [F_{-1}(\bar{\eta}_{4,4}) - F_{-1}(\bar{\theta}_{4,4})] \tag{6.15}$$

where $\bar{\eta}_{4,4} = (k_B T)^{-1}[E_{F26,4} - E_{44}]$, $\bar{\theta}_{4,4} = \left[\bar{\eta}_{4,4} - \frac{eV_D}{k_B T}\right]$, E_{44} is given by (5.20) and $E_{F26,4}$ is the Fermi energy in the present case which can be determined from

$$\frac{C_{ins}}{e}(V_T - V_{GS}) = \frac{g_v eB}{2\pi\hbar} \sum_{n=0}^{n_{\max}} \sum_{i=0}^{i_{\max}} [F_{-1}(\bar{\eta}_{4,4}) + F_{-1}(\bar{\theta}_{4,4})] \tag{6.16}$$

(c) The parabolic band model

In this case from (1.8), we can write

$$V_x = \left[\sqrt{\frac{2}{m_c} E}\right] \tag{6.17}$$

Using (6.17), (6.1)–(6.3) and (5.24), we get

$$\frac{J_D}{W} = \frac{e^2 B g_v}{\pi^2 \hbar} \sqrt{\frac{2}{m_c}} \sum_{n=0}^{n_{\max}} \sum_{i=0}^{i_{\max}} [\sqrt{E_{45}}][F_{-1}(\eta_{4,5}) - F_{-1}(\theta_{4,5})] \tag{6.18}$$

where $\eta_{4,5} = (k_B T)^{-1}[E_{F26,5} - E_{45}]$, $\theta_{4,5} = \left[\eta_{4,5} - \frac{eV_D}{k_B T}\right]$, E_{45} is given by (5.23) and $E_{F26,5}$ is the Fermi energy in the present case which can be determined from

$$\frac{C_{ins}}{e}(V_T - V_{GS}) = \frac{g_v eB}{2\pi\hbar} \sum_{n=0}^{n_{\max}} \sum_{i=0}^{i_{\max}} [F_{-1}(\eta_{4,5}) + F_{-1}(\theta_{4,5})] \tag{6.19}$$

(d) Strong electric field limit

(i) The three band Kane model

Using (6.11), (6.1)–(6.3) and (5.28), we get

$$\frac{J_D}{W} = \frac{e^2 B g_v}{\pi^2 \hbar} \sqrt{\frac{2}{m_c}} \sum_{n=0}^{n_{\max}} \sum_{i=0}^{i_{\max}} [[\sqrt{I(E_{47})}][I'(E_{47})]^{-1}][F_{-1}(\eta_{4,6}) - F_{-1}(\theta_{4,6})] \tag{6.20}$$

where $\eta_{4,6} = (k_B T)^{-1}[E_{F26,6} - E_{47}]$, $\theta_{4,6} = \left[\eta_{4,6} - \frac{eV_D}{k_B T}\right]$, E_{47} is given by (5.27) and $E_{F26,6}$ is the Fermi energy in the present case which can be determined from

$$\frac{C_{ins}}{e}(V_T - V_{GS}) = \frac{g_v eB}{2\pi\hbar} \sum_{n=0}^{n_{\max}} \sum_{i=0}^{i_{\max}} [F_{-1}(\eta_{4,6}) + F_{-1}(\theta_{4,6})] \tag{6.21}$$

(ii) The two band model of Kane

Using (6.14), (6.1)–(6.3) and (5.32), we get

$$\frac{J_D}{W} = \frac{e^2 B g_v}{\pi^2 \hbar}\sqrt{\frac{2}{m_c}}\sum_{n=0}^{n_{\max}}\sum_{i=0}^{i_{\max}}[[\sqrt{E_{48}(1+\alpha E_{48})}1 + 2\alpha E_{48}]^{-1}]$$
$$\times [F_{-1}(\eta_{4,7}) - F_{-1}(\theta_{4,7})] \tag{6.22}$$

where $\eta_{4,7} = (k_B T)^{-1}[E_{F26,7} - E_{48}]$, $\theta_{4,7} = \left[\eta_{4,7} - \frac{eV_D}{k_B T}\right]$, E_{47} is given by (5.31) and $E_{F26,7}$ is the Fermi energy in the present case which can be determined from

$$\frac{C_{ins}}{e}(V_T - V_{GS}) = \frac{g_v e B}{2\pi\hbar}\sum_{n=0}^{n_{\max}}\sum_{i=0}^{i_{\max}}[F_{-1}(\eta_{4,7}) + F_{-1}(\theta_{4,7})] \tag{6.23}$$

6.2.3 The $I_D - V_D$ equation in magneto MOSFETs of II–VI materials

From (1.51), we can write

$$V_x = \left[\sqrt{(\bar{\lambda}_0)^2 + \frac{2E\hbar^2}{m_\perp^*}}\right](\hbar)^{-1} \tag{6.24}$$

Using (6.24), (6.1)–(6.3) and (5.37), we get

$$\frac{J_D}{W} = \frac{e^2 B g_v}{2\pi^2 \hbar^2}\sum_{n=0}^{n_{\max}}\sum_{i=0}^{i_{\max}}\left[\left[\sqrt{(\bar{\lambda}_0)^2 + \frac{2E_{49}\hbar^2}{m_\perp^*}}\right][F_{-1}(\eta_{4,8}) - F_{-1}(\theta_{4,8})]\right] \tag{6.25}$$

where $\eta_{4,8} = (k_B T)^{-1}[E_{F26,8} - E_{49}]$, $\theta_{4,8} = \left[\eta_{4,8} - \frac{eV_D}{k_B T}\right]$, E_{49} is given by (5.36) and $E_{F26,8}$ is the Fermi energy in the present case which can be determined from

$$\frac{C_{ins}}{e}(V_T - V_{GS}) = \frac{g_v e B}{4\pi\hbar}\sum_{n=0}^{n_{\max}}\sum_{i=0}^{i_{\max}}[F_{-1}(\eta_{4,8}) + F_{-1}(\theta_{4,8})] \tag{6.26}$$

6.2.4 The $I_D - V_D$ equation in magneto MOSFETs of IV–VI materials

(a) Weak electric field limit

From (1.117), we can write

$$V_x = \sqrt{\frac{2}{m_1}}\left[\frac{\sqrt{E(1+\alpha E)}}{1+2\alpha E}\right] \tag{6.27}$$

Using (6.27), (6.1)–(6.3) and (5.43), we get

$$\frac{J_D}{W} = \frac{e^2 B g_v}{\pi^2 \hbar}\sqrt{\frac{2}{m_1}}\sum_{n=0}^{n_{\max}}\sum_{i=0}^{i_{\max}}\left[\left[\frac{\sqrt{E_{50}(1+\alpha E_{50})}}{(1+2\alpha E_{50})}\right][F_{-1}(\eta_{4,9}) - F_{-1}(\theta_{4,9})]\right] \tag{6.28}$$

where $\eta_{4,9} = (k_B T)^{-1}[E_{F26,9} - E_{50}]$, $\theta_{4,9} = \left[\eta_{4,9} - \frac{eV_D}{k_B T}\right]$, E_{50} is given by (5.42) and $E_{F26,9}$ is the Fermi energy in the present case which can be determined from

$$\frac{C_{ins}}{e}(V_T - V_{GS}) = \frac{g_v e B}{2\pi\hbar}\sum_{n=0}^{n_{\max}}\sum_{i=0}^{i_{\max}}[F_{-1}(\eta_{4,9}) + F_{-1}(\theta_{4,9})] \tag{6.29}$$

(b) Strong electric field limit

Using (6.27), (6.1)–(6.3) and (5.48), we get

$$\frac{J_D}{W} = \frac{e^2 B g_v}{\pi^2 \hbar}\sqrt{\frac{2}{m_1}}\sum_{n=0}^{n_{\max}}\sum_{i=0}^{i_{\max}}\left[\left[\frac{\sqrt{E_{51}(1+\alpha E_{51})}}{(1+2\alpha E_{51})}\right]\right.$$
$$\left.\times\, [F_{-1}(\eta_{4,10}) - F_{-1}(\theta_{4,10})]\right] \tag{6.30}$$

where $\eta_{4,10} = (k_B T)^{-1}[E_{F26,10} - E_{51}]$, $\theta_{4,10} = \left[\eta_{4,10} - \frac{eV_D}{k_B T}\right]$, E_{51} is given by (5.47) and $E_{F26,10}$ is the Fermi energy in the present case which can be determined from

$$\frac{C_{ins}}{e}(V_T - V_{GS}) = \frac{g_v e B}{2\pi\hbar}\sum_{n=0}^{n_{\max}}\sum_{i=0}^{i_{\max}}[F_{-1}(\eta_{4,10}) + F_{-1}(\theta_{4,10})] \tag{6.31}$$

6.2.5 The $I_D - V_D$ equation in magneto MOSFETs of stressed Kane type materials

(a) Weak electric field limit

From (1.105), we get

$$V_x = [\hbar[a^*(E)]']^{-1} \tag{6.32}$$

Using (6.32), (6.1)–(6.3) and (5.54), we get

$$\frac{J_D}{W} = \frac{e^2 B g_v}{\pi^2 \hbar^2} \sum_{n=0}^{n_{\max}} \sum_{i=0}^{i_{\max}} [[[a^*(E_{52})]']^{-1}[F_{-1}(\eta_{4,10}) - F_{-1}(\theta_{4,10})]] \tag{6.33}$$

where $\eta_{4,10} = (k_B T)^{-1}[E_{F26,10} - E_{52}]$, $\theta_{4,10} = \left[\eta_{4,10} - \frac{eV_D}{k_B T}\right]$, E_{52} is given by (5.53) and $E_{F26,10}$ is the Fermi energy in the present case which can be determined from

$$\frac{C_{ins}}{e}(V_T - V_{GS}) = \frac{g_v e B}{2\pi\hbar} \sum_{n=0}^{n_{\max}} \sum_{i=0}^{i_{\max}} [F_{-1}(\eta_{4,10}) + F_{-1}(\theta_{4,10})] \tag{6.34}$$

(b) Strong electric field limit

Using (6.32), (6.1)–(6.3) and (5.59), we get

$$\frac{J_D}{W} = \frac{e^2 B g_v}{\pi^2 \hbar^2} \sum_{n=0}^{n_{\max}} \sum_{i=0}^{i_{\max}} [[a^*(E_{53})]']^{-1}[F_{-1}(\eta_{4,11}) - F_{-1}(\theta_{4,11})] \tag{6.35}$$

where $\eta_{4,11} = (k_B T)^{-1}[E_{F26,11} - E_{53}]$, $\theta_{4,11} = [\eta_{4,11} - \frac{eV_D}{k_B T}]$, E_{53} is given by (5.58) and $E_{F26,11}$ is the Fermi energy in the present case which can be determined from

$$\frac{C_{ins}}{e}(V_T - V_{GS}) = \frac{g_v e B}{2\pi\hbar} \sum_{n=0}^{n_{\max}} \sum_{i=0}^{i_{\max}} [F_{-1}(\eta_{4,11}) + F_{-1}(\theta_{4,11})] \tag{6.36}$$

6.2.6 The $I_D - V_D$ equation in magneto MOSFETs of bismuth telluride

(a) Weak electric field limit

From (5.60), we can write

$$V_x = \left[\frac{2\sqrt{I_1 E(1+\alpha E)}}{\hbar(1+2\alpha E)}\right] \tag{6.37}$$

Using (6.37), (6.1)–(6.3) and (5.64), we get

$$\begin{aligned}\frac{J_D}{W} &= \frac{e^2 B g_v}{\pi^2\hbar^2} 2\sqrt{I_1} \sum_{n=0}^{n_{\max}} \sum_{i=0}^{i_{\max}} \\ &\times \left[\left[\frac{\sqrt{E_{54}(1+\alpha E_{54})}}{(1+2\alpha E_{54})}\right][F_{-1}(\eta_{4,11}) - F_{-1}(\theta_{4,11})]\right]\end{aligned} \tag{6.38}$$

where $\eta_{4,11} = (k_B T)^{-1}[E_{F26,11} - E_{54}]$, $\theta_{4,11} = \left[\eta_{4,11} - \frac{eV_D}{k_B T}\right]$, E_{54} is given by (5.63b) and $E_{F26,11}$ is the Fermi energy in the present case which can be determined from

$$\frac{C_{ins}}{e}(V_T - V_{GS}) = \frac{g_v e B}{2\pi\hbar} \sum_{n=0}^{n_{\max}} \sum_{i=0}^{i_{\max}} [F_{-1}(\eta_{4,11}) + F_{-1}(\theta_{4,11})] \tag{6.39}$$

(b) Strong electric field limit

Using (6.37), (6.1)–(6.3) and (5.69), we get

$$\begin{aligned}\frac{J_D}{W} &= \frac{e^2 B g_v}{\pi^2\hbar^2} 2\sqrt{I_1} \sum_{n=0}^{n_{\max}} \sum_{i=0}^{i_{\max}} \left[\left[\frac{\sqrt{E_{55}(1+\alpha E_{55})}}{(1+2\alpha E_{55})}\right]\right. \\ &\left.\times\ [F_{-1}(\eta_{4,12}) - F_{-1}(\theta_{4,12})]\right]\end{aligned} \tag{6.40}$$

where $\eta_{4,12} = (k_B T)^{-1}[E_{F26,12} - E_{55}]$, $\theta_{4,12} = \left[\eta_{4,12} - \frac{eV_D}{k_B T}\right]$, E_{55} is given by (5.68) and $E_{F26,12}$ is the Fermi energy in the present case which can be determined from

$$\frac{C_{ins}}{e}(V_T - V_{GS}) = \frac{g_v e B}{2\pi\hbar} \sum_{n=0}^{n_{\max}} \sum_{i=0}^{i_{\max}} [F_{-1}(\eta_{4,12}) + F_{-1}(\theta_{4,12})] \tag{6.41}$$

6.2.7 The $I_D - V_D$ equation in magneto MOSFETs of germanium

(a) The model of Cardona *et al.*

From (1.59), we can write

$$V_x = \left[\sqrt{\frac{2}{m_\perp^*}}\frac{\sqrt{E(1+\alpha E)}}{(1+2\alpha E)}\right] \tag{6.42}$$

Using (6.42), (6.1)–(6.3) and (5.75), we get

$$\frac{J_D}{W} = \frac{e^2 B g_v}{\pi^2\hbar}\sqrt{\frac{2}{m_\perp^*}}\sum_{n=0}^{n_{\max}}\sum_{i=0}^{i_{\max}}\left[\left[\frac{\sqrt{E_{57}(1+\alpha E_{57})}}{(1+2\alpha E_{57})}\right] \times [F_{-1}(\eta_{4,13}) - F_{-1}(\theta_{4,13})]\right] \tag{6.43}$$

where $\eta_{4,13} = (k_B T)^{-1}[E_{F26,13} - E_{57}]$, $\theta_{4,13} = \left[\eta_{4,13} - \frac{eV_D}{k_B T}\right]$, E_{57} is given by (5.74) and $E_{F26,13}$ is the Fermi energy in the present case which can be determined from

$$\frac{C_{ins}}{e}(V_T - V_{GS}) = \frac{g_v eB}{2\pi\hbar}\sum_{n=0}^{n_{\max}}\sum_{i=0}^{i_{\max}}[F_{-1}(\eta_{4,13}) + F_{-1}(\theta_{4,13})] \tag{6.44}$$

(b) The model of Wang *et al.*

From (1.63), we can write

$$V_x = [\omega'_{222}(E)]^{-1} \tag{6.45}$$

where $\omega_{222}(E) = \left[\left[\frac{1-\sqrt{1-4c_1 E}}{2c_1}\right] 2m_\perp^*\right]^{1/2}$.

Using (6.45), (6.1)–(6.3) and (5.79), we get

$$\frac{J_D}{W} = \frac{e^2 B g_v}{\pi^2\hbar}\sum_{n=0}^{n_{\max}}\sum_{i=0}^{i_{\max}}[[\omega'_{222}(E_{59})]^{-1}[F_{-1}(\eta_{4,14}) - F_{-1}(\theta_{4,14})]] \tag{6.46}$$

where $\eta_{4,14} = (k_B T)^{-1}[E_{F26,14} - E_{59}]$, $\theta_{4,14} = \left[\eta_{4,14} - \frac{eV_D}{k_B T}\right]$, E_{59} is given by (5.78) and $E_{F26,14}$ is the Fermi energy in the present case which can be determined from

$$\frac{C_{ins}}{e}(V_T - V_{GS}) = \frac{g_v eB}{2\pi\hbar}\sum_{n=0}^{n_{\max}}\sum_{i=0}^{i_{\max}}[F_{-1}(\eta_{4,14}) + F_{-1}(\theta_{4,14})] \tag{6.47}$$

6.2.8 The $I_D - V_D$ equation in magneto MOSFETs of gallium aresenide

(a) The model of Rossler

(i) Weak electric field limit

From (5.82), we can write

$$V_x = \left[\frac{2\sqrt{\gamma_{4,\pm}(E)}}{\hbar \gamma'_{4,\pm}(E)} \right] \tag{6.48}$$

Using (6.48), (6.1)–(6.3) and (5.85), we get

$$\frac{J_D}{W} = \frac{e^2 B g_v}{2\pi^2 \hbar^2} \sum_{n=0}^{n_{\max}} \sum_{i=0}^{i_{\max}} \left[\left[\frac{\sqrt{\gamma_{4,\pm}(E_{60})}}{\gamma'_{4,\pm}(E_{60})} \right] [F_{-1}(\eta_{4,15}) - F_{-1}(\theta_{4,15})] \right] \tag{6.49}$$

where $\eta_{4,15} = (k_B T)^{-1}[E_{F26,15} - E_{60}]$, $\theta_{4,15} = \left[\eta_{4,15} - \frac{eV_D}{k_B T} \right]$, E_{60} is given by (5.84) and $E_{F26,15}$ is the Fermi energy in the present case which can be determined from

$$\frac{C_{ins}}{e}(V_T - V_{GS}) = \frac{g_v e B}{4\pi\hbar} \sum_{n=0}^{n_{\max}} \sum_{i=0}^{i_{\max}} [F_{-1}(\eta_{4,15}) + F_{-1}(\theta_{4,15})] \tag{6.50}$$

(ii) Strong electric field limit

Using (6.48), (6.1)–(6.3) and (5.89), we get

$$\frac{J_D}{W} = \frac{e^2 B g_v}{2\pi^2 \hbar^2} \sum_{n=0}^{n_{\max}} \sum_{i=0}^{i_{\max}} \left[\left[\frac{\sqrt{\gamma_{4,\pm}(E_{61})}}{\gamma'_{4,\pm}(E_{61})} \right] [F_{-1}(\eta_{4,16}) - F_{-1}(\theta_{4,16})] \right] \tag{6.51}$$

where $\eta_{4,16} = (k_B T)^{-1}[E_{F26,16} - E_{61}]$, $\theta_{4,16} = \left[\eta_{4,16} - \frac{eV_D}{k_B T} \right]$, E_{61} is given by (5.88) and $E_{F26,16}$ is the Fermi energy in the present case which can be determined from

$$\frac{C_{ins}}{e}(V_T - V_{GS}) = \frac{g_v e B}{4\pi\hbar} \sum_{n=0}^{n_{\max}} \sum_{i=0}^{i_{\max}} [F_{-1}(\eta_{4,16}) + F_{-1}(\theta_{4,16})] \tag{6.52}$$

(b) The model of Stillman *et al.*

(i) Weak electric field limit

From (5.91), we can write

$$V_x = \left[\frac{2\sqrt{\gamma_5(E)}}{\hbar\gamma_5'(E)}\right] \tag{6.53}$$

Using (6.53), (21.1)–(21.3) and (5.94), we get

$$\frac{J_D}{W} = \frac{e^2 B g_v}{\pi^2\hbar^2}\sum_{n=0}^{n_{\max}}\sum_{i=0}^{i_{\max}}\left[\left[\frac{\sqrt{\gamma_5(E_{62})}}{\gamma_5'(E_{62})}\right][F_{-1}(\eta_{4,16}) - F_{-1}(\theta_{4,16})]\right] \tag{6.54}$$

where $\eta_{4,16} = (k_BT)^{-1}[E_{F26,16} - E_{62}]$, $\theta_{4,16} = \left[\eta_{4,16} - \frac{eV_D}{k_BT}\right]$, E_{62} is given by (5.93) and $E_{F26,16}$ is the Fermi energy in the present case which can be determined from

$$\frac{C_{ins}}{e}(V_T - V_{GS}) = \frac{g_v eB}{2\pi\hbar}\sum_{n=0}^{n_{\max}}\sum_{i=0}^{i_{\max}}[F_{-1}(\eta_{4,16}) + F_{-1}(\theta_{4,16})] \tag{6.55}$$

(ii) Strong electric field limit

Using (6.53), (6.1)–(6.3) and (5.98), we get

$$\frac{J_D}{W} = \frac{e^2 B g_v}{\pi^2\hbar^2}\sum_{n=0}^{n_{\max}}\sum_{i=0}^{i_{\max}}\left[\left[\frac{\sqrt{\gamma_5(E_{63})}}{\gamma_5'(E_{63})}\right][F_{-1}(\eta_{4,17}) - F_{-1}(\theta_{4,17})]\right] \tag{6.56}$$

where $\eta_{4,17} = (k_BT)^{-1}[E_{F26,17} - E_{63}]$, $\theta_{4,17} = \left[\eta_{4,17} - \frac{eV_D}{k_BT}\right]$, E_{63} is given by (5.97) and $E_{F26,17}$ is the Fermi energy in the present case which can be determined from

$$\frac{C_{ins}}{e}(V_T - V_{GS}) = \frac{g_v eB}{2\pi\hbar}\sum_{n=0}^{n_{\max}}\sum_{i=0}^{i_{\max}}[F_{-1}(\eta_{4,17}) + F_{-1}(\theta_{4,17})] \tag{6.57}$$

*6.2.9 The $I_D - V_D$ equation in magneto MOSFETs of gallium antimonide

(a) The model of Sailer *et al.*

(i) Weak electric field limit

From (5.100), we can write

$$V_x = \left[\frac{2\sqrt{\gamma_{6,\pm}(\bar{E})}}{\hbar\gamma'_{6,\pm}(\bar{E})}\right] \tag{6.58}$$

Using (6.58), (6.1)–(6.3) and (5.104), we get

$$\frac{J_D}{W} = \frac{e^2 B g_v}{2\pi^2\hbar^2}\sum_{n=0}^{n_{\max}}\sum_{i=0}^{i_{\max}}\left[\left[\frac{\sqrt{\gamma_{6\pm}(E_{64})}}{\gamma'_{6\pm}(E_{64})}\right][F_{-1}(\eta_{4,18}) - F_{-1}(\theta_{4,18})]\right] \tag{6.59}$$

where $\eta_{4,18} = (k_B T)^{-1}[E_{F26,18} - E_{64}]$, $\theta_{4,18} = \left[\eta_{4,18} - \frac{eV_D}{k_B T}\right]$, E_{64} is given by (5.103) and $E_{F26,18}$ is the Fermi energy in the present case which can be determined from

$$\frac{C_{ins}}{e}(V_T - V_{GS}) = \frac{g_v e B}{4\pi\hbar}\sum_{n=0}^{n_{\max}}\sum_{i=0}^{i_{\max}}[F_{-1}(\eta_{4,18}) + F_{-1}(\theta_{4,18})] \tag{6.60}$$

(ii) Strong electric field limit

Using (6.58), (6.1)–(6.3) and (5.108), we get

$$\frac{J_D}{W} = \frac{e^2 B g_v}{2\pi^2\hbar^2}\sum_{n=0}^{n_{\max}}\sum_{i=0}^{i_{\max}}\left[\left[\frac{\sqrt{\gamma_6(E_{65})}}{\gamma'_6(E_{65})}\right][F_{-1}(\eta_{4,19}) - F_{-1}(\theta_{4,19})]\right] \tag{6.61}$$

where $\eta_{4,19} = (k_B T)^{-1}[E_{F26,19} - E_{65}]$, $\theta_{4,19} = \left[\eta_{4,19} - \frac{eV_D}{k_B T}\right]$, E_{65} is given by (5.107) and $E_{F26,19}$ is the Fermi energy in the present case which can be determined from

$$\frac{C_{ins}}{e}(V_T - V_{GS}) = \frac{g_v e B}{4\pi\hbar}\sum_{n=0}^{n_{\max}}\sum_{i=0}^{i_{\max}}[F_{-1}(\eta_{4,19}) + F_{-1}(\theta_{4,19})] \tag{6.62}$$

(b) The model of Mathur *et al.*

(i) Weak electric field limit

From (1.97), we can write

$$V_x = \left[\frac{2\sqrt{\gamma_7(E)}}{\hbar\gamma_7'(E)}\right] \tag{6.63}$$

Using (6.63), (6.1)–(6.3) and (5.113), we get

$$\frac{J_D}{W} = \frac{e^2 B g_v}{\pi^2\hbar^2}\sum_{n=0}^{n_{\max}}\sum_{i=0}^{i_{\max}}\left[\left[\frac{\sqrt{\gamma_7(E_{66})}}{\gamma_7'(E_{66})}\right][F_{-1}(\eta_{4,20}) - F_{-1}(\theta_{4,20})]\right] \tag{6.64}$$

where $\eta_{4,20} = (k_BT)^{-1}[E_{F26,20} - E_{66}]$, $\theta_{4,20} = \left[\eta_{4,20} - \frac{eV_D}{k_BT}\right]$, E_{66} is given by (5.112) and $E_{F26,20}$ is the Fermi energy in the present case which can be determined from

$$\frac{C_{ins}}{e}(V_T - V_{GS}) = \frac{g_v eB}{2\pi\hbar}\sum_{n=0}^{n_{\max}}\sum_{i=0}^{i_{\max}}[F_{-1}(\eta_{4,20}) + F_{-1}(\theta_{4,20})] \tag{6.65}$$

(ii) Strong electric field limit

Using (6.63), (6.1)–(6.3) and (5.117), we get

$$\frac{J_D}{W} = \frac{e^2 B g_v}{\pi^2\hbar^2}\sum_{n=0}^{n_{\max}}\sum_{i=0}^{i_{\max}}\left[\left[\frac{\sqrt{\gamma_7(E_{67})}}{\gamma_7'(E_{67})}\right][F_{-1}(\eta_{4,21}) - F_{-1}(\theta_{4,21})]\right] \tag{6.66}$$

where $\eta_{4,21} = (k_BT)^{-1}[E_{F26,21} - E_{67}]$, $\theta_{4,21} = \left[\eta_{4,21} - \frac{eV_D}{k_BT}\right]$, E_{67} is given by (5.116) and $E_{F26,21}$ is the Fermi energy in the present case which can be determined from

$$\frac{C_{ins}}{e}(V_T - V_{GS}) = \frac{g_v eB}{2\pi\hbar}\sum_{n=0}^{n_{\max}}\sum_{i=0}^{i_{\max}}[F_{-1}(\eta_{4,21}) + F_{-1}(\theta_{4,21})] \tag{6.67}$$

(c) The model of Zhang

(i) Weak electric field limit

From (5.120), we can write

$$V_x = \left[\frac{2\sqrt{\gamma_8(E)}}{\hbar\gamma_8'(E)}\right] \tag{6.68}$$

Using (6.68), (6.1)–(6.3) and (5.123), we get

$$\frac{J_D}{W} = \frac{e^2 B g_v}{\pi^2\hbar^2}\sum_{n=0}^{n_{\max}}\sum_{i=0}^{i_{\max}}\left[\left[\frac{\sqrt{\gamma_8(E_{68})}}{\gamma_8'(E_{68})}\right][F_{-1}(\eta_{4,22}) - F_{-1}(\theta_{4,22})]\right] \tag{6.69}$$

where $\eta_{4,22} = (k_B T)^{-1}[E_{F26,22} - E_{68}]$, $\theta_{4,22} = \left[\eta_{4,22} - \frac{eV_D}{k_B T}\right]$, E_{68} is given by (5.122) and $E_{F26,22}$ is the Fermi energy in the present case which can be determined from

$$\frac{C_{ins}}{e}(V_T - V_{GS}) = \frac{g_v e B}{2\pi\hbar}\sum_{n=0}^{n_{\max}}\sum_{i=0}^{i_{\max}}[F_{-1}(\eta_{4,22}) + F_{-1}(\theta_{4,22})] \tag{6.70}$$

(ii) Strong electric field limit

Using (6.68), (6.1)–(6.3) and (5.127), we get

$$\frac{J_D}{W} = \frac{e^2 B g_v}{\pi^2\hbar^2}\sum_{n=0}^{n_{\max}}\sum_{i=0}^{i_{\max}}\left[\left[\frac{\sqrt{\gamma_8(E_{69})}}{\gamma_8'(E_{69})}\right][F_{-1}(\eta_{4,23}) - F_{-1}(\theta_{4,23})]\right] \tag{6.71}$$

where $\eta_{4,23} = (k_B T)^{-1}[E_{F26,23} - E_{69}]$, $\theta_{4,23} = \left[\eta_{4,23} - \frac{eV_D}{k_B T}\right]$, E_{69} is given by (5.126) and $E_{F26,23}$ is the Fermi energy in the present case which can be determined from

$$\frac{C_{ins}}{e}(V_T - V_{GS}) = \frac{g_v e B}{2\pi\hbar}\sum_{n=0}^{n_{\max}}\sum_{i=0}^{i_{\max}}[F_{-1}(\eta_{4,23}) + F_{-1}(\theta_{4,23})] \tag{6.72}$$

6.2.10 The $I_D - V_D$ equation in magneto MOSFETs of indium antimonide

(a) The model of Palik *et al.*

(i) Weak electric field limit

From (5.130), we can write

$$V_x = \left[\frac{2\sqrt{\gamma_9(E)}}{\hbar \gamma_9'(E)} \right] \tag{6.73}$$

Using (6.73), (6.1)–(6.3) and (5.133), we get

$$\frac{J_D}{W} = \frac{e^2 B g_v}{\pi^2 \hbar^2} \sum_{n=0}^{n_{\max}} \sum_{i=0}^{i_{\max}} \left[\left[\frac{\sqrt{\gamma_9(E_{70})}}{\gamma_9'(E_{70})} \right] [F_{-1}(\eta_{4,24}) - F_{-1}(\theta_{4,24})] \right] \tag{6.74}$$

where $\eta_{4,24} = (k_B T)^{-1}[E_{F26,24} - E_{70}]$, $\theta_{4,24} = \left[\eta_{4,24} - \frac{eV_D}{k_B T} \right]$, E_{70} is given by (5.132) and $E_{F26,24}$ is the Fermi energy in the present case which can be determined from

$$\frac{C_{ins}}{e}(V_T - V_{GS}) = \frac{g_v e B}{2\pi\hbar} \sum_{n=0}^{n_{\max}} \sum_{i=0}^{i_{\max}} [F_{-1}(\eta_{4,24}) + F_{-1}(\theta_{4,24})] \tag{6.75}$$

(ii) Strong electric field limit

Using (6.73), (6.1)–(6.3) and (5.137), we get

$$\frac{J_D}{W} = \frac{e^2 B g_v}{\pi^2 \hbar^2} \sum_{n=0}^{n_{\max}} \sum_{i=0}^{i_{\max}} \left[\left[\frac{\sqrt{\gamma_9(E_{71})}}{\gamma_9'(E_{71})} \right] [F_{-1}(\eta_{4,25}) - F_{-1}(\theta_{4,25})] \right] \tag{6.76}$$

where $\eta_{4,25} = (k_B T)^{-1}[E_{F26,25} - E_{71}]$, $\theta_{4,25} = \left[\eta_{4,25} - \frac{eV_D}{k_B T} \right]$, E_{71} is given by (5.136) and $E_{F26,25}$ is the Fermi energy in the present case which can be determined from

$$\frac{C_{ins}}{e}(V_T - V_{GS}) = \frac{g_v e B}{2\pi\hbar} \sum_{n=0}^{n_{\max}} \sum_{i=0}^{i_{\max}} [F_{-1}(\eta_{4,25}) + F_{-1}(\theta_{4,25})] \tag{6.77}$$

(b) The model of Johnson *et al.*

(i) Weak electric field limit
From (5.140), we can write

$$V_x = \left[\frac{2\sqrt{\gamma_{10}(E)}}{\hbar\gamma'_{10}(E)}\right] \tag{6.78}$$

Using (6.78), (6.1)–(6.3) and (5.143), we get

$$\frac{J_D}{W} = \frac{e^2Bg_v}{\pi^2\hbar^2}\sum_{n=0}^{n_{\max}}\sum_{i=0}^{i_{\max}}\left[\left[\frac{\sqrt{\gamma_{10}(E_{72})}}{\gamma'_{10}(E_{72})}\right][F_{-1}(\eta_{4,26}) - F_{-1}(\theta_{4,26})]\right] \tag{6.79}$$

where $\eta_{4,26} = (k_BT)^{-1}[E_{F26,26} - E_{72}]$, $\theta_{4,26} = \left[\eta_{4,26} - \frac{eV_D}{k_BT}\right]$, E_{72} is given by (5.142) and $E_{F26,26}$ is the Fermi energy in the present case which can be determined from

$$\frac{C_{ins}}{e}(V_T - V_{GS}) = \frac{g_veB}{2\pi\hbar}\sum_{n=0}^{n_{\max}}\sum_{i=0}^{i_{\max}}[F_{-1}(\eta_{4,26}) + F_{-1}(\theta_{4,26})] \tag{6.80}$$

(ii) Strong electric field limit

Using (6.78), (6.1)–(6.3) and (5.147), we get

$$\frac{J_D}{W} = \frac{e^2Bg_v}{\pi^2\hbar^2}\sum_{n=0}^{n_{\max}}\sum_{i=0}^{i_{\max}}\left[\left[\frac{\sqrt{\gamma_{10}(E_{73})}}{\gamma'_{10}(E_{73})}\right][F_{-1}(\eta_{4,27}) - F_{-1}(\theta_{4,27})]\right] \tag{6.81}$$

where $\eta_{4,27} = (k_BT)^{-1}[E_{F26,27} - E_{73}]$, $\theta_{4,27} = \left[\eta_{4,27} - \frac{eV_D}{k_BT}\right]$, E_{73} is given by (5.146) and $E_{F26,27}$ is the Fermi energy in the present case which can be determined from

$$\frac{C_{ins}}{e}(V_T - V_{GS}) = \frac{g_veB}{2\pi\hbar}\sum_{n=0}^{n_{\max}}\sum_{i=0}^{i_{\max}}[F_{-1}(\eta_{4,27}) + F_{-1}(\theta_{4,27})] \tag{6.82}$$

(c) The model of Agafonov *et al.*

(i) Weak electric field limit

From (5.150), we can write

$$V_x = \left[\frac{2\sqrt{\gamma_{11}(E)}}{\hbar\gamma'_{11}(E)}\right] \tag{6.83}$$

Using (6.83), (6.1)–(6.3) and (5.153), we get

$$\frac{J_D}{W} = \frac{e^2 B g_v}{\pi^2\hbar^2}\sum_{n=0}^{n_{\max}}\sum_{i=0}^{i_{\max}}\left[\left[\frac{\sqrt{\gamma_{11}(E_{74})}}{\gamma'_{11}(E_{74})}\right][F_{-1}(\eta_{4,28}) - F_{-1}(\theta_{4,28})]\right] \tag{6.84}$$

where $\eta_{4,28} = (k_BT)^{-1}[E_{F26,28} - E_{74}]$, $\theta_{4,28} = \left[\eta_{4,28} - \frac{eV_D}{k_BT}\right]$, E_{74} is given by (5.152) and $E_{F26,28}$ is the Fermi energy in the present case which can be determined from

$$\frac{C_{ins}}{e}(V_T - V_{GS}) = \frac{g_v eB}{2\pi\hbar}\sum_{n=0}^{n_{\max}}\sum_{i=0}^{i_{\max}}[F_{-1}(\eta_{4,28}) + F_{-1}(\theta_{4,28})] \tag{6.85}$$

(ii) Strong electric field limit

Using (6.83), (6.1)–(6.3) and (5.157), we get

$$\frac{J_D}{W} = \frac{e^2 B g_v}{\pi^2\hbar^2}\sum_{n=0}^{n_{\max}}\sum_{i=0}^{i_{\max}}\left[\left[\frac{\sqrt{\gamma_{11}(E_{75})}}{\gamma'_{11}(E_{75})}\right][F_{-1}(\eta_{4,29}) - F_{-1}(\theta_{4,29})]\right] \tag{6.86}$$

where $\eta_{4,29} = (k_BT)^{-1}[E_{F26,29} - E_{75}]$, $\theta_{4,29} = \left[\eta_{4,29} - \frac{eV_D}{k_BT}\right]$, E_{75} is given by (5.156) and $E_{F26,29}$ is the Fermi energy in the present case which can be determined from

$$\frac{C_{ins}}{e}(V_T - V_{GS}) = \frac{g_v eB}{2\pi\hbar}\sum_{n=0}^{n_{\max}}\sum_{i=0}^{i_{\max}}[F_{-1}(\eta_{4,29}) + F_{-1}(\theta_{4,29})] \tag{6.87}$$

6.2.11 The $I_D - V_D$ equation in magneto MOSFETs of tellurium

From (1.67), we can write

$$V_x = (\hbar)^{-1}\sqrt{\psi_4^2 + 4\psi_2 E} \tag{6.88}$$

Using (6.88), (6.1)–(6.3) and (5.163), we get

$$\frac{J_D}{W} = \frac{e^2 B g_v}{2\pi^2\hbar^2}\sum_{n=0}^{n_{\max}}\sum_{i=0}^{i_{\max}}[[\sqrt{\psi_4^2 + 4\psi_2 E_{76}}][F_{-1}(\eta_{4,30}) - F_{-1}(\theta_{4,30})]] \tag{6.89}$$

where $\eta_{4,30} = (k_B T)^{-1}[E_{F26,30} - E_{76}]$, $\theta_{4,30} = \left[\eta_{4,30} - \frac{eV_D}{k_B T}\right]$, E_{76} is given by (5.162) and $E_{F26,30}$ is the Fermi energy in the present case which can be determined from

$$\frac{C_{ins}}{e}(V_T - V_{GS}) = \frac{g_v eB}{4\pi\hbar}\sum_{n=0}^{n_{\max}}\sum_{i=0}^{i_{\max}}[F_{-1}(\eta_{4,30}) + F_{-1}(\theta_{4,30})] \tag{6.90}$$

6.2.12 The $I_D - V_D$ equation for ballistic magneto QWFETs of nonlinear optical and tetragonal materials

Using (6.1)–(6.3), (5.165) and (5.166), we get

$$\frac{J_D}{W} = \frac{e^2 B g_v}{\pi^2\hbar^2}\sum_{n=0}^{n_{\max}}\sum_{i=0}^{i_{\max}}[[g'_{21,1}(E_{77})]^{-1}[F_{-1}(\eta_{24,4,1}) - F_{-1}(\theta_{24,4,1})]] \tag{6.91}$$

where $\eta_{24,4,1} = (k_B T)^{-1}[E_{F24,1} - E_{77}]$, $\theta_{24,4,1} = \left[\eta_{24,4,1} - \frac{eV_D}{k_B T}\right]$ and $E_{F24,1}$ is the Fermi energy in the present case which can be determined from

$$\frac{C_{ins}}{e}(V_T - V_{GS}) = \frac{g_v eB}{2\pi\hbar}\sum_{n=0}^{n_{\max}}\sum_{i=0}^{i_{\max}}[F_{-1}(\eta_{24,4,1}) + F_{-1}(\theta_{24,4,1})] \tag{6.92}$$

6.2.13 The $I_D - V_D$ equation for ballistic magneto QWFETs of III–V, ternary and quaternary materials

(a) The three band model of Kane

Using (6.1)–(6.3), (5.168) and (5.169), we get

$$\frac{J_D}{W} = \frac{e^2 B g_v}{\pi^2 \hbar}\sqrt{\frac{2}{m_c}}\sum_{n=0}^{n_{\max}}\sum_{i=0}^{i_{\max}}[[\sqrt{I(E_{78})}][I'(E_{78})]^{-1} \times [F_{-1}(\eta_{24,4,3}) - F_{-1}(\theta_{24,4,3})]] \tag{6.93}$$

where $\eta_{24,4,3} = (k_B T)^{-1}[E_{F24,3} - E_{78}]$, $\theta_{24,4,3} = \left[\eta_{24,4,3} - \frac{eV_D}{k_B T}\right]$ and $E_{F24,3}$ is the Fermi energy in the present case which can be determined from

$$\frac{C_{ins}}{e}(V_T - V_{GS}) = \frac{g_v e B}{2\pi\hbar}\sum_{n=0}^{n_{\max}}\sum_{i=0}^{i_{\max}}[F_{-1}(\eta_{24,4,3}) + F_{-1}(\theta_{24,4,3})] \tag{6.94}$$

(b) The two band model of Kane

Using (6.1)–(6.3), (5.171) and (5.172), we get

$$\frac{J_D}{W} = \frac{e^2 B g_v}{\pi^2 \hbar}\sqrt{\frac{2}{m_c}}\sum_{n=0}^{n_{\max}}\sum_{i=0}^{i_{\max}}[[\sqrt{E_{79}(1+\alpha E_{79})}][1+2\alpha E_{79}]^{-1} \times [F_{-1}(\eta_{24,4,4}) - F_{-1}(\theta_{24,4,4})]] \tag{6.95}$$

where $\eta_{24,4,4} = (k_B T)^{-1}[E_{F24,4} - E_{79}]$, $\theta_{24,4,4} = \left[\eta_{24,4,4} - \frac{eV_D}{k_B T}\right]$ and $E_{F24,4}$ is the Fermi energy in the present case which can be determined from

$$\frac{C_{ins}}{e}(V_T - V_{GS}) = \frac{g_v e B}{2\pi\hbar}\sum_{n=0}^{n_{\max}}\sum_{i=0}^{i_{\max}}[F_{-1}(\eta_{24,4,4}) + F_{-1}(\theta_{24,4,4})] \tag{6.96}$$

(c) The parabolic band model

Using (6.1)–(6.3), (5.174) and (5.175), we get

$$\frac{J_D}{W} = \frac{e^2 B g_v}{\pi^2 \hbar} \sqrt{\frac{2}{m_c}} \sum_{n=0}^{n_{\max}} \sum_{i=0}^{i_{\max}} [[\sqrt{E_{80}}][F_{-1}(\eta_{24,4,5}) - F_{-1}(\theta_{24,4,5})]] \quad (6.97)$$

where $\eta_{24,4,5} = (k_B T)^{-1}[E_{F24,5} - E_{80}]$, $\theta_{24,4,5} = \left[\eta_{24,4,5} - \frac{eV_D}{k_B T}\right]$ and $E_{F24,5}$ is the Fermi energy in the present case which can be determined from

$$\frac{C_{ins}}{e}(V_T - V_{GS}) = \frac{g_v e B}{2\pi\hbar} \sum_{n=0}^{n_{\max}} \sum_{i=0}^{i_{\max}} [F_{-1}(\eta_{24,4,5}) + F_{-1}(\theta_{24,4,5})] \quad (6.98)$$

6.2.14 The $I_D - V_D$ equation for ballistic magneto QWFETs of II–VI materials

Using (6.1)–(6.3), (5.177) and (5.178), we get

$$\frac{J_D}{W} = \frac{e^2 B g_v}{2\pi^2 \hbar^2} \sum_{n=0}^{n_{\max}} \sum_{i=0}^{i_{\max}} \left[\left[\sqrt{(\bar{\lambda}_0)^2 + \frac{2E_{81}\hbar^2}{m_\perp^*}}\right] \times [F_{-1}(\eta_{24,4,6}) - F_{-1}(\theta_{24,4,6})]\right] \quad (6.99)$$

where $\eta_{24,4,6} = (k_B T)^{-1}[E_{F24,6} - E_{81}]$, $\theta_{24,4,6} = \left[\eta_{24,4,6} - \frac{eV_D}{k_B T}\right]$ and $E_{F24,6}$ is the Fermi energy in the present case which can be determined from

$$\frac{C_{ins}}{e}(V_T - V_{GS}) = \frac{g_v e B}{4\pi\hbar} \sum_{n=0}^{n_{\max}} \sum_{i=0}^{i_{\max}} [F_{-1}(\eta_{24,4,6}) + F_{-1}(\theta_{24,4,6})] \quad (6.100)$$

6.2.15 The $I_D - V_D$ equation for ballistic magneto QWFETs of IV–VI materials

(a) The McClure and Choi model

Using (6.1)–(6.3), (5.184) and (5.185), we get

$$\frac{J_D}{W} = \frac{e^2 B g_v}{\pi^2 \hbar} \sqrt{\frac{2}{m_1}} \sum_{n=0}^{n_{\max}} \sum_{i=0}^{i_{\max}} [[\sqrt{E_{82}(1 + \alpha E_{82})}][1 + 2\alpha E_{82}]^{-1} \times [F_{-1}(\eta_{24,4,7}) - F_{-1}(\theta_{24,4,7})]] \quad (6.101)$$

where $\eta_{24,4,7} = (k_BT)^{-1}[E_{F24,7} - E_{82}]$, $\theta_{24,4,7} = \left[\eta_{24,4,7} - \frac{eV_D}{k_BT}\right]$ and $E_{F24,7}$ is the Fermi energy in the present case which can be determined from

$$\frac{C_{ins}}{e}(V_T - V_{GS}) = \frac{g_v eB}{2\pi\hbar}\sum_{n=0}^{n_{\max}}\sum_{i=0}^{i_{\max}}[F_{-1}(\eta_{24,4,7}) + F_{-1}(\theta_{24,4,7})] \quad (6.102)$$

(b) The Hybrid model

Using (6.1)–(6.3), (5.187) and (5.188), we get

$$\frac{J_D}{W} = \frac{e^2Bg_v}{\pi^2\hbar}\sqrt{\frac{2}{m_1}}\sum_{n=0}^{n_{\max}}\sum_{i=0}^{i_{\max}}[[\sqrt{E_{83}(1+\alpha E_{83})}][1+2\alpha E_{83}]^{-1} \times [F_{-1}(\eta_{24,4,8}) - F_{-1}(\theta_{24,4,8})]] \quad (6.103)$$

where $\eta_{24,4,8} = (k_BT)^{-1}[E_{F24,8} - E_{83}]$, $\theta_{24,4,8} = \left[\eta_{24,4,8} - \frac{eV_D}{k_BT}\right]$ and $E_{F24,8}$ is the Fermi energy in the present case which can be determined from

$$\frac{C_{ins}}{e}(V_T - V_{GS}) = \frac{g_v eB}{2\pi\hbar}\sum_{n=0}^{n_{\max}}\sum_{i=0}^{i_{\max}}[F_{-1}(\eta_{24,4,8}) + F_{-1}(\theta_{24,4,8})] \quad (6.104)$$

(c) The Cohen model

Using (6.1)–(6.3), (5.190) and (5.191), we get

$$\frac{J_D}{W} = \frac{e^2Bg_v}{\pi^2\hbar}\sqrt{\frac{2}{m_1}}\sum_{n=0}^{n_{\max}}\sum_{i=0}^{i_{\max}}[[\sqrt{E_{84}(1+\alpha E_{84})}][1+2\alpha E_{84}]^{-1} \times [F_{-1}(\eta_{24,4,9}) - F_{-1}(\theta_{24,4,9})]] \quad (6.105)$$

where $\eta_{24,4,9} = (k_BT)^{-1}[E_{F24,9} - E_{84}]$, $\theta_{24,4,9} = \left[\eta_{24,4,9} - \frac{eV_D}{k_BT}\right]$ and $E_{F24,9}$ is the Fermi energy in the present case which can be determined from

$$\frac{C_{ins}}{e}(V_T - V_{GS}) = \frac{g_v eB}{2\pi\hbar}\sum_{n=0}^{n_{\max}}\sum_{i=0}^{i_{\max}}[F_{-1}(\eta_{24,4,9}) + F_{-1}(\theta_{24,4,9})] \quad (6.106)$$

(d) The Lax model

Using (6.1)–(6.3), (5.193) and (5.194), we get

$$\frac{J_D}{W} = \frac{e^2Bg_v}{\pi^2\hbar}\sqrt{\frac{2}{m_1}}\sum_{n=0}^{n_{\max}}\sum_{i=0}^{i_{\max}}[[\sqrt{E_{85}(1+\alpha E_{85})}][1+2\alpha E_{85}]^{-1} \times [F_{-1}(\eta_{24,4,10}) - F_{-1}(\theta_{24,4,10})]] \quad (6.107)$$

where $\eta_{24,4,10} = (k_BT)^{-1}[E_{F24,10} - E_{85}]$, $\theta_{24,4,10} = \left[\eta_{24,4,10} - \frac{eV_D}{k_BT}\right]$ and $E_{F24,10}$ is the Fermi energy in the present case which can be determined from

$$\frac{C_{ins}}{e}(V_T - V_{GS}) = \frac{g_v eB}{2\pi\hbar}\sum_{n=0}^{n_{\max}}\sum_{i=0}^{i_{\max}}[F_{-1}(\eta_{24,4,10}) + F_{-1}(\theta_{24,4,10})] \quad (6.108)$$

(e) The Dimmock model

Using (6.1)–(6.3), (5.210) and (5.211), we get

$$\frac{J_D}{W} = \frac{e^2Bg_v}{\pi^2\hbar}\sum_{n=0}^{n_{\max}}\sum_{i=0}^{i_{\max}}[[\phi'_{227}(E_{86})]^{-1}[F_{-1}(\eta_{24,4,11}) - F_{-1}(\theta_{24,4,11})]] \quad (6.109)$$

where $\eta_{24,4,11} = (k_BT)^{-1}[E_{F24,11} - E_{86}]$, $\theta_{24,4,11} = \left[\eta_{24,4,11} - \frac{eV_D}{k_BT}\right]$ and $E_{F24,11}$ is the Fermi energy in the present case which can be determined from

$$\frac{C_{ins}}{e}(V_T - V_{GS}) = \frac{g_v eB}{2\pi\hbar}\sum_{n=0}^{n_{\max}}\sum_{i=0}^{i_{\max}}[F_{-1}(\eta_{24,4,11}) + F_{-1}(\theta_{24,4,11})] \quad (6.110)$$

(f) The Ellipsoidal Parabolic model

Using (6.1)–(6.3), (5.196) and (5.197), we get

$$\frac{J_D}{W} = \frac{e^2Bg_v}{\pi^2\hbar}\sqrt{\frac{2}{m_1}}\sum_{n=0}^{n_{\max}}\sum_{i=0}^{i_{\max}}[[\sqrt{E_{87}}][F_{-1}(\eta_{24,4,12}) - F_{-1}(\theta_{24,4,12})]] \quad (6.111)$$

where $\eta_{24,4,12} = (k_BT)^{-1}[E_{F24,12} - E_{87}]$, $\theta_{24,4,12} = \left[\eta_{24,4,12} - \frac{eV_D}{k_BT}\right]$ and $E_{F24,12}$ is the Fermi energy in the present case which can be determined from

$$\frac{C_{ins}}{e}(V_T - V_{GS}) = \frac{g_v eB}{2\pi\hbar}\sum_{n=0}^{n_{\max}}\sum_{i=0}^{i_{\max}}[F_{-1}(\eta_{24,4,12}) + F_{-1}(\theta_{24,4,12})] \quad (6.112)$$

*6.2.16 The $I_D - V_D$ equation for ballistic magneto QWFETs of stressed Kane type materials

Using (6.1)–(6.3), (5.213) and (5.214), we get

$$\frac{J_D}{W} = \frac{e^2 B g_v}{\pi^2 \hbar^2} \sum_{n=0}^{n_{\max}} \sum_{i=0}^{i_{\max}} [[[\bar{a}_0(E_{88})]']^{-1}[F_{-1}(\eta_{24,4,13}) - F_{-1}(\theta_{24,4,13})]] \tag{6.113}$$

where $\eta_{24,4,13} = (k_B T)^{-1}[E_{F24,13} - E_{88}]$, $\theta_{24,4,13} = \left[\eta_{24,4,13} - \frac{eV_D}{k_B T}\right]$ and $E_{F24,13}$ is the Fermi energy in the present case which can be determined from

$$\frac{C_{ins}}{e}(V_T - V_{GS}) = \frac{g_v e B}{2\pi\hbar} \sum_{n=0}^{n_{\max}} \sum_{i=0}^{i_{\max}} [F_{-1}(\eta_{24,4,13}) + F_{-1}(\theta_{24,4,13})] \tag{6.114}$$

6.2.17 The $I_D - V_D$ equation for ballistic magneto QWFETs of bismuth telluride

Using (6.1)–(6.3), (5.218) and (5.219), we get

$$\frac{J_D}{W} = \frac{2e^2 B g_v}{\pi^2 \hbar^2} \sqrt{I_1} \sum_{n=0}^{n_{\max}} \sum_{i=0}^{i_{\max}} [[\sqrt{E_{89}(1 + \alpha E_{89})}][1 + 2\alpha E_{89}]^{-1} \times [F_{-1}(\eta_{24,4,14}) - F_{-1}(\theta_{24,4,14})]] \tag{6.115}$$

where $\eta_{24,4,14} = (k_B T)^{-1}[E_{F24,14} - E_{89}]$, $\theta_{24,4,14} = \left[\eta_{24,4,14} - \frac{eV_D}{k_B T}\right]$ and $E_{F24,14}$ is the Fermi energy in the present case which can be determined from

$$\frac{C_{ins}}{e}(V_T - V_{GS}) = \frac{g_v e B}{2\pi\hbar} \sum_{n=0}^{n_{\max}} \sum_{i=0}^{i_{\max}} [F_{-1}(\eta_{24,4,14}) + F_{-1}(\theta_{24,4,14})] \tag{6.116}$$

6.2.18 The $I_D - V_D$ equation for ballistic magneto QWFETs of germanium

(a) The model of Cardona *et al.*

Using (6.1)–(6.3), (5.224) and (5.225), we get

$$\frac{J_D}{W} = \frac{e^2 B g_v}{\pi^2 \hbar} \sqrt{\frac{2}{m_\perp^*}} \sum_{n=0}^{n_{\max}} \sum_{i=0}^{i_{\max}} [[\sqrt{E_{90}(1 + \alpha E_{90})}][1 + 2\alpha E_{90}]^{-1} \times [F_{-1}(\eta_{24,4,15}) - F_{-1}(\theta_{24,4,15})]] \tag{6.117}$$

where $\eta_{24,4,15} = (k_B T)^{-1}[E_{F24,15} - E_{90}]$, $\theta_{24,4,15} = \left[\eta_{24,4,15} - \frac{eV_D}{k_B T}\right]$ and $E_{F24,15}$ is the Fermi energy in the present case which can be determined from

$$\frac{C_{ins}}{e}(V_T - V_{GS}) = \frac{g_v eB}{2\pi\hbar}\sum_{n=0}^{n_{\max}}\sum_{i=0}^{i_{\max}}[F_{-1}(\eta_{24,4,15}) + F_{-1}(\theta_{24,4,15})] \quad (6.118)$$

(b) The model of Wang *et al.*

Using (6.1)–(6.3), (5.230) and (5.231), we get

$$\frac{J_D}{W} = \frac{e^2 B g_v}{\pi^2\hbar}\sum_{n=0}^{n_{\max}}\sum_{i=0}^{i_{\max}}[[[\omega'_{222}(E_{91})]^{-1}][F_{-1}(\eta_{24,4,16}) - F_{-1}(\theta_{24,4,16})]] \quad (6.119)$$

where $\eta_{24,4,16} = (k_B T)^{-1}[E_{F24,16} - E_{91}]$, $\theta_{24,4,16} = \left[\eta_{24,4,16} - \frac{eV_D}{k_B T}\right]$ and $E_{F24,16}$ is the Fermi energy in the present case which can be determined from

$$\frac{C_{ins}}{e}(V_T - V_{GS}) = \frac{g_v eB}{2\pi\hbar}\sum_{n=0}^{n_{\max}}\sum_{i=0}^{i_{\max}}[F_{-1}(\eta_{24,4,16}) + F_{-1}(\theta_{24,4,16})] \quad (6.120)$$

6.2.19 The $I_D - V_D$ equation for ballistic magneto QWFETs of gallium aresenide

(a) The model of Rossler

Using (6.1)–(6.3), (5.236) and (5.237), we get

$$\frac{J_D}{W} = \frac{e^2 B g_v}{2\pi^2\hbar^2}\sum_{n=0}^{n_{\max}}\sum_{i=0}^{i_{\max}}\left[\left[\frac{\sqrt{\gamma_{4,\pm}(E_{92})}}{\gamma'_{4,\pm}(E_{92})}\right][F_{-1}(\eta_{24,4,17}) - F_{-1}(\theta_{24,4,17})]\right] \quad (6.121)$$

where $\eta_{24,4,17} = (k_B T)^{-1}[E_{F24,17} - E_{92}]$, $\theta_{24,4,17} = \left[\eta_{24,4,17} - \frac{eV_D}{k_B T}\right]$ and $E_{F24,17}$ is the Fermi energy in the present case which can be determined from

$$\frac{C_{ins}}{e}(V_T - V_{GS}) = \frac{g_v eB}{4\pi\hbar}\sum_{n=0}^{n_{\max}}\sum_{i=0}^{i_{\max}}[F_{-1}(\eta_{24,4,17}) + F_{-1}(\theta_{24,4,17})] \quad (6.122)$$

(b) The model of Stillman *et al.*

Using (6.1)–(6.3), (5.240) and (5.241), we get

$$\frac{J_D}{W} = \frac{e^2 B g_v}{\pi^2 \hbar^2} \sum_{n=0}^{n_{\max}} \sum_{i=0}^{i_{\max}} \left[\left[\frac{\sqrt{\gamma_5(E_{93})}}{\gamma_5'(E_{93})} \right] [F_{-1}(\eta_{24,4,18}) - F_{-1}(\theta_{24,4,18})] \right] \tag{6.123}$$

where $\eta_{24,4,18} = (k_B T)^{-1}[E_{F24,18} - E_{93}]$, $\theta_{24,4,18} = \left[\eta_{24,4,18} - \frac{eV_D}{k_B T}\right]$ and $E_{F24,18}$ is the Fermi energy in the present case which can be determined from

$$\frac{C_{ins}}{e}(V_T - V_{GS}) = \frac{g_v e B}{2\pi\hbar} \sum_{n=0}^{n_{\max}} \sum_{i=0}^{i_{\max}} [F_{-1}(\eta_{24,4,18}) + F_{-1}(\theta_{24,4,18})] \tag{6.124}$$

*6.2.20 The $I_D - V_D$ equation for ballistic magneto QWFETs of gallium antimonide

(a) The model of Seiler *et al.*

Using (6.1)–(6.3), (5.244) and (5.245), we get

$$\frac{J_D}{W} = \frac{e^2 B g_v}{2\pi^2 \hbar^2} \sum_{n=0}^{n_{\max}} \sum_{i=0}^{i_{\max}} \left[\left[\frac{\sqrt{\gamma_6(E_{94})}}{\gamma_6'(E_{94})} \right] [F_{-1}(\eta_{24,4,19}) - F_{-1}(\theta_{24,4,19})] \right] \tag{6.125}$$

where $\eta_{24,4,19} = (k_B T)^{-1}[E_{F24,19} - E_{94}]$, $\theta_{24,4,19} = \left[\eta_{24,4,19} - \frac{eV_D}{k_B T}\right]$ and $E_{F24,19}$ is the Fermi energy in the present case which can be determined from

$$\frac{C_{ins}}{e}(V_T - V_{GS}) = \frac{g_v e B}{4\pi\hbar} \sum_{n=0}^{n_{\max}} \sum_{i=0}^{i_{\max}} [F_{-1}(\eta_{24,4,19}) + F_{-1}(\theta_{24,4,19})] \tag{6.126}$$

(b) The model of Mathur and Jain

Using (6.1)–(6.3), (5.248) and (5.249), we get

$$\frac{J_D}{W} = \frac{e^2 B g_v}{\pi^2 \hbar^2} \sum_{n=0}^{n_{\max}} \sum_{i=0}^{i_{\max}} \left[\left[\frac{\sqrt{\gamma_7(E_{95})}}{\gamma_7'(E_{95})} \right] [F_{-1}(\eta_{24,4,20}) - F_{-1}(\theta_{24,4,20})] \right] \tag{6.127}$$

where $\eta_{24,4,20} = (k_BT)^{-1}[E_{F24,20} - E_{95}]$, $\theta_{24,4,20} = \left[\eta_{24,4,20} - \frac{eV_D}{k_BT}\right]$ and $E_{F24,20}$ is the Fermi energy in the present case which can be determined from

$$\frac{C_{ins}}{e}(V_T - V_{GS}) = \frac{g_v eB}{2\pi\hbar}\sum_{n=0}^{n_{\max}}\sum_{i=0}^{i_{\max}}[F_{-1}(\eta_{24,4,20}) + F_{-1}(\theta_{24,4,20})] \quad (6.128)$$

(c) The model of Zhang

Using (6.1)–(6.3), (5.252) and (5.253), we get

$$\frac{J_D}{W} = \frac{e^2 B g_v}{\pi^2\hbar^2}\sum_{n=0}^{n_{\max}}\sum_{i=0}^{i_{\max}}\left[\left[\frac{\sqrt{\gamma_8(E_{96})}}{\gamma_8'(E_{96})}\right][F_{-1}(\eta_{24,4,21}) - F_{-1}(\theta_{24,4,21})]\right] \quad (6.129)$$

where $\eta_{24,4,21} = (k_BT)^{-1}[E_{F24,21} - E_{96}]$, $\theta_{24,4,21} = \left[\eta_{24,4,21} - \frac{eV_D}{k_BT}\right]$ and $E_{F24,21}$ is the Fermi energy in the present case which can be determined from

$$\frac{C_{ins}}{e}(V_T - V_{GS}) = \frac{g_v eB}{2\pi\hbar}\sum_{n=0}^{n_{\max}}\sum_{i=0}^{i_{\max}}[F_{-1}(\eta_{24,4,21}) + F_{-1}(\theta_{24,4,21})] \quad (6.130)$$

6.2.21 The $I_D - V_D$ equation for ballistic magneto QWFETs of indium antimonide

(a) The model of Palik *et al.*

Using (6.1)–(6.3), (5.256) and (5.257), we get

$$\frac{J_D}{W} = \frac{e^2 B g_v}{\pi^2\hbar^2}\sum_{n=0}^{n_{\max}}\sum_{i=0}^{i_{\max}}\left[\left[\frac{\sqrt{\gamma_9(E_{97})}}{\gamma_9'(E_{97})}\right][F_{-1}(\eta_{24,4,22}) - F_{-1}(\theta_{24,4,22})]\right] \quad (6.131)$$

where $\eta_{24,4,22} = (k_BT)^{-1}[E_{F24,22} - E_{97}]$, $\theta_{24,4,22} = \left[\eta_{24,4,21} - \frac{eV_D}{k_BT}\right]$ and $E_{F24,22}$ is the Fermi energy in the present case which can be determined from

$$\frac{C_{ins}}{e}(V_T - V_{GS}) = \frac{g_v eB}{2\pi\hbar}\sum_{n=0}^{n_{\max}}\sum_{i=0}^{i_{\max}}[F_{-1}(\eta_{24,4,22}) + F_{-1}(\theta_{24,4,22})] \quad (6.132)$$

(b) The model of Johnson *et al.*

Using (6.1)–(6.3), (5.260) and (5.261), we get

$$\frac{J_D}{W} = \frac{e^2 B g_v}{\pi^2 \hbar^2} \sum_{n=0}^{n_{\max}} \sum_{i=0}^{i_{\max}} \left[\left[\frac{\sqrt{\gamma_{10}(E_{98})}}{\gamma'_{10}(E_{98})} \right] [F_{-1}(\eta_{24,4,23}) - F_{-1}(\theta_{24,4,23})] \right] \tag{6.133}$$

where $\eta_{24,4,23} = (k_B T)^{-1}[E_{F24,23} - E_{98}]$, $\theta_{24,4,23} = \left[\eta_{24,4,23} - \frac{eV_D}{k_B T}\right]$ and $E_{F24,23}$ is the Fermi energy in the present case which can be determined from

$$\frac{C_{ins}}{e}(V_T - V_{GS}) = \frac{g_v e B}{2\pi\hbar} \sum_{n=0}^{n_{\max}} \sum_{i=0}^{i_{\max}} [F_{-1}(\eta_{24,4,23}) + F_{-1}(\theta_{24,4,23})] \tag{6.134}$$

(c) The model of Agafonov *et al.*

Using (6.1)–(6.3), (5.264) and (5.265), we get

$$\frac{J_D}{W} = \frac{e^2 B g_v}{\pi^2 \hbar^2} \sum_{n=0}^{n_{\max}} \sum_{i=0}^{i_{\max}} \left[\left[\frac{\sqrt{\gamma_{11}(E_{99})}}{\gamma'_{11}(E_{99})} \right] [F_{-1}(\eta_{24,4,24}) - F_{-1}(\theta_{24,4,24})] \right] \tag{6.135}$$

where $\eta_{24,4,24} = (k_B T)^{-1}[E_{F24,24} - E_{99}]$, $\theta_{24,4,24} = \left[\eta_{24,4,24} - \frac{eV_D}{k_B T}\right]$ and $E_{F24,24}$ is the Fermi energy in the present case which can be determined from

$$\frac{C_{ins}}{e}(V_T - V_{GS}) = \frac{g_v e B}{2\pi\hbar} \sum_{n=0}^{n_{\max}} \sum_{i=0}^{i_{\max}} [F_{-1}(\eta_{24,4,24}) + F_{-1}(\theta_{24,4,24})] \tag{6.136}$$

6.2.22 The $I_D - V_D$ equation for ballistic magneto QWFETs of tellurium

Using (6.1)–(6.3), (5.256) and (5.257), we get

$$\frac{J_D}{W} = \frac{e^2 B g_v}{2\pi^2 \hbar^2} \sum_{n=0}^{n_{\max}} \sum_{i=0}^{i_{\max}} [[\sqrt{\psi_4^2 + 4\psi_2 E_{100}}][F_{-1}(\eta_{24,4,25}) - F_{-1}(\theta_{24,4,25})]] \tag{6.137}$$

where $\eta_{24,4,25} = (k_B T)^{-1}[E_{F24,25} - E_{100}]$, $\theta_{24,4,25} = \left[\eta_{24,4,25} - \frac{eV_D}{k_B T}\right]$ and $E_{F24,25}$ is the Fermi energy in the present case which can be determined from

$$\frac{C_{ins}}{e}(V_T - V_{GS}) = \frac{g_v e B}{4\pi\hbar} \sum_{n=0}^{n_{\max}} \sum_{i=0}^{i_{\max}} [F_{-1}(\eta_{24,4,25}) + F_{-1}(\theta_{24,4,25})] \tag{6.138}$$

*6.2.23 The $I_D - V_D$ equation for ballistic magneto QWFETs of $PtSb_2$

Using (6.1)–(6.3), (5.274) and (5.275), we get

$$\frac{J_D}{W} = \frac{e^2 B g_v}{2\pi^2\hbar^2} \sum_{n=0}^{n_{\max}} \sum_{i=0}^{i_{\max}} [[[t'_{222}(E_{101})]^{-1}][F_{-1}(\eta_{24,4,26}) - F_{-1}(\theta_{24,4,26})]] \tag{6.139}$$

where $\eta_{24,4,26} = (k_B T)^{-1}[E_{F24,26} - E_{101}]$, $\theta_{24,4,26} = \left[\eta_{24,4,26} - \frac{eV_D}{k_B T}\right]$ and $E_{F24,26}$ is the Fermi energy in the present case which can be determined from

$$\frac{C_{ins}}{e}(V_T - V_{GS}) = \frac{g_v e B}{4\pi\hbar} \sum_{n=0}^{n_{\max}} \sum_{i=0}^{i_{\max}} [F_{-1}(\eta_{24,4,26}) + F_{-1}(\theta_{24,4,26})] \tag{6.140}$$

*6.2.24 The $I_D - V_D$ equation for ballistic magneto QWFETs of GaP

Using (6.1)–(6.3), (5.279) and (5.280), we get

$$\frac{J_D}{W} = \frac{e^2 B g_v}{2\pi^2\hbar^2} \sum_{n=0}^{n_{\max}} \sum_{i=0}^{i_{\max}} \left[\left[\sqrt{\frac{\psi_{444}(E_{102})}{\psi'_{444}(E_{102})}}\right] [F_{-1}(\eta_{24,4,27}) - F_{-1}(\theta_{24,4,27})]\right] \tag{6.141}$$

where

$$\psi_{444}(E_{102}) = \left[\frac{\begin{array}{l} c + 2a(E - |V_G|) \\ \quad \pm\sqrt{[c + 2a(E - |V_G|)a]^2 - 4a^2[(E - |V_G|)^2 - |V_G|]} \end{array}}{2a^2}\right],$$

$\eta_{24,4,27} = (k_B T)^{-1}[E_{F24,27} - E_{102}]$, $\theta_{24,4,27} = \left[\eta_{24,4,27} - \frac{eV_D}{k_B T}\right]$ and $E_{F24,27}$ is the Fermi energy in the present case which can be determined from

$$\frac{C_{ins}}{e}(V_T - V_{GS}) = \frac{g_v e B}{4\pi\hbar} \sum_{n=0}^{n_{\max}} \sum_{i=0}^{i_{\max}} [F_{-1}(\eta_{24,4,27}) + F_{-1}(\theta_{24,4,27})] \tag{6.142}$$

6.3 Summary and Conclusion

In Section 6.2.1, we have formulated the $I_D - V_D$ equations for ballistic Magneto MOSFETs made of nonlinear optical materials for both weak and strong electric field limits as given by (6.6) and (6.9) respectively and for the computation of which we need the corresponding expression of the 2D electron statistics for both the limits which are given in (6.8) and (6.10) respectively. The $I_D - V_D$ equations for Magneto MOSFETs made of III–V, ternary and quaternary materials has been investigated in Section 26.2.2 under weak electric field limit in accordance with the three and two band models of Kane together with parabolic energy bands, are given by (6.12), (6.15) and (6.18) and the corresponding 2D electron statistics are expressed through (6.13), (6.16) and (6.19) respectively. Besides the $I_D - V_D$ equations in this case for strong electric field limit in accordance with the three and two band models of Kane are given by (6.20) and (6.22) and the corresponding 2D electron statistics are expressed through (6.21) and (6.23) respectively. In Section 6.2.3, we have studied the $I_D - V_D$ equation for ballistic Magneto MOSFETs of II–VI materials which is given by (6.25) and the corresponding 2D electron statistics has been represented in (6.26). In Section 6.2.4, we have studied the $I_D - V_D$ equation in Magneto MOSFETs of IV–VI materials for both weak and strong electric field limits which are given by (6.28) and (6.30) respectively and the corresponding 2D electron statistics have been represented in (6.29) and (6.31) respectively. In Section 6.2.5, we have investigated the $I_D - V_D$ equation in ballistic Magneto MOSFETs of stressed Kane type materials for both weak and strong electric field limits which are given by (6.33) and (6.35) respectively and the corresponding 2D electron statistics have been represented in (6.34) and (6.36) respectively. In Section 6.2.6, we have studied the $I_D - V_D$ equation in ballistic Magneto MOSFETs of Bismuth Telluride for both weak and strong electric field limits which are given by (6.38) and (6.40) respectively and the corresponding 2D electron statistics have been represented in (6.39) and (6.41) respectively. In Section 6.2.7, we have studied the $I_D - V_D$ equation in ballistic Magneto MOSFETs of Germanium in accordance to the models of Cardona *et al.* and Wang *et al.* which are given by (6.43) and (6.46) and the corresponding 2D electron statistics has been represented in (6.44) and (6.47) respectively. In Section 6.2.8, we have investigated the $I_D - V_D$ equation in Magneto MOSFETs of Gallium Arsenide in accordance with the models of Rossler and Stillman *et al.* for weak electric field limits which are given by (6.49) and (6.54) and the corresponding 2D electron statistics has been represented in (6.50) and (6.55) respectively. In the same section we have also investigated the $I_D - V_D$ equation in this case in accordance with the said models for strong electric field limits which are given by (6.51) and (6.56) and the corresponding 2D electron statistics has been represented in (6.52) and (6.53) respectively. In Section 6.2.9, we have studied the $I_D - V_D$ equation in ballistic Magneto MOSFETs of Gallium Antimonide in accordance with the models of Sailer, Mathur *et al.* and Zhang for weak electric field limits which are given by (6.59), (6.64) and (6.69) and the corresponding 2D electron statistics has been represented in (6.60), (6.65) and (6.70) respectively. In the same section we have also investigated the $I_D - V_D$ equation in this case in accordance with the said models for strong electric field limits which are given by (6.61), (6.66) and (6.71) and the corresponding 2D electron statistics has been represented in (6.62), (6.67) and (6.72) respectively. In Section 6.2.10, we have studied the $I_D - V_D$ equation in ballistic Magneto MOSFETs of Indium Antimonide in accordance with the models of Palik, Johnson *et al.* and Agafonov *et al.* for weak electric field limits which are given by (6.74), (6.78) and (6.84) and the corresponding 2D electron statistics has been represented in (6.75), (6.79) and (6.85) respectively. In the same section we have also investigated the $I_D - V_D$ equation in this case in accordance with the said models for strong electric field limits which are given by (6.76), (6.81) and (6.86) and the corresponding 2D electron statistics has been represented in (6.77), (6.82) and (6.87) respectively. In Section 6.2.11, we have studied the $I_D - V_D$ equation for ballistic Magneto MOSFETs of Tellurium which is given by (6.89) and the corresponding 2D electron statistics has been represented in (6.90).

In Section 6.2.12, we have formulated $I_D - V_D$ for ballistic Magneto QWFETs made of nonlinear optical materials as given by (6.91) for the computation of which we need the expression of the 2D electron statistics which is given in (6.92). The $I_D - V_D$ equations for ballistic Magneto QWFETs made of III–V, ternary and quaternary materials has been investigated in Section 6.2.13 in accordance with the models of three band model of Kane, two band model of Kane and Parabolic energy bands are given by (6.93), (6.95) and (6.97) and the corresponding 2D electron statistics are expressed through (6.94), (6.96) and (6.98) respectively. In Section 6.2.14, we have studied the $I_D - V_D$ equation for Magneto QWFETs of II–VI materials which is given by (6.99) and the corresponding 2D electron statistics has been represented in (6.100). In Section 6.2.15, we have studied the $I_D - V_D$ equation in Magneto QWFETs of IV–VI materials in accordance with the models of McClure-Choi, Hybrid, Cohen, Lax, Dimmock and Ellipsoidal are given by (6.101), (6.103), (6.105), (6.107), (6.109) and (6.111) and the corresponding 2D electron statistics are expressed through (6.102), (6.104), (6.106), (6.108), (6.110) and (6.112) respectively. In Section 6.2.16, we have investigated the

$I_D - V_D$ equation in Magneto QWFETs of stressed Kane type materials which is given by (6.113) and the corresponding 2D electron statistics has been represented in (6.114). In Section 6.2.17, we have studied the $I_D - V_D$ equation in Magneto QWFETs of Bismuth Telluride which is given by (6.115) and the corresponding 2D electron statistics has been represented in (6.116). In Section 6.2.18, we have studied the $I_D - V_D$ equation in QWFETs of Germanium materials in accordance with the models of Cardona *et al.* and Wang *et al.* are given by (6.117) and (6.119) and the corresponding 2D electron statistics are expressed through (6.118) and (6.120) respectively. In Section 6.2.19, we have investigated the $I_D - V_D$ equation in Magneto QWFETs of Gallium Arsenide materials in accordance with the models of Rossler *et al.* and Stillman *et al.* are given by (6.121) and (6.123) and the corresponding 2D electron statistics are expressed through (6.122) and (6.124) respectively. In Section 6.2.20, we have studied the $I_D - V_D$ equation in Magneto QWFETs of Gallium Antimonide materials in accordance with the models of Sailer *et al.*, Mathur *et al.* and Zhang *et al.* are given by (6.125), (6.127) and (6.129) and the corresponding 2D electron statistics are expressed through (6.126), (6.128) and (6.130) respectively. In Section 6.2.21, we have studied the $I_D - V_D$ equation in Magneto QWFETs of Indium Antimonide materials in accordance with the models of Palik *et al.*, Johnson *et al.* and Agafonov *et al.* are given by (6.131), (6.133) and (6.135) and the corresponding 2D electron statistics are expressed through (6.132), (6.134) and (6.136) respectively. In Section 6.2.22, we have studied the $I_D - V_D$ equation in Magneto QWFETs of Tellurium materials which is given by (6.137) and the corresponding 2D electron statistics has been represented in (6.138). In Section 6.2.23, we have studied the $I_D - V_D$ equation in Magneto QWFETs of $PtSb_2$ materials which is given by (6.139) and the corresponding 2D electron statistics has been represented in (6.140). In Section 6.2.24, we have studied the $I_D - V_D$ equation in Magneto QWFETs of GaP materials which is given by (6.141) and the corresponding 2D electron statistics has been represented in (6.142). We have not given the graphs of the $I_D - V_D$ equations for ballistic Magneto MOSFETs and QWFETs made of many materials as discussed here since we feel that our readers will perform the intricate numerical computation to obtain the oscillatory plots for better assessment and joy of understanding. It is the band structure which changes in a fundamental way and consequently all the physical properties of all the electronic materials change radically leading to new physical concepts.

6.4 Open Research Problem

(R.6.1) Derive $I_D - V_D$ equations for ballistic Magneto MOSFETs made of all the materials whose DRs are given in (R.1.1). Draw the corresponding graphs and explain the results.

(R.6.2) Derive equations for ballistic Magneto QWFETs made of all the materials whose DRs are given in (R.1.1). Draw the corresponding graphs and explain the results.

Chapter 7

The MTP in Quantum Dot Superlattices

The two most important places to keep clean and tidy are my mind and heart.

7.1 Introduction

In recent years, there has been considerable interest in studying the different properties of quantum dot superlattices (QDSLs) [1–25]. In this chapter, the MTP in III–V, II–VI, IV–VI, HgTe/CdTe and strained layer QDSLs with graded interfaces has been studied in Sections 7.2.1–7.2.5. From Sections 7.2.6 to 7.2.10, the MTP in III–V, II–VI, IV–VI, HgTe/CdTe and strained layer quantum dot effective mass superlattices, respectively, has been presented. The TPSM, ΔC_{44}, and ΔC_{456} and the photoemission in the present case have also been investigated. Section 7.3 contains the results and discussion pertinent to this chapter. Section 7.4 presents 14 open research problems.

7.2 Theoretical Background

7.2.1 The MTP in III–V QDSLs with graded interfaces

The energy spectrum of the conduction electrons in bulk specimens of the constituent materials of III–V SLs whose energy band structures are defined by three-band model of Kane can be written from (1.21) as

$$(\hbar^2 k^2)/(2m_i^*) = EG(E, E_{gi}, \Delta_i)$$

$$i = 1 \text{ and } 2, G(E, E_{gi}, \Delta_i) \equiv \frac{\left(E_{gi} + \frac{2}{3}\Delta_i\right)(E + E_{gi} + \Delta_i)(E + E_{gi})}{\left[E_{gi}(E_{gi} + \Delta_i)\left(E + E_{gi} + \frac{2}{3}\Delta_i\right)\right]} \tag{7.1}$$

and m_i^* is the effective electron mass at the edge of the conduction band.

Therefore, the dispersion law of the electrons of III–V SLs with graded interfaces can be expressed, following Jiang and Lin [26], as

$$\cos(L_0 k) = \frac{1}{2}\Phi(E, k_s) \tag{7.2}$$

where $L_0(\equiv a_0 + b_0)$ is the period length, and a_0 and b_0 are the widths of the barrier and the well, respectively,

$$\begin{aligned}\Phi(E,k_s) \equiv \Bigg[&2\cosh\{\beta(E,k_s)\}\cos\{\gamma(E,k_s)\} + \varepsilon(E,k_s)\sinh\{\beta(E,k_s)\}\\ &\times \sin\{\gamma(E,k_s)\}\\ &+\Delta_0\Bigg[\left(\frac{K_1^2(E,k_s)}{K_2(E,k_s)} - 3K_2(E,k_s)\right)\cosh\{\beta(E,k_s)\}\sin\{\gamma(E,k_s)\}\\ &+\left(3K_1(E,k_s) - \frac{\{K_2(E,k_s)\}^2}{K_1(E,k_s)}\right)\sinh\{\beta(E,k_s)\}\cos\{\gamma(E,k_s)\}\Bigg]\\ &+\Delta_0\Bigg[2(\{K_1(E,k_s)\}^2 - \{K_2(E,k_s)\}^2)\cosh\{\beta(E,k_s)\}\cos\{\gamma(E,k_s)\}\\ &+\frac{1}{12}\left[\frac{5\{K_2(E,k_s)\}^3}{K_1(E,k_s)} + \frac{5\{K_1(E,k_s)\}^3}{K_2(E,k_s)} - 34K_2(E,k_s)K_1(E,k_s)\right]\\ &\times\sinh\{\beta(E,k_s)\}\sin\{\gamma(E,k_s)\}\Bigg]\Bigg]\end{aligned}$$

$$\varepsilon(E,k_s) \equiv \left[\frac{K_1(E,k_s)}{K_2(E,k_s)} - \frac{K_2(E,k_s)}{K_1(E,k_s)}\right], \beta(E,k_s) \equiv K_1(E,k_s)[a_0 - \Delta_0]$$

where Δ_0 is the interface width, $\gamma(E,k_s) = K_2(E,k_s)[b_0 - \Delta_0]$, $K_1(E,k_s) \equiv \left[\frac{2m_2^* E'}{\hbar^2}G(E - V_0, \alpha_2, \Delta_2) + k_s^2\right]^{1/2}$, $E' \equiv V_0 - E$, V_0 is the potential barrier encountered by the electron ($V_0 \equiv |E_{g_2} - E_{g_1}|$), $\alpha_i \equiv 1/E_{gi}$, $K_2(E,k_s) \equiv \left[\frac{2m_1^* E}{\hbar^2}G(E, \alpha_1, \Delta_1) - k_s^2\right]^{1/2}$, and $k_s^2 = k_x^2 + k_y^2$.

Therefore, the dispersion relation in quantum dot III–V superlattices with graded interfaces can be written from (7.2) as

$$\left(\frac{n_x\pi}{d_x}\right)^2 + \left(\frac{n_y\pi}{d_y}\right)^2 = \frac{1}{L_0^2}\left[\cos^{-1}\left[\frac{1}{2}\Phi(E_{\mathrm{SLQD1}}, n_x, n_y)\right]\right]^2 - \left(\frac{n_z\pi}{d_z}\right)^2 \tag{7.3}$$

where $n_i = 1, 2, 3 \ldots, d_i$ is the nano thickness along ith direction where $i = x, y, z$.

$$\begin{aligned}\Phi(E_{\mathrm{SLQD1}}, n_x, n_y) \equiv \Big[&2\cosh\{\beta(E_{\mathrm{SLQD1}}, n_x, n_y)\}\cos\{\gamma(E_{\mathrm{SLQD1}}, n_x, n_y)\}\\ &+\varepsilon(E_{\mathrm{SLQD1}}, n_x, n_y)\sinh\{\beta(E_{\mathrm{SLQD1}}, n_x, n_y)\}\sin\{\gamma(E_{\mathrm{SLQD1}}, n_x, n_y)\}\end{aligned}$$

$$+\Delta_0\left[\left(\frac{K_1^2(E_{\mathrm{SLQD1}},n_x,n_y)}{K_2(E_{\mathrm{SLQD1}},n_x,n_y)}-3K_2(E_{\mathrm{SLQD1}},n_x,n_y)\right)\right.$$
$$\times\cosh\{\beta(E_{\mathrm{SLQD1}},n_x,n_y)\}\sin\{\gamma(E_{\mathrm{SLQD1}},n_x,n_y)\}$$
$$+\left(3K_1(E_{\mathrm{SLQD1}},n_x,n_y)-\frac{\{K_2(E_{\mathrm{SLQD1}},n_x,n_y)\}^2}{K_1(E_{\mathrm{SLQD1}},n_x,n_y)}\right)$$
$$\left.\times\ \sinh\{\beta(E_{\mathrm{SLQD1}},n_x,n_y)\}\cos\{\gamma(E_{\mathrm{SLQD1}},n_x,n_y)\}\right]$$
$$+\Delta_0\left[2(\{K_1(E_{\mathrm{SLQD1}},n_x,n_y)\}^2-\{K_2(E_{\mathrm{SLQD1}},n_x,n_y)\}^2)\right.$$
$$\times\cosh\{\beta(E_{\mathrm{SLQD1}},n_x,n_y)\}\cos\{\gamma(E_{\mathrm{SLQD1}},n_x,n_y)\}$$
$$+\frac{1}{12}\left[\frac{5\{K_2(E_{\mathrm{SLQD1}},n_x,n_y)\}^3}{K_1(E_{\mathrm{SLQD1}},n_x,n_y)}+\frac{5\{K_1(E_{\mathrm{SLQD1}},n_x,n_y)\}^3}{K_2(E_{\mathrm{SLQD1}},n_x,n_y)}\right.$$
$$\left.-34K_2(E_{\mathrm{SLQD1}},n_x,n_y)K_1(E_{\mathrm{SLQD1}},n_x,n_y)\right]$$
$$\left.\left.\sinh\{\beta(E_{\mathrm{SLQD1}},n_x,n_y)\}\sin\{\gamma(E_{\mathrm{SLQD1}},n_x,n_y)\}\right]\right]$$

$$\varepsilon(E_{\mathrm{SLQD1}},n_x,n_y)\equiv\left[\frac{K_1(E_{\mathrm{SLQD1}},n_x,n_y)}{K_2(E_{\mathrm{SLQD1}},n_x,n_y)}-\frac{K_2(E_{\mathrm{SLQD1}},n_x,n_y)}{K_1(E_{\mathrm{SLQD1}},n_x,n_y)}\right],$$
$$\beta(E_{\mathrm{SLQD1}},n_x,n_y)\equiv K_1(E_{\mathrm{SLQD1}},n_x,n_y)[a_0-\Delta_0]$$
$$\gamma(E_{\mathrm{SLQD1}},n_x,n_y)=K_2(E_{\mathrm{SLQD1}},n_x,n_y)[b_0-\Delta_0],$$
$$K_1(E_{\mathrm{SLQD1}},n_x,n_y)\equiv\left[\frac{2m_2^*E'}{\hbar^2}G(E_{\mathrm{SLQD1}}-V_0,\alpha_2,\Delta_2)\right.$$
$$\left.+\left(\frac{n_x\pi}{d_x}\right)^2+\left(\frac{n_y\pi}{d_y}\right)^2\right]^{1/2}$$
$$K_2(E_{\mathrm{SLQD1}},n_x,n_y)\equiv\left[\frac{2m_1^*E}{\hbar^2}G(E_{\mathrm{SLQD1}},\alpha_1,\Delta_1)\right.$$
$$\left.-\left[\left(\frac{n_x\pi}{d_x}\right)^2+\left(\frac{n_y\pi}{d_y}\right)^2\right]\right]^{1/2}$$

and E_{SLQD1} is the total quantized energy in this case.

The density-of-states function is given by

$$N(E)=\frac{2}{d_xd_yd_z}\sum_{n_x=1}^{n_{x\max}}\sum_{n_y=1}^{n_{y\max}}\sum_{n_z=1}^{n_{z\max}}\delta'(E-E_{\mathrm{SLQD1}}) \tag{7.4}$$

where δ' is the Dirac's delta function.

The electron concentration can be expressed as

$$n_0 = \frac{2}{d_x d_y d_z} \sum_{n_x=1}^{n_{x\max}} \sum_{n_y=1}^{n_{y\max}} \sum_{n_z=1}^{n_{z\max}} F_{-1}(\eta_{1\text{SLQD}1}) \tag{7.5}$$

where $\eta_{1\text{SLQD}1} = (k_B T)^{-1}(E_{\text{FSLQD}} - E_{\text{SLQD}1})$ and E_{FSLQD} is the Fermi energy in this case.

By using (1.5) and (7.5), the MTP in this case can be written as

$$G = \left(\frac{\pi^2 k_B}{3e}\right) \left[\sum_{n_x=1}^{n_{x\max}} \sum_{n_y=1}^{n_{y\max}} \sum_{n_z=1}^{n_{z\max}} F_{-1}(\eta_{1\text{SLQD}1})\right]^{-1} \times \left[\sum_{n_x=1}^{n_{x\max}} \sum_{n_y=1}^{n_{y\max}} \sum_{n_z=1}^{n_{z\max}} F_{-2}(\eta_{1\text{SLQD}1})\right] \tag{7.6}$$

7.2.2 The MTP in II–VI QDSLs with graded interfaces

The energy spectrum of the conduction electrons of the constituent materials of II–VI SLs are given by

$$E = \frac{\hbar^2 k_s^2}{2m^*_{\perp,1}} + \frac{\hbar^2 k_z^2}{2m^*_{\parallel,1}} \pm \bar{\lambda}_0 k_s \tag{7.7}$$

and

$$\frac{\hbar^2 k^2}{2m_2^*} = EG(E, E_{g2}, \Delta_2) \tag{7.8}$$

where $m^*_{\perp,1}$ and $m^*_{\parallel,1}$ are the transverse and longitudinal effective electron masses, respectively, at the edge of the conduction band for the first material. The electron dispersion law in II–VI SLs with graded interfaces can be expressed as

$$\cos(L_0 k) = \frac{1}{2}\Phi_1(E, k_s) \tag{7.9}$$

where

$$\Phi_1(E, k_s) \equiv \Big[2\cosh\{\beta_1(E, k_s)\}\cos\{\gamma_1(E, k_s)\} + \varepsilon_1(E, k_s) \times \sinh\{\beta_1(E, k_s)\}\sin\{\gamma_1(E, k_s)\} + \Delta_0\left[\left(\frac{\{K_3(E, k_s)\}^2}{K_4(E, k_s)} - 3K_4(E, k_s)\right)\cosh\{\beta_1(E, k_s)\}\sin\{\gamma_1(E, k_s)\}\right.$$

$$+\left(3K_3(E,k_s)-\frac{\{K_4(E,k_s)\}^2}{K_3(E,k_s)}\right)\sinh\{\beta_1(E,k_s)\}\cos\{\gamma_1(E,k_s)\}\Bigg]$$

$$+\Delta_0\Bigg[2(\{K_3(E,k_s)\}^2-\{K_4(E,k_s)\}^2)\cosh\{\beta_1(E,k_s)\}\cos\{\gamma_1(E,k_s)\}\Bigg]\Bigg]$$

$$\varepsilon_1(E,k_s)\equiv\left[\frac{K_3(E,k_s)}{K_4(E,k_s)}-\frac{K_4(E,k_s)}{K_3(E,k_s)}\right],\beta_1(E,k_s)\equiv K_3(E,k_s)[a_0-\Delta_0],$$

$$\gamma_1(E,k_s)=K_4(E,k_s)[b_0-\Delta_0]$$

$$K_3(E,k_s)\equiv\left[\frac{2m_2^*E'}{\hbar^2}G(E-V_0,\alpha_2,\Delta_2)+k_s^2\right]^{1/2},\text{ and}$$

$$K_4(E,k_s)\equiv\left[\frac{2m_{\parallel,1}^*}{\hbar^2}\left[E-\frac{\hbar^2k_s^2}{2m_{\perp,1}^*}\mp\bar{\lambda}_0k_s\right]\right]^{1/2}$$

Therefore, the dispersion relation in quantum dot II–VI superlattices with graded interfaces can be written from (7.9) as

$$\left(\frac{n_x\pi}{d_x}\right)^2+\left(\frac{n_y\pi}{d_y}\right)^2=\frac{1}{L_0^2}\left[\cos^{-1}\left[\frac{1}{2}\Phi_1(E_{\mathrm{SLQD2}},n_x,n_y)\right]\right]^2-\left(\frac{n_z\pi}{d_z}\right)^2 \tag{7.10}$$

where

$$\Phi_1(E,n_x,n_y)\equiv\Bigg[2\cosh\{\beta_1(E_{\mathrm{SLQD2}},n_x,n_y)\}\cos\{\gamma_1(E_{\mathrm{SLQD2}},n_x,n_y)\}$$

$$+\varepsilon_1(E_{\mathrm{SLQD2}},n_x,n_y)\sinh\{\beta_1(E_{\mathrm{SLQD2}},n_x,n_y)\}\sin\{\gamma_1(E_{\mathrm{SLQD2}},n_x,n_y)\}$$

$$+\Delta_0\left[\left(\frac{\{K_3(E_{\mathrm{SLQD2}},n_x,n_y)\}^2}{K_4(E_{\mathrm{SLQD2}},n_x,n_y)}-3K_4(E_{\mathrm{SLQD2}},n_x,n_y)\right)\right.$$

$$\times\cosh\{\beta_1(E_{\mathrm{SLQD2}},n_x,n_y)\}\sin\{\gamma_1(E_{\mathrm{SLQD2}},n_x,n_y)\}$$

$$+\left(3K_3(E_{\mathrm{SLQD2}},n_x,n_y)-\frac{\{K_4(E_{\mathrm{SLQD2}},n_x,n_y)\}^2}{K_3(E_{\mathrm{SLQD2}},n_x,n_y)}\right)$$

$$\left.\times\sinh\{\beta_1(E_{\mathrm{SLQD2}},n_x,n_y)\}\cos\{\gamma_1(E_{\mathrm{SLQD2}},n_x,n_y)\}\right]$$

$$+\Delta_0\Big[2(\{K_3(E_{\mathrm{SLQD2}},n_x,n_y)\}^2-\{K_4(E_{\mathrm{SLQD2}},n_x,n_y)\}^2)$$

$$\times\cosh\{\beta_1(E_{\mathrm{SLQD2}},n_x,n_y)\}\cos\{\gamma_1(E_{\mathrm{SLQD2}},n_x,n_y)\}\Big]\Bigg]$$

$$\varepsilon_1(E_{\text{SLQD2}}, n_x, n_y) \equiv \left[\frac{K_3(E_{\text{SLQD2}}, n_x, n_y)}{K_4(E_{\text{SLQD2}}, n_x, n_y)} - \frac{K_4(E_{\text{SLQD2}}, n_x, n_y)}{K_3(E_{\text{SLQD2}}, n_x, n_y)}\right],$$

$$\beta_1(E_{\text{SLQD2}}, n_x, n_y) \equiv K_3(E_{\text{SLQD2}}, n_x, n_y)[a_0 - \Delta_0]$$

$$\gamma_1(E_{\text{SLQD2}}, n_x, n_y) = K_4(E_{\text{SLQD2}}, n_x, n_y)[b_0 - \Delta_0],$$

$$K_3(E_{\text{SLQD2}}, n_x, n_y) \equiv \left[\frac{2m_2^* E'}{\hbar^2} G(E_{\text{SLQD2}} - V_0, \alpha_2, \Delta_2) + k_s^2\right]^{1/2}$$

$$K_4(E_{\text{SLQD2}}, n_x, n_y) \equiv \left[\frac{2m_{\parallel,1}^*}{\hbar^2} E - \frac{\hbar^2}{2m_{\perp,1}^*}\left[\left(\frac{n_x\pi}{d_x}\right)^2 + \left(\frac{n_y\pi}{d_y}\right)^2\right]\right.$$

$$\left.\mp\bar{\lambda}_0\left[\left(\frac{n_x\pi}{d_x}\right)^2 + \left(\frac{n_y\pi}{d_y}\right)^2\right]^{1/2}\right]^{1/2}$$

and E_{SLQD2} is the total quantized energy in this case.

The density-of-states function is given by

$$N(E) = \frac{1}{d_x d_y d_z} \sum_{n_x=1}^{n_{x\max}} \sum_{n_y=1}^{n_{y\max}} \sum_{n_z=1}^{n_{z\max}} \delta'(E - E_{\text{SLQD2}}) \tag{7.11}$$

The electron concentration can be expressed as

$$n_0 = \frac{1}{d_x d_y d_z} \sum_{n_x=1}^{n_{x\max}} \sum_{n_y=1}^{n_{y\max}} \sum_{n_z=1}^{n_{z\max}} F_{-1}(\eta_{1\text{SLQD2}}) \tag{7.12}$$

where

$$\eta_{1\text{SLQD2}} = (k_B T)^{-1}(E_{\text{FSLQD}} - E_{\text{SLQD2}})$$

By using (1.5) and (7.12), the MTP in this case can be written as

$$G = \left(\frac{\pi^2 k_B}{3e}\right)\left[\sum_{n_x=1}^{n_{x\max}} \sum_{n_y=1}^{n_{y\max}} \sum_{n_z=1}^{n_{z\max}} F_{-1}(\eta_{1\text{SLQD2}})\right]^{-1}$$

$$\times \left[\sum_{n_x=1}^{n_{x\max}} \sum_{n_y=1}^{n_{y\max}} \sum_{n_z=1}^{n_{z\max}} F_{-2}(\eta_{1\text{SLQD2}})\right] \tag{7.13}$$

7.2.3 The MTP in IV–VI QDSLs with graded interfaces

The E–k dispersion relation of the conduction electrons of the constituent materials of the IV–VI SLs can be expressed as

$$E = a_i k_s^2 + b_i k_z^2 + \left[[c_i k_s^2 + d_i k_z^2] + \left(e_i k_s^2 + f_i k_y^2 + \frac{E_{g_i}}{2}\right)^2\right]^{1/2} - \frac{E_{g_i}}{2} \tag{7.14}$$

where $a_i \equiv [\frac{\hbar^2}{2m_{\perp,i}^-}], b_i \equiv (\frac{\hbar^2}{2m_{\|,i}^-})^2, c_i \equiv P_{\perp,i}^2, d_i \equiv P_{\perp,i}^2, e_i \equiv [\frac{\hbar^2}{2m_{\perp,i}^+}]$ and $f_i \equiv (\frac{\hbar^2}{2m_{\|,i}^+})^2$.

The electron dispersion law in IV–VI SLs with graded interfaces can be expressed as

$$\cos(L_0 k) = \frac{1}{2}\Phi_2(E, k_s) \tag{7.15}$$

where

$$\begin{aligned}
\Phi_2(E,k_s) \equiv & \Bigg[2\cosh\{\beta_2(E,k_s)\}\cos\{\gamma_2(E,k_s)\} + \varepsilon_2(E,k_s) \\
& \times \sinh\{\beta_2(E,k_s)\}\sin\{\gamma_2(E,k_s)\} \\
& + \Delta_0\Bigg[\left(\frac{\{K_5(E,k_s)\}^2}{K_6(E,k_s)} - 3K_6(E,k_s)\right)\cosh\{\beta_2(E,k_s)\}\sin\{\gamma_2(E,k_s)\} \\
& + \left(3K_5(E,k_s) - \frac{\{K_6(E,k_s)\}^2}{K_5(E,k_s)}\right)\sinh\{\beta_2(E,k_s)\}\cos\{\gamma_2(E,k_s)\}\Bigg] \\
& + \Delta_0\Bigg[2(\{K_5(E,k_s)\}^2 - \{K_6(E,k_s)\}^2)\cosh\{\beta_2(E,k_s)\}\cos\{\gamma_2(E,k_s)\} \\
& + \frac{1}{12}\left[\frac{5\{K_5(E,k_s)\}^3}{K_6(E,k_s)} + \frac{5\{K_6(E,k_s)\}^3}{K_5(E,k_s)} - 34K_6(E,k_s)K_5(E,k_s)\right] \\
& \times \sinh\{\beta_2(E,k_s)\}\sin\{\gamma_2(E,k_s)\}\Bigg]\Bigg]
\end{aligned}$$

$$\varepsilon_2(E,k_s) \equiv \left[\frac{K_5(E,k_s)}{K_6(E,k_s)} - \frac{K_6(E,k_s)}{K_5(E,k_s)}\right], \beta_2(E,k_s) \equiv K_5(E,k_s)[a_0 - \Delta_0],$$

$$\gamma_2(E,k_s) = K_6(E,k_s)[b_0 - \Delta_0]$$

$$\begin{aligned}
K_6(E,k_x,k_y) \equiv & [[EH_{11} + H_{21}(k_x,k_y)] - [H_{31}E^2 + EH_{41}(k_x,k_y) \\
& + H_{51}(k_x,k_y)]^{1/2}]^{1/2}, H_{1i} = (b_i^2 - f_i^2)^{-1}
\end{aligned}$$

and

$$H_{2i}(k_x,k_y) = [2H_{1i}]^{-1}[E_{g_i}b_i + d_i + f_iE_{g_i} + 2(e_if_i - a_ib_i)(k_x^2 + k_y^2)],$$

$$H_{3i} = \frac{f_i^2}{(b_i^2 - f_i^2)^2}$$

$$\begin{aligned}
H_{4i}(k_x,k_y) = & [4H_{1i}^2]^{-1}[4b_id_i + 4b_if_iE_{g_i} + 4f_i^2E_{g_i} + 8(k_x^2 + k_y^2) \\
& \times [e_if_ib_i - a_if_i^2]],
\end{aligned}$$

$$\begin{aligned}
H_{5i}(k_x,k_y) \equiv & [4H_{1i}^2]^{-1}\left[(k_x^2 + k_y^2)^2\,[-8a_ib_ie_if_i + 4b_i^2e_i^2 + 4f_i^2a_i^2]\right. \\
& + (k_x^2 + k_y^2)[4e_if_iE_{g_i}b_i - 4e_if_id_i + 4e_if_i^2
\end{aligned}$$

$$\times E_{g_i} - 4a_ib_i^2E_{g_i} - 4a_ib_id_i - 4a_ib_if_iE_{g_i} + 4b_i^2e_iE_{g_i}$$
$$+ 4b_i^2c_i + 4b_i^2E_{g_i}a_i - 4f_i^2e_iE_{g_i} - 4f_i^2c_i - 4f_i^2E_{g_i}a_i]$$
$$+ [E_{g_i}^2b_i^2 + d_i^2 + f_i^2g_i^2 + 2E_{g_i}b_id_i + 2E_{g_i}^2b_if_i + 2d_if_iE_{g_i}]]$$

and

$$K_5(E, k_x, k_y) \equiv [[(E - V_0)^2H_{32} + (E - V_0)H_{42}(k_x, k_y) + H_{52}(k_x, k_y)]^{1/2} - [(E - V_0)H_{12} + H_{22}(k_x, k_y)]]^{1/2}$$

Therefore, the dispersion relation in quantum well IV–VI superlattices with graded interfaces can be written from (7.15) as

$$\left(\frac{n_x\pi}{d_x}\right)^2 + \left(\frac{n_y\pi}{d_y}\right)^2 = \frac{1}{L_0^2}\left[\cos^{-1}\left[\frac{1}{2}\Phi_2(E_{\mathrm{SLQD3}}, n_x, n_y)\right]\right]^2 - \left(\frac{n_z\pi}{d_z}\right)^2 \tag{7.16}$$

where

$$\begin{aligned}\Phi_2(E_{\mathrm{SLQD3}}, n_x, n_y) \equiv \Bigg[&2\cosh\{\beta_2(E_{\mathrm{SLQD3}}, n_x, n_y)\}\cos\{\gamma_2(E_{\mathrm{SLQD3}}, n_x, n_y)\} \\ &+ \varepsilon_2(E_{\mathrm{SLQD3}}, n_x, n_y)\sinh\{\beta_2(E_{\mathrm{SLQD3}}, n_x, n_y)\}\sin\{\gamma_2(E_{\mathrm{SLQD3}}, n_x, n_y)\} \\ &+ \Delta_0\left[\left(\frac{\{K_5(E_{\mathrm{SLQD3}}, n_x, n_y)\}^2}{K_6(E_{\mathrm{SLQD3}}, n_x, n_y)} - 3K_6(E_{\mathrm{SLQD3}}, n_x, n_y)\right)\right. \\ &\times \cosh\{\beta_2(E_{\mathrm{SLQD3}}, n_x, n_y)\}\sin\{\gamma_2(E_{\mathrm{SLQD3}}, n_x, n_y)\} \\ &+ \left(3K_5(E_{\mathrm{SLQD3}}, k_s) - \frac{\{K_6(E_{\mathrm{SLQD3}}, n_x, n_y)\}^2}{K_5(E_{\mathrm{SLQD3}}, n_x, n_y)}\right) \\ &\left.\times \sinh\{\beta_2(E_{\mathrm{SLQD3}}, n_x, n_y)\}\cos\{\gamma_2(E_{\mathrm{SLQD3}}, n_x, n_y)\}\right] \\ &+ \Delta_0\Big[2(\{K_5(E_{\mathrm{SLQD3}}, n_x, n_y)\}^2 - \{K_6(3E_{\mathrm{SLQD3}}, n_x, n_y)\}^2) \\ &\times \cosh\{\beta_2(E_{\mathrm{SLQD3}}, n_x, n_y)\}\cos\{\gamma_2(E_{\mathrm{SLQD3}}, n_x, n_y)\} \\ &+ \frac{1}{12}\left[\frac{5\{K_5(E_{\mathrm{SLQD3}}, n_x, n_y)\}^3}{K_6(E_{\mathrm{SLQD3}}, n_x, n_y)} + \frac{5\{K_6(E_{\mathrm{SLQD3}}, n_x, n_y)\}^3}{K_5(E_{\mathrm{SLQD3}}, n_x, n_y)}\right. \\ &\left.- 34K_6(E_{\mathrm{SLQD3}}, n_x, n_y)K_5(E_{\mathrm{SLQD3}}, n_x, n_y)\right]\sinh\{\beta_2(E_{\mathrm{SLQD3}}, n_x, n_y)\} \\ &\times \sin\{\gamma_2(E_{\mathrm{SLQD3}}, n_x, n_y)\}\Big]\Bigg]\end{aligned}$$

$$\varepsilon_2(E_{\mathrm{SLQD3}}, n_x, n_y) \equiv \left[\frac{K_5(E_{\mathrm{SLQD3}}, n_x, n_y)}{K_6(E_{\mathrm{SLQD3}}, n_x, n_y)} - \frac{K_6(E_{\mathrm{SLQD3}}, n_x, n_y)}{K_5(E_{\mathrm{SLQD3}}, n_x, n_y)}\right],$$

$$\beta_2(E_{\text{SLQD3}}, n_x, n_y) \equiv K_5(E_{\text{SLQD3}}, n_x, n_y)[a_0 - \Delta_0]$$

$$\gamma_2(E_{\text{SLQD3}}, n_x, n_y) = K_6(E_{\text{SLQD3}}, n_x, n_y)[b_0 - \Delta_0],$$

$$K_6(E_{\text{SLQD3}}, n_x, n_y) \equiv [[EH_{11} + H_{21}(n_x, n_y)] \\ -[H_{31}E^2 + EH_{41}(E_{\text{SLQD3}}, n_x, n_y) + H_{51}(n_x, n_y)]^{1/2}]^{1/2}$$

$$H_{1i} = (b_i^2 - f_i^2)^{-1} \text{ and } H_{2i}(n_x, n_y) = [2H_{1i}]^{-1} \\ \times \left[E_{g_i}b_i + d_i + f_iE_{g_i} + 2(e_if_i - a_ib_i)\left(\left(\frac{n_x\pi}{d_x}\right)^2 + \left(\frac{n_y\pi}{d_y}\right)^2\right)\right]$$

$$H_{3i} = \frac{f_i^2}{(b_i^2 - f_i^2)^2},\ H_{4i}(n_x, n_y) = [4H_{1i}^2]^{-1}\left[4b_id_i + 4b_if_iE_{g_i} + 4f_i^2E_{g_i}\right. \\ \left. +8\left(\left(\frac{n_x\pi}{d_x}\right)^2 + \left(\frac{n_y\pi}{d_y}\right)^2\right)[e_if_ib_i - a_if_i^2]\right]$$

$$H_{5i}(n_x, n_y) \equiv [4H_{1i}^2]^{-1}\left[\left(\left(\frac{n_x\pi}{d_x}\right)^2 + \left(\frac{n_y\pi}{d_y}\right)^2\right)^2\right. \\ \times[-8a_ib_ie_if_i + 4b_i^2e_i^2 + 4f_i^2a_i^2] + \left(\left(\frac{n_x\pi}{d_x}\right)^2 + \left(\frac{n_y\pi}{d_y}\right)^2\right)[4e_if_iE_{g_i}b_i \\ - 4e_if_id_i + 4e_if_i^2E_{g_i} - 4a_ib_i^2E_{g_i} - 4a_ib_id_i - 4a_ib_if_iE_{g_i} + 4b_i^2e_iE_{g_i} \\ + 4b_i^2c_i + 4b_i^2E_{g_i}a_i - 4f_i^2e_iE_{g_i} - 4f_i^2c_i - 4f_i^2E_{g_i}a_i] \\ \left. + \left[E_{g_i}^2b_i^2 + d_i^2 + f_i^2g_i^2 + 2E_{g_i}b_id_i + 2E_{g_i}^2b_if_i + 2d_if_iE_{g_i}\right]\right],$$

$$K_5(E_{\text{SLQD3}}, n_x, n_y) \equiv [[(E - V_0)^2H_{32} + (E - V_0)H_{42}(n_x, n_y) \\ + H_{52}(n_x, n_y)]^{1/2} - [(E - V_0)H_{12} + H_{22}(n_x, n_y)]]^{1/2}$$

and E_{SLQD3} is the total quantized energy in this case.

The density-of-states function is given by

$$N(E) = \frac{2}{d_xd_yd_z}\sum_{n_x=1}^{n_{x\max}}\sum_{n_y=1}^{n_{y\max}}\sum_{n_z=1}^{n_{z\max}}\delta'(E - E_{\text{SLQD3}}) \tag{7.17}$$

The electron concentration can be expressed as

$$n_0 = \frac{2}{d_xd_yd_z}\sum_{n_x=1}^{n_{x\max}}\sum_{n_y=1}^{n_{y\max}}\sum_{n_z=1}^{n_{z\max}}F_{-1}(\eta_{1\text{SLQD3}}) \tag{7.18}$$

where

$$\eta_{1\text{SLQD}3} = (k_B T)^{-1}(E_{\text{FSLQD}} - E_{\text{SLQD}3})$$

By using (1.5) and (7.18), the MTP in this case can be written as

$$G = \left(\frac{\pi^2 k_B}{3e}\right)\left[\sum_{n_x=1}^{n_{x\max}}\sum_{n_y=1}^{n_{y\max}}\sum_{n_z=1}^{n_{z\max}} F_{-1}(\eta_{1\text{SLQD}3})\right]^{-1} \times \left[\sum_{n_x=1}^{n_{x\max}}\sum_{n_y=1}^{n_{y\max}}\sum_{n_z=1}^{n_{z\max}} F_{-2}(\eta_{1\text{SLQD}3})\right] \tag{7.19}$$

*7.2.4 The MTP in HgTe/CdTe QDSLs with graded interfaces

The dispersion relation of the conduction electrons of the constituent materials of HgTe/CdTe SLs can be expressed as

$$E = \frac{\hbar^2 k^2}{2m_1^*} + \frac{3|e|^2 k}{128\varepsilon_{sc}} \tag{7.20}$$

$$\frac{\hbar^2 k^2}{2m_2^*} = EG(E, E_{g2}, \Delta_2) \tag{7.21}$$

The electron dispersion law in HgTe/CdTe SL is given by

$$\cos(L_0 k) = \frac{1}{2}\Phi_3(E, k_s) \tag{7.22}$$

where

$$\Phi_3(E,k_s) \equiv \Big[2\cosh\{\beta_3(E,k_s)\}\cos\{\gamma_3(E,k_s)\} + \varepsilon_3(E,k_s)$$

$$\times \sinh\{\beta_3(E,k_s)\}\sin\{\gamma_3(E,k_s)\}$$

$$+\Delta_0\left[\left(\frac{\{K_7\gamma_3(E,k_s)\}^7}{K_8\gamma_3(E,k_s)} - 3K_8\gamma_3(E,k_s)\right)\cosh\{\beta_3(E,k_s)\}\sin\{\gamma_3(E,k_s)\}\right.$$

$$\left.+\left(3K_7\,(E,k_s) - \frac{\{K_8(E,k_s)\}^2}{K_7(E,k_s)}\right)\sinh\{\beta_3(E,k_s)\}\cos\{\gamma_3(E,k_s)\}\right]$$

$$+\Delta_0\,\Big[2(\{K_7(E,k_s)\}^2 - \{K_8(E,k_s)\}^2)\cosh\{\beta_3(E,k_s)\}\cos\{\gamma_3(E,k_s)\}$$

$$+\frac{1}{12}\left[\frac{5\{K_8(E,k_s)\}^3}{K_7(E,k_s)}+\frac{5\{K_7(E,k_s)\}^3}{K_8(E,k_s)}-34K_7(E,k_s)K_8(E,k_s)\right]$$

$$\times \sinh\{\beta_3(E,k_s)\}\sin\{\gamma_3(E,k_s)\}\Big]\Big]$$

$$\varepsilon_3(E,k_s)\equiv\left[\frac{K_7(E,k_s)}{K_8(E,k_s)}-\frac{K_8(E,k_s)}{K_7(E,k_s)}\right],\beta_3(E,k_s)\equiv K_7(E,k_s)[a_0-\Delta_0],$$

$$\gamma_3(E,k_s)=K_8(E,k_s)[b_0-\Delta_0]$$

$$K_8(E,k_s)\equiv\left[\frac{B_0^2+2AE-B_0\sqrt{B_0^2+4AE}}{2A^2}-k_s^2\right]^{1/2},$$

$$B_0=\frac{3|e|^2}{128\varepsilon_{sc}},A=\frac{\hbar^2}{2m_1^*}$$

and

$$K_7(E,k_s)\equiv\left[\frac{2m_2^*E'}{\hbar^2}G(E-V_0,E_{g_2},\Delta_2)+k_s^2\right]^{1/2}$$

Therefore, the dispersion relation in quantum dot HgTe/CdTe superlattices with graded interfaces can be written from (7.22) as

$$\left(\frac{n_x\pi}{d_x}\right)^2+\left(\frac{n_y\pi}{d_y}\right)^2=\frac{1}{L_0^2}\left[\cos^{-1}\left[\frac{1}{2}\Phi_3(E_{\mathrm{SLQD4}},n_x,n_y)\right]\right]^2-\left(\frac{n_z\pi}{d_z}\right)^2 \tag{7.23}$$

where

$$\Phi_3(E_{\mathrm{SLQD4}},n_x,n_y)\equiv\Big[2\cosh\{\beta_3(E_{\mathrm{SLQD4}},n_x,n_y)\}\cos\{\gamma_3(E_{\mathrm{SLQD4}},n_x,n_y)\}$$

$$+\varepsilon_3(E_{\mathrm{SLQD4}},n_x,n_y)\sinh\{\beta_3(E_{\mathrm{SLQD4}},n_x,n_y)\}\sin\{\gamma_3(E_{\mathrm{SLQD4}},n_x,n_y)\}$$

$$+\Delta_0\left[\left(\frac{\{K_7\gamma_3(E_{\mathrm{SLQD4}},n_x,n_y)\}^7}{K_8\gamma_3(E_{\mathrm{SLQD4}},n_x,n_y)}-3K_8\gamma_3(E_{\mathrm{SLQD4}},n_x,n_y)\right)\right.$$

$$\times\cosh\{\beta_3(E_{\mathrm{SLQD4}},n_x,n_y)\}\sin\{\gamma_3(E_{\mathrm{SLQD4}},n_x,n_y)\}$$

$$+\left(3K_7(E_{\mathrm{SLQD4}},n_x,n_y)-\frac{\{K_8(E_{\mathrm{SLQD4}},n_x,n_y)\}^2}{K_7(E_{\mathrm{SLQD4}},n_x,n_y)}\right)$$

$$\times\sinh\{\beta_3(E_{\mathrm{SLQD4}},n_x,n_y)\}\cos\{\gamma_3(E_{\mathrm{SLQD4}},n_x,n_y)\}\Big]$$

$$+\Delta_0\left[2(\{K_7(E_{\text{SLQD4}},n_x,n_y)\}^2-\{K_8(E_{\text{SLQD4}},n_x,n_y)\}^2)\right.$$

$$\times\cosh\{\beta_3(E_{\text{SLQD4}},n_x,n_y)\}\cos\{\gamma_3(E_{\text{SLQD4}},n_x,n_y)\}$$

$$+\frac{1}{12}\left[\frac{5\{K_8(E_{\text{SLQD4}},n_x,n_y)\}^3}{K_7(E_{\text{SLQD4}},n_x,n_y)}+\frac{5\{K_7(E_{\text{SLQD4}},n_x,n_y)\}^3}{K_8(E_{\text{SLQD4}},n_x,n_y)}\right.$$

$$\left.-34K_7(E_{\text{SLQD4}},n_x,n_y)K_8(E_{\text{SLQD4}},n_x,n_y)\right]\sinh\{\beta_3(E_{\text{SLQD4}},n_x,n_y)\}$$

$$\left.\left.\times\sin\{\gamma_3(E_{\text{SLQD4}},n_x,n_y)\}\right]\right]$$

$$\varepsilon_3(E_{\text{SLQD4}},n_x,n_y)\equiv\left[\frac{K_7(E_{\text{SLQD4}},n_x,n_y)}{K_8(E_{\text{SLQD4}},n_x,n_y)}-\frac{K_8(E_{\text{SLQD4}},n_x,n_y)}{K_7(E_{\text{SLQD4}},n_x,n_y)}\right],$$

$$\beta_3(E_{\text{SLQD4}},n_x,n_y)\equiv K_7(E_{\text{SLQD4}},n_x,n_y)[a_0-\Delta_0],$$

$$\gamma_3(E_{\text{SLQD4}},n_x,n_y)=K_8(E_{\text{SLQD4}},n_x,n_y)[b_0-\Delta_0],$$

$$K_8(E_{\text{SLQD4}},n_x,n_y)\equiv\left[\frac{B_0^2+2AE-B_0\sqrt{B_0^2+4AE}}{2A^2}\right.$$

$$\left.-\left[\left(\frac{n_x\pi}{d_x}\right)^2+\left(\frac{n_y\pi}{d_y}\right)^2\right]\right]^{1/2}$$

$$K_7(E_{\text{SLQD4}},n_x,n_y)\equiv\left[\frac{2m_2^*E'}{\hbar^2}G(E-V_0,E_{g_2},\Delta_2)\right.$$

$$\left.+\left[\left(\frac{n_x\pi}{d_x}\right)^2+\left(\frac{n_y\pi}{d_y}\right)^2\right]\right]^{1/2}$$

and E_{SLQD4} is the total quantized energy in this case.

The density-of-states function is given by

$$N(E)=\frac{2}{d_xd_yd_z}\sum_{n_x=1}^{n_{x\max}}\sum_{n_y=1}^{n_{y\max}}\sum_{n_z=1}^{n_{z\max}}\delta'(E-E_{\text{SLQD4}}) \tag{7.24}$$

The electron concentration can be expressed as

$$n_0=\frac{2}{d_xd_yd_z}\sum_{n_x=1}^{n_{x\max}}\sum_{n_y=1}^{n_{y\max}}\sum_{n_z=1}^{n_{z\max}}F_{-1}(\eta_{1\text{SLQD4}}) \tag{7.25}$$

where

$$\eta_{1\text{SLQD4}}=(k_BT)^{-1}(E_{\text{FSLQD}}-E_{\text{SLQD4}})$$

By using (1.5) and (7.25), the MTP in this case can be written as

$$G = \left(\frac{\pi^2 k_B}{3e}\right)\left[\sum_{n_x=1}^{n_{x\max}}\sum_{n_y=1}^{n_{y\max}}\sum_{n_z=1}^{n_{z\max}} F_{-1}(\eta_{1\mathrm{SLQD4}})\right]^{-1} \times \left[\sum_{n_x=1}^{n_{x\max}}\sum_{n_y=1}^{n_{y\max}}\sum_{n_z=1}^{n_{z\max}} F_{-2}(\eta_{1\mathrm{SLQD4}})\right] \quad (7.26)$$

*7.2.5 The MTP in strained layer QDSLs with graded interfaces

The dispersion relation of the conduction electrons of the constituent materials of the strained layer superlattices can be expressed following (1.105) as

$$[E - T_{1i}]k_x^2 + [E - T_{2i}]k_y^2 + [E - T_{3i}]k_z^2 = q_1 E^3 - R_1 E^2 + V_i E + \xi_i \quad (7.27)$$

where

$$T_{1i} = \bar{\theta}_i,\ \bar{\theta}_i = \left[E_{gi} - C_{1i}^c \varepsilon_i - (a_i + C_{1i}^c)\varepsilon_i + \frac{3}{2} b_i \varepsilon_{xxi} - \frac{b_i \varepsilon_i}{2} + \frac{\sqrt{3} d_i \varepsilon_{xyi}}{2}\right]$$

$$T_{2i} = \omega_i,\ \omega_i = \left[E_{gi} - C_{1i}^c \varepsilon_i - (a_i + C_{1i}^c)\varepsilon_i + \frac{3}{2} b_i \varepsilon_{xxi} - \frac{b_i \varepsilon_i}{2} + \frac{\sqrt{3} d_i \varepsilon_{xyi}}{2}\right]$$

$$T_{3i} = \delta_i,\ \delta_i = \left[E_{gi} - C_{1i}^c \varepsilon_i - (a_i + C_{1i}^c)\varepsilon_i + \frac{3}{2} b_i \varepsilon_{xxi} - \frac{b_i \varepsilon_i}{2}\right],$$

$$R_i = q_i[2A_i + C_{1i}^c \varepsilon_i],\ q_i = \frac{3}{2B_{2i}^2},\ A_i = E_{gi} - C_{1i}^c \varepsilon_i$$

$$V_i = q_i\left[A_i^2 - \frac{2C_{2i}^2 \varepsilon_{xyi}}{3} + 2A_i C_{1i}^c \varepsilon_i\right],\ \xi_i = q_i\left[\frac{2C_{2i}^2 \varepsilon_{xyi}}{3} - C_{1i}^c \varepsilon_i A_i^2\right]$$

Therefore, (2.18) can be expressed as

$$\bar{P}_i(E)k_x^2 + \bar{Q}_i(E)k_y^2 + \bar{S}_i(E)k_z^2 = 1 \quad (7.28)$$

where

$$\bar{P}_i(E) = [(E - T_{1i})/\omega_i(E)],\ \omega_i(E) = [q_i E^3 - R_i E^2 + V_i E + \xi_i],\ \bar{Q}_i(E) = [(E - T_{2i})/\omega_i(E)]$$

and

$$\bar{S}_i(E) = [(E - T_{3i})/\omega_i(E)]$$

Therefore, the dispersion relation in strained layer SLs with graded interfaces can be written as

$$\cos(L_0 k) = \frac{1}{2}\Phi_4(E, k_s) \tag{7.29}$$

where

$$\begin{aligned}
\Phi_4(E, k_s) &= [2\cosh[T_4(E, k_s)]\cos[T_5(E, k_s)]] + [[T_6(E, k_s)] \\
&\times \sinh[T_4(E, k_s)]\sin[T_5(E, k_s)] \\
&+ \Delta_0\left[\left(\frac{k_0^2(E, k_s)}{k'(E, k_s)} - 3k'(E, k_s)\right)\cosh[T_4(E, k_s)]\sin[T_5(E, k_s)]\right. \\
&+ \left.\left(3k_0(E, k_s) - \frac{k'^2(E, k_s)}{k_0(E, k_s)}\right)\sinh[T_4(E, k_s)]\sin[T_5(E, k_s)]\right] \\
&+ \Delta_0[2(k_0^2(E, k_s) - k'^2(E, k_s))\cosh[T_4(E, k_s)]\cos[T_5(E, k_s)] \\
&+ \frac{1}{12}\left(\frac{5k_0^3(E, k_s)}{k'(E, k_s)} + \frac{5k'^3(E, k_s)}{k_0^3(E, k_s)} - 34k_0(E, k_s)k'(E, k_s)\right) \\
&\times \sinh[T_4(E, k_s)]\sin[T_5(E, k_s)]
\end{aligned}$$

$$\begin{aligned}
[T_4(E, k_s)] &= k_0(E, k_s)[a_0 - \Delta_0] \\
k_0(E, k_s) &= [\bar{S}_2(E - V_0)]^{-1/2}[\bar{P}_2(E - V_0)k_x^2 + \bar{Q}_2(E - V_0)k_y^2 - 1]^{1/2} \\
T_5(E, k_s) &= k'(E, k_s)[b_0 - \Delta_0],\ k'(E, k_s) = [\bar{S}_1(E)]^{1/2} \\
&\times [1 - \bar{P}_1(E)k_x^2 - \bar{Q}_1(E)k_y^2]^{1/2}
\end{aligned}$$

and

$$T_6(E, k_s) = \left[\frac{k_0(E, k_s)}{k'(E, k_s)} - \frac{k'(E, k_s)}{k_0(E, k_s)}\right]$$

Therefore, the dispersion relation in quantum dot strained layer superlattices with graded interfaces can be written from (7.29) as

$$\left(\frac{n_x\pi}{d_x}\right)^2 + \left(\frac{n_y\pi}{d_y}\right)^2 = \frac{1}{L_0^2}\left[\cos^{-1}\left[\frac{1}{2}\Phi_4(E_{\text{SLQD5}}, n_x, n_y)\right]\right]^2 - \left(\frac{n_z\pi}{d_z}\right)^2 \tag{7.30}$$

$$\cos(L_0 k) = \frac{1}{2}\Phi_4(E_{\text{SLQD5}}, k_s) \tag{7.31}$$

where

$$\Phi_4(E_{\mathrm{SLQD5}},n_x,n_y) = [2\cosh[T_4(E_{\mathrm{SLQD5}},n_x,n_y)]\cos[T_5(E_{\mathrm{SLQD5}},n_x,n_y)]]$$

$$+ [[T_6(E_{\mathrm{SLQD5}},n_x,n_y)]\mathrm{sinh}[T_4(E_{\mathrm{SLQD5}},n_x,n_y)]\mathrm{sin}[T_5(E_{\mathrm{SLQD5}},n_x,n_y)]]$$

$$+ \Delta_0\left[\left(\frac{k_0^2(E_{\mathrm{SLQD5}},n_x,n_y)}{k'(E_{\mathrm{SLQD5}},n_x,n_y)} - 3k'(E_{\mathrm{SLQD5}},n_x,n_y)\right)\right.$$

$$\times \cosh[T_4(E_{\mathrm{SLQD5}},n_x,n_y)]\sin[T_5(E_{\mathrm{SLQD5}},n_x,n_y)]$$

$$+ \left(3k_0(E_{\mathrm{SLQD5}},n_x,n_y) - \frac{k'^2(E_{\mathrm{SLQD5}},n_x,n_y)}{k_0(E_{\mathrm{SLQD5}},n_x,n_y)}\right)$$

$$\left.\times \sinh[T_4(E_{\mathrm{SLQD5}},n_x,n_y)]\sin[T_5(E_{\mathrm{SLQD5}},n_x,n_y)]\right]$$

$$+ \Delta_0[2(k_0^2(E_{\mathrm{SLQD5}},n_x,n_y) - k'^2(E_{\mathrm{SLQD5}},n_x,n_y))]$$

$$\times \cosh[T_4(E_{\mathrm{SLQD5}},n_x,n_y)]\cos[T_5(E_{\mathrm{SLQD5}},n_x,n_y)]$$

$$+ \frac{1}{12}\left(\frac{5k_0^3(E_{\mathrm{SLQD5}},n_x,n_y)}{k'(E_{\mathrm{SLQD5}},n_x,n_y)} + \frac{5k'^3(E_{\mathrm{SLQD5}},n_x,n_y)}{k_0^3(E_{\mathrm{SLQD5}},n_x,n_y)}\right.$$

$$\left. - 34k_0\ (E_{\mathrm{SLQD5}},n_x,n_y)k'(E_{\mathrm{SLQD5}},n_x,n_y)\right)$$

$$\times \sinh[T_4(E_{\mathrm{SLQD5}},n_x,n_y)]\sin[T_5(E_{\mathrm{SLQD5}},n_x,n_y)]$$

$$[T_4(E_{\mathrm{SLQD5}},n_x,n_y)] = k_0(E_{\mathrm{SLQD5}},n_x,n_y)[a_0 - \Delta_0]$$

$$k_0(E_{\mathrm{SLQD5}},n_x,n_y) = [\bar{S}_2(E-V_0)]^{-1/2}\left[\bar{P}_2(E-V_0)\left[\left(\frac{n_x\pi}{d_x}\right)^2\right]\right.$$

$$\left. + \bar{Q}_2(E-V_0)\left[\left(\frac{n_y\pi}{d_y}\right)^2\right] - 1\right]^{1/2}$$

$$T_5(E_{\mathrm{SLQD5}},n_x,n_y) = k'(E_{\mathrm{SLQD5}},n_x,n_y)[b_0 - \Delta_0]$$

$$k'(E_{\mathrm{SLQD5}},n_x,n_y) = [\bar{S}_1(E_{\mathrm{SLQD5}})]^{1/2}\left[1 - \bar{P}_1(E_{\mathrm{SLQD5}})\left[\left(\frac{n_x\pi}{d_x}\right)^2\right]\right.$$

$$\left. - \bar{Q}_1(E_{\mathrm{SLQD5}})\left[\left[\left(\frac{n_y\pi}{d_y}\right)^2\right]\right]\right]^{1/2}$$

$$T_6(E_{\mathrm{SLQD5}},n_x,n_y) = \left[\frac{k_0(E_{\mathrm{SLQD5}},n_x,n_y)}{k'(E_{\mathrm{SLQD5}},n_x,n_y)} - \frac{k'(E_{\mathrm{SLQD5}},n_x,n_y)}{k_0(E_{\mathrm{SLQD5}},n_x,n_y)}\right]$$

and E_{SLQD5} is the total quantized energy in this case.

The density-of-states function is given by

$$N(E) = \frac{2}{d_x d_y d_z} \sum_{n_x=1}^{n_{x\max}} \sum_{n_y=1}^{n_{y\max}} \sum_{n_z=1}^{n_{z\max}} \delta'(E - E_{\mathrm{SLQD5}}) \tag{7.32}$$

The electron concentration can be expressed as

$$n_0 = \frac{2}{d_x d_y d_z} \sum_{n_x=1}^{n_{x\max}} \sum_{n_y=1}^{n_{y\max}} \sum_{n_z=1}^{n_{z\max}} F_{-1}(\eta_{1\mathrm{SLQD5}}) \tag{7.33}$$

where

$$\eta_{1\mathrm{SLQD5}} = (k_B T)^{-1}(E_{\mathrm{FSLQD}} - E_{\mathrm{SLQD5}})$$

By using (1.5) and (7.33), the MTP in this case can be written as

$$G = \left(\frac{\pi^2 k_B}{3e}\right) \left[\sum_{n_x=1}^{n_{x\max}} \sum_{n_y=1}^{n_{y\max}} \sum_{n_z=1}^{n_{z\max}} F_{-1}(\eta_{1\mathrm{SLQD5}})\right]^{-1} \times \left[\sum_{n_x=1}^{n_{x\max}} \sum_{n_y=1}^{n_{y\max}} \sum_{n_z=1}^{n_{z\max}} F_{-2}(\eta_{1\mathrm{SLQD5}})\right] \tag{7.34}$$

7.2.6 The MTP in III–V QD effective mass superlattices

Following Sasaki [27], the electron dispersion law in III–V effective mass superlattices (EMSLs) can be written as

$$k_x^2 = \left[\frac{1}{L_0^2}\{\cos^{-1}(f(E, k_y, k_z))\}^2 - k_\perp^2\right] \tag{7.35}$$

in which

$$f(E, k_y, k_z) = a_1 \cos[a_0 C_1(E, k_\perp) + b_0 D_1(E, k_\perp)] - a_2 \cos[a_0 C_1(E, k_\perp) - b_0 D_1(E, k_\perp)], k_\perp^2 = k_y^2 + k_z^2$$

$$a_1 = \left[\sqrt{\frac{m_2^*}{m_1^*}} + 1\right]^2 \left[4\left(\frac{m_2^*}{m_1^*}\right)^{1/2}\right]^{-1}, a_2 = \left[-1 + \sqrt{\frac{m_2^*}{m_1^*}}\right]^2 \left[4\left(\frac{m_2^*}{m_1^*}\right)^{1/2}\right]^{-1},$$

$$C_1(E, k_\perp) \equiv \left[\left(\frac{2m_1^* E}{\hbar^2}\right) G(E, E_{g_1}, \Delta_1) - k_\perp^2\right]^{1/2}$$

and

$$D_1(E, k_\perp) \equiv \left[\left(\frac{2m_2^* E}{\hbar^2}\right) G(E, E_{g_2}, \Delta_2) - k_\perp^2\right]^{1/2}$$

Therefore, the dispersion relation in III–V quantum dot EMSLs can be written as

$$\left(\frac{n_x\pi}{d_x}\right)^2 = \left[\frac{1}{L_0^2}\{\cos^{-1}(f(E_{\mathrm{SLQD6}}, n_y, n_z))\}^2 - \left[\left(\frac{n_y\pi}{d_y}\right)^2 + \left(\frac{n_z\pi}{d_z}\right)^2\right]\right] \tag{7.36}$$

where

$$f(E_{\mathrm{SLQD6}}, n_y, n_z) = a_1\cos[a_0C_1(E_{\mathrm{SLQD6}}, n_y, n_z) + b_0D_1(E_{\mathrm{SLQD6}}, n_y, n_z)]$$
$$- a_2\cos[a_0C_1(E_{\mathrm{SLQD6}}, n_y, n_z) - b_0D_1(E_{\mathrm{SLQD6}}, n_y, n_z)]$$

$$C_1(E_{\mathrm{SLQD6}}, n_y, n_z) \equiv \left[\left(\frac{2m_1^*E}{\hbar^2}\right)G(E_{\mathrm{SLQD6}}, E_{g_1}, \Delta_1)\right.$$
$$\left.-\left[\left(\frac{n_y\pi}{d_y}\right)^2 + \left(\frac{n_z\pi}{d_z}\right)^2\right]\right]^{1/2}$$

$$D_1(E_{\mathrm{SLQD6}}, n_y, n_z) \equiv \left[\left(\frac{2m_2^*E}{\hbar^2}\right)G(E_{\mathrm{SLQD6}}, E_{g_2}, \Delta_2)\right.$$
$$\left.-\left[\left(\frac{n_y\pi}{d_y}\right)^2 + \left(\frac{n_z\pi}{d_z}\right)^2\right]\right]^{1/2}$$

and E_{SLQD6} is the total quantized energy in this case.

The density-of-states function is given by

$$N(E) = \frac{2}{d_xd_yd_z}\sum_{n_x=1}^{n_{x\max}}\sum_{n_y=1}^{n_{y\max}}\sum_{n_z=1}^{n_{z\max}}\delta'(E - E_{\mathrm{SLQD6}}) \tag{7.37}$$

The electron concentration can be expressed as

$$n_0 = \frac{2}{d_xd_yd_z}\sum_{n_x=1}^{n_{x\max}}\sum_{n_y=1}^{n_{y\max}}\sum_{n_z=1}^{n_{z\max}}F_{-1}(\eta_{1\mathrm{SLQD6}}) \tag{7.38}$$

where

$$\eta_{1\mathrm{SLQD6}} = (k_BT)^{-1}(E_{\mathrm{FSLQD}} - E_{\mathrm{SLQD6}})$$

By using (1.5) and (7.38), the MTP in this case can be written as

$$G = \left(\frac{\pi^2k_B}{3e}\right)\left[\sum_{n_x=1}^{n_{x\max}}\sum_{n_y=1}^{n_{y\max}}\sum_{n_z=1}^{n_{z\max}}F_{-1}(\eta_{1\mathrm{SLQD6}})\right]^{-1}$$
$$\times\left[\sum_{n_x=1}^{n_{x\max}}\sum_{n_y=1}^{n_{y\max}}\sum_{n_z=1}^{n_{z\max}}F_{-2}(\eta_{1\mathrm{SLQD6}})\right] \tag{7.39}$$

7.2.7 The MTP in II–VI QD effective mass superlattices

Following Sasaki [27], the electron dispersion law in II–VI EMSLs can be written as

$$k_z^2 = \left[\frac{1}{L_0^2}\{\cos^{-1}(f_1(E, k_x, k_y))\}^2 - k_s^2\right] \tag{7.40}$$

in which

$$f_1(E, k_x, k_y) = a_3 \cos\left[a_0 C_2(E, k_s) + b_0 D_2(E, k_s)\right]$$
$$-a_4 \cos\left[a_0 C_2(E, k_s) - b_0 D_2(E, k_s)\right], k_s^2 = k_x^2 + k_y^2$$

$$a_3 = \left[\sqrt{\frac{m_2^*}{m_{\parallel,1}^*}} + 1\right]^2 \left[4\left(\frac{m_2^*}{m_{\parallel,1}^*}\right)^{1/2}\right]^{-1},$$

$$a_4 = \left[-1 + \sqrt{\frac{m_2^*}{m_{\parallel,1}^*}}\right]^2 \left[4\left(\frac{m_2^*}{m_{\parallel,1}^*}\right)^{1/2}\right]^{-1},$$

$$C_2(E, k_s) \equiv \left(\frac{2m_{\parallel,1}^*}{\hbar^2}\right)^{1/2} \left[E - \frac{\hbar^2 k_s^2}{2m_{\perp,1}^*} \mp \bar{\lambda}_0 k_s\right]^{1/2}$$

and

$$D_2(E, k_s) \equiv \left[\left(\frac{2m_2^*}{\hbar^2}\right) EG(E, E_{g_2}, \Delta_2) - k_s^2\right]^{1/2}$$

Therefore, the dispersion relation in quantum dot II–VI superlattices with graded interfaces can be written as

$$\left(\frac{n_z\pi}{d_z}\right)^2 = \left[\frac{1}{L_0^2}\{\cos^{-1}(f_1(E_{\text{SLQD7}}, n_x, n_y))\}^2 - \left[\left(\frac{n_x\pi}{d_x}\right)^2 + \left(\frac{n_y\pi}{d_y}\right)^2\right]\right] \tag{7.41}$$

where

$$f_1(E_{\text{SLQD7}}, n_x, n_y) = a_3 \cos[a_0 C_2(E_{\text{SLQD7}}, n_x, n_y) + b_0 D_2(E_{\text{SLQD7}}, n_x, n_y)]$$
$$- a_4 \cos[a_0 C_2(E_{\text{SLQD7}}, n_x, n_y) - b_0 D_2(E_{\text{SLQD7}}, n_x, n_y)]$$

$$C_2(E_{\text{SLQD7}}, n_x, n_y) \equiv \left(\frac{2m^*_{\|,1}}{\hbar^2}\right)^{1/2}$$

$$\times \left[E - \frac{\hbar^2\left[\left(\frac{n_x\pi}{d_x}\right)^2 + \left(\frac{n_y\pi}{d_y}\right)^2\right]}{2m^*_{\perp,1}} \mp \bar{\lambda}_0\sqrt{\left[\left(\frac{n_x\pi}{d_x}\right)^2 + \left(\frac{n_y\pi}{d_y}\right)^2\right]}\right]^{1/2}$$

$$D_2(E_{\text{SLQD7}}, n_x, n_y) \equiv \left[\left(\frac{2m^*_2}{\hbar^2}\right)EG(E, E_{g_2}, \Delta_2)\right.$$

$$\left. - \left[\left(\frac{n_x\pi}{d_x}\right)^2 + \left(\frac{n_y\pi}{d_y}\right)^2\right]\right]^{1/2}$$

and E_{SLQD7} is the total quantized energy in this case.

The density-of-states function is given by

$$N(E) = \frac{1}{d_x d_y d_z} \sum_{n_x=1}^{n_{x\max}} \sum_{n_y=1}^{n_{y\max}} \sum_{n_z=1}^{n_{z\max}} \delta'(E - E_{\text{SLQD7}}) \tag{7.42}$$

The electron concentration can be expressed as

$$n_0 = \frac{1}{d_x d_y d_z} \sum_{n_x=1}^{n_{x\max}} \sum_{n_y=1}^{n_{y\max}} \sum_{n_z=1}^{n_{z\max}} F_{-1}(\eta_{1\text{SLQD7}}) \tag{7.43}$$

where

$$\eta_{1\text{SLQD7}} = (k_B T)^{-1}(E_{\text{FSLQD}} - E_{\text{SLQD7}})$$

By using (1.5) and (7.43), the MTP in this case can be written as

$$G = \left(\frac{\pi^2 k_B}{3e}\right)\left[\sum_{n_x=1}^{n_{x\max}} \sum_{n_y=1}^{n_{y\max}} \sum_{n_z=1}^{n_{z\max}} F_{-1}(\eta_{1\text{SLQD7}})\right]^{-1}$$

$$\times \left[\sum_{n_x=1}^{n_{x\max}} \sum_{n_y=1}^{n_{y\max}} \sum_{n_z=1}^{n_{z\max}} F_{-2}(\eta_{1\text{SLQD7}})\right] \tag{7.44}$$

7.2.8 The MTP in IV–VI QDHD effective mass superlattices

Following Sasaki [27], the electron dispersion law in IV–VI, EMSLs can be written as

$$k_x^2 = \left[\frac{1}{L_0^2}\{\cos^{-1}(f_2(E,k_y,k_z))\}^2 - k_\perp^2\right] \tag{7.45}$$

in which

$$f_2(E,k_y,k_z) = a_5\cos[a_0C_3(E,k_y,k_z) + b_0D_3(E,k_y,k_z)]$$
$$- a_6\cos[a_0C_3(E,k_y,k_z) - b_0D_3(E,k_y,k_z)],$$
$$a_5 = \left[\sqrt{\frac{m_2^*}{m_1^*}} + 1\right]^2\left[4\left(\frac{m_2^*}{m_1^*}\right)^{1/2}\right]^{-1},$$
$$m_i^* = \left[\frac{2\hbar^2}{\{a_i^2 - b_i^2\}}\right][a_i + [a_iC_i + a_ie_iE_{g_i} - e_i^2E_{g_i}][E_{g_i}^2a_i^2 + C_i^2$$
$$+ e_i^2E_{g_i}^2 + 2C_ie_iE_{g_i} - 2E_{g_i}a_iC_i - 2e_ia_iE_{g_i}^2]^{1/2}]$$
$$C_3(E,k_y,k_z) \equiv [[EH_{11} + H_{21}(k_y,k_z)]$$
$$- [E^2H_{31} + EH_{41}(k_y,k_z) + H_{51}(k_y,k_z)]^{1/2}]^{1/2}$$
$$D_3(E,k_y,k_z) \equiv [[EH_{12} + H_{22}(k_y,k_z)]$$
$$- [E^2H_{32} + EH_{42}(k_y,k_z) + H_{52}(k_y,k_z)]^{1/2}]^{1/2}$$

and

$$a_6 = \left[-1 + \sqrt{\frac{m_2^*}{m_1^*}}\right]^2\left[4\left(\frac{m_2^*}{m_1^*}\right)^{1/2}\right]^{-1}$$

Therefore, the dispersion relation in quantum dot IV–VI effective mass superlattices can be written as

$$\left(\frac{n_x\pi}{d_x}\right)^2 = \left[\frac{1}{L_0^2}\{\cos^{-1}(f_2(E_{\text{SLQD8}},n_y,n_z))\}^2 - \left[\left(\frac{n_y\pi}{d_y}\right)^2 + \left(\frac{n_z\pi}{d_z}\right)^2\right]\right] \tag{7.46}$$

where

$$f_2(E_{\text{SLQD8}},n_y,n_z) = a_5\cos[a_0C_3(E_{\text{SLQD8}},n_y,n_z) + b_0D_3(E_{\text{SLQD8}},n_y,n_z)]$$
$$- a_6\cos[a_0C_3(E_{\text{SLQD8}},n_y,n_z) - b_0D_3(E_{\text{SLQD8}},n_y,n_z)]$$

$$C_3(E_{\mathrm{SLQD8}}, n_y, n_z) \equiv [[EH_{11} + H_{21}(n_y, n_z)] - [E^2H_{31} + EH_{41}(n_y, n_z) + H_{51}(n_y, n_z)]^{1/2}]^{1/2}$$

$$D_3(E_{\mathrm{SLQD8}}, n_y, n_z) \equiv [[EH_{12} + H_{22}(n_y, n_z)] - [E^2H_{32} + EH_{42}(n_y, n_z) + H_{52}(n_y, n_z)]^{1/2}]^{1/2}$$

and E_{SLQD8} is the total quantized energy in this case.

The density-of-states function is given by

$$N(E) = \frac{2}{d_x d_y d_z} \sum_{n_x=1}^{n_{x\max}} \sum_{n_y=1}^{n_{y\max}} \sum_{n_z=1}^{n_{z\max}} \delta'(E - E_{\mathrm{SLQD8}}) \tag{7.47}$$

The electron concentration can be expressed as

$$n_0 = \frac{2}{d_x d_y d_z} \sum_{n_x=1}^{n_{x\max}} \sum_{n_y=1}^{n_{y\max}} \sum_{n_z=1}^{n_{z\max}} F_{-1}(\eta_{1\mathrm{SLQD8}}) \tag{7.48}$$

where

$$\eta_{1\mathrm{SLQD8}} = (k_B T)^{-1}(E_{\mathrm{FSLQD}} - E_{\mathrm{SLQD8}})$$

By using (1.5) and (7.48), the MTP in this case can be written as

$$G = \left(\frac{\pi^2 k_B}{3e}\right) \left[\sum_{n_x=1}^{n_{x\max}} \sum_{n_y=1}^{n_{y\max}} \sum_{n_z=1}^{n_{z\max}} F_{-1}(\eta_{1\mathrm{SLQD8}})\right]^{-1} \times \left[\sum_{n_x=1}^{n_{x\max}} \sum_{n_y=1}^{n_{y\max}} \sum_{n_z=1}^{n_{z\max}} F_{-2}(\eta_{1\mathrm{SLQD8}})\right] \tag{7.49}$$

7.2.9 The MTP in HgTe/CdTe QD effective mass superlattices

Following Sasaki [27], the electron dispersion law in HgTe/CdTe EMSLs can be written as

$$k_x^2 = \left[\frac{1}{L_0^2}\{\cos^{-1}(f_3(E, k_y, k_z))\}^2 - k_\perp^2\right] \tag{7.50}$$

in which

$$f_3(E,k_\perp) = a_7\cos\left[a_0C_4(E,k_\perp)+b_0D_4(E,k_\perp)\right] - a_8\cos\left[a_0C_4(E,k_\perp) - b_0D_4(E,k_\perp)\right]$$

$$a_1 = \left[\sqrt{\frac{m_2^*}{m_1^*}}+1\right]^2\left[4\left(\frac{m_2^*}{m_1^*}\right)^{1/2}\right]^{-1}, a_2 = \left[-1+\sqrt{\frac{m_2^*}{m_1^*}}\right]^2\left[4\left(\frac{m_2^*}{m_1^*}\right)^{1/2}\right]^{-1},$$

$$C_4(E,k_\perp) \equiv \left[\frac{B_0^2+2AE-B_0\sqrt{B_0^2+4AE}}{2A^2} - k_\perp^2\right]^{1/2}$$

and

$$D_4(E,k_\perp) \equiv \left[\left(\frac{2m_2^*E}{\hbar^2}\right)G(E,E_{g_2},\Delta_2) - k_\perp^2\right]^{1/2}$$

Therefore, the dispersion relation in quantum dot HgTe/CdTe effective mass superlattices can be written as

$$\left(\frac{n_x\pi}{d_x}\right)^2 = \left[\frac{1}{L_0^2}\{\cos^{-1}(f_3(E_{\mathrm{SLQD9}},n_y,n_z))\}^2 - \left[\left(\frac{n_y\pi}{d_y}\right)^2+\left(\frac{n_z\pi}{d_z}\right)^2\right]\right] \tag{7.51}$$

where

$$f_3(E_{\mathrm{SLQD9}},n_y,n_z) = a_7\cos[a_0C_4(E_{\mathrm{SLQD9}},n_y,n_z)+b_0D_4(E_{\mathrm{SLQD9}},n_y,n_z)] - a_8\cos[a_0C_4(E_{\mathrm{SLQD9}},n_y,n_z)-b_0D_4(E_{\mathrm{SLQD9}},n_y,n_z)]$$

$$C_4(E_{\mathrm{SLQD9}},n_y,n_z) \equiv \left[\frac{B_0^2+2AE-B_0\sqrt{B_0^2+4AE}}{2A^2} - \left[\left(\frac{n_y\pi}{d_y}\right)^2+\left(\frac{n_z\pi}{d_z}\right)^2\right]\right]^{1/2}$$

$$D_4(E_{\mathrm{SLQD9}},n_y,n_z) \equiv \left[\left(\frac{2m_2^*E}{\hbar^2}\right)G(E,E_{g_2},\Delta_2) - \left[\left(\frac{n_y\pi}{d_y}\right)^2+\left(\frac{n_z\pi}{d_z}\right)^2\right]\right]^{1/2}$$

and E_{SLQD9} is the total quantized energy in this case.

The density-of-states function is given by

$$N(E) = \frac{2}{d_x d_y d_z} \sum_{n_x=1}^{n_{x\max}} \sum_{n_y=1}^{n_{y\max}} \sum_{n_z=1}^{n_{z\max}} \delta'(E - E_{\mathrm{SLQD9}}) \tag{7.52}$$

The electron concentration can be expressed as

$$n_0 = \frac{2}{d_x d_y d_z} \sum_{n_x=1}^{n_{x\max}} \sum_{n_y=1}^{n_{y\max}} \sum_{n_z=1}^{n_{z\max}} F_{-1}(\eta_{1\mathrm{SLQD9}}) \tag{7.53}$$

where

$$\eta_{1\mathrm{SLQD9}} = (k_B T)^{-1}(E_{\mathrm{FSLQD}} - E_{\mathrm{SLQD9}})$$

By using (1.5) and (7.53), the MTP in this case can be written as

$$G = \left(\frac{\pi^2 k_B}{3e}\right) \left[\sum_{n_x=1}^{n_{x\max}} \sum_{n_y=1}^{n_{y\max}} \sum_{n_z=1}^{n_{z\max}} F_{-1}(\eta_{1\mathrm{SLQD9}})\right]^{-1} \times \left[\sum_{n_x=1}^{n_{x\max}} \sum_{n_y=1}^{n_{y\max}} \sum_{n_z=1}^{n_{z\max}} F_{-2}(\eta_{1\mathrm{SLQD9}})\right] \tag{7.54}$$

*7.2.10 The MTP in strained layer QD effective mass superlattices

The electron energy spectrum in heavily doped strained layer effective mass superlattices can be written as

$$k_z^2 = \left[\frac{1}{L_0^2}\{\cos^{-1}(f_4(E, k_x, k_y))\}^2 - k_s^2\right] \tag{7.55}$$

where

$$f_4(E, k_x, k_y) = a_{20}\cos[a_0 C_{40}(E, k_x, k_y) + b_0 D_{40}(E, k_x, k_y)] - a_{21}\cos[a_0 C_{40}(E, k_x, k_y) - b_0 D_{40}(E, k_x, k_y)],$$

$$a_{20} = \left[\sqrt{\frac{M_{s2}(0)}{M_{s1}(0)}} + 1\right]^2 \left[4\left(\frac{M_{s2}(0)}{M_{s1}(0)}\right)^{1/2}\right]^{-1}$$

$$M_{si}(0) = \frac{\hbar^2}{2}\left[\frac{T_{3i}V_i + \xi_i}{T_{3i}^2}\right], a_{20} = \left[\sqrt{\frac{M_{s2}(0)}{M_{s1}(0)}} + 1\right]^2 \left[4\left(\frac{M_{s2}(0)}{M_{s1}(0)}\right)^{1/2}\right]^{-1},$$

$$a_{21} = \left[\sqrt{\frac{M_{s2}(0)}{M_{s1}(0)}} + 1\right]^2 \left[4\left(\frac{M_{s2}(0)}{M_{s1}(0)}\right)^{1/2}\right]^{-1}$$

$$C_{40}(E, k_x, k_y) = [1 - \bar{P}_1(E)k_x^2 - \bar{Q}_1(E)k_y^2]^{1/2}[\bar{S}_1(E)]^{-1/2}$$

$$D_{40}(E, k_x, k_y) = [1 - \bar{P}_2(E)k_x^2 - \bar{Q}_2(E)k_y^2]^{1/2}[\bar{S}_2(E)]^{-1/2}$$

Therefore, the dispersion relation in strained layer effective mass quantum dot superlattices can be expressed as

$$\left(\frac{n_z\pi}{d_z}\right)^2 = \left[\frac{1}{L_0^2}\{\cos^{-1}(f_4(E_{\mathrm{SLQD10}}, n_x, n_y))\}^2 - \left[\left(\frac{n_x\pi}{d_x}\right)^2 + \left(\frac{n_y\pi}{d_y}\right)^2\right]\right] \tag{7.56}$$

where

$$\begin{aligned} f_4(E_{\mathrm{SLQD10}}, n_x, n_y) &= a_{20}\cos[a_0C_{40}(E_{\mathrm{SLQD10}}, n_x, n_y) \\ &\quad + b_0D_{40}(E_{\mathrm{SLQD10}}, n_x, n_y)] \\ &\quad - a_{21}\cos[a_0C_{40}(E_{\mathrm{SLQD10}}, n_x, n_y) - b_0D_{40}(E_{\mathrm{SLQD10}}, n_x, n_y)] \end{aligned}$$

$$\begin{aligned} &C_{40}(E_{\mathrm{SLQD10}}, n_x, n_y) \\ &= \left[1 - \bar{P}_1(E)\left[\left(\frac{n_x\pi}{d_x}\right)^2\right] - \bar{Q}_1(E)\left[\left(\frac{n_y\pi}{d_y}\right)^2\right]\right]^{1/2}[\bar{S}_1(E)]^{-1/2} \end{aligned}$$

$$\begin{aligned} &D_{40}(E_{\mathrm{SLQD10}}, n_x, n_y) \\ &= \left[1 - \bar{P}_2(E)\left[\left(\frac{n_x\pi}{d_x}\right)^2\right] - \bar{Q}_2(E)\left[\left(\frac{n_y\pi}{d_y}\right)^2\right]\right]^{1/2}[\bar{S}_2(E)]^{-1/2} \end{aligned}$$

and E_{SLQD10} is the total quantized energy in this case.

The density-of-states function is given by

$$N(E) = \frac{2}{d_xd_yd_z}\sum_{n_x=1}^{n_{x\max}}\sum_{n_y=1}^{n_{y\max}}\sum_{n_z=1}^{n_{z\max}}\delta'(E - E_{\mathrm{SLQD10}}) \tag{7.57}$$

The electron concentration can be expressed as

$$n_0 = \frac{2}{d_xd_yd_z}\sum_{n_x=1}^{n_{x\max}}\sum_{n_y=1}^{n_{y\max}}\sum_{n_z=1}^{n_{z\max}}F_{-1}(\eta_{1\mathrm{SLQD10}}) \tag{7.58}$$

where

$$\eta_{1\text{SLQD}10} = (k_B T)^{-1}(E_{\text{FSLQD}} - E_{\text{SLQD}10})$$

By using (1.5) and (7.58), the MTP in this case can be written as

$$G = \left(\frac{\pi^2 k_B}{3e}\right)\left[\sum_{n_x=1}^{n_{x\max}}\sum_{n_y=1}^{n_{y\max}}\sum_{n_z=1}^{n_{z\max}} F_{-1}(\eta_{1\text{SLQD}10})\right]^{-1} \times \left[\sum_{n_x=1}^{n_{x\max}}\sum_{n_y=1}^{n_{y\max}}\sum_{n_z=1}^{n_{z\max}} F_{-2}(\eta_{1\text{SLQD}10})\right] \quad (7.59)$$

7.3 Results and Discussion

Using the band constants from Table 1.1 and the appropriate equations, the normalized MTP in this case in $HgTe/Hg_{1-x}Cd_xTe$, CdS/ZnSe, PbSe/PbTe and HgTe/CdTe quantum dot superlattices with graded interfaces have been plotted as a function of film thickness as shown by curves (a)–(d), respectively, in Fig. 7.1. Figure 7.2 demonstrates the normalized

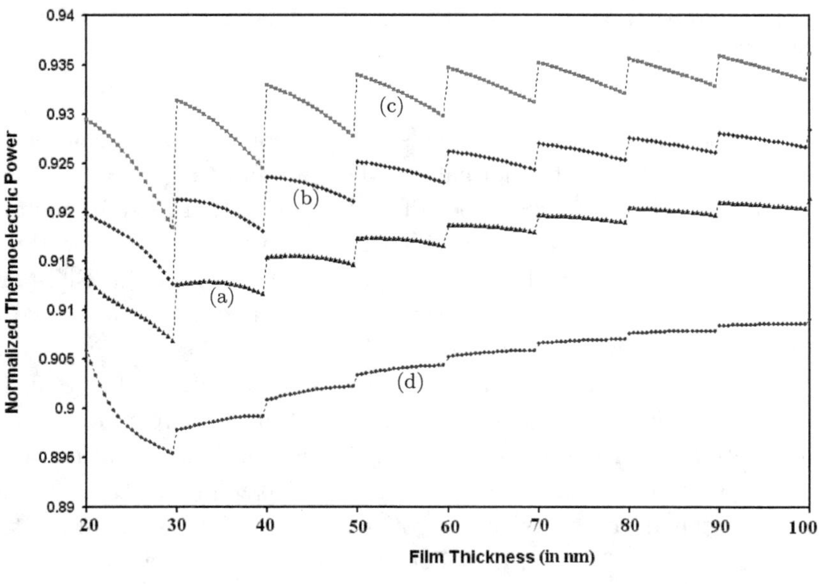

Fig. 7.1. Plot of the MTP in (a) $HgTe/Hg_{1-x}Cd_xTe$, (b) CdS/ZnSe, (c) PbSe/PbTe and (d) HgTe/CdTe quantum dot superlattices with graded interfaces as a function of normalized film thickness.

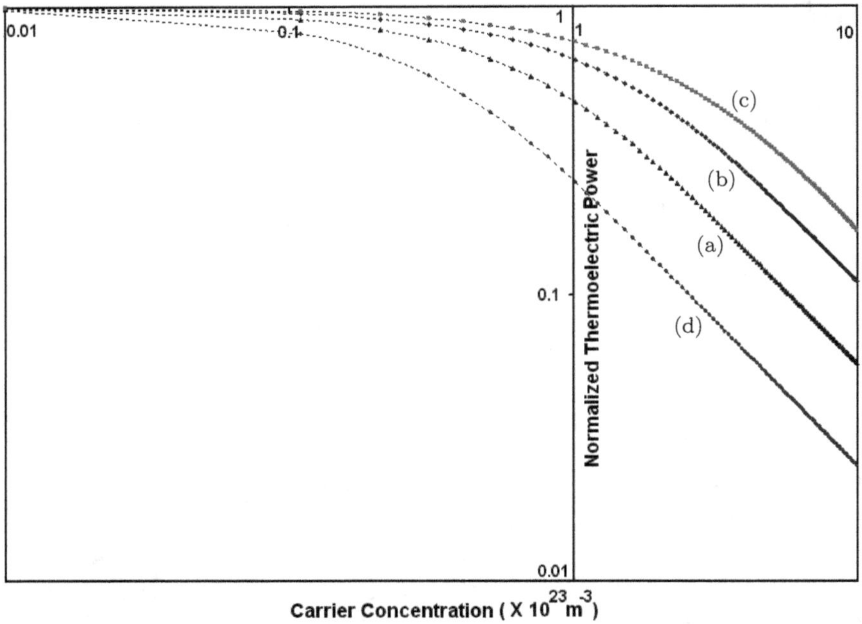

Fig. 7.2. Plot of the MTP in (a) $HgTe/Hg_{1-x}Cd_xTe$, (b) CdS/ZnSe, (c) PbSe/PbTe and (d) HgTe/CdTe quantum dot superlattices with graded interfaces as a function of carrier concentration.

MTP for the said quantized structures as a function of impurity concentration. Figure 7.3 exhibits the normalized MTP as a function of film thickness in $HgTe/Hg_{1-x}Cd_xTe$, CdS/ZnSe, PbSe/PbTe and HgTe/CdTe quantum dot effective mass superlattices as shown by curves (a)–(d), respectively. The normalized MTP in $HgTe/Hg_{1-x}Cd_xTe$, CdS/ZnSe, PbSe/PbTe and HgTe/CdTe quantum dot effective mass superlattices has been plotted as a function of electron concentration in Fig. 7.4.

It appears from Fig. 7.1 that the MTP in $HgTe/Hg_{1-x}Cd_xTe$, CdS/ZnSe, PbSe/PbTe and HgTe/CdTe quantum dot superlattices with graded interfaces increases with increasing film thickness exhibiting quantum jumps for fixed values of film thickness depending on the values of the energy band constants of the particular quantized structures. It is observed from Fig. 7.2 that the MTP in quantum dots of aforementioned superlattices decreases with increasing carrier degeneracy and differ widely for large values of the same, whereas for relatively small values of electron concentration, the MTP exhibits a converging behavior. From Fig. 7.3, it

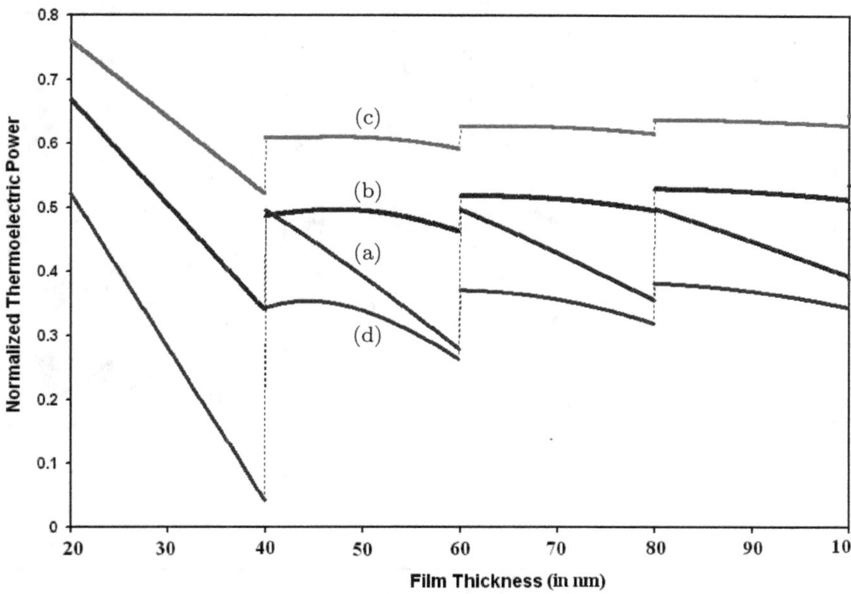

Fig. 7.3. Plot of the MTP in (a) $HgTe/Hg_{1-x}Cd_xTe$, (b) CdS/ZnSe, (c) PbSe/PbTe and (d) HgTe/CdTe quantum dot effective mass superlattices as a function of normalized film thickness.

is observed that the TPSM in $HgTe/Hg_{1-x}Cd_xTe$, CdS/ZnSe, PbSe/PbTe and HgTe/CdTe quantum dot effective mass superlattices oscillates with increasing film thickness. From Fig. 7.4, it appears that the MTP of the aforementioned superlattices decreases with increasing concentration. It should be noted that all types of variations of MTP with respect to thickness and concentration are basically band structure-dependent.

It may further be noted that the MTP of a two-dimensional electron gas in the presence of a periodic potential has already been formulated in the literature. The SL is a three-dimensional system under periodic potential. There is a radical difference in the dispersion relations of the 3D quantized structures and the corresponding carrier energy spectra of the 2D systems.

From the dispersion relations of various superlattices as discussed in this chapter, the energy spectra of the various other types of low-dimensional systems can be formulated and the corresponding MTPs can also be investigated. The results will be fundamentally different in all cases due to system asymmetry together with the change in the respective wave functions exhibiting new physical features in the respective cases.

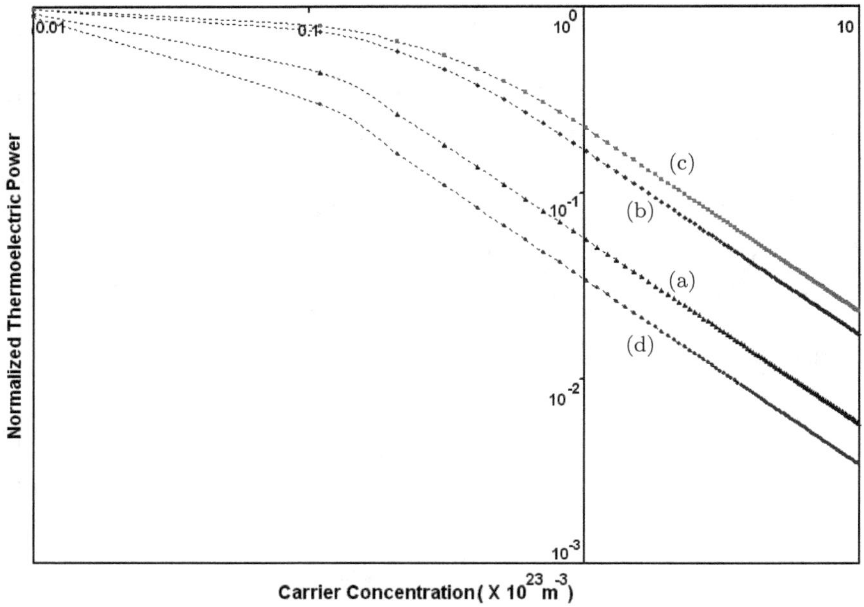

Fig. 7.4. Plot of the MTP in (a) HgTe/$Hg_{1-x}Cd_xTe$, (b) CdS/ZnSe, (c) PbSe/PbTe and (d) HgTe/CdTe quantum dot effective mass superlattices as a function of carrier concentration.

Therefore, it appears that the dispersion law and the corresponding wave function play a cardinal role in formulating any electronic property of any electronic material, since they change in a fundamental way in the presence of dimension reduction. Consequently, the derivations and the respective physical interpretations of the different transport quantities change radically.

7.4 Open Research Problems

(R.7.1) Investigate the thermoelectric power in the absence of magnetic field by considering all types of scattering mechanisms for III–V, II–VI, IV–VI and HgTe/CdTe quantum dot superlattices with graded interfaces and also the effective mass superlattices of the aforementioned materials with the appropriate dispersion relations, as formulated in this chapter.

(R.7.2) Investigate the thermoelectric power in the absence of magnetic field by considering all types of scattering mechanisms for strained layer, random, short period, polytype and saw-toothed quantum dot superlattices, respectively.

(R.7.3) Investigate the thermoelectric power in the absence of magnetic field by considering all types of scattering mechanisms for (R.7.1) and (R.7.2) under an arbitrarily oriented (a) non-uniform electric field and (b) alternating electric field for all types of quantum dot superlattices of Chapter 7, respectively.

(R.7.4) Investigate the thermoelectric power by considering all types of scattering mechanisms for (R.7.1) and (R.7.2) under an arbitrarily oriented alternating magnetic field by including broadening and the electron spin for all types of quantum dot superlattices of Chapter 7, respectively.

(R.7.5) Investigate the thermoelectric power by considering all types of scattering mechanisms for (R.7.1) and (R.7.2) under an arbitrarily oriented alternating magnetic field and crossed alternating electric field by including broadening and the electron spin for all types of quantum dot superlattices of Chapter 7, respectively.

(R.7.6) Investigate the thermoelectric power by considering all types of scattering mechanisms for (R.7.1) and (R.7.2) under an arbitrarily oriented alternating magnetic field and crossed alternating non-uniform electric field by including broadening and the electron spin for all types of quantum dot superlattices of Chapter 7, respectively.

(R.7.7) Investigate the thermoelectric power in the absence of magnetic field for all types of superlattices as considered in this chapter under exponential, Kane, Halperin, Lax and Bonch-Bruevich band tails, respectively.

(R.7.8) Investigate the thermoelectric power in the absence of magnetic field for the problem as defined in (R.7.7) under an arbitrarily oriented (a) non-uniform electric field and (b) alternating electric field for all types of quantum dot superlattices of Chapter 7, respectively.

(R.7.9) Investigate the thermoelectric power for the problem as defined in Chapter 7 under an arbitrarily oriented alternating magnetic field by including broadening and the electron spin, respectively.

(R.7.10) Investigate the thermoelectric power for the problem as defined in Chapter 7 under an arbitrarily oriented alternating magnetic field and crossed alternating electric field by including broadening and the electron spin, respectively.

(R.7.11) Investigate the problems as defined in (R.7.1)–(R.7.10) for all types of quantum dot superlattices, as discussed in this chapter.
(R.7.12) Investigate the problems as defined in (R.7.1)–(R.7.10) for all types of quantum dot superlattices as discussed in this chapter in the presence of strain.
(R.7.13) Introducing new theoretical formalisms, investigate all the problems of this chapter in the presence of hot electron effects.
(R.7.14) Investigate the influence of deep traps and surface states separately for all the appropriate problems of this chapter after proper modifications.

References

[1] S. Rakshit, S. N. Biswas, A. N. Chakravarti, *Phys. Lett.*, 37A, 64 **(1971)**; S. Rakshit, S. N. Biswas, A. N. Chakravarti, *Sol. Stat. Electron.*, 5, 702, **(1972)**; A. N. Chakravarti, D. P. Parui, *Phys. Lett.*, 40A, 113 **(1972)**.
[2] A. N. Chakravarti, D. P. Parui, *Phys. Stat. Sol. (a)*, 14, K23 **(1972)**; A. N. Chakravarti, D. P. Parui, *Phys. Stat. Sol. (a)*, 14, K139 **(1972)**; A. N. Chakravarti, D. P. Parui, *Phys. Stat. Sol. (a)*, 14, K55 **(1972)**.
[3] A. N. Chakravarti, D. P. Parui, *Czech. J. Phys. B*, 23, 548 **(1973)**; A. N. Chakravarti, D. P. Parui, *Phys. Lett.*, 43A, 60 **(1973)**.
[4] A. N. Chakravarti, D. P. Parui, *Int. J. Electron.*, 34, 709 **(1973)**; A. N. Chakravarti, D. P. Parui, *Czech. J. Phys. B*, 23, 923 **(1973)**; B. R. Nag and A. N. Chakravarti, *Phys. Stat. Sol. (b)*, 60, K49 **(1973)**.
[5] A. N. Chakravarti, D. Mukherjee, *Phys. Lett.*, 53A, 403 **(1975)**; A. N. Chakravarti and S. Swaminathan, *Phys. Stat. Sol. (a)*, 23, K191 **(1974)**.
[6] A. N. Chakravarti, K. P. Ghatak, K. K. Ghosh and A. Dhar, *Phys. Stat. Sol. (b)*, 103, K55 **(1981)**; A. N. Chakravarti, D. P. Parui, *Int. J. Electron.*, 36, 275 **(1974)**; A. N. Chakravarti, B. R. Nag, *Phys. Stat. Sol. (a)*, 22, K153 **(1974)**; A. N. Chakravarti, S. Swaminathan, *Phys. Stat. Sol. (a)*, 25, K47 **(1974)**.
[7] B. R. Nag, A. N. Chakravarti and P. K. Basu, *Phys. Stat. Sol. (a)*, 68, K75 **(1981)**; B. R. Nag and A. N. Chakravarti, *Phys. Stat. Sol. (a)*, 67, K113 **(1981)**; A. N. Chakravarti and D. P. Parui, *Phys. Lett.*, 40A, 113 **(1972)**; A. N. Chakravarti, D. Mukherjee, *Phys. Lett.*, 53A, 57 **(1975)**; B. R. Nag and A. N. Chakravarti, *Sol. Stat. Sol. B*, 18, 109 **(1975)**; A. N. Chakravarti, S. Swaminathan, *Czech. J. Phys. B*, 25, 712 **(1975)**.
[8] A. N. Chakravarti and D. P. Parui, *Canad. J. Phys.*, 51, 451 **(1973)**; D. Mukherjee, A. N. Chakravarti and B. R. Nag, *Phys. Stat. Sol. (a)*, 26, K27 **(1974)**; A. N. Chakravarti, *Czech. J. Phys. B*, 25, 778 **(1975)**.
[9] S. Ghosh and A. N. Chakravarti, *Phys. Stat. Sol. (b)*, 147, 355 **(1988)**; A. N. Chakravarti, K. P. Ghatak, K. K. Ghosh, H. M. Mukherjee, *Phys. Stat. Sol. (b)*, 118, 843 **(1983)**.

[10] A. N. Chakravarti, K. P. Ghatak, A. Dhar, K. K. Ghosh, *Czech. J. Phys.*, 33 (1), 65–69 **(1983)**; B. R. Nag, A. N. Chakravarti, *Phys. Sol. Stat. Sol. (b)*, 1, K45 **(1975)**.
[11] A. N. Chakravarti, K. P. Ghatak, K. K. Ghosh, S. Ghosh, A. Dhar, *Zeitschrift für Physik B Condensed Matter*, 47, 149 **(1983)**; D. R. Choudhury, P. K. Basu, A. N. Chakravarti, *Phys. Stat. Sol. (a)*, 36, K65 **(1976)**.
[12] A. N. Chakravarti, K. P. Ghatak, K. K. Ghosh, G. B. Rao, *Phys. Stat. Sol. (b)*, 111, K61 **(1983)**; D. R. Choudhury, P. K. Basu, A. N. Chakravarti, *Phys. Stat. Sol. (a)*, 38, K35 **(1976)**.
[13] A. N. Chakravarti, K. P. Ghatak, S. Ghosh, A. K. Chowdhury, *Phys. Stat. Sol. (b)*, 109, 705 **(1982)**; A. K. Chowdhury, A. N. Chakravarti, *Czech. J. Phys. B*, 28, 228 **(1978)**; A. N. Chakravarti, A. K. Chowdhury, *Phys. Stat. Sol. (b)*, 1, K31 **(1977)**.
[14] A. N. Chakravarti, K. P. Ghatak, K. K. Ghosh, S. Ghosh, H. M. Mukherjee, *Phys. Stat. Sol. (b)*, 108, 609 **(1981)**; A. N. Chakravarti, A. K. Chowdhury, S. Ghosh, K. P. Ghatak, A. Dhar, *Czech. J. Phys. B*, 31, 1037 **(1981)**; A. N. Chakravarti, A. K. Chowdhury, K. P. Ghatak, A. Dhar, D. R. Choudhury, *Czech. J. Phys. B*, 31, 905 **(1981)**.
[15] A. N. Chakravarti, K. P. Ghatak, A. Dhar, K. K. Ghosh, S. Ghosh, *Acta Phys. Polonica A*, 60, 151 **(1981)**; A. N. Chakravarti, K. P. Ghatak, A. Dhar, S. Ghosh, *Phys. Stat. Sol. (b)*, 105, 691 **(1981)**; K. P. Ghatak, A. K. Chowdhury, S. Ghosh, A. N. Chakravarti, *Appl. Phys.*, 23, 241 **(1980)**.
[16] A. N. Chakravarti, A. K. Chowdhury, K. P. Ghatak, D. R. Choudhury, *Czech. J. Phys. B*, 30, 1161 **(1980)**; K. P. Ghatak, A. K. Chowdhury, S. Ghosh, A. N. Chakravarti, *Acta Phys. Polonica A*, 58, 389 **(1980)**; K. P. Ghatak, A. K. Chowdhury, S. Ghosh, A. N. Chakravarti, *Czech. J. Phys. B*, 30, 925 **(1980)**.
[17] A. N. Chakravarti, A. K. Chowdhury, K. P. Ghatak, D. R. Choudhury, *Phys. Stat. Sol. (a)*, 59, K211 **(1980)**; A. N. Chakravarti, A. K. Chowdhury, K. P. Ghatak, D. R. Choudhury, *Acta Phys. Polonica A*, 58, 251 **(1980)**.
[18] C. Majumdar, A. B. Maity, A. N. Chakravarti, *Phys. Stat. Sol. (b)*, 140, K7 **(1987)**; C. Majumdar, A. B. Maity, A. N. Chakravarti, *Phys. Stat. Sol. (b)*, 141, K35 **(1987)**; N. R. Das, K. K. Ghosh, D. Ghoshal, *Phys. Stat. Sol. (b)*, 197, 97 **(1996)**; A. N. Chakravarti, A. K. Chowdhury, *Czech. J. Phys.*, 27, 1161 **(1977)**; A. K. Chowdhury, A. N. Chakravarti, *Phys. Stat. Sol. (a)*, 44, K111 **(1977)**.
[19] C. Majumdar, A. B. Maity, A. N. Chakravarti, *Phys. Stat. Sol. (b)*, 144, K13 **(1987)**; N. R. Das, A. N. Chakravarti, *Phys. Stat. Sol. (b)*, 176, 335 **(1993)**; S. Sen, N. R. Das and A. N. Chakravarti, *J. Phys. Conden. Mat.*, 19, 186205 **(2007)**; K. P. Ghatak, A. K. Chowdhury, S. Ghosh, A. N. Chakravarti, *Phys. Stat. Sol. (b)*, 99, K55 **(1980)**.
[20] N. R. Das, S. Ghosh, A. N. Chakravarti, *Phys. Stat. Sol. (b)*, 174, 45 **(1992)**; A. B. Maity, C. Majumdar, A. N. Chakravarti, *Phys. Stat. Sol. (b)*, 144, K93 **(1987)**; A. B. Maity, C. Majumdar, A. N. Chakravarti, *Phys. Stat. Sol. (b)*, 149, 565 **(1988)**; K. P. Ghatak, A. K. Chowdhury, S. Ghosh, A. N. Chakravarti, *Appl. Phys.*, 23, 241 **(1980)**.

[21] S. Bhattacharya, D. De, S. M. Adhikari, K. P. Ghatak, *Superlatt. Microstruct.*, 51, 203 **(2012)**; S. Bhattacharya, D. De, S. M. Adhikari, S. Saha, K. M. Chatterjee, S. Choudhury, K. P. Ghatak, *Superlatt. Microstruct.*, 50, 609 **(2011)**; S. Bhattacharya, S. Choudhury, K. P. Ghatak, *Superlatt. Microstruct.*, 48, 257 **(2010)**.

[22] D. De, A. Kumar, S. M. Adhikari, S. Pahari, N. Islam, P. Banerjee, S. K. Biswas, S. Bhattacharya, K. P. Ghatak, *Superlatt. Microstruct.*, 47, 277 **(2010)**; S. Pahari, S. Bhattacharya, S. Roy, A. Saha, D. De, K. P. Ghatak, *Superlatt. Microstruct.*, 46, 760 **(2009)**; K. P. Ghatak, S. Bhattacharya, S. Pahari, D. De, R. Benedictus, *Superlatt. Microstruct.*, 46, 387 **(2009)**.

[23] S. Debbarma, K. P. Ghatak, *Nanosci. Nanotech.*, 16, 1095 **(2016)**; M. Chakraborty, K. P. Ghatak, *Quant. Matt.*, 4, 104 **(2015)**; S. M. Adhikari, A. Sakar, K. P. Ghatak, *Quant. Matt.*, 2, 455 **(2013)**.

[24] P. K. Bose, N. Paitya, S. Bhattacharya, D. De, S. Saha, K. M. Chatterjee, S. Pahari, K. P. Ghatak, *Quant. Matt.*, 1, 89 **(2012)**; S. Pahari, S. Bhattacharya, K. P. Ghatak, *Comput. Theor. Nanosci.*, 6, 2088 **(2009)**; S. K. Biswas, A. R. Ghatak, A. Neogi, A. Sharma, S. Bhattacharya, K. P. Ghatak, *Phys. E: Low-dimen. Syst. Nanostruct.*, 36, 263 **(2007)**.

[25] L. J. Singh, S. Choudhury, D. Baruah, S. K. Biswas, S. Pahari, K. P. Ghatak, *Phys. B: Conden. Matter*, 368, 188 **(2005)**; S. Chowdhary, L. J. Singh, K. P. Ghatak, *Phys. B: Conden. Matter*, 365, 5 **(2005)**; L. J. Singh, S. Choudhary, A. Mallik, K. P. Ghatak, *Comput. Theor. Nanosci.*, 2, 287 **(2005)**.

[26] H. X. Jiang, J. Y. Lin, *J. Appl. Phys.*, 61, 624 **(1987)**.

[27] H. Sasaki, *Phys. Rev. B*, 30, 7016 **(1984)**.

Chapter 8

The Einstein's Photoemission from QWSLs Under Magnetic Quantization

To know is to know that to know is to know and not to know is to know.

8.1 Introduction

In recent years there has been considerable interest in investigating various physical properties of quantum well superlattices (QWSLs) [1–11]. In this chapter, the magneto EP from III–V, II–VI, IV–VI, HgTe/CdTe and strained layer quantum well superlattices (QWSLs) with graded interfaces has been studied in Sections 8.2.1–8.2.5. From Sections 8.2.6 to 8.2.10, the magneto EP from III–V, II–VI, IV–VI, HgTe/CdTe and strained layer quantum well effective mass superlattices respectively has been presented. Section 8.3 contains the result and discussions pertinent to this chapter. The last Section 8.4 presents 14 open research problems.

8.2 Theoretical Background

8.2.1 The EP from III–V QWSLs with graded interfaces under magnetic quantization

In the presence of a quantizing magnetic field B along z-direction, the simplified dispersion relation in this case can be written using (7.1) as

$$\frac{n_z\pi}{d_z} = \frac{1}{L_0}\left[\rho(n, E_{6\text{SLMSQ}}) - \left\{\frac{2|e|B}{\hbar}L_0^2\left(n+\frac{1}{2}\right)\right\}\right]^{1/2} \tag{8.1}$$

where

$$\rho(n, E_{6SLMSQ}) = \left[\cos^{-1}\left[\frac{1}{2}\Phi(n, E_{6SLMSQ})\right]\right]^2$$

$$\Phi(E_{6SLMSQ}, n) \equiv \Big[2\cosh\{\beta(E_{6SLMSQ}, n)\}\cos\{\gamma(E_{6SLMSQ}, n)\}$$
$$+ \varepsilon(E_{6SLMSQ}, n)\sinh\{\beta(E_{6SLMSQ}, n)\}\sin\{\gamma(E_{6SLMSQ}, n)\}$$
$$+ \Delta_0\left[\left(\frac{K_1^2(E_{6SLMSQ}, n)}{K_2(E_{6SLMSQ}, n)} - 3K_2(E_{6SLMSQ}, n)\right)\right.$$
$$\times \cosh\{\beta(E_{6SLMSQ}, n)\}\sin\{\gamma(E_{6SLMSQ}, n)\} + \Big(3K_1(E_{6SLMSQ}, n)$$
$$\left. - \frac{\{K_2(E_{6SLMSQ}, n)\}^2}{K_1(E_{6SLMSQ}, n)}\Big)\sinh\{\beta(E_{6SLMSQ}, n)\}\cos\{\gamma(E_{6SLMSQ}, n)\}\right]$$
$$+ \Delta_0\Big[2(\{K_1(E_{6SLMSQ}, n)\}^2 - \{K_2(E_{6SLMSQ}, n)\}^2)$$
$$\times \cosh\{\beta(E_{6SLMSQ}, n)\}\cos\{\gamma(E_{6SLMSQ}, n)\}$$
$$+ \frac{1}{12}\left[\frac{5\{K_2(E_{6SLMSQ}, n)\}^3}{K_1(E_{6SLMSQ}, n)} + \frac{5\{K_1(E_{6SLMSQ}, n)\}^3}{K_2(E_{6SLMSQ}, n)}\right.$$
$$\left. - 34K_2(E_{6SLMSQ}, n)K_1(E_{6SLMSQ}, n)\right]\sinh\{\beta(E_{6SLMSQ}, n)\}$$
$$\times \sin\{\gamma(E_{6SLMSQ}, n)\}\Big]\Big]$$

$$\varepsilon(E_{6SLMSQ}, n) \equiv \left[\frac{K_1(E_{6SLMSQ}, n)}{K_2(E_{6SLMSQ}, n)} - \frac{K_2(E_{6SLMSQ}, n)}{K_1(E_{6SLMSQ}, n)}\right],$$

$$\beta(E_{6SLMSQ}, n) \equiv K_1(E_{6SLMSQ}, n)[a_0 - \Delta_0]$$

$$\gamma(E_{6SLMSQ}, n) = K_2(E_{6SLMSQ}, n)[b_0 - \Delta_0]$$

$$K_1(E_{6SLMSQ}, n) \equiv \left[\frac{2m_2^* E'}{\hbar^2}G(E_{6SLMSQ} - V_0, \alpha_2, \Delta_2)\right.$$
$$\left. + \left(\frac{2eB}{\hbar}\right)\left(n + \frac{1}{2}\right)\right]^{1/2},$$

$$K_2(E_{6SLMSQ}, n) \equiv \left[\frac{2m_1^* E}{\hbar^2}G(E_{6SLMSQ}, \alpha_1, \Delta_1) - \left(\frac{2eB}{\hbar}\right)\left(n + \frac{1}{2}\right)\right]^{1/2}$$

and $E_{6\text{SLMSQ}}$ is the total quantized energy in this case.

The DOS function in this case can be written as

$$N(E) = \frac{eB}{\pi\hbar}\sum_{n=0}^{n_{\max}}\sum_{n_z=1}^{n_{z\max}}\delta'(E - E_{6\text{SLMSQ}}) \tag{8.2}$$

The electron concentration assumes the form

$$n_0 = \frac{eB}{\pi\hbar}\sum_{n=0}^{n_{\max}}\sum_{n_z=1}^{n_{z\max}}F_{-1}(\eta_{6\text{SLMSQ}}) \tag{8.3}$$

where $\eta_{6\text{SLMSQ}} = (k_BT)^{-1}[E_{F\text{SLMSQ}} - E_{6\text{SLMSQ}}]$ in which $E_{F\text{SLMSQ}}$ is the Fermi energy in this case.

Thus, the photo-emitted current density from III–V quantum well superlattices with graded interfaces under magnetic quantization can be written as

$$J_{\text{PHOTO}} = \frac{\alpha_0 e^2 BL_0 g_v}{2\pi\hbar^2 d_z}\sum_{n=0}^{n_{\max}}\sum_{n_{z_{\min}}}^{n_{z\max}}F_{-1}(\eta_{6\text{SLMSQ}})R_{31}(E_{n_{z1}}, n_z) \tag{8.4}$$

where

$$R_{31}(E_{n_{z1}}, n_z) = \left|\frac{\sin\left(\frac{L_0\pi n_z}{d_z}\right)}{\Phi'(E_{n_{z1}}, 0)}\right|$$

in which

$$\Phi'(E_{n_{z1}}, 0) = \frac{\partial}{\partial E_{n_{z1}}}[\Phi(E_{n_{z1}}, 0)],$$

$$\begin{aligned}
\Phi(E_{n_{z1}}, 0) \equiv \Bigg[&2\cosh\{\beta(E_{n_{z1}}, 0)\}\cos\{\gamma(E_{n_{z1}}, 0)\} + \varepsilon(E_{n_{z1}}, 0)\\
&\times \sinh\{\beta(E_{n_{z1}}, 0)\}\sin\{\gamma(E_{n_{z1}}, 0)\}\\
&+ \Delta_0\left[\left(\frac{K_1^2(E_{n_{z1}}, 0)}{K_2(E_{n_{z1}}, 0)} - 3K_2(E_{n_{z1}}, 0)\right)\cosh\{\beta(E_{n_{z1}}, 0)\}\right.\\
&\times \sin\{\gamma(E_{n_{z1}}, 0)\} + \left(3K_1(E_{n_{z1}}, 0) - \frac{\{K_2(E_{n_{z1}}, 0)\}^2}{K_1(E_{n_{z1}}, 0)}\right)\\
&\left.\times \sinh\{\beta(E_{n_{z1}}, 0)\}\cos\{\gamma(E_{n_{z1}}, 0)\}\right] + \Delta_0\Big[2(\{K_1(E_{n_{z1}}, 0)\}^2\\
&- \{K_2(E_{n_{z1}}, 0)\}^2)\cosh\{\beta(E_{n_{z1}}, 0)\}\cos\{\gamma(E_{n_{z1}}, 0)\}
\end{aligned}$$

$$+\frac{1}{12}\left[\frac{5\{K_2(E_{n_{z1}},0)\}^3}{K_1(E_{n_{z1}},0)}+\frac{5\{K_1(E_{n_{z1}},0)\}^3}{K_2(E_{n_{z1}},0)}\right.$$

$$\left.\left.-\;34K_2(E_{n_{z1}},0)K_1(E_{n_{z1}},0)\right]\sinh\{\beta(E_{n_{z1}},0)\}\sin\{\gamma(E_{n_{z1}},0)\}\right]\Bigg]$$

$$\varepsilon(E_{n_{z1}},0)\equiv\left[\frac{K_1(E_{n_{z1}},0)}{K_2(E_{n_{z1}},0)}-\frac{K_2(E_{n_{z1}},0)}{K_1(E_{n_{z1}},0)}\right],$$

$$\beta(E_{n_{z1}},0)\equiv K_1(E_{n_{z1}},0)[a_0-\Delta_0]\ \gamma(E_{n_{z1}},0)=K_2(E_{n_{z1}},0)[b_0-\Delta_0],$$

$$K_1(E_{n_{z1}},0)\equiv\left[\frac{2m_2^*(V_0-E_{n_{z1}})}{\hbar^2}G(E_{n_{z1}}-V_0,\alpha_2,\Delta_2)\right]^{1/2},$$

$$K_2(E_{n_{z1}},0)\equiv\left[\frac{2m_1^*E_{n_{z1}}}{\hbar^2}G(E_{n_{z1}},\alpha_1,\Delta_1)\right]^{1/2}$$

$E_{n_{z1}}$ can be determine from the equation

$$\cos(L_0n_z\pi/d_z)=\frac{1}{2}\Phi(E_{n_{z1}},0)\tag{8.5}$$

and $n_{z_{\min}}$ is the nearest integer of the following inequality

$$n_{z_{\min}}\geq\left(\frac{d_z}{\pi L_0}\right)\cos^{-1}\left[\frac{1}{2}\Phi((W-hv),0)\right]\tag{8.6}$$

8.2.2 The EP from II–VI QWSLs with graded interfaces under magnetic quantization

In the presence of a quantizing magnetic field B along z-direction, the simplified dispersion relation can be written using (7.9) as

$$\left(\frac{n_z\pi}{d_z}\right)^2=\frac{1}{L_0^2}\left[\rho_1(n,E_{7\text{SLMSQ}})-\left\{\frac{2|e|B}{\hbar}L_0^2\left(n+\frac{1}{2}\right)\right\}\right]\tag{8.7}$$

where

$$\rho_1(n,E_{7\text{SLMSQ}})=\left[\cos^{-1}\left\{\frac{1}{2}\psi_1(n,E_{7\text{SLMSQ}})\right\}\right]^2,$$

$$\psi_1(n,E_{7\text{SLMSQ}})=\Big[2\cosh\{\beta_1(n,E_{7\text{SLMSQ}})\}\cos\{\gamma_1(n,E_{7\text{SLMSQ}})\}$$

$$+\varepsilon_1(n,E)\sinh\{\beta_1(n,E_{7\text{SLMSQ}})\}\sin\{\gamma_1(n,E_{7\text{SLMSQ}})\}$$

$$
\begin{aligned}
&+\Delta_0\left[\left(\frac{\{K_3(n,E_{7\mathrm{SLMSQ}})\}^2}{K_4(n,E_{7\mathrm{SLMSQ}})}-3K_4(n,E_{7\mathrm{SLMSQ}})\right)\right.\\
&\times\cosh\{\beta_1(n,E_{7\mathrm{SLMSQ}})\}\sin\{\gamma_1(n,E_{7\mathrm{SLMSQ}})\}\\
&+\left(3K_3(n,E_{7\mathrm{SLMSQ}})-\frac{\{K_4(n,E_{7\mathrm{SLMSQ}})\}^2}{K_3(n,E_{7\mathrm{SLMSQ}})}\right)\sinh\{\beta_1(n,E_{7\mathrm{SLMSQ}})\}\\
&\left.\times\cos\{\gamma_1(n,E_{7\mathrm{SLMSQ}})\}\right]+\Delta_0\left[2(K_3(n,E_{7\mathrm{SLMSQ}})\right.\\
&-\{K_4(n,E_{7\mathrm{SLMSQ}})\}^2)\cosh\{\beta_1(n,E_{7\mathrm{SLMSQ}})\}\cos\{\gamma_1(n,E_{7\mathrm{SLMSQ}})\}\\
&+\frac{1}{12}\left[\frac{5\{K_3(n,E_{7\mathrm{SLMSQ}})\}^3}{K_4(n,E_{7\mathrm{SLMSQ}})}+\frac{5\{K_4(n,E_{7\mathrm{SLMSQ}})\}^3}{K_3(n,E_{7\mathrm{SLMSQ}})}\right.\\
&\left.-\{34K_4(n,E_{7\mathrm{SLMSQ}})K_3(n,E_{7\mathrm{SLMSQ}})\}\right]\sinh\{\beta_1(n,E_{7\mathrm{SLMSQ}})\}\\
&\left.\left.\times\sin\{\gamma_1(n,E_{7\mathrm{SLMSQ}})\}\right]\right]
\end{aligned}
$$

$$
\varepsilon_1(n,E_{7\mathrm{SLMSQ}})\equiv\left[\frac{K_3(n,E_{7\mathrm{SLMSQ}})}{K_4(n,E_{7\mathrm{SLMSQ}})}-\frac{K_4(n,E_{7\mathrm{SLMSQ}})}{K_3(n,E_{7\mathrm{SLMSQ}})}\right],
$$

$$
\beta_1(n,E_{7\mathrm{SLMSQ}})\equiv K_3(n,E_{7\mathrm{SLMSQ}})[a_0-\Delta_0],
$$

$$
\gamma_1(n,E_{7\mathrm{SLMSQ}})=K_4(n,E_{7\mathrm{SLMSQ}})[b_0-\Delta_0],
$$

$$
\begin{aligned}
K_3(n,E_{7\mathrm{SLMSQ}})\equiv&\left[\frac{2m_2^*}{\hbar^2}E'G(E_{7\mathrm{SLMSQ}}-V_0,\alpha_2,\Delta_2)\right.\\
&\left.+\frac{2|e|B}{\hbar}\left(n+\frac{1}{2}\right)\right]^{1/2},
\end{aligned}
$$

$$
\begin{aligned}
K_4(n,E_{7\mathrm{SLMSQ}})\equiv&\left[\frac{2m_{\|,1}^*}{\hbar^2}\left[E_{7\mathrm{SLMSQ}}-\frac{\hbar|e|B}{m_{\perp,1}^*}\left(n+\frac{1}{2}\right)\right.\right.\\
&\left.\left.\mp\bar{\lambda}_0\left\{\frac{2|e|B}{\hbar}\left(n+\frac{1}{2}\right)\right\}^{1/2}\right]\right]^{1/2}
\end{aligned}
$$

and $E_{7\mathrm{SLMSQ}}$ is the total quantized energy in this case.

The DOS function in this case can be written as

$$
N(E)=\frac{eB}{h}\sum_{n=0}^{n_{\max}}\sum_{n_z=1}^{n_{z\max}}\delta'(E-E_{7\mathrm{SLMSQ}}) \tag{8.8}
$$

The electron concentration assumes the form

$$n_0 = \frac{eB}{h} \sum_{n=0}^{n_{\max}} \sum_{n_z=1}^{n_{z\max}} F_{-1}(\eta_{7\mathrm{SLMSQ}}) \tag{8.9}$$

where $\eta_{7\mathrm{SLMSQ}} = (k_B T)^{-1}[E_{F\mathrm{SLMSQ}} - E_{7\mathrm{SLMSQ}}]$.

Thus, the photo-emitted current density from II–VI quantum well superlattices with graded interfaces under magnetic quantization can be written as

$$J_{\mathrm{PHOTO}} = \frac{\alpha_0 e^2 B L_0 g_v}{4\pi\hbar^2 d_z} \sum_{n=0}^{n_{\max}} \sum_{n_{z_{\min}}}^{n_{z\max}} F_{-1}(\eta_{7\mathrm{SLMSQ}}) R_{32}(E_{n_{z2}}, n_z) \tag{8.10}$$

where

$$R_{32}(E_{n_{z2}}, n_z) = \left| \frac{\sin\left(\frac{L_0 \pi n_z}{d_z}\right)}{\Phi_1'(E_{n_{z2}}, 0)} \right|$$

in which

$$\Phi_1'(E_{n_{z2}}, 0) = \frac{\partial}{\partial E_{n_{z2}}}[\Phi_1(E_{n_{z2}}, 0)],$$

$$\Phi_1(E_{n_{z2}}, 0) \equiv \Big[2\cosh\{\beta_1(E_{n_{z2}}, 0)\}\cos\{\gamma_1(E_{n_{z2}}, 0)\}$$
$$+ \varepsilon_1(E_{n_{z2}}, 0)\sinh\{\beta_1(E_{n_{z2}}, 0)\}\sin\{\gamma_1(E_{n_{z2}}, 0)\}$$
$$+ \Delta_0\left[\left(\frac{\{K_3(E_{n_{z2}}, 0)\}^2}{K_4(E_{n_{z2}}, 0)} - 3K_4(E_{n_{z2}}, 0)\right)\cosh\{\beta_1(E_{n_{z2}}, 0)\}\right.$$
$$\times \sin\{\gamma_1(E_{n_{z2}}, 0)\} + \left(3K_3(E_{n_{z2}}, 0) - \frac{\{K_4(E_{n_{z2}}, 0)\}^2}{K_3(E_{n_{z2}}, 0)}\right)$$
$$\left.\times \sinh\{\beta_1(E_{n_{z2}}, 0)\}\cos\{\gamma_1(E_{n_{z2}}, 0)\}\right] + \Delta_0\Big[2(\{K_3(E_{n_{z2}}, 0)\}^2$$
$$- \{K_4(E_{n_{z2}}, 0)\}^2)\cosh\{\beta_1(E_{n_{z2}}, 0)\}\cos\{\gamma_1(E_{n_{z2}}, 0)\}\Big]\Big]$$

$$\varepsilon_1(E_{n_{z2}}, 0) \equiv \left[\frac{K_3(E_{n_{z2}}, 0)}{K_4(E_{n_{z2}}, 0)} - \frac{K_4(E_{n_{z2}}, 0)}{K_3(E_{n_{z2}}, 0)}\right],$$

$$\beta_1(E_{n_{z2}}, 0) \equiv K_3(E_{n_{z2}}, 0)[a_0 - \Delta_0],$$

$$\gamma_1(E_{n_{z2}},0) = K_4(E_{n_{z2}},0)[b_0 - \Delta_0]$$

$$K_3(E_{n_{z2}},0) \equiv \left[\frac{2m_2^*(V_0 - E_{n_{z2}})}{\hbar^2} G(E_{n_{z2}} - V_0, \alpha_2, \Delta_2)\right]^{1/2} \quad \text{and}$$

$$K_4(E_{n_{z2}},0) \equiv \left[\frac{2m_{\parallel,1}^*}{\hbar^2}[E_{n_{z2}}]\right]^{1/2}$$

$E_{n_{z2}}$ can be determine from the equation

$$\cos(L_0 n_z \pi/d_z) = \frac{1}{2}\Phi_1(E_{n_{z2}},0) \tag{8.11}$$

and $n_{z_{\min}}$ is the nearest integer of the following inequality

$$n_{z_{\min}} \geq \left(\frac{d_z}{\pi L_0}\right) \cos^{-1}\left[\frac{1}{2}\Phi_1((W - hv),0)\right] \tag{8.12}$$

8.2.3 The EP from IV–VI QWSLs with graded interfaces under magnetic quantization

The simplified dispersion relations in this case can be written using (7.15) as

$$\frac{n_z \pi}{d_z} = \frac{1}{L_0}\left[\rho_2(n, E_{8\mathrm{SLMSQ}}) - \left\{\frac{2|e|B}{\hbar} L_0^2 \left(n + \frac{1}{2}\right)\right\}\right]^{1/2} \tag{8.13}$$

where

$$\rho_2(n, E_{8\mathrm{SLMSQ}}) = \left[\cos^{-1}\left\{\frac{1}{2}\psi_2(n, E_{8\mathrm{SLMSQ}})\right\}\right]^2$$

$$\begin{aligned}
\psi_2(n, E_{8\mathrm{SLMSQ}}) = \Big[& 2\cosh\{\beta_2(n, E_{8\mathrm{SLMSQ}})\}\cos\{\gamma_2(n, E_{8\mathrm{SLMSQ}})\} \\
& + \varepsilon_2(n, E_{8\mathrm{SLMSQ}})\sinh\{\beta_2(n, E_{8\mathrm{SLMSQ}})\}\sin\{\gamma_2(n, E_{8\mathrm{SLMSQ}})\} \\
& + \Delta_0\left[\left(\frac{\{K_5(n, E_{8\mathrm{SLMSQ}})\}^2}{K_6(n, E_{8\mathrm{SLMSQ}})} - 3K_6(n, E_{8\mathrm{SLMSQ}})\right)\right. \\
& \times \cosh\{\beta_2(n, E_{8\mathrm{SLMSQ}})\}\sin\{\gamma_2(n, E_{8\mathrm{SLMSQ}})\} + \Bigg(3K_5(n, E_{8\mathrm{SLMSQ}}) \\
& \left. - \frac{\{K_6(n, E_{8\mathrm{SLMSQ}})\}^2}{K_5(n, E_{8\mathrm{SLMSQ}})}\Bigg)\sinh\{\beta_2(n, E_{8\mathrm{SLMSQ}})\}\cos\{\gamma_2(n, E_{8\mathrm{SLMSQ}})\}\right] \\
& + \Delta_0\Big[2(K_5(n, E_{8\mathrm{SLMSQ}}) - \{K_6(n, E_{8\mathrm{SLMSQ}})\}^2)
\end{aligned}$$

$$\times \cosh\{\beta_2(n, E_{8\mathrm{SLMSQ}})\}\cos\{\gamma_2(n, E_{8\mathrm{SLMSQ}})\}$$

$$+\frac{1}{12}\left[\frac{5\{K_5(n, E_{8\mathrm{SLMSQ}})\}^3}{K_6(n, E_{8\mathrm{SLMSQ}})} + \frac{5\{K_6(n, E_{8\mathrm{SLMSQ}})\}^3}{K_5(n, E_{8\mathrm{SLMSQ}})}\right.$$

$$\left. - \{34K_6(n, E_{8\mathrm{SLMSQ}})K_5(n, E_{8\mathrm{SLMSQ}})\}\right]$$

$$\left.\left. \times \sinh\{\beta_2(n, E_{8\mathrm{SLMSQ}})\}\sin\{\gamma_2(n, E_{8\mathrm{SLMSQ}})\}\right]\right]$$

$$\varepsilon_2(n, E_{8\mathrm{SLMSQ}}) \equiv \left[\frac{K_5(n, E_{8\mathrm{SLMSQ}})}{K_6(n, E_{8\mathrm{SLMSQ}})} - \frac{K_6(n, E_{8\mathrm{SLMSQ}})}{K_5(n, E_{8\mathrm{SLMSQ}})}\right],$$

$$\beta_2(n, E_{8\mathrm{SLMSQ}}) \equiv K_5(n, E_{8\mathrm{SLMSQ}})[a_0 - \Delta_0],$$

$$K_6(n, E_{8\mathrm{SLMSQ}}) \equiv [[EH_{11} + H_{21}(n)] - [E^2H_{31} + EH_{41}(n)$$

$$+ H_{51}(n)]^{1/2}]^{1/2}, \quad H_{1i} = (b_i^2 - f_i^2)^{-1},$$

$$H_{2i}(n) = [2H_{1i}]^{-1}\left[E_{g_i}b_i + d_i + f_iE_{g_i} + 2(e_if_i - a_ib_i)\right.$$

$$\left.\times\left\{\frac{2|e|B}{\hbar}\left(n+\frac{1}{2}\right)\right\}\right], \quad H_{3i} = \frac{f_i^2}{(b_i^2 - f_i^2)^2}$$

$$H_{4i}(n) = [4H_{1i}^2]^{-1}\left[4b_id_i + 4b_if_iE_{g_i} + 4f_i^2E_{g_i} + 8\left\{\frac{2|e|B}{\hbar}\left(n+\frac{1}{2}\right)\right\}\right.$$

$$\left.\times [e_if_ib_i - a_if_i^2]\right]$$

$$H_{5i}(n) \equiv [4H_{1i}^2]^{-1}\left[\left\{\frac{2|e|B}{\hbar}\left(n+\frac{1}{2}\right)\right\}^2[-8a_ib_ie_if_i + 4b_i^2e_i^2 + 4f_i^2a_i^2]\right.$$

$$+\left\{\frac{2|e|B}{\hbar}\left(n+\frac{1}{2}\right)\right\}[4e_if_iE_{g_i}b_i - 4e_if_id_i + 4e_if_i^2E_{g_i} - 4a_ib_i^2E_{g_i}$$

$$- 4a_ib_id_i - 4a_ib_if_iE_{g_i} + 4b_i^2e_iE_{g_i} + 4b_i^2c_i + 4b_i^2E_{g_i}a_i - 4f_i^2e_iE_{g_i}$$

$$- 4f_i^2c_i - 4f_i^2E_{g_i}a_i] + [E_{g_i}^2b_i^2 + d_i^2 + f_i^2g_i^2 + 2E_{g_i}b_id_i + 2E_{g_i}^2b_if_i$$

$$\left. + 2d_if_iE_{g_i}]\right]$$

$$K_5(n, E_{8\mathrm{SLMSQ}}) \equiv [[(E_{8\mathrm{SLMSQ}} - V_0)^2H_{32} + (E_{8\mathrm{SLMSQ}} - V_0)H_{42}(n)$$

$$+ H_{52}(n)]^{1/2} - [(E_{8\mathrm{SLMSQ}} - V_0)H_{12} + H_{22}(n)]]^{1/2}$$

and $E_{8\mathrm{SLMSQ}}$ is the total quantized energy in this case.

The DOS function in this case can be written as

$$N(E) = \frac{eB}{\pi\hbar} \sum_{n=0}^{n_{\max}} \sum_{n_z=1}^{n_{z\max}} \delta'(E - E_{8\mathrm{SLMSQ}}) \tag{8.14}$$

The electron concentration assumes the form

$$n_0 = \frac{eB}{\pi\hbar} \sum_{n=0}^{n_{\max}} \sum_{n_z=1}^{n_{z\max}} F_{-1}(\eta_{8\mathrm{SLMSQ}}) \tag{8.15}$$

where $\eta_{8\mathrm{SLMSQ}} = (k_B T)^{-1}[E_{F\mathrm{SLMSQ}} - E_{8\mathrm{SLMSQ}}]$.

Thus, the photo-emitted current density from IV–VI quantum well superlattices with graded interfaces under magnetic quantization can be written as

$$J_{\mathrm{PHOTO}} = \frac{\alpha_0 e^2 B L_0 g_v}{2\pi\hbar^2 d_z} \sum_{n=0}^{n_{\max}} \sum_{n_{z\min}}^{n_{z\max}} F_{-1}(\eta_{8\mathrm{SLMSQ}}) R_{33}(E_{n_{z3}}, n_z) \tag{8.16}$$

where

$$R_{33}(E_{n_{z3}}, n_z) = \left| \frac{\sin\left(\frac{L_0 \pi n_z}{d_z}\right)}{\Phi_2'(E_{n_{z3}}, 0)} \right|$$

in which

$$\Phi_2'(E_{n_{z3}}, 0) = \frac{\partial}{\partial E_{n_{z3}}} [\Phi_2(E_{n_{z3}}, 0)]$$

where

$$\begin{aligned}
\Phi_2(E_{n_{z3}}, 0) \equiv & \Big[2\cosh\{\beta_2(E_{n_{z3}}, 0)\}\cos\{\gamma_2(E_{n_{z3}}, 0)\} + \varepsilon_2(E_{n_{z3}}, 0) \\
& \times \sinh\{\beta_2(E_{n_{z3}}, 0)\}\sin\{\gamma_2(E_{n_{z3}}, 0)\} + \Delta_0 \Bigg[\left(\frac{\{K_5(E_{n_{z3}}, 0)\}^2}{K_6(E_{n_{z3}}, 0)} \right. \\
& \left. - 3K_6(E_{n_{z3}}, 0) \right) \cosh\{\beta_2(E_{n_{z3}}, 0)\}\sin\{\gamma_2(E_{n_{z3}}, 0)\} + \left(3K_5(E_{n_{z3}}, 0) \right. \\
& \left. - \frac{\{K_6(E_{n_{z3}}, 0)\}^2}{K_5(E_{n_{z3}}, 0)} \right) \sinh\{\beta_2(E_{n_{z3}}, 0)\}\cos\{\gamma_2(E_{n_{z3}}, 0)\} \Bigg] \\
& + \Delta_0 \Big[2(\{K_5(E_{n_{z3}}, 0)\}^2 - \{K_6(E_{n_{z3}}, 0)\}^2)\cosh\{\beta_2(E_{n_{z3}}, 0)\} \\
& \times \cos\{\gamma_2(E_{n_{z3}}, 0)\} + \frac{1}{12}\left[\frac{5\{K_5(E_{n_{z3}}, 0)\}^3}{K_6(E_{n_{z3}}, 0)} + \frac{5\{K_6(E_{n_{z3}}, 0)\}^3}{K_5(E_{n_{z3}}, 0)} \right. \\
& \left. - 34K_6(E_{n_{z3}}, 0)K_5(E_{n_{z3}}, 0) \right] \sinh\{\beta_2(E_{n_{z3}}, 0)\}\sin\{\gamma_2(E_{n_{z3}}, 0)\} \Big] \Big]
\end{aligned}$$

$$\varepsilon_2(E_{n_{z3}},0) \equiv \left[\frac{K_5(E_{n_{z3}},0)}{K_6(E_{n_{z3}},0)} - \frac{K_6(E_{n_{z3}},0)}{K_5(E_{n_{z3}},0)}\right],$$

$$\beta_2(E_{n_{z3}},0) \equiv K_5(E_{n_{z3}},0)[a_0 - \Delta_0], \quad \gamma_2(E_{n_{z3}},0) = K_6(E_{n_{z3}},0)[b_0 - \Delta_0],$$

$$K_6(E_{n_{z3}},0) \equiv [[E_{n_{z3}}H_{11} + H_{21}(0)] - [H_{31}(E_{n_{z3}})^2 + E_{n_{z3}}H_{41}(0) + H_{51}(0)]^{1/2}]^{1/2}, \quad H_{1i} = (b_i^2 - f_i^2)^{-1}$$

and

$$H_{2i}(0) = [2H_{1i}]^{-1}[E_{g_i}b_i + d_i + f_iE_{g_i}], \quad H_{3i} = \frac{f_i^2}{(b_i^2 - f_i^2)^2},$$

$$H_{4i}(0) = [4H_{1i}^2]^{-1}[4b_id_i + 4b_if_iE_{g_i} + 4f_i^2E_{g_i}],$$

$$H_{5i}(0) \equiv [4H_{1i}^2][E_{g_i}^2b_i^2 + d_i^2 + f_i^2g_i^2 + 2E_{g_i}b_id_i + 2E_{g_i}^2b_if_i + 2d_if_iE_{g_i}]$$

and

$$K_5(E_{n_{z3}},0) \equiv [[(E_{n_{z3}} - V_0)^2H_{32} + (E_{n_{z3}} - V_0)H_{42}(0) + H_{52}(0)]^{1/2} - [(E_{n_{z3}} - V_0)H_{12} + H_{22}(0)]]^{1/2}$$

$E_{n_{z3}}$ can be determine from the equation

$$\cos(L_0n_z\pi/d_z) = \frac{1}{2}\Phi_2(E_{n_{z3}},0) \tag{8.17}$$

and $n_{z_{\min}}$ is the nearest integer of the following inequality

$$n_{z_{\min}} \geq \left(\frac{d_z}{\pi L_0}\right)\cos^{-1}\left[\frac{1}{2}\Phi_2((W - hv),0)\right] \tag{8.18}$$

*8.2.4 The EP from HgTe/CdTe QWSLs with graded interfaces under magnetic quantization

The dispersion relation in this case can be written using (7.22) as

$$\frac{n_z\pi}{d_z} = \frac{1}{L_0}\left[\rho_3(n,E_{9\text{SLMSQ}}) - \left\{\frac{2|e|B}{\hbar}L_0^2\left(n + \frac{1}{2}\right)\right\}\right]^{1/2} \tag{8.19}$$

where

$$\rho_3(n,E_{9\text{SLMSQ}}) = \left[\cos^{-1}\left\{\frac{1}{2}\psi_3(n,E_{9\text{SLMSQ}})\right\}\right]^2$$

$$\psi_3(n,E_{9\text{SLMSQ}}) = \Big[2\cosh\{\beta_3(n,E_{9\text{SLMSQ}})\}\cos\{\gamma_3(n,E_{9\text{SLMSQ}})\}$$

$$
\begin{aligned}
&+\varepsilon_3(n, E_{9\mathrm{SLMSQ}}) \sinh\{\beta_3(n, E_{9\mathrm{SLMSQ}})\} \sin\{\gamma_3(n, E_{9\mathrm{SLMSQ}})\} \\
&+\Delta_0 \left[\left(\frac{\{K_7(n, E_{9\mathrm{SLMSQ}})\}^2}{K_8(n, E_{9\mathrm{SLMSQ}})} - 3K_8(n, E_{9\mathrm{SLMSQ}})\right)\right. \\
&\times \cosh\{\beta_3(n, E_{9\mathrm{SLMSQ}})\} \sin\{\gamma_3(n, E_{9\mathrm{SLMSQ}})\} + \left(3K_7(n, E_{9\mathrm{SLMSQ}})\right. \\
&\left.\left. - \frac{\{K_8(n, E_{9\mathrm{SLMSQ}})\}^2}{K_7(n, E_{9\mathrm{SLMSQ}})}\right) \sinh\{\beta_3(n, E_{9\mathrm{SLMSQ}})\} \cos\{\gamma_3(n, E_{9\mathrm{SLMSQ}})\}\right] \\
&+\Delta_0 \left[2(\{K_7(n, E_{9\mathrm{SLMSQ}})\}^2 - \{K_8(n, E_{9\mathrm{SLMSQ}})\}^2)\right. \\
&\times \cosh\{\beta_3(n, E_{9\mathrm{SLMSQ}})\} \cos\{\gamma_3(n, E_{9\mathrm{SLMSQ}})\} \\
&+\frac{1}{12}\left[\frac{5\{K_7(n, E_{9\mathrm{SLMSQ}})\}^3}{K_8(n, E_{9\mathrm{SLMSQ}})} + \frac{5\{K_8(n, E_{9\mathrm{SLMSQ}})\}^3}{K_7(n, E_{9\mathrm{SLMSQ}})}\right. \\
&\left. -\{34K_8(n, E_{9\mathrm{SLMSQ}})K_7(n, E_{9\mathrm{SLMSQ}})\}\right] \sinh\{\beta_3(n, E_{9\mathrm{SLMSQ}})\} \\
&\left.\left.\times \sin\{\gamma_3(n, E_{9\mathrm{SLMSQ}})\}\right]\right]
\end{aligned}
$$

$$
\varepsilon_3(n, E_{9\mathrm{SLMSQ}}) \equiv \left[\frac{K_7(n, E_{9\mathrm{SLMSQ}})}{K_8(n, E_{9\mathrm{SLMSQ}})} - \frac{K_8(n, E_{9\mathrm{SLMSQ}})}{K_7(n, E_{9\mathrm{SLMSQ}})}\right],
$$

$$
\beta_3(n, E_{9\mathrm{SLMSQ}}) \equiv K_7(n, E_{9\mathrm{SLMSQ}})[a_0 - \Delta_0],
$$

$$
\gamma_3(n, E_{9\mathrm{SLMSQ}}) = K_8(n, E_{9\mathrm{SLMSQ}})[b_0 - \Delta_0],
$$

$$
K_8(n, E_{9\mathrm{SLMSQ}}) \equiv \left[\frac{B_0^2 + 2AE - B_0\sqrt{B_0^2 + 4AE}}{2A^2} - \left\{\frac{2|e|B}{\hbar}\left(n + \frac{1}{2}\right)\right\}\right]^{1/2}
$$

$$
\begin{aligned}
K_7(n, E_{9\mathrm{SLMSQ}}) \equiv &\left[\left(\frac{2m_2^* E'}{\hbar^2}\right) G(E_{9\mathrm{SLMSQ}} - V_0, \alpha_2, \Delta_2)\right. \\
&\left. + \left\{\frac{2|e|B}{\hbar}\left(n + \frac{1}{2}\right)\right\}\right]^{1/2}
\end{aligned}
$$

and $E_{9\mathrm{SLMSQ}}$ is the total quantized energy in this case.

The DOS function in this case can be written as

$$
N(E) = \frac{eB}{\pi\hbar} \sum_{n=0}^{n_{\max}} \sum_{n_z=1}^{n_{z\max}} \delta'(E - E_{9\mathrm{SLMSQ}}) \tag{8.20}
$$

The electron concentration assumes the form

$$n_0 = \frac{eB}{\pi\hbar} \sum_{n=0}^{n_{\max}} \sum_{n_z=1}^{n_{z\max}} F_{-1}(\eta_{9\text{SLMSQ}}) \tag{8.21}$$

where $\eta_{9\text{SLMSQ}} = (k_B T)^{-1}[E_{F\text{SLMSQ}} - E_{9\text{SLMSQ}}]$.

Thus, the photo-emitted current density from HgTe/CdTe quantum well superlattices with graded interfaces under magnetic quantization can be written as

$$J_{\text{PHOTO}} = \frac{\alpha_0 e^2 B L_0 g_v}{2\pi\hbar^2 d_z} \sum_{n=0}^{n_{\max}} \sum_{n_{z_{\min}}}^{n_{z\max}} F_{-1}(\eta_{9\text{SLMSQ}}) R_{34}(E_{n_{z4}}, n_z) \tag{8.22}$$

where

$$R_{34}(E_{n_{z4}}, n_z) = \left| \frac{\sin\left(\frac{L_0 \pi n_z}{d_z}\right)}{\Phi_3'(E_{n_{z4}}, 0)} \right|$$

in which

$$\Phi_3'(E_{n_{z4}}, 0) = \frac{\partial}{\partial E_{n_{z4}}}[\Phi_3(E_{n_{z4}}, 0)],$$

where

$$\begin{aligned}
\Phi_3(E_{n_{z4}}, 0) \equiv &\Bigg[2\cosh\{\beta_3(E_{n_{z4}}, 0)\}\cos\{\gamma_3(E_{n_{z4}}, 0)\} + \varepsilon_3(E_{n_{z4}}, 0) \\
&\times \sinh\{\beta_3(E_{n_{z4}}, 0)\}\sin\{\gamma_3(E_{n_{z4}}, 0)\} + \Delta_0 \Bigg[\Bigg(\frac{\{K_7\gamma_3(E_{n_{z4}}, 0)\}^7}{K_8\gamma_3(E_{n_{z4}}, 0)} \\
&- 3K_8\gamma_3(E_{n_{z4}}, 0)\Bigg)\cosh\{\beta_3(E_{n_{z4}}, 0)\}\sin\{\gamma_3(E_{n_{z4}}, 0)\} \\
&+ \left(3K_7(E_{n_{z4}}, 0) - \frac{\{K_8(E_{n_{z4}}, 0)\}^2}{K_7(E_{n_{z4}}, 0)}\right)\sinh\{\beta_3(E, k_s)\}\cos\{\gamma_3(E, k_s)\}\Bigg] \\
&+ \Delta_0 \Bigg[2(\{K_7(E_{n_{z4}}, 0)\}^2 - \{K_8(E_{n_{z4}}, 0)\}^2)\cosh\{\beta_3(E_{n_{z4}}, 0)\} \\
&\times \cos\{\gamma_3(E_{n_{z4}}, 0)\} + \frac{1}{12}\Bigg[\frac{5\{K_8(E_{n_{z4}}, 0)\}^3}{K_7(E_{n_{z4}}, 0)} + \frac{5\{K_7(E_{n_{z4}}, 0)\}^3}{K_8(E_{n_{z4}}, 0)} \\
&-34K_7(E_{n_{z4}}, 0)K_8(E_{n_{z4}}, 0)\Bigg]\sinh\{\beta_3(E_{n_{z4}}, 0)\}\sin\{\gamma_3(E_{n_{z4}}, 0)\}\Bigg]\Bigg]
\end{aligned}$$

$$\varepsilon_3(E_{n_{z4}}, 0) \equiv \left[\frac{K_7(E_{n_{z4}}, 0)}{K_8(E_{n_{z4}}, 0)} - \frac{K_8(E_{n_{z4}}, 0)}{K_7(E_{n_{z4}}, 0)}\right],$$

$$\beta_3(E_{n_{z4}},0) \equiv K_7(E_{n_{z4}},0)[a_0 - \Delta_0], \quad \gamma_3(E_{n_{z4}},0) = K_8(E_{n_{z4}},0)[b_0 - \Delta_0],$$

$$K_8(E_{n_{z4}},0) \equiv \left[\frac{B_0^2 + 2AE_{n_{z4}} - B_0\sqrt{B_0^2 + 4AE_{n_{z4}}}}{2A^2}\right]^{1/2},$$

and

$$K_7(E_{n_{z4}},0) \equiv \left[\frac{2m_2^*(V_0 - E_{n_{z4}})}{\hbar^2} G(E_{n_{z4}} - V_0, E_{g_2}, \Delta_2)\right]^{1/2}$$

$E_{n_{z4}}$ can be determine from the equation

$$\cos(L_0 n_z \pi/d_z) = \frac{1}{2}\Phi_3(E_{n_{z4}},0) \tag{8.23}$$

and $n_{z_{\min}}$ is the nearest integer of the following inequality

$$n_{z_{\min}} \geq \left(\frac{d_z}{\pi L_0}\right)\cos^{-1}\left[\frac{1}{2}\Phi_3((W - hv),0)\right] \tag{8.24}$$

8.2.5 The EP from strained layer QWSLs with graded interfaces under magnetic quantization

The dispersion relation in this case can be written using (7.29) as

$$\left(\frac{n_z\pi}{d_z}\right)^2 = \omega_{60}(E_{10\mathrm{SLMSQ}}, n) \tag{8.25}$$

where

$$\omega_{60}(E_{10\mathrm{SLMSQ}}, n) = \frac{1}{L_0^2}\left[\cos^{-1}\left(\frac{1}{2}\phi_4(E,k_s)\right)\right]^2 - \frac{2eB}{\hbar}\left(n + \frac{1}{2}\right)$$

$$\Phi_4(E_{10\mathrm{SLMSQ}}, n) = [2\cosh[T_4(E_{10\mathrm{SLMSQ}}, n)]\cos[T_5(E_{10\mathrm{SLMSQ}}, n)]]$$

$$+ \Bigg[[T_6(E_{10\mathrm{SLMSQ}}, n)]\sinh[T_4(E_{10\mathrm{SLMSQ}}, n)]\sin[T_5(E_{10\mathrm{SLMSQ}}, n)]$$

$$+\Delta_0\left[\left(\frac{k_0^2(E_{10\mathrm{SLMSQ}}, n)}{k'(E_{10\mathrm{SLMSQ}}, n)} - 3k'(E_{10\mathrm{SLMSQ}}, n)\right)\cosh[T_4(E_{10\mathrm{SLMSQ}}, n)]\right.$$

$$\times \sin[T_5(E_{10\mathrm{SLMSQ}}, n)] + \left(3k_0(E_{10\mathrm{SLMSQ}}, n) - \frac{k'^2(E_{10\mathrm{SLMSQ}}, n)}{k_0(E_{10\mathrm{SLMSQ}}, n)}\right)$$

$$\left.\times \sinh[T_4(E_{10\mathrm{SLMSQ}}, n)]\sin[T_5(E_{10\mathrm{SLMSQ}}, n)]\right] + \Delta_0\Big[2(k_0^2(E_{10\mathrm{SLMSQ}}, n)$$

$$- k'^2(E_{10\mathrm{SLMSQ}}, n))\cosh[T_4(E_{10\mathrm{SLMSQ}}, n)]\cos[T_5(E_{10\mathrm{SLMSQ}}, n)]$$

$$+\frac{1}{12}\left(\frac{5k_0^3(E_{10\mathrm{SLMSQ}},n)}{k'(E_{10\mathrm{SLMSQ}},n)}+\frac{5k'^3(E_{10\mathrm{SLMSQ}},n)}{k_0^3(E_{10\mathrm{SLMSQ}},n)}-34k_0(E_{10\mathrm{SLMSQ}},n)\right.$$

$$\left.\left.\left.\times\ k'(E_{10\mathrm{SLMSQ}},n)\right)\right]\sinh[T_4(E_{10\mathrm{SLMSQ}},n)]\sin[T_5(E_{10\mathrm{SLMSQ}},n)]\right]$$

$$[T_4(E_{10\mathrm{SLMSQ}},n)]=k_0(E_{10\mathrm{SLMSQ}},n)[a_0-\Delta_0]$$

$$k_0(E_{10\mathrm{SLMSQ}},n)=[\bar{S}_2(E_{10\mathrm{SLMSQ}}-V_0)]^{-1/2}\left[\left(n+\frac{1}{2}\right)\hbar\omega_{08}-1\right]^{1/2},$$

$$\omega_{08}=\frac{eB}{\sqrt{m_{18}m_{28}}},\quad m_{18}=\frac{\hbar}{\sqrt{2\bar{P}_2(E_{10\mathrm{SLMSQ}}-V_0)}}$$

$$m_{28}=\frac{\hbar}{\sqrt{2\bar{Q}_2(E_{10\mathrm{SLMSQ}}-V_0)}},$$

$$T_5(E_{10S\mathrm{LMSQ}},n)=k'(E_{10\mathrm{SLMSQ}},n)[b_0-\Delta_0],$$

$$k'(E_{10\mathrm{SLMSQ}},n)=[\bar{S}_1(E_{10\mathrm{SLMSQ}}-V_0)]^{-1/2}\left[1-\left(n+\frac{1}{2}\right)\hbar\omega_{081}\right]^{1/2},$$

$$\omega_{081}=\frac{eB}{\sqrt{m_{181}m_{281}}},\quad m_{181}=\frac{\hbar}{\sqrt{2\bar{P}_1(E_{10\mathrm{SLMSQ}})}}$$

$$m_{281}=\frac{\hbar}{\sqrt{2\bar{Q}_1(E_{10\mathrm{SLMSQ}})}},$$

$$T_5(E_{10\mathrm{SLMSQ}},n)=k'(E_{10\mathrm{SLMSQ}},n)[b_0-\Delta_0],$$

and

$$T_6(E_{10\mathrm{SLMSQ}},n)=\left[\frac{k_0(E_{10\mathrm{SLMSQ}},n)}{k'(E_{10\mathrm{SLMSQ}},n)}-\frac{k'(E_{10\mathrm{SLMSQ}},n)}{k_0(E_{10\mathrm{SLMSQ}},n)}\right]$$

and $E_{10\mathrm{SLMSQ}}$ is the total quantized energy in this case.

The DOS function in this case can be written as

$$N(E)=\frac{eB}{\pi\hbar}\sum_{n=0}^{n_{\max}}\sum_{n_z=1}^{n_{z\max}}\delta'(E-E_{10\mathrm{SLMSQ}}) \tag{8.26}$$

The electron concentration assumes the form

$$n_0=\frac{eB}{\pi\hbar}\sum_{n=0}^{n_{\max}}\sum_{n_z=1}^{n_{z\max}}F_{-1}(\eta_{10\mathrm{SLMSQ}}) \tag{8.27}$$

where $\eta_{10\mathrm{SLMSQ}}=(k_BT)^{-1}[E_{F\mathrm{SLMSQ}}-E_{10\mathrm{SLMSQ}}]$.

Thus, the photo-emitted current density from strained layer quantum well superlattices with graded interfaces under magnetic quantization can be written as

$$J_{\mathrm{PHOTO}} = \frac{\alpha_0 e^2 B L_0 g_v}{2\pi\hbar^2 d_z} \sum_{n=0}^{n_{\max}} \sum_{n_{z_{\min}}}^{n_{z_{\max}}} F_{-1}(\eta_{10\mathrm{SLMSQ}}) R_{35}(E_{n_{z5}}, n_z) \tag{8.28}$$

where

$$R_{35}(E_{n_{z4}}, n_z) = \left| \frac{\sin\left(\frac{L_0 \pi n_z}{d_z}\right)}{\Phi_4'(E_{n_{z5}}, 0)} \right|$$

in which

$$\Phi_4'(E_{n_{z5}}, 0) = \frac{\partial}{\partial E_{n_{z5}}}[\Phi_4(E_{n_{z5}}, 0)],$$

where

$$\begin{aligned}
\Phi_4(E_{n_{z5}}, 0) &= [2\cosh[T_4(E_{n_{z5}}, 0)]\cos[T_5(E_{n_{z5}}, 0)]] + \Bigg[[T_6(E_{n_{z5}}, 0)] \\
&\times \sinh[T_4(E_{n_{z5}}, 0)]\sin[T_5(E_{n_{z5}}, 0)] + \Delta_0 \left[\left(\frac{k_0^2(E_{n_{z5}}, 0)}{k'(E_{n_{z5}}, 0)} - 3k'(E_{n_{z5}}, 0)\right)\right. \\
&\times \cosh[T_4(E_{n_{z5}}, 0)]\sin[T_5(E_{n_{z5}}, 0)] + \left(3k_0(E_{n_{z5}}, 0) - \frac{k'^2(E_{n_{z5}}, 0)}{k_0(E_{n_{z5}}, 0)}\right) \\
&\left.\times \sinh[T_4(E_{n_{z5}}, 0)]\sin[T_5(E_{n_{z5}}, 0)]\right] + \Delta_0 \Bigg[2(k_0^2(E_{n_{z5}}, 0) - k'^2(E_{n_{z5}}, 0)) \\
&\times \cosh[T_4(E_{n_{z5}}, 0)]\cos[T_5(E_{n_{z5}}, 0)] + \frac{1}{12}\left(\frac{5k_0^3(E_{n_{z5}}, 0)}{k'(E_{n_{z5}}, 0)} + \frac{5k'^3(E_{n_{z5}}, 0)}{k_0^3(E_{n_{z5}}, 0)}\right. \\
&\left.\left. - 34k_0(E_{n_{z5}}, 0)k'(E_{n_{z5}}, 0)\right)\right] \sinh[T_4(E_{n_{z5}}, 0)]\sin[T_5(E_{n_{z5}}, 0)]\Bigg]
\end{aligned}$$

$$[T_4(E_{n_{z5}}, 0)] = k_0(E_{n_{z5}}, 0)[a_0 - \Delta_0]$$

$$k_0(E_{n_{z5}}, 0) = [(\omega_2, E_{n_{z5}})]^{1/2}[V + T_{32} - E_{n_{z5}}]^{-1/2},$$

$$T_5(E_{n_{z5}}, 0) = k'(E_{n_{z5}}, 0)[b_0 - \Delta_0], \quad k'(E_{n_{z5}}, 0) = [\bar{S}_1(E_{n_{z5}})]^{1/2}$$

and

$$T_6(E_{n_{z5}},0) = \left[\frac{k_0(E_{n_{z5}},0)}{k'(E_{n_{z5}},0)} - \frac{k'(E_{n_{z5}},0)}{k_0(E_{n_{z5}},0)}\right]$$

$E_{n_{z5}}$ can be determine from the equation

$$\cos(L_0 n_z \pi / d_z) = \frac{1}{2}\Phi_4(E_{n_{z5}},0) \tag{8.29}$$

and $n_{z_{\min}}$ is the nearest integer of the following inequality

$$n_{z_{\min}} \geq \left(\frac{d_z}{\pi L_0}\right)\cos^{-1}[\Phi_4((W - h\upsilon),0)] \tag{8.30}$$

8.2.6 The EP from III–V QW effective mass superlattices under magnetic quantization

In the presence of an external magnetic field along x-direction, the dispersion relation in this case can be written using (7.35) as follows.

$$\left(\frac{n_x\pi}{d_x}\right)^2 = [\rho_4(n, E_{11\mathrm{SLMSQ}})] \tag{8.31}$$

where

$$\rho_4(n, E_{11\mathrm{SLMSQ}}) = \frac{1}{L_0^2}[\cos^{-1}(f(n, E_{11\mathrm{SLMSQ}}))]^2 - \left\{\frac{2|e|B}{\hbar}\left(n+\frac{1}{2}\right)\right\},$$

$$f(n, E_{11\mathrm{SLMSQ}}) = a_1\cos[a_0C_1(n, E_{11\mathrm{SLMSQ}}) + b_0D_1(n, E_{11\mathrm{SLMSQ}})]$$
$$- a_2\cos[a_0C_1(n, E_{11\mathrm{SLMSQ}}) - b_0D_1(n, E_{11\mathrm{SLMSQ}})],$$

$$C_1(n, E_{11\mathrm{SLMSQ}}) \equiv \left[\left(\frac{2m_1^*E}{\hbar^2}\right)G(E_{11\mathrm{SLMSQ}}, E_{g_1}, \Delta_1)\right.$$
$$\left.- \left\{\frac{2|e|B}{\hbar}\left(n+\frac{1}{2}\right)\right\}\right]^{1/2},$$

$$D_1(n, E_{11\mathrm{SLMSQ}}) \equiv \left[\left(\frac{2m_2^*E}{\hbar^2}\right)G(E_{11\mathrm{SLMSQ}}, E_{g_2}, \Delta_2)\right.$$
$$\left.- \left\{\frac{2|e|B}{\hbar}\left(n+\frac{1}{2}\right)\right\}\right]^{1/2}$$

and $E_{11\mathrm{SLMSQ}}$ is the total quantized energy in this case.

The DOS function in this case can be written as

$$N(E) = \frac{eB}{\pi\hbar} \sum_{n=0}^{n_{\max}} \sum_{n_x=1}^{n_{x\max}} \delta'(E - E_{11\mathrm{SLMSQ}}) \tag{8.32}$$

The electron concentration assumes the form

$$n_0 = \frac{eB}{\pi\hbar} \sum_{n=0}^{n_{\max}} \sum_{n_x=1}^{n_{x\max}} F_{-1}(\eta_{11\mathrm{SLMSQ}}) \tag{8.33}$$

where $\eta_{11\mathrm{SLMSQ}} = (k_B T)^{-1}[E_{F\mathrm{SLMSQ}} - E_{11\mathrm{SLMSQ}}]$.

Thus, the photo-emitted current density from III–V effective mass superlattices under magnetic quantization can be written as

$$J_{\mathrm{PHOTO}} = \frac{\alpha_0 g_v e^2 B}{2\pi\hbar^2 L_0 d_x} \sum_{n=0}^{n_{\max}} \sum_{n_{x_{\min}}}^{n_{x\max}} F_{-1}(\eta_{11\mathrm{SLMSQ}}) R_{36}(E_{n_{x6}}, n_x) \tag{8.34}$$

where

$$R_{36}(E_{n_{x6}}, n_x) = \left| \frac{\sin\left(\frac{L_0 \pi n_x}{d_x}\right)}{f'(E_{n_{x6}}, 0)} \right|$$

in which

$$f'(E_{n_{x6}}, 0) = \frac{\partial}{\partial E_{n_{x6}}}[f(E_{n_{x6}}, 0)],$$

$$f(E_{n_{x6}}, 0) = a_1 \cos[a_0 C_1(E_{n_{x6}}, 0) + b_0 D_1(E_{n_{x6}}, 0)] - a_2 \cos[a_0 C_1(E_{n_{x6}}, 0) - b_0 D_1(E_{n_{x6}}, 0)],$$

$$C_1(E_{n_{x6}}, 0) \equiv \left[\left(\frac{2m_1^* E_{n_{x6}}}{\hbar^2}\right) G(E_{n_{x6}}, E_{g_1}, \Delta_1)\right]^{1/2} \quad \text{and}$$

$$D_1(E_{n_{x6}}, 0) \equiv \left[\left(\frac{2m_2^* E_{n_{x6}}}{\hbar^2}\right) G(E_{n_{x6}}, E_{g_2}, \Delta_2)\right]^{1/2}$$

$E_{n_{z1}}$ can be determine from the equation

$$\cos(L_0 n_x \pi / d_x) = f(E_{n_{x6}}, 0) \tag{8.35}$$

and $n_{x_{\min}}$ is the nearest integer of the following inequality

$$n_{x_{\min}} \geq \left(\frac{d_x}{\pi L_0}\right) \cos^{-1}[f((W - h\upsilon), 0)] \tag{8.36}$$

8.2.7 The EP from II–VI QW effective mass superlattices under magnetic quantization

Under magnetic quantization along z-direction, the dispersion relation can be expressed using (7.40) as

$$\left(\frac{n_z\pi}{d_z}\right)^2 = [\rho_5(n, E_{12\mathrm{SLMSQ}})] \tag{8.37}$$

where

$$\rho_5(n, E_{12\mathrm{SLMSQ}}) = \frac{1}{L_0^2}[\cos^{-1}(f_1(n, E_{12\mathrm{SLMSQ}}))]^2 - \left\{\frac{2|e|B}{\hbar}\left(n+\frac{1}{2}\right)\right\},$$

$$f_1(n, E_{12\mathrm{SLMSQ}}) = a_3\cos[a_0C_2(n, E_{12\mathrm{SLMSQ}}) + b_0D_2(n, E_{12\mathrm{SLMSQ}})]$$
$$- a_4\cos[a_0C_2(n, E_{12\mathrm{SLMSQ}}) - b_0D_2(n, E_{12\mathrm{SLMSQ}})],$$

$$C_2(n, E_{12\mathrm{SLMSQ}}) \equiv \left[\left(\frac{2m_1^*E}{\hbar^2}\right)^{1/2}\left[E_{12\mathrm{SLMSQ}} - \frac{\hbar|e|B}{m_{\perp,1}^*}\left(n+\frac{1}{2}\right)\right.\right.$$
$$\left.\left.\mp \bar{\lambda}_0\left\{\frac{2|e|B}{\hbar}\left(n+\frac{1}{2}\right)\right\}\right]^{1/2}\right],$$

$$D_2(n, E_{12\mathrm{SLMSQ}}) \equiv \left[\left(\frac{2m_2^*E}{\hbar^2}\right)E_{12\mathrm{SLMSQ}}G(E_{12\mathrm{SLMSQ}}, E_{g_2}, \Delta_2)\right.$$
$$\left. - \left\{\frac{2|e|B}{\hbar}\left(n+\frac{1}{2}\right)\right\}\right]^{1/2}$$

and $E_{12\mathrm{SLMSQ}}$ is the total quantized energy in this case.

The DOS function in this case can be written as

$$N(E) = \frac{eB}{h}\sum_{n=0}^{n_{\max}}\sum_{n_z=1}^{n_{z\max}}\delta'(E - E_{12\mathrm{SLMSQ}}) \tag{8.38}$$

The electron concentration assumes the form

$$n_0 = \frac{eB}{h}\sum_{n=0}^{n_{\max}}\sum_{n_z=1}^{n_{z\max}}F_{-1}(\eta_{12\mathrm{SLMSQ}}) \tag{8.39}$$

where $\eta_{12\mathrm{SLMSQ}} = (k_BT)^{-1}[E_{F\mathrm{SLMSQ}} - E_{12\mathrm{SLMSQ}}]$.

Thus, the photo-emitted current density from II–VI effective mass superlattices under magnetic quantization can be written as

$$J_{\text{PHOTO}} = \frac{\alpha_0 g_v e^2 B}{4\pi\hbar^2 L_0 d_z} \sum_{n=0}^{n_{\max}} \sum_{n_{z_{\min}}}^{n_{z_{\max}}} F_{-1}(\eta_{12\text{SLMSQ}}) R_{37}(E_{n_{z7}}, n_z) \quad (8.40)$$

where

$$R_{37}(E_{n_{z7}}, n_z) = \left| \frac{\sin\left(\frac{L_0 \pi n_z}{d_z}\right)}{f_1'(E_{n_{z7}}, 0)} \right|$$

in which

$$f_1'(E_{n_{z7}}, 0) = \frac{\partial}{\partial E_{n_{z7}}}[f_1(E_{n_{z7}}, 0)],$$

$$f_1(E_{n_{z7}}, 0) = a_3 \cos[a_0 C_2(E_{n_{z7}}, 0) + b_0 D_2(E_{n_{z7}}, 0)] - a_4 \cos[a_0 C_2(E_{n_{z7}}, 0) - b_0 D_2(E_{n_{z7}}, 0)],$$

$$C_2(E_{n_{z7}}, 0) \equiv \left(\frac{2m_{\parallel,1}^*}{\hbar^2}\right)^{1/2} [E_{n_{z7}}]^{1/2} \quad \text{and}$$

$$D_2(E_{n_{z7}}, 0) \equiv \left[\left(\frac{2m_2^*}{\hbar^2}\right) E_{n_{z7}} G(E_{n_{z7}}, E_{g2}, \Delta_2)\right]^{1/2}$$

$E_{n_{z7}}$ can be determine from the equation

$$\cos(L_0 n_z \pi / d_z) = f_1(E_{n_{z7}}, 0) \quad (8.41)$$

and $n_{z_{\min}}$ is the nearest integer of the following inequality

$$n_{z_{\min}} \geq \left(\frac{d_z}{\pi L_0}\right) \cos^{-1}[f_1((W - h\upsilon), 0)] \quad (8.42)$$

*8.2.8 The EP from IV–VI QW effective mass superlattices under magnetic quantization

Thus, in the presence of a quantizing magnetic field along x-direction, the dispersion relation in this case can be written using (7.45) as

$$\left(\frac{n_x \pi}{d_x}\right)^2 = [\rho_6(n, E_{13\text{SLMSQ}})] \quad (8.43)$$

where

$$\rho_6(n, E_{13SLMSQ}) = \frac{1}{L_0^2}[\cos^{-1}(f_2(n, E_{13SLMSQ}))]^2 - \left\{\frac{2|e|B}{\hbar}\left(n+\frac{1}{2}\right)\right\},$$

$$f_2(n, E_{13SLMSQ}) = a_5 \cos[a_0 C_3(n, E_{13SLMSQ}) + b_0 D_3(n, E_{13SLMSQ})]$$
$$- a_6 \cos[a_0 C_3(n, E_{13SLMSQ}) - b_0 D_3(n, E_{13SLMSQ})],$$

$$C_3(n, E_{13SLMSQ}) \equiv [[E_{13SLMSQ} H_{11} + H_{21}(n)] - [(E_{13SLMSQ})^2 H_{31}$$
$$+ E_{13SLMSQ} H_{41}(n) + H_{51}(n)]^{1/2}]^{1/2},$$

$$D_3(n, E_{13SLMSQ}) \equiv [[E_{13SLMSQ} H_{12} + H_{22}(n)] - [(E_{13SLMSQ})^2 H_{32}$$
$$+ E_{13SLMSQ} H_{42}(n) + H_{52}(n)]^{1/2}]^{1/2},$$

and $E_{13SLMSQ}$ is the total quantized energy in this case.

The DOS function in this case can be written as

$$N(E) = \frac{eB}{\pi\hbar} \sum_{n=0}^{n_{\max}} \sum_{n_x=1}^{n_{x\max}} \delta'(E - E_{13SLMSQ}) \tag{8.44}$$

The electron concentration assumes the form

$$n_0 = \frac{eB}{\pi\hbar} \sum_{n=0}^{n_{\max}} \sum_{n_x=1}^{n_{x\max}} F_{-1}(\eta_{13SLMSQ}) \tag{8.45}$$

where $\eta_{13SLMSQ} = (k_B T)^{-1}[E_{FSLMSQ} - E_{13SLMSQ}]$.

Thus, the photo-emitted current density from IV–VI effective mass superlattices under magnetic quantization can be written as

$$J_{PHOTO} = \frac{\alpha_0 g_v e^2 B}{2\pi\hbar^2 L_0 d_x} \sum_{n=0}^{n_{\max}} \sum_{n_{x_{\min}}}^{n_{x\max}} F_{-1}(\eta_{13SLMSQ}) R_{38}(E_{n_{x8}}, n_x) \tag{8.46}$$

where

$$R_{38}(E_{n_{x8}}, n_x) = \left| \frac{\sin\left(\frac{L_0 \pi n_x}{d_x}\right)}{f_2'(E_{n_{x8}}, 0)} \right|$$

In which

$$f_2'(E_{n_{z8}}, 0) = \frac{\partial}{\partial E_{n_{z8}}}[f_2(E_{n_{z8}}, 0)],$$

$$f_2(E_{n_{x8}}, 0) = a_5 \cos[a_0 C_3(E_{n_{x8}}, 0) + b_0 D_3(E_{n_{x8}}, 0)]$$
$$- a_6 \cos[a_0 C_3(E_{n_{x8}}, 0) - b_0 D_3(E_{n_{x8}}, 0)],$$

$$C_3(E_{n_{x8}},0) \equiv [[E_{n_{x8}}H_{11}+H_{21}(0)]-[(E_{n_{x8}})^2H_{31}+E_{n_{x8}}H_{41}(0)$$
$$+H_{51}(0)]^{1/2}]^{1/2}$$
$$D_3(E_{n_{x8}},0) \equiv [[E_{n_{x8}}H_{12}+H_{22}(0)]-[(E_{n_{x8}})^2H_{32}+E_{n_{x8}}H_{42}(0)$$
$$+H_{52}(0)]^{1/2}]^{1/2}$$

$E_{n_{z8}}$ can be determine from the equation

$$\cos(L_0 n_x \pi/d_x) = f_2(E_{n_{x8}},0) \tag{8.47}$$

and $n_{x_{\min}}$ is the nearest integer of the following inequality

$$n_{x_{\min}} \geq \left(\frac{d_x}{\pi L_0}\right)\cos^{-1}[f_2((W-h\upsilon),0)] \tag{8.48}$$

*8.2.9 The EP from HgTe/CdTe QW effective mass superlattices under magnetic quantization

In the presence of an external magnetic field along x-direction, the dispersion relation in this case can be written using (7.50) as

$$\left(\frac{n_x\pi}{d_x}\right)^2 = [\rho_7(n,E_{14\mathrm{SLMSQ}})] \tag{8.49}$$

where

$$\rho_7(n,E_{14\mathrm{SLMSQ}}) = \left\{\frac{1}{L_0^2}[\cos^{-1}(f_3(n,E_{14\mathrm{SLMSQ}}))]^2\right\}-\left\{\frac{2|e|B}{\hbar}\left(n+\frac{1}{2}\right)\right\},$$

$$f_3(n,E_{14\mathrm{SLMSQ}}) = a_7\cos[a_0C_4(n,E_{14\mathrm{SLMSQ}})+b_0D_4(n,E_{14\mathrm{SLMSQ}})]$$
$$-a_8\cos[a_0C_4(n,E_{14\mathrm{SLMSQ}})-b_0D_4(n,E_{14\mathrm{SLMSQ}})],$$

$$C_4(n,E_{14\mathrm{SLMSQ}}) \equiv \left[\frac{B_0^2+2AE_{14\mathrm{SLMSQ}}-B_0\sqrt{B_0^2+4AE_{14\mathrm{SLMSQ}}}}{2A^2}\right.$$
$$\left.-\left\{\frac{2|e|B}{\hbar}\left(n+\frac{1}{2}\right)\right\}\right]^{1/2},$$

$$D_4(n,E_{14\mathrm{SLMSQ}}) \equiv \left[\left(\frac{2m_2^*E_{14\mathrm{SLMSQ}}}{\hbar^2}\right)G(E_{14\mathrm{SLMSQ}},E_{g_2},\Delta_2)\right.$$
$$\left.-\left\{\frac{2|e|B}{\hbar}\left(n+\frac{1}{2}\right)\right\}\right]^{1/2}$$

and $E_{14\mathrm{SLMSQ}}$ is the total quantized energy in this case.

The DOS function in this case can be written as

$$N(E) = \frac{eB}{\pi\hbar}\sum_{n=0}^{n_{\max}}\sum_{n_x=1}^{n_{x\max}} \delta'(E - E_{14\mathrm{SLMSQ}}) \tag{8.50}$$

The electron concentration assumes the form

$$n_0 = \frac{eB}{\pi\hbar}\sum_{n=0}^{n_{\max}}\sum_{n_x=1}^{n_{x\max}} F_{-1}(\eta_{14\mathrm{SLMSQ}}) \tag{8.51}$$

where $\eta_{14\mathrm{SLMSQ}} = (k_B T)^{-1}[E_{F\mathrm{SLMSQ}} - E_{14\mathrm{SLMSQ}}]$.

Thus, the photo-emitted current density from HgTe/CdTe effective mass superlattices under magnetic quantization can be written as

$$J_{\mathrm{PHOTO}} = \frac{\alpha_0 g_v e^2 B}{2\pi\hbar^2 L_0 d_x}\sum_{n=0}^{n_{\max}}\sum_{n_{x\min}}^{n_{x\max}} F_{-1}(\eta_{14\mathrm{SLMSQ}}) R_{39}(E_{n_{x9}}, n_x) \tag{8.52}$$

where

$$R_{39}(E_{n_{x9}}, n_x) = \left|\frac{\sin\left(\frac{L_0\pi n_x}{d_x}\right)}{f_3'(E_{n_{x9}}, 0)}\right|$$

In which

$$f_3'(E_{n_{z9}}, 0) = \frac{\partial}{\partial E_{n_{z7}}}[f_3(E_{n_{z9}}, 0)],$$

$$f_3(E_{n_{x9}}, 0) = a_7\cos[a_0 C_4(E_{n_{x9}}, 0) + b_0 D_4(E_{n_{x9}}, 0)]$$
$$- a_8\cos[a_0 C_4(E_{n_{x9}}, 0) - b_0 D_4(E_{n_{x9}}, 0)],$$

$$C_4(E_{n_{x9}}, 0) \equiv \left[\frac{B_0^2 + 2AE_{n_{x9}} - B_0\sqrt{B_0^2 + 4AE_{n_{x9}}}}{2A^2}\right]^{1/2} \quad \text{and}$$

$$D_4(E_{n_{x9}}, 0) \equiv \left[\left(\frac{2m_2^* E_{n_{x9}}}{\hbar^2}\right) G(E_{n_{x9}}, E_{g_2}, \Delta_2)\right]^{1/2}$$

$E_{n_{z9}}$ can be determine from the equation

$$\cos(L_0 n_x \pi/d_x) = f_3(E_{n_{x9}}, 0) \tag{8.53}$$

and $n_{x_{\min}}$ is the nearest integer of the following inequality

$$n_{x_{\min}} \geq \left(\frac{d_x}{\pi L_0}\right)\cos^{-1}[f_3((W - hv), 0)] \tag{8.54}$$

*8.2.10 The EP from strained layer QW effective mass superlattices under magnetic quantization

The electron energy spectrum in heavily doped strained layer EMSL can be written using (7.55) as

$$\left(\frac{n_z\pi}{d_z}\right)^2 = \omega_{61}(E_{15\text{SLMSQ}}, n) \tag{8.55}$$

$$\left(\frac{n_z\pi}{d_z}\right)^2 = \omega_{61}(E_{15\text{SLMSQ}}, n) \tag{8.56}$$

where

$$\omega_{61}(E_{15\text{SLMSQ}}, n) = \left[\frac{1}{L_0^2}\{\cos^{-1}(f_4(E_{15\text{SLMSQ}}, n))\}^2 - \frac{2eB}{\hbar}\left(n+\frac{1}{2}\right)\right]$$

$$f_4(E_{15\text{SLMSQ}}, n) = a_{20}\cos[a_0C_{40}(E_{15\text{SLMSQ}}, n) + b_0D_{40}(E_{15\text{SLMSQ}}, n)] - a_{21}\cos[a_0C_{40}(E_{15\text{SLMSQ}}, n) - b_0D_{40}(E_{15\text{SLMSQ}}, n)],$$

$$C_{40}(E_{15\text{SLMSQ}}, n) = [\bar{S}_1(E_{15\text{SLMSQ}})]^{-1/2}\left[1-\left(n+\frac{1}{2}\right)\hbar\omega_{0810}\right]^{1/2},$$

$$\omega_{0810} = \frac{eB}{\sqrt{m_{1810}m_{2810}}}, \quad m_{1810} = \frac{\hbar}{\sqrt{2\bar{P}_1(E_{15\text{SLMSQ}})}}$$

$$m_{2810} = \frac{\hbar}{\sqrt{2\bar{Q}_1(E_{15\text{SLMSQ}})}},$$

$$D_{40}(E_{15\text{SLMSQ}}, n) = \left[1-\left(n+\frac{1}{2}\right)\hbar\omega_{0812}\right]^{1/2}[\bar{S}_2(E_{15\text{SLMSQ}})]^{-1/2}$$

$$\omega_{0812} = \frac{eB}{\sqrt{m_{1812}m_{2812}}}, \quad m_{1812} = \frac{\hbar}{\sqrt{2\bar{P}_2(E_{15\text{SLMSQ}})}},$$

$$m_{2812} = \frac{\hbar}{\sqrt{2\bar{Q}_2(E_{15\text{SLMSQ}})}},$$

and $E_{15\text{SLMSQ}}$ is the total quantized energy in this case.

The DOS function in this case can be written as

$$N(E) = \frac{eB}{\pi\hbar}\sum_{n=0}^{n_{\max}}\sum_{n_z=1}^{n_{z\max}}\delta'(E - E_{15\text{SLMSQ}}) \tag{8.57}$$

The electron concentration assumes the form

$$n_0 = \frac{eB}{\pi\hbar} \sum_{n=0}^{n_{\max}} \sum_{n_z=1}^{n_{z\max}} F_{-1}(\eta_{15\text{SLMSQ}}) \tag{8.58}$$

where $\eta_{15\text{SLMSQ}} = (k_B T)^{-1}[E_{F\text{SLMSQ}} - E_{15\text{SLMSQ}}]$.

Thus, the photo-emitted current density from strained layer effective mass superlattices under magnetic quantization can be written as

$$J_{\text{PHOTO}} = \frac{\alpha_0 e^2 B L_0 g_v}{2\pi\hbar^2 d_z} \sum_{n=0}^{n_{\max}} \sum_{n_{z\min}}^{n_{z\max}} F_{-1}(\eta_{15\text{SLMSQ}}) R_{35}(E_{n_z 10}, n_z) \tag{8.59}$$

where

$$R_{40}(E_{n_z 10}, n_z) = \left| \frac{\sin\left(\frac{L_0 \pi n_z}{d_z}\right)}{f_4'(E_{n_z 10}, 0)} \right|$$

in which

$$f_4'(E_{n_z 10}, 0) = \frac{\partial}{\partial E_{n_z 10}}[f_4(E_{n_z 10}, 0)],$$

$$f_4(E_{n_z 10}, 0) = a_{20}\cos[a_0 C_{40}(E_{n_z 10}, 0) + b_0 D_{40}(E_{n_z 10}, 0)]$$

$$- a_{21}\cos[a_0 C_{40}(E_{n_z 10}, 0) - b_0 D_{40}(E_{n_z 10}, 0)],$$

$$C_{40}(E_{n_z 10}, 0) = [\bar{S}_1(E_{n_z 10})]^{-1/2} \quad \text{and} \quad D_{40}(E_{n_z 10}, 0) = [\bar{S}_2(E_{n_z 10})]^{-1/2}$$

$E_{n_z 10}$ can be determine from the equation

$$\cos(L_0 n_z \pi / d_z) = \frac{1}{2} f_4(E_{n_z 10}, 0) \tag{8.60}$$

and $n_{z_{\min}}$ is the nearest integer of the following inequality

$$n_{z_{\min}} \geq \left(\frac{d_z}{\pi L_0}\right) \cos^{-1}[f_4((W - h\upsilon), 0)] \tag{8.61}$$

8.3 Result and Discussions

In Figs. 8.1–8.3, we have plotted the normalized photocurrent density from QW III–V SLs (taking GaAs/Ga$_{1-x}$Al$_x$As and In$_x$Ga$_{1-x}$As/InP QW SLs) with graded interfaces under the magnetic quantization has been plotted as functions of the inverse quantizing magnetic field,

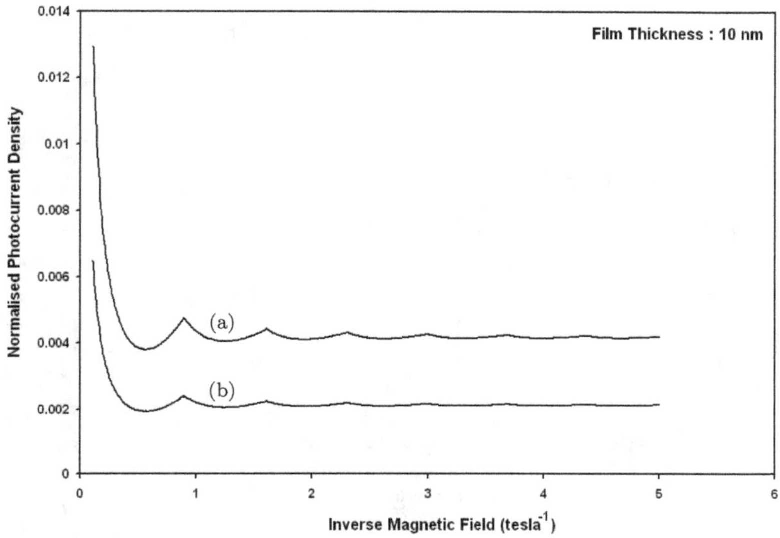

Fig. 8.1. Plot of the normalized photocurrent density from (a) GaAs/Ga$_{1-x}$Al$_x$As and (b) In$_x$Ga$_{1-x}$As/InP quantum well superlattices with graded interfaces as a function of inverse quantizing magnetic field.

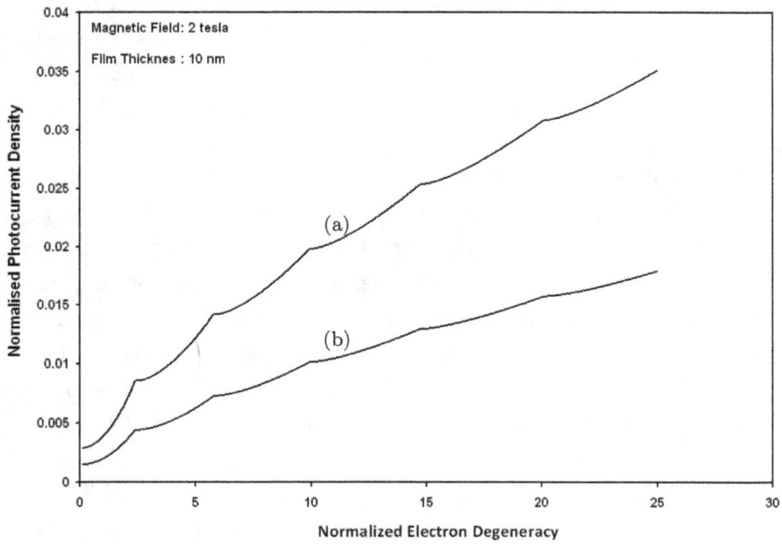

Fig. 8.2. Plot of the normalized photocurrent density from (a) GaAs/Ga$_{1-x}$Al$_x$As and (b) In$_x$Ga$_{1-x}$As/InP quantum well superlattices with graded interfaces under quantizing magnetic field as a function of normalized electron degeneracy.

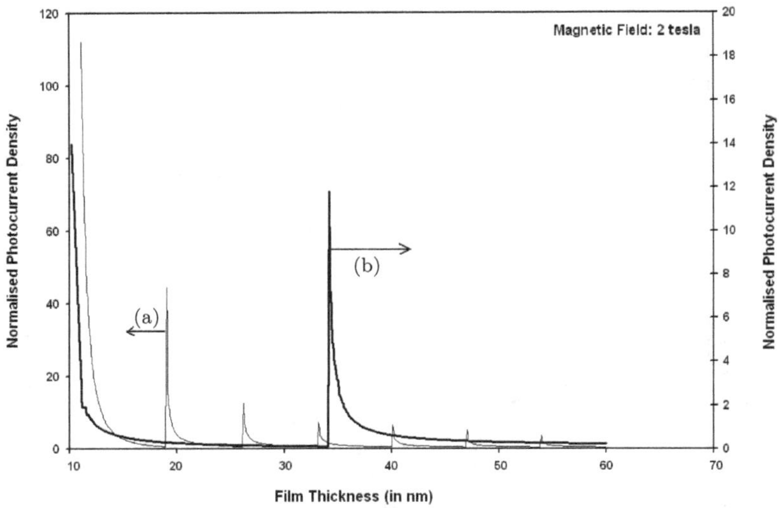

Fig. 8.3. Plot of the normalized photocurrent density from (a) $GaAs/Ga_{1-x}Al_xAs$ and (b) $In_xGa_{1-x}As/InP$ quantum well superlattices with graded interfaces under quantizing magnetic field as a function of film thickness.

normalized electron degeneracy, film thickness and the normalized incident photon energy respectively. It appears from Fig. 8.1 that the photoemission in this case oscillates with inverse quantizing magnetic field due to SdH effect. Figure 8.2 exhibits the fact that the photocurrent density increases with increasing carrier degeneracy in an oscillatory way and the nature of oscillations is different as compared with Fig. 8.1. From Fig. 8.3, it can be inferred that the photoemission oscillates with film thickness and for certain values of film thickness the photoemission exhibits very large values. From Fig. 8.4, it appears that photoemission increases with increasing photon energy in quantum steps. The plot of the normalized magneto-photoemission from II–VI QWSLs with graded interfaces as a function of inverse quantizing magnetic field as shown in the curve (b) of Fig. 8.5, where the plot (a) has been drawn with $\bar{\lambda}_0 = 0$ for the purpose of assessing the splitting of the two spin states by the spin orbit coupling and the crystalline field on the magneto photoemission in this case. The plot (c) of Fig. 8.5 has been drawn for HgTe/CdTe QWSL whereas the plot (d) is valid for IV–VI QWSL (using PbSe/PbTe as an example).

Figures 8.6–8.8 demonstrate the plots of the photoemission as functions of normalized electron degeneracy, film thickness and normalized incident photon energy respectively, for all the cases of Fig. 8.5. The plot of the

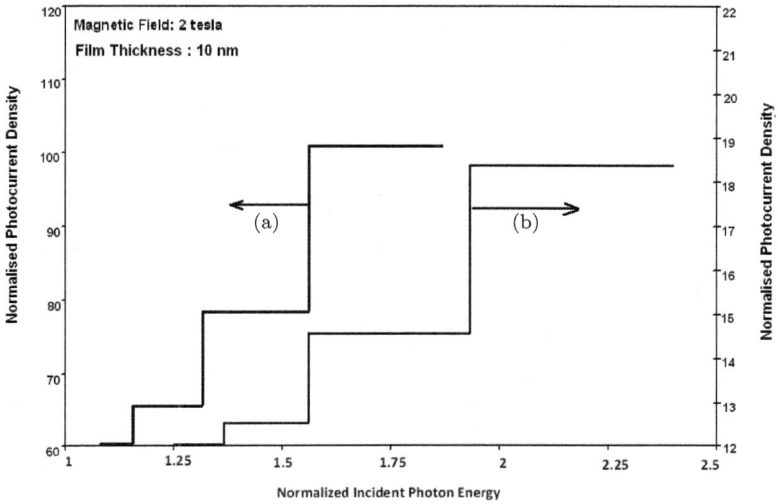

Fig. 8.4. Plot of the normalized photocurrent density from (a) $GaAs/Ga_{1-x}Al_xAs$ and (b) $In_xGa_{1-x}As/InP$ quantum well superlattices with graded interfaces under quantizing magnetic field as a function of normalized incident photon energy.

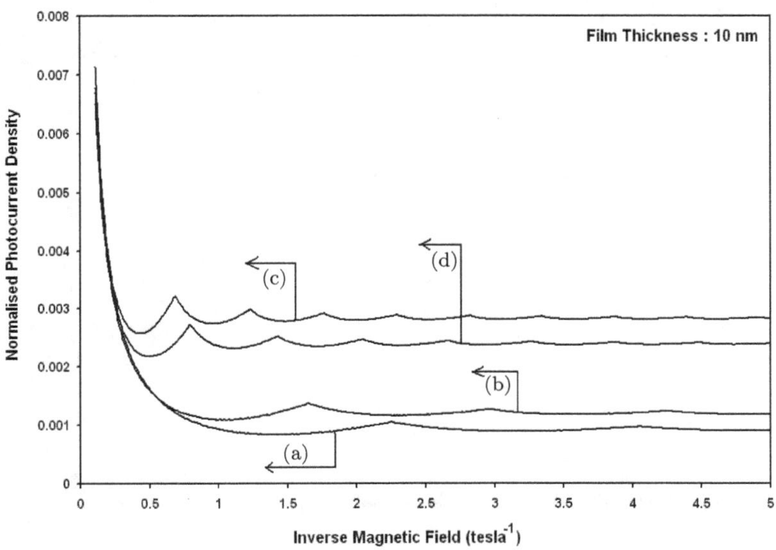

Fig. 8.5. Plot of the normalized photocurrent density from (a) CdS/ZnSe with $C_0 = 0$, (b) CdS/ZnSe with $\bar{\lambda}_0 \neq 0$ (c) HgTe/CdTe and (d) PbSe/PbTe quantum well superlattices with graded interfaces as a function of inverse magnetic field.

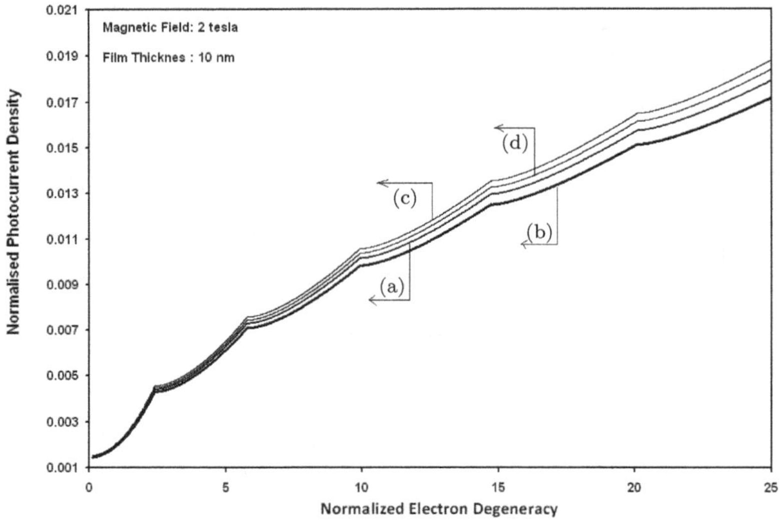

Fig. 8.6. Plot of the normalized photocurrent density from (a) CdS/ZnSe with $C_0 = 0$, (b) CdS/ZnSe with $\bar{\lambda}_0 \neq 0$ (c) HgTe/CdTe and (d) PbSe/PbTe quantum well superlattices with graded interfaces under quantizing magnetic field as a function of normalized electron degeneracy.

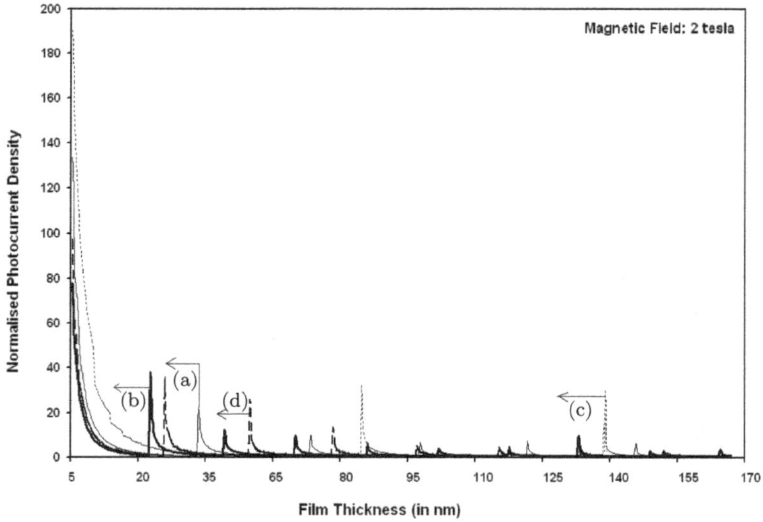

Fig. 8.7. Plot of the normalized photocurrent density from (a) CdS/ZnSe with $C_0 = 0$, (b) CdS/ZnSe with $\bar{\lambda}_0 \neq 0$ (c) HgTe/CdTe and (d) PbSe/PbTe quantum well superlattices with graded interfaces under quantizing magnetic field as a function of film thickness.

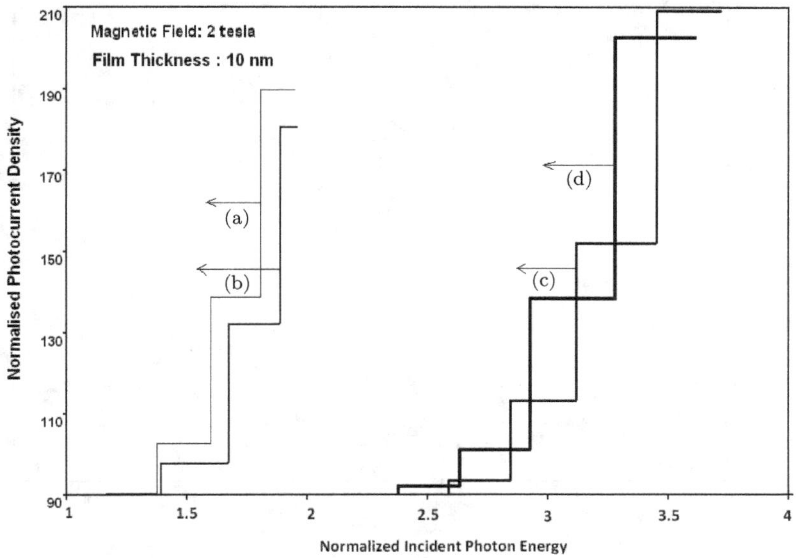

Fig. 8.8. Plot of the normalized photocurrent density from (a) CdS/ZnSe with $C_0 = 0$, (b) CdS/ZnSe with $\bar{\lambda}_0 \neq 0$ (c) HgTe/CdTe and (d) PbSe/PbTe quantum well superlattices with graded interfaces under quantizing magnetic field as a function of normalized incident photon energy.

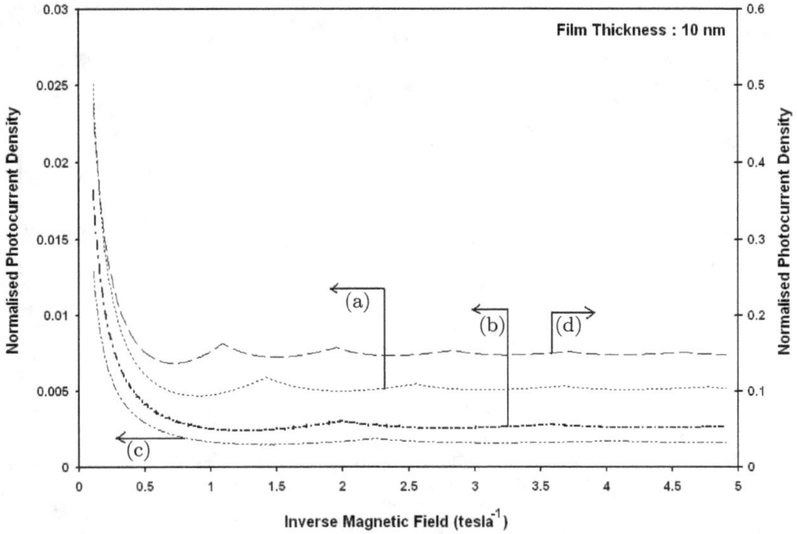

Fig. 8.9. Plot of the normalized photocurrent density from (a) $GaAs/Ga_{1-x}Al_xAs$, (b) CdS/ZnSe with $\bar{\lambda}_0 \neq 0$, (c) PbSe/PbTe and (d) HgTe/CdTe quantum well effective mass superlattices as a function of inverse quantizing magnetic field.

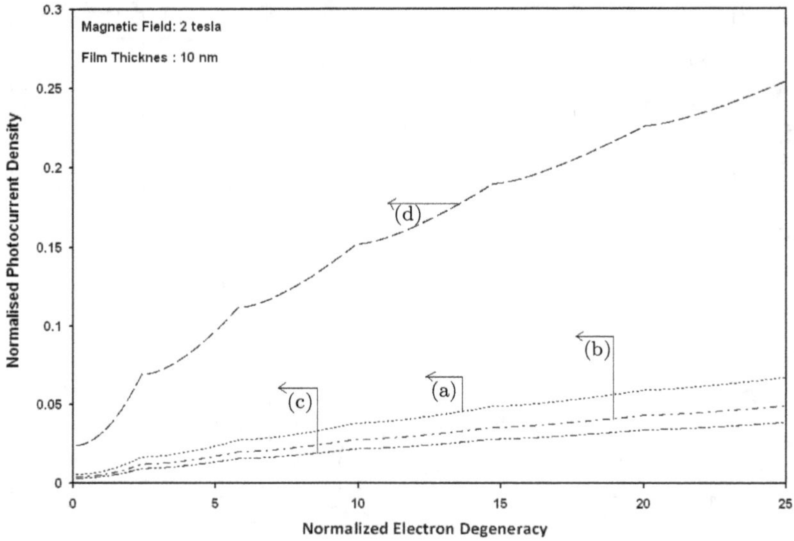

Fig. 8.10. Plot of the normalized photocurrent density from (a) GaAs/$Ga_{1-x}Al_xAs$, (b) CdS/ZnSe with $\bar{\lambda}_0 \neq 0$, (c) PbSe/PbTe and (d) HgTe/CdTe quantum well effective mass superlattices under quantizing magnetic field as a function of normalized electron degeneracy.

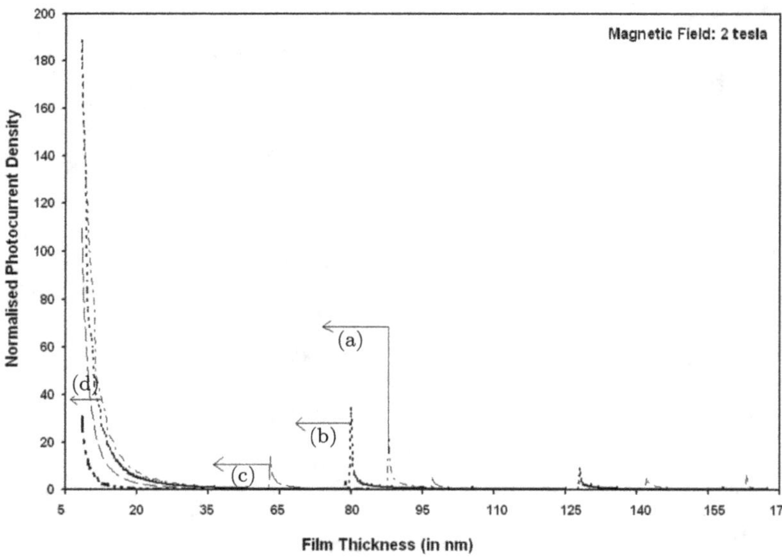

Fig. 8.11. Plot of the normalized photocurrent density from (a) GaAs/$Ga_{1-x}Al_xAs$, (b) CdS/ZnSe with $\bar{\lambda}_0 \neq 0$, (c) PbSe/PbTe and (d) HgTe/CdTe quantum well effective mass superlattices under quantizing magnetic field as a function of film thickness.

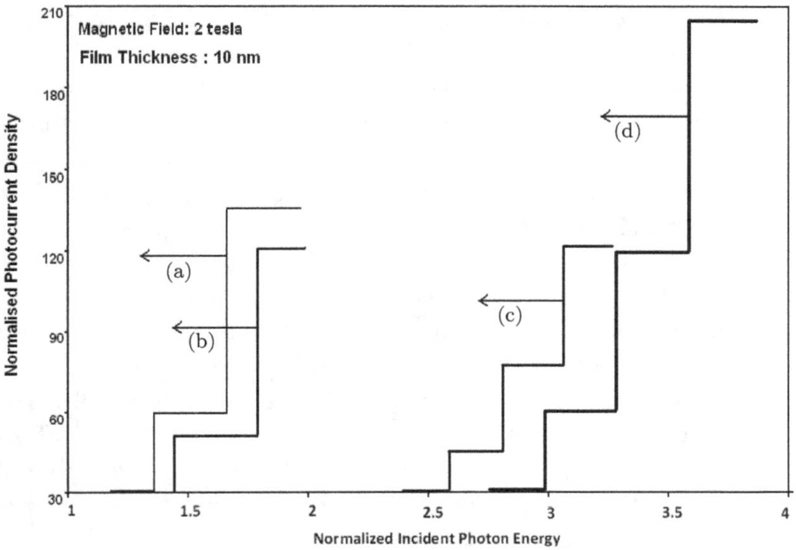

Fig. 8.12. Plot of the normalized photocurrent density from (a) $GaAs/Ga_{1-x}Al_xAs$, (b) CdS/ZnSe with $\bar{\lambda}_0 \neq 0$, (c) PbSe/PbTe and (d) HgTe/CdTe quantum well effective mass superlattices under quantizing magnetic field as a function of normalized incident photon energy.

normalized magneto-photoemission for III–V QW effective mass SLs (taking $GaAs/Ga_{1-x}Al_xAs$ as an example) as a function of inverse quantizing magnetic field shown in the curve (a) of Fig. 8.8. The plots (b), (c) and (d) in the same figure have been drawn for II–VI QW effective mass SL (taking CdS/ZnSe as an example), IV–VI QW effective mass SL (taking PbSe/PbTe as an example) and HgTe/CdTe QW effective mass SL respectively. In Figs. 8.9–8.12, we have plotted normalized photocurrent density as functions of inverse quantizing magnetic field, electron degeneracy, film thickness and incident photon energy respectively. Finally, it is logical to conclude that the numerical values of the photo-emitted current density are totally different in all cases which exhibit the signature of the respective band structure under different physical conditions and the rates of variation are again energy spectrum dependent.

8.4 Open Research Problems

(R.8.1) Investigate the photoemission in the absence of magnetic field by considering all types of scattering mechanisms for III–V,

II–VI, IV–VI and HgTe/CdTe quantum well and quantum wire superlattices with graded interfaces and also the effective mass superlattices of the aforementioned materials.

(R.8.2) Investigate the photoemission in the absence of magnetic field by considering all types of scattering mechanisms for strained layer, random, short period and Fibonacci, polytype and saw-tooth quantum well and quantum wire superlattices.

(R.8.3) Investigate the photoemission in the presence of an arbitrarily oriented quantizing magnetic field in the presence of spin and broadening by considering all types of scattering mechanisms for (R.8.1) and (R.8.2) under an arbitrarily oriented (a) non-uniform electric field and (b) alternating electric field respectively.

(R.8.4) Investigate the photoemission by considering all types of scattering mechanisms for (R.8.1) and (R.8.2) under an arbitrarily oriented alternating magnetic field by including broadening and the electron spin respectively.

(R.8.5) Investigate the photoemission by considering all types of scattering mechanisms for (R.8.1) and (R.8.2) under an arbitrarily oriented quantizing alternating magnetic field and crossed alternating electric field by including broadening and the electron spin respectively.

(R.8.6) Investigate the photoemission by considering all types of scattering mechanisms for (R.8.1) and (R.8.2) under an arbitrarily oriented alternating quantizing magnetic field and crossed alternating non-uniform electric field by including broadening and the electron spin respectively.

(R.8.7) Investigate the photoemission in the absence of magnetic field for all types of quantum well and quantum wire superlattices as considered in this chapter under exponential, Kane, Halperin, Lax and Bonch-Bruevich band tails [2] respectively.

(R.8.8) Investigate the photoemission in the presence of quantizing magnetic field including spin and broadening for the problem as defined in (R.8.7) under an arbitrarily oriented (a) non-uniform electric field and (b) alternating electric field respectively.

(R.8.9) Investigate the photoemission for the problem as defined in (R.8.7) under an arbitrarily oriented alternating quantizing magnetic field by including broadening and the electron spin respectively.

(R.8.10) Investigate the photoemission for the problem as defined in (R.8.7) under an arbitrarily oriented alternating quantizing magnetic field and crossed alternating electric field by including broadening and the electron spin respectively.

(R.8.11) Investigate all the appropriate problems as defined in (R.8.1) to (R.8.10) for all types of quantum dot superlattices.

(R.8.12) Investigate all the appropriate problems as defined in (R.8.1) to (R.8.10) for all types of quantum dot superlattices in the presence of strain.

(R.8.13) Introducing new theoretical formalisms, investigate all the problems of this chapter in the presence of hot electron effects.

(R.8.14) Investigate the influence of deep traps and surface states separately for all the appropriate problems of this chapter after proper modifications.

References

[1] N. Paitya and K. P. Ghatak, *Quant. Matt.*, 5, 191 **(2016)**.

[2] S. Debbarma and K. P. Ghatak, *J. Nanosci. Nanotech.*, 16, 1095 **(2016)**.

[3] M. Chakraborty and K. P. Ghatak, *Quant. Matt.*, 4, 10 **(2015)**.

[4] S. Chakrabarti, S. K. Sen, S. Chakraborty, L. S. Singh and K. P. Ghatak, *Adv. Sci. Engg. Medi.*, 6, 1042 **(2014)**.

[5] S. M. Adhikari, A. Sakar and K. P. Ghatak, *Quant. Matt.*, 2, 455 **(2013)**.

[6] P. K. Bose, N. Paitya, S. Bhattacharya, D. De, S. Saha, K. M. Chatterjee, S. Pahari and K. P. Ghatak, *Quant. Matt.*, 1, 89 **(2012)**.

[7] N. Paitya, S. Bhattacharya, D. De and K. P. Ghatak, *Adv. Sci. Engg. Medi.*, 4, 96 **(2012)**.

[8] S. Bhattacharya, D. De, S. M. Adhikari and K. P. Ghatak, *Superlatt. Microstruct.*, 51, 203 **(2012)**.

[9] S. Bhattacharya, D. De, S. M. Adhikari, S. Saha, K. M. Chatterjee, S. Choudhury and K. P. Ghatak, *Superlatt. Microstruct.*, 50, 609 **(2011)**.

[10] D. De, S. Bhattacharya, S. M. Adhikari, A. Kumar, P. K. Bose and K. P. Ghatak, *Beilstein J. Nanotech.*, 2, 339 **(2011)**.

[11] D. De, A. Kumar, S. M. Adhikari, S. Pahari, N. Islam, P. Banerjee, S. K. Biswas, S. Bhattacharya and K. P. Ghatak, *Superlatt. Microstruct.*, 47, 377 **(2010)**.

Chapter 9

The Heat Capacity in QWSLs Under Magnetic Quantization

Believe in yourself. If you don't, who will?

9.1 Introduction

In recent years there has been considerable interest in investigating the heat capacity (HC) in nano-structured materials under different physical conditions [1–10]. In this chapter, the magneto HC in III–V, II–VI, IV–VI, HgTe/CdTe and strained layer quantum well superlattices (QWSLs) with graded interfaces has been studied in Sections 9.2.1–9.2.5. From Sections 9.2.6 to 9.2.10, the magneto HC in III–V, II–VI, IV–VI, HgTe/CdTe and strained layer quantum well effective mass superlattices respectively has been presented. Section 8.3 contains the summary and conclusion pertinent to this chapter. The last Section 8.4 presents 4 open research problems.

9.2 Theoretical Background

9.2.1 The HC in III–V QWSLs with graded interfaces under magnetic quantization

Using (1.197) and (8.3), we can study the magneto HC in this case.

9.2.2 The HC in II–VI QWSLs with graded interfaces under magnetic quantization

Using (1.197) and (8.9), we can study the magneto HC in this case.

9.2.3 The HC in IV–VI QWSLs with graded interfaces under magnetic quantization

Using (1.197) and (8.15), we can study the magneto HC in this case.

9.2.4 The HC in HgTe/CdTe QWSLs with graded interfaces under magnetic quantization

Using (1.197) and (8.21), we can study the magneto HC in this case.

9.2.5 The HC in strained layer QWSLs with graded interfaces under magnetic quantization

Using (1.197) and (8.27), we can study the magneto HC in this case.

9.2.6 The HC in III–V QW effective mass superlattices under magnetic quantization

Using (1.197) and (8.33), we can study the magneto HC in this case.

9.2.7 The HC in II–VI QW effective mass superlattices under magnetic quantization

Using (1.197) and (8.39), we can study the magneto HC in this case.

9.2.8 The HC in IV–VI QW effective mass superlattices under magnetic quantization

Using (1.197) and (8.45), we can study the magneto HC in this case.

*9.2.9 The HC in HgTe/CdTe QW effective mass superlattices under magnetic quantization

Using (1.197) and (8.51), we can study the magneto HC in this case.

*9.2.10 The HC in strained layer QW effective mass superlattices under magnetic quantization

Using (1.197) and (8.58), we can study the magneto HC in this case.

9.3 Summary and Conclusion

The DRs in magneto-size quantized of HD materials exhibit the fact that the total energy is quantized since the corresponding wave vector space is totally quantized by quantizing magnetic field and size quantization along z-direction.

- The DOS functions for all the QWSL in this case are series of non-uniformly distributed Dirac's Delta functions at specified quantized points in the respective energy axis. The spacing between the consecutive Delta functions are functions of energy band constants and quantization of the wave vector space of a particular material. The DOS function needs two summations namely one summation over the Landau quantum number and the other one is due to size quantization.
- It may be noted that the energy levels in magneto-size quantized QWSL lead to the discrete energy levels, somewhat like atomic energy levels, which produce very large changes. This follows from the inherent nature of the quantum confinement of the carrier gas dealt with here. Consequently, the crossing of the Fermi level by the size-quantized levels in this case would have much greater impact on the redistribution of the carriers among the allowed levels, as compared to that found for QWs, NWs and QDs respectively for the electronic properties in general.
- By using the appropriate equations we can plot the HC with respect to many variables in all the cases of this chapter. We have not given any plot of the HC in this case since we believe our able readers will plot the same and enjoy intricate physics, computational analysis, advanced mathematics, and will create new problems both theoretical and experimental.
- It is the band structure which changes in a fundamental way and consequently all the physical properties of all the electronic materials changes radically leading to new physical concepts.

9.4 Open Research Problems

(R.9.1) Investigate the magneto HC in the presence of arbitrarily oriented quantizing magnetic field for III–V, II–VI, IV–VI and HgTe/CdTe quantum well with graded interfaces and also the effective mass superlattices of the aforementioned materials.

(R.9.2) Investigate the magneto HC in the presence of arbitrarily oriented quantizing magnetic field for strained layer, random, short period and Fibonacci, polytype and saw-tooth quantum well superlattices.

(R.9.3) Investigate the magneto HC in the presence of arbitrarily oriented quantizing magnetic field in the presence of spin and broadening respectively.

(R.9.4) Investigate the magneto HC under an arbitrarily oriented quantizing magnetic alternating field and crossed alternating electric field by including broadening and the electron spin respectively.

References

[1] K. P. Ghatak and S. N. Biswas, A. Fizika, *J. Exp. Theor. Phys.*, 3, 7 **(1994)**.
[2] K. P. Ghatak and S. N. Biswas, *Reflective and Refractive Optical Materials for Earth and Space Applications*, *Proceeding of SPIE*, 1485, 206 **(1991)**.
[3] K. P. Ghatak, A. Goshal, S. Bhattacharya, M. Mondal, *Electro-Optical Materials for Switches, Coatings, Sensor Optics, and Detectors, Proceeding of SPIE*, 1307, 414 **(1990)**.
[4] K. P. Ghatak, A. Goshal, S. Bhattacharya, M. Mondal, *Thermosense XII: An International Conference on Thermal Sensing and Imaging Diagnostic Applications*, *Proceeding of SPIE*, 1313, 225 **(1990)**.
[5] M. Mondal, K. P. Ghatak, *J. Phys. Stat. Sol. (b)*, 128, K21 **(1985)**.
[6] M. Mondal, K. P. Ghatak, *J. Acta Phys. Polonica A*, 66, 539 **(1984)**.
[7] M. Mondal, K. P. Ghatak, *J. Phys. Stat. Sol. (b)*, 120, K63 **(1983)**.
[8] K. P. Ghatak, K. K. Ghosh, H. M. Mukherjee, A. N. Chakravarti, *J. Phys. Stat. Sol. (b)*, 110, 323 **(1982)**.
[9] K. P. Ghatak, K. K. Ghosh, H. M. Mukherjee, A. N. Chakravarti, *J. Phys. Stat. Sol. (b)*, 109, 705 **(1982)**.
[10] A. N. Chakravarti, A. K. Chowdhury, K. P. Ghatak, *J. Phys. Stat. Sol. (b)*, 63, K97 **(1981)**.

Chapter 10

The Carrier Contribution to the Elastic Constants in QWSLs Under Magnetic Quantization

A bad attitude cancels all other good skills.

10.1 Introduction

In recent years there has been considerable interest in investigating the second and third order carrier contribution to the elastic constants (CEC) in non-parabolic materials under different physical conditions [1–14]. In this chapter, the magneto CEC in III–V, II–VI, IV–VI, HgTe/CdTe and strained layer quantum well superlattices (QWSLs) with graded interfaces has been studied in Sections 10.2.1–10.2.5. From Sections 10.2.6 to 10.2.10, the magneto EC in III–V, II–VI, IV–VI, HgTe/CdTe and strained layer quantum well effective mass superlattices respectively has been presented. Section 10.3 contains the summary and conclusion pertinent to this chapter. The last Section 10.4 presents 4 open research problems.

10.2 Theoretical Background

10.2.1 The CEC in III–V QWSLs with graded interfaces under magnetic quantization

Using (1.169a), (1.169b) and (8.3), the ΔC_{44} and ΔC_{456} can respectively be expressed in this case as

$$\Delta C_{44} = \left(\frac{-G_0^2}{9k_BTd_z}\right)\frac{eB}{\pi\hbar}\sum_{n=0}^{n_{\max}}\sum_{n_z=1}^{n_{z\max}} F_{-2}(\eta_{6\mathrm{SLMSQ}}) \tag{10.1}$$

$$\Delta C_{456} = \left(\frac{G_0^3}{27(k_BT)^2d_z}\right)\frac{eB}{\pi\hbar}\sum_{n=0}^{n_{\max}}\sum_{n_z=1}^{n_{z\max}} F_{-3}(\eta_{6\mathrm{SLMSQ}}) \tag{10.2}$$

10.2.2 The CEC in II–VI QWSLs with graded interfaces under magnetic quantization

Using (1.169a), (1.169b) and (8.9), the ΔC_{44} and ΔC_{456} can respectively be expressed in this case as

$$\Delta C_{44} = \left(\frac{-G_0^2}{9k_BTd_z}\right)\frac{eB}{\pi\hbar}\sum_{n=0}^{n_{\max}}\sum_{n_z=1}^{n_{z\max}} F_{-2}(\eta_{7\mathrm{SLMSQ}}) \tag{10.3}$$

$$\Delta C_{456} = \left(\frac{G_0^3}{27(k_BT)^2d_z}\right)\frac{eB}{\pi\hbar}\sum_{n=0}^{n_{\max}}\sum_{n_z=1}^{n_{z\max}} F_{-3}(\eta_{7\mathrm{SLMSQ}}) \tag{10.4}$$

10.2.3 The CEC in IV–VI QWSLs with graded interfaces under magnetic quantization

Using (1.169a), (1.169b) and (8.15), the ΔC_{44} and ΔC_{456} can respectively be expressed in this case as

$$\Delta C_{44} = \left(\frac{-G_0^2}{9k_BTd_z}\right)\frac{eB}{\pi\hbar}\sum_{n=0}^{n_{\max}}\sum_{n_z=1}^{n_{z\max}} F_{-2}(\eta_{8\mathrm{SLMSQ}}) \tag{10.5}$$

$$\Delta C_{456} = \left(\frac{G_0^3}{27(k_BT)^2d_z}\right)\frac{eB}{\pi\hbar}\sum_{n=0}^{n_{\max}}\sum_{n_z=1}^{n_{z\max}} F_{-3}(\eta_{8\mathrm{SLMSQ}}) \tag{10.6}$$

10.2.4 The CEC in HgTe/CdTe QWSLs with graded interfaces under magnetic quantization

Using (1.169a), (1.169b) and (8.21), the ΔC_{44} and ΔC_{456} can respectively be expressed in this case as

$$\Delta C_{44} = \left(\frac{-G_0^2}{9k_BTd_z}\right)\frac{eB}{\pi\hbar}\sum_{n=0}^{n_{\max}}\sum_{n_z=1}^{n_{z\max}} F_{-2}(\eta_{9\mathrm{SLMSQ}}) \tag{10.7}$$

$$\Delta C_{456} = \left(\frac{G_0^3}{27(k_BT)^2d_z}\right)\frac{eB}{\pi\hbar}\sum_{n=0}^{n_{\max}}\sum_{n_z=1}^{n_{z\max}} F_{-3}(\eta_{9\mathrm{SLMSQ}}) \tag{10.8}$$

10.2.5 The CEC in strained layer QWSLs with graded interfaces under magnetic quantization

Using (1.169a), (1.169b) and (8.27), the ΔC_{44} and ΔC_{456} can respectively be expressed in this case as

$$\Delta C_{44} = \left(\frac{-G_0^2}{9k_BTd_z}\right)\frac{eB}{\pi\hbar}\sum_{n=0}^{n_{\max}}\sum_{n_z=1}^{n_{z\max}} F_{-2}(\eta_{10\mathrm{SLMSQ}}) \tag{10.9}$$

$$\Delta C_{456} = \left(\frac{G_0^3}{27(k_BT)^2d_z}\right)\frac{eB}{\pi\hbar}\sum_{n=0}^{n_{\max}}\sum_{n_z=1}^{n_{z\max}} F_{-3}(\eta_{10\mathrm{SLMSQ}}) \tag{10.10}$$

10.2.6 The CEC in III–V QW effective mass superlattices under magnetic quantization

Using (1.169a), (1.169b) and (8.33), the ΔC_{44} and ΔC_{456} can respectively be expressed in this case as

$$\Delta C_{44} = \left(\frac{-G_0^2}{9k_BTd_z}\right)\frac{eB}{\pi\hbar}\sum_{n=0}^{n_{\max}}\sum_{n_z=1}^{n_{z\max}} F_{-2}(\eta_{11\mathrm{SLMSQ}}) \tag{10.11}$$

$$\Delta C_{456} = \left(\frac{G_0^3}{27(k_BT)^2d_z}\right)\frac{eB}{\pi\hbar}\sum_{n=0}^{n_{\max}}\sum_{n_z=1}^{n_{z\max}} F_{-3}(\eta_{11\mathrm{SLMSQ}}) \tag{10.12}$$

10.2.7 The CEC in II–VI QW effective mass superlattices under magnetic quantization

Using (1.169a), (1.169b) and (8.39), the ΔC_{44} and ΔC_{456} can respectively be expressed in this case as

$$\Delta C_{44} = \left(\frac{-G_0^2}{9k_BTd_z}\right)\frac{eB}{\pi h}\sum_{n=0}^{n_{\max}}\sum_{n_z=1}^{n_{z\max}} F_{-2}(\eta_{12\mathrm{SLMSQ}}) \tag{10.13}$$

$$\Delta C_{456} = \left(\frac{G_0^3}{27(k_BT)^2d_z}\right)\frac{eB}{\pi h}\sum_{n=0}^{n_{\max}}\sum_{n_z=1}^{n_{z\max}} F_{-3}(\eta_{12\mathrm{SLMSQ}}) \tag{10.14}$$

10.2.8 The CEC in IV–VI QW effective mass superlattices under magnetic quantization

Using (1.169a), (1.169b) and (8.45), the ΔC_{44} and ΔC_{456} can respectively be expressed in this case as

$$\Delta C_{44} = \left(\frac{-G_0^2}{9k_BTd_z}\right)\frac{eB}{\pi\hbar}\sum_{n=0}^{n_{\max}}\sum_{n_z=1}^{n_{z\max}} F_{-2}(\eta_{13\mathrm{SLMSQ}}) \tag{10.15}$$

$$\Delta C_{456} = \left(\frac{G_0^3}{27(k_BT)^2d_z}\right)\frac{eB}{\pi\hbar}\sum_{n=0}^{n_{\max}}\sum_{n_z=1}^{n_{z\max}} F_{-3}(\eta_{13\mathrm{SLMSQ}}) \tag{10.16}$$

*10.2.9 The CEC in HgTe/CdTe QW effective mass superlattices under magnetic quantization

Using (1.169a), (1.169b) and (8.51), the ΔC_{44} and ΔC_{456} can respectively be expressed in this case as

$$\Delta C_{44} = \left(\frac{-G_0^2}{9k_BTd_z}\right)\frac{eB}{\pi\hbar}\sum_{n=0}^{n_{\max}}\sum_{n_z=1}^{n_{z\max}} F_{-2}(\eta_{14\mathrm{SLMSQ}}) \tag{10.17}$$

$$\Delta C_{456} = \left(\frac{G_0^3}{27(k_BT)^2d_z}\right)\frac{eB}{\pi\hbar}\sum_{n=0}^{n_{\max}}\sum_{n_z=1}^{n_{z\max}} F_{-3}(\eta_{14\mathrm{SLMSQ}}) \tag{10.18}$$

*10.2.10 The CEC in strained layer QW effective mass superlattices under magnetic quantization

Using (1.169a), (1.169b) and (8.58), the ΔC_{44} and ΔC_{456} can respectively be expressed in this case as

$$\Delta C_{44} = \left(\frac{-G_0^2}{9k_BTd_z}\right)\frac{eB}{\pi\hbar}\sum_{n=0}^{n_{\max}}\sum_{n_z=1}^{n_{z\max}} F_{-2}(\eta_{15\mathrm{SLMSQ}}) \tag{10.19}$$

$$\Delta C_{456} = \left(\frac{G_0^3}{27(k_BT)^2d_z}\right)\frac{eB}{\pi\hbar}\sum_{n=0}^{n_{\max}}\sum_{n_z=1}^{n_{z\max}} F_{-3}(\eta_{15\mathrm{SLMSQ}}) \tag{10.20}$$

10.3 Summary and Conclusion

The DRs in magneto-size quantized of HD materials exhibit the fact that the total energy is quantized since the corresponding wave vector space is totally quantized by quantizing magnetic field and size quantization along z-direction.

- The DOS functions for all the QWSL in this case are series of non-uniformly distributed Dirac's Delta functions at specified quantized points in the respective energy axis. The spacing between the consecutive Delta functions are functions of energy band constants and quantization of the wave vector space of a particular material. The DOS function needs two summations namely one summation over the Landau quantum number and the other one is due to size quantization.
- It may be noted that the energy levels in magneto-size quantized QWSL lead to the discrete energy levels, somewhat like atomic energy levels, which produce very large changes. This follows from the inherent nature of the quantum confinement of the carrier gas dealt with here. Consequently, the crossing of the Fermi level by the size-quantized levels in this case would have much greater impact on the redistribution of the carriers among the allowed levels, as compared to that found for QWs, NWs and QDs respectively for the electronic properties in general.
- By using the appropriate equations we can plot the CEC with respect to many variables in all the cases of this chapter. We have not given any plot of the CEC in this case since we believe our able readers will plot the same and enjoy intricate physics, computational analysis, advanced mathematics, and will create new problems both theoretical and experimental.
- It is the band structure which changes in a fundamental way and consequently all the physical properties of all the electronic materials changes radically leading to new physical concepts.

10.4 Open Research Problems

(R.10.1) Investigate the magneto CEC in the presence of arbitrarily oriented quantizing magnetic field for III–V, II–VI, IV–VI and HgTe/CdTe quantum well with graded interfaces and also the effective mass superlattices of the aforementioned materials.

(R.10.2) Investigate the magneto CEC in the presence of arbitrarily oriented quantizing magnetic field for strained layer, random, short period and Fibonacci, polytype and saw-tooth quantum well superlattices.

(R.10.3) Investigate the magneto CEC in the presence of arbitrarily oriented quantizing magnetic field in the presence of spin and broadening respectively.

(R.10.4) Investigate the magneto CEC under an arbitrarily oriented quantizing magnetic alternating field and crossed alternating electric field by including broadening and the electron spin respectively.

References

[1] K. P. Ghatak, B. Mitra, *Phys. Scr.*, 46, 182 **(1992)**; B. Nag, K. P. Ghatak, *J. Phys. Chem. Solids*, 58, 427 **(1997)**; K. P. Ghatak, B. Nag, *Phys. Stat. Sol. (b)*, 205, 519 **(1998)**.

[2] K. P. Ghatak, J. Y. Siddiqui, B. Nag, *Phys. Lett. A*, 282, 428 **(2001)**.

[3] K. P. Ghatak, S. N. Biswas, *MRS Proc.*, 308, 445 **(1993)**; K. P. Ghatak, J. P. Banerjee, P. K. Chakrabarty, B. Nag, *J. Wave Mater. Interact.*, 11, 166 **(1996)**; L. J. Singh, S. Choudhary, A. Mallik, K. P. Ghatak, *J. Comput. Theor. Nanosci.*, 2, 287 **(2005)**.

[4] A. Mallik, K. P. Ghatak, S. Choudhary, L. J. Singh, *J. Comput. Theor. Nanosci.*, 2(2), 287–292 **(2005)**; K. P. Ghatak, *Int. J. Electron.*, 71, 239 **(1991)**; K. P. Ghatak, *Acta Phys. Hungarica*, 68, 253 **(1990)**.

[5] B. Mitra, K. P. Ghatak, *Phys. Stat. Sol. (b)*, 154, K35 **(1989)**; K. P. Ghatak, P. K. Bose, *J. Wave Mater. Interact.*, 12, 53 **(1997)**; S. M. Adhikari, K. P. Ghatak, *J. Nanoeng. Nanomanuf.*, 3, 48 **(2013)**.

[6] B. Chatterjee, S. Chakrabarti, S. K. Sen, M. Mitra, K. P. Ghatak, *Quant. Matt.*, 5, 85 **(2016)**; K. P. Ghatak, S. Bhattacharya, S. S. Roy, L. J. Singh, *Nonlinear Opt.*, 32, 307 **(2010)**; D. Baruah, S. Choudhury, K. M. Singh, K. P. Ghatak, *J. Phys. Conf. Ser.*, 61, 80 **(2007)**.

[7] K. P. Ghatak, S. N. Biswas, *Acta Phys. Slovaca*, 43, 425 **(1993)**. K. P. Ghatak, M. Mondal, *Phys. Stat. Sol. (b)*, 170, 57 **(1992)**; K. P. Ghatak, D. K. Basu, B. Nag, *J. Phys. Chem. Solids*, 58, 133 **(1997)**.

[8] K. P. Ghatak, *Phys. Stat. Sol. (b)*, 154, K29 **(1989)**; K. P. Ghatak, J. P. Banerjee, B. Goswami, B. Nag, *Nonlinear Optics-Reading*, 16, 241 **(1996)**; B. Nag, K. P. Ghatak, *Nonlinear Optics-Reading*, 19, 1 **(1998)**.

[9] M. Mitra, M. Chakraborty, S. Debbarma, S. Chakraborty, S. K. Sen, *Quant. Matt.*, 5, 58 **(2016)**.

[10] K. P. Ghatak, B. Nag, *Nanostruct. Mater.*, 10, 923 **(1998)**; M. Mitra, M. Chakraborty, S. Debbarma, S. Chakraborty, S. K. Sen, *Quant. Matt.*, 5, 58 **(2016)**; B. Nag, K. P. Ghatak, *Phys. Scr.*, 54, 657 **(1996)**.

[11] K. P. Ghatak, S. Dutta, D. K. Basu, B. Nag, *Il Nuovo Cimento D*, 20, 227 **(1998)**; S. Bhattacharya, S. Chowdhury, S. Ghoshal, S. K. Biswas, D. De, K. P. Ghatak, *J. Comput. Theor. Nanosci.*, 3, 423 **(2006)**.

[12] K. P. Ghatak, S. Bhattacharya, S. Pahari, S. N. Mitra, P. K. Bose, D. De, *J. Phys. Chem. Solids*, 70, 122 **(2009)**; K. P. Ghatak, S. Karmakar, D. De, S. Pahari, S. K. Charaborty, S. K. Biswas, *J. Comput. Theor. Nanosci.*, 3,

153 **(2006)**; K. P. Ghatak, J. P. Banerjee, B. Nag, *J. Appl. Phys.*, 83, 1420 **(1998)**.

[13] K. P. Ghatak, S. K. Biswas, D. De, S. Ghosal, S. Chatterjee, *Phys. B*, 353, 127 **(2004)**; K. P. Ghatak, J. P. Banerjee, D. Bhattacharyya, B. Nag, *Nanotechnology*, 7, 110 **(1996)**.

[14] K. P. Ghatak, B. Nag, G. Mazumder, *MRS Proc.*, 379, 109 **(1995)**; R. W. Keyes, *IBM J. Res. Dev.*, 5, 266 **(1961)**; A. K. Sreedhar and S. C. Gupta, *Phys. Rev.*, 5B, 1360 **(1972)**.

Chapter 11

The CEC in Heavily Doped Quantum Dots

I must bow down my head to the dust of the all people's feet so that I can hear God's laugh from every heart.

11.1 Introduction

In Sections 11.2.1–11.2.15, we have investigated the CEC in QDs of HD nonlinear optical and tetragonal, III–V, II–VI, GaP, Ge, Te, Graphite, $PtSb_2$, II–V, GaSb, stressed Kane type, IV–VI, Lead Germanium Telluride, Zinc and Cadmium Diphosphides and Bi_2Te_3 respectively. Section 11.3 contains the summary and conclusion pertaining to this chapter. Section 11.4 presents 22 open research problems.

11.2 Theoretical Background

*11.2.1 The CEC in HDQD of nonlinear optical and tetragonal materials

It is well known that the band tails are being formed in the forbidden zone of HDS and can be explained by the overlapping of the impurity band with the conduction and valence bands [1, 2]. Kane [3] and Bonch Bruevich [4] have independently derived the theory of band tailing for semiconductors having unperturbed parabolic energy bands. Kane's model [3] was used to explain the experimental results on tunneling [5] and the optical absorption edges [6,7] in this context. Halperin and Lax [8] developed a model for band tailing applicable only to the deep tailing states. Although Kane's concept is often used in the literature for the investigation of band tailing [9,10], it may be noted that this model [3,11] suffers from serious assumptions in the sense that the local impurity potential is assumed to be small and slowly varying in space coordinates [10]. In this respect, the local impurity potential may be assumed to be a constant. In order to avoid these approximations, we are presenting here, the electron energy spectra for HDS for studying the EP based on the concept of the variation of the kinetic energy [2, 9] of the electron with the local point in space coordinates. This kinetic energy is then averaged over the entire region of variation using a Gaussian type potential energy. It may be noted that, a more general treatment of many-body theory for the DOS of HDS merges with one-electron theory under macroscopic conditions [2]. Also, the experimental results for the Fermi energy and others are the average effect of this macroscopic case. So, the present treatment of the one-electron system is more applicable to the experimental point of view and it is also easy to understand the overall effect in such a case [12]. In a HDS, each impurity atom is surrounded by the electrons, assuming a regular distribution of atoms, and it is screened independently [8, 11, 13]. The interaction energy between electrons and impurities is known as the impurity screening potential. This energy is determined by the inter-impurity distance and the screening radius, which is known as the screening length. The screening radius changes with the electron concentration and the effective mass. Furthermore, these entities are important for HDS in characterizing the semiconductor properties [14,15] and the modern electronic devices [9,16]. The works on Fermi energy and the screening length in an n-type GaAs have already been initiated in the literature [17,18], based on Kane's model. Incidentally, the limitations of Kane's model [3,10], as mentioned above, are also present in their studies.

The Gaussian distribution $F(V)$ of the impurity potential is given by [3,10]

$$F(V) = (\pi\eta_g^2)^{-1/2}\exp(-V^2/\eta_g^2) \tag{11.1}$$

where η_g is the impurity screening potential. It appears from (11.1) that the variance parameter η_g is not equal to zero, but the mean value is zero. Further, the impurities are assumed to be uncorrelated and the band mixing effect has been neglected in this simplified theoretical formalism.

The generalized unperturbed electron energy spectrum for the bulk specimens of the nonlinear optical materials in the absence of band tails can be expressed following (2.2) as

$$\frac{\hbar^2 k_z^2}{2m_\parallel^*} + \left(\frac{b_\parallel}{b_\perp}\frac{c_\perp}{c_\parallel}\right)\frac{\hbar^2 k_s^2}{2m_\perp^*}$$

$$= \left\{\frac{E(\alpha E+1)\left(b_\parallel E+1\right)}{\left(c_\parallel E+1\right)} + \frac{\alpha b_\parallel}{c_\parallel}\left[\delta E + \frac{2}{9}\left(\Delta_\parallel^2 - \Delta_\perp^2\right)\right]\right.$$

$$-\left(\frac{2}{9}\right)\frac{\alpha b_{\|}}{c_{\|}}\frac{\left(\Delta_{\|}^2-\Delta_{\perp}^2\right)}{(c_{\|}E+1)}\Bigg\}-\left(\frac{\hbar^2k_s^2}{2m_{\perp}^*}\right)\Bigg\{\left(\frac{b_{\|}}{b_{\perp}}\frac{c_{\perp}}{c_{\|}}\right)$$

$$\times\left[\left(\frac{\delta}{2}+\frac{\Delta_{\|}^2-\Delta_{\perp}^2}{6\Delta_{\|}}\right)\frac{\alpha_{\|}}{\alpha_{\|}E+1}+\left(\frac{\delta}{2}-\left\{\frac{\Delta_{\|}^2-\Delta_{\perp}^2}{6\Delta_{\|}}\right\}\right)\frac{c_{\|}}{c_{\|}E+1}\right]\Bigg\} \tag{11.2}$$

where $b_{\|} \equiv \left(E_g+\Delta_{\|}\right)^{-1}$, $c_{\perp} \equiv \left(E_g+\frac{2}{3}\Delta_{\perp}\right)^{-1}, b_{\perp} \equiv (E_g+\Delta_{\perp})^{-1}$, $c_{\|} = \left(E_g+\frac{2}{3}\Delta_{\|}\right)^{-1}$ and $\alpha = (E_g)^{-1}$. We have to consider the average of the kinetic energy in the order to obtain the EM in nonlinear optical materials in the presence of band tails. Using the (11.1) and (11.2), we get

$$\left[\frac{\hbar^2k_z^2}{2m_{\|}^*}\int_{-\infty}^{E}F(V)dV\right]+\left[\left(\frac{b_{\|}}{b_{\perp}}\frac{c_1}{c_{\|}}\right)\frac{\hbar^2k_s^2}{2m_{\perp}^*}\int_{-\infty}^{E}F(V)dV\right]$$

$$=\Bigg\{\int_{-\infty}^{E}\frac{(E-V)[\alpha(E-V)+1]\left[b_{\|}(E-V)+1\right]}{\left[c_{\|}(E-V)+1\right]}F(V)dV$$

$$+\frac{\alpha b_{\|}}{c_{\|}}\left[\delta\int_{-\infty}^{E}(E-V)F(V)dV+\frac{2}{9}\left(\Delta_{\|}^2-\Delta_{\perp}^2\right)\int_{-\infty}^{E}F(V)dV\right]$$

$$-\left(\frac{2}{9}\right)\frac{\alpha b_{\|}}{c_{\|}}\left(\Delta_{\|}^2-\Delta_{\perp}^2\right)\int_{-\infty}^{E}\frac{F(V)dV}{\left[c_{\|}(E-V)+1\right]}\Bigg\}-\left(\frac{\hbar^2k_s^2}{2m_{\perp}^*}\right)$$

$$\times\Bigg\{\left(\frac{b_{\|}}{b_{\perp}}\frac{c_{\perp}}{c_{\|}}\right)\left[\left(\frac{\delta}{2}+\frac{\Delta_{\|}^2-\Delta_{\perp}^2}{6\Delta_{\|}}\right)\alpha\int_{-\infty}^{E}\frac{F(V)dV}{[\alpha(E-V)+1]}\right.$$

$$\left.+\left(\frac{\delta}{2}-\frac{\Delta_{\|}^2-\Delta_{\perp}^2}{6\Delta_{\|}}\right)c_{\|}\int_{-\infty}^{E}\frac{F(V)dV}{\left[c_{\|}(E-V)+1\right]}\right]\Bigg\} \tag{11.3}$$

Equation (3.5) can be rewritten as

$$\frac{\hbar^2k_z^2}{2m_{\|}^*}I(1)+\left(\frac{b_{\|}}{b_{\perp}}\frac{c_1}{c_{\|}}\right)\frac{\hbar^2k_s^2}{2m_{\perp}^*}I(1)=\Bigg\{I_3(c_{\|})+\frac{\alpha b_{\|}}{c_{\|}}$$

$$\times\left[\delta I(4)+\frac{2}{9}(\Delta_{\|}^2-\Delta_{\perp}^2)I(1)\right]-\left(\frac{2}{9}\right)\frac{\alpha b_{\|}}{c_{\|}}(\Delta_{\|}^2-\Delta_{\perp}^2)I_6(c_{\|})\Bigg\}$$

$$-\left(\frac{\hbar^2 k_s^2}{2m_\perp^*}\right)\times\left\{\left(\frac{b_\parallel}{b_\perp}\frac{c_\perp}{c_\parallel}\right)\left[\left(\frac{\delta}{2}+\frac{\Delta_\parallel^2-\Delta_\perp^2}{6\Delta_\parallel}\right)\alpha I(\alpha)\right.\right.$$
$$\left.\left.+\left(\frac{\delta}{2}\left\{\frac{\Delta_\parallel^2-\Delta_\perp^2}{6\Delta_\parallel}\right\}\right)c_\parallel I(c_\parallel)\right]\right\} \tag{11.4}$$

where

$$I(1)=\int_{-\infty}^{E}F(V)dV \tag{11.5}$$

$$I_3(c_\parallel)=\int_{-\infty}^{E}\frac{(E-V)[\alpha(E-V)+1]\left[b_\parallel(E-V)+1\right]}{\left[c_\parallel(E-V)+1\right]}F(V)dV \tag{11.6}$$

$$I(4)=\int_{-\infty}^{E}(E-V)F(V)dV \tag{11.7}$$

$$I(\alpha)=\int_{-\infty}^{E}\frac{F(V)dV}{[\alpha(E-V)+1]} \tag{11.8}$$

Substituting $E-V\equiv x$ and, $\frac{x}{\eta_g}\equiv t_0$, we get from (3.7)

$$I(1)=\left(\exp\left(-\frac{E^2}{\eta_g^2}\right)/\sqrt{\pi}\right)\int_0^{\infty}\exp\left[-t_0^2+\left(\frac{2Et_0}{\eta_g}\right)\right]dt_0 \tag{11.9}$$

Thus,

$$I(1)=\left[\frac{1+Erf(E/\eta_g)}{2}\right] \tag{11.10}$$

where $Erf(E/\eta_g)$ is the error function.

From (11.7), one can write

$$I(4)=\left(1/\eta_g\sqrt{\pi}\right)\int_{-\infty}^{E}(E-V)\exp\left(-V^2/\eta_g^2\right)dV$$
$$=\frac{E}{2}\left[1+Erf(E/\eta_g)\right]-\left\{\frac{1}{\sqrt{\pi\eta_g^2}}\int_{-\infty}^{E}V\exp(-V^2/\eta_g^2)dV\right\} \tag{11.11}$$

After computing this simple integration, one obtains

Thus,

$$I(4) = \eta_g \exp(-E^2/\eta_g^2)(2\sqrt{\pi}) + \frac{E}{2}(1 + Erf(E/\eta_g)) = \gamma_0(E, \eta_g) \tag{11.12}$$

From (11.8), we can write

$$I(\alpha) = \frac{1}{\sqrt{\pi\eta_g^2}} \int_{-\infty}^{E} \frac{\exp(-V^2/\eta_g^2)dV}{[\alpha(E-V)+1]} \tag{11.13}$$

where $V \to \pm\infty$, $\frac{1}{[\alpha(E-V)+1]} \to 0$ and $\exp(-V^2/\eta_g^2) \to 0$.

Thus, (11.13) can be expressed as

$$I(\alpha) = (1/\alpha\eta_g\sqrt{\pi}) \int_{-\infty}^{\infty} \exp\left(-t^2\right)(u-t)^{-1}dt \tag{11.14}$$

where

$$\frac{V}{\eta_g} = t \quad \text{and} \quad u \equiv \left(\frac{1+\alpha E}{\alpha\eta}\right)$$

It is well known that [19, 20]

$$W(Z) = (i/\pi) \int_{-\infty}^{\infty} (Z-t)^{-1} \exp(-t^2)dt \tag{11.15}$$

In which $i = \sqrt{-1}$ and Z, in general, is a complex number.

We also know that [19, 20]

$$W(Z) = \exp(-Z^2)Erfc(-iZ) \tag{11.16}$$

where $Erfc(Z) \equiv 1 - Erf(Z)$.

Thus,

$$Erfc(-iu) = 1 - Erf(-iu) \tag{11.17}$$

Since, $Erf(-iu) = -Erf(iu)$

Therefore,

$$I(\alpha) = [-i\sqrt{\pi}/\alpha\eta_g] \exp(-u^2)[1 + Erf(iu)] \tag{11.18}$$

It is well-know that [19, 20]

$$\begin{aligned} Erf(x+iy) = {} & Erf(x) + \left(\frac{e^{-x^2}}{2\pi x}\right) \\ & \times \left[(1-\cos(2xy)) + i\sin(2xy) + \frac{2}{\pi}e^{-x^2}\sum_{p=1}^{\infty}\frac{\exp(-p^2/4)}{(p^2+4x^2)}\right] \\ & \times [f_p(x,y) + ig_p(x,y) + \varepsilon(x,y)] \end{aligned} \tag{11.19}$$

where

$$
\begin{aligned}
f_p(x,y) &= [2x - 2x\cosh(py)\cos(2xy) + p\sinh(py)\sin(2xy)]g_p(x,y) \\
&\equiv [2x\cosh(py)\sin(2xy) + p\sinh(py)\cos(2xy)], |\varepsilon(x,y)| \\
&\approx 10^{-15}|Erf(x+iy)|
\end{aligned}
$$

Substituting $x = 0$ and $y = u$ in (11.19), one obtains,

$$
Erf(iu) = \left(\frac{2i}{\pi}\right)\sum_{p=1}^{\infty}\left\{\frac{\exp\left(-p^2/4\right)}{p}\sinh(pu)\right\} \tag{11.20}
$$

Therefore, one can write

$$
I(\alpha) = C_{21}(\alpha, E, \eta_g) - iD_{21}(\alpha, E, \eta_g) \tag{11.21}
$$

where

$$
C_{21}(\alpha, E, \eta_g) = \left[\frac{2}{\alpha\eta_g\sqrt{\pi}}\right]\exp(-u^2)\left[\sum_{p=1}^{\infty}\left\{\frac{\exp(-p^2/4)}{p}\sinh(pu)\right\}\right]
$$

and

$$
D_{21}(\alpha, E, \eta_g) = \left[\frac{\sqrt{\pi}}{\alpha\eta_g}\exp(-u^2)\right]
$$

Equation (11.21) consists of both real and imaginary parts and therefore, $I(\alpha)$ is complex, which can also be proved by using the method of analytic continuation of Complex Analysis.

The integral $I_3(c_{||})$ in (11.6) can be written as

$$
\begin{aligned}
I_3(c_{||}) &= \left(\frac{\alpha b_{||}}{c_{||}}\right)I(5) + \left(\frac{\alpha c_{||} + b_{||}c_{||} - \alpha b_{||}}{c_{||}^2}\right)I(4) + \frac{1}{c_{||}}\left(1 - \frac{\alpha}{c_{||}}\right) \\
&\times \left(1 - \frac{b_{||}}{c_{||}}\right)I(1) - \left\{\frac{1}{c_{||}}\left(1 - \frac{\alpha}{c_{||}}\right)\left(1 - \frac{b_{||}}{c_{||}}\right)I\left(c_{||}\right)\right\}
\end{aligned} \tag{11.22}
$$

where

$$
I(5) = \int_{-\infty}^{E}(E - V)^2 F(V)dV \tag{11.23}
$$

From (11.23), we can write

$$I(5) = \frac{1}{\sqrt{\pi\eta_g^2}}\left[E^2\int_{-\infty}^{E}\exp\left(\frac{-V^2}{\eta_g^2}\right)dV - 2E\int_{-\infty}^{E}V\exp\left(\frac{-V^2}{\eta_g^2}\right)dV + \int_{-\infty}^{E}V^2\exp\left(\frac{-V^2}{\eta_g^2}\right)dV\right]$$

The evaluations of the component integrals lead us to write

$$\begin{aligned} I(5) &= \frac{\eta_g E}{2\sqrt{\pi}}\exp\left(\frac{-E^2}{\eta_g^2}\right) + \frac{1}{4}\left(\eta_g^2 + 2E^2\right)\left[1 + Erf\left(\frac{E}{\eta_g}\right)\right] \\ &= \theta_0\left(E, \eta_g\right) \end{aligned} \tag{11.24}$$

Thus, combining the aforementioned equations, $I_3(c_{\parallel})$ can be expressed as

$$I_3\left(c_{\parallel}\right) = A_{21}\left(E, \eta_g\right) + iB_{21}\left(E, \eta_g\right) \tag{11.25}$$

where

$$\begin{aligned} A_{21}(E,\eta) = &\left[\frac{\alpha b_{\parallel}}{c_{\parallel}}\left[\frac{\eta_g E}{2\sqrt{\pi}}\exp\left(\frac{-E^2}{\eta_g^2}\right) - \frac{1}{4}\left(\eta_g^2 + 2E^2\right)\right.\right. \\ &\left.\times\left\{1 + Erf\left(\frac{E}{\eta_g}\right)\right\}\right] + \left[\frac{\alpha c_{\parallel} + b_{\parallel}c_{\parallel} - \alpha b_{\parallel}}{c_{\parallel}^2}\right] \\ &\times\left\{\frac{E}{2}\left[1 + Erf\left(\frac{E}{\eta_g}\right)\right] + \frac{\eta\exp(-E^2/\eta_g^2)}{2\sqrt{\pi}}\right\} \\ &+ \frac{1}{c_{\parallel}}\left(1 - \frac{\alpha}{c_{\parallel}}\right)\left(1 - \frac{b_{\parallel}}{c_{\parallel}}\right)\frac{1}{2}\left[1 + Erf\left(\frac{E}{\eta_g}\right)\right] \\ &- \left\{\frac{2}{c_{\parallel}^2\eta_g\sqrt{\pi}}\left(1 - \frac{\alpha}{c_{\parallel}}\right)\left(1 - \frac{b_{\parallel}}{c_{\parallel}}\right)\exp\left(-u_1^2\right)\right\} \\ &\left.\times\left[\sum_{p=1}^{\infty}\left\{\frac{\exp(-p^2/4)}{p}\sinh(pu_1)\right\}\right]\right], \quad u_1 \equiv \left[\frac{1 + c_{\parallel}E}{c_{\parallel}\eta_g}\right] \end{aligned}$$

and

$$B_{21}(E,\eta_g) \equiv \frac{\sqrt{\pi}}{c_{\parallel}^2\eta_g}\left(1 - \frac{\alpha}{c_{\parallel}}\right)\left(1 - \frac{b_{\parallel}}{c_{\parallel}}\right)\exp(-u_1^2)$$

Therefore, the combination of all the appropriate integrals together with algebraic manipulations leads to the expression of the dispersion relation of the conduction electrons of HD nonlinear optical materials forming Gaussian band tails as

$$\frac{\hbar^2 k_z^2}{2m_{\|}^* T_{21}(E,\eta_g)} + \frac{\hbar^2 k_s^2}{2m_{\perp}^* T_{22}(E,\eta_g)} = 1 \tag{11.26}$$

where $T_{21}(E,\eta_g)$ and $T_{22}(E,\eta_g)$ have both real and complex parts and are given by

$$T_{21}(E,\eta_g) = [T_{27}(E,\eta_g) + iT_{28}(E,\eta_g)],$$

$$T_{27}(E,\eta_g) = \left[\frac{T_{23}(E,\eta_g)}{T_5(E,\eta_g)}\right]$$

$$T_{23}(E,\eta_g) \equiv \left[A_{21}(E,\eta_g) + \frac{\alpha b_{\|}}{c_{\|}}\left[\delta\gamma_0(E,\eta_g) + \frac{1}{9}(\Delta_{\|}^2 - \Delta_{\perp}^2)\right.\right.$$
$$\left.\left.\times\ [1 + Erf(E/\eta_g)]\right] - \left\{\frac{2}{9}\left(\frac{\alpha b_{\|}}{c_{\|}}\right)\left(\Delta_{\|}^2 - \Delta_{\perp}^2\right) G_{21}(c_{\|}, E, \eta_g)\right\}\right],$$

$$G_{21}(E,\eta_g) \equiv \frac{2}{c_{\|}\eta_g\sqrt{\pi}} \exp\left(-u_1^2\right) \sum_{p=1}^{\infty} \left\{\frac{\exp\left(-p^2/4\right)}{p} \sinh(pu_1)\right\},$$

$$T_5(E,\eta_g) \equiv \frac{1}{2}[1 + Erf(E/\eta_g)]$$

$$T_{28}(E,\eta_g) \equiv \left[\frac{T_{24}(E,\eta_g)}{T_5(E,\eta_g)}\right],$$

$$T_{24}(E,\eta_g) \equiv \left[B_{21}(E,\eta_g) + \frac{2}{9}\frac{\alpha b_{\|}}{c_{\|}}\left(\Delta_{\|}^2 - \Delta_{\perp}^2\right) H_{21}(c_{\|}, E, \eta_g)\right]$$

$$H_{21}(c_{\|}, E, \eta_g) \equiv \left[\frac{\sqrt{\pi}}{\eta_g c_{\|}} \exp\left(-u_1^2\right)\right],$$

$$T_{22}(E,\eta_g) \equiv [T_{29}(E,\eta_g) + iT_{30}(E,\eta_g)],$$

$$T_{29}(E,\eta_g) \equiv \frac{T_3(E,\eta_g)T_{25}(E,\eta_g) - T_{24}(E,\eta_g)T_{26}(E,\eta_g)}{[(T_{25}(E,\eta_g))^2 + (T_{26}(E,\eta_g))^2]}$$

$$T_{25}(E,\eta_g) \equiv \left[\left(\frac{b_{\|}}{b_{\perp}}\frac{c_{\perp}}{c_{\|}}\right)\frac{1}{2}\left[1 + Erf\left(\frac{E}{\eta_g}\right)\right] + \left(\frac{b_{\|}}{b_{\perp}}\frac{c_{\|}}{c_{\perp}}\right)\right.$$

$$\times\left(\frac{\delta}{2}+\left[\frac{\Delta_{\|}^{2}-\Delta_{\perp}^{2}}{6\Delta_{\|}}\right]\right)\alpha_{\|}C_{21}(\alpha_{\|},E,\eta_g)+\left(\frac{b_{\|}c_{\perp}}{b_{\perp}}\right)$$

$$\times\left(\frac{\delta}{2}-\left[\frac{\Delta_{\|}^{2}-\Delta_{\perp}^{2}}{6\Delta_{\|}}\right]\right)G_{21}\left(\alpha_{\|},E,\eta_g\right)\Bigg],$$

$$C_{21}\left(\alpha,E,\eta_g\right)\equiv\left[\frac{2}{\alpha\sqrt{\pi}\eta_g}\exp(-u^2)\left[\sum_{p=1}^{\infty}\frac{\exp(-p^2/4)}{p}\sinh(pu)\right]\right],$$

$$T_{26}\left(E,\eta_g\right)\equiv\left(\frac{b_{\|}}{b_{\perp}}\frac{c_{\perp}}{c_{\|}}\right)\left(\frac{\delta}{2}-\frac{\Delta_{\|}^{2}-\Delta_{\perp}^{2}}{6\Delta_{\|}}\right)\alpha D_{21}\left(\alpha,E,\eta_g\right)$$

$$+\frac{b_{\|}c_{\perp}}{b_{\perp}}\left(\frac{\delta}{2}-\frac{\Delta_{\|}^{2}-\Delta_{\perp}^{2}}{6\Delta_{\|}}\right)H_{21}\left(c_{\|},E,\eta_g\right)$$

and

$$T_{30}(E,\eta_g)\equiv\frac{T_{24}(E,\eta_g)T_{25}(E,\eta_g)+T_{23}(E,\eta_g)T_{26}(E,\eta_g)}{[(T_{25}(E,\eta_g))^2+(T_{26}(E,\eta_g))^2]}$$

From (11.26), it appears that the energy spectrum in HD nonlinear optical and tetragonal materials is complex.

The complex nature of the electron dispersion law in HD materials occurs from the existence of the essential poles in the corresponding electron energy spectrum in the absence of band tails. It may be noted that the complex band structures have already been studied for bulk materials and superlattices without heavy doping and bears no relationship with the complex electron dispersion law as indicated by (11.26). The physical picture behind the formulation of the complex energy spectrum in HDS is the interaction of the impurity atoms in the tails with the splitting constants of the valance bands. More is the interaction; more is the prominence of the complex part than the other case.

The DR in HDQDs in this case can be written following (11.26) as

$$\frac{\hbar^2\left(n_z\pi/d_z\right)^2}{2m_{\|}^{*}T_{21}\left(E_{1\mathrm{QBHD}},\eta_g\right)}+\frac{\hbar^2\left(n_y\pi/d_y\right)^2}{2m_{\perp}^{*}T_{22}\left(E_{1\mathrm{QBHD}},\eta_g\right)}$$
$$+\frac{\hbar^2\left(n_x\pi/d_x\right)^2}{2m_{\perp}^{*}T_{22}\left(E_{1\mathrm{QBHD}},\eta_g\right)}=1\tag{11.27}$$

where $E_{1\mathrm{QBHD}}$ is the total quantized energy in this case.

The total DOS function in this case is given by

$$N_{0\mathrm{DT}}(E) = \frac{2g_v}{d_x d_y d_z} \sum_{n_x=1}^{n_{x\max}} \sum_{n_y=1}^{n_{y\max}} \sum_{n_z=1}^{n_{z\max}} \delta'\left(E - E_{1\mathrm{QBHD}}\right) \tag{11.28}$$

The electron concentration can be expressed as

$$n_0 = \text{Real part of } \bar{n}_0 \tag{11.29}$$

where

$$\bar{n}_0 = \frac{2g_v}{d_x d_y d_z} \sum_{n_x=1}^{n_{x\max}} \sum_{n_y=1}^{n_{y\max}} \sum_{n_z=1}^{n_{z\max}} F_{-1}(\eta_{1\mathrm{QD}}) \tag{11.30}$$

in which $\eta_{1\mathrm{QD}} = (E_{\mathrm{FQD}} - E_{1\mathrm{QBHD}})(k_B T)^{-1}$ and E_{FQD} is the Fermi energy in this case.

Using (1.170a), (1.170b) and (11.29), we can find the ΔC_{44} and ΔC_{456} in this case, where $n_{0\mathrm{HD}}$ is replaced by n_0.

11.2.2 The CEC in HDQD of III–V materials

(a) The three band model of Kane [21]

Under the conditions, $\delta = 0, \Delta_{\|} = \Delta_{\perp} = \Delta$ (isotropic spin orbit splitting constant) and $m_{\|}^* = m_{\perp}^* = m_c$ (isotropic effective electron mass at the edge of the conduction band), the HD electron dispersion law in this case can be written as

$$\frac{\hbar^2 k^2}{2m_c} = T_{31}\left(E, \eta_g\right) + iT_{32}\left(E, \eta_g\right) \tag{11.31}$$

where

$$T_{31}(E, \eta_g) \equiv \left(\frac{2}{1 + Erf(E/\eta_g)}\right)$$

$$\left[\frac{ab}{c}\theta_0(E, \eta_g) + \left[\frac{\alpha c + bc - \alpha b}{c^2}\right]\gamma_0(E, \eta_g)\right.$$

$$+ \frac{1}{c}\left(1 - \frac{\alpha}{c}\right)\left(1 - \frac{b}{c}\right)\frac{1}{2}\left[1 + Erf\left(\frac{E}{\eta_g}\right)\right] - \frac{1}{c}\left(1 - \frac{\alpha}{c}\right)$$

$$\left.\times\left(1 - \frac{b}{c}\right)\frac{2}{c\eta_g\sqrt{\pi}}\exp\left(-u_2^2\right)\left[\sum_{p=1}^{\infty}\frac{\exp\left(-p^2/4\right)}{p}\sinh\left(pu_2\right)\right]\right],$$

$$b \equiv (E_g + \Delta)^{-1}, \quad c \equiv \left(E_g + \frac{2}{3}\Delta\right)^{-1}$$

$$u_2 \equiv \frac{1 + cE}{c\eta_g} \quad \text{and} \quad T_{32}(E, \eta_g) \equiv \left(\frac{2}{1 + Erf\,(E/\eta_g)}\right)\frac{1}{c}$$

$$\times \left(1 - \frac{\alpha}{c}\right)\left(1 - \frac{b}{c}\right)\frac{\sqrt{\pi}}{c\eta_g}\frac{\sqrt{\pi}}{c\eta_g}\exp\left(-u_2^2\right)$$

Thus, the complex energy spectrum occurs due to the term $T_{32}(E, \eta_g)$ and this imaginary band is quite different from the forbidden energy band.

The DR in HDQD in this case can be written following (11.31) as

$$\frac{\hbar^2 (n_z\pi/d_z)^2}{2m_c} + \frac{\hbar^2 (n_y\pi/d_y)^2}{2m_c} + \frac{\hbar^2 k_x{}^2}{2m_c} = T_{44}\left(E_{2\text{QBHD}}, \eta_g\right) \quad (11.32)$$

where $T_{44}\left(E_{2\text{QBHD}}, \eta_g\right) = T_{31}\left(E_{2\text{QBHD}}, \eta_g\right) + iT_{31}\left(E_{2\text{QBHD}}, \eta_g\right)$ and $E_{2\text{QBHD}}$ is the total quantized energy in this case.

The DOS function is given by

$$N_{\text{0DT}}(E) = \frac{2g_v}{d_x d_y d_z}\sum_{n_x=1}^{n_{x\max}}\sum_{n_y=1}^{n_{y\max}}\sum_{n_z=1}^{n_{z\max}}\delta'\left(E - E_{2\text{QBHD}}\right) \quad (11.33)$$

The electron concentration can be expressed as

$$n_0 = \text{Real part of } \bar{n}_0 \quad (11.34)$$

where

$$\bar{n}_0 = \frac{2g_v}{d_x d_y d_z}\sum_{n_x=1}^{n_{x\max}}\sum_{n_y=1}^{n_{y\max}}\sum_{n_z=1}^{n_{z\max}}F_{-1}\left(\eta_{2\text{QD}}\right) \quad (11.35)$$

In which $\eta_{2\text{QD}} = \left(E_{\text{FQD}} - E_{2\text{QBHD}}\right)(k_B T)^{-1}$.

Using (1.170a), (1.170b) and (11.34), we can find the ΔC_{44} and ΔC_{456} in this case, where $n_{0\text{HD}}$ is replaced by n_0.

(b) The HD two band model of Kane [21]

The dispersion relation in HD III–V, ternary and quaternary materials whose energy spectrum in the absence of band tails obeys the two band model of Kane can be written as

$$\frac{\hbar^2 k^2}{2m_c} = \gamma_2 (E, \eta_g) \tag{11.36}$$

where $\gamma_2 (E, \eta_g) \equiv \left[\frac{2}{1+Erf(E/\eta_g)}\right] [\gamma_0 (E, \eta_g) + \alpha\theta_0 (E, \eta_g)]$.

The DR in HDQD in this case can be written as

$$\frac{\hbar^2 (n_z\pi/d_z)^2}{2m_c} + \frac{\hbar^2 (n_y\pi/d_y)^2}{2m_c} + \frac{\hbar^2 (n_x\pi/d_x)^2}{2m_c} = \gamma_2 (\mathrm{E}_{3\mathrm{QBHD}}, \eta_g) \tag{11.37}$$

where $E_{3\mathrm{QBHD}}$ is the total quantized energy in this case.

The DOS function is given by

$$N_{0\mathrm{DT}}(E) = \frac{2g_v}{d_x d_y d_z} \sum_{n_x=1}^{n_{x\max}} \sum_{n_y=1}^{n_{y\max}} \sum_{n_z=1}^{n_{z\max}} \delta' (E - E_{3\mathrm{QBHD}}) \tag{11.38}$$

The electron concentration can be expressed as

$$n_0 = \frac{2g_v}{d_x d_y d_z} \sum_{n_x=1}^{n_{x\max}} \sum_{n_y=1}^{n_{y\max}} \sum_{n_z=1}^{n_{z\max}} F_{-1} (\eta_{3\mathrm{QD}}) \tag{11.39}$$

where $\eta_{3\mathrm{QD}} = (E_{\mathrm{FQD}} - E_{3\mathrm{QBHD}}) (k_B T)^{-1}$.

Using (1.170a), (1.170b) and (11.39), we can find the ΔC_{44} and ΔC_{456} in this case, where $n_{0\mathrm{HD}}$ is replaced by n_0.

(c) The HD parabolic band model

The dispersion relation in HDS whose energy spectrum in the absence of band tails obeys the parabolic energy bands is given by

$$\frac{\hbar^2 k^2}{2m_c} = \gamma_3 (E, \eta_g) \tag{11.40}$$

where $\gamma_3(E, \eta_g) \equiv \left[\frac{2}{(1+Erf(E/\eta_g))}\right] \gamma_0(E, \eta_g)$.

Since the dispersion relation in accordance with the said model is an all zero function with no pole in the finite complex plane, therefore the HD counterpart will be totally real, which is also apparent form the expression (11.40).

The DR in HDQD in this case can be written as

$$\frac{\hbar^2 (n_z\pi/d_z)^2}{2m_c} + \frac{\hbar^2 (n_y\pi/d_y)^2}{2m_c} + \frac{\hbar^2 (n_x\pi/d_x)^2}{2m_c} = \gamma_3 (\mathrm{E}_{4\mathrm{QBHD}}, \eta_g) \tag{11.41}$$

where $E_{4\mathrm{QBHD}}$ is the total quantized energy in this case.

The DOS function is given by

$$N_{0\mathrm{DT}}(E) = \frac{2g_v}{d_x d_y d_z} \sum_{n_x=1}^{n_{x\max}} \sum_{n_y=1}^{n_{y\max}} \sum_{n_z=1}^{n_{z\max}} \delta' (E - E_{4\mathrm{QBHD}}) \tag{11.42}$$

The electron concentration can be expressed as

$$n_0 = \frac{2g_v}{d_x d_y d_z} \sum_{n_x=1}^{n_{x\max}} \sum_{n_y=1}^{n_{y\max}} \sum_{n_z=1}^{n_{z\max}} F_{-1} (\eta_{4\mathrm{QD}}) \tag{11.43}$$

where $\eta_{4\mathrm{QD}} = (E_{\mathrm{FQD}} - E_{4\mathrm{QBHD}})(k_B T)^{-1}$.

Using (1.170a), (1.170b) and (11.43), we can find the ΔC_{44} and ΔC_{456} in this case, where $n_{0\mathrm{HD}}$ is replaced by n_0.

(d) The HD model of Stillman *et al.* [22]

In the presence of band tails, (1.26) gets transformed as

$$\frac{\hbar^2 k^2}{2m_c} = \theta_4 (E, \eta_g) \tag{11.44}$$

where $\theta_4(E, \eta_g) = a_{11}[1 - (1 - a_{12}\gamma_3(E, \eta_g))^{1/2}]$.

The DR in HDQD in this case can be written as

$$\frac{\hbar^2 (n_z\pi/d_z)^2}{2m_c} + \frac{\hbar^2 (n_y\pi/d_y)^2}{2m_c} + \frac{\hbar^2 (n_x\pi/d_x)^2}{2m_c} = \theta_4 (\mathrm{E}_{5\mathrm{QBHD}}, \eta_g) \tag{11.45}$$

where $\mathrm{E}_{5\mathrm{QBHD}}$ is the total quantized energy in this case.

The DOS function is given by

$$N_{0\mathrm{DT}}(E) = \frac{2g_v}{d_x d_y d_z} \sum_{n_x=1}^{n_{x\max}} \sum_{n_y=1}^{n_{y\max}} \sum_{n_z=1}^{n_{z\max}} \delta' (E - E_{5\mathrm{QBHD}}) \tag{11.46}$$

The electron concentration can be expressed as

$$n_0 = \frac{2g_v}{d_x d_y d_z} \sum_{n_x=1}^{n_{x\max}} \sum_{n_y=1}^{n_{y\max}} \sum_{n_z=1}^{n_{z\max}} F_{-1}\left(\eta_{5\mathrm{QD}}\right) \tag{11.47}$$

where $\eta_{5\mathrm{QD}} = (E_{\mathrm{FQD}} - E_{5\mathrm{QBHD}})(k_B T)^{-1}$.

Using (1.170a), (1.170b) and (11.47), we can find the ΔC_{44} and ΔC_{456} in this case, where $n_{0\mathrm{HD}}$ is replaced by n_0.

(e) The HD model of Palik *et al.* [23]

Under the condition of heavy doping forming Gaussian band tails, (1.39) assumes the form

$$\frac{\hbar^2 k^2}{2m_c} = I_{13}\left(E, \eta_g\right) \tag{11.48}$$

where $I_{13}\ (E, \eta_g) = b_{12}[a_{12} - ((a_{12})^2 - 4b_{11}\gamma_3(E, \eta_g))^{1/2}]$.

The DR in HDQD in this case can be written as

$$\frac{\hbar^2 \left(n_z \pi / d_z\right)^2}{2m_c} + \frac{\hbar^2 \left(n_y \pi / d_y\right)^2}{2m_c} + \frac{\hbar^2 \left(n_x \pi / d_x\right)^2}{2m_c} = \theta_5\left(\mathrm{E}_{6\mathrm{QBHD}}, \eta_g\right) \tag{11.49}$$

where $\theta_5\left(\mathrm{E}_{6\mathrm{QBHD}}, \eta_g\right) \equiv I_{13}\left(\mathrm{E}_{6\mathrm{QBHD}}, \eta_g\right)$ and $\mathrm{E}_{6\mathrm{QBHD}}$ is the total quantized energy in this case.

The DOS function is given by

$$N_{0\mathrm{DT}}(E) = \frac{2g_v}{d_x d_y d_z} \sum_{n_x=1}^{n_{x\max}} \sum_{n_y=1}^{n_{y\max}} \sum_{n_z=1}^{n_{z\max}} \delta'\left(E - E_{6\mathrm{QBHD}}\right) \tag{11.50}$$

The electron concentration can be expressed as

$$n_0 = \frac{2g_v}{d_x d_y d_z} \sum_{n_x=1}^{n_{x\max}} \sum_{n_y=1}^{n_{y\max}} \sum_{n_z=1}^{n_{z\max}} F_{-1}\left(\eta_{6\mathrm{QD}}\right) \tag{11.51}$$

$$\text{where } \eta_{6\mathrm{QD}} = (E_{\mathrm{FQD}} - E_{6\mathrm{QBHD}})(k_B T)^{-1}. \tag{11.52}$$

Using (1.170a), (1.170b) and (11.51), we can find the ΔC_{44} and ΔC_{456} in this case, where $n_{0\mathrm{HD}}$ is replaced by n_0.

(f) The HD model of Rossler [24]

The HDDR in QDHD in this case can be expressed using (1.34) as

$$\begin{aligned}\gamma_3\left(\bar{E}_{8\mathrm{QBHD}},\eta_g\right) \equiv \frac{\hbar^2\pi^2}{2m_c}\left[\left(\frac{n_x}{d_x}\right)^2+\left(\frac{n_y}{d_y}\right)^2+\left(\frac{n_z}{d_z}\right)^2\right] \\ + \left[\alpha_{11}+\alpha_{12}\left[\left(\frac{\pi n_x}{d_x}\right)^2+\left(\frac{\pi n_y}{d_y}\right)^2+\left(\frac{\pi n_z}{d_z}\right)^2\right]^{1/2}\right] \\ \times\left[\left(\frac{\pi n_x}{d_x}\right)^2+\left(\frac{\pi n_y}{d_y}\right)^2+\left(\frac{\pi n_z}{d_z}\right)^2\right]^2 \\ + \left[\beta_{11}+\beta_{12}\left[\left(\frac{\pi n_x}{d_x}\right)^2+\left(\frac{\pi n_y}{d_y}\right)^2+\left(\frac{\pi n_z}{d_z}\right)^2\right]^{1/2}\right] \\ \times\left[\pi^4\left(\frac{n_x n_y}{d_x d_y}\right)^2+\pi^4\left(\frac{n_y n_z}{d_y d_z}\right)^2+\pi^4\left(\frac{n_z n_x}{d_z d_x}\right)^2\right] \\ \pm\left[\gamma_{11}+\gamma_{12}\left[\left(\frac{\pi n_x}{d_x}\right)^2+\left(\frac{\pi n_y}{d_y}\right)^2+\left(\frac{\pi n_z}{d_z}\right)^2\right]^{1/2}\right] \\ \times\left[\left[\left(\frac{\pi n_x}{d_x}\right)^2+\left(\frac{\pi n_y}{d_y}\right)^2+\left(\frac{\pi n_z}{d_z}\right)^2\right]\right. \\ \times\left[\pi^4\left(\frac{n_x n_y}{d_x d_y}\right)^2+\pi^4\left(\frac{n_y n_z}{d_y d_z}\right)^2+\pi^4\left(\frac{n_z n_x}{d_z d_x}\right)^2\right] \\ \left.-9\pi^6\left(n_x n_y n_z/d_x d_y d_z\right)^6\right]\end{aligned} \tag{11.53}$$

The DOS function can be expressed as

$$N_{\mathrm{ODT}}(E)=\frac{g_v}{d_x d_y d_z}\sum_{n_x=1}^{n_{x\max}}\sum_{n_y=1}^{n_{y\max}}\sum_{n_z=1}^{n_{z\max}}\delta'\left(E-\bar{E}_{8\mathrm{QBHD}}\right) \tag{11.54}$$

The electron concentration can be written as

$$n_0=\frac{g_v}{d_x d_y d_z}\sum_{n_x=1}^{n_{x\max}}\sum_{n_y=1}^{n_{y\max}}\sum_{n_z=1}^{n_{z\max}}F_{-1}\left(\bar{\eta}_{8\mathrm{QD}}\right) \tag{11.55}$$

where $\bar{\eta}_{8\mathrm{QD}}=(E_{\mathrm{FQD}}-\bar{E}_{8\mathrm{QBHD}})(k_BT)^{-1}$.

Using (1.170a), (1.170b) and (11.55), we can find the ΔC_{44} and ΔC_{456} in this case, where $n_{0\mathrm{HD}}$ is replaced by n_0.

(g) The HD model of Agafonov *et al.* [25]

The HD electron dispersion law can be written following (1.47) as

$$\gamma_3(E,\eta_g) = \frac{\bar{y}-\bar{E}_{g_0}}{2}\left[1 - T_5\left(\frac{k_x^4+k_y^4+k_z^4}{\bar{y}k^2}\right)\right] \tag{11.56}$$

The totally quantized energy can be written as

$$\gamma_3(\overline{\mathrm{E}}_{8\mathrm{QBHD1}},\eta_g) = \left(\frac{\Psi_{30}-E_{g_0}}{2}\right)\left[1 - \left[T_5\left[\left(\frac{\pi n_x}{d_x}\right)^4 + \left(\frac{\pi n_y}{d_y}\right)^4 + \left(\frac{\pi n_z}{d_z}\right)^4\right]\right]\right] \times \left[\psi_{30}\left[\left(\frac{\pi n_x}{d_x}\right)^4 + \left(\frac{\pi n_y}{d_y}\right)^4 + \left(\frac{\pi n_z}{d_z}\right)^4\right]\right]^{-1} \tag{11.57}$$

where $\psi_{30} = \left[{E_{g_0}}^2 + \frac{8}{3}P^2\left[\left(\frac{\pi n_x}{d_x}\right)^2 + \left(\frac{\pi n_y}{d_y}\right)^2 + \left(\frac{\pi n_z}{d_z}\right)^2\right]\right]^{1/2}$ and $\overline{\mathrm{E}}_{8\mathrm{QBHD1}}$ is the total quantized energy in this case.

The DOS function can be expressed as

$$N_{0\mathrm{DT}}(E) = \frac{2g_v}{d_x d_y d_z}\sum_{n_x=1}^{n_{x\max}}\sum_{n_y=1}^{n_{y\max}}\sum_{n_z=1}^{n_{z\max}} \delta'\left(E - \bar{E}_{8\mathrm{QBHD1}}\right) \tag{11.58}$$

The electron concentration can be expressed as

$$n_0 = \frac{2g_v}{d_x d_y d_z}\sum_{n_x=1}^{n_{x\max}}\sum_{n_y=1}^{n_{y\max}}\sum_{n_z=1}^{n_{z\max}} F_{-1}\left(\bar{\eta}_{8\mathrm{QD3}}\right) \tag{11.59a}$$

where $\bar{\eta}_{8\mathrm{QD3}} = \left(E_{\mathrm{FQD}} - \bar{E}_{8\mathrm{QBHD1}}\right)(k_B T)^{-1}$.

Using (1.170a), (1.170b) and (11.59a), we can find the ΔC_{44} and ΔC_{456} in this case, where $n_{0\mathrm{HD}}$ is replaced by n_0.

(h) The HD model of Newson and Kurobe [26]

The HD electron dispersion law can be written following (1.30) as

$$\gamma_3(E,\eta_g) = a_{13}k_z^4\left[\frac{\hbar^2}{2m_c} + a_{14}k_s^2\right]k_z^2 + \frac{\hbar^2}{2m_c}k_s^2 + a_{14}k_x^2k_y^2 + a_{13}\left(k_x^4 + k_y^4\right) \tag{11.59b}$$

The totally quantized energy can be written as

$$\begin{aligned}\gamma_3(E_{8\mathrm{QBHD1N}},\eta_g) &= a_{13}(n_z\pi/d_z)^4 \\ &+ \left[\frac{\hbar^2}{2m_c} + a_{14}\left\{\left(\frac{n_x\pi}{d_x}\right)^2 + \left(\frac{n_y\pi}{d_y}\right)^2\right\}\right]\left(\frac{n_z\pi}{d_z}\right)^2 \\ &+ \frac{\hbar^2}{2m_c}\left[\left(\frac{n_x\pi}{d_x}\right)^2 + \left(\frac{n_y\pi}{d_y}\right)^2\right] + a_{14}\left(\frac{n_xn_y\pi^2}{d_xd_y}\right)^2 \\ &+ a_{13}\left[\left(\frac{n_x\pi}{d_x}\right)^4 + \left(\frac{n_y\pi}{d_y}\right)^4\right]\end{aligned} \tag{11.59c}$$

where $E_{8\mathrm{QBHD1N}}$ is the total quantized energy in this case.

The DOS function can be expressed as

$$N_{0\mathrm{DT}}(E) = \frac{2g_v}{d_xd_yd_z}\sum_{n_x=1}^{n_{x\max}}\sum_{n_y=1}^{n_{y\max}}\sum_{n_z=1}^{n_{z\max}}\delta'(E - E_{8\mathrm{QBHD1N}}) \tag{11.59d}$$

The electron concentration can be expressed as

$$n_0 = \frac{2g_v}{d_xd_yd_z}\sum_{n_x=1}^{n_{x\max}}\sum_{n_y=1}^{n_{y\max}}\sum_{n_z=1}^{n_{z\max}}F_{-1}(\eta_{\mathrm{QD3N}}) \tag{11.59e}$$

where $\bar{\eta}_{\mathrm{QD3N}} = (k_BT)^{-1}(E_{\mathrm{FQD}} - E_{8\mathrm{QBHD1N}})$.

Using (11.59e), (1.170a) and (1.170b), we can find the ΔC_{44} and ΔC_{456} in this case, where $n_{0\mathrm{HD}}$ is replaced by n_0.

(i) The HD model of Johnson and Dicley [27]

Equation (1.43) can be expressed under Gaussian band tails in this case as

$$I(5) + (E_{g_0} - ax)I(4) + \alpha_1d_5k^2I(\alpha_1) = (D_6k^2 + E_{g_0}ax - a^2x^2)I(1) \tag{11.59f}$$

where $\alpha_1 = (E_{g_0} + \Delta)^{-1}$, $d_5 = \frac{1}{3}\Delta D_6$, $D_6 = \frac{E_{g0}^2}{4}\left(\frac{E_{g0}+\Delta}{E_{g0}+\frac{2}{3}\Delta}\right)$, $a = \frac{\hbar^2}{2}\left[\frac{1}{m_c} - \frac{1}{m_0}\right]$ and $x = k^2 s$ and the other notations are defined in Section (11.2.1).

After algebraic manipulations, (11.59f) can be written as

$$\theta_5(E, \eta_g) - k^2[\theta_6(\alpha_1, E, \eta_g) + i\theta_3(\alpha_1, E, \eta_g)] + a^2k^4 = 0 \tag{11.59g}$$

where

$$\begin{aligned}
\theta_5 &= \theta_1(E, \eta_g) + E_{g_0}\gamma_3(E, \eta_g), \\
\theta_1(E, \eta_g) &= 2[1 + Erf(E, \eta_g)]^{-1}\theta_0(E, \eta_g), \\
\theta_6(\alpha_1, E, \eta_g) &= [D_6 + aE_{g_0} + a\gamma_3(E, \eta_g) - \theta_2(\alpha_1, E, \eta_g)], \\
\theta_2(\alpha_1, E, \eta_g) &= [2\alpha_1 d_5[1 + Erf(E/\eta_g)c_{21}(\alpha_1, E, \eta_g)]], \\
\theta_3(\alpha_1, E, \eta_g) &= [2\alpha_1 d_5[1 + Erf(E/\eta_g)D_{21}(\alpha_1, E, \eta_g)]]
\end{aligned}$$

The electron dispersion law HDQD assumes the form

$$\begin{aligned}
&\theta_5(E_{\mathrm{QBHD1J}}, \eta_g) - k_1^2[\theta_6(\alpha_1, E_{\mathrm{QBHD1J}}, \eta_g) \\
&\quad + i\theta_3(\alpha_1, E_{\mathrm{QBHD1J}}, \eta_g)] + a^2k_1^4 = 0
\end{aligned} \tag{11.59h}$$

where E_{QBHD1J} is the total quantized energy in this case and $k_1^2 = \left[\left(\frac{\pi n_x}{d_x}\right)^2 + \left(\frac{\pi n_y}{d_y}\right)^2 + \left(\frac{\pi n_z}{d_z}\right)^2\right]$.

The DOS function can be expressed as

$$N_{\mathrm{0DT}}(E) = \frac{2g_v}{d_x d_y d_z}\sum_{n_x=1}^{n_{x\max}}\sum_{n_y=1}^{n_{y\max}}\sum_{n_z=1}^{n_{z\max}}\delta'(E - E_{\mathrm{QBHD1J}}) \tag{11.59i}$$

The electron concentration can be expressed as

$$n_0 = \frac{2g_v}{d_x d_y d_z}\text{ Real part of }\sum_{n_x=1}^{n_{x\max}}\sum_{n_y=1}^{n_{y\max}}\sum_{n_z=1}^{n_{z\max}}F_{-1}(\eta_{\mathrm{QD3N}}) \tag{11.59j}$$

where $\eta_{\mathrm{QD3J}} = (k_BT)^{-1}(E_{\mathrm{FQD}} - E_{\mathrm{QBHD1J}})$.

Using (1.170a), (1.170b) and (11.59j), we can find the ΔC_{44} and ΔC_{456} in this case, where $n_{0\mathrm{HD}}$ is replaced by n_0.

(j) The HD model of Kolodziejczak *et al.*

In accordance with the model of Kolodziejczak *et al.* [28], the HD electron energy spectrum of III–V compounds can be expressed, taking into account the interaction of the higher bands as

$$\gamma_3(E,\eta_g) = \frac{\hbar^2k^2}{2m_0} + \frac{\chi_0 - E_g}{2} + a\left[\frac{\chi_0 + E_g}{2\chi_0}\right]\frac{\hbar^2k^2}{2m_0} + \left[\frac{\chi_0 - E_g}{2}\right]\frac{\hbar^2k^2}{2m_0}\left[b + \frac{c}{5}\right] \quad (11.59\text{k})$$

The HDQDDR assumes the form

$$\gamma_3(E_{70},\eta_g) = \frac{\hbar^2k_1^2}{2m_0} + \frac{\bar{\chi}_{0,15} - E_g}{2} + a\left[\frac{\bar{\chi}_{0,15} + E_g}{2\chi_0}\right]\frac{\hbar^2k_1^2}{2m_0} + \left[\frac{\bar{\chi}_{0,15} - E_g}{2}\right]\frac{\hbar^2k_1^2}{2m_0}\left[b + \frac{c}{5}\right] \quad (11.59\text{l})$$

where $\bar{\chi}_{0,15} \equiv \left[E_g^2 - 4FE_g\left(\frac{E_g+\frac{2}{3}\Delta}{E_g+\Delta}\right)\left(\frac{\hbar^2k_1^2}{2m_c}\right)\right]^{1/2}$ and E_{70} is the total quantized energy in this case.

The DOS function can be expressed as

$$N_{\mathrm{ODT}}(E) = \frac{2g_v}{d_xd_yd_z}\sum_{n_x=1}^{n_{x\max}}\sum_{n_y=1}^{n_{y\max}}\sum_{n_z=1}^{n_{z\max}}\delta'(E - E_{70}) \quad (11.59\text{m})$$

The electron concentration can be expressed as

$$n_0 = \frac{2g_v}{d_xd_yd_z}\sum_{n_x=1}^{n_{x\max}}\sum_{n_y=1}^{n_{y\max}}\sum_{n_z=1}^{n_{z\max}}F_{-1}(\eta_{70}) \quad (11.59\text{n})$$

where $\eta_{70} = (k_BT)^{-1}(E_{\mathrm{FQD}} - E_{70})$.

Using (1.170a), (1.170b) and (11.59m), we can find the ΔC_{44} and ΔC_{456} in this case, where $n_{0\mathrm{HD}}$ is replaced by n_0.

(k) The HD model of Chaikovskii

The HDDR in the presence of a strong oscillatory electric field along z-axis for a semiconductor having parabolic energy band is given by [29]

$$\gamma_3(E,\eta_g) = \frac{\hbar^2k^2}{2m_c} + \frac{e^2E_0^2}{4(\omega_0)^2m_c} \quad (11.59\text{o})$$

The QDDR can be expressed as

$$\gamma_3\left(E_{153},\eta_g\right)=\frac{\hbar^2k_1^2}{2m_c}+\frac{e^2E_0^2}{4\left(\omega_0\right)^2m_c} \tag{11.59p}$$

where E_{153} is the total quantized energy in this case.

The electron concentration is given by

$$n_0=\frac{2g_v}{d_xd_yd_z}\sum_{n_x=1}^{n_{x\max}}\sum_{n_y=1}^{n_{y\max}}\sum_{n_z=1}^{n_{z\max}}F_{-1}\left(\eta_{1,9,153}\right) \tag{11.59q}$$

where $\eta_{1,9,153}=(k_BT)^{-1}(E_{\mathrm{FQD}}-E_{153})$.

Using (1.5) and (11.59q), we get

$$G_0=\frac{\pi^2k_B}{3e}\left[\sum_{n_x=1}^{n_{x\max}}\sum_{n_y=1}^{n_{y\max}}\sum_{n_z=1}^{n_{z\max}}F_{-1}\left(\eta_{1,9,153}\right)\right]^{-1}$$
$$\times\left[\sum_{n_x=1}^{n_{x\max}}\sum_{n_y=1}^{n_{y\max}}\sum_{n_z=1}^{n_{z\max}}F_{-2}\left(\eta_{1,9,153}\right)\right] \tag{11.59r}$$

*11.2.3 The CEC in HDQD of II–VI materials

(a) The HD model of Hopfield

The dispersion relation of the carriers in HD II–VI materials in the presence of Gaussian band tails can be expressed following (1.51) as [30]

$$\gamma_3\left(E,\eta_g\right)=a_0'k_s^2+b_0'k_z^2\pm\bar{\lambda}_0k_s \tag{11.60}$$

Thus, the energy spectrum in this case is real since the corresponding E–k relation in the absence of band tails as given by (11.60) is a no pole function in the finite complex plane.

The 0D electron dispersion law in QD of HD II–VI materials can be written following (11.60) as

$$\gamma_3(\bar{E}_{9\mathrm{QBHD2}},\eta_g)=a_0'\left[\left(\frac{n_x\pi}{d_x}\right)^2+\left(\frac{n_y\pi}{d_y}\right)^2\right]\pm\bar{\lambda}_0\left(\frac{n_x\pi}{d_x}\right)^2$$
$$+\left(\frac{n_y\pi}{d_y}\right)^{1/2}+\frac{\hbar^2\left(n_z\pi/d_z\right)^2}{2m_{\|}^*} \tag{11.61}$$

where $\bar{E}_{9\mathrm{QBHD2}}$ is the total quantized energy in this case.

The DOS function is given by

$$N_{0DT}(E) = \frac{g_v}{d_x d_y d_z} \sum_{n_x=1}^{n_{x\max}} \sum_{n_y=1}^{n_{y\max}} \sum_{n_z=1}^{n_{z\max}} \delta'\left(E - \bar{E}_{9QBHD2}\right) \tag{11.62}$$

The electron concentration can be expressed as

$$n_0 = \frac{g_v}{d_x d_y d_z} \sum_{n_x=1}^{n_{x\max}} \sum_{n_y=1}^{n_{y\max}} \sum_{n_z=1}^{n_{z\max}} F_{-1}\left(\bar{\eta}_{9QD}\right) \tag{11.63}$$

where $\bar{\eta}_{9QD} = (E_{FQD} - \bar{E}_{9QBHD2})(k_B T)^{-1}$.

Using (1.170a), (1.170b) and (11.63), we can find the ΔC_{44} and ΔC_{456} in this case, where n_{0HD} is replaced by n_0.

11.2.4 The CEC in HDQD of n-gallium phosphide

(a) The HD model of Rees

The dispersion relation of the conduction electrons in HD n-GaP can be expressed following (1.55) as [31]

$$\gamma_3(E, \eta_g) = \frac{\hbar^2 k_s^2}{2m_\perp^*} + \frac{\hbar^2}{2m_\parallel^*}\left[k_s^2 + k_z^2\right] - \left[\frac{\hbar^4 k_0^2}{m_\parallel^{*2}}\left(k_s^2 + k_z^2\right) + |V_G|^2\right]^{1/2} + |V_G| \tag{11.64}$$

The electron energy spectrum in HDQD of n-GaP can be expressed using (11.64) as

$$\begin{aligned}
\gamma_3\left(\bar{E}_{14QBHD14}, \eta_g\right) &= \frac{\hbar^2}{2m_\perp^*}\left[\left(\frac{n_x \pi}{d_x}\right)^2 + \left(\frac{n_y \pi}{d_y}\right)^2\right] \\
&+ \frac{\hbar^2}{2m_\parallel^*}\left[\left(\frac{n_x \pi}{d_x}\right)^2 + \left(\frac{n_y \pi}{d_y}\right)^2 + \left(\frac{n_z \pi}{d_z}\right)^2\right] \\
&- \left[\frac{\hbar^4 k_0^2}{m_\parallel^{*2}}\left[\left(\frac{n_x \pi}{d_x}\right)^2 + \left(\frac{n_y \pi}{d_y}\right)^2 + \left(\frac{n_z \pi}{d_z}\right)^2\right]\right] + |V_G|^2 \\
&- |V_G|
\end{aligned} \tag{11.65}$$

where $\bar{E}_{14QBHD14}$ is the total quantized energy in this case.

The DOS function is given by

$$N_{0\mathrm{DT}}(E)=\frac{2g_v}{d_xd_yd_z}\sum_{n_x=1}^{n_{x\max}}\sum_{n_y=1}^{n_{y\max}}\sum_{n_z=1}^{n_{z\max}}\delta'\left(E-\bar{E}_{14\mathrm{QBHD}14}\right) \tag{11.66}$$

The electron concentration can be expressed as

$$n_0=\frac{2g_v}{d_xd_yd_z}\sum_{n_x=1}^{n_{x\max}}\sum_{n_y=1}^{n_{y\max}}\sum_{n_z=1}^{n_{z\max}}F_{-1}\left(\bar{\eta}_{15\mathrm{QD}14}\right) \tag{11.67a}$$

where $\bar{\eta}_{15\mathrm{QD}14}=\left(E_{\mathrm{FQD}}-\bar{E}_{14\mathrm{QBHD}14}\right)(k_BT)^{-1}$.

Using (1.170a), (1.170b) and (11.67a), we can find the ΔC_{44} and ΔC_{456} in this case, where $n_{0\mathrm{HD}}$ is replaced by n_0.

(b) The HD model of Ivchenko and Pikus

The dispersion relation of the conduction electrons in bulk specimens of HD n-GaP can be written by following Ivchenko and Pikus [32] as given by (1.58b)

$$\gamma_3(E,\eta_g)=\frac{\hbar^2k_z^2}{2m^*}+\frac{\hbar^2k_s^2}{2m_\perp^*}\mp\frac{\bar{\Delta}_0}{2}\pm\left[\left(\frac{\bar{\Delta}_0}{2}\right)^2+P_1k_z^2+D_1k_x^2k_y^2\right]^{1/2} \tag{11.67b}$$

The electron energy spectrum of the HDQD in this case can be written following (1.67b) as

$$\gamma_3\left(\bar{E}_{14\mathrm{QBHD}141},\eta_g\right)=\frac{\hbar^2}{2m^*}\left(\frac{n_z\pi}{d_z}\right)^2+\frac{\hbar^2}{2m_\perp^*}\left[\left(\frac{n_x\pi}{d_x}\right)^2+\left(\frac{n_y\pi}{d_y}\right)^2\right]\mp\frac{\bar{\Delta}_0}{2}\pm\left[\left(\frac{\bar{\Delta}_0}{2}\right)^2+P_1\left(\frac{n_z\pi}{d_z}\right)^2+D_1\left(\frac{n_x\pi}{d_x}\right)^2\left(\frac{n_y\pi}{d_y}\right)^2\right]^{1/2} \tag{11.67c}$$

where $\bar{E}_{14\mathrm{QBHD}141}$ is the total quantized energy in this case.

The DOS function is given by

$$N_{0\mathrm{DT}}(E)=\frac{g_v}{d_xd_yd_z}\sum_{n_x=1}^{n_{x\max}}\sum_{n_y=1}^{n_{y\max}}\sum_{n_z=1}^{n_{z\max}}\delta'\left(E-\bar{E}_{141\mathrm{QBHD}141}\right) \tag{11.67d}$$

The electron concentration can be expressed as

$$n_0 = \frac{g_v}{d_x d_y d_z} \sum_{n_x=1}^{n_{x\max}} \sum_{n_y=1}^{n_{y\max}} \sum_{n_z=1}^{n_{z\max}} F_{-1}\left(\bar{\eta}_{15\text{QD}141}\right) \tag{11.67e}$$

where $\bar{\eta}_{15\text{QD}141} = \left(E_{\text{FQD}} - \bar{E}_{14\text{QBHD}141}\right)\left(k_B T\right)^{-1}$.

Using (1.170a), (1.170b) and (11.67e), we can find the ΔC_{44} and ΔC_{456} in this case, where $n_{0\text{HD}}$ is replaced by n_0.

11.2.5 The CEC in HDQD of *n*-germanium

(a) The model Cardona *et al.*

The DR under the condition of heavy doping can be expressed from (1.59) as [33]

$$\gamma_3\left(E, \eta_g\right) = -\frac{E_{g0}}{2} + \frac{\hbar^2 k_z^2}{2m_\parallel^*} + \left[\frac{E_{g0}^2}{4} + E_{g0} k_s^2 \left(\frac{\hbar^2}{2m_\perp^*}\right)\right]^{1/2} \tag{11.68}$$

The DR in HDQD of n-Ge in the present case can be written as

$$\gamma_3\left(\bar{E}_{20\text{QBHD}}, \eta_g\right) = -\frac{E_{g0}}{2} + \frac{\hbar^2}{2m_\parallel^*}\left(\frac{n_z \pi}{d_z}\right)^2 + \left[\frac{E_{g0}^2}{4} + E_{g0}\left[\left(\frac{n_x \pi}{d_x}\right)^2 + \left(\frac{n_y \pi}{d_y}\right)^2\right]\left(\frac{\hbar^2}{2m_\perp^*}\right)\right]^{1/2} \tag{11.69}$$

where $\bar{E}_{20\text{QBHD}}$ is the total quantized energy in this case.

The DOS function is given by

$$N_{0\text{DT}}(E) = \frac{2g_v}{d_x d_y d_z} \sum_{n_x=1}^{n_{x\max}} \sum_{n_y=1}^{n_{y\max}} \sum_{n_z=1}^{n_{z\max}} \delta'\left(E - \bar{E}_{20\text{QBHD}}\right) \tag{11.70}$$

The electron concentration can be expressed as

$$n_0 = \frac{2g_v}{d_x d_y d_z} \sum_{n_x=1}^{n_{x\max}} \sum_{n_y=1}^{n_{y\max}} \sum_{n_z=1}^{n_{z\max}} F_{-1}\left(\bar{\eta}_{18\text{QD}20}\right) \tag{11.71}$$

$$\text{where } \bar{\eta}_{18\text{QD}20} = \left(E_{\text{FQD}} - \bar{E}_{20\text{QBHD}}\right)\left(k_B T\right)^{-1}. \tag{11.72}$$

Using (1.170a), (1.170b) and (11.72), we can find the ΔC_{44} and ΔC_{456} in this case, where $n_{0\mathrm{HD}}$ is replaced by n_0.

(b) The model of Wang and Ressler

The DR in HD n-Ge can be written following (1.63) as [34]

$$\gamma_3(E,\eta_g) = \frac{\hbar^2 k_z^2}{2m_\parallel^*} + \frac{\hbar^2 k_s^2}{2m_\perp^*} - c_1\left(\frac{\hbar^2 k_s^2}{2m_\perp^*}\right)^2 - d_1\left(\frac{\hbar^2 k_s^2}{2m_\perp^*}\right)\left(\frac{\hbar^2 k_z^2}{2m_\parallel^*}\right) - e_1\left(\frac{\hbar^2 k_z^2}{2m_\parallel^*}\right)^2 \tag{11.73}$$

The DR in HDQD of n-Ge can be expressed in this case following (11.73) as

$$\begin{aligned}\gamma_3(\bar{E}_{24\mathrm{QBHD}24},\eta_g) &= \left(\frac{\hbar^2}{2m_\parallel^*}\right)\left[\frac{n_z\pi}{d_z}\right]^2 \\ &+ \left(\frac{\hbar^2}{2m_\perp^*}\right)\left[\left[\frac{n_x\pi}{d_x}\right]^2 + \left[\frac{n_y\pi}{d_y}\right]^2\right] - c_1\left(\frac{\hbar^2}{2m_\perp^*}\right)^2 \\ &\times \left[\left[\frac{n_x\pi}{d_x}\right]^2 + \left[\frac{n_y\pi}{d_y}\right]^2\right]^2 - d_1\left(\frac{\hbar^2}{2m_\perp^*}\right)\left[\left[\frac{n_x\pi}{d_x}\right]^2\right. \\ &+ \left.\left[\frac{n_y\pi}{d_y}\right]^2\right]\left(\frac{\hbar^2}{2m_\parallel^*}\right)\left[\frac{n_z\pi}{d_z}\right]^2 - e_1\left(\frac{\hbar^2}{2m_\parallel^*}\right)^2\left[\frac{n_z\pi}{d_z}\right]^4\end{aligned} \tag{11.74}$$

where $\bar{E}_{24\mathrm{QBHD}24}$ is the total quantized energy in this case.

The DOS function is given by

$$N_{0\mathrm{DT}}(E) = \frac{2g_v}{d_x d_y d_z}\sum_{n_x=1}^{n_{x\max}}\sum_{n_y=1}^{n_{y\max}}\sum_{n_z=1}^{n_{z\max}}\delta'\left(E - \bar{E}_{24\mathrm{QBHD}24}\right) \tag{11.75}$$

The electron concentration can be expressed as

$$n_0 = \frac{2g_v}{d_x d_y d_z}\sum_{n_x=1}^{n_{x\max}}\sum_{n_y=1}^{n_{y\max}}\sum_{n_z=1}^{n_{z\max}} F_{-1}(\bar{\eta}_{18\mathrm{QD}22}) \tag{11.76}$$

where $\bar{\eta}_{18\mathrm{QD}22} = (E_{\mathrm{FQD}} - \bar{E}_{24\mathrm{QBHD}24})(k_B T)^{-1}$.

Using (1.170a), (1.170b) and (11.76), we can find the ΔC_{44} and ΔC_{456} in this case, where $n_{0\mathrm{HD}}$ is replaced by n_0.

11.2.6 The CEC in HDQD of tellurium

(a) The HD model of Bouat *et al.*

The dispersion relation of the conduction electrons of HD Te can be written following (1.67) as [35]

$$\gamma_3(E,\eta_g) = A_b k_z^2 + A_z k_s^2 \pm \left[A_g k_z^2 + A_g k_s^2\right]^{1/2} \tag{11.77}$$

The DR in HDQD of n-Te can be expressed in this case following (11.77) as

$$\gamma_3(E_{11\mathrm{QBHD}16},\eta_g) = A_6\left(\frac{n_z\pi}{d_z}\right)^2 + A_7\left[\left(\frac{n_x\pi}{d_x}\right)^2 + \left(\frac{n_y\pi}{d_y}\right)^2\right]$$
$$\pm\left[A_8\left(\frac{n_z\pi}{d_z}\right)^2 + A_9\left[\left(\frac{n_x\pi}{d_x}\right)^2 + \left(\frac{n_y\pi}{d_y}\right)^2\right]\right]^{1/2} \tag{11.78}$$

where $E_{11\mathrm{QBHD}16}$ is the total quantized energy in this case.

The DOS function is given by

$$N_{0\mathrm{DT}}(E) = \frac{g_v}{d_x d_y d_z}\sum_{n_x=1}^{n_{x\max}}\sum_{n_y=1}^{n_{y\max}}\sum_{n_z=1}^{n_{z\max}}\delta'\left(E - E_{11\mathrm{QBHD}16}\right) \tag{11.79}$$

The electron concentration can be expressed as

$$n_0 = \frac{g_v}{d_x d_y d_z}\sum_{n_x=1}^{n_{x\max}}\sum_{n_y=1}^{n_{y\max}}\sum_{n_z=1}^{n_{z\max}}F_{-1}\left(\bar{\eta}_{14\mathrm{QD}12}\right) \tag{11.80}$$

where $\bar{\eta}_{14\mathrm{QD}12} = \left(E_{\mathrm{FQD}} - E_{11\mathrm{QBHD}16}\right)(k_B T)^{-1}$.

Using (1.170a), (1.170b) and (11.80), we can find the ΔC_{44} and ΔC_{456} in this case, where $n_{0\mathrm{HD}}$ is replaced by n_0.

(b) HD model of Ortenberg and Button

The dispersion relation of the conduction electrons of HD Te can be written in this case following (1.71) as [36]

$$\gamma_3(E,\eta_g) = \mp t_1 + t_2 k_z^2 + t_3 k_s^2 + t_4 k_s^4 + t_5 k_s^2 k_z^2$$
$$\pm\left[\left(t_1 + t_6 k_s^2\right)^2 + t_7 k_z^2\right]^{1/2} \tag{11.81}$$

The DR in HDQD of n-Te can be expressed in this case following (11.82) as

$$\gamma_3\left(\bar{E}_{11\mathrm{QBHD}16},\eta_g\right)=\mp t_1+t_2\left(\frac{n_z\pi}{d_z}\right)^2+t_3\left[\left(\frac{n_x\pi}{d_x}\right)^2+\left(\frac{n_y\pi}{d_y}\right)^2\right]$$
$$+t_4\left[\left(\frac{n_x\pi}{d_x}\right)^2+\left(\frac{n_y\pi}{d_y}\right)^2\right]^2+t_5\left[\left(\frac{n_x\pi}{d_x}\right)^2+\left(\frac{n_y\pi}{d_y}\right)^2\right]$$
$$\times\left(\frac{n_z\pi}{d_z}\right)^2\pm\left[\left(t_1+t_6\left[\left(\frac{n_x\pi}{d_x}\right)^2+\left(\frac{n_y\pi}{d_y}\right)^2\right]\right)^2\right.$$
$$\left.+t_7\left(\frac{n_z\pi}{d_z}\right)^2\right]^{1/2} \tag{11.82}$$

where $\bar{E}_{11\mathrm{QBHD}16}$ is the total quantized energy in this case.

The DOS function is given by

$$N_{0\mathrm{DT}}(E)=\frac{g_v}{d_xd_yd_z}\sum_{n_x=1}^{n_{x\max}}\sum_{n_y=1}^{n_{y\max}}\sum_{n_z=1}^{n_{z\max}}\delta'\left(E-\bar{E}_{11\mathrm{QBHD}16}\right) \tag{11.83}$$

The electron concentration can be expressed as

$$n_0=\frac{g_v}{d_xd_yd_z}\sum_{n_x=1}^{n_{x\max}}\sum_{n_y=1}^{n_{y\max}}\sum_{n_z=1}^{n_{z\max}}F_{-1}\left(\bar{\eta}_{14\mathrm{QD}121}\right) \tag{11.84}$$

where $\bar{\eta}_{14\mathrm{QD}121}=\left(E_{\mathrm{FQD}}-\bar{E}_{11\mathrm{QBHD}16}\right)(k_BT)^{-1}$.

Using (1.170a), (1.170b) and (11.84), we can find the ΔC_{44} and ΔC_{456} in this case, where $n_{0\mathrm{HD}}$ is replaced by n_0.

*11.2.7 The CEC in HDQD of graphite

(a) The HD model of Ushio *et al.*

The totally quantized energy in HDQD Graphite can be expressed following (1.76) as [37]

$$\gamma_3\left(\bar{E}_{16,\pm},\eta_g\right)=\frac{1}{2}\left[\Delta_1-2\gamma_1\cos\left(\frac{\bar{c}(\pi n_z)}{2d_z}\right)\right.$$
$$\left.+2\gamma_5\cos^2\left(\frac{\bar{c}(\pi n_z)}{2d_z}\right)+2\gamma_2\cos^2\left(\frac{\bar{c}(\pi n_z)}{2d_z}\right)\right]$$

$$\pm\left[\frac{1}{4}\left(\Delta_1-2\gamma_1\cos^2\left(\frac{\bar{c}(\pi n_z)}{2d_z}\right)+2\gamma_5\cos^2\left(\frac{\bar{c}(\pi n_z)}{2d_z}\right)\right.\right.$$
$$\left.-2\gamma_2\cos^2\left(\frac{\bar{c}(\pi n_z)}{2d_z}\right)\right)^2+\frac{3}{4}(\bar{a})^2\left(\gamma_0+2\gamma_4\cos\left(\frac{n_z\pi\bar{c}}{2d_z}\right)\right)^2$$
$$\left.\times\left[\left(\frac{\pi n_x}{d_x}\right)^2+\left(\frac{\pi n_y}{d_y}\right)^2+\left(\frac{\pi n_z}{d_z}\right)^2\right]\right]^{1/2} \tag{11.85}$$

where $\bar{E}_{16,\pm}$ is the total quantized energy in this case.

The DOS function is given by

$$N_{0\mathrm{DT}}(E)=\frac{g_v}{d_xd_yd_z}\sum_{n_x=1}^{n_{x\max}}\sum_{n_y=1}^{n_{y\max}}\sum_{n_z=1}^{n_{z\max}}\delta'\left(E-\bar{E}_{16,\pm}\right) \tag{11.86}$$

The electron concentration can be expressed as

$$n_0=\frac{g_v}{d_xd_yd_z}\sum_{n_x=1}^{n_{x\max}}\sum_{n_y=1}^{n_{y\max}}\sum_{n_z=1}^{n_{z\max}}F_{-1}\left(\bar{\eta}_{14\mathrm{QD}122}\right) \tag{11.87}$$

where $\bar{\eta}_{14\mathrm{QD}122}=\left(E_{\mathrm{FQD}}-\bar{E}_{16,\pm}\right)(k_BT)^{-1}$.

Using (1.170a), (1.170b) and (11.87), we can find the ΔC_{44} and ΔC_{456} in this case, where $n_{0\mathrm{HD}}$ is replaced by n_0.

*11.2.8 The CEC in HDQD of platinum antimonide

(a) The HD model of Emtage

The carrier dispersion law in HD $PtSb_2$ can be written following (1.80) as [38]

$$T_{11}k_s^4-k_s^2\left[T_{21}(E,\eta_g)-T_{31}k_z^2\right]$$
$$+\left[T_{41}k_z^4-T_{51}(E,\eta_g)k_z^2-T_{61}(E,\eta_g)\right]=0 \tag{11.88}$$

where

$$T_{11}=(I_1-\omega_2\omega_3),T_{21}(E,\eta_g)=[\omega_1\delta_0+\omega_1\gamma_3(E,\eta_g)$$
$$+\omega_3\gamma_3(E,\eta_g)],$$

$$T_{31} = [2I_1 + \omega_2\omega_4 - \omega_2\omega_3], \quad T_{41} = [2I_1 + \omega_2\omega_4], \quad T_{51}(E, \eta_g)$$
$$= [\omega_2\gamma_0 - \omega_4\gamma_3(E, \eta_g) + \omega_2\gamma_3(E, \eta_g)], T_{61}(E', \eta_g)$$
$$= [\gamma_8(E, \eta_g) + \gamma_0\gamma_3(E, \eta_g)] \quad \text{and}$$
$$\gamma_8(E, \eta_g) = 2\theta_0(E, \eta_g)[1 + Erf(E/\eta_g)]^{-1}$$

The DR in HDQD of $PtSb_2$ can be written as

$$T_{11}\left[\left(\frac{\pi n_x}{d_x}\right)^2 + \left(\frac{\pi n_y}{d_y}\right)^2\right]^2 - \left[\left(\frac{\pi n_x}{d_x}\right)^2 + \left(\frac{\pi n_y}{d_y}\right)^2\right]$$
$$\times \left[T_{21}\left(\bar{E}_{15\mathrm{QBHD}15}, \eta_g\right) - T_{31}\left(\frac{\pi n_z}{d_z}\right)^2\right]$$
$$+ \left[T_{41}\left[\left(\frac{\pi n_x}{d_x}\right)^2 + \left(\frac{\pi n_y}{d_y}\right)^2\right]^2 - T_{51}\left(\bar{E}_{15\mathrm{QBHD}15}, \eta_g\right)\right.$$
$$\left.\times \left(\frac{\pi n_z}{d_z}\right)^2\right] - \left[T_{61}\left(\bar{E}_{15\mathrm{QBHD}15}, \eta_g\right)\right] = 0 \quad (11.89)$$

where

$$T_{21}\left(\bar{E}_{15\mathrm{QBHD}15}, \eta_g\right) = \left[\omega_1\delta_0 + \omega_1\gamma_3\left(\bar{E}_{15\mathrm{QBHD}15}, \eta_g\right)\right.$$
$$\left. + \omega_3\gamma_3\left(\bar{E}_{15\mathrm{QBHD}15}, \eta_g\right)\right],$$
$$T_{51}\left(\bar{E}_{15\mathrm{QBHD}15}, \eta_g\right) = \left[\omega_2\gamma_0 - \omega_4\gamma_3\left(\bar{E}_{15\mathrm{QBHD}15}, \eta_g\right)\right.$$
$$\left. + \omega_2\gamma_3\left(\bar{E}_{15\mathrm{QBHD}15}, \eta_g\right)\right],$$
$$T_{61}\left(\bar{E}_{15\mathrm{QBHD}15}, \eta_g\right) = \left[\gamma_8\left(\bar{E}_{15\mathrm{QBHD}15}, \eta_g\right)\right.$$
$$\left. + \gamma_0\gamma_3\left(\bar{E}_{15\mathrm{QBHD}15}, \eta_g\right)\right] \quad \text{and}$$
$$\gamma_8\left(\bar{E}_{15\mathrm{QBHD}15}, \eta_g\right) = 2\theta_0\left(\bar{E}_{15\mathrm{QBHD}15}, \eta_g\right)$$
$$\left[1 + Erf\left(\bar{E}_{15\mathrm{QBHD}15}/\eta_g\right)\right]^{-1}$$

The DOS function is given by

$$N_{0\mathrm{DT}}(E) = \frac{2g_v}{d_x d_y d_z} \sum_{n_x=1}^{n_{x\max}} \sum_{n_y=1}^{n_{y\max}} \sum_{n_z=1}^{n_{z\max}} \delta'\left(E - \bar{E}_{15\mathrm{QBHD}15}\right) \quad (11.90)$$

The electron concentration can be expressed as

$$n_0 = \frac{2g_v}{d_x d_y d_z} \sum_{n_x=1}^{n_{x\max}} \sum_{n_y=1}^{n_{y\max}} \sum_{n_z=1}^{n_{z\max}} F_{-1}\left(\bar{\eta}_{16\text{QD}16}\right) \tag{11.91}$$

where $\bar{\eta}_{16\text{QD}16} = \left(E_{\text{FQD}} - \bar{E}_{15\text{QBHD}15}\right)(k_B T)^{-1}$.

Using (1.170a), (1.170b) and (11.91), we can find the $\Delta C_{,44}$ and ΔC_{456} in this case, where $n_{0\text{HD}}$ is replaced by n_0.

*11.2.9 The CEC in HDQD of II–V materials

(a) The HD model of Yamada

The DR of the holes in QDs of HD II–V compounds can be expressed following (1.88) as [39]

$$\gamma_3\left(\bar{E}_{150\text{QBHD},\pm}, \eta_g\right) = A_{10}\left(\frac{n_x\pi}{d_x}\right)^2 + A_{11}\left(\frac{n_y\pi}{d_y}\right)^2 + A_{12}\left(\frac{n_z\pi}{d_z}\right)^2 + A_{13}\left(\frac{n_x\pi}{d_x}\right)^2 \pm \left[A_{14}\left(\frac{n_x\pi}{d_x}\right)^2 + A_{15}\left(\frac{n_y\pi}{d_y}\right)^2 + A_{16}\left(\frac{n_z\pi}{d_z}\right)^2 + A_{17}\left(\frac{n_x\pi}{d_x}\right)^2 + A_{18}\left(\frac{n_y\pi}{d_y}\right)^2 + A_{19}^2\right]^{1/2} \mp A_{19} \tag{11.92}$$

where $\bar{E}_{150\text{QBHD},\pm}$ is the total quantized energy in this case.

The DOS function is given by

$$N_{\text{ODT}}(E) = \frac{g_v}{d_x d_y d_z} \sum_{n_x=1}^{n_{x\max}} \sum_{n_y=1}^{n_{y\max}} \sum_{n_z=1}^{n_{z\max}} \delta'\left(E - \bar{E}_{150\text{QBHD},\pm}\right) \tag{11.93}$$

The electron concentration can be expressed as

$$n_0 = \frac{g_v}{d_x d_y d_z} \sum_{n_x=1}^{n_{x\max}} \sum_{n_y=1}^{n_{y\max}} \sum_{n_z=1}^{n_{z\max}} F_{-1}\left(\bar{\eta}_{22\text{QD}30}\right) \tag{11.94}$$

where $\bar{\eta}_{22\text{QD}30} = \left(E_{\text{FQD}} - \bar{E}_{150\text{QBHD},\pm}\right)(k_B T)^{-1}$.

Using (1.170a), (1.170b) and (11.94), we can find the ΔC_{44} and ΔC_{456} in this case, where $n_{0\text{HD}}$ is replaced by n_0.

11.2.10 The CEC in HDQD of gallium antimonide

(a) HD model of Seiler *et al.*

The DR in HDQD of GaSb in accordance with the model of Seiler *et al.* [40] can be written following (1.92) as

$$\gamma_3\left(\bar{E}_{30\text{QBHD}11}, \eta_g\right) = \left[-\frac{E_{g_0}}{2} + \frac{E_{g_0}}{2}\left[1+\alpha_4 k_1^2\right]^{1/2} + \frac{\bar{\zeta}_0 \hbar^2 k_1^2}{2m_0} + \frac{\bar{v}_0 \theta_1(k_1)\hbar^2}{2m_0} \pm \frac{\bar{w}_0 \theta_2(k_1)\hbar^2}{2m_0}\right] \tag{11.95}$$

where $\bar{E}_{30\text{QBHD}11}$ is the total quantized energy in this case,

$$\theta_1(k_1) = k_1^{-2}\left[\left(\frac{n_x\pi}{d_x}\right)^2\left(\frac{n_y\pi}{d_y}\right)^2 + \left(\frac{n_y\pi}{d_y}\right)^2\left(\frac{n_z\pi}{d_z}\right)^2 + \left(\frac{n_z\pi}{d_z}\right)^2\left(\frac{n_x\pi}{d_x}\right)^2\right]$$

and

$$\theta_2(k_1) = \left[\left\{k_1^2\left[\left(\frac{n_x\pi}{d_x}\right)^2\left(\frac{n_y\pi}{d_y}\right)^2 + \left(\frac{n_y\pi}{d_y}\right)^2\left(\frac{n_z\pi}{d_z}\right)^2 + \left(\frac{n_z\pi}{d_z}\right)^2 \times \left(\frac{n_x\pi}{d_x}\right)^2\right] - 9\left(\frac{n_x\pi}{d_x}\right)^2\left(\frac{n_y\pi}{d_y}\right)^2\left(\frac{n_z\pi}{d_z}\right)^2\right\}^{1/2} \cdot k_1^{-1}\right] \tag{11.96}$$

The DOS function is given by

$$N_{0\text{DT}}(E) = \frac{g_v}{d_x d_y d_z}\sum_{n_x=1}^{n_{x\max}}\sum_{n_y=1}^{n_{y\max}}\sum_{n_z=1}^{n_{z\max}} \delta'\left(E - \bar{E}_{30\text{QBHD}11}\right) \tag{11.97}$$

The electron concentration can be expressed as

$$n_0 = \frac{g_v}{d_x d_y d_z}\sum_{n_x=1}^{n_{x\max}}\sum_{n_y=1}^{n_{y\max}}\sum_{n_z=1}^{n_{z\max}} F_{-1}\left(\bar{\eta}_{19\text{QD}24}\right) \tag{11.98}$$

where $\bar{\eta}_{19\text{QD}24} = \left(E_{\text{FQD}} - \bar{E}_{30\text{QBHD}11}\right)(k_B T)^{-1}$.

Using (1.170a), (1.170b) and (11.98), we can find the ΔC_{44} and ΔC_{456} in this case, where $n_{0\mathrm{HD}}$ is replaced by n_0.

(b) HD model of Mathur *et al.*

The totally quantized energy $\bar{E}_{30\mathrm{QBHD}}$ in this case can be written following (1.96) as [41]

$$\gamma_3\left(\bar{E}_{30\mathrm{QBHD}}, \eta_g\right) = \alpha_9\left[\left(\frac{\pi n_x}{d_x}\right)^2 + \left(\frac{\pi n_y}{d_y}\right)^2 + \left(\frac{\pi n_z}{d_z}\right)^2\right] + \frac{E_{g1}}{2}\left[\sqrt{1 + \alpha_{10}\left[\left(\frac{\pi n_x}{d_x}\right)^2 + \left(\frac{\pi n_y}{d_y}\right)^2 + \left(\frac{\pi n_z}{d_z}\right)^2\right]} - 1\right] \quad (11.99)$$

where $\bar{E}_{30\mathrm{QBHD}}$ is the total quantized energy in this case.

The DOS function is given by

$$N_{0\mathrm{DT}}(E) = \frac{2g_v}{d_x d_y d_z}\sum_{n_x=1}^{n_{x\max}}\sum_{n_y=1}^{n_{y\max}}\sum_{n_z=1}^{n_{z\max}} \delta'\left(E - \bar{E}_{30\mathrm{QBHDM}}\right) \quad (11.100)$$

The electron concentration can be expressed as

$$n_0 = \frac{2g_v}{d_x d_y d_z}\sum_{n_x=1}^{n_{x\max}}\sum_{n_y=1}^{n_{y\max}}\sum_{n_z=1}^{n_{z\max}} F_{-1}\left(\bar{\eta}_{20\mathrm{QD}26}\right) \quad (11.101)$$

where $\bar{\eta}_{20\mathrm{QD}26} = \left(E_{\mathrm{FQD}} - \bar{E}_{30\mathrm{QBHD}M}\right)(k_B T)^{-1}$.

Using (1.170a), (1.170b) and (11.101), we can find the ΔC_{44} and ΔC_{456} in this case, where $n_{0\mathrm{HD}}$ is replaced by n_0.

(c) HD model of Zhang *et al.*

The DR in HDQD of GaSb in accordance with this model, can be written following (1.101) as [42]

$$\gamma_3\left(\bar{E}_{30\mathrm{QBHD}22}, \eta_g\right) = \left[\overline{E_1} + \overline{E_2} Z_1(k_1)\right] k_1^2 + \left[\overline{E_3} + \overline{E_4} Z_1(k_1)\right] k_1^4 + k_1^6\left[\overline{E_5} + \overline{E_6} Z_1(k_1) + \overline{E_7} Z_2(k_1)\right] \quad (11.102)$$

where $\bar{E}_{30\text{QBHD}22}$ is the total quantized energy in this case.

$$Z_1(k_1) = \frac{5\sqrt{21}}{4}\left[\frac{\left(\frac{n_x\pi}{d_x}\right)^4 + \left(\frac{n_y\pi}{d_y}\right)^4 + \left(\frac{n_z\pi}{d_z}\right)^4}{k_1^4} - \frac{3}{5}\right] \quad \text{and}$$

$$Z_2(k_1) = \left[\frac{639639}{32}\right]^{1/2}\left[\frac{\left(\frac{n_x\pi}{d_x}\right)^2\left(\frac{n_y\pi}{d_y}\right)^2\left(\frac{n_z\pi}{d_z}\right)^2}{k_1^6}\right.$$
$$\left. + \frac{1}{12}\left(\frac{\left(\frac{n_x\pi}{d_x}\right)^4 + \left(\frac{n_y\pi}{d_y}\right)^4 + \left(\frac{n_z\pi}{d_z}\right)^4}{k_1^4} - \frac{3}{5}\right) - \frac{1}{105}\right]$$

The DOS function is given by

$$N_{\text{0DT}}(E) = \frac{2g_v}{d_x d_y d_z}\sum_{n_x=1}^{n_{x\max}}\sum_{n_y=1}^{n_{y\max}}\sum_{n_z=1}^{n_{\max}} \delta'\left(E - \bar{E}_{30\text{QBHD}36}\right) \tag{11.103}$$

The electron concentration can be expressed as

$$n_0 = \frac{g_v}{d_x d_y d_z}\sum_{n_x=1}^{n_{x\max}}\sum_{n_y=1}^{n_{y\max}}\sum_{n_z=1}^{n_{z\max}} F_{-1}\left(\bar{\eta}_{21\text{QD}28}\right) \tag{11.104}$$

where $\bar{\eta}_{21\text{QD}28} = \left(E_{\text{FQD}} - \bar{E}_{30\text{QBHD}36}\right)(k_B T)^{-1}$.

Using (1.170a), (1.170b) and (11.104), we can find the ΔC_{44} and ΔC_{456} in this case, where $n_{0\text{HD}}$ is replaced by n_0.

11.2.11 The CEC in HDQD of stressed materials

(a) The HD model of Seiler *et al.* [40]

Using (5.50), we get

$$Ek^2 - T_{17}k_x^2 - T_{27}k_y^2 - T_{37}k_z^2 = \left[q_{67}E^3 - R_{67}E^2 + V_{67}E + \rho_{67}\right] \tag{11.105}$$

where

$$\{T_{17} = \alpha_1, \quad T_{27} = \alpha_2, \quad T_{37} = \alpha_3, \quad t_1 = q_{67}, \quad t_2 = R_{67}, \\ t_3 = V_{67} \quad \text{and} \quad t_4 = \rho_{67}\} \tag{11.106}$$

Under the condition of heavy doping, (11.105) can be written as

$$I(4)k^2 - T_{17}I(1)k_x^2 - T_{27}I(1)k_y^2 - T_{37}k_z^2 I(1) \\ = [q_{67}I(6) - R_{67}I(5) + V_{67}I(4) + \rho_{67}I(1)] \tag{11.107}$$

where

$$I(6) = \int_{-\infty}^{E} (E - V)^3 F(V)dV \tag{11.108}$$

Equation (11.108) can be written as

$$I(6) = E^3 I(1) - 3E^2 I(7) + 3EI(8) - I(9) \tag{11.109}$$

In which

$$I(7) = \int_{-\infty}^{E} VF(V)dV \tag{11.110}$$

$$I(8) = \int_{-\infty}^{E} V^2 F(V)dV \tag{11.111}$$

$$I(9) = \int_{-\infty}^{E} V^3 F(V)dV \tag{11.112}$$

Using (11.112), together with simple algebraic manipulations, one obtains

$$I(7) = \frac{-\eta_g}{2\sqrt{\pi}} \exp\left(\frac{-E^2}{\eta_g^2}\right) \tag{11.113}$$

$$I(8) = \frac{\eta_g^2}{4}\left[1 + Erf\left(\frac{E}{\eta_g}\right)\right] \tag{11.114}$$

and

$$I(9) = \frac{-\eta_g^3}{2\sqrt{\pi}} \exp\left(\frac{-E^2}{\eta_g^2}\right)\left[1 + \frac{E^2}{\eta_g^2}\right] \tag{11.115}$$

Thus, (11.115) can be written as

$$I(6) = \left[\frac{E}{2}\left[1 + Erf\left(\frac{E}{\eta_g}\right)\right]\left[E^2 + \frac{3}{2}\eta_g^2\right] + \frac{\eta_g}{2\sqrt{\pi}}\exp\left(\frac{-E^2}{\eta_g^2}\right)\left[4E^2 + \eta_g^2\right]\right] \tag{11.116}$$

Thus, combining the appropriate equations, the dispersion relations of the conduction electrons in HD stressed materials can be expressed as

$$P_{11}(E,\eta_g)\,k_x^2 + Q_{11}(E,\eta_g)\,k_y^2 + S_{11}(E,\eta_g)\,k_z^2 = 1 \tag{11.117}$$

where

$$P_{11}(E,\eta_g) \equiv \left[\frac{\gamma_0(E,\eta_g) - (T_{17}/2)\left[1 + Erf(E/\eta_g)\right]}{\Delta_{14}(E,\eta_g)}\right]$$

$$\Delta_{14}(E,\eta_g) \equiv \left[q_{67}\left\{\frac{E}{2}\left[1 + Erf(E/\eta_g)\right]\left[E^2 + \frac{3}{2}\eta_g^2\right] + \frac{\eta_g}{2\sqrt{\pi}}\exp\left(-E^2/\eta_g^2\right)\left[4E^2 + \eta_g^2\right]\right\} - R_{67}\theta_0(E,\eta_g) + V_{67}\gamma_0(E,\eta_g) + \frac{\rho_{67}}{2}\left[1 + Erf(E/\eta_g)\right]\right],$$

$$Q_{11}(E,\eta_g) \equiv \left[\frac{\gamma_0(E,\eta_g) - (T_{27}/2)\left[1 + Erf(E/\eta_g)\right]}{\Delta_{14}(E,\eta_g)}\right] \quad \text{and}$$

$$S_{11}(E,\eta_g) \equiv \left[\frac{\gamma_0(E,\eta_g) - (T_{37}/2)\left[1 + Erf(E/\eta_g)\right]}{\Delta_{14}(E,\eta_g)}\right]$$

Thus, the energy spectrum in this case is real since the dispersion relation of the corresponding materials in the absence of band tails as given by (11.116) has no poles in the finite complex plane.

The dispersion relations of the conduction electrons in HDQD of stressed materials can be expressed as

$$P_{11}(E_{10QBHD14},\eta_g)\left(\frac{n_x\pi}{d_x}\right)^2 + Q_{11}(E_{10QBHD14},\eta_g)\left(\frac{n_y\pi}{d_y}\right)^2 + S_{11}(E_{10QBHD14},\eta_g)\left(\frac{n_z\pi}{d_z}\right)^2 = 1 \tag{11.118a}$$

where

$$P_{11}\left(E_{10\mathrm{QBHD}14},\eta_g\right) \equiv \left[\frac{\gamma_0\left(E_{10\mathrm{QBHD}14},\eta_g\right)-\left(T_{17}/2\right)\left[1+Erf\left(E_{10\mathrm{QBHD}14}/\eta_g\right)\right]}{\Delta_{14}\left(E,\eta_g\right)}\right]$$

$$\Delta_{14}\left(E_{10\mathrm{QBHD}14},\eta_g\right) \equiv \left[q_{67}\left\{\frac{E_{10\mathrm{QBHD}14}}{2}\left[1+Erf\left(E_{10\mathrm{QBHD}14}/\eta_g\right)\right]\right.\right.$$
$$\left[\left(E_{10\mathrm{QBHD}14}\right)^2+\frac{3}{2}\eta_g^2\right]$$
$$\left.+\frac{\eta_g}{2\sqrt{\pi}}\exp\left(-\left(E_{10\mathrm{QBHD}14}\right)^2/\eta_g^2\right)\left[4\left(E_{10\mathrm{QBHD}14}\right)^2+\eta_g^2\right]\right\}$$
$$-R_{67}\theta_0\left(E_{10\mathrm{QBHD}14},\eta_g\right)+V_{67}\gamma_0\left(E_{10\mathrm{QBHD}14},\eta_g\right)$$
$$\left.+\frac{\rho_{67}}{2}\left[1+Erf\left(E_{10\mathrm{QBHD}14}/\eta_g\right)\right]\right],$$

$$S_{11}\left(E_{10\mathrm{QBHD}14},\eta_g\right) \equiv \left[\frac{\gamma_0\left(E_{10\mathrm{QBHD}14},\eta_g\right)-\left(T_{37}/2\right)\left[1+Erf\left(E_{10\mathrm{QBHD}14}/\eta_g\right)\right]}{\Delta_{14}\left(E_{10\mathrm{QBHD}14},\eta_g\right)}\right]$$

and $E_{10\mathrm{QBHD}14}$ is the total quantized energy in this case.

$$Q_{11}\left(E_{10\mathrm{QBHD}14},\eta_g\right) \equiv \left[\frac{\gamma_0\left(E_{10\mathrm{QBHD}14},\eta_g\right)-\left(T_{27}/2\right)\left[1+Erf\left(E_{10\mathrm{QBHD}14}/\eta_g\right)\right]}{\Delta_{14}\left(E_{10\mathrm{QBHD}14},\eta_g\right)}\right]$$

The DOS function is given by

$$N_{0\mathrm{DT}}(E)=\frac{2g_v}{d_x d_y d_z}\sum_{n_x=1}^{n_{x\max}}\sum_{n_y=1}^{n_{y\max}}\sum_{n_z=1}^{n_{z\max}}\delta'\left(E-E_{10\mathrm{QBHD}14}\right) \tag{11.118b}$$

The electron concentration can be expressed as

$$n_0=\frac{2g_v}{d_x d_y d_z}\sum_{n_x=1}^{n_{x\max}}\sum_{n_y=1}^{n_{y\max}}\sum_{n_z=1}^{n_{z\max}}F_{-1}\left(\bar{\eta}_{13\mathrm{QD}10}\right) \tag{11.119}$$

where $\bar{\eta}_{13\text{QD}10} = (E_{\text{FQD}} - E_{10\text{QBHD}14}) (k_B T)^{-1}$. (11.120)

Using (1.170a), (1.170b) and (11.119), we can find the ΔC_{44} and ΔC_{456} in this case, where $n_{0\text{HD}}$ is replaced by n_0.

11.2.12 The CEC in HDQD of IV–VI materials

(a) The HD model of Dimmock

From (5.199), the DR assumes the form [22]

$$\frac{\alpha\hbar^4 k_s^4}{4m_t^+ m_t^-} + \hbar^2 k_s^2 \left[\left(\frac{1}{2m_t^*} - \frac{1}{2m_t^-}\right) + \alpha E\left(\frac{1}{2m_t^-} - \frac{1}{2m_t^+}\right) + \frac{\alpha\hbar^2 k_z^2}{4m_l^- m_t^+}\right] + \left[\left(\frac{\hbar^2 k_z^2}{2m_l^*} + \frac{\hbar^2 k_z^2}{2m_l^-}\right) + \frac{\alpha E}{2}\hbar^2 k_z^2\left(\frac{1}{m_l^-} - \frac{1}{m_l^+}\right) + \frac{\alpha\hbar^4 k_z^4}{4m_l^+ m_t^-} - E(1+\alpha E)\right] = 0 \tag{11.121}$$

Using (11.121), the dispersion relation of the conduction electrons in HD IV–VI materials can be expressed as

$$\frac{\alpha\hbar^4 k_s^4}{4m_t^+ m_l^-} Z_0(E,\eta_g) + \hbar^2 k_s^2 \left[\lambda_{11}(E,\eta_g) k_z^2 + \lambda_{72}(E,\eta_g)\right] + \left[\lambda_{73}(E,\eta_g) k_z^2 + \lambda_{74}(E,\eta_g) k_z^4 - \lambda_{75}(E,\eta_g)\right] = 0 \tag{11.122}$$

where

$$Z_0(E,\eta_g) \equiv \frac{1}{2}\left[1 + Erf\left(\frac{E}{\eta_g}\right)\right],$$

$$\lambda_{70}(E,\eta_g) \equiv \frac{\alpha}{4m_t^+ m_t^-} Z_0(E,\eta_g),$$

$$\lambda_{71}(E,\eta_g) \equiv \left[\frac{\alpha\hbar^2}{4m_t^- m_l^+} Z_0(E,\eta_g) + \frac{\alpha\hbar^2}{4m_l^- m_t^+} Z_0(E,\eta_g)\right],$$

$$\lambda_{72}(E,\eta_g) \equiv \left[\left(\frac{1}{2m_t^*} - \frac{1}{2m_t^-}\right) Z_0(E,\eta_g) + \alpha\left(\frac{1}{2m_t^-} - \frac{1}{2m_t^+}\right)\gamma_0(E,\eta_g)\right],$$

$$\lambda_{73}(E,\eta_g) \equiv \left[\left(\frac{\hbar^2}{2m_l^*}+\frac{\hbar^2}{2m_l^-}\right)Z_0(E,\eta_g) + \frac{\alpha\hbar^2}{2}\left(\frac{1}{m_l^-}-\frac{1}{2m_l^+}\right)\gamma_0(E,\eta_g)\right],$$

$$\lambda_{74}(E,\eta_g) = \frac{\alpha\hbar^4 Z_0(E,\eta_g)}{4m_l^+m_l^-} \quad \text{and}$$

$$\lambda_{75}(E,\eta_g) \equiv [\gamma_0(E,\eta_g)+\alpha\theta_0(E,\eta_g)]$$

Using (11.122), the DR in HDQD of IV–VI materials in this case can be written as

$$\frac{\alpha\hbar^4k_s^4}{4m_t^+m_l^-}Z_0(E_{81\text{QBHD}},\eta_g) + \left[\hbar^2\left(\frac{n_x\pi}{d_x}\right)^2+\left(\frac{n_y\pi}{d_y}\right)^2\right] \times \left[\lambda_{11}(E_{81\text{QBHD}},\eta_g)\left(\frac{n_z\pi}{d_z}\right)^2+\lambda_{72}(E_{81\text{QBHD}},\eta_g)\right] + [\lambda_{73}(E_{81\text{QBHD}},\eta_g)k_z^2+\lambda_{74}(E_{81\text{QBHD}},\eta_g)k_z^4 - \lambda_{75}(E_{81\text{QBHD}},\eta_g)] = 0 \tag{11.123}$$

where

$$Z_0(E_{81\text{QBHD}},\eta_g) \equiv \frac{1}{2}\left[1+Erf\left(\frac{E_{81QBHD}}{\eta_g}\right)\right],$$

$$\lambda_{70}(E_{81\text{QBHD}},\eta_g) \equiv \frac{\alpha}{4m_t^+m_t^-}Z_0(E_{81\text{QBHD}},\eta_g),$$

$$\lambda_{71}(E_{81\text{QBHD}},\eta_g) \equiv \left[\frac{\alpha\hbar^2}{4m_t^-m_l^+}Z_0(E_{81\text{QBHD}},\eta_g)+\frac{\alpha\hbar^2}{4m_l^-m_t^+}Z_0(E_{81\text{QBHD}},\eta_g)\right],$$

$$\lambda_{72}(E_{81\text{QBHD}},\eta_g) \equiv \left[\left(\frac{1}{2m_t^*}-\frac{1}{2m_t^-}\right)Z_0(E_{81\text{QBHD}},\eta_g) + \alpha\left(\frac{1}{2m_t^-}-\frac{1}{2m_t^+}\right)\gamma_0(E_{81\text{QBHD}},\eta_g)\right],$$

$$\lambda_{73}\left(E_{81\text{QBHD}}\eta_g\right) \equiv \left[\left(\frac{\hbar^2}{2m_l^*}+\frac{\hbar^2}{2m_l^-}\right) Z_0\left(E_{81\text{QBHD}},\eta_g\right)\right.$$
$$\left.+\frac{\alpha\hbar^2}{2}\left(\frac{1}{m_l^-}-\frac{1}{2m_l^+}\right)\gamma_0\left(E_{81\text{QBHD}},\eta_g\right)\right],$$
$$\lambda_{74}\left(E_{81\text{QBHD}},\eta_g\right) = \frac{\alpha\hbar^4 Z_0\left(E_{81\text{QBHD}},\eta_g\right)}{4m_l^+ m_l^-},$$
$$\lambda_{75}\left(E_{81\text{QBHD}},\eta_g\right) \equiv \left[\gamma_0\left(E_{81\text{QBHD}},\eta_g\right)+\alpha\theta_0\left(E_{81\text{QBHD}},\eta_g\right)\right]$$

and $E_{81\text{QBHD}}$ is the total quantized energy in this case.

The DOS function is given by

$$N_{\text{0DT}}(E) = \frac{2g_v}{d_x d_y d_z}\sum_{n_x=1}^{n_{x\max}}\sum_{n_y=1}^{n_{y\max}}\sum_{n_z=1}^{n_{z\max}}\delta'\left(E-E_{81\text{QBHD}}\right) \tag{11.124}$$

The electron concentration can be expressed as

$$n_0 = \frac{2g_v}{d_x d_y d_z}\sum_{n_x=1}^{n_{x\max}}\sum_{n_y=1}^{n_{y\max}}\sum_{n_z=1}^{n_{z\max}} F_{-1}\left(\bar{\eta}_{10\text{QD4}}\right) \tag{11.125}$$

where $\bar{\eta}_{10\text{QD4}} = (E_{\text{FQD}} - E_{81\text{QBHD}})(k_B T)^{-1}$.

Using (1.170a), (1.170b) and (11.125), we can find the ΔC_{44} and ΔC_{456} in this case, where $n_{0\text{HD}}$ is replaced by n_0.

(b) The HD model of Bangert and Kastner

The dispersion relation of the conduction electrons in bulk specimens of IV–VI semiconductors in this case can be written using (1.132) as [43]

$$\omega_1(E)k_s^2 + \omega_2(E)k_z^2 = 1 \tag{11.126}$$

where

$$\omega_1(E) = (2E)^{-1}\left[\frac{(\overline{R})^2}{E_{g_0}\left(1+\alpha_1 E\right)}+\frac{(\overline{S})^2}{\Delta_c'\left(1+\alpha_3 E\right)}+\frac{(\overline{Q})^2}{\Delta_c''\left(1+\alpha_3 E\right)}\right] \quad \text{and}$$

$$\omega_2(E) = (2E)^{-1}\left[\frac{(\overline{A})^2}{E_{g0}\left(1+\alpha_1 E\right)}+\frac{(\overline{S}+\overline{Q})^2}{\Delta_c''\left(1+\alpha_3 E\right)}\right],$$

$$(\bar{R})^2 = 2.3 \times 10^{-10}(eVm)^2$$

$$(\bar{S})^2 = 4.6(\bar{R})^2, \quad \alpha_1 = \frac{1}{E_{g0}}, \quad \alpha_2 = \frac{1}{\Delta_c{}'}, \quad \alpha_3 = \frac{1}{\Delta_c{}''}$$

The electron energy spectrum in heavily doped IV–VI materials in this case can be expressed as

$$\begin{aligned}2I(4) = k_s^2 \Big[&\{c_1(\alpha_1 E, E_g) - iD_1(\alpha_1, E, E_g)\}\frac{(\bar{R})^2}{E_{g_0}} \\ &+ \{c_2(\alpha_2 E, E_g) - iD_2(\alpha_2, E, E_g)\}\frac{(\bar{S})^2}{\Delta_c'} + \{c_3(\alpha_3 E, E_g) \\ &- iD_3(\alpha_3 E, E_g)\}\frac{(\bar{Q})^2}{\Delta_c''}\Big] + k_z^2\left[\frac{2(\bar{A})^2}{E_{g_0}}\{c_1(\alpha_1 E, E_g)\right. \\ &\left. - iD_1(\alpha_1 E, E_g)\} + \frac{(\bar{S}+\bar{Q})^2}{\Delta_c''}\{c_3(\alpha_3 E, E_g) - iD_3(\alpha_3 E, E_g)\}\right] \end{aligned} \quad (11.127)$$

where

$$u_i = \frac{1+\alpha_i E}{\eta_g \alpha_i}, c_i(\alpha_i E, \eta_g) = \left[\frac{2}{\alpha_i \eta_g \sqrt{\pi}}\right]\exp(-u_i^2)$$
$$\times \left[\sum_{p=1}^{\infty}\{\exp(-p^2/4)(\sinh(pu_i))\}p^{-1}\right], \quad i = 1, 2, 3, 4, \ldots$$

and

$$D_i(\alpha_i, E, \eta_g) = \left[\frac{\sqrt{\pi}}{\alpha_i \eta_g}\right]\exp(-u_i^2)$$

Therefore, (11.127) can be written as,

$$F_1(E, \eta_g)\,k_s^2 + F_2(E, \eta_g)\,k_z^2 = 1 \quad (11.128)$$

where

$$\begin{aligned}F_1(E, \eta_g) &= [2\gamma_0(E, \eta_g)]^{-1} \\ &\times \left[\frac{(\bar{R})^2}{E_g}\{c_1(\alpha_1 E, E_g) - iD_1(\alpha_1, E, E_g)\} + \frac{(\bar{S})^2}{\Delta_c'}\{c_2(\alpha_2, E, E_g)\right. \\ &\left. - iD_2(\alpha_2, E, E_g)\} + \frac{(\bar{Q})^2}{\Delta_c''}\{c_3(\alpha_3, E, E_g) - iD_3(\alpha_3, E, E_g)\}\right]\end{aligned}$$

and

$$F_2(E,\eta_g) = [2\gamma_0(E,\eta_g)]^{-1} \times \left[\frac{2(\bar{A})^2}{E_g}\{c_1(\alpha_1,E,\eta_g) - iD_1(\alpha_1,E,\eta_g)\} + \frac{(\bar{S}+\bar{Q})^2}{\Delta_c''}\{c_3(\alpha_3,E,\eta_g) - iD_3(\alpha_3,E,\eta_g)\}\right]$$

Since $F_1(E,\eta_g)$ and $F_2(E,\eta_g)$ are complex, the energy spectrum is also complex in the presence of Gaussian band tails.

From (11.128), the HDDR in this case assumes the form

$$F_1(E_{9\mathrm{QBHD4}},\eta_g)\left[\left(\frac{n_x\pi}{d_x}\right)^2 + \left(\frac{n_x\pi}{d_x}\right)^2\right] + F_2(E_{9\mathrm{QBHD4}},\eta_g)\left(\frac{n_z\pi}{d_z}\right)^2 = 1 \tag{11.129}$$

where

$$F_1(E_{9\mathrm{QBHD4}},\eta_g) = [2\gamma_0(E_{9\mathrm{QBHD4}},\eta_g)]^{-1} \times \left[\frac{(\bar{R})^2}{E_g}\{c_1(\alpha_1,E_{9\mathrm{QBHD4}},E_g) - iD_1(\alpha_1,E_{9\mathrm{QBHD4}},E_g)\} + \frac{(\bar{S})^2}{\Delta_c'}\{c_2(\alpha_2,E_{9\mathrm{QBHD4}},E_g) - iD_2(\alpha_2,E_{9\mathrm{QBHD4}},E_g)\} + \frac{(\bar{Q})^2}{\Delta_c''}\{c_3(\alpha_3,E_{9\mathrm{QBHD4}},E_g) - iD_3(\alpha_3,E_{9\mathrm{QBHD4}},E_g)\}\right],$$

$$F_2(E_{9\mathrm{QBHD4}},\eta_g) = [2\gamma_0(E_{9\mathrm{QBHD4}},\eta_g)]^{-1} \times \left[\frac{2(\bar{A})^2}{E_g}\{c_1(\alpha_1,E_{9\mathrm{QBHD4}},\eta_g) - iD_1(\alpha_1,E_{9\mathrm{QBHD4}},\eta_g)\} + \frac{(\bar{S}+\bar{Q})^2}{\Delta_c''}\{c_3(\alpha_3,E_{9\mathrm{QBHD4}},\eta_g) - iD_3(\alpha_3,E_{9\mathrm{QBHD4}},\eta_g)\}\right],$$

$$\bar{u}_i = \frac{1+\alpha_i E_{9\mathrm{QBHD4}}}{\eta_g\alpha_i}, \quad c_i(\alpha_i E_{9\mathrm{QBHD4}},\eta_g) = \left[\frac{2}{\alpha_i\eta_g\sqrt{\pi}}\right]\exp\left(-\bar{u}_i^2\right) \times \left[\sum_{p=1}^{\infty}\{\exp(-p^2/4)(\sinh(p\bar{u}_i))\}p^{-1}\right], \quad i = 1,2,3,4,\ldots,$$

and

$$D_i\left(\alpha_i, E_{9\mathrm{QBHD}4}, \eta_g\right) = \left[\frac{\sqrt{\pi}}{\alpha_i \eta_g}\right] \exp\left(-\bar{u}_i^2\right)$$

The DOS function is given by

$$N_{\mathrm{0DT}}(E) = \frac{2g_v}{d_x d_y d_z} \sum_{n_x=1}^{n_{x\max}} \sum_{n_y=1}^{n_{y\max}} \sum_{n_z=1}^{n_{z\max}} \delta'\left(E - E_{9\mathrm{QBHD}4}\right) \quad (11.130)$$

The electron concentration can be expressed as

$$n_0 = \text{Real part of } \bar{n}_0 \quad (11.131)$$

where

$$\bar{n}_0 = \frac{2g_v}{d_x d_y d_z} \sum_{n_x=1}^{n_{x\max}} \sum_{n_y=1}^{n_{y\max}} \sum_{n_z=1}^{n_{z\max}} F_{-1}\left(\bar{\eta}_{11\mathrm{QD}6}\right) \quad (11.132)$$

in which $\bar{\eta}_{11\mathrm{QD}6} = \left(E_{\mathrm{FQD}} - E_{9\mathrm{QBHD}4}\right)\left(k_B T\right)^{-1}$.

Using (1.170a), (1.170b) and (11.131), we can find the ΔC_{44} and ΔC_{456} in this case, where $n_{0\mathrm{HD}}$ is replaced by n_0.

(c) The HD model of Foley *et al.*

The total quantized energy $\bar{E}_{10\mathrm{QBHD}12}$ in QD for this model can be expressed using (1.136) as [44]

$$\begin{aligned}
\gamma_3\left(\bar{E}_{10\mathrm{QBHD}12}, \eta_g\right) &= -\frac{E_{g_0}}{2} + \frac{\hbar^2}{2m_\perp^-}\left[\left(\frac{n_\chi \pi}{d_x}\right)^2 + \left(\frac{n_y \pi}{d_y}\right)^2\right] \\
&+ \frac{\hbar^2}{2m_\parallel^-}\left(\frac{n_z \pi}{d_z}\right)^2 + \left[\left[\frac{E_{g_0}}{2} + \frac{\hbar^2}{2m_\parallel^+}\left(\frac{n_z \pi}{d_z}\right)^2\right.\right. \\
&\left.+ \frac{\hbar^2}{2m_\perp^+}\left\{\left(\frac{n_x \pi}{d_x}\right)^2 + \left(\frac{n_y \pi}{d_y}\right)^2\right\}\right]^2 \\
&\left.+ P_\parallel^2\left(\frac{n_z \pi}{d_z}\right)^2 + P_\perp^2\left\{\left(\frac{n_x \pi}{d_x}\right)^2 + \left(\frac{n_y \pi}{d_y}\right)^2\right\}\right]^{1/2}
\end{aligned} \quad (11.133)$$

The DOS function is given by

$$N_{\mathrm{0DT}}(E) = \frac{2g_v}{d_x d_y d_z} \sum_{n_x=1}^{n_{x\max}} \sum_{n_y=1}^{n_{y\max}} \sum_{n_z=1}^{n_{z\max}} \delta'\left(E - \bar{E}_{10\mathrm{QBHD}12}\right) \quad (11.134)$$

The electron concentration can be expressed as

$$n_0 = \frac{2g_v}{d_x d_y d_z} \sum_{n_x=1}^{n_{x\max}} \sum_{n_y=1}^{n_{y\max}} \sum_{n_z=1}^{n_{z\max}} F_{-1}\left(\bar{\eta}_{12\mathrm{QD8}}\right) \tag{11.135a}$$

where $\bar{\eta}_{12\mathrm{QD8}} = \left(E_{\mathrm{FQD}} - \bar{E}_{10\mathrm{QBHD12}}\right)\left(k_B T\right)^{-1}$.

Using (1.170a), (1.170b) and (11.135a), we can find the ΔC_{44} and ΔC_{456} in this case, where $n_{0\mathrm{HD}}$ is replaced by n_0.

(d) The HD model of McClure and Choi

Using (1.109) the HDDR of the conduction electrons for the McClure and Choi model of IV–VI materials can be expressed as [45]

$$\gamma_2(E,\eta_g) - \gamma_3(E,\eta_g)\frac{\alpha p_y^2}{2m_2}\left(1-\frac{m_2}{m_2'}\right) = \frac{p_x^2}{2m_1} + \frac{p_y^2}{2m_2} + \frac{p_z^2}{2m_3} + \frac{\alpha p_y^4}{4m_2 m_2'} - \frac{\alpha p_x^2 p_y^2}{4m_1 m_2} - \frac{\alpha p_y^2 p_z^2}{4m_2 m_3} \tag{11.135b}$$

The electron dispersion law in HDQD in this case can be written following (11.135b)

$$\gamma_2\left(\bar{E}_{10\mathrm{QBHD121}},\eta_g\right) - \gamma_3\left(\bar{E}_{10\mathrm{QBHD121}},\eta_g\right)\frac{\alpha}{2m_2}\left[\frac{\hbar\pi n_y}{d_y}\right]^2 \times\left(1-\frac{m_2}{m_2'}\right) = \frac{1}{2m_1}\left[\frac{\hbar\pi n_x}{d_x}\right]^2 + \frac{1}{2m_2}\left[\frac{\hbar\pi n_y}{d_y}\right]^2 + \frac{1}{2m_3}\left[\frac{\hbar\pi n_z}{d_z}\right]^2 + \frac{\alpha}{4m_2 m_2'}\left[\frac{\hbar\pi n_y}{d_y}\right]^4 - \frac{\alpha}{4m_1 m_2} \times\left[\frac{\hbar\pi n_x}{d_x}\right]^2\left[\frac{\hbar\pi n_y}{d_y}\right]^2 - \frac{\alpha}{4m_2 m_3}\left[\frac{\hbar\pi n_y}{d_y}\right]^2\left[\frac{\hbar\pi n_z}{d_z}\right]^2 \tag{11.135c}$$

where $\bar{E}_{10\mathrm{QBHD121}}$ is the total quantized energy in this case.

The DOS function is given by

$$N_{0\mathrm{DT}}(E) = \frac{2g_v}{d_x d_y d_z} \sum_{n_x=1}^{n_{x\max}} \sum_{n_y=1}^{n_{y\max}} \sum_{n_z=1}^{n_{z\max}} \delta'\left(E - \bar{E}_{10\mathrm{QBHD121}}\right) \tag{11.135d}$$

The electron concentration can be expressed as

$$n_0 = \frac{2g_v}{d_x d_y d_z} \sum_{n_x=1}^{n_{x\max}} \sum_{n_y=1}^{n_{y\max}} \sum_{n_z=1}^{n_{z\max}} F_{-1}\left(\bar{\eta}_{12QD81}\right) \tag{11.135e}$$

where $\bar{\eta}_{12QD81} = \left(E_{FQD} - \bar{E}_{10QBHD121}\right)\left(k_B T\right)^{-1}$.

Using (1.170a), (1.170b) and (11.135e), we can find the ΔC_{44} and ΔC_{456} in this case, where n_{0HD} is replaced by n_0.

(e) The HD model of Hybrid

The HD DR in this case can be written following (1.113) as [46]

$$\gamma_3(E, \eta_g) - \alpha(1-\gamma_0)\frac{\hbar^2 k_y^2}{2m_2}\gamma_3(E, \eta_g) = \left(1+\bar{\delta}_0\right)\frac{\hbar^2 k_y^2}{2M_2} + \frac{\alpha\gamma_0 \hbar^4 k_y^4}{4M_2^2} + \frac{\hbar^2 k_y^2}{2M_2} + \frac{\hbar^2 k_z^2}{2M_3} \tag{11.135f}$$

The electron dispersion law in HDQD in this case can be written following (11.135f)

$$\gamma_3\left(\bar{E}_{10QBHD122}, \eta_g\right) - \alpha(1-\gamma_0)\frac{\hbar^2}{2m_2}\left[\frac{\pi n_y}{d_y}\right]^2$$
$$\gamma_3\left(\bar{E}_{10QBHD122}, \eta_g\right)\left(1-\bar{\delta}_0\right)\frac{\hbar^2}{2M_2}\left[\frac{\pi n_y}{d_y}\right]^2$$
$$+\frac{\alpha\gamma_0\hbar^4}{4M_2^2}\left[\frac{\pi n_y}{d_y}\right]^4 + \frac{\hbar^2}{2m_2}\left[\frac{\pi n_y}{d_y}\right]^2 + \frac{\hbar^2}{2M_3}\left[\frac{\pi n_z}{d_z}\right]^2 \tag{11.135g}$$

where $\bar{E}_{10QBHD122}$ is the total quantized energy in this case.

The DOS function is given by

$$N_{0DT}(E) = \frac{2g_v}{d_x d_y d_z} \sum_{n_x=1}^{n_{x\max}} \sum_{n_y=1}^{n_{y\max}} \sum_{n_z=1}^{n_{z\max}} \delta'\left(E - \bar{E}_{10QBHD122}\right) \tag{11.135h}$$

The electron concentration can be expressed as

$$n_0 = \frac{2g_v}{d_x d_y d_z} \sum_{n_x=1}^{n_{x\max}} \sum_{n_y=1}^{n_{y\max}} \sum_{n_z=1}^{n_{z\max}} F_{-1}\left(\bar{\eta}_{12QD82}\right) \tag{11.135i}$$

where $\bar{\eta}_{12QD82} = \left(E_{FQD} - \bar{E}_{10QBHD122}\right)\left(k_B T\right)^{-1}$.

Using (1.170a), (1.170b) and (11.135i), we can find the ΔC_{44} and ΔC_{456} in this case, where $n_{0\mathrm{HD}}$ is replaced by n_0.

(f) The HD model of Cohen

The HD DR in this case can be written following (1.117) as [47]

$$\gamma_2(E,\eta_g)+\frac{\alpha p_y^2}{2}\left(\frac{1}{m_2'}-\frac{1}{m_2}\right)\gamma_3(E,\eta_g)=\frac{p_x^2}{2m_1}+\frac{p_y^2}{2m_2}+\frac{p_z^2}{2m_3}+\frac{\alpha d_y^4}{4m_2m_2'} \tag{11.135j}$$

The electron dispersion law in HDQD in this case can be written following (11.135j).

$$\gamma_2(\bar{E}_{10\mathrm{QBHD123}},\eta_g)+\frac{\alpha}{2}\left[\frac{\hbar\pi n_y}{d_y}\right]^2\left(\frac{1}{m_2'}-\frac{1}{m_2}\right)\gamma_3(\bar{E}_{10\mathrm{QBHD123}},\eta_g)=\frac{1}{2m_1}\left[\frac{\hbar\pi n_x}{d_x}\right]^2+\frac{1}{2m_2}\left[\frac{\hbar\pi n_y}{d_y}\right]^2+\frac{1}{2m_3}\left[\frac{\hbar\pi n_z}{d_z}\right]^2+\frac{\alpha d_y^4}{4m_2m_2'} \tag{11.135k}$$

where $\bar{E}_{10\mathrm{QBHD123}}$ is the total quantized energy in this case.

The DOS function is given by

$$N_{0\mathrm{DT}}(E)=\frac{2g_v}{d_xd_yd_z}\sum_{n_x=1}^{n_{x\max}}\sum_{n_y=1}^{n_{y\max}}\sum_{n_z=1}^{n_{z\max}}\delta'\left(E-\bar{E}_{10\mathrm{QBHD123}}\right) \tag{11.135l}$$

The electron concentration can be expressed as

$$n_0=\frac{2g_v}{d_xd_yd_z}\sum_{n_x=1}^{n_{x\max}}\sum_{n_y=1}^{n_{y\max}}\sum_{n_z=1}^{n_{z\max}}F_{-1}\left(\bar{\eta}_{12\mathrm{QD83}}\right) \tag{11.135m}$$

where $\bar{\eta}_{12\mathrm{QD83}}=\left(E_{\mathrm{FQD}}-\bar{E}_{10\mathrm{QBHD123}}\right)(k_BT)^{-1}$.

Using (1.170a), (1.170b) and (11.135m), we can find the ΔC_{44} and ΔC_{456} in this case, where $n_{0\mathrm{HD}}$ is replaced by n_0.

(g) The HD model of Lax

The HD DR in this case can be written following (1.121) as [48]

$$\gamma_2(E,\eta_g)=\frac{p_x^2}{2m_1}+\frac{p_y^2}{2m_2}+\frac{p_z^2}{2m_3} \tag{11.135n}$$

The electron dispersion law in HDQD in this case can be written following (11.122)

$$\gamma_2\left(\bar{E}_{10QBHD124}, \eta_g\right) = \frac{1}{2m_1}\left[\frac{\hbar\pi n_x}{d_x}\right]^2 + \frac{1}{2m_2}\left[\frac{\hbar\pi n_y}{d_y}\right]^2 + \frac{1}{2m_3}\left[\frac{\hbar\pi n_z}{d_z}\right]^2 \tag{11.135o}$$

where $\bar{E}_{10QBHD124}$ is the total quantized energy in this case.

The DOS function is given by

$$N_{0DT}(E) = \frac{2g_v}{d_x d_y d_z} \sum_{n_x=1}^{n_{x\max}} \sum_{n_y=1}^{n_{y\max}} \sum_{n_z=1}^{n_{z\max}} \delta'\left(E - \bar{E}_{10QBHD124}\right) \tag{11.135p}$$

The electron concentration can be expressed as

$$n_0 = \frac{2g_v}{d_x d_y d_z} \sum_{n_x=1}^{n_{x\max}} \sum_{n_y=1}^{n_{y\max}} \sum_{n_z=1}^{n_{z\max}} F_{-1}\left(\bar{\eta}_{12QD84}\right) \tag{11.135q}$$

where $\bar{\eta}_{12QD84} = \left(E_{FQD} - \bar{E}_{10QBHD124}\right)\left(k_B T\right)^{-1}$.

Using (1.170a), (1.170b) and (11.135q), we can find the ΔC_{44} and ΔC_{456} in this case, where n_{0HD} is replaced by n_0.

(h) The HD model of ellipsoidal parabolic energy bands

The HD DR in this case can be written following (1.121) as [49]

$$\gamma_3\left(E, \eta_g\right) = \frac{p_x^2}{2m_1} + \frac{p_y^2}{2m_2} + \frac{p_z^2}{2m_3} \tag{11.135r}$$

The electron dispersion law in HDQD in this case can be written following (11.121)

$$\gamma_3\left(\bar{E}_{10QBHD125}, \eta_g\right) = \frac{1}{2m_1}\left[\frac{\hbar\pi n_x}{d_x}\right]^2 + \frac{1}{2m_2}\left[\frac{\hbar\pi n_y}{d_x}\right]^2 + \frac{1}{2m_3}\left[\frac{\hbar\pi n_z}{d_x}\right]^2 \tag{11.135s}$$

where $\bar{E}_{10QBHD125}$ is the total quantized energy in this case.

The DOS function is given by

$$N_{0\mathrm{DT}}(E) = \frac{2g_v}{d_x d_y d_z} \sum_{n_x=1}^{n_{x\max}} \sum_{n_y=1}^{n_{y\max}} \sum_{n_z=1}^{n_{z\max}} \delta'\left(E - \bar{E}_{10\mathrm{QBHD}125}\right) \tag{11.135t}$$

The electron concentration can be expressed as

$$n_0 = \frac{2g_v}{d_x d_y d_z} \sum_{n_x=1}^{n_{x\max}} \sum_{n_y=1}^{n_{y\max}} \sum_{n_z=1}^{n_{z\max}} F_{-1}\left(\bar{\eta}_{12\mathrm{QD}85}\right) \tag{11.135u}$$

where $\bar{\eta}_{12\mathrm{QD}85} = \left(E_{\mathrm{FQD}} - \bar{E}_{10\mathrm{QBHD}125}\right)(k_B T)^{-1}$.

Using (1.170a), (1.170b) and (11.135u), we can find the ΔC_{44} and ΔC_{456} in this case, where $n_{0\mathrm{HD}}$ is replaced by n_0.

*11.2.13 The CEC in HDQD of lead germanium telluride

(a) The HD model of Vassilev

The 0D electron energy spectrum in n-type $\mathrm{Pb}_{1-x}\mathrm{Ge}_x\mathrm{Te}$ under the condition of formation of band tails can be written following (1.140) as [50]

$$\begin{aligned}
&\left[\frac{2}{1+Erf\left(\frac{\bar{E}_{101\mathrm{QBHD},\pm}}{\eta_g}\right)}\right]\theta_0\left(\bar{E}_{101\mathrm{QBHD},\pm},\eta_g\right)+\gamma_3\left(\bar{E}_{101\mathrm{QBHD},\pm},\eta_g\right)\\
&\left[\bar{E}_{g_0}-0.195\left[\left(\frac{n_x\pi}{d_x}\right)^2+\left(\frac{n_y\pi}{d_y}\right)^2\right]-0.345\left(\frac{n_z\pi}{d_z}\right)^2\right]\\
&\quad=\left[0.23\left[\left(\frac{n_x\pi}{d_x}\right)^2+\left(\frac{n_y\pi}{d_y}\right)^2\right]+0.02\left(\frac{n_z\pi}{d_z}\right)^2\right]\\
&\qquad\pm\left[0.06\bar{E}_{g_0}+0.061\left[\left(\frac{n_x\pi}{d_x}\right)^2+\left(\frac{n_y\pi}{d_y}\right)^2\right]+0.0066\left(\frac{n_x\pi}{d_x}\right)^2\right]\\
&\qquad\times\left[\left(\frac{n_x\pi}{d_x}\right)^2+\left(\frac{n_y\pi}{d_y}\right)^2\right]^{1/2}+\left[\bar{E}_{g_0}+0.411\left[\left(\frac{n_x\pi}{d_x}\right)^2+\left(\frac{n_y\pi}{d_y}\right)^2\right]\right.\\
&\qquad\left.+0.377\left(\frac{n_z\pi}{d_z}\right)^2\right]\left[0.606\left(\frac{n_x\pi}{d_x}\right)^2+\left(\frac{n_y\pi}{d_y}\right)^2\right]+0.722\left(\frac{n_z\pi}{d_z}\right)^2
\end{aligned} \tag{11.136}$$

where $\bar{E}_{101\mathrm{QBHD},\pm}$ is the total quantized energy in this case.

The DOS function is given by

$$N_{0\mathrm{DT}}(E) = \frac{g_v}{d_x d_y d_z} \sum_{n_x=1}^{n_{x\max}} \sum_{n_y=1}^{n_{y\max}} \sum_{n_z=1}^{n_{z\max}} \delta'\left(E - \bar{E}_{101\mathrm{QBHD},\pm}\right) \tag{11.137}$$

The electron concentration can be expressed as

$$n_0 = \frac{g_v}{d_x d_y d_z} \sum_{n_x=1}^{n_{x\max}} \sum_{n_y=1}^{n_{y\max}} \sum_{n_z=1}^{n_{z\max}} F_{-1}\left(\bar{\eta}_{23\mathrm{QD}32}\right) \tag{11.138}$$

where $\bar{\eta}_{23\mathrm{QD}32} = \left(E_{\mathrm{FQD}} - \bar{E}_{101\mathrm{QBHD},\pm}\right)(k_B T)^{-1}$.

Using (1.170a), (1.170b) and (11.138), we can find the ΔC_{44} and ΔC_{456} in this case, where $n_{0\mathrm{HD}}$ is replaced by n_0.

*11.2.14 The CEC in HDQD of zinc and cadmium diphosphides

(a) The HD model of Chuiko

The DR in HD QDs of Zinc and Cadmium diphosphides can be written following (1.144) as [51]

$$\begin{aligned}
\gamma_3\left(\bar{E}_{102\mathrm{QBHD},\pm}, \eta_g\right) &= \left[\beta_1 + \frac{\beta_2 \beta_{31}(n_x, n_y, n_z)}{8\beta_4}\right] \\
&\times \left[\left(\frac{n_x \pi}{d_x}\right)^2 + \left(\frac{n_y \pi}{d_y}\right)^2 + \left(\frac{n_z \pi}{d_z}\right)^2\right] \\
&\pm \left\{\left[\beta_4 \beta_{31}(n_x, n_y, n_z)\left(\beta_5 - \frac{\beta_2 \beta_{31}(n_x, n_y, n_z)}{8\beta_4}\right)\right.\right. \\
&\times \left.\left[\left(\frac{n_x \pi}{d_x}\right)^2 + \left(\frac{n_y \pi}{d_y}\right)^2 + \left(\frac{n_z \pi}{d_z}\right)^2\right]\right] \\
&+ 8{\beta_4}^2\left(1 - \frac{{\beta_{31}}^2(n_x, n_y, n_z)}{4}\right) - \beta_2\left(1 - \frac{{\beta_{31}}^2(n_x, n_y, n_z)}{4}\right) \\
&\times \left.\left[\left(\frac{n_x \pi}{d_x}\right)^2 + \left(\frac{n_y \pi}{d_y}\right)^2 + \left(\frac{n_z \pi}{d_z}\right)^2\right]\right\}^{1/2}
\end{aligned} \tag{11.139}$$

where

$$\beta_{31}(n_x, n_y, n_z) = \left[\frac{\left(\frac{n_x \pi}{d_x}\right)^2 + \left(\frac{n_y \pi}{d_y}\right)^2 - 2\left(\frac{n_z \pi}{d_z}\right)^2}{\left[\left(\frac{n_x \pi}{n_x}\right)^2 + \left(\frac{n_y \pi}{d_y}\right)^2 + \left(\frac{n_z \pi}{d_z}\right)^2\right]}\right]$$

and $\bar{E}_{102\mathrm{QBHD},\pm}$ is the total quantized energy in this case.

The DOS function is given by

$$N_{0DT}(E) = \frac{g_v}{d_x d_y d_z} \sum_{n_x=1}^{n_{x\max}} \sum_{n_y=1}^{n_{y\max}} \sum_{n_z=1}^{n_{z\max}} \delta'\left(E - \bar{E}_{102QBHD,\pm}\right) \quad (11.140)$$

The electron concentration can be expressed as

$$n_0 = \frac{g_v}{d_x d_y d_z} \sum_{n_x=1}^{n_{x\max}} \sum_{n_y=1}^{n_{y\max}} \sum_{n_z=1}^{n_{z\max}} F_{-1}\left(\bar{\eta}_{24QD34}\right) \quad (11.141)$$

where $\bar{\eta}_{24QD34} = (E_{FQD} - \bar{E}_{102QBHD,\pm})(k_B T)^{-1}$.

Using (1.170a), (1.170b) and (11.141), we can find the ΔC_{44} and ΔC_{456} in this case, where n_{0HD} is replaced by n_0.

11.2.15 The CEC in HDQD of bismuth telluride

(a) The HD model of Kohler *et al.*

Following (1.148), the DR in HD Bi_2Te_3 can be written as [52]

$$\gamma_2(E, \eta_g) = \frac{\hbar^2}{2m_0}\left(\left(\bar{\alpha}_{11}\right) k_x{}^2 + \alpha_{22} k_y{}^2 + \alpha_{33} k_z{}^2 + 2\alpha_{23} k_y k_z\right) \quad (11.142a)$$

The totally quantized energy $\bar{E}_{18QBHD}$ in this case can be written following (11.142a) as

$$\gamma_2\left(\bar{E}_{18QBHD}, \eta_g\right) = \frac{\hbar^2}{2m_0}\left[\left(\bar{\alpha}_{111}\right)\left(\frac{n_x \pi}{d_x}\right)^2 + \alpha_{22}\left(\frac{n_y \pi}{d_y}\right)^2 + \alpha_{33}\left(\frac{n_z \pi}{d_z}\right)^2 + 2\alpha_{23}\left(\frac{\pi^2 n_y n_z}{d_y d_z}\right)\right] \quad (11.142b)$$

The DOS function is given by

$$N_{0DT}(E) = \frac{2g_v}{d_x d_y d_z} \sum_{n_x=1}^{n_{x\max}} \sum_{n_y=1}^{n_{y\max}} \sum_{n_z=1}^{n_{z\max}} \delta'\left(E - \bar{E}_{18QBHD}\right) \quad (11.143)$$

The electron concentration can be expressed as

$$n_0 = \frac{2g_v}{d_x d_y d_z} \sum_{n_x=1}^{n_{x\max}} \sum_{n_y=1}^{n_{y\max}} \sum_{n_z=1}^{n_{z\max}} F_{-1}\left(\bar{\eta}_{17\mathrm{QD}18}\right) \tag{11.144}$$

where $\bar{\eta}_{17\mathrm{QD}18} = (E_{\mathrm{FQD}} - \bar{E}_{18\mathrm{QBHD}})(k_B T)^{-1}$.

Using (1.170a), (1.170b) and (11.144), we can find the ΔC_{44} and ΔC_{456} in this case, where $n_{0\mathrm{HD}}$ is replaced by n_0.

11.3 Summary and Conclusion

- The DRs for QDs of HD materials exhibit the fact that the total energy is quantized since the corresponding wave vector space is totally quantized.
- The DOS functions for all the materials in this case are series of non-uniformly distributed Dirac's Delta functions at specified quantized points in the respective energy axes. The spacing between the consecutive Delta functions are functions of energy band constants and quantization of the wave vector space of a particular material.
- It may be noted that the HD QDs lead to the discrete energy levels, somewhat like atomic energy levels, which produce very large changes. This follows from the inherent nature of the quantum confinement of the carrier gas dealt with here. In QDs, there remain no free carrier states in between any two allowed sets of size-quantized levels unlike that found for QWs and NWs where the quantum confinements are 1D and 2D, respectively. Consequently, the crossing of the Fermi level by the size-quantized levels in HD QDs would have much greater impact on the redistribution of the carriers among the allowed levels, as compared to that found for QWs and NWs respectively. The quantum signature of HD QDs for the DR is rather prominent as compared to the same from QWs and NWs.
- Any property of QDs, the determination of which needs the DOS function will exhibit the non-uniformly distributed Dirac's Delta functions dependence along y-axis at the specified points of the independent variable which is plotted along x-axis.
- It is the band structure which changes in a fundamental way and consequently all the physical properties of all the electronic materials changes radically leading to new physical concepts.

We have not given any plot of CECs in HDQDs of all the materials as considered in this chapter since we believe our able readers will plot the same and enjoy intricate physics, computational analysis and advanced mathematics and will create new problems both theoretical and experimental.

11.4 Open Research Problems

(R.11.1) Investigate the CECs' for QDs of the HDS in the presences of exponential, Kane, Halperian, Lax and Bonch-Burevich types of band tails for all systems whose unperturbed carrier energy spectra are defined in Chapter 1.

(R.11.2) Investigate the CECs for QDs of all the HD materials as considered in Chapter 1 under non uniform strain.

(R.11.3) Investigate the CECs' of all the materials of Chapter 1 in the presence of non-uniform strain for QDs of HD negative refractive index, organic, magnetic and other advanced optical materials in the presence of an alternating electric field.

(R.11.4) Investigate the CECs' for the QDs of all the HD materials as considered in Chapter 1 HD negative refractive index, organic, magnetic and other advanced optical materials in the presence of an arbitrarily oriented alternating magnetic field.

(R.11.5) Investigate the CECs' for the multiple QDs of HD materials whose unperturbed carrier energy spectra are defined in Chapter 1.

(R.11.6) Investigate the CECs' for all the appropriate HD zero dimensional systems of this chapter in the presence of finite potential wells.

(R.11.7) Investigate the CECs for all the appropriate HD zero dimensional systems of this chapter in the presence of parabolic potential wells.

(R.11.8) Investigate the CECs' for all the above appropriate problems in the presence of elliptical Hill and quantum square rings in the presence of strain.

(R.11.9) Investigate the CECs' for parabolic cylindrical HD zero dimensional systems in the presence of an arbitrarily oriented alternating electric field for all the HD materials whose unperturbed carrier energy spectra are defined in Chapter 1 in the presence of strain.

(R.11.10) Investigate the CECs' for HD zero dimensional systems of the negative refractive index and other advanced optical materials in the presence of an arbitrarily oriented alternating electric field and non-uniform light waves and in the presence of strain.

(R.11.11) Investigate the CECs' for HDQD materials of Chapter 1 in the presence of triangular potential well zero dimensional systems of the negative refractive index, organic, magnetic and other advanced optical materials in the presence of an arbitrarily oriented alternating electric field in the presence of strain.

(R.11.12) Investigate the CECs' for all the problems of (R.11.11) in the presence of arbitrarily oriented magnetic field.

(R.11.13) Investigate the CECs' for all the problems of (R.11.11) in the presence of alternating electric field.

(R.11.14) Investigate the CECs' for all the problems of (R.11.11) in the presence of alternating magnetic field.

(R.11.15) Investigate the CECs' for all the problems of (R.11.11) in the presence of crossed electric field and quantizing magnetic fields.

(R.11.16) Investigate the CECs' for all the problems of (R.11.11) in the presence of crossed alternating electric field and alternating quantizing magnetic fields.

(R.11.17) Investigate the CECs' for HD QDs of the negative refractive index, organic and magnetic materials.

(R.11.18) Investigate the CECs' for HD QDs of the negative refractive index, organic and magnetic materials in the presence of alternating time dependent magnetic field.

(R.11.19) Investigate the CECs' for HD QDs of the negative refractive index, organic and magnetic materials in the presence of in the presence of crossed alternating electric field and alternating quantizing magnetic fields.

(R.11.20) (a) Investigate the CECs' for HD QDs of the negative refractive index, organic, magnetic and other advanced optical materials in the presence of an arbitrarily oriented alternating electric field considering many body effects.

(b) Investigate all the appropriate problems of this chapter for a Dirac electron.

(R.11.21) Investigate all the appropriate problems of this chapter by including the many body, image force, broadening and hot carrier effects respectively.

(R.11.22) Investigate all the appropriate problems of this chapter by removing all the mathematical approximations and establishing the respective appropriate uniqueness conditions.

References

[1] J. Bodnar, in *Proceedings of the International Conference on Physics of Narrow-gap Semiconductors*, (Polish Science Publishers, Warsaw, **1978**).
[2] R. K. Willardson, Ed. by A. C. Beer, 1, *Semiconductors and Semimetals*, (Academic Press, New York, **1966**), p. 102.
[3] E. O. Kane, *Phys. Rev.*, 131, 79 **(1963)**.
[4] V. L. Bonch Bruevich, *Sov. Phys. Sol. Stat.*, 4, 53 **(1963)**.
[5] R. A. Logan, A. G. Chynoweth, *Phys. Rev.*, 131, 89 **(1963)**.
[6] C. J. Hwang, *J. Appl. Phys.*, 40, 3731 **(1969)**.
[7] J. I. Pankove, *Phys. Rev. A*, 130, 2059 **(1965)**.
[8] B. I. Halperin, M. Lax, *Phys. Rev.*, 148, 722 **(1966)**.
[9] R. A. Abram, G. J. Rees, B. L. H. Wilson, *Adv. Phys.*, 27, 799 **(1978)**.
[10] B. I. Shklovskii, A. L. Efros, *Electronic Properties of Doped Semiconductors*, Springer Series in Solid State Sciences, Vol. 45, (Springer, Berlin, **1984**).
[11] E. O. Kane, *Sol. Stat. Electron.*, 28, 3 **(1985)**.
[12] P. K. Chakraborty, J. C. Biswas, *J. Appl. Phys.*, 82, 3328 **(1997)**.
[13] B. R. Nag, *Electron Transport in Compound Semiconductors*, Springer Series in Solid State Sciences, Vol. 11, (Springer, Heidelberg, **1980**).
[14] P. E. Schmid, *Phys. Rev. B*, 23, 5531 **(1981)**.
[15] Jr. G. E. Jellison, F. A. Modine, C. W. White, R. F. Wood, R. T. Young, *Phys. Rev. Lett.*, 46, 1414 **(1981)**.
[16] V. I. Fistul, *Heavily Doped Semiconductors*, (Plenum, New York, **1969**), Ch. 7.
[17] C. J. Hwang, *J. Appl. Phys.*, 41, 2668 **(1970)**; W. Sritrakool, H. R. Glyde, V. Sa Yakanit, *Can. J. Phys.*, 60, 373 **(1982)**; H. Ikoma, *J. Phys. Soc. Jap.*, 27, 514 **(1969)**.
[18] M. H. Chen, C. C. Wu, C. J. Lin, *J. Low Temp. Phys.*, 55, 127 **(1984)**.
[19] M. Abramowitz and I. A. Stegun, *Handbook of Mathematical Functions with Formulas, Graphs and Mathematical Tables*, (New York: Wiley, **1964**).
[20] I. S. Gradshteyn and I. M. Ryzhik, *Tables of Integrals, Series and Products*, (New York: Academic, **1965**).
[21] E. O. Kane, in *Semiconductors and Semimetals*, Vol. 1, Ed. by R. K. Willardson, A. C. Beer, (Academic Press, New York, **1966**), p. 75.
[22] G. E. Stillman, C. M. Wolfe, J. O. Dimmock, in *Semiconductors and Semimetals*, 12, Ed. by R. K. Willardson, A. C. Beer, (Academic Press, New York, 169, **1977**).

[23] E. D. Palik, G. S. Picus, S. Teither, R. E. Wallis, *Phys. Rev.*, 475 **(1961)**.
[24] U. Rossler, *Sol. Stat. Commun.*, 49, 943 **(1984)**.
[25] V. G. Agafonov, P. M. Valov, B. S. Ryvkin and I. D. Yaroshetskii, *Sov. Phys. Semicond.*, 12, 1182 **(1978)**.
[26] D. J. Newson, A. Kurobe, *Semicond. Sci. Technol.*, 3, 786 **(1988)**.
[27] E. J. Johnson, D. H. Dicley, *Phys. Rev.*, 1, 2676 **(1970)**.
[28] J. Kolodziejczak, S. Zukotynski and H. Stramska, *Phys. Stat. Sol.*, 14, 471 **(1966)**.
[29] I. A. Chaikovskii, *Sov. Phys. Semicond.*, 6, 1 **(1972)**.
[30] J. J. Hopfield, *J. Appl. Phys.*, 32, 2277 **(1961)**.
[31] G. J. Rees, *Physics of Compounds*, in Proc. of the 13th Inter. Nat. Conf. Ed. by F. G. Fumi, 1166, (North Holland Company, The Netherlands, **1976**).
[32] E. L. Ivechenko, G. E. Pikus, *Sov. Phys. Semicond.*, 13, 579 **(1979)**.
[33] M. Cardona, W. Paul, H. Brooks, *Helv. Acta Phys.*, 33, 329 **(1960)**.
[34] C. C. Wang, N. W. Ressler, *Phys. Rev.*, 2, 1827 **(1970)**.
[35] J. Bouat, J. C. Thuillier, *Surface Sci.*, 73, 528 **(1978)**.
[36] M. V. Ortenberg, K. J. Button, *Phys. Rev. B*, 16, 2618 **(1977)**.
[37] H. Ushio, T. Dau, Y. Uemura, *J. Phys. Soc. Jap.*, 33, 1551 **(1972)**.
[38] P. R. Emtage, *Phys. Rev.*, 138, A246 **(1965)**.
[39] Y. Yamada, *J. Phys. Soc. Jap.*, 35, 1600 **(1973)**.
[40] D. G. Seiler, B. D. Bajaj and A. E. Stephens, *Phys. Rev. B*, 16, 2822 **(1977)**; A. V. Germaneko and G. M. Minkov, *Phys. Stat. Sol. (b)*, 184, 9 **(1994)**; G. L. Bir and G. E. Pikus, *Symmetry and Strain–Induced Effects in Semiconductors Nauka, Russia* **(1972)** (in Russian); M. Mondal and K. P. Ghatak, *Phys. Stat. Sol. (b)*, 135, K21 **(1986)**.
[41] P. C. Mathur and S. Jain, *Phys. Rev.*, 19, 3159 **(1979)**.
[42] H. I. Zhang, *Phys. Rev. B*, 1, 3450 **(1970)**.
[43] E. Bangert, P. Kastner, *Phys. Stat. Sol. (b)*, 61, 503 **(1974)**.
[44] G. M. T. Foley, P. N. Langenberg, *Phys. Rev. B*, 15, 4850 **(1977)**.
[45] J. W. McClure, K. H. Choi, *Sol. Stat. Commun.*, 21, 1015 **(1977)**.
[46] S. Takaoka, H. Kawamura, K. Murase, S. Takano, *Phys. Rev. B*, 13, 1428 **(1976)**; C. C. Wu, C. J. Lin, *J. Low. Temp. Phys.*, 57, 469 **(1984)**.
[47] M. H. Cohen, *Phys. Rev.*, 121, 387 **(1961)**.
[48] B. Lax, J. G. Mavroides, H. J. Zieger, R. J. Keyes, *Phys. Rev. Lett.*, 5, 241 **(1960)**.
[49] D. Shoenberg, *Proc. Roy. Soc. (London)*, 170, 341 **(1939)**; B. Abeles, S. Meiboom, *Phys. Rev.*, 101, 544 **(1956)**.
[50] L. A. Vassilev, *Phys. Stat. Sol. (b)*, 121, 203 **(1984)**; S. Takaoka, K. Murase, *Phys. Rev. B*, 20, 2823 **(1979)**.
[51] G. P. Chuiko, *Sov. Phys. Semicond.*, 19, 1381 **(1985)**.
[52] H. Köhler, *Phys. Stat. Sol. (b)*, 74, 591 **(1976)**.

Chapter 12

Einstein's Photoemission from HDQD of Various Non-Parabolic Materials

Education is the best investment in life.

12.1 Introduction

In Sections 12.2.1–12.2.15, we have investigated the EP from QDs of HD nonlinear optical and tetragonal, III–V, II–VI, IV–VI, stressed Kane type, Te, GaP, $PtSb_2, Bi_2Te_3$, Ge, GaSb, Graphite, II–V, Lead Germanium Telluride and Zinc and Cadmium diphosphides respectively. Section 12.3 contains the summary and conclusion pertaining to this chapter. Section 12.4 presents 23 open research problems.

12.2 Theoretical Background

*12.2.1 The EP from HDQD of nonlinear optical and tetragonal materials

Following (11.29), the electron concentration per sub-band is given by

$$\Delta n_{0\text{D}} = \frac{2g_v}{d_x d_y d_z} \text{ Real Part of } [F_{-1}(\eta_{1\text{QD}})] \tag{12.1}$$

The quantized energy along Z direction the $E_{n_z\text{HD1}}$ in this case is given by

$$\frac{\hbar^2 (n_z\pi/d_z)^2}{2m_{\|}^* T_{21}(E_{n_z\text{HD1}}, \eta_g)} = 1 \tag{12.2}$$

The expression of the total photo-emitted current density in this case is

$$J_{0\text{D}} = \frac{e\alpha_0 g_v}{2} \text{ Real Part of } \sum_{n_x=1}^{n_{x\max}} \sum_{n_y=1}^{n_{y\max}} \sum_{n_{z\min}}^{n_{z\max}} (\Delta n_{0\text{D}}) v_z(E_{n_z\text{HD1}}) \tag{12.3a}$$

where

$$v_z\left(E_{n_z\mathrm{HD1}}\right) = [n_z\pi\hbar]\left[m_{\|}^{*}d_z T_{21}'\left(E_{n_z\mathrm{HD1}},\eta_g\right)\right]^{-1}$$

and $n_{z_{\min}}$ is the nearest integer of the following equations

$$\frac{\hbar^2\left(n_{z_{\min}}\pi/d_z\right)^2}{2m_{\|}^{*}T_{21}((W-hv),\eta_g)} = 1 \tag{12.3b}$$

12.2.2 The EP from HDQD of III–V materials

(a) The three band model of Kane

Following (11.35), the electron concentration per sub-band is given by

$$\Delta n_{0\mathrm{D}} = \frac{2g_v}{d_x d_y d_z}\ \text{Real Part of}\ [F_{-1}(\eta_{2\mathrm{QD}})] \tag{12.4}$$

The quantized energy along Z direction the $E_{n_z\mathrm{HD2}}$ in this case is given by

$$\frac{\hbar^2\left(n_z\pi/d_z\right)^2}{2m_c T_{44}\left(E_{n_z\mathrm{HD2}},\eta_g\right)} = 1 \tag{12.5}$$

The expression of the total photo-emitted current density in this case is

$$J_{0\mathrm{D}} = \frac{e\alpha_0 g_v}{2}\ \text{Real Part of}\ \sum_{n_x=1}^{n_{x\max}}\sum_{n_y=1}^{n_{y\max}}\sum_{n_{z_{\min}}}^{n_{z\max}}(\Delta n_{0\mathrm{D}})v_{z2}(E_{n_z\mathrm{HD2}}) \tag{12.6a}$$

where

$$v_{z2}(E_{n_z\mathrm{HD2}}) = [n_z\pi\hbar][m_c d_z T_{44}'(E_{n_z\mathrm{HD2}},\eta_g)]^{-1}$$

and $n_{z_{\min}}$ is the nearest integer of the following equation

$$\frac{\hbar^2\left(n_{z_{\min}}\pi/d_z\right)^2}{2m_c T_{44}\left((W-hv),\eta_g\right)} = 1 \tag{12.6b}$$

(b) The two band model of Kane

Following (11.39), the electron concentration per sub-band is given by

$$\Delta n_{0\mathrm{D}} = \frac{2g_v}{d_x d_y d_z}\left[F_{-1}\left(\eta_{3\mathrm{QD}}\right)\right] \tag{12.7}$$

The quantized energy along Z direction the $E_{n_z\mathrm{HD3}}$ in this case is given by

$$\frac{\hbar^2(n_z\pi/d_z)^2}{2m_c\gamma_2(E_{n_z\mathrm{HD3}},\eta_g)} = 1 \tag{12.8}$$

The expression of the total photo-emitted current density in this case is

$$J_{\mathrm{0D}} = \frac{e\alpha_0 g_v}{2} \sum_{n_x=1}^{n_{x\max}} \sum_{n_y=1}^{n_{y\max}} \sum_{n_{z\min}}^{n_{z\max}} (\Delta n_{\mathrm{0D}})\, v_{z3}\left(E_{n_z\mathrm{HD3}}\right) \tag{12.9a}$$

where

$$v_{z3}\left(E_{n_z\mathrm{HD3}}\right) = \left[n_z \pi \hbar\right]\left[m_c d_z \gamma_2'\left(E_{n_z\mathrm{HD3}}, \eta_g\right)\right]^{-1}$$

and $n_{z_{\min}}$ is the nearest integer of the following equations

$$\frac{\hbar^2 \left(n_{z_{\min}} \pi / d_z\right)^2}{2m_c \gamma_2\left((W - hv), \eta_g\right)} = 1 \tag{12.9b}$$

(c) The Parabolic band model

Following (11.43), the electron concentration per sub-band is given by

$$\Delta n_{\mathrm{0D}} = \frac{2g_v}{d_x d_y d_z}\left[F_{-1}\left(\eta_{4\mathrm{QD}}\right)\right] \tag{12.10}$$

The quantized energy along Z direction the $E_{n_z\mathrm{HD4}}$ in this case is given by

$$\frac{\hbar^2 \left(n_z \pi / d_z\right)^2}{2m_c \gamma_3\left(E_{n_z\mathrm{HD4}}, \eta_g\right)} = 1 \tag{12.11}$$

The expression of the total photo-emitted current density in this case is

$$J_{\mathrm{0D}} = \frac{e\alpha_0 g_v}{2} \sum_{n_x=1}^{n_{x\max}} \sum_{n_y=1}^{n_{y\max}} \sum_{n_{z\min}}^{n_{z\max}} (\Delta n_{\mathrm{0D}})\, v_{z4}\left(E_{n_z\mathrm{HD4}}\right) \tag{12.12a}$$

where

$$v_{z4}(E_{n_z\mathrm{HD4}}) = \left[n_z \pi \hbar\right]\left[m_c d_z \gamma_3'(E_{n_z\mathrm{HD4}}, \eta_g)\right]^{-1}$$

and $n_{z_{\min}}$ is the nearest integer of the following equations

$$\frac{\hbar^2 \left(n_{z_{\min}} \pi / d_z\right)^2}{2m_c \gamma_3\left((W - hv), \eta_g\right)} = 1 \tag{12.12b}$$

(d) The model of Stillman *et al.*

Following (11.47), the electron concentration per sub-band is given by

$$\Delta n_{\mathrm{0D}} = \frac{2g_v}{d_x d_y d_z}\left[F_{-1}\left(\eta_{5\mathrm{QD}}\right)\right] \tag{12.13}$$

The quantized energy along Z direction the $E_{n_z\mathrm{HD4}}$ in this case is given by

$$\frac{\hbar^2 \left(n_z \pi / d_z\right)^2}{2m_c \theta_4\left(E_{n_z\mathrm{HD5}}, \eta_g\right)} = 1 \tag{12.14}$$

The expression of the total photo-emitted current density in this case is

$$J_{0D} = \frac{e\alpha_0 g_v}{2} \sum_{n_x=1}^{n_{x\max}} \sum_{n_y=1}^{n_{y\max}} \sum_{n_{z\min}}^{n_{z\max}} (\Delta n_{0D})\, v_{z5}\left(E_{n_z\mathrm{HD5}}\right) \tag{12.15a}$$

where

$$v_{z5}\left(E_{n_z\mathrm{HD5}}\right) = \left[n_z \pi \hbar\right]\left[m_c d_z \theta_4'\left(E_{n_z\mathrm{HD5}}, \eta_g\right)\right]^{-1}$$

and $n_{z_{\min}}$ is the nearest integer of the following equations

$$\frac{\hbar^2\left(n_{z_{\min}}\pi/d_z\right)^2}{2m_c\theta_4\left((W - hv), \eta_g\right)} = 1 \tag{12.15b}$$

(e) The HD Palik model

Following (11.51), the electron concentration per sub-band is given by

$$\Delta n_{0D} = \frac{2g_v}{d_x d_y d_z}\left[F_{-1}\left(\eta_{6QD}\right)\right] \tag{12.16}$$

The quantized energy along Z direction the $E_{n_z\mathrm{HD6}}$ in this case is given by

$$\frac{\hbar^2\left(n_z\pi/d_z\right)^2}{2m_c\theta_5\left(E_{n_z\mathrm{HD6}}, \eta_g\right)} = 1 \tag{12.17}$$

The expression of the total photo-emitted current density in this case is

$$J_{0D} = \frac{e\alpha_0 g_v}{2} \sum_{n_x=1}^{n_{x\max}} \sum_{n_y=1}^{n_{y\max}} \sum_{n_{z\min}}^{n_{z\max}} (\Delta n_{0D}) v_{z6}(E_{n_z\mathrm{HD6}}) \tag{12.18a}$$

where

$$v_{z6}\left(E_{n_z\mathrm{HD6}}\right) = \left[n_z \pi \hbar\right]\left[m_c d_z \theta_5'\left(E_{n_z\mathrm{HD6}}, \eta_g\right)\right]^{-1}$$

and $n_{z_{\min}}$ is the nearest integer of the following equations

$$\frac{\hbar^2\left(n_{z_{\min}}\pi/d_z\right)^2}{2m_c\theta_5\left((W - hv), \eta_g\right)} = 1 \tag{12.18b}$$

(f) The HD Rossler model

Following (11.55), the electron concentration per sub-band is given by

$$n_0 = \frac{g_v}{d_x d_y d_z}\left[F_{-1}\left(\bar{\eta}_{8QD}\right)\right] \tag{12.18c}$$

The quantized energy along z-axis the $E_{n_z\mathrm{HD61}}$ in this case is given by

$$\gamma_3\left(E_{n_z\mathrm{HD61}},\eta_g\right)=\frac{\hbar^2}{2m_c}\left(\pi n_z/d_z\right)^2+(\alpha_{11}+\alpha_{12}(\pi n_z/d_z))(\pi n_z/d_z)^4 \tag{12.18d}$$

The expression of the total photo-emitted current density in this case is

$$J_{\mathrm{0D}}=\frac{e\alpha_0 g_v}{2}\sum_{n_x=1}^{n_{x\max}}\sum_{n_y=1}^{n_{y\max}}\sum_{n_{z\min}}^{n_{z\max}}(\Delta n_{\mathrm{0D}})v_{z61}(E_{n_z\mathrm{HD61}}) \tag{12.18e}$$

where

$$v_{z61}(E_{n_z\mathrm{HD61}})=\frac{1}{\hbar}[\gamma_3'(E_{n_z\mathrm{HD61}},\eta_g)]^{-1}$$
$$\times\left[\frac{\hbar^2}{m_c}(\pi n_z/d_z)+\alpha_{12}(\pi n_z/d_z)^4+4(\pi n_z/d_z)^3(\alpha_{11}+\alpha_{12}(\pi n_z/d_z))\right]$$

and $n_{z_{\min}}$ is the nearest integer of the following equation

$$\gamma_3\left((W-hv),\eta_g\right)=\frac{\hbar^2}{2m_c}\left(\pi n_{z_{\min}}/d_z\right)^2+(\alpha_{11}+\alpha_{12}(\pi n_{z_{\min}}/d_z))\times(\pi n_{z_{\min}}/d_z)^4 \tag{12.18f}$$

(g) The HD model of Agafonov *et al.*

Following (11.59a), the electron concentration per sub-band is given by

$$n_0=\frac{2g_v}{d_x d_y d_z}\left[F_{-1}\left(\bar{\eta}_{8\mathrm{QD3}}\right)\right] \tag{12.18g}$$

The quantized energy along z-axis the $E_{n_z\mathrm{HD62}}$ in this case is given by

$$\gamma_3\left(E_{n_z\mathrm{HD62}},\eta_g\right)=\left[\frac{\left(\sqrt{E_{g_0}^2+\frac{8}{3}P_0^2\left(\pi n_z/d_z\right)^2}\right)-E_{g_0}}{2}\right]\times\left[1-\frac{T_3\left(\pi n_z/d_z\right)^2}{\sqrt{E_{g_0}^2+\frac{8}{3}P_0^2\left(\pi n_z/d_z\right)^2}}\right] \tag{12.18h}$$

The expression of the total photo-emitted current density in this case is

$$J_{\mathrm{0D}} = \frac{e\alpha_0 g_v}{2} \sum_{n_x=1}^{n_{x\max}} \sum_{n_y=1}^{n_{y\max}} \sum_{n_{z_{\min}}}^{n_{z\max}} (\Delta n_{\mathrm{0D}})\, v_{z61}\left(E_{n_z\mathrm{HD62}}\right) \tag{12.18i}$$

where

$$v_{z62}\left(E_{n_z\mathrm{HD62}}\right) = \frac{1}{\hbar}\left[\gamma_3'\left(E_{n_z\mathrm{HD62}}, \eta_g\right)\right]^{-1}\left[\frac{4}{3}\left(\delta_{n_z}\right)^{-1/2} P_0^2\left(\pi n_z/d_z\right) \times \left[1 - \frac{T_3\left(\pi n_z/d_z\right)^2}{\sqrt{\delta_{n_z}}}\right] + \left[\frac{\left(\sqrt{\delta_{n_z}}\right) - E_{g_0}}{2}\right] \times \left[\frac{-2T_3\left(\pi n_z/d_z\right)}{\sqrt{\delta_{n_z}}} + \frac{4}{3}T_3 P_0^2\left(\pi n_z/d_z\right)^3\left(\delta_{n_z}\right)^{-3/2}\right]\right],$$

$$\delta_{n_z} = E_{g_0}^2 + \frac{8}{3}P_0^2\left(\pi n_z/d_z\right)^2$$

and $n_{z_{\min}}$ is the nearest integer of the following equation

$$\gamma_3\left((W - hv), \eta_g\right) = \left[\frac{\left(\sqrt{E_{g_0}^2 + \frac{8}{3}P_0^2\left(\pi n_{z_{\min}}/d_z\right)^2}\right) - E_{g_0}}{2}\right] \times \left[1 - \frac{T_3\left(\pi n_{z_{\min}}/d_z\right)^2}{\sqrt{E_{g_0}^2 + \frac{8}{3}P_0^2\left(\pi n_{z_{\min}}/d_z\right)^2}}\right] \tag{12.18j}$$

(h) The HD model of Newson and Kurobe

Following (11.59d), the electron concentration per sub-band is given by

$$n_0 = \frac{2g_v}{d_x d_y d_z}\left[F_{-1}\left(\eta_{\mathrm{QD3N}}\right)\right] \tag{12.18k}$$

The quantized energy along z-axis the $E_{n_z\mathrm{HD63}}$ in this case is given by

$$\gamma_3\left(E_{n_z\mathrm{HD63}}, \eta_g\right) = \frac{\hbar^2}{2m_c}\left(\pi n_z/d_z\right)^2 + (a_{13})\left(\pi n_z/d_z\right)^4 \tag{12.18l}$$

The expression of the total photo-emitted current density in this case is

$$J_{\mathrm{0D}} = \frac{e\alpha_0 g_v}{2} \sum_{n_x=1}^{n_{x\max}} \sum_{n_y=1}^{n_{y\max}} \sum_{n_{z_{\min}}}^{n_{z\max}} (\Delta n_{\mathrm{0D}})\, v_{z63}\left(E_{n_z\mathrm{HD63}}\right) \tag{12.18m}$$

where

$$v_{z63}\left(E_{n_z\mathrm{HD63}}\right) = \frac{1}{\hbar}\left[\gamma_3'\left(E_{n_z\mathrm{HD63}}, \eta_g\right)\right]^{-1}\left[\frac{\hbar^2}{m_c}\left(\pi n_z/d_z\right) + 4a_{13}\left(\pi n_z/d_z\right)^3\right]$$

and $n_{z_{\min}}$ is the nearest integer of the following equation

$$\gamma_3\left((W-hv), \eta_g\right) = \frac{\hbar^2}{2m_c}\left(\pi n_{z_{\min}}/d_z\right)^2 + (a_{13})\left(\pi n_{z_{\min}}/d_z\right)^4 \tag{12.18n}$$

(i) HD model of Johnson and Dickey

Following (11.59i), the electron concentration per sub-band is given by

$$n_0 = \frac{2g_v}{d_x d_y d_z}\ \text{Real part of}\ \left[F_{-1}(\eta_{\mathrm{QD3N}})\right] \tag{12.18o}$$

The quantized energy along z-axis the $E_{n_z\mathrm{HD64}}$ in this case is given by

$$\theta_5\left(E_{n_z\mathrm{HD64}}, \eta_g\right) - \left(\pi n_z/d_z\right)^2\left[\theta_6\left(\alpha_1, E_{n_z\mathrm{HD64}}, \eta_g\right) + i\theta_3\left(\alpha_1, E_{n_z\mathrm{HD64}}, \eta_g\right)\right] + a^2\left(\pi n_z/d_z\right)^4 = 0 \tag{12.18p}$$

The expression of the total photo-emitted current density in this case is

$$J_{0\mathrm{D}} = \frac{e\alpha_0 g_v}{2}\ \text{Real part of}\ \sum_{n_x=1}^{n_{x\max}}\sum_{n_y=1}^{n_{y\max}}\sum_{n_{z_{\min}}}^{n_{z\max}}\left(\Delta n_{0\mathrm{D}}\right) v_{z64}\left(E_{n_z\mathrm{HD64}}\right) \tag{12.18q}$$

where

$$v_{z64}(E_{n_z\mathrm{HD64}}) = \frac{1}{\hbar}\frac{\begin{array}{c}[2((\pi n_z/d_z)(\theta_6(\alpha_1, E_{n_z\mathrm{HD64}}, \eta_g)\\ + i\theta_3(\alpha_1, E_{n_z\mathrm{HD64}}, \eta_g)) - 4a^2(\pi n_z/d_z)^3]\end{array}}{\begin{array}{c}[[\theta_5(\alpha_1, E_{n_z\mathrm{HD64}}, \eta_g)]' - (\pi n_z/d_z)^2[\theta_6(\alpha_1, E_{n_z\mathrm{HD64}}, \eta_g)\\ + i\theta_3(\alpha_1, E_{n_z\mathrm{HD64}}, \eta_g)]']\end{array}}$$

and $n_{z_{\min}}$ is the nearest integer of the following equation

$$\left[\theta_5\left((W-hv), \eta_g\right) - \left(\pi n_{z_{\min}}/d_z\right)^2\left[\theta_6\left(\alpha_1, (W-hv), \eta_g\right) + i\theta_3\left(\alpha_1, (W-hv), \eta_g\right)\right] + a^2\left(\pi n_{z_{\min}}/d_z\right)^4\right] = 0 \tag{12.18r}$$

(j) The HD model of Kolodziejczak *et al.*

Following (11.59i), the electron concentration per sub-band is given by

$$n_0 = \frac{2g_v}{d_x d_y d_z} F_{-1}(\eta_{70}) \tag{12.18s}$$

The quantized energy along z-axis the $\bar{E}_{n_{z\text{HD}64}}$ in this case is given by

$$\begin{aligned}\gamma_3(\bar{E}_{n_{z\text{HD}64}}, \eta_g) &= \frac{\hbar^2}{2m_0}(\pi n_z/d_z)^2 + \frac{1}{2}\left[\sqrt{t_7 - t_8(\pi n_z/d_z)^2} - E_{g_0}\right] \\ &+ a\left[\frac{\sqrt{t_7 - t_8(\pi n_z/d_z)^2} + E_{g_0}}{2}\right] \times \frac{\hbar^2}{2m_0}(\pi n_z/d_z)^2 \\ &+ \left[\frac{\sqrt{t_7 - t_8(\pi n_z/d_z)^2} - E_{g_0}}{2}\right]\left(b + \frac{c}{5}\right)\frac{\hbar^2}{2m_0}(\pi n_z/d_z)^2\end{aligned} \tag{12.18t}$$

where

$$t_7 = E_{g0}^2, t_8 = 4FE_{g_0}\left[\frac{E_{g_0} + \frac{2}{3}\Delta}{E_{g_0} + \Delta}\right]$$

The expression of the total photo-emitted current density in this case is

$$J_{0\text{D}} = \frac{e\alpha_0 g_v}{2} \sum_{n_x=1}^{n_{x\max}} \sum_{n_y=1}^{n_{y\max}} \sum_{n_{z\min}}^{n_{z\max}} (\Delta n_{0\text{D}})\, \bar{v}_{z64}(\bar{E}_{n_{z\text{HD}64}}) \tag{12.18u}$$

where

$$\begin{aligned}\bar{v}_{z64}(\bar{E}_{n_{z\text{HD}64}}) &= [\hbar\gamma_3'(\bar{E}_{n_{z\text{HD}64}}, \eta_g)]^{-1}\left[\frac{\hbar^2}{m_0}(\pi n_z/d_z)\right. \\ &- \frac{t_8(\pi n_z/d_z)}{2}\left(t_7 - t_8(\pi n_z/d_z)^2\right)^{-1/2} + \frac{a\hbar^2(\pi n_z/d_z)}{m_0} \\ &\times \left[\frac{\sqrt{t_7 - t_8(\pi n_z/d_z)^2} + E_{g_0}}{2}\right] - \frac{a\hbar^2(\pi n_z/d_z)^3 t_8}{2m_0} \\ &\times [t_7 - t_8(\pi n_z/d_z)^2]^{-1/2} + \frac{\hbar^2}{2m_0}\left(b + \frac{c}{5}\right)(\pi n_z/d_z) \\ &\times \left[\sqrt{t_7 - t_8(\pi n_z/d_z)^2} - E_{g_0}\right] - \frac{\hbar^2(\pi n_z/d_z)^3 t_8}{2m_0}\left(b + \frac{c}{5}\right) \\ &\left.\times [t_7 - t_8(\pi n_z/d_z)^2]^{-1/2}\right]\end{aligned}$$

and $n_{z_{\min}}$ is the nearest integer of the following equation

$$\gamma_3\left((W-hv),\eta_g\right)=\frac{\hbar^2}{2m_0}\left(\pi n_{z_{\min}}/d_z\right)^2+\frac{1}{2}\left[\sqrt{t_7-t_8\left(\pi n_{z_{\min}}/d_z\right)^2}-E_{g0}\right]$$
$$+a\left[\frac{\sqrt{t_7-t_8\left(\pi n_{z_{\min}}/d_z\right)^2}+E_{g0}}{2}\right]\frac{\hbar^2}{2m_0}\left(\pi n_{z_{\min}}/d_z\right)^2$$
$$+\left[\frac{\sqrt{t_7-t_8\left(\pi n_{z_{\min}}/d_z\right)^2}-E_{g0}}{2}\right]\left(b+\frac{c}{5}\right)\frac{\hbar^2}{2m_0}\left(\pi n_{z_{\min}}/d_z\right)^2 \tag{12.18v}$$

(k) The HD model of Chaikovskii

Following (11.59p), the electron concentration per sub-band is given by

$$n_0=\frac{2g_v}{d_xd_yd_z}F_{-1}\left(\eta_{1,9,153}\right) \tag{12.19a}$$

The quantized energy along z-axis the $\bar{E}_{n_z\mathrm{HD65}}$ in this case is given by

$$\gamma_3(\bar{E}_{n_z\mathrm{HD65}},\eta_g)=\frac{\hbar^2}{2m_c}\left(\pi n_z/d_z\right)^2+\frac{e^2E_0^2}{\omega_0^2m_c} \tag{12.19b}$$

The expression of the total photo-emitted current density in this case is

$$J_{\mathrm{OD}}=\frac{e\alpha_0g_v}{2}\sum_{n_x=1}^{n_{x\max}}\sum_{n_y=1}^{n_{y\max}}\sum_{n_{z_{\min}}}^{n_{z\max}}\left(\Delta n_{\mathrm{OD}}\right)\bar{v}_{z65}\left(\bar{E}_{n_z\mathrm{HD65}}\right) \tag{12.19c}$$

where

$$\bar{v}_{z65}\left(\bar{E}_{n_z\mathrm{HD65}}\right)=\left[\hbar\gamma_3'\left(\bar{E}_{n_z\mathrm{HD65}},\eta_g\right)\right]^{-1}\left[\frac{\hbar}{m_c}\left(\pi n_z/d_z\right)\right]$$

and $n_{z_{\min}}$ is the nearest integer of the following equation

$$\gamma_3\left((W-hv),\eta_g\right)=\frac{\hbar^2}{2m_c}\left(\pi n_{z_{\min}}/d_z\right)^2+\frac{e^2E_0^2}{\omega_0^2m_c} \tag{12.19d}$$

12.2.3 The EP from HDQD of II–VI materials

Following (11.63), the electron concentration per sub-band is given by

$$\Delta n_{\mathrm{OD}}=\frac{g_v}{d_xd_yd_z}\left[F_{-1}\left(\bar{\eta}_{9\mathrm{QD}}\right)\right] \tag{12.19e}$$

The quantized energy along Z direction the $E_{n_z\text{HD7}}$ in this case is given by

$$\frac{\hbar^2(n_z\pi/d_z)^2}{2m_{\|}^*\gamma_3(E_{n_z\text{HD7}},\eta_g)}=1 \tag{12.20}$$

The expression of the total photo-emitted current density in this case is

$$J_{0\text{D}}=\frac{e\alpha_0 g_v}{4}\sum_{n_x=1}^{n_{x\max}}\sum_{n_y=1}^{n_{y\max}}\sum_{n_{z\min}}^{n_{z\max}}(\Delta n_{0\text{D}})\,v_{z7}\left(E_{n_z\text{HD7}}\right) \tag{12.21a}$$

where

$$v_{z7}\left(E_{n_z\text{HD7}}\right)=\left[n_z\pi\hbar\right]\left[m_{\|}^*d_z\gamma_3'\left(E_{n_z\text{HD7}},\eta_g\right)\right]^{-1}$$

and $n_{z_{\min}}$ is the nearest integer of the following equation

$$\frac{\hbar^2\left(n_{z_{\min}}\pi/\mathrm{d}_z\right)^2}{2m_{\|}^*\gamma_3\left((W-hv),\eta_g\right)}=1 \tag{12.21b}$$

*12.2.4 The EP from HDQD of IV–VI materials

(a) The Dimmock model

Following (11.124), the electron concentration per sub-band is given by

$$\Delta n_{0\text{D}}=\frac{g_v}{d_xd_yd_z}\left[F_{-1}\left(\bar{\eta}_{10\text{QD4}}\right)\right] \tag{12.22}$$

The quantized energy along Z direction the $E_{n_z\text{HD8}}$ in this case is given by

$$\lambda_{73}\left(E_{n_z\text{HD8}},\eta_g\right)\left(\frac{n_z\pi}{d_z}\right)^2+\lambda_{74}\left(E_{n_z\text{HD8}},\eta_g\right)\left(\frac{n_z\pi}{d_z}\right)^4 -\lambda_{75}\left(E_{n_z\text{HD8}},\eta_g\right)=0 \tag{12.23}$$

The expression of the total photo-emitted current density in this case is

$$J_{0\text{D}}=\frac{e\alpha_0 g_v}{4}\sum_{n_x=1}^{n_{x\max}}\sum_{n_y=1}^{n_{y\max}}\sum_{n_{z\min}}^{n_{z\max}}(\Delta n_{0\text{D}})\,v_{z8}\left(E_{n_z\text{HD8}}\right) \tag{12.24a}$$

where

$$v_{z8}\left(E_{n_z\mathrm{HD8}}\right) = \frac{1}{\hbar}\left[\frac{\left[2\left(\pi n_z/d_z\right)\lambda_{73}\left(E_{n_z\mathrm{HD8}},\eta_g\right)+4\left(\pi n_z/d_z\right)^3\lambda_{74}\left(E_{n_z\mathrm{HD8}},\eta_g\right)\right]}{\begin{array}{c}[[\lambda_{75}(E_{n_z\mathrm{HD8}},\eta_g)]' - [\lambda_{76}(E_{n_z\mathrm{HD8}},\eta_g)]'(\pi n_z/d_z)^4 \\ -[\lambda_{73}(E_{n_z\mathrm{HD8}},\eta_g)]'(\pi n_z/d_z)^2]\end{array}}\right]$$

and $n_{z_{\min}}$ is the nearest integer of the following equation

$$\lambda_{73}\left((W-hv),\eta_g\right)\left(\frac{n_{z_{\min}}\pi}{d_z}\right)^2+\lambda_{74}\left((W-hv),\eta_g\right)\left(\frac{n_{z_{\min}}\pi}{d_z}\right)^4 - \lambda_{75}\left((W-hv),\eta_g\right)=0 \tag{12.24b}$$

(b) The Bangert and Kastner model

Following (11.131), the electron concentration per sub-band is given by

$$\Delta n_{0\mathrm{D}} = \frac{2g_v}{d_x d_y d_z}\ \text{Real Part of}\ [F_{-1}(\bar{\eta}_{11\mathrm{QD6}})] \tag{12.25}$$

The quantized energy along Z direction the $E_{n_z\mathrm{HD9}}$ in this case is given by

$$F_2\left(E_{n_z\mathrm{HD9}},\eta_g\right)\left(\frac{n_z\pi}{d_z}\right)^2=1 \tag{12.26}$$

The expression of the total photo-emitted current density in this case is

$$J_{0\mathrm{D}} = \frac{e\alpha_0 g_v}{2}\ \text{Real Part of}\ \sum_{n_x=1}^{n_{x\max}}\sum_{n_y=1}^{n_{y\max}}\sum_{n_{z_{\min}}}^{n_{z\max}}\left(\Delta n_{0\mathrm{D}}\right)v_{z9}\left(E_{n_z\mathrm{HD9}}\right) \tag{12.27a}$$

where

$$v_{z9}(E_{n_z\mathrm{HD9}}) = \left[-2\left(\frac{n_z\pi}{d_z}\right)F_2(E_{n_z\mathrm{HD9}},\eta_g)\right]\left[\hbar F_2'(E_{n_z\mathrm{HD9}},\eta_g)\right]^{-1}$$

and $n_{z_{\min}}$ is the nearest integer of the following equation

$$F_2\left((W-hv),\eta_g\right)\left(\frac{\pi n_{z_{\min}}}{d_z}\right)^2=1 \tag{12.27b}$$

(c) The HD model of Foley *et al.*

Following (11.134a), the electron concentration per sub-band is given by

$$\Delta n_{0\mathrm{D}} = \frac{2g_v}{d_x d_y d_z}\left[F_{-1}\left(\bar{\eta}_{12\mathrm{QD}8}\right)\right] \tag{12.27c}$$

The quantized energy along Z direction the $E_{n_z\mathrm{HD}91}$ in this case is given by

$$\gamma_3\left(E_{n_z\mathrm{HD}91}, \eta_g\right) = \left[\left[-\frac{E_{g_0}}{2} + \frac{\hbar^2}{2m_{\|}^{-}}\left(\pi n_z/d_z\right)^2\right]\right.$$
$$\left. + \left[\left[\frac{E_{g_0}}{2} + \frac{\hbar^2}{2m_{\|}^{+}}\left(\pi n_z/d_z\right)^2\right]^2 + P_{\|}^2\left(\pi n_z/d_z\right)^2\right]^{1/2}\right] \tag{12.27d}$$

The expression of the total photo-emitted current density in this case is

$$J_{0\mathrm{D}} = \frac{e\alpha_0 g_v}{2}\sum_{n_x=1}^{n_{x\max}}\sum_{n_y=1}^{n_{y\max}}\sum_{n_{z\min}}^{n_{z\max}}\left(\Delta n_{0\mathrm{D}}\right) v_{z91}\left(E_{n_z\mathrm{HD}91}\right) \tag{12.27e}$$

where

$$v_{z91}(E_{n_z\mathrm{HD}91}) = \left[\frac{1}{\hbar}\left[\gamma_3'\left(E_{n_z\mathrm{HD}91}, \eta_g\right)\right]^{-1}\left[\frac{\hbar^2}{m_{\|}^{-}}\left(\pi n_z/d_z\right)\right.\right.$$
$$+\frac{1}{2}\left[\left[\frac{E_{g_0}}{2} + \frac{\hbar^2}{2m_{\|}^{+}}\left(\pi n_z/d_z\right)^2\right]^2 + P_{\|}^2\left(\pi n_z/d_z\right)^2\right]^{-1/2}$$
$$\left.\left.\times\left[\frac{2\hbar^2}{m_{\|}^{+}}\left[\frac{E_{g_0}}{2} + \frac{\hbar^2}{2m_{\|}^{+}}\left(\pi n_z/d_z\right)^2\right]\left(\pi n_z/d_z\right) + 2P_{\|}^2\left(\pi n_z/d_z\right)\right]\right]\right]$$

and $n_{z_{\min}}$ is the nearest integer of the following equation

$$\gamma_3\left((W - hv), \eta_g\right) = \left[\left[-\frac{E_{g_0}}{2} + \frac{\hbar^2}{2m_{\|}^{-}}\left(\pi n_{z_{\min}}/d_z\right)^2\right]\right.$$
$$\left. + \left[\left[\frac{E_{g_0}}{2} + \frac{\hbar^2}{2m_{\|}^{+}}\left(\pi n_{z_{\min}}/d_z\right)^2\right]^2 + P_{\|}^2\left(\pi n_{z_{\min}}/d_z\right)^2\right]^{1/2}\right] \tag{12.27f}$$

(d) The HD McClure and Choi model

Following (11.134e), the electron concentration per sub-band is given by

$$n_0 = \frac{2g_v}{d_x d_y d_z}[F_{-1}(\bar{\eta}_{12QD81})] \quad (12.27g)$$

The quantized energy along Z direction the $E_{n_{zHD92}}$ in this case is given by

$$\gamma_2(E_{n_{zHD92}}, \eta_g) = \frac{\hbar^2}{2m_3}(\pi n_z/d_z)^2 \quad (12.27h)$$

The expression of the total photo-emitted current density in this case is

$$J_{0D} = \frac{e\alpha_0 g_v}{2} \sum_{n_x=1}^{n_{x\max}} \sum_{n_y=1}^{n_{y\max}} \sum_{n_{z\min}}^{n_{z\max}} (\Delta n_{0D})\, v_{z92}(E_{n_{zHD92}}) \quad (12.27i)$$

where

$$v_{z92}(E_{n_{zHD92}}) = [\hbar[\gamma_2'(E_{n_{zHD92}}, \eta_g)]^{-1}[(\pi n_z/d_z)]]$$

and $n_{z_{\min}}$ is the nearest integer of the following equation

$$\gamma_2((W - hv), \eta_g) = \frac{\hbar^2}{2m_3}(\pi n_z/d_z)^2 \quad (12.27j)$$

(e) The HD Hybrid model

Following (11.134i), the electron concentration per sub-band is given by

$$n_0 = \frac{2g_v}{d_x d_y d_z}[F_{-1}(\bar{\eta}_{12QD82})] \quad (12.27k)$$

The quantized energy along Z direction the $E_{n_{zHD93}}$ in this case is given by

$$\gamma_2(E_{n_{zHD93}}, \eta_g) = \frac{\hbar^2}{2m_3}(\pi n_z/d_z)^2 \quad (12.27l)$$

The expression of the total photo-emitted current density in this case is

$$J_{0D} = \frac{e\alpha_0 g_v}{2} \sum_{n_x=1}^{n_{x\max}} \sum_{n_y=1}^{n_{y\max}} \sum_{n_{z\min}}^{n_{z\max}} (\Delta n_{0D})\, v_{z93}(E_{n_{zHD93}}) \quad (12.27m)$$

where

$$v_{z93}(E_{n_{zHD93}}) = [\hbar[\gamma_2'(E_{n_{zHD93}}, \eta_g)]^{-1}[(\pi n_z/d_z)]]$$

and $n_{z_{\min}}$ is the nearest integer of the following equation

$$\gamma_2((W - hv), \eta_g) = \frac{\hbar^2}{2m_3}(\pi n_z/d_z)^2 \quad (12.27n)$$

(f) The HD Cohen model

Following (11.134m), the electron concentration per sub-band is given by

$$n_0 = \frac{2g_v}{d_x d_y d_z}\left[F_{-1}\left(\bar{\eta}_{12\mathrm{QD}83}\right)\right] \tag{12.27o}$$

The quantized energy along Z direction the $E_{n_z\mathrm{HD}94}$ in this case is given by

$$\gamma_2\left(E_{n_z\mathrm{HD}94}, \eta_g\right) = \frac{\hbar^2}{2m_3}\left(\pi n_z/d_z\right)^2 \tag{12.27p}$$

The expression of the total photo-emitted current density in this case is

$$J_{0\mathrm{D}} = \frac{e\alpha_0 g_v}{2}\sum_{n_x=1}^{n_{x\max}}\sum_{n_y=1}^{n_{y\max}}\sum_{n_{z\min}}^{n_{z\max}}\left(\Delta n_{0\mathrm{D}}\right) v_{z94}\left(E_{n_z\mathrm{HD}94}\right) \tag{12.27q}$$

where

$$v_{z94}(E_{n_z\mathrm{HD}94}) = [\hbar[\gamma_2'(E_{n_z\mathrm{HD}94}, \eta_g)]^{-1}[(\pi n_z/d_z)]]$$

and $n_{z_{\min}}$ is the nearest integer of the following equation

$$\gamma_2((W - hv), \eta_g) = \frac{\hbar^2}{2m_3}(\pi n_z/d_z)^2 \tag{12.27r}$$

(g) The HD model of Lax

Following (11.134q), the electron concentration per sub-band is given by

$$n_0 = \frac{2g_v}{d_x d_y d_z}\left[F_{-1}\left(\bar{\eta}_{12\mathrm{QD}84}\right)\right] \tag{12.27s}$$

The quantized energy along Z direction the $E_{n_z\mathrm{HD}95}$ in this case is given by

$$\gamma_2\left(E_{n_z\mathrm{HD}95}, \eta_g\right) = \frac{\hbar^2}{2m_3}\left(\pi n_z/d_z\right)^2 \tag{12.27t}$$

The expression of the total photo-emitted current density in this case is

$$J_{0\mathrm{D}} = \frac{e\alpha_0 g_v}{2}\sum_{n_x=1}^{n_{x\max}}\sum_{n_y=1}^{n_{y\max}}\sum_{n_{z\min}}^{n_{z\max}}\left(\Delta n_{0\mathrm{D}}\right) v_{z95}\left(E_{n_z\mathrm{HD}95}\right) \tag{12.27u}$$

where

$$v_{z95}\left(E_{n_z\mathrm{HD}95}\right) = \left[\hbar\left[\gamma_2'\left(E_{n_z\mathrm{HD}95}, \eta_g\right)\right]^{-1}\left[(\pi n_z/d_z)\right]\right]$$

and $n_{z_{\min}}$ is the nearest integer of the following equation

$$\gamma_2\left((W - hv), \eta_g\right) = \frac{\hbar^2}{2m_3}\left(\pi n_z/d_z\right)^2 \tag{12.27v}$$

(h) The HD ellipsoidal parabolic model

Following (11.134q), the electron concentration per sub-band is given by

$$n_0 = \frac{2g_v}{d_x d_y d_z}\left[F_{-1}\left(\bar{\eta}_{12\mathrm{QD}85}\right)\right] \tag{12.27w}$$

The quantized energy along Z direction the $E_{n_z\mathrm{HD}96}$ in this case is given by

$$\gamma_3\left(E_{n_z\mathrm{HD}96}, \eta_g\right) = \frac{\hbar^2}{2m_3}\left(\pi n_z/d_z\right)^2 \tag{12.27x}$$

The expression of the total photo-emitted current density in this case is

$$J_{0\mathrm{D}} = \frac{e\alpha_0 g_v}{2}\sum_{n_x=1}^{n_{x\max}}\sum_{n_y=1}^{n_{y\max}}\sum_{n_{z_{\min}}}^{n_{z\max}}\left(\Delta n_{0\mathrm{D}}\right) v_{z96}\left(E_{n_z\mathrm{HD}96}\right) \tag{12.27y}$$

where

$$v_{z96}(E_{n_z\mathrm{HD}96}) = [\hbar[\gamma_3'(E_{n_z\mathrm{HD}96}, \eta_g)]^{-1}[(\pi n_z/d_z)]]$$

and $n_{z_{\min}}$ is the nearest integer of the following equation

$$\gamma_2((W - hv), \eta_g) = \frac{\hbar^2}{2m_3}(\pi n_z/d_z)^2 \tag{12.27z}$$

12.2.5 The EP from HDQD of stressed Kane type materials

Following (11.118), the electron concentration per sub-band is given by

$$\Delta n_{0\mathrm{D}} = \frac{g_v}{d_x d_y d_z}\left[F_{-1}\left(\bar{\eta}_{13\mathrm{QD}10}\right)\right] \tag{12.28}$$

The quantized energy along Z direction the $E_{n_z\mathrm{HD}10}$ in this case is given by

$$S_{11}\left(E_{n_z\mathrm{HD}10}, \eta_g\right)\left(\frac{n_z\pi}{d_z}\right)^2 = 1 \tag{12.29}$$

The expression of the total photo-emitted current density in this case is

$$J_{0\mathrm{D}} = \frac{e\alpha_0 g_v}{2}\sum_{n_x=1}^{n_{x\max}}\sum_{n_y=1}^{n_{y\max}}\sum_{n_{z_{\min}}}^{n_{z\max}}\left(\Delta n_{0\mathrm{D}}\right) v_{z10}\left(E_{n_z\mathrm{HD}10}\right) \tag{12.30a}$$

where

$$v_{z10}(E_{n_z\mathrm{HD}10}) = \left[-2\left(\frac{n_z\pi}{d_z}\right)\mathrm{S}_{11}(E_{n_z\mathrm{HD}10}, \eta_g)\right][\hbar S_{11}{}'(E_{n_z\mathrm{HD}10}\eta_g)]^{-1}$$

and $n_{z_{\min}}$ is the nearest integer of the following equation

$$S_{11}\left((W - hv), \eta_g\right)\left(\frac{n_z\pi}{d_z}\right)^2 = 1 \tag{12.30b}$$

12.2.6 The EP from HDQD of Te

(a) HD model of Bouat *et al.*

Following (11.80), the electron concentration per sub-band is given by

$$\Delta n_{\mathrm{0D}} = \frac{g_v}{d_x d_y d_z}\left[F_{-1}\left(\bar{\eta}_{14\mathrm{QD}12}\right)\right] \tag{12.31}$$

The quantized energy along Z direction the $E_{n_z\mathrm{HD}11}$ in this case is given by

$$\gamma_3(E_{n_z\mathrm{HD}11}, \eta_g) = A_6(\pi n_z/d_z)^2 \pm (\sqrt{A_8})(\pi n_z/d_z) \tag{12.32}$$

The expression of the total photo-emitted current density in this case is

$$J_{\mathrm{0D}} = \frac{e\alpha_0 g_v}{4}\sum_{n_x=1}^{n_{x\max}}\sum_{n_y=1}^{n_{y\max}}\sum_{n_{z\min}}^{n_{z\max}}\left(\Delta n_{\mathrm{0D}}\right) v_{z11}\left(E_{n_z\mathrm{HD}11}\right) \tag{12.33a}$$

where

$$\left[v_{z11}\left(E_{n_z\mathrm{HD}11}\right)\right] = \frac{1}{\hbar}\left[\gamma_3'\left(E_{n_z\mathrm{HD}11}, \eta_g\right)\right]^{-1}\left[2A_6\left(\pi n_z/d_z\right) \pm \sqrt{A_8}\right]$$

and $n_{z_{\min}}$ is the nearest integer of the following equation

$$\gamma_3\left((W - hv), \eta_g\right) = A_6\left(\pi n_{z_{\min}}/d_z\right)^2 \pm (\sqrt{A_8})\left(\pi n_{z_{\min}}/d_z\right) \tag{12.33b}$$

(b) The HD model of Ortenberg and Button

Following (11.83), the electron concentration per sub-band is given by

$$\Delta n_{\mathrm{0D}} = \frac{g_v}{d_x d_y d_z}\left[F_{-1}\left(\bar{\eta}_{14\mathrm{QD}121}\right)\right] \tag{12.33c}$$

The quantized energy along Z direction the $E_{n_z\mathrm{HD}111}$ in this case is given by

$$\gamma_3(E_{n_z\mathrm{HD}111}, \eta_g) = \left[\mp t_1 + t_2(\pi n_z/d_z)^2 \pm \sqrt{t_1^2 + t_7(\pi n_z/d_z)^2}\right] \tag{12.33d}$$

The expression of the total photo-emitted current density in this case is

$$J_{\mathrm{0D}} = \frac{e\alpha_0 g_v}{4}\sum_{n_x=1}^{n_{x\max}}\sum_{n_y=1}^{n_{y\max}}\sum_{n_{\min}}^{n_{z\max}}\left(\Delta n_{\mathrm{0D}}\right) v_{z111}\left(E_{n_z\mathrm{HD}111}\right) \tag{12.33e}$$

where

$$\left[v_{z111}(E_{n_z\mathrm{HD}111})\right] = \left[\frac{1}{\hbar}[\gamma_3'(E_{n_z\mathrm{HD}111}, \eta_g)]^{-1} \times \left[2t_2(\pi n_z/d_z) \pm [t_1^2 + t_7(\pi n_z/d_z)^2]^{-1/2}[t_7(\pi n_z/d_z)]\right]\right]$$

and $n_{z_{\min}}$ is the nearest integer of the following equation

$$\gamma_3\left((W-hv),\eta_g\right)=\left[\mp t_1+t_2\left(\pi n_z/d_z\right)^2\pm\sqrt{t_1^2+t_7\left(\pi n_z/d_z\right)^2}\right] \tag{12.33f}$$

*12.2.7 The EP from HDQD of gallium phosphides

(a) The HD model of Rees

Following (11.67), the electron concentration per sub-band is given by

$$\Delta n_{0\mathrm{D}}=\frac{g_v}{d_x d_y d_z}\left[F_{-1}\left(\bar{\eta}_{15\mathrm{QD}14}\right)\right] \tag{12.34}$$

The quantized energy along Z direction the $E_{n_z\mathrm{HD}14}$ in this case is given by

$$\gamma_3\left(E_{n_z\mathrm{HD}14},\eta_g\right)=\left[\frac{\hbar^2}{2m_\parallel^*}\left(\pi n_z/d_z\right)^2-\left[\frac{\hbar^4k_0^2}{\left(m_\parallel^*\right)^2}\left(\pi n_z/d_z\right)^2+\left(|V_G|\right)^2\right]^{1/2}+\left(|V_G|\right)\right] \tag{12.35}$$

The expression of the total photo-emitted current density in this case is

$$J_{0\mathrm{D}}=\frac{e\alpha_0 g_v}{4}\sum_{n_x=1}^{n_{x\max}}\sum_{n_y=1}^{n_{y\max}}\sum_{n_{z\min}}^{n_{z\max}}\left(\Delta n_{0\mathrm{D}}\right)v_{z14}\left(E_{n_{2\mathrm{FD}14}}\right) \tag{12.36a}$$

where

$$\left[v_{z14}\left(E_{n_z\mathrm{HD}14}\right)\right]=\left[\frac{1}{\hbar}\left[\gamma_3'\left(E_{n_z\mathrm{HD}14},\eta_g\right)\right]^{-1}\left[\frac{\hbar^2}{m_\parallel^*}\left(\pi n_z/d_z\right)-\frac{\hbar^4k_0^2}{(m_\parallel^*)^2}\left(\pi n_z/d_z\right)\left[\frac{\hbar^4k_0^2}{(m_\parallel^*)^2}\left(\pi n_z/d_z\right)^2+\left(|V_G|\right)^2\right]^{-1/2}\right]\right]$$

and $n_{z_{\min}}$ is the nearest integer of the following equation

$$\gamma_3\left((W-hv),\eta_g\right)=\left[\frac{\hbar^2}{2m_\parallel^*}\left(\pi n_{z_{\min}}/d_z\right)^2-\left[\frac{\hbar^4k_0^2}{(m_\parallel^*)^2}\left(\pi n_{z_{\min}}/d_z\right)^2+\left(|V_G|\right)^2\right]^{1/2}+\left(|V_G|\right)\right] \tag{12.36b}$$

(b) The HD model of Ivchenko and Pikus

Following (11.67e), the electron concentration per sub-band is given by

$$n_0 = \frac{g_v}{d_x d_y d_z}\left[F_{-1}\left(\bar{\eta}_{15\mathrm{QD}141}\right)\right] \tag{12.36c}$$

The quantized energy along Z direction the $E_{n_z\mathrm{HD}141}$ in this case is given by

$$\gamma_3(E_{n_z\mathrm{HD}141}, \eta_g) = \left[\frac{\hbar^2}{2m_\parallel^*}\left(\pi n_z/d_z\right)^2 \mp \frac{\bar{\Delta}_0}{2} \pm \left[\left(\frac{\bar{\Delta}_0}{2}\right)^2 + P_1\left(\pi n_z/d_z\right)^2\right]^{1/2}\right] \tag{12.36d}$$

The expression of the total photo-emitted current density in this case is

$$J_{0\mathrm{D}} = \frac{e\alpha_0 g_v}{4}\sum_{n_x=1}^{n_{x\max}}\sum_{n_y=1}^{n_{y\max}}\sum_{n_{z\min}^*}^{n_{z\max}}\left(\Delta n_{0\mathrm{D}}\right) v_{z141}\left(E_{n_z\mathrm{HD}141}\right) \tag{12.36e}$$

where

$$\left[v_{z141}\left(E_{n_z\mathrm{HD}141}\right)\right] = \left[\frac{1}{\hbar}\left[\gamma_3'\left(E_{n_z\mathrm{HD}141}, \eta_g\right)\right]^{-1}\right.$$

$$\left.\times\left[\frac{\hbar^2}{m_\parallel^*}\left(\pi n_z/d_z\right) - P_1\left(\pi n_z/d_z\right)\left[P_1\left(\pi n_z/d_z\right)^2 + \left(\frac{\bar{\Delta}_0}{2}\right)^2\right]^{-1/2}\right]\right]$$

and $n_{z_{\min}}$ is the nearest integer of the following equation

$$\gamma_3\left((W - hv), \eta_g\right) = \left[\frac{\hbar^2}{2m_\parallel^*}\left(\pi n_{z_{\min}}/d_z\right)^2 \mp \frac{\bar{\Delta}_0}{2}\right.$$

$$\left.\pm\left[\left(\frac{\bar{\Delta}_0}{2}\right)^2 + P_1\left(\pi n_{z_{\min}}/d_z\right)^2\right]^{1/2}\right] \tag{12.36f}$$

*12.2.8 The EP from HDQD of platinum antimonide

Following (11.90), the electron concentration per sub-band is given by

$$\Delta n_{0\mathrm{D}} = \frac{g_v}{d_x d_y d_z}\left[F_{-1}\left(\bar{\eta}_{16\mathrm{QD}16}\right)\right] \tag{12.37}$$

The quantized energy along Z direction the $E_{n_z\mathrm{HD15}}$ in this case is given by

$$T_{41}\left(\frac{n_z\pi}{d_z}\right)^4 - T_{51}\left(E_{n_z\mathrm{HD15}},\eta_g\right)\left(\frac{n_z\pi}{d_z}\right)^2 - T_{61}\left(E_{n_z\mathrm{HD15}},\eta_g\right) = 0 \tag{12.38}$$

The expression of the total photo-emitted current density in this case is

$$J_{\mathrm{0D}} = \frac{e\alpha_0 g_v}{4}\sum_{n_x=1}^{n_{x\max}}\sum_{n_y=1}^{n_{y\max}}\sum_{n_{z\min}}^{n_{z\max}}\left(\Delta n_{\mathrm{0D}}\right)v_{z15}\left(E_{n_z\mathrm{HD15}}\right) \tag{12.39a}$$

where

$$\left[v_{z15}\left(E_{n_z\mathrm{HD15}}\right)\right] = \frac{1}{\hbar}\left[\frac{\left[4T_{41}\left(\frac{n_z\pi}{d_z}\right)^3 - 2T_{51}\left(E_{n_z\mathrm{HD15}},\eta_g\right)\left(\frac{n_z\pi}{d_z}\right)\right]}{\left[\mathrm{T}'_{51}\left(E_{n_z\mathrm{HD15}},\eta_g\right)\left(\frac{n_z\pi}{d_z}\right)^2 + \mathrm{T}'_{61}\left(E_{n_z\mathrm{HD15}},\eta_g\right)\right]}\right]$$

and $n_{z_{\min}}$ is the nearest integer of the following equation

$$T_{41}\left(\frac{n_{z_{\min}}\pi}{d_z}\right)^4 - T_{51}\left((W-hv),\eta_g\right)\left(\frac{n_{z_{\min}}\pi}{d_z}\right)^2 - T_{61}((W-hv),\eta_g) = 0 \tag{12.39b}$$

12.2.9 The EP from HDQD of bismuth telluride

Following (11.143), the electron concentration per sub-band is given by

$$\Delta n_{\mathrm{0D}} = \frac{2g_v}{d_x d_y d_z}\left[F_{-1}\left(\bar{\eta}_{17\mathrm{QD}18}\right)\right] \tag{12.40}$$

The quantized energy along Z direction the $E_{n_z\mathrm{HD20}}$ in this case is given by

$$\gamma_2\left(E_{n_z\mathrm{HD20}},\eta_g\right) = \frac{\hbar^2}{2m_0}\bar{\alpha}_{11}\left(\frac{n_x\pi}{d_x}\right)^2 \tag{12.41}$$

The expression of the total photo-emitted current density in this case is

$$J_{\mathrm{0D}} = \frac{e\alpha_0 g_v}{2}\sum_{n_{x\min}}^{n_{x\max}}\sum_{n_y=1}^{n_{y\max}}\sum_{n_z=1}^{n_{z\max}}\left(\Delta n_{\mathrm{0D}}\right)v_{x20}\left(E_{n_z\mathrm{HD20}}\right) \tag{12.42a}$$

where

$$[v_{x20}(E_{n_z\text{HD20}}] = [\gamma_2'(E_{n_z\text{HD20}}, \eta_g)]^{-1}\frac{\hbar\bar{\alpha}_{11}}{m_0}(\pi n_x/d_x)^2$$

and $n_{x_{\min}}$ is the nearest integer of the following equation

$$\gamma_2\left((W - hv), \eta_g\right) = \frac{\hbar^2}{2m_0}\bar{\alpha}_{11}\left(\frac{n_{x_{\min}}\pi}{d_x}\right)^2 \tag{12.42b}$$

12.2.10 The EP from HDQD of n-germanium

(a) The model Cardona *et al.*

Following (11.72), the electron concentration per sub-band is given by

$$\Delta n_{0\text{D}} = \frac{2g_v}{d_x d_y d_z}\left[F_{-1}\left(\bar{\eta}_{18\text{QD}20}\right)\right] \tag{12.43}$$

The quantized energy along Z direction the $E_{n_z\text{HD22}}$ in this case is given by

$$\gamma_2\left(E_{n_z\text{HD22}}, \eta_g\right) = \left[\frac{\hbar^2}{2m_\parallel^*}\left(\frac{n_z\pi}{d_z}\right)^2\right] \tag{12.44}$$

The expression of the total photo-emitted current density in this case is

$$J_{0\text{D}} = \frac{e\alpha_0 g_v}{2}\sum_{n_x=1}^{n_{x\max}}\sum_{n_y=1}^{n_{y\max}}\sum_{n_{z_{\min}}}^{n_{z\max}}\left(\Delta n_{0\text{D}}\right)v_{z22}\left(E_{n_z\text{HD22}}\right) \tag{12.45a}$$

where

$$[v_{z22}\left(E_{n_z\text{HD22}}\right)] = \frac{\hbar\pi n_z}{d_z m_\parallel^*\gamma_2'\left(E_{n_z\text{HD22}}, \eta_g\right)}$$

and $n_{z_{\min}}$ is the nearest integer of the following equation

$$\gamma_2\left((W - hv), \eta_g\right) = \left[\frac{\hbar^2}{2m_\parallel^*}\left(\frac{n_{z_{\min}}\pi}{d_z}\right)^2\right] \tag{12.45b}$$

(b) The model of Wang and Ressler

Following (11.76), the electron concentration per sub-band is given by

$$\Delta n_{0\text{D}} = \frac{2g_v}{d_x d_y d_z}\left[F_{-1}\left(\bar{\eta}_{18\text{QD}22}\right)\right] \tag{12.46}$$

The quantized energy along Z direction the $E_{n_z\text{HD24}}$ in this case is given by

$$\gamma_3\left(E_{n_z\text{HD24}}, \eta_g\right) = \frac{\hbar^2}{2m_\parallel^*}\left(\frac{n_z\pi}{d_z}\right)^2 - e_1\left(\frac{\hbar^2}{2m_\parallel^*}\right)^2\left(\frac{n_z\pi}{d_z}\right)^4 \tag{12.47}$$

The expression of the total photo-emitted current density in this case is

$$J_{0D} = \frac{e\alpha_0 g_v}{2} \sum_{n_x=1}^{n_{x\max}} \sum_{n_y=1}^{n_{y\max}} \sum_{n_{z\min}}^{n_{z\max}} (\Delta n_{0D})\, v_{z24}\left(E_{n_z\mathrm{HD24}}\right) \qquad (12.48a)$$

where

$$v_{z24}\left(E_{n_z\mathrm{HD24}}\right) = \frac{1}{\hbar}\left[\gamma_3'\left(E_{n_z\mathrm{HD24}}, \eta_g\right)\right]^{-1} \times \left[\frac{\hbar^2}{m_\parallel^*}(\pi n_z/d_z) - e_1\left(\frac{\hbar^2}{m_\parallel^*}\right)^2 (\pi n_z/d_z)^3\right]$$

and $n_{z_{\min}}$ is the nearest integer of the following equation

$$\gamma_3\left((W - hv), \eta_g\right) = \frac{\hbar^2}{2m_\parallel^*}\left(\frac{n_{z_{\min}}\pi}{d_z}\right)^2 - e_1\left(\frac{\hbar^2}{2m_\parallel^*}\right)^2\left(\frac{n_{z_{\min}}\pi}{d_z}\right)^4 \qquad (12.48b)$$

12.2.11 The EP from HDQD of gallium antimonide

(a) The HD model of Sailer *et al.*

Following (11.97), the electron concentration per sub-band is given by

$$\Delta n_{0D} = \frac{2g_v}{d_x d_y d_z}\left[F_{-1}\left(\bar{\eta}_{19\mathrm{QD24}}\right)\right] \qquad (12.49)$$

The quantized energy along Z direction the $E_{n_z\mathrm{HD301}}$ in this case is given by

$$\gamma_3\left(E_{n_z\mathrm{HD301}}, \eta_g\right) = -\frac{E_{g0}}{2} + \frac{E_{g0}}{2}\left(\sqrt{1 + \alpha_4\left(\frac{n_z\pi}{d_z}\right)^2}\right) + \bar{\xi}_0\frac{\hbar^2}{2m_0}\left(\frac{n_z\pi}{d_z}\right)^2 \qquad (12.50)$$

The expression of the total photo-emitted current density in this case is

$$J_{0D} = \frac{e\alpha_0 g_v}{2} \sum_{n_x=1}^{n_{x\max}} \sum_{n_y=1}^{n_{y\max}} \sum_{n_{z\min}}^{n_{z\max}} (\Delta n_{0D})\, v_{z30}\left(E_{n_z\mathrm{HD301}}\right) \qquad (12.51)$$

where

$$v_{z30}(E_{n_z\mathrm{HD301}}) = \left[\frac{1}{\hbar}[\gamma_3'(E_{n_z\mathrm{HD301}}, \eta_g)]^{-1} \times \left[\frac{\alpha_4 E_{g0}}{2}(\pi n_z/d_z)\left[1 + \alpha_4(\pi n_z/d_z)^2\right]^{-1/2} + \frac{\bar{\xi}_0\hbar^2}{m_0}(\pi n_z/d_z)\right]\right]$$

and $n_{z_{\min}}$ is the nearest integer of the following equation

$$\gamma_3\left((W-hv),\eta_g\right)=-\frac{E_{g_0}}{2}+\frac{E_{g_0}}{2}\left(\sqrt{1+\alpha_4\left(\frac{\pi n_{z_{\min}}}{d_z}\right)^2}\right)+\bar{\xi}_0\frac{\hbar^2}{2m_0}\left(\frac{\pi n_{z_{\min}}}{d_z}\right)^2 \tag{12.52a}$$

(b) HD model of Mathur *et al.*

Following (11.100), the electron concentration per sub-band is given by

$$\Delta n_{0\mathrm{D}}=\frac{2g_v}{d_x d_y d_z}\left[F_{-1}\left(\bar{\eta}_{20\mathrm{QD}26}\right)\right] \tag{12.52b}$$

The quantized energy along Z direction the $E_{n_z\mathrm{HD}302}$ in this case is given by

$$\gamma_3\left(E_{n_z\mathrm{HD}302},\eta_g\right)=\left[\alpha_9\left(\pi n_z/d_z\right)^2+\frac{E_{g_1}}{2}\left[\left(\sqrt{1+\alpha_{10}\left(\pi n_z/d_z\right)^2}\right)-1\right]\right] \tag{12.52c}$$

The expression of the total photo-emitted current density in this case is

$$J_{0\mathrm{D}}=\frac{e\alpha_0 g_v}{2}\sum_{n_x=1}^{n_{x\max}}\sum_{n_y=1}^{n_{y\max}}\sum_{n_{z_{\min}}}^{n_{z\max}}\left(\Delta n_{0\mathrm{D}}\right)v_{z302}\left(E_{n_z\mathrm{HD}302}\right) \tag{12.52d}$$

where

$$v_{z302}(E_{n_z\mathrm{HD}302})=\left[\frac{1}{\hbar}\left[\gamma_3'\left(E_{n_z\mathrm{HD}302},\eta_g\right)\right]^{-1}\times\left[2\alpha_9(\pi n_z/d_z)+\frac{E_{g_1}}{2}\alpha_{10}(\pi n_z/d_z)[1+\alpha_{10}(\pi n_z/d_z)^2]^{-1/2}\right]\right]$$

and $n_{z_{\min}}$ is the nearest integer of the following equation

$$\gamma_3((W-hv),\eta_g)=\left[\alpha_9\left(\pi n_{z_{\min}}/d_z\right)^2+\frac{E_{g_1}}{2}\left[\left(\sqrt{1+\alpha_{10}\left(\pi n_{z_{\min}}/d_z\right)^2}\right)-1\right]\right] \tag{12.52e}$$

(c) HD model of Zhang *et al.*

Following (11.103), the electron concentration per sub-band is given by

$$\Delta n_{0\mathrm{D}} = \frac{2g_v}{d_x d_y d_z}\left[F_{-1}\left(\bar{\eta}_{21\mathrm{QD}28}\right)\right] \tag{12.53}$$

The quantized energy along Z direction the $E_{n_z\mathrm{HD}303}$ in this case is given by

$$\gamma_3(E_{n_z\mathrm{HD}303}, \eta_g) = C_{11}(\pi n_z/d_z)^2 + C_{12}(\pi n_z/d_z)^4 + C_{13}(\pi n_z/d_z)^6 \tag{12.54}$$

where

$$C_{11} = \left[\bar{E}_1 + \bar{E}_2(\sqrt{21}/2)\right], \quad C_{12} = \left[\bar{E}_3 + \bar{E}_4(\sqrt{21}/2)\right],$$

$$C_{13} = \left[\left[\bar{E}_5 + \bar{E}_6(\sqrt{21}/2) + \bar{E}_7\left(\frac{639639}{32}\right)^{1/2}\left[\frac{1}{60} - \frac{1}{105}\right]\right]\right]$$

The expression of the total photo-emitted current density in this case is

$$J_{0\mathrm{D}} = \frac{e\alpha_0 g_v}{2}\sum_{n_x=1}^{n_{x\max}}\sum_{n_y=1}^{n_{y\max}}\sum_{n_{z\min}}^{n_{z\max}}\left(\Delta n_{0\mathrm{D}}\right) v_{z303}\left(E_{n_z\mathrm{HD}303}\right) \tag{12.55}$$

where

$$v_{z303}\left(E_{n_z\mathrm{HD}303}\right) = \left[\frac{1}{\hbar}\left[\gamma_3'\left(E_{n_z\mathrm{HD}303}, \eta_g\right)\right]^{-1}\right.$$
$$\left.\times\left[2C_{11}\left(\pi n_z/d_z\right) + 4C_{12}\left(\pi n_z/d_z\right)^3 + 6C_{13}\left(\pi n_z/d_z\right)^5\right]\right]$$

and $n_{z_{\min}}$ is the nearest integer of the following equation

$$\gamma_3\left((W - hv), \eta_g\right) = C_{11}\left(\pi n_{z_{\min}}/d_z\right)^2 + C_{12}\left(\pi n_{z_{\min}}/d_z\right)^4 + C_{13}\left(\pi n_{z_{\min}}/d_z\right)^6 \tag{12.56}$$

*12.2.12 The EP from HDQD of graphite

Following (11.86), the electron concentration per sub-band is given by

$$\Delta n_{0\mathrm{D}} = \frac{g_v}{d_x d_y d_z}\left[F_{-1}\left(\bar{\eta}_{14\mathrm{QD}122}\right)\right] \tag{12.57}$$

The quantized energy along Z direction the $E_{n_z\text{HD304}}$ in this case is given by

$$\gamma_3\left(E_{n_z\text{HD304}},\eta_g\right)=\left[\frac{1}{2}\left[E_{20}+E_{30}\right]\pm\left[\frac{1}{4}\left[E_{20}-E_{30}\right]^2+\left(\eta_{46}\right)^2\left(\pi n_z/d_z\right)^2\right]^{1/2}\right] \quad (12.58)$$

where $E_{20}=\left(\Delta_1-2\gamma_1\cos\bar{\phi}_0+2\gamma_5\cos^2\bar{\phi}_0\right),\bar{\phi}_0=\left(\bar{c}\left(\pi n_z/2d_z\right)\right)$, $E_{30}=\left(2\gamma_2\cos^2\bar{\phi}_0\right)$.

The expression of the total photo-emitted current density in this case is

$$J_{0\text{D}}=\frac{e\alpha_0 g_v}{4}\sum_{n_x=1}^{n_{x\max}}\sum_{n_y=1}^{n_{y\max}}\sum_{n_{z\min}}^{n_{z\max}}\left(\Delta n_{0\text{D}}\right)v_{z304}\left(E_{n_z\text{HD304}}\right) \quad (12.59)$$

where

$$v_{z304}\left(E_{n_z\text{HD304}}\right)=\left[\frac{1}{\hbar}\left[\gamma_3'\left(E_{n_z\text{HD304}},\eta_g\right)\right]^{-1}\left[\frac{\bar{c}}{2}\gamma_1\left(\sin\bar{\phi}_0\right)-\frac{\bar{c}}{2}\gamma_5\left(\sin 2\bar{\phi}_0\right)-\frac{\bar{c}}{2}\gamma_2\left(\sin 2\bar{\phi}_0\right)\pm\frac{1}{2}\left[\frac{1}{4}\left(E_{20}-E_{30}\right)^2+\left(\eta_{46}\right)^2\left(\pi n_z/d_z\right)^2\right]^{-1/2}\times\left[\frac{1}{2}\left(E_{20}-E_{30}\right)\left\{\bar{c}\gamma_1\left(\sin\bar{\phi}_0\right)-\bar{c}\gamma_5\left(\sin 2\bar{\phi}_0\right)+\bar{c}\gamma_2\left(\sin 2\bar{\phi}_0\right)\right\}+2\left(\eta_{46}\right)^2\left(\pi n_z/d_z\right)^2\right]\right]\right]$$

and $n_{z_{\min}}$ is the nearest integer of the following equation

$$\gamma_3\left(\left(W-hv\right),\eta_g\right)=\left[\frac{1}{2}\left[\bar{E}_{20}+\bar{E}_{30}\right]\pm\left[\frac{1}{4}\left[\bar{E}_{20}-\bar{E}_{30}\right]^2+\left(\eta_{46}\right)^2\left(\pi n_{z_{\min}}/d_z\right)^2\right]^{1/2}\right] \quad (12.60)$$

where $\bar{E}_{20}=\left(\Delta_1-2\gamma_1\cos\bar{\phi}_{0_{\min}}+2\gamma_5\cos^2\bar{\phi}_{0_{\min}}\right)$, $\bar{\phi}_{0_{\min}}=(\bar{c}(\pi n_{2_{\min}}/2d_z))$, $\bar{E}_{30}=(2\gamma_2\cos^2\bar{\phi}_{0_{\min}})$.

12.2.13 The EP from HDQD of II–V materials

Following (11.93), the electron concentration per sub-band is given by

$$\Delta n_{0\text{D}}=\frac{g_v}{d_x d_y d_z}\left[F_{-1}\left(\bar{\eta}_{22\text{QD}30}\right)\right] \quad (12.61)$$

The quantized energy along Z direction the $E_{n_{z\text{HD305}}}$ in this case is given by

$$\gamma_3(E_{n_{z\text{HD305}}},\eta_g) = [A_{12}(\pi n_z/d_z)^2 \pm [A_{16}^2(\pi n_z/d_z)^4 + A_{19}^2]^{1/2} \mp A_{19}] \tag{12.62}$$

The expression of the total photo-emitted current density in this case is

$$J_{0\text{D}} = \frac{e\alpha_0 g_v}{4} \sum_{n_x=1}^{n_{x\max}} \sum_{n_y=1}^{n_{y\max}} \sum_{n_{z_{\min}}}^{n_{z\max}} (\Delta n_{0\text{D}})\, v_{z305}(E_{n_{z\text{HD305}}}) \tag{12.63}$$

where

$$v_{z305}(E_{n_{z\text{HD305}}}) = \left[\frac{1}{\hbar}[\gamma_3'(E_{n_{z\text{HD305}}},\eta_g)]^{-1}\right.$$
$$\left.\times\left[2A_{12}(\pi n_z/d_z) + 2A_{16}^2(\pi n_z/d_z)^3\left[A_{16}^2(\pi n_z/d_z)^4 + A_{19}^2\right]^{-1/2}\right]\right]$$

and $n_{z_{\min}}$ is the nearest integer of the following equation

$$\gamma_3((W-hv),\eta_g) = [A_{12}(\pi n_{z_{\min}}/d_z)^2 \pm [A_{16}^2(\pi n_{z_{\min}}/d_z)^4 + A_{19}^2]^{1/2} \mp A_{19}] \tag{12.64}$$

*12.2.14 The EP from HDQD of lead germanium telluride

Following (11.93), the electron concentration per sub-band is given by

$$\Delta n_{0\text{D}} = \frac{g_v}{d_x d_y d_z}[F_{-1}(\bar{\eta}_{23\text{QD}32})] \tag{12.65}$$

The quantized energy along Z direction the $E_{n_{z\text{HD306}}}$ in this case is given by

$$\beta_{30}(E_{n_{z\text{HD306}}},\eta_g) + \gamma_3(E_{n_{z\text{HD306}}},\eta_g)[\bar{E}_{g_0} - 0.345(\pi n_z/d_z)^2]$$
$$= [0.02(\pi n_z/d_z)^2 + [\bar{E}_{g_0} + 0.377(\pi n_z/d_z)^2](0.0722)(\pi n_z/d_z)^2] \tag{12.66}$$

where

$$\beta_{30}\left(E_{n_z\mathrm{HD306}},\eta_g\right)=\left[2\theta_0\left(E_{n_z\mathrm{HD306}},\eta_g\right)\left[1+Erf\left(\frac{E_{n_z\mathrm{HD306}}}{\eta_g}\right)\right]^{-1}\right]$$

The expression of the total photo-emitted current density in this case is

$$J_{\mathrm{0D}}=\frac{e\alpha_0 g_v}{4}\sum_{n_x=1}^{n_{x\max}}\sum_{n_y=1}^{n_{y\max}}\sum_{n_{z_{\min}}}^{n_{z\max}}\left(\Delta n_{\mathrm{0D}}\right)v_{z306}\left(E_{n_z\mathrm{HD306}}\right) \tag{12.67}$$

where

$$\begin{aligned}v_{z306}(E_{n_z\mathrm{HD306}})=&\left[\frac{1}{\hbar}[0.690\gamma_3\left(E_{n_z\mathrm{HD306}},\eta_g\right)\left(\pi n_z/d_z\right)\right.\\&+0.04\left(\pi n_z/d_z\right)+0.1414\left(\pi n_z/d_z\right)[\bar{E}_{g_0}+0.377\left(\pi n_z/d_z\right)^2]\\&+C_{20}\left(\pi n_z/d_z\right)^3][\beta_{30}'\left(E_{n_z\mathrm{HD306}},\eta_g\right)+\gamma_3'\left(E_{n_z\mathrm{HD306}},\eta_g\right)\\&\left.\times[\bar{E}_{g_0}-0.345\left(\pi n_z/d_z\right)^2]]^{-1}\right]\end{aligned}$$

and $C_{20}=[0.07722\times 0.754]$ and $n_{z_{\min}}$ is the nearest integer of the following equation

$$\begin{aligned}&\beta_{30}\left((W-hv),\eta_g\right)+\gamma_3\left((W-hv),\eta_g\right)[\bar{E}_{g_0}-0.345\left(\pi n_{z_{\min}}/d_z\right)^2]\\&\quad=[0.02\left(\pi n_{z_{\min}}/d_z\right)^2+[\bar{E}_{g_0}+0.377(\pi n_{z_{\min}}/d_z)^2]\\&\qquad\times(0.0722)\left(\pi n_{z_{\min}}/d_z\right)^2]\end{aligned} \tag{12.68}$$

*12.2.15 The EP from HDQD of zinc and cadmium diphosphides

Following (11.140), the electron concentration per sub-band is given by

$$\Delta n_{\mathrm{0D}}=\frac{g_v}{d_x d_y d_z}\left[F_{-1}\left(\bar{\eta}_{24\mathrm{QD}34}\right)\right] \tag{12.69}$$

The quantized energy along Z direction the $E_{n_z\mathrm{HD307}}$ in this case is given by

$$\gamma_3(E_{n_z\mathrm{HD307}},\eta_g)=\left[\beta_1-\frac{\beta_2}{4\beta_4}\right]\left(\pi n_z/d_z\right)^2 \tag{12.70}$$

The expression of the total photo-emitted current density in this case is

$$J_{\mathrm{OD}} = \frac{e\alpha_0 g_v}{4} \sum_{n_x=1}^{n_{x\max}} \sum_{n_y=1}^{n_{y\max}} \sum_{n_{z\min}}^{n_{z\max}} (\Delta n_{\mathrm{OD}})\, v_{z307}\left(E_{n_z\mathrm{HD}307}\right) \tag{12.71}$$

where

$$v_{z307}\left(E_{n_z\mathrm{HD}307}\right) = \left[\frac{1}{\hbar}\left[\gamma_3'\left(E_{n_z\mathrm{HD}307}, \eta_g\right)\right]^{-1} \times \left[C_{90} + 2\left(\beta_1 - \frac{\beta_2}{4\beta_4}\right)(\pi n_z/d_z)^2\right]\right]$$

and $C_{90} = \left[\left[2\beta_4\left(-\beta_5 - \frac{\beta_2}{4\beta_4}\right)\right]^{1/2}\right]$ and $n_{z_{\min}}$ is the nearest integer of the following equation

$$\gamma_3\left((W - hv), \eta_g\right) = \left[\beta_1 - \frac{\beta_2}{4\beta_4}\right](\pi n_{z_{\min}}/d_z)^2 \tag{12.72}$$

12.3 Summary and Conclusion

It may be noted that the HD QDs lead to the discrete energy levels, somewhat like atomic energy levels, which produce very large changes. This follows from the inherent nature of the quantum confinement of the carrier gas dealt with here. In QDs, there remain no free carrier states in between any two allowed sets of size-quantized levels unlike that found for QWs and NWs where the quantum confinements are 1D and 2D, respectively. Consequently, the crossing of the Fermi level by the size-quantized levels in HD QDs would have much greater impact on the redistribution of the carriers among the allowed levels, as compared to that found for QWs and NWs respectively. Although the EP varies in various manners with all the variables the rates of variations are totally band-structure dependent. The quantum signature of HD QDs for the EP is rather prominent as compared to the same from QWs and NWs. It is the band structure which changes in a fundamental way and consequently alters the photoemission together with the fact that in general all the physical properties of all the electronic materials changes radically leading to new physical concepts. We have not given any plot of EPs' in QDs since we believe our able readers will plot the same and enjoy intricate physics, computational analysis and advanced mathematics and will create new problems both theoretical and experimental.

12.4 Open Research Problems

(R.12.1) Investigate the EP for bulk specimens of the HDS under non uniform strain in the presence of exponential, Kane, Halperian, Lax and Bonch-Burevich types of band tails for all systems whose unperturbed carrier energy spectra are defined in Chapter 1.

(R.12.2) Investigate the EP for QDs of all the HD semiconductors as considered in (R.12.1).

(R.12.3) Investigate the EP in the presence of non uniform strain for HD bulk specimens of the negative refractive index, organic, magnetic and other advanced optical materials in the presence of an arbitrarily oriented alternating electric field.

(R.12.4) Investigate the EP for the QDs of HD negative refractive index, organic, magnetic and other advanced optical materials in the presence of an arbitrarily oriented alternating electric field.

(R.12.5) Investigate the EP for the multiple QDs of HD materials whose unperturbed carrier energy spectra are defined in Chapter 1.

(R.12.6) Investigate the EP for all the appropriate HD zero dimensional systems of this chapter in the presence of finite potential wells.

(R.12.7) Investigate the EP for all the appropriate HD zero dimensional systems of this chapter in the presence of parabolic potential wells.

(R.12.8) Investigate the EP for all the appropriate HD totally quantized systems of this chapter forming quantum rings.

(R.12.9) Investigate the EP for all the above appropriate problems in the presence of elliptical Hill and quantum square rings in the presence of strain.

(R.12.10) Investigate the EP for parabolic cylindrical HD low dimensional systems in the presence of an arbitrarily oriented alternating electric field for all the HD materials whose unperturbed carrier energy spectra are defined in Chapter 1 in the presence of strain.

(R.12.11) Investigate the EP for HD zero dimensional systems of the negative refractive index and other advanced optical materials in the presence of an arbitrarily oriented alternating electric field and non-uniform light waves and in the presence of strain.

(R.12.12) Investigate the EP for triangular HD zero dimensional systems of the negative refractive index, organic, magnetic and other advanced optical materials in the presence of an arbitrarily oriented alternating electric field in the presence of strain.

(R.12.13) Investigate the EP for all the problems of (R.12.12) in the presence of arbitrarily oriented magnetic field.

(R.12.14) Investigate the EP for all the problems of (R.12.12) in the presence of alternating electric field.

(R.12.15) Investigate the EP for all the problems of (R.12.12) in the presence of alternating magnetic field.

(R.12.16) Investigate the EP for all the problems of (R.12.14) in the presence of crossed electric field and quantizing magnetic fields.

(R.12.17) Investigate the EP for all the problems of Chapter 1 in the presence of crossed alternating electric field and alternating quantizing magnetic fields.

(R.12.18) Investigate the EP for HD QDs of the negative refractive index, organic and magnetic materials.

(R.12.19) Investigate the EP for HD QDs of the negative refractive index, organic and magnetic materials in the presence of alternating time dependent magnetic field.

(R.12.20) Investigate the EP for HD QDs of the negative refractive index, organic and magnetic materials in the presence of crossed alternating electric field and alternating quantizing magnetic fields.

(R.12.21) (a) Investigate the EP for HD QDs of the negative refractive index, organic, magnetic and other advanced optical materials in the presence of an arbitrarily oriented alternating electric field considering many body effects.

(b) Investigate all the appropriate problems of this chapter for a Dirac electron.

(R.12.22) Investigate all the appropriate problems of this chapter by including the many body, image force, broadening and hot carrier effects respectively.

(R.12.23) Investigate all the appropriate problems of this chapter by removing all the mathematical approximations and establishing the respective appropriate uniqueness conditions.

Chapter 13

The Activity Coefficient in Heavily Doped of Doping Superlattices of Non-Parabolic Materials Under Magnetic Quantization

The smallest deeds always exceeds the grandest of intentions.

13.1 Introduction

The importance of studying the Activity Coefficient (AC) has already been discussed in Chapter 1 and magneto doping superlattices find applications in this regard [1–29]. In Section 13.2.1, of the theoretical background, the DOS and the AC in doping superlattices of HD nonlinear optical and tetragonal semiconductors have been investigated under magnetic quantization. Section 13.2.2 contains the results for doping superlattices under magnetic quantization of HD III–V, ternary and quaternary semiconductors in accordance with the three and the two band models of Kane together with parabolic energy bands and they form the special cases of Section 13.2.1. Sections 13.2.3–13.2.5 contain the study of the DOS and the AC under magnetic quantization for doping superlattices of HD II–VI, IV–VI and stressed Kane type semiconductors respectively. Sections 13.3 and 13.4 contain the summary and conclusion and the five open research problems for this chapter.

13.2 Theoretical Background

13.2.1 The AC in doping superlattices of HD nonlinear optical and tetragonal materials under magnetic quantization

The DR of the conduction electrons in HD doping superlattices of nonlinear optical and tetragonal materials can be expressed by using (11.26) and following the method as given in [1,2] as

$$\frac{\left(n_i+\frac{1}{2}\right)}{\hbar T_{21}(E,\eta_g)}\omega_{8\mathrm{HD}}(E,\eta_g)+\frac{\hbar^2k_s^2}{2m_\perp^* T_{22}(E,\eta_g)}=1 \tag{13.1}$$

where

$$\omega_{8\mathrm{HD}}(E,\eta_g)\equiv\left(\frac{n_0\,|e|^2}{\varepsilon_{sc}[m_\parallel^* T_{21}'(E,\eta_g)]}\right)^{1/2}$$

n_i (=0, 1, 2...) is the mini-band index for NIPI structures, d_0 is the superlattice period and $\bar{n}_0=\frac{n_0}{d_0}$.

The DR of the conduction electrons in doping superlattices under magnetic quantization of HD nonlinear optical and tetragonal materials can be expressed by using (13.1) as

$$\frac{\left(n_i+\frac{1}{2}\right)}{\hbar T_{21}(E_{11,1},\eta_g)}\omega_{8\mathrm{HD}}(E_{11,1},\eta_g)+\frac{\hbar eB\left(n_i+\frac{1}{2}\right)}{m_\perp^* T_{22}(E_{11,1},\eta_g)}=1 \tag{13.2}$$

where $E_{11,1}$ is the total quantized energy in this case.

The DOS function is given by

$$N_{Bn_z}=\frac{g_veB}{\pi\hbar}\sum_{n=0}^{n_{\max}}\sum_{n_z=1}^{n_{z\max}}\delta'(E-E_{11,1}) \tag{13.3}$$

The electron concentration can be written as

$$n_0=\frac{g_veB}{\pi\hbar}\text{ Real part of }\sum_{n=0}^{n_{\max}}\sum_{n_z=1}^{n_{z\max}}F_{-1}(\eta_{11,1}) \tag{13.4}$$

where $\eta_{11,1}=(k_BT)^{-1}[E_{F11,1}-E_{11,1}]$ and $E_{F11,1}$ is the Fermi energy in this case.

Using (1.181) and (13.3), we can write the AC as

$$f_e^*=\frac{4}{n_0}\frac{g_veB}{\pi\hbar}\text{ Real part of }\sum_{n=0}^{n_{\max}}\sum_{n_z=1}^{n_{z\max}}F_{-2}(\eta_{11,1}) \tag{13.5}$$

13.2.2 The AC in doping superlattices of HD III–V, ternary and quaternary materials under magnetic quantization

(a) The electron energy spectrum in accordance with heavily doped three band model of Kane in doping superlattices under magnetic quantization of

HD III–V, ternary and quaternary materials can be expressed from (13.2) under the conditions $\delta = 0$, $\Delta_{\parallel} = \Delta_{\perp} = \Delta$ and $m_{\parallel}^* = m_{\perp}^* = m_c$ as

$$\left(n+\frac{1}{2}\right)\hbar\omega_0 = \left[T_{31}(E_{11,3},\eta_g) + iT_{32}(E_{11,3},\eta_g) - \left(n_i+\frac{1}{2}\right)\hbar\omega_{9\mathrm{HD}}(E_{11,3},\eta_g)\right] \quad (13.6)$$

where $E_{11,3}$ is the total quantized energy.

The DOS function is given by

$$N_{Bn_z} = \frac{g_v eB}{\pi\hbar}\sum_{n=0}^{n_{\max}}\sum_{n_z=1}^{n_{z\max}} \delta'(E - E_{11,3}) \quad (13.7)$$

The electron concentration can be written as

$$n_0 = \frac{g_v eB}{\pi\hbar}\ \text{Real part of}\ \sum_{n=0}^{n_{\max}}\sum_{n_z=1}^{n_{z\max}} F_{-1}(\eta_{11,3}) \quad (13.8)$$

where $\eta_{11,3} = (k_B T)^{-1}[E_{F11,3} - E_{11,3}]$ and $E_{F11,3}$ is the Fermi energy in this case.

Using (1.181) and (13.7), we can write the AC as

$$f_e^* = \frac{4}{n_0}\frac{g_v eB}{\pi\hbar}\ \text{Real part of}\ \sum_{n=0}^{n_{\max}}\sum_{n_z=1}^{n_{z\max}} F_{-2}(\eta_{11,3}) \quad (13.9)$$

(b) The DR in doping superlattics of HD III–V, ternary and quaternary materials under magnetic quantization whose energy band structures in the absence of band tails are described by the two band model of Kane can be expressed from (13.6) under the conditions $\Delta \gg E_g$ or $\Delta \ll E_g$, as

$$\left(n+\frac{1}{2}\right)\hbar\omega_0 = \left[\gamma_2(E_{11,5},\eta_g) - \left(n_i+\frac{1}{2}\right)\hbar\omega_{10\mathrm{HD}}(E_{11,5},\eta_g,\eta_g)\right] \quad (13.10)$$

where $E_{11,5}$ is the total quantized energy.

The DOS function is given by

$$N_{Bn_z} = \frac{g_v eB}{\pi\hbar}\sum_{n=0}^{n_{\max}}\sum_{n_z=1}^{n_{z\max}} \delta'(E - E_{11,5}) \quad (13.11)$$

The electron concentration can be written as

$$n_0 = \frac{g_v eB}{\pi\hbar}\sum_{n=0}^{n_{\max}}\sum_{n_z=1}^{n_{z\max}} F_{-1}(\eta_{11,5}) \quad (13.12)$$

where $\eta_{11,5} = (k_B T)^{-1}[E_{F11,5} - E_{11,5}]$ and $E_{F11,5}$ is the Fermi energy in this case.

Using (1.181) and (13.11), we can write the AC as

$$f_e^* = \frac{4}{n_0}\frac{g_v eB}{\pi\hbar}\sum_{n=0}^{n_{\max}}\sum_{n_z=1}^{n_{z\max}} F_{-2}(\eta_{11,5}) \tag{13.13}$$

(c) The electron energy spectrum in NIPI structures of HD III–V, ternary and quaternary materials under magnetic quantization whose energy band structures in the absence of band tails are described by the parabolic energy bands can be expressed as

$$\left(n+\frac{1}{2}\right)\hbar\omega_0 = \left[\gamma_3(E_{11,7},\eta_g) - \left(n+\frac{1}{2}\right)\hbar\omega_{11\mathrm{HD}}(E_{11,7},\eta_g)\right] \tag{13.14}$$

where $E_{11,7}$ is the total quantized energy.

The DOS function is given by

$$N_{Bn_z} = \frac{g_v eB}{\pi\hbar}\sum_{n=0}^{n_{\max}}\sum_{n_z=1}^{n_{z\max}} \delta'(E - E_{11,7}) \tag{13.15}$$

The electron concentration can be written as

$$n_0 = \frac{g_v eB}{\pi\hbar}\sum_{n=0}^{n_{\max}}\sum_{n_z=1}^{n_{z\max}} F_{-1}(\eta_{11,7}) \tag{13.16}$$

where $\eta_{11,7} = (k_B T)^{-1}[E_{F11,7} - E_{11,7}]$ and $E_{F11,7}$ is the Fermi energy in this case.

Using (1.181) and (13.15), we can write the AC as

$$f_e^* = \frac{4}{n_0}\frac{g_v eB}{\pi\hbar}\sum_{n=0}^{n_{\max}}\sum_{n_z=1}^{n_{z\max}} F_{-2}(\eta_{11,7}) \tag{13.17}$$

13.2.3 The AC in doping superlattices of HD II–VI materials under magnetic quantization

The 2D electron dispersion law in HD doping superlattices of II–VI semiconductors can be expressed following (11.60) as

$$\gamma_3(E,\eta_g) = a_0' k_s^2 + \left(n_i + \frac{1}{2}\right)\hbar\omega_{30}(E,\eta_g) \pm \bar{\lambda}_0 k_s \tag{13.18}$$

where

$$\omega_{30}(E,\eta_g) \equiv \left(\frac{\bar{n}_0|e|^2}{\gamma_3'(E,\eta_g)\varepsilon_{sc} m_{||}^*}\right)^{1/2}$$

The 2D DR in doping superlattices of HD II–VI semiconductors under magnetic quantization can be expressed from (13.18) as

$$\gamma_3(E_{11,9},\eta_g) = a_0'\left[\frac{2eB}{\hbar}\left(n+\frac{1}{2}\right)\right] + \left(n_i+\frac{1}{2}\right)\hbar\omega_{30\mathrm{HD}}(E_{11,9},\eta_g)$$
$$\pm\bar{\lambda}_0\sqrt{\left[\frac{2eB}{\hbar}\left(n+\frac{1}{2}\right)\right]} \tag{13.19}$$

where $E_{11,9}$ is the total quantized energy in this case.

The DOS function is given by

$$N_{Bn_z} = \frac{g_veB}{2\pi\hbar}\sum_{n=0}^{n_{\max}}\sum_{n_z=1}^{n_{z\max}}\delta'(E-E_{11,9}) \tag{13.20}$$

The electron concentration can be written as

$$n_0 = \frac{g_veB}{2\pi\hbar}\sum_{n=0}^{n_{\max}}\sum_{n_z=1}^{n_{z\max}}F_{-1}(\eta_{11,9}) \tag{13.21}$$

where $\eta_{11,9} = (k_BT)^{-1}[E_{F11,9}-E_{11,9}]$ and $E_{F11,9}$ is the Fermi energy in this case.

Using (1.181) and (13.20), we can write the AC as

$$f_e^* = \frac{4}{n_0}\frac{g_veB}{2\pi\hbar}\sum_{n=0}^{n_{\max}}\sum_{n_z=1}^{n_{z\max}}F_{-2}(\eta_{11,9}) \tag{13.22}$$

13.2.4 The AC in doping superlattices of HD IV–VI materials under magnetic quantization

The dispersion relation in HD doping superlattices made of IV–VI materials can be written from (11.121) as

$$k_s^2 = \delta_{15}(E,\eta_g,n_i) \tag{13.23}$$

where

$$\delta_{15}(E,\eta_g,n_i) = [2\delta_{12}(E,\eta_g)]^{-1}[-\delta_{13}(E,\eta_g,n_i)$$
$$+\sqrt{\delta_{13}^2(E,\eta_g,n_i)-4\delta_{12}(E,\eta_g)\delta_{14}(E,\eta_g,n_i)}],$$

$$\delta_{12}(E,\eta_g) = \frac{\alpha\hbar^4 Z_0(E,\eta_g)}{4m_t^+ m_l^-}, \delta_{13}(E,\eta_g,n_i)$$
$$= \hbar^2[\lambda_{71}(E,\eta_g)\delta_{11}(E,\eta_g,n_i) + \lambda_{12}(E,\eta_g)],$$
$$\delta_{14}(E,\eta_g,n_i) = [\lambda_{73}(E,\eta_g)\delta_{11}^2(E,\eta_g,n_i) + \lambda_{74}(E,\eta_g)\delta_{11}^4(E,\eta_g,n_i) - \lambda_{74}(E,\eta_g)],$$
$$\delta_{11}(E,\eta_g,n_i) = \frac{2}{\hbar}m_{\mathrm{HD}}^*(0,\eta_g)\left(n_i+\frac{1}{2}\right)\left[\frac{e^2 n_0}{\varepsilon_{\mathrm{sc}} m_{\mathrm{HD}}^*(E,\eta_g)}\right]^{1/2}$$

and

$$m_{\mathrm{HD}}^*(E,\eta_g) = \frac{\hbar^2}{4\lambda_{76}^2(E,\eta_g)}\left[2\lambda_{74}(E,\eta_g) - \left\{-\lambda_{73}'(E,\eta_g) + \frac{\lambda_3(E,\eta_g)\lambda_{73}'(E,\eta_g) + 2\lambda_{74}'(E,\eta_g)\lambda_{75}(E,\eta_g) + 2\lambda_{74}(E,\eta_g)\lambda_{75}'(E,\eta_g)}{\sqrt{\lambda_{73}^2(E,\eta_g) + 4\lambda_{74}(E,\eta_g)\lambda_{75}(E,\eta_g)}}\right\} - 2\lambda_{74}'(E,\eta_g)\{-\lambda_{73}(E,\eta_g) + \sqrt{\lambda_{73}^2(E,\eta_g) + 4\lambda_{74}(E,\eta_g)\lambda_{75}(E,\eta_g)}\}\right]$$

The 2D magneto DR from (13.23) in this case is given by

$$\frac{2eB}{\hbar}\left(n+\frac{1}{2}\right) = \delta_{15}(E_{11,11},\eta_g,n_i) \tag{13.24}$$

where $E_{11,11}$ is the total quantized energy,

The DOS function is given by

$$N_{Bn_z} = \frac{g_v eB}{\pi\hbar}\sum_{n=0}^{n_{\max}}\sum_{n_z=1}^{n_{z\max}}\delta'(E - E_{11,11}) \tag{13.25}$$

The electron concentration can be written as

$$n_0 = \frac{g_v eB}{\pi\hbar}\sum_{n=0}^{n_{\max}}\sum_{n_z=1}^{n_{z\max}}F_{-1}(\eta_{11,11}) \tag{13.26}$$

where $\eta_{11,11} = (k_B T)^{-1}[E_{F11,11} - E_{11,11}]$ and $E_{F11,11}$ is the Fermi energy in this case.

Using (1.181) and (13.25), we can write the AC as

$$f_e^* = \frac{4}{n_0}\frac{g_v eB}{\pi\hbar}\sum_{n=0}^{n_{\max}}\sum_{n_z=1}^{n_{z\max}} F_{-2}(\eta_{11,11}) \tag{13.27}$$

*13.2.5 The DOS function in doping superlattices of HD stressed Kane type materials under magnetic quantization and the AC

The 2D dispersion relation in this case can be written following (11.116) as

$$P_{11}(E,\eta_g)k_z^2 + Q_{11}(E,\eta_g)k_y^2 + S_{11}(E,\eta_g)\delta_{19}(E,\eta_g,n_i) = 1 \tag{13.28}$$

where

$$\delta_{19}(E,\eta_g,n_i) = \frac{2}{\hbar}m_{zz}^*(0,\eta_g)\left(n_i+\frac{1}{2}\right)\left[\frac{\bar{n}_0 e^2}{\varepsilon_{\mathrm{sc}} m_{\mathrm{zz}}(E,\eta_g)}\right]^{1/2}$$

From (13.28), the 2D DR in this case is given by

$$\left(n+\frac{1}{2}\right)\hbar\omega_{90\mathrm{HD}}(E_{11,13},\eta_g) + S_{11}(E_{11,13},\eta_g)\delta_{19}(E_{11,13},\eta_g,n_i) \tag{13.29}$$

where $E_{11,13}$ is the total quantized energy in this case.

$$\omega_{90}(E_{11,13},\eta_g) = \frac{eB}{\sqrt{m_1^*(E_{11,13},\eta_g)m_2^*(E_{11,13},\eta_g)}},$$

$$m_1^*(E_{10,13},\eta_g) = \frac{\hbar^2}{2P_{11}(E_{11,13},\eta_g)}$$

and

$$m_2^*(E_{11,13},\eta_g) = \frac{\hbar^2}{2Q_{11}(E_{11,13},\eta_g)}$$

The DOS function is given by

$$N_{Bn_z} = \frac{g_\nu eB}{\pi\hbar}\sum_{n=0}^{n_{\max}}\sum_{n_z=1}^{n_{z\max}} \delta'(E - E_{11,13}) \tag{13.30}$$

The electron concentration can be written as

$$n_0 = \frac{g_v eB}{\pi\hbar}\sum_{n=0}^{n_{\max}}\sum_{n_z=1}^{n_{z\max}} F_{-1}(\eta_{11,13}) \tag{13.31}$$

where $\eta_{11,13} = (k_B T)^{-1}[E_{F11,13} - E_{11,13}]$ and $E_{F11,13}$ is the Fermi energy in this case.

Using (1.181) and (13.30), we can write the AC as

$$f_e^* = \frac{4}{n_0}\frac{g_v eB}{\pi\hbar}\sum_{n=0}^{n_{\max}}\sum_{n_z=1}^{n_{z\max}} F_{-2}(\eta_{11,14}) \tag{13.32}$$

13.3 Summary and Conclusion

From the 2D DR in doping superlattices of HD nonlinear optical and tetragonal materials as given by (13.1) under magnetic quantization, we observe that the electron energy is totally quantized in 3D wave vector surfaces in the complex energy plane which is the consequence of non-removable poles in the corresponding DR in the absence of band tails. Equations (13.3)–(13.5) represent the DOS function, the electron concentration and the corresponding AC in this case respectively. Equations (13.6)–(13.9) represent the DR, DOS function, electron concentration and corresponding AC in this case respectively for doping superlattices of HD III–V materials under magnetic quantization whose un-perturbed conduction electrons obey the three band model of Kane, which contains one non removal pole in energy axis.

The 0D electrons in HD doping superlattices of III–V materials under magnetic quantization are also described by two band model of Kane and parabolic energy bands with the DRs as given by (13.10) and (13.14) respectively.

Equations (13.19)–(13.22) represent the DR, DOS function, electron concentration and the corresponding AC respectively for HD superlattices of II–VI materials under magnetic quantization. Equations (13.24)–(13.27) represent the DR, DOS function, electron concentration and the corresponding AC respectively for HD superlattices of IV–VI materials under magnetic quantization. Since all the DRs in this case possess no poles in the finite energy planes, the constant quantized energies in this case exist in the real plane. The DR (13.29) in doping superlattices of HD stressed Kane type semiconductors under magnetic quantization reflects the identical conclusion. Besides in all the cases the DOS functions are non-uniformly placed Dirac delta functions in the energy axis.

It is the band structure which changes in a fundamental way and consequently all the physical properties of all the electronic materials change radically leading to new physical concepts.

We have not given any plot of AC in any case of this chapter since we believe our creative readers will plot the same and enjoy intricate physics, computational analysis and advanced mathematics and will create new problems both theoretical and experimental.

13.4 Open Research Problems

(R.13.1) Investigate the DOS and the AC in the presence of an arbitrarily oriented quantizing magnetic field for NIPI structures of HD nonlinear optical and tetragonal semiconductors by including the electron spin. Study all the special cases for HD III–V, ternary and quaternary materials in this context.

(R.13.2) Investigate the DOS and the AC in NIPI structures of HD IV–VI, II–VI and stressed Kane type compounds in the presence of an arbitrarily oriented quantizing magnetic field by including the electron spin.

(R.13.3) Investigate the DOS and the AC for HD NIPI structures of all the materials as stated in Chapter 1 under alternating quantizing magnetic field.

(R.13.4) Investigate the DOS and the AC for all the problems of (R.13.3) in the presence of an additional arbitrarily oriented electric field.

(R.13.5) Investigate the DOS and the AC for all the problems of (R.13.4) in the presence of arbitrarily oriented crossed alternating electric and magnetic fields.

References

[1] N. G. Anderson, W. D. Laidig, R. M. Kolbas, Y. C. Lo, *J. Appl. Phys.*, 60, 2361 **(1986)**.

[2] F. Capasso, *Semicond. Semimetals*, 22, 2 **(1985)**.

[3] F. Capasso, K. Mohammed, A. Y. Cho, R. Hull, A. L. Hutchinson, *Appl. Phys. Lett.*, 47, 420 **(1985)**.

[4] F. Capasso, R. A. Kiehl, *J. Appl. Phys.*, 58, 1366 **(1985)**.

[5] K. Ploog, G. H. Doheler, *Adv. Phys.*, 32, 285 **(1983)**.

[6] F. Capasso, K. Mohammed, A. Y. Cho, *Appl. Phys. Lett.*, 48, 478 **(1986)**.

[7] R. Grill, C. Metzner, G. H. Döhler, *Phys. Rev. B*, 63, 235316 **(2001)**.

[8] A. R. Kost, M. H. Jupina, T. C. Hasenberg, E. M. Garmire, *J. Appl. Phys.*, 99, 023501 **(2006)**.

[9] A. G. Smirnov, D. V. Ushakov, V. K. Kononenko, *Proc. SPIE*, 4706, 70 **(2002)**.

[10] D. V. Ushakov, V. K. Kononenko, I. S. Manak, *Proc. SPIE*, 4358, 171 **(2001)**.

[11] J. Z. Wang, Z. G. Wang, Z. M. Wang, S. L. Feng, Z. Yang, *Phys. Rev. B*, 62, 6956 **(2000)**.
[12] A. R. Kost, L. West, T. C. Hasenberg, J. O. White, M. Matloubian, G. C. Valley, *Appl. Phys. Lett.*, 63, 3494 **(1993)**.
[13] A. S. Bastola, S. J. Chua, S. J. Xu, *J. Appl. Phys.*, 83, 1476 **(1998)**.
[14] Z. J. Yang, E. M. Garmire, D. Doctor, *J. Appl. Phys.*, 82, 3874 **(1997)**.
[15] G. H. Avetisyan, V. B. Kulikov, I. D. Zalevsky, P. V. Bulaev, *Proc. SPIE*, 2694, 216 **(1996)**.
[16] U. Pfeiffer, M. Kneissl, B. Knüpfer, N. Müller, P. Kiesel, G. H. Döhler, J. S. Smith, *Appl. Phys. Lett.*, 68, 1838 **(1996)**.
[17] H. L. Vaghjiani, E. A. Johnson, M. J. Kane, R. Grey, C. C. Phillips, *J. Appl. Phys.*, 76, 4407 **(1994)**.
[18] P. Kiesel, K. H. Gulden, A. Hoefler, M. Kneissl, B. Knuepfer, S. U. Dankowski, P. Riel, X. X. Wu, J. S. Smith, G. H. Doehler, *Proc. SPIE*, 1985, 278 **(1993)**.
[19] G. H. Doheler, *Phys. Script.*, 24, 430 **(1981)**.
[20] S. Mukherjee, S. N. Mitra, P. K. Bose, A. R. Ghatak, A. Neoigi, J. P. Banerjee, A. Sinha, M. Pal, S. Bhattacharya, K. P. Ghatak, *J. Comput. Theor. Nanosci.*, 4, 550 **(2007)**; N. Paitya, K. P. Ghatak, *J. Adv. Phys.*, 1, 161 **(2012)**; N. Paitya, S. Bhattacharya, D. De, K. P. Ghatak, *Adv. Sci. Engg. Medi.*, 4, 96 **(2012)**.
[21] S. Bhattacharya, D. De, S. M. Adhikari, K. P. Ghatak, *Superlatt. Microstruct.*, 51, 203 **(2012)**; D. De, S. Bhattacharya, S. M. Adhikari, A. Kumar, P. K. Bose, K. P. Ghatak, *Beilstein J. Nanotech.*, 2, 339 **(2012)**; D. De, A. Kumar, S. M. Adhikari, S. Pahari, N. Islam, P. Banerjee, S. K. Biswas, S. Bhattacharya, K. P. Ghatak, *Superlatt. Microstruct.*, 47, 377 **(2010)**.
[22] S. Pahari, S. Bhattacharya, S. Roy, A. Saha, D. De, K. P. Ghatak, *Superlatt. Microstruct.*, 46, 760 **(2009)**; S. Pahari, S. Bhattacharya, K. P. Ghatak, *J. Comput. Theor. Nanosci.*, 6, 2088 **(2009)**; S. K. Biswas, A. R. Ghatak, A. Neogi, A. Sharma, S. Bhattacharya, K. P. Ghatak, *Phys. E: Low-dimen. Syst. Nanostruct.*, 36, 163 **(2007)**.
[23] L. J. Singh, S. Choudhury, D. Baruah, S. K. Biswas, S. Pahari, K. P. Ghatak, *Phys. B: Conden. Matter*, 368, 188 **(2005)**; S. Chowdhary, L. J. Singh, K. P. Ghatak, *Phys. B: Conden. Matter*, 365, 5 **(2005)**; L. J. Singh, S. Choudhary, A. Mallik, K. P. Ghatak, *J. Comput. Theor. Nanosci.*, 2, 287 **(2005)**.
[24] K. P. Ghatak, J. Mukhopadhyay, J. P. Banerjee, *SPIE Proc. Ser.*, 4746, 1292 **(2002)**; K. P. Ghatak, S. Dutta, D. K. Basu, B. Nag, *Il Nuovo Cimento D*, 20, 227 **(1998)**; K. P. Ghatak, D. K. Basu, B. Nag, *J. Phys. Chem. Solids*, 58, 133 **(1997)**; K. P. Ghatak, B. De, *Mat. Res. Soc. Proc.*, 300, 513 **(1993)**.
[25] K. P. Ghatak, B. Mitra, *Il Nuovo Cimento D*, 15, 97 **(1993)**; K. P. Ghatak, *Inter. Soc. Opt. Photon. Proc. Soc. Photo Opt. Instru. Engg.*, 1626, 115 **(1992)**; K. P. Ghatak, A. Ghoshal, *Phys. Stat. Sol. (b)*, 170, K27 **(1992)**.
[26] K. P. Ghatak, S. Bhattacharya, S. N. Biswas, *Proc. Soc. Photo Opt. Instru. Engg.*, 836, 72 **(1988)**; K. P. Ghatak, A. Ghoshal, S. N. Biswas, M. Mondal,

Proc. Soc. Photo Opt. Instru. Engg., 1308, 356 **(1990)**; K. P. Ghatak, G. Mazumder, *Proc. Mat. Res. Soc.*, 484, 679 **(1998)**.

[27] K. P. Ghatak, B. De, *Defect Engg. Semi. Growth, Process. Device Tech. Symp., Mat. Res. Soc.*, 262, 911 **(1992)**; S. N. Biswas, K. P. Ghatak, *Int. J. Electron. Theor. Exp.*, 70, 125 **(1991)**; B. Mitra, K. P. Ghatak, *Phys. Lett. A*, 146, 357 **(1990)**; B. Mitra, K. P. Ghatak, *Phys. Lett. A*, 142, 401 **(1989)**.

[28] K. P. Ghatak, B. Mitra, A. Ghoshal, *Phys. Stat. Sol. (b)*, 154, K121 **(1989)**; B. Mitra, K. P. Ghatak, *Phys. Stat. Sol. (b)*, 149, K117 **(1988)**; K. P. Ghatak, S. N. Biswas, *Proc. Soc. Photo Opt. Instru. Engg.*, 792, 239 **(1987)**.

[29] S. Bhattacharyya, K. P. Ghatak, S. Biswas, *OE/Fibers' 87*, *Inter. Soc. Opt. Photon.*, 836, 73 **(1988)**; M. Mondal, K. P. Ghatak, *Czech. J. Phys. B*, 36, 1389 **(1986)**; K. P. Ghatak, A. N. Chakravarti, *Phys. Stat. Sol. (b)*, 117, 707 **(1983)**.

Chapter 14

The Heat Capacity in Doping Superlattices of HD Non-Parabolic Materials Under Magnetic Quantization

Pray as if everything depends on LORD. Work as if everything depends on you.

14.1 Introduction

In recent years there has been considerable interest in investigating the Heat Capacity (HC) in nano-structured materials under different physical conditions [1–10]. In this chapter, the magneto HC in doping superlattices of nonlinear optical and tetragonal, III–V, II–VI, IV–VI, and strained Kane type materials has been studied in Sections 14.2.1–14.2.5. Section 14.3 contains the summary and conclusion pertinent to this chapter. The last Section 14.4 presents five open research problems.

14.2 Theoretical Background

14.2.1 The HC in doping superlattices of HD nonlinear optical and tetragonal materials under magnetic quantization

Using (1.197) and (13.3), we can study the magneto HC in this case.

14.2.2 The HC in doping superlattices of HD III–V, ternary and quaternary materials under magnetic quantization

Using (1.197) and (13.7), we can study the magneto HC in this case.

14.2.3 The HC in doping superlattices of HD II–VI materials under magnetic quantization

Using (1.197) and (13.20), we can study the magneto HC in this case.

14.2.4 The HC in doping superlattices of HD IV–VI materials under magnetic quantization

Using (1.197) and (13.25), we can study the magneto HC in this case.

*14.2.5 The HC in doping superlattices of HD stressed Kane type materials under magnetic quantization

Using (1.197) and (13.30), we can study the magneto HC in this case.

14.3 Summary and Conclusion

The DRs in magneto-size quantized magneto HD doping superlattices of non-parabolic materials exhibit the fact that the total energy is quantized since the corresponding wave vector space is totally quantized by quantizing magnetic field and size quantization along z-direction.

- The DOS functions for all the quantized magneto HD doping superlattices of non-parabolic materials in this case are series of non-uniformly distributed Dirac's Delta functions at specified quantized points in the respective energy axis. The spacing between the consecutive Delta functions are functions of energy band constants and quantization of the wave vector space of a particular material. The DOS function needs two summations namely one summation over the Landau quantum number and the other one is due to size quantization.
- It may be noted that the energy levels in this case lead to the discrete energy levels, somewhat like atomic energy levels, which produce very large changes. This follows from the inherent nature of the quantum confinement of the carrier gas dealt with here. Consequently, the crossing of the Fermi level by the size-quantized levels in this case would have much greater impact on the redistribution of the carriers among the allowed levels, as compared to that found for QWs, NWs and QDs respectively for the electronic properties in general.
- By using the appropriate equations we can plot the HC with respect to many variables in all the cases of this chapter. We have not given

any plot of the HC in this case since we believe our able readers will plot the same and enjoy intricate physics, computational analysis, advanced mathematics, and will create new problems both theoretical and experimental.

It is the band structure which changes in a fundamental way and consequently all the physical properties of all the electronic materials changes radically leading to new physical concepts.

14.4 Open Research Problems

(R.14.1) Investigate the HC in the presence of an arbitrarily oriented quantizing magnetic field for doping superlattices of nonlinear optical semiconductors by including the electron spin. Study all the special cases for III–V, ternary and quaternary materials in this context.

(R.14.2) Investigate the HC in doping superlattices of IV–VI, II–VI and stressed Kane type compounds in the presence of an arbitrarily oriented quantizing magnetic field by including the electron spin.

(R.14.3) Investigate the HC for doping superlattices of all the materials as stated in Chapter 1.

(R.14.4) Investigate the HC for all the problems of (R.14.3) in the presence of an additional arbitrarily oriented electric field.

(R.14.5) Investigate the HC for all the problems of (R.14.3) in the presence of arbitrarily oriented crossed electric and magnetic fields.

References

[1] K. P. Ghatak and S. N. Biswas, A. Fizika, *J. Exp. Theor. Phys.*, 3, 7 **(1994)**.

[2] K. P. Ghatak and S. N. Biswas, *Reflective and Refractive Optical Materials for Earth and Space Applications*, *Proceeding of SPIE*, 1485, 206 **(1991)**.

[3] K. P. Ghatak, A. Goshal, S. Bhattacharya, M. Mondal, *Electro-Optical Materials for Switches, Coatings, Sensor Optics, and Detectors*, *Proceeding of SPIE*, 1307, 414 **(1990)**.

[4] K. P. Ghatak, A. Goshal, S. Bhattacharya, M. Mondal, *Thermosense XII: An International Conference on Thermal Sensing and Imaging Diagnostic Applications*, *Proceeding of SPIE*, 1313, 225 **(1990)**.

[5] M. Mondal, K. P. Ghatak, *J. Phys. Stat. Sol. (b)*, 128, K21 **(1985)**.

[6] M. Mondal, K. P. Ghatak, *J. Acta Phys. Polonica A*, 66, 539 **(1984)**.

[7] M. Mondal, K. P. Ghatak, *J. Phys. Stat. Sol. (b)*, 120, K63 **(1983)**.

[8] K. P. Ghatak, K. K. Ghosh, H. M. Mukherjee, A. N. Chakravarti, *J. Phys. Stat. Sol. (b)*, 110, 323 **(1982)**.
[9] K. P. Ghatak, K. K. Ghosh, H. M. Mukherjee, A. N. Chakravarti, *J. Phys. Stat. Sol. (b)*, 109, 705 **(1982)**.
[10] A. N. Chakravarti, A. K. Chowdhury, K. P. Ghatak, *J. Phys. Stat. Sol. (b)*, 63, K97 **(1981)**.

Chapter 15

The QC in Heavily Doped Quantized Transistors Under Magnetic Quantization

I will never find a rainbow if I am looking down.

15.1 Introduction

In recent years there has been considerable interest in studying the effect of magnetic quantization on the different electronic properties of quantized transistors [1–14]. In Section 15.2.1, of the theoretical background, the QC has been investigated in HDQWTs of nonlinear optical and tetragonal materials in the presence of a quantizing magnetic field. Section 15.2.2 contains the results for HDQWTs of III–V, ternary and quaternary compounds in accordance with the three and the two band models of Kane along with the parabolic model. In the same section the QC in accordance with the models of Stillman *et al.* [15], Palik *et al.* [16], Rossler, Johnson and Dickey, Agafonov *et al.* and Zhang have also been studied for the purpose of relative comparison. Section 15.2.3 contains the study of the QC for HDQWTs of II–VI semiconductors under magnetic quantization. In Section 15.2.4, the QC in HDQWTs of IV–VI materials has been discussed in accordance with the models of Cohen, Lax, Dimmock, Bangert and Kastner, Foley and Landenberg and McClure and Choi respectively. In Section 15.2.5, the magneto-QC for the stressed HDQWTs of Kane type semiconductors has been investigated. In Section 15.2.6, the QC in HDQWTs of Te has been studied under magnetic quantization. In Section 15.2.7, the magneto-QC in HDQWTs of GaP has been studied. In Section 15.2.8, the QC in HDQWTs of $PtSb_2$ has been explored under magnetic quantization. In Section 15.2.9, the magneto-QC in HDQWTs of Bi_2Te_3 has been studied. In Section 15.2.10, the QC in HDQWTs of Ge has been studied under magnetic quantization in accordance with the models of Cardona *et al.* and Wang and Ressler respectively. In Section 15.2.11, the magneto-QC in HDQWTs of GaSb has been studied in accordance with the models of Mathur *et al.* and Seiler respectively. In Sections 15.2.12 and 15.2.13, the magneto-QC in HDQWTs of II–V compounds and $Pb_{1-x}Ge_xTe$ has respectively been discussed. In Section 15.2.14, the magneto-QC in HDQWTs of Zinc and Cadmium Diphosphides has been studied. It may also be noted from the recent literature that the magneto QC in 2D QMOSFETs of important materials operated under accumulation mode has yet to be investigated. In this chapter, we study the magneto QC in 2D QMOSFETs of tetragonal and nonlinear optical materials in Section 15.2.15 of theoretical background 15.2. The magneto QC in 2D QMOSFETs of III–V, ternary and quaternary materials under accumulation mode has been studied in Section 15.2.16 in accordance with the models of Three band model of Kane, Two band model of Kane and Parabolic Energy Bands. From Sections 15.2.17 to 15.2.19 we shall investigate the same in 2D QMOSFETs of II–VI, IV–VI, and stressed Kane type materials under accumulation mode respectively. Section 15.3 explores the summary and conclusion and the last Section 15.4 contains 21 open research problems for this chapter.

15.2 Theoretical Background

15.2.1 The QC in HDQWTs of nonlinear optical and tetragonal materials under magnetic quantization

The DR under magnetic quantization in tetragonal and nonlinear optical materials can be written as

$$\gamma(E) = \frac{\hbar^2 k_s^2}{2m_\perp^*} f_3(E) + \frac{\hbar^2 k_z^2}{2m_{\parallel}^*} f_4(E) \pm \frac{eB\hbar E_g}{6}\left[\frac{(E_{g_0}+\Delta_\perp)}{(E_{g_0}+\frac{2}{3}\Delta_\perp)}\right]\left[E+E_{g_0}+\delta+\frac{\Delta_{\parallel}^2-\Delta_\perp^2}{3\Delta_{\parallel}}\right] \quad (15.1)$$

where

$$f_3(E) = \frac{E_g(E_g+\Delta_\perp)}{(E_g+\frac{2}{3}\Delta_\perp)}\left[(E+E_g)\left(E+E_g+\frac{2}{3}\Delta_{\parallel}\right)+\delta\left(E+E_g+\frac{1}{3}\Delta_{\parallel}\right) + \frac{1}{9}(\Delta_{\parallel}^2-\Delta_\perp^2)\right]$$

and

$$f_4(E) = \frac{E_g(E_g + \Delta_{||})}{(E_g + \frac{2}{3}\Delta_{||})}\left[(E + E_g)(E + E_g + \frac{2}{3}\Delta_{||})\right]$$

Equation (15.1) can be expressed as

$$\frac{\hbar^2 k_z^2}{2m_{||}^*} + \left(\frac{b_{||}c_\perp}{b_\perp c_{||}}\right)\left(\frac{\hbar^2 k_s^2}{2m_\perp^*}\right) = \left\{\left[\frac{ab_{||}}{c_{||}}E^2 + \frac{(ac_{||} + b_{||} - ab_{||})}{c_{||}^2}E\right.\right.$$
$$+\frac{1}{c_{||}}\left(1 - \frac{a}{c_{||}}\right)\left(1 - \frac{b_{||}}{c_{||}}\right) - \frac{1}{c_{||}}\left(1 - \frac{a}{c_{||}}\right)\left(1 - \frac{b_{||}}{c_{||}}\right)\frac{1}{(c_{||}E + 1)}\Bigg]$$
$$+\frac{ab_{||}}{c_{||}}\left[\delta E + \frac{2}{9}(\Delta_{||}^2 - \Delta_\perp^2)\right] - \frac{2}{9}\frac{ab_{||}}{c_{||}}\frac{(\Delta_{||}^2 - \Delta_\perp^2)}{(c_{||}E + 1)}\Bigg\} - \left(\frac{\hbar^2 k_s^2}{2m_\perp^*}\right)\left\{\left(\frac{b_{||}c_\perp}{b_\perp c_{||}}\right)\right.$$
$$\times\left[\left(\frac{\delta}{2} + \frac{\Delta_{||}^2 - \Delta_\perp^2}{6\Delta_{||}}\right)\frac{a}{(aE + 1)} + \left(\frac{\delta}{2} - \frac{\Delta_{||}^2 - \Delta_\perp^2}{6\Delta_{||}}\right)\frac{c_{||}}{(c_{||}E + 1)}\right]\Bigg\}$$
$$\pm e_1\left[\frac{\rho_1}{E + E_g} + \frac{\rho_2}{E + E_g + \frac{2}{3}\Delta_{||}}\right] \tag{15.2}$$

where

$$e_1 = \left[\frac{eB\hbar\,(E_g + \frac{2}{3}\Delta_{||})}{6(E_g + \Delta_\perp)}\frac{(E_g + \Delta_\perp)}{(E_g + \frac{2}{3}\Delta_{||})}\right], \quad \rho_1 = \frac{(-E_g + G_1)}{(\frac{2}{3}\Delta_{||})},$$

$$G_1 = \left[E_{g0} + \delta + \frac{\Delta_{||}^2 - \Delta_\perp^2}{3\Delta_{||}}\right] \quad \text{and} \quad \rho_2 = \frac{3}{2\Delta_{||}}\left[\frac{\Delta_{||}}{3} + \frac{\Delta_\perp^2}{3\Delta_{||}} - \delta\right]$$

Therefore, the DR of the conduction electrons in HDQWTs of tetragonal and nonlinear optical materials in the presence of a quantizing magnetic field B can be written following the methods as developed in Chapter 11 as

$$\frac{\hbar^2 k_z^2}{2m_{||}^*} = U_{1,\pm}(E, n, \eta_g) + iU_{2,\pm}(E, n, \eta_g) \tag{15.3}$$

where

$$U_{1,\pm}(E, n, \eta_g) = \left[\frac{-eB}{m_\perp^*}\left(n + \frac{1}{2}\right)\left(\frac{b_{||}c_\perp}{c_{||}b_\perp}\right) + \left[\frac{1 + Erf(E/\eta_g)}{2}\right]^{-1}\right.$$

$$\times\left[\left[\frac{ab_{\|}}{c_{\|}}\theta_0(E,\eta_g)+\frac{ac_{\|}+b_{\|}c_{\|}-ab_{\|}}{c_{\|}^2}\gamma_0(E,\eta_g)+\frac{1}{c_{\|}}\left(1-\frac{a}{c_{\|}}\right)\left(1-\frac{b_{\|}}{c_{\|}}\right)\right.\right.$$

$$\times\left[\frac{1+Erf(E/\eta_g)}{2}\right]-\frac{1}{c_{\|}}\left(1-\frac{a}{c_{\|}}\right)\left(1-\frac{b_{\|}}{c_{\|}}\right)c(\beta_1,E,\eta_g)$$

$$+\frac{ab_{\|}}{c_{\|}}\left[\delta\gamma_0(E,\eta_g)+\frac{2}{9}(\Delta_{\|}^2-\Delta_{\perp}^2)\left[\frac{1+Erf(E/\eta_g)}{2}\right]\right]$$

$$\left.-\frac{2}{9}\frac{ab_{\|}}{c_{\|}}(\Delta_{\|}^2-\Delta_{\perp}^2)c(\beta_1,E,\eta_g)\right]-\frac{\hbar eB}{m_{\perp}^*}\left(n+\frac{1}{2}\right)\left\{\left(\frac{b_{\|}c_{\perp}}{c_{\|}b_{\perp}}\right)\right.$$

$$\left.\times\left[\left(\frac{\delta}{2}+\frac{\Delta_{\|}^2-\Delta_{\perp}^2}{6\Delta_{\|}}\right)ac(\beta_2,E,\eta_g)+\left(\frac{\delta}{2}+\frac{\Delta_{\|}^2-\Delta_{\perp}^2}{6\Delta_{\|}}\right)c_{\|}c(\beta_1,E,\eta_g)\right\}$$

$$\left.\left.\times\pm\frac{\rho_1e_1}{E_g}(\beta_2,E,\eta_g)+\frac{\rho_2e_1}{(E_g+\frac{2}{3}\Delta_{\|})}c(\beta_3,E,\eta_g)\right]\right]$$

$$U_{2,\pm}(E,n,\eta_g)=\left[\left[\frac{1+Erf(E/\eta_g)}{2}\right]^{-1}\left[\frac{1}{c_{\|}}\left(1-\frac{a}{c_{\|}}\right)\left(1-\frac{b_{\|}}{c_{\|}}\right)\right.\right.$$

$$\times D(\beta_1,E,\eta_g)+\frac{2}{9}\frac{ab_{\|}}{c_{\|}}(\Delta_{\|}^2-\Delta_{\perp}^2)D(\beta_1,E,\eta_g)-\frac{\hbar eB}{m_{\perp}^*}\left(n+\frac{1}{2}\right)$$

$$\times\left\{\left(\frac{b_{\|}c_{\perp}}{c_{\|}b_{\perp}}\right)\left[\left(\frac{\delta}{2}+\frac{\Delta_{\|}^2-\Delta_{\perp}^2}{6\Delta_{\|}}\right)aD(\beta_2,E,\eta_g)+\left(\frac{\delta}{2}-\frac{\Delta_{\|}^2-\Delta_{\perp}^2}{6\Delta_{\|}}\right)\right.\right.$$

$$\left.\left.\left.\times c_{\|}D(\beta_1,E,\eta_g)\right\}\right]\mp\frac{\rho_1e_1}{E_g}D(\beta_2,E,\eta_g)\pm\frac{\rho_2e_1}{(E_g+\frac{2}{3}\Delta_{\|})}D(\beta_3,E,\eta_g)\right]$$

$$c(\beta_i,E,\eta_g)=\left[\frac{2}{\beta_i\eta_g\sqrt{\pi}}\right]\exp(-u_i^2)\times\left[\sum_{p=1}^{\infty}\left\{\frac{\exp\left(\frac{-p^2}{4}\right)}{p}\right\}\sinh(pu_i)\right],$$

$$u_i=\frac{1+\beta_iE}{\beta_i\eta_g}\quad\text{and}\quad D(\beta_i,E,\eta_g)=\left[\left[\frac{\sqrt{\pi}}{\beta_i\eta_g}\exp(-u_i^2)\right]\right]$$

The DR of the conduction electrons in HDQWTs of tetragonal and nonlinear optical semiconductors in the presence of a quantizing magnetic field B can be written following (15.3) as

$$\frac{\hbar^2\left(\frac{n_z\pi}{d_z}\right)^2}{2m_{\|}^*}=U_{1,\pm}(e_{81},n,\eta_g)+iU_{2,\pm}(e_{81},n,\eta_g)\tag{15.4}$$

where e_{81} is the total quantized energy in this case.

The DOS function is given by

$$N_{Bn_z} = \frac{g_v eB}{2\pi\hbar} \sum_{n=0}^{n_{\max}} \sum_{n_z=1}^{n_{z\max}} \delta'(E - e_{81}) \tag{15.5}$$

The electron concentration per unit area can be written as

$$N_s = \frac{g_v eB}{2\pi\hbar} \text{ Real part of } \sum_{n=0}^{n_{\max}} \sum_{n_z=1}^{n_{z\max}} F_{-1}(\eta_{12,1}) \tag{15.6}$$

where $\eta_{12,1} = (k_B T)^{-1}[E_{F12\text{HD}} - e_{81}]$ and $E_{F12\text{HD}}$ is the Fermi energy in the present case. The QC can be expressed as

$$C_G = e\frac{\partial}{\partial V_g}(N_s) \tag{15.7}$$

Thus, by using (15.7) and other appropriate equations we can study the QC in this case.

15.2.2 The QC in HDQWTs of III–V, ternary and quaternary materials under magnetic quantization

(a) The three band model of Kane

Substituting $\delta = 0$, $\Delta_{\parallel} = \Delta_{\perp} = \Delta$ and $m_{\parallel}^* = m_{\perp}^* = m_c$ in (15.1), the electron energy spectrum in III–V materials under magnetic quantization is given by

$$\frac{E(E+E_g)(E+E_g+\Delta)\left(E_g+\frac{2}{3}\Delta\right)}{E_g(E_g+\Delta)\left(E+E_g+\frac{2}{3}\Delta\right)} = \left(n+\frac{1}{2}\right)\hbar\omega_0 + \frac{\hbar^2 k_z^2}{2m_c} \pm \frac{eB\hbar\Delta}{6m_c\left(E+E_g+\frac{2}{3}\Delta\right)} \tag{15.8a}$$

Under the condition of heavy doping, the (15.8a) can be written as

$$\left[\frac{ab}{c}E^2 + \left(\frac{ac+bc-ab}{c^2}\right)E + \frac{1}{c}\left(1-\frac{a}{c}\right)\left(1-\frac{b}{c}\right) - \frac{1}{c}\left(1-\frac{a}{c}\right)\left(1-\frac{b}{c}\right)\frac{1}{(1+cE)}\right] = \left(n+\frac{1}{2}\right)\hbar\omega_0 + \frac{\hbar^2 k_z^2}{2m_c} + \frac{eB\hbar\Delta}{6m_c(1+cE)\left(E_g+\frac{2}{3}\Delta\right)} \tag{15.8b}$$

where

$$a = \frac{1}{E_g}, \quad b = \frac{1}{E_g + \Delta} \quad \text{and} \quad c = \frac{1}{E_g + \frac{2}{3}\Delta}$$

Therefore,

$$\frac{ab}{c} I(5) + \left(\frac{ac + bc - ab}{c^2}\right) I(4) + \left[\frac{1}{c}\left(1 - \frac{a}{c}\right)\left(1 - \frac{b}{c}\right) - \left(n + \frac{1}{2}\right)\hbar\omega_0\right]$$

$$I(1) - g_{\pm}[G(c, E, \eta_g) - iH(c, E, \eta_g)] = \frac{\hbar^2 k_z^2}{2m_c} I(1) \tag{15.9}$$

where

$$g_{\pm} = \left[\frac{1}{c}\left(1 - \frac{a}{c}\right)\left(1 - \frac{b}{c}\right) \pm \frac{eB\hbar\Delta}{6m_c\left(E_g + \frac{2}{3}\Delta\right)}\right]$$

$$G(c, E, \eta_g) = \left[\frac{2}{c\eta_g\sqrt{\pi}}\right] \exp(-u^2) \times \left[\sum_{p=1}^{\infty} \left\{\frac{\exp\left(\frac{-p^2}{4}\right)}{p}\right\} \sinh(pu)\right],$$

$$u = \frac{1 + cE}{c\eta_g}, \quad H(c, E, \eta_g) = \left[\frac{\sqrt{\pi}}{c\eta_g} \exp(-u^2)\right]$$

Therefore,

$$\begin{aligned}\frac{\hbar^2 k_z^2}{2m_c} &= \left[\frac{ab}{c}\right] \theta_0(E, \eta_g) \left[\frac{1 + Erf(E/\eta_g)}{2}\right]^{-1} \\ &+ \left(\frac{ac + bc - ab}{c^2}\right) \gamma_0(E, \eta_g) \left[\frac{1 + Erf(E/\eta_g)}{2}\right]^{-1} \\ &+ \left[\frac{1}{c}\left(1 - \frac{a}{c}\right)\left(1 - \frac{b}{c}\right) - \left(n + \frac{1}{2}\right)\hbar\omega_0\right] - g_{\pm}\left[\frac{1 + Erf(E/\eta_g)}{2}\right]^{-1} \\ &\times [G(c, E, \eta_g) - iH(c, E, \eta_g)]\end{aligned} \tag{15.10}$$

Therefore, the DR is given by

$$\frac{\hbar^2 k_z^2}{2m_c} = U_{3,\pm}(E, n, \eta_g) + iU_{4,\pm}(E, n, \eta_g) \tag{15.11}$$

where

$$\begin{aligned}U_{3,\pm}(E, n, \eta_g) = &\left[\frac{ab}{c}\theta_0(E, \eta_g) \times \left[\frac{1 + Erf(E/\eta_g)}{2}\right]^{-1}\right. \\ &+ \left(\frac{ac + bc - ab}{c^2}\right) \gamma_0(E, \eta_g) \times \left[\frac{1 + Erf(E/\eta_g)}{2}\right]^{-1}\end{aligned}$$

$$+\frac{1}{c}\left(1-\frac{a}{c}\right)\left(1-\frac{b}{c}\right)-\left(n+\frac{1}{2}\right)\hbar\omega_0+g_\pm\left[\frac{1+Erf(E/\eta_g)}{2}\right]^{-1}$$

$$\left.\times G(c,E,\eta_g)\right] \quad \text{and} \quad U_{4,\pm}(E,\eta_g)=g_\pm\left[\frac{1+Erf(E/\eta_g)}{2}\right]^{-1}H(c,E,\eta_g)$$

In accordance with three band model of Kane, the DR in the present case can be written following (15.11) as

$$\frac{\hbar^2\left(\frac{n_z\pi}{d_z}\right)^2}{2m_c}=U_{3,\pm}(e_{82},n,\eta_g)+iU_{4,\pm}(e_{82},n,\eta_g) \tag{15.12a}$$

where e_{82} is the total quantized energy in this case.

The DOS function is given by

$$N_{Bn_z}=\frac{g_v eB}{2\pi\hbar}\sum_{n=0}^{n_{\max}}\sum_{n_z=1}^{n_{z\max}}\delta'(E-e_{82}) \tag{15.12b}$$

The electron concentration per unit area in this case can be written as

$$N_s=\frac{g_v eB}{2\pi\hbar}\text{ Real part of }\sum_{n=0}^{n_{\max}}\sum_{n_z=1}^{n_{z\max}}F_{-1}(\eta_{12,2}) \tag{15.13}$$

where $\eta_{12,2}=(k_BT)^{-1}[E_{F12\text{HD}}-e_{82}]$.

Thus, by using (15.7), (15.13) and other appropriate equations we can study the QC in this case.

(b) The two band model of Kane

The DR in this case is given by

$$\left(\hbar^2\left(\frac{n_z\pi}{d_z}\right)^2/2m_c\right)=\gamma_2(e_{83},\eta_g)-\left(n+\frac{1}{2}\right)\hbar\omega_0\mp\frac{1}{2}g^*\mu_0B \tag{15.14}$$

where e_{83} is the total quantized energy in this case.

The DOS function is given by

$$N_{Bn_z}=\frac{g_v eB}{2\pi\hbar}\sum_{n=0}^{n_{\max}}\sum_{n_z=1}^{n_{z\max}}\delta'(E-e_{83}) \tag{15.15}$$

The electron concentration per unit area in this case can be written as

$$N_s=\frac{g_v eB}{2\pi\hbar}\sum_{n=0}^{n_{\max}}\sum_{n_z=1}^{n_{z\max}}F_{-1}(\eta_{12,3}) \tag{15.16}$$

where $\eta_{12,3}=(k_BT)^{-1}[E_{F12\text{HD}}-e_{83}]$.

Thus, by using (15.7), (15.16) and other appropriate equations we can study the QC in this case.

(c) The Parabolic energy bands

The DR in this case is given by

$$\frac{\hbar^2 \left(\frac{n_z \pi}{d_z}\right)^2}{2m_c} = \gamma_3(e_{84}, \eta_g) - \left(n + \frac{1}{2}\right) \hbar\omega_0 \mp \frac{1}{2} g^* \mu_0 B \tag{15.17}$$

where e_{84} is the total quantized energy in this case.

The DOS function is given by

$$N_{Bn_z} = \frac{g_v eB}{2\pi\hbar} \sum_{n=0}^{n_{\max}} \sum_{n_z=1}^{n_{z\max}} \delta'(E - e_{84}) \tag{15.18}$$

The electron concentration per unit area in this case can be written as

$$N_s = \frac{g_v eB}{2\pi\hbar} \sum_{n=0}^{n_{\max}} \sum_{n_z=1}^{n_{z\max}} F_{-1}(\eta_{12,4}) \tag{15.19}$$

where $\eta_{12,4} = (k_B T)^{-1}[E_{F12\text{HD}} - e_{84}]$.

Thus, by using (15.7), (15.19) and other appropriate equations we can study the QC in this case.

(d) The model of Stillman *et al.* [15]

From (11.23), we get

$$k^2 = U_7(E, n, \eta_g) \tag{15.20}$$

where

$$U_7(E, n, \eta_g) = \left[\frac{2m_c}{\hbar^2} I_{12}(E, \eta_g) - \frac{2eB}{\hbar}\left(n + \frac{1}{2}\right)\right] \quad \text{and} \quad I_{12}(E, \eta_g) = \theta_4(E, \eta_g)$$

Following (15.20), the DR in the present case can be written as

$$\left(\frac{n_z \pi}{d_z}\right)^2 = U_7(e_{85}, n, \eta_g) \tag{15.21}$$

where e_{85} is the total quantized energy in this case.

The DOS function is given by

$$N_{Bn_z} = \frac{g_v eB}{\pi\hbar} \sum_{n=0}^{n_{\max}} \sum_{n_z=1}^{n_{z\max}} \delta'(E - e_{85}) \tag{15.22}$$

The electron concentration per unit area in this case can be written as

$$N_s = (g_v eB/\pi\hbar) \sum_{n=0}^{n_{\max}} \sum_{n_z=1}^{n_{z\max}} F_{-1}(\eta_{12,5}) \tag{15.23}$$

where $\eta_{12,5} = (k_B T)^{-1}[E_{F12\mathrm{HD}} - e_{85}]$.

Thus, by using (15.7), (15.23) and other appropriate equations we can study the QC in this case.

(e) The model of Palik *et al.* [16]

To the fourth order in effective mass theory and taking into account the interactions of the conduction, light hole, heavy-hole and split-off hole bands, the electron energy spectrum in III–V semiconductors in the presence of a quantizing magnetic field $\vec{B}$ can be written [16] as

$$\begin{aligned} E = J_{31} + \left(n + \frac{1}{2}\right)\hbar\omega_0 + \frac{\hbar^2 k_z^2}{2m_c} \pm \frac{1}{4}\left(\frac{m_c}{m_0}\right)\hbar\omega_0 g_0^* \pm k_{30}\alpha\left(n + \frac{1}{2}\right)(\hbar\omega_0)^2 \\ \pm k_{31}\alpha\hbar\omega_0 \frac{\hbar^2 k_z^2}{2m_c} + k_{32}\alpha\left[\hbar\omega_0\left(n + \frac{1}{2}\right) + \frac{\hbar^2 k_z^2}{2m_c}\right]^2 \end{aligned} \tag{15.24}$$

where

$$J_{31} = -\frac{1}{2}\alpha\hbar\omega_0[(1 - y_{11})/(2 + x_{11})^2] \cdot J_{32}$$

$$\begin{aligned} J_{32} = \left\{\left[\frac{1}{3}(1 - x_{11})^2 - (2 + x_{11})^2\right](2 + x_{11}) \cdot y_{11} \right. \\ \left. + \frac{1}{2}(1 - x_{11}^2)(1 + x_{11})(1 + y_{11})\right\} \end{aligned}$$

$$g_0^* = 2\left\{1 - \left[\frac{(1 - x_{11})}{(2 + x_{11})}\right]\left[\frac{(1 - y_{11})}{y_{11}}\right]\right\},$$

$$k_{30} = (1 - y_{11})(1 - x_{11})\left\{\left[\left(2 + \frac{3}{2}x_{11} + x_{11}^2\right) \cdot \frac{(1 - x_{11})}{(2 + x_{11})^2}\right] - \frac{2}{3}y_{11}\right\}$$

$$k_{31} = (1 - y_{11})\left[\frac{(1-x_{11})}{(2+x_{11})}\right] \cdot \left\{\left[\left(2 + \frac{3}{2}x_{11} + x_{11}^2\right) \cdot \frac{(1-y_{11})}{(2+x_{11})^2}\right] - \frac{2}{3}(1-x_{11})y_{11}\right\}$$

$$k_{32} = \left[\left(1 + \frac{1}{2}x_{11}^2\right) \bigg/ \left(1 + \frac{1}{2}x_{11}\right)(1-y_{11})^2,\right.$$

$$x_{11} = \left[1 + \left(\frac{\Delta}{E_g}\right)\right]^{-1} \quad \text{and} \quad y_{11} = \left.\frac{m_c}{m_0}\right]$$

Under the condition of heavy doping, (15.24) assumes the form

$$J_{34}k_z^4 + J_{35,\pm}(n)k_z^2 + J_{36,\pm}(n) - \gamma_3(E, \eta_g) = 0 \tag{15.25}$$

where

$$J_{34} = \alpha k_{32}\left(\frac{\hbar^2}{2m_c}\right)^2,$$

$$J_{35,\pm}(n) = \left[\frac{\hbar^2}{2m_c} \pm \alpha k_{31}\hbar\omega_0 \cdot \frac{\hbar^2}{2m_c} + \alpha k_{32}\hbar\omega_0 \cdot \frac{\hbar^2}{2m_c}\left(n + \frac{1}{2}\right)\right]$$

$$J_{36,\pm}(n) = \left[J_{31} \pm \frac{1}{4}\left(\frac{m_c}{m_0}\right)\hbar\omega_0 g_0^* \pm k_{30}\alpha(\hbar\omega_0)^2\left(n + \frac{1}{2}\right) + k_{32}\alpha\left[(\hbar\omega_0)\left(n + \frac{1}{2}\right)\right]^2\right]$$

Equation (15.25) can be written as

$$k_z^2 = A_{35,\pm}(E, n, \eta_g) \tag{15.26}$$

where

$$A_{35,\pm}(E, n, \eta_g) = (2J_{34})^{-1}[-J_{35,\pm}(n) + \sqrt{(J_{35,\pm}(n))^2 - 4J_{34}[J_{36,\pm}(n) - \gamma_3(E, \eta_g)]}]$$

Following (15.26), the DR in the present case can be written as

$$\left(\frac{n_z\pi}{d_z}\right)^2 = A_{35,\pm}(e_{86}, n, \eta_g) \tag{15.27}$$

where e_{86} is the total quantized energy in this case.

The DOS function is given by

$$N_{Bn_z} = \frac{g_v eB}{2\pi\hbar} \sum_{n=0}^{n_{\max}} \sum_{n_z=1}^{n_{z\max}} \delta'(E - e_{86}) \tag{15.28}$$

The electron concentration per unit area in this case can be written as

$$N_s = \frac{g_v eB}{2\pi\hbar} \sum_{n=0}^{n_{\max}} \sum_{n_z=1}^{n_{z\max}} F_{-1}(\eta_{12,6}) \tag{15.29a}$$

where $\eta_{12,6} = (k_B T)^{-1}[E_{F12\mathrm{HD}} - e_{86}]$.

Thus, by using (15.7), (15.29a) and other appropriate equations we can study the QC in this case.

(f) The model of Rossler

Equation (5.82) under the condition of heavy doping assumes the form

$$k^2 = \bar{\gamma}_{15\pm}(E, \eta_g) \tag{15.29b}$$

where

$$\bar{\gamma}_{15\pm}(E, \eta_g) = (2e_2)^{-1}[+e_1 - \sqrt{e_1^2 - 4e_2[\gamma_3(E, \eta_g) \pm \beta_{15}(E, \eta_g)]}] \quad \text{and}$$

$$\beta_{15}(E, \eta_g) = \left(e_3 \left(\frac{1}{2e_2}\right)^{3/2}\right) [+e_1 - \sqrt{e_1^2 - 4\gamma_3(E, \eta_g)e_2}]^{3/2}$$

The DR in HDQD in the present case can be written as

$$\left(\frac{\pi n_z}{d_z}\right)^2 + \frac{2eB}{\hbar}\left(n + \frac{1}{2}\right) = \bar{\gamma}_{15\pm}(E_{86,15}, \eta_g) \tag{15.29c}$$

where $E_{86,15}$ is the total quantized energy in this case.

The DOS function is given by

$$N_{Bn_z} = \frac{g_v eB}{2\pi\hbar} \sum_{n=0}^{n_{\max}} \sum_{n_z=1}^{n_{z\max}} \delta'(E - E_{86,15}) \tag{15.29d}$$

The electron concentration per unit area in this case can be written as

$$N_s = \frac{g_v eB}{2\pi\hbar} \sum_{n=0}^{n_{\max}} \sum_{n_z=1}^{n_{z\max}} F_{-1}(\eta_{15,\pm}) \tag{15.29e}$$

where $\eta_{15,\pm} = (k_B T)^{-1}[E_{F12\mathrm{HD}} - E_{86,15}]$.

Thus, by using (15.7), (15.29e) and other appropriate equations we can study the QC in this case.

(g) The HD model of Johnson and Dickey

The DR in HDQD in the present case can be written from (11.59f) as

$$\theta_5(E_{151},\eta_g) - k_g^2[\theta_6(\alpha_1,E_{151},\eta_g) + i\theta_3(\alpha_1,E_{151},\eta_g)] + a^2k_g^4 = 0 \tag{15.29f}$$

where

$$k_g^2 = \left[\left(\frac{\pi n_z}{d_z}\right)^2 + \frac{2eB}{\hbar}\left(n+\frac{1}{2}\right)\right]$$

and E_{151} is the total quantized energy in this case.

The DOS function is given by

$$N_{Bn_z} = \frac{g_v eB}{\pi\hbar}\sum_{n=0}^{n_{\max}}\sum_{n_z=1}^{n_{z\max}}\delta'(E-E_{151}) \tag{15.29g}$$

The electron concentration per unit area in this case can be written as

$$N_s = \frac{g_v eB}{\pi\hbar}\sum_{n=0}^{n_{\max}}\sum_{n_z=1}^{n_{z\max}}F_{-1}(\eta_{16}) \tag{15.29h}$$

where $\eta_{16} = (k_BT)^{-1}[E_{F12\mathrm{HD}} - E_{151}]$.

Thus, by using (15.7), (15.29h) and other appropriate equations we can study the QC in this case.

(h) The HD model of Agafonov *et al.*

The HD DR in HDQD in the present case can be written from (5.150) as

$$k_g^2 = \gamma_{152}(E_{152},\eta_g) \tag{15.29i}$$

E_{152} is the totally quantized energy in this case,

$$\begin{aligned}
&\gamma_{152}(E_{152},\eta_g)\\
&= \left[\frac{8(\gamma_3(E_{152},\eta_g))^2N_{01}^2}{(2(\gamma_3(E_{152},\eta_g))+E_{g_0})^2}\right]^{-1}\left[-\left[4(\gamma_3(E_{152},\eta_g))N_{01} - \frac{8P_0^2}{3}\right]\right.\\
&\quad + \left[\left(4(\gamma_3(E_{152},\eta_g))N_{01} - \frac{8P_0^2}{3}\right)^2\right.\\
&\quad \left.\left. - \frac{64(\gamma_3(E_{152},\eta_g))^2N_{01}^2(\gamma_3(E_{152},\eta_g))^2 + (\gamma_3(E_{152},\eta_g)E_{g_0})}{(2(\gamma_3(E_{152},\eta_g))+E_{g_0})^2}\right]^{1/2}\right]
\end{aligned}$$

The DOS function is given by

$$N_{Bn_z} = \frac{g_v eB}{\pi\hbar} \sum_{n=0}^{n_{\max}} \sum_{n_z=1}^{n_{z\max}} \delta'(E - E_{152}) \tag{15.29j}$$

The electron concentration per unit area in this case can be written as

$$N_s = \frac{g_v eB}{\pi\hbar} \sum_{n=0}^{n_{\max}} \sum_{n_z=1}^{n_{z\max}} F_{-1}(\eta_{17}) \tag{15.29k}$$

where $\eta_{17} = (k_B T)^{-1}[E_{F12\text{HD}} - E_{152}]$.

Thus, by using (15.7), (15.29k) and other appropriate equations we can study the QC in this case.

(i) The HD model of Zhang

The HD DR in HDQD in the present case can be written from (5.150) as

$$k_g^2 = \gamma_{153}(E_{153}, \eta_g) \tag{15.29l}$$

where

$$\gamma_{153}(E_{153}, \eta_g) = (2\delta_2)^{-1}[\delta_1 - \sqrt{\delta_1^2 - 4\delta_2[\gamma_3(E_{153}, \eta_g) - \beta_{20}(E_{153}, \eta_g)]}]$$

$$\beta_{20}(E_{153}, \eta_g) = \delta_3 \left[\frac{\delta_1 - \sqrt{\delta_1^2 - 4\gamma_3(E_{153}, \eta_g)\delta_3}}{2\delta_3}\right]^3$$

and E_{153} is the total quantized energy in this case.

The DOS function is given by

$$N_{Bn_z} = \frac{g_v eB}{\pi\hbar} \sum_{n=0}^{n_{\max}} \sum_{n_z=1}^{n_{z\max}} \delta'(E - E_{153}) \tag{15.29m}$$

The electron concentration per unit area in this case can be written as

$$N_s = \frac{g_v eB}{\pi\hbar} \sum_{n=0}^{n_{\max}} \sum_{n_z=1}^{n_{z\max}} F_{-1}(\eta_{18}) \tag{15.29n}$$

where $\eta_{18} = (k_B T)^{-1}[E_{F12\text{HD}} - E_{153}]$.

Thus, by using (15.7), (15.29n) and other appropriate equations we can study the QC in this case.

15.2.3 The QC in HDQWTs of II–VI materials under magnetic quantization

The magneto dispersion relation of the carriers in heavily doped II–VI semiconductors can be expressed using (11.60) as

$$\gamma_3(E,\eta_g) = a_0' \frac{2eB}{\hbar}\left(n+\frac{1}{2}\right) + b_0' k_z^2 \pm \bar{\lambda}_0 \left[\frac{2eB}{\hbar}\left(n+\frac{1}{2}\right)\right]^{1/2} \quad (15.29o)$$

(15.29o) can be written as

$$k_z^2 = U_{8,\pm}(E,n,\eta_g) \quad (15.29p)$$

where

$$U_{8,\pm}(E,n,\eta_g) = (b_0')^{-1}\left[\gamma_3(E,\eta_g) - \frac{2eBa_0'}{\hbar}\left(n+\frac{1}{2}\right) \mp \bar{\lambda}_0 \left[\frac{2eB}{\hbar}\left(n+\frac{1}{2}\right)\right]^{1/2}\right]$$

Following (15.29p), the DR in the present case can be written as

$$\left(\frac{n_z\pi}{d_z}\right)^2 = U_{8,\pm}(e_{87},n,\eta_g) \quad (15.30)$$

where e_{87} is the total quantized energy in this case.

The DOS function is given by

$$N_{Bn_z} = \frac{g_v eB}{2\pi\hbar}\sum_{n=0}^{n_{\max}}\sum_{n_z=1}^{n_{z\max}} \delta'(E-e_{87}) \quad (15.31)$$

The electron concentration per unit area in this case can be written as

$$N_s = \frac{g_v eB}{2\pi\hbar}\sum_{n=0}^{n_{\max}}\sum_{n_z=1}^{n_{z\max}} F_{-1}(\eta_{12,7}) \quad (15.32)$$

where $\eta_{12,7} = (k_B T)^{-1}[E_{F12\mathrm{HD}} - e_{87}]$.

Thus, by using (15.7), (15.32) and other appropriate equations we can study the QC in this case.

15.2.4 The QC in HDQWTs of IV–VI materials under magnetic quantization

The electron energy spectrum in IV–VI semiconductors are defined by the models of Cohen, Lax, Dimmock, Bangert and Kastner, Foley and

Langenberg and McClure and Choi model respectively. The magneto DR in HD IV–VI semiconductors is discussed in accordance with the said models for the purpose of relative comparison.

(a) The Cohen model

Following (1.109), the magneto electron energy spectrum in IV–VI semiconductors in the presence of quantizing magnetic field B along z-direction can be written as

$$E(1+\alpha E) = \left[\left(n+\frac{1}{2}\right)\hbar\omega_{16}(E) \pm \frac{1}{2}g_0^*\mu_0 B + \frac{3\alpha}{8}\left(n^2+n+\frac{1}{2}\right)[\hbar\omega_{16}(E)]^2 + \frac{\hbar^2 k_z^2}{2m_c}\right] \tag{15.33}$$

where

$$\omega(E) = \frac{|e|B}{\sqrt{m_1 m_2}}\left[1+\alpha E\left(1-\frac{m_2}{m_2'}\right)\right]^{1/2}$$

Therefore, the magneto dispersion law in HDQWTs of IV–VI materials can be expressed as

$$\frac{\hbar^2 k_z^2}{2m_3} = U_{16,\pm}(E,n,\eta_g) \tag{15.34}$$

where

$$\begin{aligned}
U_{16,\pm}&(E,n,\eta_g) \\
&= \left[\gamma_2(E,\eta_g) - \left(n+\frac{1}{2}\right)\frac{\hbar eB}{\sqrt{m_1 m_2}} \mp \frac{1}{2}g^*\mu_0 B - \frac{3\alpha}{8}\left(n^2+n+\frac{1}{2}\right)\right. \\
&\quad \times \left(\frac{\hbar eB}{\sqrt{m_1 m_2}}\right)^2 - \gamma_3(E,\eta_g)\left[\frac{\alpha}{2}\left(n+\frac{1}{2}\right)\frac{\hbar eB}{\sqrt{m_1 m_2}}\left(1-\frac{m_2}{m_2'}\right)\right. \\
&\quad \left.\left. + \frac{3\alpha^2}{8}\left(n^2+n+\frac{1}{2}\right)\left(\frac{\hbar eB}{\sqrt{m_1 m_2}}\right)^2\left(1-\frac{m_2}{m_2'}\right)\right]\right]
\end{aligned}$$

The DR in the present case can be written as

$$\left(\frac{n_z\pi}{d_z}\right)^2 = U_{16,\pm}(e_{88},n,\eta_g) \tag{15.35}$$

where e_{88} is the total quantized energy in this case.

The DOS function is given by

$$N_{Bn_z} = \frac{g_v eB}{2\pi\hbar}\sum_{n=0}^{n_{\max}}\sum_{n_z=1}^{n_{z\max}} \delta'(E - e_{88}) \tag{15.36}$$

The electron concentration per unit area in this case can be written as

$$N_s = \frac{g_v eB}{2\pi\hbar}\sum_{n=0}^{n_{\max}}\sum_{n_z=1}^{n_{z\max}} F_{-1}(\eta_{12,8}) \tag{15.37}$$

where
$$\eta_{12,8} = (k_B T)^{-1}[E_{F12\mathrm{HD}} - e_{88}]. \tag{15.38}$$

Thus, by using (15.7), (15.37) and other appropriate equations we can study the QC in this case.

(b) The Lax model

In accordance with this model, the magneto DR assumes the form using (1.121) as

$$E(1+\alpha E) = \left(n+\frac{1}{2}\right)\hbar\omega_{03}(E) + \frac{\hbar^2 k_z^2}{2m_3} \pm \frac{1}{2}\mu_0 g^* B \tag{15.39}$$

where

$$\omega_{03}(E) \equiv \frac{eB}{\sqrt{m_1 m_2}}$$

The magneto DR in HDQWTs of IV–VI materials, can be written following (15.39) as

$$\gamma_2(E,\eta_g) = \left(n+\frac{1}{2}\right) + \hbar\omega_{03}(E) + \frac{\hbar^2 k_z^2}{2m_3} \pm \frac{1}{2}g^*\mu_0 B \tag{15.40}$$

Equation (15.40) can be written as

$$\frac{\hbar^2 k_z^2}{2m_3} = U_{17,\pm}(E,n,\eta_g) \tag{15.41}$$

where

$$U_{17,\pm}(E,n,\eta_g) = \gamma_2(E,\eta_g) - \left(n+\frac{1}{2}\right)\hbar\omega_{03}(E) \pm \frac{1}{2}g^*\mu_0 B$$

The DR in the present case can be written as

$$\left(\frac{n_z\pi}{d_z}\right)^2 = U_{17,\pm}(e_{89},n,\eta_g) \tag{15.42}$$

where e_{89} is the total quantized energy in this case.

The DOS function is given by

$$N_{Bn_z} = \frac{g_v eB}{2\pi\hbar} \sum_{n=0}^{n_{\max}} \sum_{n_z=1}^{n_{z\max}} \delta'(E - e_{89}) \tag{15.43}$$

The electron concentration per unit area in this case can be written as

$$N_s = \frac{g_v eB}{2\pi\hbar} \sum_{n=0}^{n_{\max}} \sum_{n_z=1}^{n_{z\max}} F_{-1}(\eta_{12,9}) \tag{15.44}$$

where
$$\eta_{12,9} = (k_B T)^{-1}[E_{F12\text{HD}} - e_{89}]. \tag{15.45}$$

Thus, by using (15.7), (15.44) and other appropriate equations we can study the QC in this case.

(c) The Dimmock model

Following (11.120), the dispersion relation under magnetic quantization in HDQWTs of IV–VI materials can be expressed in accordance with the Dimmock model as

$$\begin{aligned}
&\gamma_2(E,\eta_g) + \alpha\gamma_3(E,\eta_g)\frac{2eB}{\hbar}\left(n+\frac{1}{2}\right)\frac{\hbar^2}{2}\left(\frac{1}{m_t^+} - \frac{1}{m_t^-}\right) \\
&\quad + \alpha\gamma_3(E,\eta_g)x\frac{\hbar^2}{2}\left(\frac{1}{m_l^+} - \frac{1}{m_l^-}\right) \\
&= \frac{\hbar^2 k_s^2}{2m_t^*} + \frac{\hbar^2 k_z^2}{2m_l^*} + \frac{\hbar^2 k_s^2}{2m_t^-} + \frac{\hbar^2 k_z^2}{2m_l^-} \\
&\quad + \alpha\left[\frac{\hbar^4 k_s^4}{4m_t^- m_t^+} + \frac{\hbar^2 k_s^2 k_z^2}{4m_l^- m_l^+} + \frac{\hbar^2 k_z^2 k_s^2}{4m_t^+ m_t^-} + \frac{\hbar^4 k_z^4}{4m_l^- m_l^+}\right] \\
&= \frac{2eB}{\hbar}\left(n+\frac{1}{2}\right)\frac{\hbar^2}{2}\left(\frac{1}{m_t^*} - \frac{1}{m_t^-}\right) + x\frac{\hbar^2}{2}\left(\frac{1}{m_l^*} - \frac{1}{m_l^-}\right) \\
&\quad + \alpha\left[\frac{\hbar^4}{4m_t^- m_t^+}\left(\frac{2eB}{\hbar}\left(n+\frac{1}{2}\right)\right)^2 + x\left[\frac{\hbar^4 eB}{2m_l^+ m_t^- \hbar} + \frac{\hbar^4 eB}{2m_t^+ m_l^- \hbar}\right]\right. \\
&\quad \left.\times\left(n+\frac{1}{2}\right) + \frac{\hbar^4}{4m_l^- m_l^+}x^2\right]
\end{aligned} \tag{15.46}$$

where $x = k_z^2$.

Therefore, the magneto dispersion relation in HDQWTs of IV–VI materials, whose unperturbed carriers obey the Dimmock Model can be expressed as

$$k_z^2 = U_{170}(E, n, \eta_g) \tag{15.47a}$$

where

$$U_{170}(E, n, \eta_g) = [2p_9]^{-1}[-q_9(E, n, \eta_g) + [q_9^2(E, n, \eta_g) + 4p_9 R_9(E, n, \eta_g)]^{1/2}],$$

$$p_9 = \frac{\alpha\hbar^4}{4m_l^- m_l^+}, \quad q_9(E, n, \eta_g) = \left[\frac{\hbar^2}{2}\left(\frac{1}{m_l^*} + \frac{1}{m_l^-}\right) + \frac{\alpha\hbar^3 eB}{2}\left(n + \frac{1}{2}\right) \times \left(\frac{1}{m_t^- m_l^+} + \frac{1}{m_l^- m_t^+}\right) - \alpha\gamma_3(E, \eta_g)\left(\frac{1}{m_l^+} + \frac{1}{m_l^-}\right)\right]$$

and

$$R_9(E, n, \eta_g) = \gamma_3(E, \eta_g) + \alpha eB\gamma_3(E, \eta_g)\left(n + \frac{1}{2}\right)\hbar\left(\frac{1}{m_t^+} - \frac{1}{m_t^-}\right) - \frac{\alpha\hbar^2}{m_t^- m_t^+}\left[eB\left(n + \frac{1}{2}\right)^2\right]$$

The DR in the present case can be written as

$$\left(\frac{n_z\pi}{d_z}\right)^2 = U_{170}(e_{90}, n, \eta_g) \tag{15.47b}$$

where e_{90} is the total quantized energy in this case.

The DOS function is given by

$$N_{Bn_z} = \frac{g_v eB}{\pi\hbar} \sum_{n=0}^{n_{\max}} \sum_{n_z=1}^{n_{z\max}} \delta'(E - e_{90}) \tag{15.48a}$$

The electron concentration per unit area in this case can be written as

$$N_s = \frac{g_v eB}{\pi\hbar} \sum_{n=0}^{n_{\max}} \sum_{n_z=1}^{n_{z\max}} F_{-1}(\eta_{12,10}) \tag{15.48b}$$

where $\eta_{12,10} = (k_B T)^{-1}[E_{F12\mathrm{HD}} - e_{90}]$.

Thus, by using (15.7), (15.48b) and other appropriate equations we can study the QC in this case.

(d) The model of Bangert and Kastner

In accordance with the model of Bangert and Kastner [8], the carrier energy spectrum in HD IV–VI semiconductors can be written following (11.127) as

$$\frac{k_s^2}{\rho_{11}^2(E,\eta_g)}+\frac{k_y^2}{\rho_{12}^2(E,\eta_g)}=1 \tag{15.49}$$

where

$$\rho_{11}(E,\eta_g)=\frac{1}{\sqrt{S_1(E,\eta_g)}},\quad \rho_{12}(E,\eta_g)=\frac{1}{\sqrt{S_2(E,\eta_g)}}$$

$$S_1(E,\eta_g)=[2\gamma_0(E,\eta_g)]^{-1}\left[\frac{(\bar{R})^2}{E_g}\{c_1(\alpha_1E,E_g)-iD_1(\alpha_1E,E_g)\}\frac{(\bar{S})^2}{\Delta_c'}\right.$$
$$\times\{c_2(\alpha_2E,E_g)-iD_2(\alpha_2E,E_g)\}+\frac{(\bar{Q})^2}{\Delta_c''}\{c_3(\alpha_3E,E_g)$$
$$\left.-iD_3(\alpha_3E,E_g)\}\right]$$

and

$$S_2(E,\eta_g)+[2\gamma_0(E,\eta_g)]^{-1}\left[\frac{(2\bar{A})^2}{E_g}\{c_1(\alpha_1E,E_g)-iD_1(\alpha_1E,E_g)\}\right.$$
$$\left.+\frac{(\bar{S}+\bar{Q})^2}{\Delta_c''}\{c_3(\alpha_3E,E_g)-iD_3(\alpha_3E,E_g)\}\right]$$

Since $S_1(E,\eta_g)$ and $S_2(E,\eta_g)$ are complex, the energy spectrum is also complex in the presence of Gaussian band tails.
Therefore, the magneto dispersion law in this case can be written as

$$k_z^2=U_{18}(E,n,\eta_g) \tag{15.50}$$

where

$$U_{18}(E,n,\eta_g)=[\rho_{12}^2(E,\eta_g)]$$
$$-\left[\frac{2eB}{\hbar}\left(n+\frac{1}{2}\right)[(\rho_{11}^2(E,\eta_g)\rho_{12}^2(E,\eta_g))^{-1}\{\rho_{12}^2(E,\eta_g)\}^{3/2}]\right]$$

Using (15.50), the DR in the present case can be written as

$$\left(\frac{n_z\pi}{d_z}\right)^2=U_{18}(e_{91},n,\eta_g) \tag{15.51}$$

where e_{91} is the total quantized energy in this case.

The DOS function is given by

$$N_{Bn_z} = \frac{g_v eB}{\pi\hbar} \sum_{n=0}^{n_{\max}} \sum_{n_z=1}^{n_{z\max}} \delta'(E - e_{91}) \tag{15.52}$$

The electron concentration per unit area in this case can be written as

$$N_s = \frac{g_v eB}{\pi\hbar} \text{ Real part of } \sum_{n=0}^{n_{\max}} \sum_{n_z=1}^{n_{z\max}} F_{-1}(\eta_{12,11}) \tag{15.53}$$

where $\eta_{12,11} = (k_B T)^{-1}[E_{F12\text{HD}} - e_{91}]$.

Thus, by using (15.7), (15.53) and other appropriate equations we can study the QC in this case.

(e) The model of Foley and Langenberg

Using (1.136), and following the methods as given in Chapter 11, the DR in HDQWTs of IV–VI materials in the present case is given by

$$\left[\left[\gamma_3(E,\eta_g) + \frac{E_{g_0}}{2}\right] - \left[\frac{\hbar^2 k_s^2}{2m_\perp^-} + \frac{\hbar^2 k_z^2}{2m_{||}^-}\right]\right]^2$$
$$= \left[\frac{\hbar^2 k_s^2}{2m_\perp^-} + \frac{\hbar^2 k_z^2}{2m_{||}^-}\right] + \frac{E_{g_0}^2}{4} + E_{g_0}\left[\frac{\hbar^2 k_s^2}{2m_\perp^-} + \frac{\hbar^2 k_z^2}{2m_{||}^-}\right] + P_{||}^2 k_z^2 + P_\perp^2 k_s^2 \tag{15.54}$$

Therefore, the magneto-DR in HDQWTs of IV–VI materials can be written as

$$\gamma_3^2(E,\eta_g) + \frac{E_g^2}{4} + E_g\gamma_3(E,\eta_g) + \left[\frac{\hbar eB}{m_\perp^-}\left(n+\frac{1}{2}\right) + \frac{\hbar^2 x}{2m_{||}^-}\right]^2$$
$$-2\left[\gamma_3(E,\eta_g) + \frac{E_g}{2}\right]\left[\frac{\hbar eB\left(n+\frac{1}{2}\right)}{m_\perp^-} + \frac{\hbar^2 x}{2m_{||}^-}\right]$$
$$= \left[\frac{\hbar eB\left(n+\frac{1}{2}\right)}{m_\perp^+} + \frac{\hbar^2 x}{2m_{||}^+}\right] + \left[\frac{\hbar eB}{m_\perp^+}\left(n+\frac{1}{2}\right) + \frac{\hbar^2 x}{2m_{||}^+}\right]^2$$
$$+ P_{||}^2 x + P_\perp^2 \frac{2eB}{\hbar}\left(n+\frac{1}{2}\right) \tag{15.55}$$

where $k_z^2 = x$.

Therefore, the magneto DR in HDQWTs of IV–VI materials, where unperturbed carriers follow the model of Foley *et al.* can be expressed as

$$k_z^2 = U_{19}(E, n, \eta_g) \tag{15.56}$$

where

$$U_{19}(E, n, \eta_g) = [[2p_{91}]^{-1}[-q_{91}(E, n, \eta_g) + \{q_{91}^2(E, n, \eta_g) + 4p_{91}R_{91}(E, n, \eta_g)\}^{1/2}]], \quad p_{91} = \frac{\hbar^4}{4}\left[\frac{1}{(m_{||}^{2})^2} - \frac{1}{(m_{||}^{-})^2}\right],$$

$$q_{91}(E, n, \eta_g) = \left[\frac{\hbar^2 eB}{m_\perp^- m_{||}^+}\left(n + \frac{1}{2}\right) + P_{||}^2 + \frac{\hbar^2 E_g}{2m_{||}^+} - \frac{\hbar^2 eB\left(n + \frac{1}{2}\right)}{m_\perp^- m_{||}^-} + \frac{\hbar^2}{m_{||}^-}\left(\gamma_3(E, \eta_g) + \frac{E_g}{2}\right)\right]$$

$$R_{91}(E, n, \eta_g) = \left[\gamma_3^2(E, \eta_g) + E_g\gamma_3(E, \eta_g) - \frac{2\hbar eB}{m_\perp^-}\left(\gamma_3(E, \eta_g) + \frac{E_g}{2}\right) \times \left(n + \frac{1}{2}\right) - E_g\frac{\hbar eB}{m_\perp^+}\left(n + \frac{1}{2}\right) - P_\perp^2 \cdot \frac{2eB}{\hbar}\left(n + \frac{1}{2}\right)\right]$$

The DR in the present case can be written as

$$\left(\frac{n_z\pi}{d_z}\right)^2 = U_{19}(e_{911}, n, \eta_g) \tag{15.57}$$

where e_{911} is the total quantized energy in this case.

The DOS function is given by

$$N_{Bn_z} = \frac{g_v eB}{\pi\hbar}\sum_{n=0}^{n_{\max}}\sum_{n_z=1}^{n_{z\max}}\delta'(E - e_{92}) \tag{15.58}$$

The electron concentration per unit area in this case can be written as

$$N_s = \frac{g_v eB}{\pi\hbar}\sum_{n=0}^{n_{\max}}\sum_{n_z=1}^{n_{z\max}}F_{-1}(\eta_{12,12}) \tag{15.59a}$$

where $\eta_{12,12} = (k_BT)^{-1}[E_{F12\text{HD}} - e_{92}]$.

Thus, by using (15.7), (15.59a) and other appropriate equations we can study the QC in this case.

(f) The HD model of McClure and Choi model

Using (1.109), and following the methods as given in Chapter 11, the magneto DR in HDQWTs of IV–VI materials in the present case can be written as

$$\left(n+\frac{1}{2}\right)\hbar\omega_{15} = \left[\gamma_3(E_{54},\eta_g) - \gamma_3(E_{54},\eta_g)\frac{\alpha\hbar^2}{2m_2}\left(\frac{\pi n_y}{d_y}\right)^2\left(1-\frac{m_2}{m_2'}\right) - \frac{\alpha\hbar^4}{4m_2m_2'}\left(\frac{\pi n_y}{d_y}\right)^4 - \frac{\hbar^2}{2m_2}\left(\frac{\pi n_y}{d_y}\right)^2\right] \tag{15.59b}$$

where

$$\omega_{15} = \left(\frac{eB}{\sqrt{m_1m_3}}\right)\left[1-\frac{\alpha\hbar^2}{2m_2}\left(\frac{\pi n_y}{d_y}\right)^2\right]$$

and E_{54} is the total quantized energy in this case.

The DOS function is given by

$$N_{Bn_z} = \frac{g_v eB}{\pi\hbar}\sum_{n=0}^{n_{\max}}\sum_{n_z=1}^{n_{z\max}}\delta'(E-E_{54}) \tag{15.59c}$$

The electron concentration per unit area in this case can be written as

$$N_s = \frac{g_v eB}{\pi\hbar}\sum_{n=0}^{n_{\max}}\sum_{n_z=1}^{n_{z\max}}F_{-1}(\eta_{54}) \tag{15.59d}$$

where $\eta_{54} = (k_BT)^{-1}[E_{F12\text{HD}} - E_{54}]$.

Thus, by using (15.7), (15.59d) and other appropriate equations we can study the QC in this case.

15.2.5 The QC in HDQWTs of stressed Kane type materials under magnetic quantization

The DR of the conduction electrons in heavily doped stressed Kane type semiconductors can be written following (11.116) of Chapter 11 as

$$[k_x/a_{11}(E,\eta_g)]^2 + [k_y/b_{11}(E,\eta_g)]^2 + [k_z/c_{11}(E,\eta_g)]^2 = 1 \tag{15.60a}$$

where

$$a_{11}(E,\eta_g) = \frac{1}{\sqrt{P_{11}(E,\eta_g)}},\quad b_{11}(E,\eta_g) = \frac{1}{\sqrt{Q_{11}(E,\eta_g)}}\quad \text{and}$$

$$c_{11}(E,\eta_g) = \frac{1}{\sqrt{S_{11}(E,\eta_g)}}$$

The electron energy spectrum in heavily doped Kane type semiconductors in the presence of an arbitrarily oriented quantizing magnetic field B along z-axis can be written as

$$(k_z)^2 = U_{41}(E, n, \eta_g) \tag{15.60b}$$

where

$$U_{41}(E, n, \eta_g) = I_2(E, \eta_g)[1 - I_3(E, n, \eta_g)], \quad I_2(E, \eta_g) = [c_{11}(E, \eta_g)]^2$$

and

$$I_3(E, n, \eta_g) = \frac{2eB}{\hbar}\left(n + \frac{1}{2}\right)[a_{11}(E, \eta_g)b_{11}(E, \eta_g)c_{11}(E, \eta_g)]^{-1}[I_2(E, \eta_g)]^{1/2}$$

Following (15.60b), the DR in this case is given by

$$\left(\frac{n_z\pi}{d_z}\right)^2 = U_{41}(e_{93}, n, \eta_g) \tag{15.61}$$

where e_{93} is the total quantized energy in this case.

The DOS function is given by

$$N_{Bn_z} = \frac{g_v eB}{\pi\hbar}\sum_{n=0}^{n_{\max}}\sum_{n_z=1}^{n_{z\max}} \delta'(E - e_{93}) \tag{15.62}$$

The electron concentration per unit area in this case can be written as

$$N_s = \frac{g_v eB}{\pi\hbar}\sum_{n=0}^{n_{\max}}\sum_{n_z=1}^{n_{z\max}} F_{-1}(\eta_{12,13}) \tag{15.63}$$

where $\eta_{12,13} = (k_B T)^{-1}[E_{F12\mathrm{HD}} - e_{93}]$.

Thus, by using (15.7), (15.63) and other appropriate equations we can study the QC in this case.

*15.2.6 The QC in HDQWTs of Te under magnetic quantization

Following (1.67), the magneto DR in this case can be written as

$$E = A_6 k_z^2 + A_7\left(\frac{2eB}{\hbar}\right)\left(n + \frac{1}{2}\right) \pm \left[A_8 k_z^2 + A_9\left(\frac{2eB}{\hbar}\right)\left(n + \frac{1}{2}\right)\right]^{1/2} \tag{15.64}$$

The magneto sized DR can be expressed using (1.67) as

$$\bar{e}_{94} = A_6\left(\frac{n_z\pi}{d_z}\right)^2 + A_7\left(\frac{2eB}{\hbar}\right)\left(n+\frac{1}{2}\right) \pm \left[A_8\left(\frac{n_z\pi}{d_z}\right)^2 + A_9\left(\frac{2eB}{\hbar}\right)\left(n+\frac{1}{2}\right)\right]^{1/2} \quad (15.65)$$

where $\bar{e}_{94}$ is the total quantized energy in this case.

The DOS function is given by

$$N_{Bn_z} = \frac{g_v eB}{2\pi\hbar}\sum_{n=0}^{n_{\max}}\sum_{n_z=1}^{n_{z\max}}\delta'(E-\bar{e}_{94}) \quad (15.66)$$

The electron concentration per unit area in this case can be written as

$$N_s = \frac{g_v eB}{2\pi\hbar}\sum_{n=0}^{n_{\max}}\sum_{n_z=1}^{n_{z\max}}F_{-1}(\eta_{12,14}) \quad (15.67)$$

where $\eta_{12,14} = (k_BT)^{-1}[E_{F12\mathrm{HD}} - \bar{e}_{94}]$.

Thus, by using (15.7), (15.67) and other appropriate equations we can study the QC in this case.

*15.2.7 The QC in HDQWTs of gallium phosphide under magnetic quantization

The magneto dispersion relation of the conduction electrons in HD size quantized n-GaP can be expressed following (11.64) as

$$\gamma_3(\bar{e}_{941},\eta_g) = [[\hbar eB/m_\perp^*](n+2^{-1}) + [\hbar^2k_9^2/2m_\parallel^*] - [\hbar^4k_0^2k_9^2/m_\parallel^{*2}] + |V_G|^2]^{1/2} + |V_G| \quad (15.68)$$

where $\bar{e}_{941}$ is the total quantized energy in this case.

The DOS function is given by

$$N_{Bn_z} = \frac{g_v eB}{\pi\hbar}\sum_{n=0}^{n_{\max}}\sum_{n_z=1}^{n_{z\max}}\delta'(E-\bar{e}_{941}) \quad (15.69)$$

The electron concentration per unit area in this case can be written as

$$N_s = \frac{g_v eB}{\pi\hbar}\sum_{n=0}^{n_{\max}}\sum_{n_z=1}^{n_{z\max}}F_{-1}(\eta_{12,15}) \quad (15.70)$$

where $\eta_{12,15} = (k_BT)^{-1}[E_{F12\mathrm{HD}} - \bar{e}_{941}]$.

Thus, by using (15.7), (15.70) and other appropriate equations we can study the QC in this case.

*15.2.8 The QC in HDQWTs of platinum antimonide under magnetic quantization

The magneto DR in HDQWTs of $PtSb_2$ can be written following (11.87) as

$$T_{11}\left[\left(\frac{2eB}{\hbar}\right)\left(n+\frac{1}{2}\right)\right]^2-\left(\frac{2eB}{\hbar}\right)\left(n+\frac{1}{2}\right)\left[T_{21}(e_{95},\eta_g)-T_{31}\left(\frac{n_z\pi}{d_z}\right)^2\right]$$
$$+\left[T_{41}\left[\left(\frac{n_z\pi}{d_z}\right)^4\right]-T_{51}(e_{95},\eta_g)\left(\frac{n_z\pi}{d_z}\right)^2-T_{61}(e_{95},\eta_g)\right]=0 \quad (15.71)$$

where e_{95} is the total quantized energy in this case.

The DOS function is given by

$$N_{Bn_z}=\frac{g_veB}{\pi\hbar}\sum_{n=0}^{n_{\max}}\sum_{n_z=1}^{n_{z\max}}\delta'(E-e_{95}) \quad (15.72)$$

The electron concentration per unit area in this case can be written as

$$N_s=\frac{g_veB}{\pi\hbar}\sum_{n=0}^{n_{\max}}\sum_{n_z=1}^{n_{z\max}}F_{-1}(\eta_{12,16}) \quad (15.73)$$

where
$$\eta_{12,16}=(k_BT)^{-1}[E_{F12\mathrm{HD}}-e_{95}]. \quad (15.74)$$

Thus, by using (15.7), (15.73) and other appropriate equations we can study the QC in this case.

15.2.9 The QC in HDQWTs of bismuth telluride under magnetic quantization

The magneto DR of the valance bands in bulk p-type HD Bismuth telluride can be written as

$$k_x^2=U_{45}(E,\eta_g,n) \quad (15.75)$$

where

$$U_{45}(E,\eta_g,n)=\frac{\gamma_2(E,\eta_g)-\left(n+\frac{1}{2}\right)\frac{e\hbar B}{M_{31}}}{\bar{\omega}_1} \quad \text{and}$$

$$M_{31}=\frac{m_0}{\left(\bar{\alpha}_{22}\bar{\alpha}_{33}-\frac{(\bar{\alpha}_{23})^2}{4}\right)^{1/2}}$$

Therefore, in the present case the hole energy spectrum can be expressed as

$$\left(\frac{n_z\pi}{d_z}\right)^2 = U_{45}(e_{96}, n, \eta_g) \tag{15.76}$$

where e_{96} is the total quantized energy in this case.

The DOS function is given by

$$N_{Bn_z} = \frac{g_v eB}{\pi\hbar}\sum_{n=0}^{n_{\max}}\sum_{n_z=1}^{n_{z\max}}\delta'(E - e_{96}) \tag{15.77}$$

The electron concentration per unit area in this case can be written as

$$N_s = \frac{g_v eB}{\pi\hbar}\sum_{n=0}^{n_{\max}}\sum_{n_z=1}^{n_{z\max}}F_{-1}(\eta_{12,17}) \tag{15.78}$$

where $\eta_{12,17} = (k_BT)^{-1}[E_{F12\mathrm{HD}} - e_{96}]$.

Thus, by using (15.7), (15.78) and other appropriate equations we can study the QC in this case.

15.2.10 The QC in HDQWTs of germanium under magnetic quantization

(a) The model of Cardona *et al.*

The magneto DR in HDQWT in this case can be expressed from (11.68) as

$$\gamma_3(e_{97}, \eta_g) = -\frac{E_{g_0}}{2} + \frac{\hbar^2}{2m_\parallel^*}\left(\frac{n_z\pi}{d_z}\right)^2 + \left[\frac{E_{g_0}^2}{4} + E_{g_0}\left(\frac{2eB}{\hbar}\right)\left(n+\frac{1}{2}\right)\left(\frac{\hbar^2}{2m_\perp^*}\right)\right]^{1/2} \tag{15.79}$$

where e_{97} is the total quantized energy in this case.

The DOS function is given by

$$N_{Bn_z} = \frac{g_v eB}{\pi\hbar}\sum_{n=0}^{n_{\max}}\sum_{n_z=1}^{n_{z\max}}\delta'(E - e_{97}) \tag{15.80}$$

The electron concentration per unit area in this case can be written as

$$N_s = \frac{g_v eB}{\pi\hbar}\sum_{n=0}^{n_{\max}}\sum_{n_z=1}^{n_{z\max}}F_{-1}(\eta_{12,18}) \tag{15.81}$$

where $$\eta_{12,18} = (k_B T)^{-1}[E_{F12\text{HD}} - e_{97}]. \tag{15.82}$$

Thus, by using (15.7), (15.81) and other appropriate equations we can study the QC in this case.

(b) The model of Wang and Ressler

The magneto DR in HD Ge can be written following (11.73) as

$$\begin{aligned}\gamma_3(e_{98}, \eta_g) &= \frac{\hbar^2}{2m_{\|}^*}\left(\frac{n_z \pi}{d_z}\right)^2 + \frac{\hbar^2}{2m_{\perp}^*}\left[\left(\frac{2eB}{\hbar}\right)\left(n+\frac{1}{2}\right)\right] \\ &- c_1\left(\frac{\hbar^2}{2m_{\perp}^*}\right)^2\left[\left(\frac{2eB}{\hbar}\right)\left(n+\frac{1}{2}\right)\right] \\ &- d_1\left(\frac{\hbar^2}{2m_{\perp}^*}\left(\frac{2eB}{\hbar}\right)\left(n+\frac{1}{2}\right)\right)\left(\frac{\hbar^2}{2m_{\|}^*}\left[\left(\frac{n_z \pi}{d_z}\right)^2\right]\right) \\ &- e_1\left(\frac{\hbar^2}{2m_{\|}^*}\left[\left(\frac{n_z \pi}{d_z}\right)^2\right]\right)^2 \end{aligned} \tag{15.83}$$

where e_{98} is the total quantized energy in this case.

The DOS function is given by

$$N_{Bn_z} = \frac{g_v eB}{\pi\hbar} \sum_{n=0}^{n_{\max}} \sum_{n_z=1}^{n_{z\max}} \delta'(E - e_{98}) \tag{15.84}$$

The electron concentration per unit area in this case can be written as

$$N_s = \frac{g_v eB}{\pi\hbar} \sum_{n=0}^{n_{\max}} \sum_{n_z=1}^{n_{z\max}} F_{-1}(\eta_{12,19}) \tag{15.85}$$

where $$\eta_{12,19} = (k_B T)^{-1}[E_{F12\text{HD}} - e_{98}]. \tag{15.86}$$

Thus, by using (15.7), (15.85) and other appropriate equations we can study the QC in this case.

15.2.11 The QC in HDQWTs of gallium antimonide under magnetic quantization

(a) The HD model of Mathur *et al.*

The heavily doped magneto DR in the present case can be written from (1.96) as

$$\gamma_3(e_{99},\eta_g)=\alpha_9\left[\left(\frac{n_z\pi}{d_z}\right)^2+\left(\frac{2eB}{\hbar}\right)\left(n+\frac{1}{2}\right)\right]$$
$$+\frac{E_{g1}}{2}\left[\sqrt{1+\alpha_{10}\left[\left(\frac{n_z\pi}{d_z}\right)^2+\left(\frac{2eB}{\hbar}\right)\left(n+\frac{1}{2}\right)\right]}-1\right] \tag{15.87}$$

where e_{99} is the total quantized energy in this case.

The DOS function is given by

$$N_{Bn_z}=\frac{g_veB}{\pi\hbar}\sum_{n=0}^{n_{\max}}\sum_{n_z=1}^{n_{z\max}}\delta'(E-e_{99}) \tag{15.88}$$

The electron concentration per unit area in this case can be written as

$$N_s=\frac{g_veB}{\pi\hbar}\sum_{n=0}^{n_{\max}}\sum_{n_z=1}^{n_{z\max}}F_{-1}(\eta_{12,20}) \tag{15.89}$$

where
$$\eta_{12,20}=(k_BT)^{-1}[E_{\mathrm{F12HD}}-e_{99}]. \tag{15.90}$$

Thus, by using (15.7), (15.89) and other appropriate equations we can study the QC in this case.

(b) The HD model of Seiler

The heavily doped magneto DR in the present case can be written extending (5.100) as

$$k_9^2=\phi_{67}(E_{67},\eta_g)$$

where

$$\phi_{67}(E_{67},\eta_g)$$
$$=(2S_{1,\pm}^2)^{-1}[S_{1,\pm}(1+2\alpha\gamma_3(E_{67},\eta_g))-\alpha_4-[\alpha_4^2-4S_{1,\pm}\alpha_4$$

$$\times (1 + 2\alpha\gamma_3(E_{67}, \eta_g)) + 4S_{1,\pm}^2(1 - 20\alpha\gamma_3(E_{67}, \eta_g)$$
$$- 28\alpha^2(\gamma_3(E_{67}, \eta_g))^2)]^{1/2}]$$

and E_{67} is the total quantized energy in this case.

The DOS function is given by

$$N_{Bn_z} = \frac{g_v eB}{2\pi\hbar} \sum_{n=0}^{n_{\max}} \sum_{n_z=1}^{n_{z\max}} \delta'(E - E_{67}) \tag{15.91}$$

The electron concentration per unit area in this case can be written as

$$N_s = \frac{g_v eB}{\pi\hbar} \sum_{n=0}^{n_{\max}} \sum_{n_z=1}^{n_{z\max}} F_{-1}(\eta_{67}) \tag{15.92}$$

where
$$\eta_{67} = (k_B T)^{-1}[E_{F12\text{HD}} - E_{67}]. \tag{15.93}$$

Thus, by using (15.7), (15.92) and other appropriate equations we can study the QC in this case.

15.2.12 The QC in HDQWTs of II–V materials under magnetic quantization

The hole energy spectrum in HDQWTs of II–V materials can be written following (1.88) as

$$y_3(E, n_g) = \theta_1 k_x^2 + \theta_2 k_y^2 + \theta_3 k_z^2 + \delta_4 k_x \pm [\{\theta_5 k_x^2 + \theta_6 k_y^2 + \theta_7 k_z^2 + \delta_5 k_x\}^2$$
$$+ G_3^2 k_y^2 + \Delta_3^2]^{\frac{1}{2}} \pm \Delta_3 \tag{15.94}$$

where

$$\theta_1 = A_{10}, \quad \theta_2 = A_{11}, \quad \theta_3 = A_{12}, \quad \delta_4 = A_{13}, \quad \theta_5 = A_{14}, \quad \theta_6 = A_{15},$$
$$\theta_7 = A_{16}, \quad \delta_5 = A_{17}, \quad G_3^2 = A_{18} \quad \text{and} \quad \Delta_3 = A_{19}$$

The magneto dispersion law in HDQWTs of II–V materials assumes the form

$$k_y^2 = U_{49,\pm}(E, \eta_g, n) \tag{15.95a}$$

where

$$U_{49,\pm}(E,\eta_g,n) = [I_{35}\gamma_3(E,\eta_g) + I_{36,\pm}(n) \pm [\gamma_3^2(E,\eta_g) + \gamma_3(E,\eta_g)I_{38,\pm}(n) + I_{39,\pm}(n)]^{1/2}]$$

$$I_{35} = \frac{\theta_2}{(\theta_2^2-\theta_5^2)}, \quad I_{36,\pm}(n) = \frac{I_{33,\pm}(n)}{2(\theta_2^2-\theta_5^2)},$$

$$I_{38,\pm}(n) = (4\theta_5^2)^{-1}[4\theta_2 I_{33,\pm}(n) + 8\theta_2^2 I_{31,\pm}(n) - \theta_5^2 I_{31,\pm}(n)]$$

$$I_{39,\pm} = (4\theta_5^2)^{-1}[I_{33,\pm}^2(n) + 4\theta_2^3 I_{34,\pm}(n) - 4\theta_5^3 I_{34,\pm}(n)],$$

$$I_{33,\pm}(n) = [G_3^2 + 2\theta_5 I_{32}(n) - 2\theta_2 I_{31,\pm}(n)]$$

$$I_{34,\pm}(n) = [I_{32,\pm}^2(n) + \Delta_3^2 - I_{31,\pm}(n)]$$

$$I_{31,\pm}(n) = \left[\left(n+\frac{1}{2}\right)\hbar\omega_{31} - \frac{\delta_4^2}{4\theta_1} \pm \Delta_3\right],$$

$$I_{32}(n) = \left[\left(n+\frac{1}{2}\right)\hbar\omega_{32} - \frac{\delta_5^2}{4\theta_5}\right]$$

$$\omega_{31} = \frac{eB}{\sqrt{M_{31}M_{32}}}, \quad \omega_{32} = \frac{eB}{\sqrt{M_{33}M_{34}}}, \quad M_{31} = \frac{\hbar^2}{2\theta_1}, \quad M_{32} = \frac{\hbar^2}{2\theta_3},$$

$$M_{33} = \frac{\hbar^2}{2\theta_5} \quad \text{and} \quad M_{34} = \frac{\hbar^2}{2\theta_7}$$

The DR in the present case can be written as

$$\left(\frac{n_y\pi}{d_y}\right)^2 = U_{49,\pm}(e_{100}, n, \eta_g) \tag{15.95b}$$

where e_{100} is the total quantized energy in this case.

The DOS function is given by

$$N_{Bn_z} = \frac{g_v eB}{2\pi\hbar}\sum_{n=0}^{n_{\max}}\sum_{n_z=1}^{n_{z\max}}\delta'(E-e_{100}) \tag{15.95c}$$

The electron concentration per unit area in this case can be written as

$$N_s = \frac{g_v eB}{2\pi\hbar}\sum_{n=0}^{n_{\max}}\sum_{n_z=1}^{n_{y\max}}F_{-1}(\eta_{12,21}) \tag{15.96}$$

where $\eta_{12,21} = (k_B T)^{-1}[E_{F12\text{HD}} - e_{100}]$.

Thus, by using (15.7), (15.96) and other appropriate equations we can study the QC in this case.

15.2.13 The QC in HDQWTs of lead germanium telluride under magnetic quantization

The magneto dispersion law in HDQWTs of $Pb_{1-x}Ge_xTe$ can be expressed as

$$\left[\frac{2\theta_0(E,\eta_g)}{1+Erf(E/\eta_g)}\right]+\gamma_3(E,\eta_g)\left[\overline{E_{g0}}-0.345x-0.390\frac{eB}{\hbar}\left(n+\frac{1}{2}\right)\right]$$
$$=\frac{0.46eB}{\hbar}\left(n+\frac{1}{2}\right)+0.02x\pm\left[0.06\overline{E_{g0}}+0.122\frac{eB}{\hbar}\left(n+\frac{1}{2}\right)+0.0066x\right]$$
$$\times\left(\frac{2eB}{\hbar}\left(n+\frac{1}{2}\right)\right)^{1/2}+\left[\overline{E_{g0}}+\frac{0.822eB}{\hbar}\left(n+\frac{1}{2}\right)+0.377x\right]$$
$$\times\left[\frac{1.212eB}{\hbar}\left(n+\frac{1}{2}\right)+0.722x\right] \tag{15.97}$$

Equation (15.97) assumes the form

$$k_z^2=U_{50,\pm}(E,\eta_g,n) \tag{15.98}$$

where

$$U_{50,\mp}(E,\eta_g,n)$$
$$=(2p_{10})^{-1}[q_{10}(E,\eta_g,n)-[q_{10}^2(E,\eta_g,n)+4p_{10}R_{10,\mp}(E,\eta_g,n)]^{1/2}]$$
$$p_{10}(0.377\times 0.722),q_{10}(E,\eta_g,n)$$
$$=\left[0.02+0.345\gamma_3(E,\eta_g)\pm 0.0066\left(\frac{2eB}{\hbar}\left(n+\frac{1}{2}\right)\right)^{1/2}\right.$$
$$\left.+0.377\times\frac{1.212eB}{\hbar}\left(n+\frac{1}{2}\right)+0.722\left[\bar{E}_{g0}+0.822\frac{eB}{\hbar}\left(n+\frac{1}{2}\right)\right]\right]$$

and

$$R_{10,\mp}(E,\eta_g,n)$$
$$=\left[\frac{2\theta_0(E,\eta_g)}{1+Eof(E/\eta_g)}+\gamma_3(E,\eta_g)\left[\bar{E}_{g0}-0.390\frac{eB}{\hbar}\left(n+\frac{1}{2}\right)\right]\right]$$

$$\mp\left(0.06\bar{E}_{g_0}+0.122\frac{eB}{\hbar}\left(n+\frac{1}{2}\right)\right)\left(\frac{2eB}{\hbar}\left(n+\frac{1}{2}\right)\right)^{1/2}$$
$$-\left(\bar{E}_{g_0}+0.822\frac{eB}{\hbar}\left(n+\frac{1}{2}\right)\right)\frac{1.212eB}{\hbar}\left(n+\frac{1}{2}\right)-\frac{0.46eB}{\hbar}\left(n+\frac{1}{2}\right)\Bigg]$$

The DR in the present case can be written as

$$\left(\frac{n_z\pi}{d_z}\right)^2=U_{50\pm}(e_{101},n,\eta_g) \tag{15.99}$$

where e_{101} is the total quantized energy in this case.

The DOS function is given by

$$N_{Bn_z}=\frac{g_v eB}{2\pi\hbar}\sum_{n=0}^{n_{\max}}\sum_{n_z=1}^{n_{z\max}}\delta'(E-e_{101}) \tag{15.100}$$

The electron concentration per unit area in this case can be written as

$$N_s=\frac{g_v eB}{2\pi\hbar}\sum_{n=0}^{n_{\max}}\sum_{n_z=1}^{n_{z\max}}F_{-1}(\eta_{12,22}) \tag{15.101a}$$

where $\eta_{12,22}=(k_BT)^{-1}[E_{\mathrm{F12HD}}-e_{101}]$.

Thus, by using (15.7), (15.101a) and other appropriate equations we can study the QC in this case.

15.2.14 The QC in HDQWTs of zinc and cadmium Diphosphides under magnetic quantization

From (1.44), the HDDR in this case assumes the form

$$\gamma_3(E_{15,15},\eta_g)$$
$$=\left[\beta_1+\frac{\beta_2\theta_{15}}{8\beta_4}\right]\left[(\pi n_z/d_z)^2+\frac{2eB}{\hbar}\left(n+\frac{1}{2}\right)\right]$$
$$\pm\Bigg\{\Bigg[\beta_4\theta_{15}\left(\beta_5-\frac{\beta_2\theta_{15}}{8\beta_4}\right)\left[(\pi n_z/d_z)^2+\frac{2eB}{\hbar}\left(n+\frac{1}{2}\right)\right]$$
$$+8\beta_4^2\left(1-\frac{\theta_{15}^2}{4}\right)-\beta_2\left(1-\frac{\theta_{15}^2}{4}\right)\left\{(\pi n_z/d_z)^2+\frac{2eB}{\hbar}\left(n+\frac{1}{2}\right)\right\}^{1/2}\Bigg]\Bigg\} \tag{15.101b}$$

where

$$\theta_{15} = \frac{(2eB/\hbar)\left(n+\frac{1}{2}\right) - 2(\pi n_z/d_z)^2}{\left[(\pi n_z/d_z)^2 + (2eB/\hbar)\left(n+\frac{1}{2}\right)\right]}$$

and $E_{15,15}$ is the total quantized energy in this case.

The DOS function is given by

$$N_{Bn_z} = \frac{g_v eB}{2\pi\hbar} \sum_{n=0}^{n_{\max}} \delta'(E - E_{15,15}) \tag{15.101c}$$

The electron concentration per unit area in this case can be written as

$$N_s = \frac{g_v eB}{2\pi\hbar} \sum_{n=0}^{n_{\max}} \sum_{n_z=1}^{n_{z\max}} F_{-1}(\eta_{103}) \tag{15.101d}$$

where $\eta_{103} = (k_B T)^{-1}[E_{F12\mathrm{HD}} - E_{15,15}]$.

Thus, by using (15.7), (15.101d) and other appropriate equations we can study the QC in this case.

15.2.15 The QC in magneto 2D QMOSFETs of nonlinear optical and tetragonal materials operated under accumulation mode

The DR under magnetic quantization in nonlinear optical materials can be written from (11.26) as

$$\frac{\hbar^2 k_z^2}{2m_\parallel^*} = U_1(E, n, \eta_g) \tag{15.102}$$

where

$$U_1(E, n, \eta_g) = \left[[T_{21}(E, \eta_g)]\left[1 - \left[\hbar eB\left(n + \frac{1}{2}\right)/m_\perp^* T_{22}(E, \eta_g)\right]\right]\right]$$

The electron concentration can be expressed as

$$n_0 = \text{Real part of } \bar{n}_0 \tag{15.103}$$

where

$$\bar{n}_0 = \frac{eB}{\pi^2\hbar^2}\sqrt{2m_\parallel^*} \sum_{n=0}^{n_{\max}} \left[[\theta_1(E_{\mathrm{FBHD}}, n, \eta_g)] + \sum_{r=1}^{r_s} L(r)[\theta_1(E_{\mathrm{FBHD}}, n, \eta_g)]\right] \tag{15.104}$$

in which

$$\theta_1(E_{\mathrm{FBHD}}, n, \eta_g) = [U_1(E_{\mathrm{FBHD}}, n, \eta_g)]^{1/2},$$

$$L(r) \equiv [2(k_B T)^{2r}(1-2^{1-2r})\xi(2r)]\left[\frac{\partial^{2r}}{\partial E_F^{2r}}\right]$$

$\xi(2r)$ is the zeta function of order $(2r)$ and in general E_F is Fermi energy which in this case is $E_F = E_{\mathrm{FBHD}}$ In the presence of a surface electric field F_s along z-direction and perpendicular to the surface, (11.26) assumes the form

$$\frac{\hbar^2 k_z^2}{2m_{\|}^*} + \frac{\hbar^2 k_s^2}{2m_{\perp}^*}\frac{T_{21}(E-|e|F_s z, \eta_g)}{T_{22}(E-|e|F_s z, \eta_g)} = T_{21}(E-|e|F_s z, \eta_g) \quad (15.105)$$

where, for this chapter, E represents the electron energy as measured from the edge of the conduction band at the surface in the vertically upward direction.

The use of (5.5) and (15.105) leads to the DR of the 2D electrons in QMOSFETs of tetragonal and nonlinear optical materials operated under accumulation mode can be written under weak electric field limit as

$$\frac{\hbar^2 k_s^2}{2m_{\|}^*} = L_6(E, i, \eta_g) \quad (15.106)$$

where

$$L_6(E, i, \eta_g) = \frac{T_{21}(E, \eta_g) - L_3(E, i, \eta_g)}{L_4(E, i, \eta_g)},$$

$$L_3(E, i, \eta_g) = S_i[T_{21}'(E, \eta_g)]^{2/3}\left[\frac{\hbar|e|F_s}{\sqrt{2m_{\|}^*}}\right]^{2/3}$$

and

$$L_4(E, i, \eta_g) = \left[\frac{T_{21}(E, \eta_g)}{T_{22}(E, \eta_g)} + L_3(E, i, \eta_g)\frac{T_{21}(E, \eta_g)}{T_{21}'(E, \eta_g)T_{22}(E, \eta_g)} \times \frac{2}{3}\left\{\frac{T_{21}'(E, \eta_g)}{T_{21}(E, \eta_g)} - \frac{T_{22}'(E, \eta_g)}{T_{22}(E, \eta_g)}\right\}\right]$$

The DR of the 2D electrons in QMOSFETs of HD tetragonal and nonlinear optical materials operated in accumulation mode under the condition of weak electric field limit under magnetic quantization can be written from (15.106) as

$$\frac{heB\left(n+\frac{1}{2}\right)}{m_{\|}^{*}} = L_6(E_{14,1}, i, \eta_g) \tag{15.107}$$

where $E_{14,1}$ is the total quantized energy in this case.

The DOS function is given by

$$N_{Bi} = \frac{g_v eB}{\pi\hbar} \sum_{n=0}^{n_{\max}} \sum_{i=0}^{n_{i\max}} \delta'(E - E_{14,1}) \tag{15.108}$$

The electron concentration per unit area can be written as

$$n_s = \text{Real Part of } \bar{n}_s \tag{15.109}$$

where

$$\bar{n}_s = \frac{g_v eB}{\pi\hbar} \sum_{n=0}^{n_{\max}} \sum_{i=0}^{n_{i\max}} F_{-1}(\eta_{14,1}) \tag{15.110}$$

in which $\eta_{14,1} = (E_{\text{FHD2D14}} - E_{14,1})(k_B T)^{-1}$ and E_{FHD2D14} is the Fermi energy in this case and is given by $E_{\text{FHD2D14}} \equiv eV_g - \frac{e^2 N_s}{C_{\text{OX}}}$.

The total surface concentration in QMOSFETs of tetragonal and nonlinear optical materials in this case can be written as

$$N_s = \text{Real Part of } g_v \left[\sum_{n=0}^{n_{\max}} \sum_{i=0}^{n_{i\max}} \left[\frac{eB}{\pi\hbar}[F_{-1}(\eta_{14,1})] + \frac{eBt_{i1\text{B}}}{\pi^2\hbar^2}\sqrt{2m_{\|}^{*}} \times \left[[\theta_1(E_{\text{FBHD}}, n, \eta_g)] + \sum_{r=1}^{r_s} L(r)[\theta_1(E_{\text{FBHD}}, n, \eta_g)]\right]\right]\right] \tag{15.111}$$

$t_{i1\text{B}} = \frac{E_{14,1\max}}{eF_s(1+i_{\max})}$, $E_{14,1\max}$, is the maximum value of the real part of the equation

$$\frac{heB\left(n+\frac{1}{2}\right)}{m_{\|}^{*}} = L_6(E_{14,1\max}, i_{\max}, \eta_g) \tag{15.112}$$

Thus, combining the appropriate equations with (5.4), we can study the QC in this case.

15.2.16 The QC in magneto 2D QMOSFETs of III–V, ternary and quaternary materials operated under accumulation mode

(a) The three band model of Kane

The electron energy spectrum in HD III–V materials under magnetic quantization can be expressed using (11.31) as

$$\left(n+\frac{1}{2}\right)\hbar\omega_0+\frac{\hbar^2k_z^2}{2m_c}=T_{31}(E,\eta_g)+iT_{32}(E,\eta_g) \tag{15.113}$$

where

$$\omega_0=\frac{eB}{m_c}$$

Therefore, the DR is given by

$$\frac{\hbar^2k_z^2}{2m_c}=U_3(E,n,\eta_g) \tag{15.114}$$

where

$$U_3(E,n,\eta_g)=\left[T_{90}(E,\eta_g)-\left(n+\frac{1}{2}\right)\hbar\omega_0\right] \quad \text{and}$$

$$T_{90}(E,\eta_g)=[T_{31}(E,\eta_g)+iT_{32}(E,\eta_g)]$$

The electron concentration can be expressed as

$$n_0=\text{Real part of }\bar{n}_0 \tag{15.115}$$

where

$$\bar{n}_0=\frac{eBg_v}{\pi^2\hbar^2}\sqrt{2m_c}\left[\sum_{n=0}^{n_{\max}}\left[[\theta_3(E_{\text{FBHD}},n,\eta_g)]+\sum_{r=1}^{r_s}L(r)[\theta_3(E_{\text{FBHD}},n,\eta_g)]\right]\right] \tag{15.116}$$

in which $\theta_3(E_{\text{FBHD}},n,\eta_g)=[U_3(E_{\text{FBHD}},n,\eta_g)]^{1/2}$.

The DR of the 2D electrons in QMOSFETs of HD III–V, ternary and quaternary materials operated in accumulation mode under the condition of weak electric field limit under magnetic quantization can be written using (5.5) and (15.113) as

$$T_{90}(E_{14,3},\eta_g)=\left(n+\frac{1}{2}\right)\hbar\omega_0+S_i\left[\frac{\hbar|e|F_s[T_{90}(E_{14,3},\eta_g)]'}{\sqrt{2m_c}}\right]^{2/3} \tag{15.117}$$

where $E_{14,3}$ is the total quantized energy in this case.

The DOS function is given by

$$N_{Bi} = \frac{g_v eB}{\pi\hbar} \sum_{n=0}^{n_{\max}} \sum_{i=0}^{n_{i\max}} \delta'(E - E_{14,3}) \tag{15.118}$$

Equation (15.117) represents the DR of the 2D electrons in accumulation layers of HD QMOSFETs of III–V, ternary and quaternary materials under the weak electric field limit under magnetic quantization whose bulk electrons obey the HD three band model of Kane. Since the electron energy spectrum in accordance with the HD three-band model of Kane is complex in nature, (15.117) will also be complex. The both complexities occur due to the presence of poles in the finite complex plane of the dispersion relation of the materials in the absence of band tails.

The electron concentration per unit area can be written as

$$n_s = \text{Real Part of } \bar{n}_s \tag{15.119}$$

where

$$\bar{n}_s = \frac{g_v eB}{\pi\hbar} \sum_{n=0}^{n_{\max}} \sum_{i=0}^{n_{i\max}} F_{-1}(\eta_{14,3}) \tag{15.120}$$

in which $\eta_{14,3} = (E_{\mathrm{FHD2D14}} - E_{14,3})(k_B T)^{-1}$.

The total surface concentration in QMOSFETs in this case can be written as

$$N_s = g_v \text{ Real Part of } \left[\sum_{n=0}^{n_{\max}} \sum_{i=0}^{n_{i\max}} \left[\frac{eB}{\pi\hbar}[F_{-1}(\eta_{14,3})] + \frac{eBt_{i3\mathrm{B}}}{\pi^2\hbar^2}\sqrt{2m_c}\left[\left[\theta_3(E_{\mathrm{FBHD}}, n, \eta_g)\right] + \sum_{r=1}^{r_s} L(r)[\theta_3(E_{\mathrm{FBHD}}, n, \eta_g)\right]\right]\right] \tag{15.121}$$

$t_{i3\mathrm{B}} = \frac{E_{14,3\max}}{eF_s(1+i_{\max})}$, $E_{14,3\max}$, is the maximum value of the real part of the equation

$$T_{90}(E_{14,3\max}, \eta_g) = \left(n + \frac{1}{2}\right)\hbar\omega_0 + S_{i\max}\left[\frac{\hbar|e|F_s[T_{90}(E_{14,3\max}, \eta_g)]'}{\sqrt{2m_c}}\right]^{2/3} \tag{15.122}$$

Thus, combining the appropriate equations with (5.4), we can study the QC in this case.

(b) The two band model of Kane

The magneto-dispersion law in this case can be expressed using (11.36) as

$$\frac{\hbar^2 k_z^2}{2m_c} = \gamma_2(E,\eta_g) - \left(n+\frac{1}{2}\right)\hbar\omega_0 = U_5(E,n,\eta_g) \tag{15.123}$$

The electron concentration can be expressed as

$$n_0 = \frac{eB}{\pi^2\hbar^2}\sqrt{2m_c}\left[\sum_{n=0}^{n_{\max}}[\theta_5(E_{\mathrm{FHD2D14}},n,\eta_g)] + \sum_{r=1}^{r_s} L(r)[\theta_5(E_{\mathrm{FBHD}},n,\eta_g)]\right] \tag{15.124}$$

where $\theta_5(E_{\mathrm{FHD2D14}},n,\eta_g) = [U_5(E_{\mathrm{FHD2D14}},n,\eta_g)]^{1/2}$.

Using (5.5) and (15.123), the DR of the 2D electrons in QMOSFETs of HD nonlinear optical materials whose energy band structures are defined by two band model of Kane operated in accumulation mode under the condition of weak electric field limit under magnetic quantization can be written as

$$\gamma_2(E_{14,5},\eta_g) = \left(n+\frac{1}{2}\right)\hbar\omega_0 + S_i\left[\frac{\hbar|e|F_s[\gamma_2(E_{14,5},\eta_g)]'}{\sqrt{2m_c}}\right]^{2/3} \tag{15.125}$$

where $E_{14,5}$ is the total quantized energy in this case.

The DOS function is given by

$$N_{Bi} = \frac{g_v eB}{\pi\hbar}\sum_{n=0}^{n_{\max}}\sum_{i=0}^{n_{i\max}}\delta'(E - E_{14,5}) \tag{15.126}$$

The electron concentration per unit area can be written as

$$n_s = \bar{n}_s \tag{15.127}$$

where

$$\bar{n}_s = \frac{g_v eB}{\pi\hbar}\sum_{n=0}^{n_{\max}}\sum_{i=0}^{n_{i\max}} F_{-1}(\eta_{14,5}) \tag{15.128}$$

in which $\eta_{14,5} = (E_{\mathrm{FHD2D14}} - E_{14,5})(k_BT)^{-1}$.

The total surface concentration in QMOSFETs operated under accumulation layer mode of nonlinear optical material can be written as

$$N_s = g_v \left[\sum_{n=0}^{n_{\max}} \sum_{i=0}^{n_{i\max}} \left[\frac{eB}{\pi\hbar}[F_{-1}(\eta_{14,5})] + \frac{eBt_{i5B}}{\pi^2\hbar^2}\sqrt{2m_c}\left[[\theta_5(E_{\mathrm{FBHD}}, n, \eta_g)] \right.\right.\right.$$
$$\left.\left.\left. + \sum_{r=1}^{r_s} L(r)[\theta_5(E_{\mathrm{FBHD}}, n, \eta_g)] \right]\right]\right] \tag{15.129}$$

$t_{i5\mathrm{B}} = \frac{E_{14,5\,\max}}{eF_s(1+i_{\max})}$, $E_{14,5\,\max}$, is the maximum value of the real part of the equation

$$\gamma_2(E_{14,5\,\max}, \eta_g) = \left(n + \frac{1}{2}\right)\hbar\omega_0 + S_i\left[\frac{\hbar|e|F_s[\gamma_2(E_{14,5\,\max}, \eta_g)]'}{\sqrt{2m_c}}\right]^{2/3} \tag{15.130}$$

Thus, combining the appropriate equations with (1.4), we can study the QC in this case.

(c) The Parabolic energy bands

The magneto-dispersion law in this case can be expressed using (11.40) as

$$\frac{\hbar^2 k_z^2}{2m_c} = \gamma_3(E, \eta_g) - \left(n + \frac{1}{2}\right)\hbar\omega_0 = U_6(E, n, \eta_g) \tag{15.131}$$

The electron concentration can be expressed as

$$n_0 = \frac{eB}{\pi^2\hbar^2}\sqrt{2m_c}\left[\sum_{n=0}^{n_{\max}} [\theta_7(E_{\mathrm{FHD2D14}}, n, \eta_g)] + \sum_{r=1}^{r_s} L(r)[\theta_7(E_{\mathrm{FBHD}}, n, \eta_g)]\right] \tag{15.132}$$

where $\theta_7(E_{\mathrm{FHD2D14}}, n, \eta_g) = [U_6(E_{\mathrm{FHD2D14}}, n, \eta_g)]^{1/2}$.

Using (5.5) and (15.131), the DR of the 2D electrons in QMOSFETs of HD nonlinear optical materials whose energy band structures are defined by parabolic energy bands operated in accumulation mode under the condition of weak electric field limit under magnetic quantization can be written as

$$\gamma_3(E_{14,7}, \eta_g) = \left(n + \frac{1}{2}\right)\hbar\omega_0 + S_i\left[\frac{\hbar|e|F_s[\gamma_3(E_{14,7}, \eta_g)]'}{\sqrt{2m_c}}\right]^{2/3} \tag{15.133}$$

where $E_{14,7}$ is the total quantized energy in this case.

The DOS function is given by

$$N_{Bi} = \frac{g_v eB}{\pi\hbar} \sum_{n=0}^{n_{\max}} \sum_{i=0}^{n_{i_{\max}}} \delta'(E - E_{14,7}) \tag{15.134}$$

The electron concentration per unit area can be written as

$$n_s = \bar{n}_s \tag{15.135}$$

where

$$\bar{n}_s = \frac{g_v eB}{\pi\hbar} \sum_{n=0}^{n_{\max}} \sum_{i=0}^{n_{i_{\max}}} F_{-1}(\eta_{14,7}) \tag{15.136}$$

in which $\eta_{14,7} = (E_{\mathrm{FHD2D14}} - E_{14,7})(k_B T)^{-1}$.

The total surface concentration in QMOSFETs of in this case can be written as

$$N_s = g_v \left[\sum_{n=0}^{n_{\max}} \sum_{i=0}^{n_{i_{\max}}} \left[\frac{eB}{\pi\hbar}[F_{-1}(\eta_{14,7})] + \frac{eBt_{i7\mathrm{B}}}{\pi^2\hbar^2}\sqrt{2m_c} \left[[\theta_7(E_{\mathrm{FHD2D14}}, n, \eta_g)] \right.\right.\right.$$
$$\left.\left.\left. + \sum_{r=1}^{r_s} L(r)[\theta_7(E_{\mathrm{FBHD}}, n, \eta_g)] \right]\right]\right] \tag{15.137}$$

$t_{i7\mathrm{B}} = \frac{E_{14,7\,\max}}{eF_s(1+i_{\max})}$, $E_{14,7\,\max}$, is the maximum value of the equation

$$\gamma_3(E_{14,7\,\max}, \eta_g) = \left(n + \frac{1}{2}\right)\hbar\omega_0 + S_{i_{\max}} \left[\frac{\hbar|e|F_s[\gamma_3(E_{14,7\,\max}, \eta_g)]'}{\sqrt{2m_c}} \right]^{2/3} \tag{15.138}$$

Thus, combining the appropriate equations with (5.4), we can study the QC in this case.

15.2.17 The QC in magneto 2D QMOSFETs of II–VI materials operated under accumulation mode

The electron concentration can be expressed as

$$n_0 = \frac{eB}{2\pi^2\hbar} \left[\sum_{n=0}^{n_{\max}} [\theta_{13,\pm}(E_{\mathrm{FBHD}}, n, \eta_g)] + \sum_{r=1}^{r_s} L(r)[\theta_{13,\pm}(E_{\mathrm{FBHD}}, n, \eta_g)] \right] \tag{15.139}$$

where $\theta_{13,\pm}(E_{\mathrm{FBHD}}, n, \eta_g) = [U_{8,\pm}(E_{\mathrm{FBHD}}, n, \eta_g)]^{1/2}$.

Using (5.5) and (15.139), the DR of the 2D electrons in QMOSFETs of HD QMOSFETs of II–VI materials whose unperturbed energy band structures are defined by Hopfield model operated in accumulation mode under the condition of weak electric field limit under magnetic quantization can be written as

$$\gamma_3(E_{14,9},\eta_g) = a_0' \left[\frac{2eB}{\hbar}\left(n+\frac{1}{2}\right)\right] \pm (\bar{\lambda}_0)\left[\frac{2eB}{\hbar}\left(n+\frac{1}{2}\right)\right]^{1/2} + S_i\left[\frac{\hbar|e|F_s\gamma_3'(E_{14,9},\eta_g)}{\sqrt{2m_{\parallel}^*}}\right]^{2/3} \tag{15.140}$$

where $E_{14,9}$ is the total quantized energy in this case.

The DOS function is given by

$$N_{Bi} = \frac{g_v eB}{2\pi\hbar}\sum_{n=0}^{n_{\max}}\sum_{i=0}^{n_{i_{\max}}}\delta'(E-E_{14,9}) \tag{15.141}$$

The electron concentration per unit area can be written as

$$n_s = \bar{n}_s \tag{15.142}$$

where

$$\bar{n}_s = \frac{g_v eB}{2\pi\hbar}\sum_{n=0}^{n_{\max}}\sum_{i=0}^{n_{i_{\max}}}F_{-1}(\eta_{14,9}) \tag{15.143}$$

in which $\eta_{14,9} = (E_{\text{FBHD}} - E_{14,9})(k_BT)^{-1}$.

The total surface concentration in QMOSFETs in this case can be written as

$$N_s = g_v\left[\sum_{n=0}^{n_{\max}}\sum_{i=0}^{n_{i_{\max}}}\left[\frac{eB}{2\pi\hbar}[F_{-1}(\eta_{14,9})] + \frac{eBt_{i9\text{B}}}{2\pi^2\hbar^2}\sqrt{2m_c}\left[[\theta_{13,\pm}(E_{\text{FBHD}},n,\eta_g)] + \sum_{r=1}^{r_s}L(r)[\theta_{13,\pm}(E_{\text{FBHD}},n,\eta_g)]\right]\right]\right] \tag{15.144}$$

$t_{i9\text{B}} = \frac{E_{14,9\,\max}}{eF_s(1+i_{\max})}$, $E_{14,9\,\max}$, is the maximum value of the equation

$$\gamma_3(E_{14,9\,\max},\eta_g) = a_0' \left[\frac{2eB}{\hbar}\left(n+\frac{1}{2}\right)\right] \pm (\bar{\lambda}_0)\left[\frac{2eB}{\hbar}\left(n+\frac{1}{2}\right)\right]^{1/2} + S_{i_{\max}}\left[\frac{\hbar|e|F_s\gamma_3'(E_{14,9\,\max},\eta_g)}{\sqrt{2m_{\parallel}^*}}\right]^{2/3} \tag{15.145}$$

Thus, combining the appropriate equations with (5.4), we can study the QC in this case.

*15.2.18 The QC in magneto 2D QMOSFETs of IV–VI materials operated under accumulation mode

The electron concentration can be expressed as

$$n_0 = \text{Real part of } \bar{n}_0 \tag{15.146}$$

where

$$\bar{n}_0 = \frac{eB}{\pi^2 \hbar}\left[\sum_{n=0}^{n_{\max}}[\theta_{21}(E_{\mathrm{FBHD}}, n, \eta_g)] + \sum_{r=1}^{r_s} L(r)[\theta_{21}(E_{\mathrm{FBHD}}, n, \eta_g)]\right] \tag{15.147}$$

in which $\theta_{21}(E_{\mathrm{FBHD}}, n, \eta_g) = [U_{18}(E_{\mathrm{FBHD}}, n, \eta_g)]^{1/2}$.

The 2D electron DR in accumulation layers of IV–VI semiconductors under weak electric field limit can be written as

$$\theta_1(E, i, \eta_g)k_x^2 + \theta_2(E, i, \eta_g)k_y^2 = \theta_3(E, i, \eta_g) \tag{15.148}$$

where

$$\theta_1(E, i, \eta_g) = [F_1(E, \eta_g)S_i(eF_s a_1(E, \eta_g))^{2/3} F_2(E, \eta_g)]$$

$$a_1(E, \eta_g) = \frac{1}{F_2(E, \eta_g)}\left[\frac{F_2'(E, \eta_g)}{F_2(E, \eta_g)} F_1(E, \eta_g) - F_1'(E, \eta_g)\right]$$

$$\theta_2(E, i, \eta_g) = \left[\left[F_1(E, \eta_g) + \frac{2a_2(E, \eta_g)}{3a_1(E, \eta_g)}(eF_s a_1(E, \eta_g))^{2/3} S_i F_1(E, \eta_g)\right]\right]$$

$$a_2(E, \eta_g) = \frac{1}{F_2(E, \eta_g)}\left[\frac{F_2'(E, \eta_g)}{F_2(E, \eta_g)} F_1(E, \eta_g) - F_1'(E, \eta_g)\right]$$

$$\theta_3(E, i, \eta_g) = \left[1 + \frac{2C(E, \eta_g)}{3a_1(E, \eta_g)} S_i(eF_s a_1(E, \eta_g))^{2/3} F_2(E, \eta_g)\right] \quad \text{and}$$

$$C(E, \eta_g) = \left[\frac{F_2'(E, \eta_g)}{F_2^2(E, \eta_g)}\right]$$

The DR of the 2D electrons in QMOSFETs of HD IV–VI materials operated in accumulation mode under the condition of weak electric field limit under magnetic quantization can be written using (15.148) as

$$\left(n+\frac{1}{2}\right)\hbar\frac{eB}{M_{9.4}(E_{14,11},i,\eta_g)}=\theta_3(E_{14,11},i,\eta_g) \tag{15.149}$$

where

$$M_{9.4}(E_{14,11},i,\eta_g)=\frac{\hbar^2}{2}[\theta_1(E_{14,11},i,\eta_g)\theta_2(E_{14,11},i,\eta_g)]^{-1/2}$$

and $E_{14,11}$ is the total quantized energy in this case.

The DOS function is given by

$$N_{Bi}=\frac{g_v eB}{\pi\hbar}\sum_{n=0}^{n_{\max}}\sum_{i=0}^{n_{i\max}}\delta'(E-E_{14,11}) \tag{15.150}$$

The electron concentration per unit area can be written as

$$n_s=\text{Real part of }\bar{n}_s \tag{15.151}$$

where

$$\bar{n}_s=\frac{g_v eB}{\pi\hbar}\sum_{n=0}^{n_{\max}}\sum_{i=0}^{n_{i\max}}F_{-1}(\eta_{14,11}) \tag{15.152}$$

in which $\eta_{14,11}=(E_{\mathrm{FHD2D14}}-E_{14,11})(k_BT)^{-1}$.

The total surface concentration in QMOSFETs in this case can be written as

$$N_s=g_v\text{ Real Part of }\left[\sum_{n=0}^{n_{\max}}\sum_{i=0}^{n_{i\max}}\left[\frac{eB}{\pi\hbar}[F_{-1}(\eta_{14,11})]\right.\right.$$
$$\left.\left.+\frac{eBt_{i11\mathrm{B}}}{\pi^2\hbar}\left[[\theta_{21}(E_{\mathrm{FBHD}},n,\eta_g)]+\sum_{r=1}^{r_s}L(r)[\theta_{21}(E_{\mathrm{FBHD}},n,\eta_g)]\right]\right]\right] \tag{15.153}$$

$t_{i11\mathrm{B}}=\frac{E_{14,11\,\max}}{eF_s(1+i_{\max})}$, $E_{14,11\,\max}$, is the maximum value of the real part of the equation

$$\left(n+\frac{1}{2}\right)\hbar\frac{eB}{M_{9.4}(E_{14,11\,\max},i_{\max},\eta_g)}=\theta_3(E_{14,11\,\max},i_{\max},\eta_g) \tag{15.154}$$

Thus, combining the appropriate equations with (5.4), we can study the QC in this case.

*15.2.19 The QC in magneto 2D QMOSFETs of stressed Kane type materials operated under accumulation mode

The electron concentration can be expressed as

$$n_0 = \frac{eB}{\pi^2\hbar}\left[\sum_{n=0}^{n_{\max}}\left[[\theta_{25}(E_{\mathrm{FBHD}}, n, \eta_g)] + \sum_{r=1}^{r_s} L(r)[\theta_{25}(E_{\mathrm{FBHD}}, n, \eta_g)]\right]\right] \tag{15.155}$$

where $\theta_{25}(E_{\mathrm{FBHD}}, n, \eta_g) = [U_{41}(E_{\mathrm{FBHD}}, n, \eta_g)]^{1/2}$.

The 2D electron DR in accumulation layers of stressed III–V semiconductors can be written using (11.116) and (5.5) as

$$\theta_{13}(E, i, \eta_g)k_x^2 + \theta_{23}(E, i, \eta_g)k_y^2 = \theta_{33}(E, i, \eta_g) \tag{15.156}$$

where

$$\theta_{13}(E, i, \eta_g) = [P_{11}(E, \eta_g) + S_i(eF_s a_{13}(E, \eta_g))^{2/3} S_{11}(E, \eta_g)]$$

$$a_{13}(E, \eta_g) = \frac{1}{S_{11}(E, \eta_g)}\left[\frac{S'_{11}(E, \eta_g)}{S_{11}(E, \eta_g)} P_{11}(E, \eta_g) - P'_{11}(E, \eta_g)\right]$$

$$\theta_{23}(E, i, \eta_g) = \left[\left[Q_{11}(E, \eta_g) + \frac{2a_{23}(E, \eta_g)}{3a_{13}(E, \eta_g)}(eF_s a_{13}(E, \eta_g))^{2/3} S_i Q_{11}(E, \eta_g)\right]\right]$$

$$a_{23}(E, \eta_g) = \frac{1}{S_{11}(E, \eta_g)}\left[\frac{S'_{11}(E, \eta_g)}{S_{11}(E, \eta_g)} Q_{11}(E, \eta_g) - Q'_{11}(E, \eta_g)\right]$$

$$\theta_{33}(E, i, \eta_g) = \left[1 + \frac{2C_3(E, \eta_g)}{3a_{13}(E, \eta_g)} S_i (eF_s a_{13}(E, \eta_g))^{2/3} S_{11}(E, \eta_g)\right] \quad \text{and}$$

$$C_3(E, \eta_g) = \left[\frac{S'_{11}(E, \eta_g)}{S_{11}^2(E, \eta_g)}\right]$$

The DR of the 2D electrons in QMOSFETs of HD stressed Kane type materials operated in accumulation mode under the condition of weak electric field limit under magnetic quantization can be written as

$$\left(n + \frac{1}{2}\right)\hbar \frac{eB}{M_{9.5}(E_{14,13}, i, \eta_g)} = \theta_{33}(E_{14,13}, i, \eta_g) \tag{15.157}$$

where

$$M_{9.5}(E_{14,13}, i, \eta_g) = \frac{\hbar^2}{2}[\theta_{13}(E_{14,13}, i, \eta_g)\theta_{23}(E_{14,13}, i, \eta_g)]^{-1/2}$$

and $E_{14,13}$ is the total quantized energy in this case.

The DOS function is given by

$$N_{Bi} = \frac{g_v eB}{\pi\hbar} \sum_{n=0}^{n_{\max}} \sum_{i=0}^{n_{i\max}} \delta'(E - E_{14,13}) \tag{15.158}$$

The electron concentration per unit area can be written as

$$n_s = \bar{n}_s \tag{15.159}$$

where

$$\bar{n}_s = \frac{g_v eB}{\pi\hbar} \sum_{n=0}^{n_{\max}} \sum_{i=0}^{n_{i\max}} F_{-1}(\eta_{14,13}) \tag{15.160}$$

in which $\eta_{14,13} = (E_{\mathrm{FHD2D14}} - E_{14,13})(k_B T)^{-1}$.

The total surface concentration in QMOSFETs in this case can be written as

$$N_s = g_v \left[\sum_{n=0}^{n_{\max}} \sum_{i=0}^{n_{i\max}} \left[\frac{eB}{\pi\hbar}[F_{-1}(\eta_{14,13})] + \frac{eBt_{i13\mathrm{B}}}{\pi^2\hbar} \left[[\theta_{25}(E_{\mathrm{FBHD}}, n, \eta_g)] + \sum_{r=1}^{r_s} L(r)[\theta_{25}(E_{\mathrm{FBHD}}, n, \eta_g)] \right] \right] \right] \tag{15.161}$$

$t_{i13\mathrm{B}} = \frac{E_{14,13\,\max}}{eF_s(1+i_{\max})}$, $E_{14,13\,\max}$, is the maximum value of the real part of the equation

$$\left(n + \frac{1}{2}\right)\hbar \frac{eB}{M_{9.5\max}(E_{14,13\,\max}, i_{\max}, \eta_g)} = \theta_3(E_{14,13\,\max}, i_{\max}, \eta_g) \tag{15.162}$$

where

$$M_{9.5\max}(E_{14,13\,\max}, i, \eta_g) = \frac{\hbar^2}{2}[\theta_{13}(E_{14,13\,\max}, i_{\max}, \eta_g)\theta_{23}(E_{14,13\,\max}, i_{\max}, \eta_g)]^{-1/2}$$

Thus, combining the appropriate equations with (5.4), we can study the QC in this case.

15.3 Summary and Conclusion

The DRs in HDQWTs made of the said HD materials exhibit the fact that the total energy is quantized since the corresponding wave vector space is totally quantized by quantizing magnetic field and size quantization along z-direction. We note the following two points in this context.

1. The Density-of-States (DOS) functions for all the materials in this case are series of non-uniformly distributed Dirac's Delta functions at specified quantized points in the respective energy axis. The spacing between the consecutive Delta functions are functions of energy band constants and quantization of the wave vector space of a particular material. The DOS function needs two summations namely one summation over the Landau quantum number and the other one is due to size quantization.
2. It may be noted that the energy levels in magneto HD QWTs' lead to the discrete energy levels, somewhat like atomic energy levels, which produce very large changes. This follows from the inherent nature of the quantum confinement of the carrier gas dealt with here. In this case there remain no free carrier states in between any two allowed sets of size-quantized levels unlike that found for QWs and NWs where the quantum confinements are 1D and 2D respectively. Consequently, the crossing of the Fermi level by the size-quantized levels in this case would have much greater impact on the redistribution of the carriers among the allowed levels, as compared to that found for QWs and NWs respectively.

In Section 15.2.1, we have formulated the electron statistics in HDQWTs of nonlinear optical and tetragonal materials under magnetic quantization as given by (15.6) and by combining it with (15.7) we can study the QC in this case. In Section 15.2.2, we have formulated the electron statistics in HDQWTs of III–V, ternary and quaternary materials under magnetic quantization in accordance with the HD three band model of Kane, two band model of Kane, isotropic parabolic energy bands together with the models of Stillman *et al.* [16], Palik *et al.* [15], Rossler, Johnson and Dickey, Agafonov *et al.* and Zhang as given by (15.13), (15.16), (15.19), (15.23), (15.29a), (15.29e), (15.29h), (15.29k) and (15.29m) and by combining each of them with (15.7) we can study the QC for each model in this case. In Section 15.2.3, we have formulated the electron statistics in HDQWTs of II–VI materials under magnetic quantization as given by (15.32) and by combining it with (15.7) we can study the QC in this case.

In Section 15.2.4, we have formulated the electron statistics in HDQWTs of IV–VI materials under magnetic quantization in accordance with the models of Cohen, Lax, Dimmock, Bangert and Kastner, Foley and Langenberg and McClure and Choi as given by (15.37), (15.44), (15.48b), (15.53), (15.59a) and (15.59d) and by combining them with (15.7) we can study the QC for each model in this case. In Section 15.2.5, we have formulated the electron statistics in HDQWTs of stressed Kane type under magnetic quantization as given by (15.63) and by combining it with (15.7) we can study the QC in this case. In Section 15.2.6, we have formulated the electron statistics in HDQWTs of Te under magnetic quantization as given by (15.67) and by combining it with (15.7) we can study the QC in this case. In Section 15.2.7, we have formulated the electron statistics in HDQWTs of GaP under magnetic quantization as given by (15.70) and by combining it with (15.7) we can study the QC in this case. In Section 15.2.8, we have formulated the electron statistics in HDQWTs of $PtSb_2$ under magnetic quantization as given by (15.73) and by combining it with (15.7) we can study the QC in this case. In Section 15.2.9, we have formulated the electron statistics in HDQWTs of Bi_2Te_3 under magnetic quantization as given by (15.78) and by combining it with (15.7) we can study the QC in this case.

In Section 15.2.10, we have formulated the electron statistics in HDQWTs of Ge under magnetic quantization in accordance with the models of together with the models of Cardona *et al.* and Wang and Ressler as given by (15.81) and (15.85) and by combining them with (15.7) we can study the QC for each model in this case. In Section 15.2.11, we have formulated the electron statistics in HDQWTs of GaSb in accordance with the models of Mathur *et al.* and Seiler under magnetic quantization as given by (15.89) and (15.92) respectively and by combining it with (15.7) we can study the QC in this case. In Section 15.2.12, we have formulated the electron statistics in HDQWTs of II–V materials under magnetic quantization as given by (15.96) and by combining it with (15.7) we can study the QC in this case. In Section 15.2.13, we have formulated the electron statistics in HDQWTs of Lead Germanium Telluride under magnetic quantization as given by (15.101a) and by combining it with (15.7) we can study the QC in this case. In Section 15.2.14, we have formulated the electron statistics in HDQWTs of Zinc and Cadmium Diphosphides under magnetic quantization as given by (15.101d) and by combining it with (15.7) we can study the QC in this case.

From theoretical background we note that the expression for surface electron concentration under magnetic quantization of 2D QMOSFETs operated under accumulation mode of nonlinear optical and tetragonal, III–V (three and two band models of Kane together with parabolic energy bands), II–VI, IV–VI and stressed Kane type materials are given by (15.111), (15.121), (15.129), (15.137), (15.144), (15.153) and (15.161) respectively. By combining each of them with the (5.4) we can study the QC in each case.

We have not given the graphs of the magneto QCs' for various 2D QMOSFETs made of many materials as discussed here operated under accumulation layer mode since we feel that our readers will perform the intricate numerical computation to obtain the oscillatory plots for better assessment and joy of understanding.

15.4 Open Research Problems

(R.15.1) Investigate the QC in the presence of an arbitrarily oriented quantizing magnetic field for all the HD QWTs' made by using the materials as given in problems of Chapter 1. Draw the corresponding graphs and explain the results.

(R.15.2) Investigate the QC in the presence of an arbitrarily oriented quantizing magnetic field in HD QWTs' of the nonlinear optical semiconductors by including broadening and the electron spin. Study all the special cases for HD QWTs' made by using the III–V, ternary and quaternary materials in this context. Draw the corresponding graphs and explain the results.

(R.15.3) Investigate the QC for HD QWTs' made by using the IV–VI, II–VI and stressed Kane type compounds in the presence of an arbitrarily oriented quantizing magnetic field by including broadening and electron spin. Draw the corresponding graphs and explain the results.

(R.15.4) Investigate the QC for all the HD QWTs' made by using the materials as stated in of Chapter 1 in the presence of an arbitrarily oriented quantizing magnetic field by including broadening and electron spin. Draw the corresponding graphs and explain the results.

(R.15.5) Investigate the QC in the presence of an arbitrarily oriented quantizing magnetic field and crossed electric fields in HD QWTs' made by using the nonlinear optical and tetragonal materials by including broadening and the electron spin. Study all the special cases for HD III–V, ternary and quaternary materials in this context. Draw the corresponding graphs and explain the results.

(R.15.6) Investigate the QC for HD QWTs' made by using the IV–VI, II–VI and stressed Kane type compounds in the presence of an arbitrarily oriented quantizing magnetic field and crossed electric field by including broadening and electron spin. Draw the corresponding graphs and explain the results.

(R.15.7) Investigate the QC for all the HD QWTs' made by using the materials as stated in of Chapter 1 in the presence of an arbitrarily oriented quantizing magnetic field and crossed electric field by including broadening and electron spin. Draw the corresponding graphs and explain the results.

(R.15.8) Investigate the QC in HD QWTs' made by using the nonlinear optical, III–V, II–VI, IV–VI and stressed Kane type semiconductors. Draw the corresponding graphs and explain the results.

(R.15.9) Investigate the 2D QC for HD QWTs' made by using for all the materials as considered in problems of Chapter 1. Draw the corresponding graphs and explain the results.

(R.15.10) Investigate the 2D QC in the presence of an arbitrarily oriented non-quantizing magnetic field for the HD QWTs' made by using the nonlinear optical and tetragonal semiconductors by including the electron spin. Study all the special cases for III–V, ternary and quaternary materials in this context. Draw the corresponding graphs and explain the results.

(R.15.11) Investigate the DRs in QWs HD IV–VI, II–VI and stressed Kane type compounds in the presence of an arbitrarily oriented classical magnetic field by including the electron spin. Draw the corresponding graphs and explain the results.

(R.15.12) Investigate the 2D QC for HD QWTs' made by using for all the materials as stated in Chapter 1 in the presence of an arbitrarily oriented magnetic field by including electron spin and broadening. Draw the corresponding graphs and explain the results.

(R.15.13) Investigate all the problems of this chapter by removing all the mathematical approximations and establishing the uniqueness conditions in each case. Draw the corresponding graphs and explain the results.

(R.15.14) Investigate the QC in all the HD QWTs' made by using the semiconductors as considered in this chapter in the presence of defects and magnetic quantization. Draw the corresponding graphs and explain the results.

(R.15.15) Investigate the magneto QCs' in the presence of an arbitrarily oriented magnetic field in 2D QMOSFETs of nonlinear optical and tetragonal materials operated under accumulation mode as stated in this chapter. Draw the corresponding graphs and explain the results.

(R.15.16) Investigate the magneto QCs' in the presence of an arbitrarily oriented magnetic field in 2D QMOSFETs of IV–VI, II–VI and stressed Kane type compounds operated under accumulation mode as stated in this chapter. Draw the corresponding graphs and explain the results.

(R.15.17) Investigate the magneto QCs' in the presence of an arbitrarily oriented magnetic field in 2D QMOSFETs operated under accumulation mode of all the materials as stated in Chapter 1. Draw the corresponding graphs and explain the results.

(R.15.18) Investigate the QCs' in the presence of an arbitrarily oriented magnetic field in 2D QMOSFETs operated under accumulation mode of all the materials as studied in this chapter in the presence of an arbitrarily oriented non-uniform magnetic field. Draw the corresponding graphs and explain the results.

(R.15.19) Investigate the QCs' in the presence of an arbitrarily oriented magnetic field in 2D QMOSFETs operated under accumulation mode of IV–VI, II–VI and stressed Kane type compounds in the presence of an arbitrarily oriented non-uniform magnetic field. Draw the corresponding graphs and explain the results.

(R.15.20) Investigate the QCs' in the presence of an arbitrarily oriented magnetic field in 2D QMOSFETs operated under accumulation mode of all the materials as stated in Chapter 1 in the presence of an arbitrarily oriented non-uniform magnetic field. Draw the corresponding graphs and explain the results.

(R.15.21) Investigate the QCs' in 2D QMOSFETs operated under accumulation mode of all the problems in Chapter 2 in the presence of an arbitrarily oriented non-uniform magnetic field. Draw the corresponding graphs and explain the results.

References

[1] N. Miura, *Physics of Semiconductors in High Magnetic Fields*, Series on Semiconductor Science and Technology, (Oxford University Press, USA, **2007**); K. H. J. Buschow, F. R. de Boer, *Physics of Magnetism and Magnetic Materials*, (Springer, New York, **2003**); D. Sellmyer, R. Skomski (Eds.), *Advanced Magnetic Nanostructures*, (Springer, New York, **2005**); J. A. C. Bland, B. Heinrich (Eds.), *Ultrathin Magnetic Structures III: Fundamentals of Nanomagnetism (Pt. 3)*, (Springer-Verlag, Germany, **2005**); B. K. Ridley, *Quantum Processes in Semiconductors*, Fourth Edition, (Oxford publications, Oxford, **1999**); J. H. Davies, *Physics of Low Dimensional Semiconductors*, (Cambridge University Press, UK, **1998**); S. Blundell, *Magnetism in Condensed Matter*, Oxford Master Series in Condensed Matter Physics, (Oxford University Press, USA, **2001**); C. Weisbuch, B. Vinter, *Quantum Semiconductor Structures: Fundamentals and Applications*, (Academic Publishers, USA, **1991**); D. Ferry, *Semiconductor Transport*, (CRC, USA, **2000**); M. Reed (Ed.), *Semiconductors and Semimetals: Nanostructured Systems*, (Academic Press, USA, **1992**); T. Dittrich, *Quantum Transport and Dissipation*, (Wiley-VCH Verlag GmbH, Germany, **1998**); A. Y. Shik, *Quantum Wells: Physics & Electronics of Two Dimensional Systems*, (World Scientific, USA, **1997**).

[2] K. P. Ghatak, M. Mondal, *Zietschrift fur Naturforschung A*, 41a, 881 **(1986)**; K. P. Ghatak, M. Mondal, *J. Appl. Phys.*, 62, 922 **(1987)**; K. P. Ghatak, S. N. Biswas, *Phys. Stat. Sol. (b)*, 140, K107 **(1987)**; K. P. Ghatak, M. Mondal, *J. Mag. and Mag. Mat.*, 74, 203 **(1988)**; K. P. Ghatak, M. Mondal, *Phys. Stat. Sol. (b)*, 139, 195 **(1987)**; K. P. Ghatak, M. Mondal, *Phys. Stat. Sol. (b)*, 148, 645 **(1988)**; K. P. Ghatak, B. Mitra, A. Ghoshal, *Phys. Stat. Sol. (b)*, 154, K121 **(1989)**; K. P. Ghatak, S. N. Biswas, *J. Low Temp. Phys.*, 78, 219 **(1990)**; K. P. Ghatak, M. Mondal, *Phys. Stat. Sol. (b)*, 160, 673 **(1990)**; K. P. Ghatak, B. Mitra, *Phys. Lett. A*, 156, 233 **(1991)**; K. P. Ghatak, A. Ghoshal, B. Mitra, *Nuovo Cimento D*, 13, 867 **(1991)**; K. P. Ghatak, M. Mondal, *Phys. Stat. Sol. (b)*, 148, 645 **(1989)**; K. P. Ghatak, B. Mitra, *Int. J. Electron.*, 70, 345 **(1991)**; K. P. Ghatak, S. N. Biswas, *J. Appl. Phys.*, 70, 299 **(1991)**; K. P. Ghatak, A. Ghoshal, *Phys. Stat. Sol. (b)*, 170, K27 **(1992)**; K. P. Ghatak, *Nuovo Cimento D*, 13, 1321 **(1992)**; K. P. Ghatak, B. Mitra, *Int. J. Electron.*, 72, 541 **(1992)**; K. P. Ghatak, S. N. Biswas, *Nonlinear Opt.*, 4, 347 **(1993)**; K. P. Ghatak, M. Mondal, *Phys. Stat. Sol. (b)*, 175, 113 **(1993)**; K. P. Ghatak, S. N. Biswas, *Nonlinear Opt.*, 4, 39 **(1993)**; K. P. Ghatak, B. Mitra, *Nuovo Cimento D*, 15, 97 **(1993)**; K. P. Ghatak, S. N. Biswas, *Nanostruct. Mater.*, 2, 91 **(1993)**; K. P. Ghatak, M. Mondal, *Phys. Stat. Sol. (b)*, 185, K5 **(1994)**; K. P. Ghatak, B. Goswami, M. Mitra, B. Nag, *Nonlinear Opt.*, 16, 9 **(1996)**; K. P. Ghatak, M. Mitra, B. Goswami, B. Nag, *Nonlinear Opt.*, 16, 167 **(1996)**; K. P. Ghatak, B. Nag, *Nanostruct. Mater.*, 10, 923 **(1998)**.

[3] D. Roy Choudhury, A. K. Choudhury, K. P. Ghatak, A. N. Chakravarti, *Phys. Stat. Sol. (b)*, 98, K141 **(1980)**; A. N. Chakravarti, K. P. Ghatak, A. Dhar, S. Ghosh, *Phys. Stat. Sol. (b)*, 105, K55 **(1981)**; A. N. Chakravarti, A. K. Choudhury, K. P. Ghatak, *Phys. Stat. Sol. (a)*, 63, K97 **(1981)**; A. N. Chakravarti, A. K. Choudhury, K. P. Ghatak, S. Ghosh, A. Dhar, *Appl. Phys.*, 25, 105 **(1981)**; A. N. Chakravarti, K. P. Ghatak, G. B. Rao, K. K. Ghosh, *Phys. Stat. Sol. (b)*, 112, 75 **(1982)**; A. N. Chakravarti, K. P. Ghatak, K. K. Ghosh, H. M. Mukherjee, *Phys. Stat. Sol. (b)*, 116, 17 **(1983)**; M. Mondal, K. P. Ghatak, *Phys. Stat. Sol. (b)*, 133, K143 **(1984)**; M. Mondal, K. P. Ghatak, *Phys. Stat. Sol. (b)*, 126, K47 **(1984)**; M. Mondal, K. P. Ghatak, *Phys. Stat. Sol. (b)*, 126, K41 **(1984)**; M. Mondal, K. P. Ghatak, *Phys. Stat. Sol. (b)*, 129, K745 **(1985)**; M. Mondal, K. P. Ghatak, *Phys. Scr.*, 31, 615 **(1985)**; M. Mondal, K. P. Ghatak, *Phys. Stat. Sol. (b)*, 135, 239 **(1986)**; M. Mondal, K. P. Ghatak, *Phys. Stat. Sol. (b)*, 93, 377 **(1986)**; M. Mondal, K. P. Ghatak, *Phys. Stat. Sol. (b)*, 135, K21 **(1986)**; M. Mondal, S. Bhattacharyya, K. P. Ghatak, *Appl. Phys. A*, 42, 331 **(1987)**; S. N. Biswas, N. Chattopadhyay, K. P. Ghatak, *Phys. Stat. Sol. (b)*, 141, K47 **(1987)**; B. Mitra, K. P. Ghatak, *Phys. Stat. Sol. (b)*, 149, K117 **(1988)**; B. Mitra, A. Ghoshal, K. P. Ghatak, *Phys. Stat. Sol. (b)*, 150, K67 **(1988)**; M. Mondal, K. P. Ghatak, *Phys. Stat. Sol. (b)*, 147, K179 **(1988)**; M. Mondal, K. P. Ghatak, *Phys. Stat. Sol. (b)*, 146, K97 **(1988)**; B. Mitra, A. Ghoshal, K. P. Ghatak, *Phys. Stat. Sol. (b)*, 153, K209 **(1989)**; B. Mitra, K. P. Ghatak, *Phys. Lett. A*, 142, 401 **(1989)**; B. Mitra, A. Ghoshal, K. P. Ghatak, *Phys. Stat. Sol. (b)*, 154, K147 **(1989)**; B. Mitra, K. P. Ghatak, *Sol. Stat. Electron.*, 32, 515 **(1989)**; B. Mitra, A. Ghoshal, K. P. Ghatak, *Phys. Stat. Sol. (b)*, 155, K23 **(1989)**; B. Mitra, K. P. Ghatak, *Phys. Lett. A*, 135, 397 **(1989)**; B. Mitra, K. P. Ghatak, *Phys. Lett. A*, 146, 357 **(1990)**; B. Mitra, K. P. Ghatak, *Phys. Stat. Sol. (b)*, 164, K13 **(1991)**; S. N. Biswas, K. P. Ghatak, *Int. J. Electron.*, 70, 125 **(1991)**.

[4] P. R. Wallace, *Phys. Stat. Sol. (b)*, 92, 49 **(1979)**.

[5] B. R. Nag, *Electron Transport in Compound Semiconductors*, Springer Series in Solid State Sciences, Vol. 11, (Springer-Verlag, Germany, **1980**).

[6] K. P. Ghatak, S. Bhattacharya, D. De, *Einstein Relation in Compound Semiconductors and Their Nanostructures*, Springer Series in Materials Science, Vol. 116, (Springer-Verlag, Germany, **2009**).

[7] C. C. Wu and C. J. Lin, *J. Low Temp. Phys.*, 57, 469 **(1984)**; M. H. Chen, C. C. Wu, C. J. Lin, *J. Low Temp. Phys.*, 55, 127 **(1984)**.

[8] E. Bangert and P. Kastner, *Phys. Stat. Sol. (b)*, 61, 503 **(1974)**.

[9] G. M. T. Foley, P. N. Langenberg, *Phys. Rev. B*, 15, 4850 **(1977)**.

[10] M. Singh, P. R. Wallace, S. D. Jog, E. Arushanov, *J. Phys. Chem. Solids*, 45, 409 **(1984)**; Y. Yamada, *Phys. Soc. Jap.*, 35, 1600 **(1973)**, 37, 606 **(1974)**.

[11] L. A. Vassilev, *Phys. Stat. Sol. (b)*, 121, 203 **(1984)**.

[12] R. W. Cunningham, *Phys. Rev.*, 167, 761 **(1968)**.

[13] A. I. Yekimov, A. A. Onushchenko, A. G. Plyukhin, A. L. Efros, *J. Exp. Theor. Phys.*, 88, 1490 **(1985)**.

[14] B. J. Roman, A. W. Ewald, *Phys. Rev. B*, 5, 3914 **(1972)**.
[15] G. E. Stillman, C. M. Wolfe and J. C. Dinmockin, in *Semiconductors and Semimetals*, Ed. R. K. Willardson and A. C. Beer, 12, 169 **(1977)**.
[16] E. D. Palik, G. S. Picus, S. Teither and R. F. Wallis, *Phys. Rev.*, 122, 475 **(1961)**.

Chapter 16

The $I_D - V_D$ Equation for Ballistic HDQWFETs Under Magnetic Quantization

Begin at the beginning and go on till you come to an end remembering that the end is where we start from.

16.1 Introduction

This is completely a new chapter where from Sections 16.2.1 to 16.2.11, we study the $I_D - V_D$ equation in Ballistic magneto HDQWFETs of nonlinear optical and tetragonal, III–V, ternary and quaternary, II–VI, IV–VI, stressed Kane type, Te, GaP, $PtSb_2$, Bi_2Te_3, Ge and GaSb respectively. In Section 16.3, we present summary and conclusion and in the last Section 16.4, we offer one multidimensional open research problem.

16.2 Theoretical Background

16.2.1 The $I_D - V_D$ equation for ballistic magneto HDQWFETs of nonlinear optical and tetragonal materials

Using (11.26), we get

$$V_x = \sqrt{\frac{2}{m_\perp^*} \frac{\sqrt{T_{22}(E,\eta_g)}}{T'_{22}(E,\eta_g)}} \tag{16.1}$$

Using (6.1)–(6.4), (15.4) and (15.5), we get

$$\frac{J_D}{W} = \frac{e^2 B g_v}{2\pi^2\hbar}\sqrt{\frac{2}{m_\perp^*}}\text{Real part of}\sum_{n=0}^{n_{\max}}\sum_{n_z=1}^{n_{z\max}}\left[\frac{\sqrt{T_{22}(e_{81},\eta_g)}}{T'_{22}(e_{81},\eta_g)}\right]$$
$$\times\left[F_{-1}(\bar{\eta}_{25,1}) - F_{-1}(\bar{\theta}_{25,1})\right] \tag{16.2}$$

where $\bar{\eta}_{25,1} = (k_B T)^{-1}[E_{F25,1} - e_{81}]$, $\bar{\theta}_{25,1} = \left[\bar{\eta}_{25,1} - \frac{eV_D}{k_B T}\right]$ and $E_{F25,1}$ is the Fermi energy in the present case which can be determined from

$$\frac{C_{ins}}{e}(V_T - V_{GS}) = \frac{g_v eB}{4\pi\hbar} \text{ Real part of} \sum_{n=0}^{n_{\max}} \sum_{i=0}^{i_{\max}} [F_{-1}(\bar{\eta}_{25,1}) + F_{-1}(\bar{\theta}_{25,1})] \tag{16.3}$$

16.2.2 The $I_D - V_D$ equation for ballistic magneto HDQWFETs of III–V, ternary and quaternary materials

(a) Three band model of Kane

From (11.31), we can write

$$V_x = \sqrt{\frac{2}{m_c}} \sum_{n=0}^{n_{\max}} \sum_{i=0}^{i_{\max}} \left[\frac{\sqrt{T_{31}(E,\eta_g) + iT_{32}(E,\eta_g)}}{T'_{31}(E,\eta_g) + iT'_{32}(E,\eta_g)}\right] \tag{16.4}$$

Using (6.1)–(6.4), (15.12a), (15.12b) and (16.4), we get

$$\frac{J_D}{W} = \frac{e^2 B g_v}{2\pi^2\hbar}\sqrt{\frac{2}{m_c}} \text{ Real part of} \sum_{n=0}^{n_{\max}} \sum_{i=0}^{i_{\max}} \left[\left[\frac{\sqrt{T_{31}(e_{82},\eta_g) + iT_{32}(e_{82},\eta_g)}}{T'_{31}(e_{82},\eta_g) + iT'_{32}(e_{82},\eta_g)}\right]\right.$$

$$\left. \times\ [F_{-1}(\bar{\eta}_{25,2}) - F_{-1}(\bar{\theta}_{25,2})]\right] \tag{16.5}$$

where $\bar{\eta}_{25,2} = (k_B T)^{-1}[E_{F25,2} - e_{82}]$, $\bar{\theta}_{25,2} = \left[\bar{\eta}_{25,2} - \frac{eV_D}{k_B T}\right]$ and $E_{F25,2}$ is the Fermi energy in the present case which can be determined from

$$\frac{C_{ins}}{e}(V_T - V_{GS}) = \frac{g_v eB}{4\pi\hbar} \text{ Real part of} \sum_{n=0}^{n_{\max}} \sum_{i=0}^{i_{\max}} [F_{-1}(\bar{\eta}_{25,2}) + F_{-1}(\bar{\theta}_{25,2})] \tag{16.6}$$

(b) Two band model of Kane

From (11.36), we can write

$$V_x = \sqrt{\frac{2}{m_c}} \left[\frac{\sqrt{\gamma_2(E,\eta_g)}}{\gamma'_2(E,\eta_g)}\right] \tag{16.7}$$

Using (6.1)–(6.4), (15.15), (15.16) and (16.7), we get

$$\frac{J_D}{W} = \frac{e^2 B g_v}{2\pi^2\hbar}\sqrt{\frac{2}{m_c}}\sum_{n=0}^{n_{\max}}\sum_{i=0}^{i_{\max}}\left[\left[\frac{\sqrt{\gamma_2(e_{83},\eta_g)}}{\gamma_2'(e_{83},\eta_g)}\right][F_{-1}(\bar{\eta}_{25,3}) - F_{-1}(\bar{\theta}_{25,3})]\right] \tag{16.8}$$

where $\bar{\eta}_{25,3} = (k_B T)^{-1}[E_{F25,3} - e_{83}]$, $\bar{\theta}_{25,3} = \left[\bar{\eta}_{25,3} - \frac{eV_D}{k_B T}\right]$ and $E_{F25,3}$ is the Fermi energy in the present case which can be determined from

$$\frac{C_{ins}}{e}(V_T - V_{GS}) = \frac{g_v e B}{4\pi\hbar}\sum_{n=0}^{n_{\max}}\sum_{i=0}^{i_{\max}}[F_{-1}(\bar{\eta}_{25,3}) + F_{-1}(\bar{\theta}_{25,3})] \tag{16.9}$$

(c) Parabolic band model

From (11.40), we can write

$$V_x = \sqrt{\frac{2}{m_c}}\left[\frac{\sqrt{\gamma_3(E,\eta_g)}}{\gamma_3'(E,\eta_g)}\right] \tag{16.10}$$

Using (6.1)–(6.4), (15.17), (15.18) and (16.10), we get

$$\frac{J_D}{W} = \frac{e^2 B g_v}{2\pi^2\hbar}\sqrt{\frac{2}{m_c}}\sum_{n=0}^{n_{\max}}\sum_{i=0}^{i_{\max}}\left[\left[\frac{\sqrt{\gamma_2(e_{84},\eta_g)}}{\gamma_2'(e_{84},\eta_g)}\right][F_{-1}(\bar{\eta}_{25,4}) - F_{-1}(\bar{\theta}_{25,4})]\right] \tag{16.11}$$

where $\bar{\eta}_{25,4} = (k_B T)^{-1}[E_{F25,4} - e_{84}]$, $\bar{\theta}_{25,4} = \left[\bar{\eta}_{25,4} - \frac{eV_D}{k_B T}\right]$ and $E_{F25,4}$ is the Fermi energy in the present case which can be determined from

$$\frac{C_{ins}}{e}(V_T - V_{GS}) = \frac{g_v e B}{4\pi\hbar}\sum_{n=0}^{n_{\max}}\sum_{i=0}^{i_{\max}}[F_{-1}(\bar{\eta}_{25,4}) + F_{-1}(\bar{\theta}_{25,4})] \tag{16.12}$$

(d) Model of Stillman *et al.*

From (11.44), we can write

$$V_x = \sqrt{\frac{2}{m_c}}\left[\frac{\sqrt{\theta_4(E,\eta_g)}}{\theta_4'(E,\eta_g)}\right] \tag{16.13}$$

Using (6.1)–(6.4), (15.21), (15.22) and (16.13), we get

$$\frac{J_D}{W} = \frac{e^2 B g_v}{\pi^2 \hbar}\sqrt{\frac{2}{m_c}}\sum_{n=0}^{n_{\max}}\sum_{i=0}^{i_{\max}}\left[\left[\frac{\sqrt{\theta_4(e_{85},\eta_g)}}{\theta_4'(e_{85},\eta_g)}\right][F_{-1}(\bar{\eta}_{25,5}) - F_{-1}(\bar{\theta}_{25,5})]\right] \tag{16.14}$$

where $\bar{\eta}_{25,5} = (k_BT)^{-1}[E_{F25,5} - e_{85}]$, $\bar{\theta}_{25,5} = \left[\bar{\eta}_{25,5} - \frac{eV_D}{k_BT}\right]$ and $E_{F25,5}$ is the Fermi energy in the present case which can be determined from

$$\frac{C_{ins}}{e}(V_T - V_{GS}) = \frac{g_v e B}{4\pi\hbar}\sum_{n=0}^{n_{\max}}\sum_{i=0}^{i_{\max}}[F_{-1}(\bar{\eta}_{25,5}) + F_{-1}(\bar{\theta}_{25,5})] \tag{16.15}$$

(e) Model of Palik *et al.*

From (11.44), we can write

$$V_x = \sqrt{\frac{2}{m_c}}\left[\frac{\sqrt{\theta_5(E,\eta_g)}}{\theta_5'(E,\eta_g)}\right] \tag{16.16}$$

Using (6.1)–(6.4), (15.27), (15.28) and (16.16), we get

$$\frac{J_D}{W} = \frac{e^2 B g_v}{\pi^2 \hbar}\sqrt{\frac{2}{m_c}}\sum_{n=0}^{n_{\max}}\sum_{i=0}^{i_{\max}}\left[\left[\frac{\sqrt{\theta_5(e_{86},\eta_g)}}{\theta_5'(e_{86},\eta_g)}\right][F_{-1}(\bar{\eta}_{25,6}) - F_{-1}(\bar{\theta}_{25,6})]\right] \tag{16.17}$$

where $\bar{\eta}_{25,6} = (k_BT)^{-1}[E_{F25,6} - e_{86}]$, $\bar{\theta}_{25,6} = \left[\bar{\eta}_{25,6} - \frac{eV_D}{k_BT}\right]$ and $E_{F25,6}$ is the Fermi energy in the present case which can be determined from

$$\frac{C_{ins}}{e}(V_T - V_{GS}) = \frac{g_v e B}{4\pi\hbar}\sum_{n=0}^{n_{\max}}\sum_{i=0}^{i_{\max}}[F_{-1}(\bar{\eta}_{25,6}) + F_{-1}(\bar{\theta}_{25,6})] \tag{16.18}$$

16.2.3 The $I_D - V_D$ equation for ballistic magneto HDQWFETs of II–VI materials

From (11.60), we get

$$V_x = \frac{1}{\hbar}\left[\frac{\sqrt{(\bar{\lambda}_0)^2 + 4a_0'\gamma_3(E,\eta_g)}}{\gamma_3'(E,\eta_g)}\right] \tag{16.19}$$

Using (6.1)–(6.4), (15.30), (15.31) and (16.19), we get

$$\frac{J_D}{W} = \frac{e^2 B g_v}{2\pi^2\hbar^2}\sqrt{\frac{2}{m_c}}\sum_{n=0}^{n_{\max}}\sum_{i=0}^{i_{\max}}\left[\left[\frac{\sqrt{(\bar{\lambda}_0)^2 + 4a_0'\gamma_3(e_{87},\eta_g)}}{\gamma_3'(e_{87},\eta_g)}\right]\right.$$

$$\left.\times\ [F_{-1}(\bar{\eta}_{25,7}) - F_{-1}(\bar{\theta}_{25,7})]\right] \tag{16.20}$$

where $\bar{\eta}_{25,7} = (k_B T)^{-1}[E_{F25,7} - e_{87}]$, $\bar{\theta}_{25,7} = \left[\bar{\eta}_{25,7} - \frac{eV_D}{k_B T}\right]$ and $E_{F25,7}$ is the Fermi energy in the present case which can be determined from

$$\frac{C_{ins}}{e}(V_T - V_{GS}) = \frac{g_v e B}{4\pi\hbar}\sum_{n=0}^{n_{\max}}\sum_{i=0}^{i_{\max}}[F_{-1}(\bar{\eta}_{25,7}) + F_{-1}(\bar{\theta}_{25,7})] \tag{16.21}$$

16.2.4 The $I_D - V_D$ equation for ballistic magneto HDQWFETs of IV–VI materials

From (11.127), we get

$$V_x = \frac{2}{\hbar}\left[\frac{[F_{1\mathrm{HD}}(E,\eta_g)]^{3/2}}{F_{1\mathrm{HD}}'(E,\eta_g)}\right] \tag{16.22}$$

Using (6.1)–(6.4), (15.51), (15.52) and (16.22), we get

$$\frac{J_D}{W} = \frac{2e^2 B g_v}{\pi^2\hbar^2}\ \text{Real part of}\sum_{n=0}^{n_{\max}}\sum_{i=0}^{i_{\max}}\left[\left[\frac{[F_{1\mathrm{HD}}(e_{91},\eta_g)]^{3/2}}{F_{1\mathrm{HD}}'(e_{91},\eta_g)}\right]\right.$$

$$\left.\times\ [F_{-1}(\bar{\eta}_{25,8}) - F_{-1}(\bar{\theta}_{25,8})]\right] \tag{16.23}$$

where $\bar{\eta}_{25,8} = (k_B T)^{-1}[E_{F25,8} - e_{91}]$, $\bar{\theta}_{25,8} = \left[\bar{\eta}_{25,8} - \frac{eV_D}{k_B T}\right]$ and $E_{F25,8}$ is the Fermi energy in the present case which can be determined from

$$\frac{C_{ins}}{e}(V_T - V_{GS}) = \frac{g_v e B}{2\pi\hbar}\ \text{Real part of}\sum_{n=0}^{n_{\max}}\sum_{i=0}^{i_{\max}}[F_{-1}(\bar{\eta}_{25,8}) + F_{-1}(\bar{\theta}_{25,8})] \tag{16.24}$$

16.2.5 The $I_D - V_D$ equation for ballistic magneto HDQWFETs of stressed Kane type materials

From (11.116), we get

$$V_x = \frac{2}{\hbar}\left[\frac{[P_{11}(E,\eta_g)]^{3/2}}{P'_{11}(E,\eta_g)}\right] \tag{16.25}$$

Using (6.1)–(6.4), (15.61), (15.62) and (16.25), we get

$$\frac{J_D}{W} = \frac{2e^2Bg_v}{\pi^2\hbar^2}\sum_{n=0}^{n_{\max}}\sum_{i=0}^{i_{\max}}\left[\left[\frac{[P_{11}(e_{93},\eta_g)]^{3/2}}{P'_{11}(e_{93},\eta_g)}\right][F_{-1}(\bar{\eta}_{25,9}) - F_{-1}(\bar{\theta}_{25,9})]\right] \tag{16.26}$$

where $\bar{\eta}_{25,9} = (k_BT)^{-1}[E_{F25,9} - e_{93}]$, $\bar{\theta}_{25,9} = \left[\bar{\eta}_{25,9} - \frac{eV_D}{k_BT}\right]$ and $E_{F25,9}$ is the Fermi energy in the present case which can be determined from

$$\frac{C_{ins}}{e}(V_T - V_{GS}) = \frac{g_veB}{2\pi\hbar}\sum_{n=0}^{n_{\max}}\sum_{i=0}^{i_{\max}}[F_{-1}(\bar{\eta}_{25,9}) + F_{-1}(\bar{\theta}_{25,9})] \tag{16.27}$$

16.2.6 The $I_D - V_D$ equation for ballistic magneto HDQWFETs of tellurium

From (11.71), we get

$$V_x = \frac{1}{\hbar}\left[\frac{\sqrt{A_9 + 4A_7\gamma_3(E,\eta_g)}}{\gamma'_3(E,\eta_g)}\right] \tag{16.28}$$

Using (6.1)–(6.4), (15.65), (15.66) and (16.28), we get

$$\frac{J_D}{W} = \frac{e^2Bg_v}{2\pi^2\hbar^2}\sum_{n=0}^{n_{\max}}\sum_{i=0}^{i_{\max}}\left[\left[\frac{\sqrt{A_9 + 4A_7\gamma_3(\bar{e}_{94},\eta_g)}}{\gamma'_3(\bar{e}_{94},\eta_g)}\right]\right.$$
$$\left.\times [F_{-1}(\bar{\eta}_{25,10}) - F_{-1}(\bar{\theta}_{25,10})]\right] \tag{16.29}$$

where $\bar{\eta}_{25,10} = (k_BT)^{-1}[E_{F25,10} - \bar{e}_{94}]$, $\bar{\theta}_{25,10} = \left[\bar{\eta}_{25,10} - \frac{eV_D}{k_BT}\right]$ and $E_{F25,10}$ is the Fermi energy in the present case which can be determined from

$$\frac{C_{ins}}{e}(V_T - V_{GS}) = \frac{g_veB}{4\pi\hbar}\sum_{n=0}^{n_{\max}}\sum_{i=0}^{i_{\max}}[F_{-1}(\bar{\eta}_{25,10}) + F_{-1}(\bar{\theta}_{25,10})] \tag{16.30}$$

*16.2.7 The $I_D - V_D$ equation for ballistic magneto HDQWFETs of GaP

From (11.77), we get

$$V_x = \frac{1}{\hbar}[F'_{15}(E, \eta_g)]^{-1} \tag{16.31}$$

where

$$F'_{15}(E, \eta_g) = \left[(\bar{C}_{15})^1 \left[\left[|V_G|\bar{C}_{15} + \frac{C}{2} + \gamma_3(E, \eta_g)\bar{C}_{15}\right]\right.\right.$$
$$\left.\left.\pm \sqrt{\left[|V_G|\bar{C}_{15} + \frac{C}{2} + \gamma_3(E, \eta_g)\bar{C}_{15}\right]^2 - \bar{C}_{15}[\gamma_3^2 - 2\gamma_3(E, \eta_g)|V_G|]}\right]\right]^{1/2},$$

$$\bar{C}_{15} = \frac{\hbar^2}{2}\left[\frac{1}{m_{\|}^*} + \frac{1}{m_{\perp}^*}\right] \quad \text{and} \quad C = \frac{\hbar^4 k_0^2}{m_{\|}^{*2}}$$

using (6.1)–(6.4), (15.68), (15.69) and (16.31), we get

$$\frac{J_D}{W} = \frac{2e^2 B g_v}{\pi^2 \hbar^2} \sum_{n=0}^{n_{\max}} \sum_{i=0}^{i_{\max}} [[F'_{15}(\bar{e}_{941}, \eta_g)]^{-1}[F_{-1}(\bar{\eta}_{25,11}) - F_{-1}(\bar{\theta}_{25,11})]] \tag{16.32}$$

where $\bar{\eta}_{25,11} = (k_B T)^{-1}[E_{F25,11} - \bar{e}_{941}]$, $\bar{\theta}_{25,11} = \left[\bar{\eta}_{25,11} - \frac{eV_D}{k_B T}\right]$ and $E_{F25,11}$ is the Fermi energy in the present case which can be determined from

$$\frac{C_{ins}}{e}(V_T - V_{GS}) = \frac{g_v e B}{4\pi\hbar} \sum_{n=0}^{n_{\max}} \sum_{i=0}^{i_{\max}} [F_{-1}(\bar{\eta}_{25,11}) + F_{-1}(\bar{\theta}_{25,11})] \tag{16.33}$$

*16.2.8 The $I_D - V_D$ equation for ballistic magneto HDQWFETs of $PtSb_2$

From (11.87), we get

$$V_x = \frac{2}{\hbar}\left[\frac{\sqrt{\beta_{23}(E, \eta_g)}}{\beta'_{23}(E, \eta_g)}\right] \tag{16.34}$$

where

$$\beta_{23}(E, \eta_g) = [2T_{11}]^{-1}\left[T_{21}(E, \eta_g) \pm \sqrt{T_{21}^2(E, \eta_g) + 4T_{11}T_{61}(E, \eta_g)}\right]$$

Using (6.1)–(6.4), (15.71), (15.72) and (16.34), we get

$$\frac{J_D}{W} = \frac{2e^2Bg_v}{\pi^2\hbar^2}\sum_{n=0}^{n_{\max}}\sum_{i=0}^{i_{\max}}\left[\left[\frac{\sqrt{\beta_{23}(e_{95},\eta_g)}}{\beta'_{23}(e_{95},\eta_g)}\right][F_{-1}(\bar{\eta}_{25,12}) - F_{-1}(\bar{\theta}_{25,12})]\right] \tag{16.35}$$

where $\bar{\eta}_{25,12} = (k_BT)^{-1}[E_{F25,12} - e_{95}]$, $\bar{\theta}_{25,12} = \left[\bar{\eta}_{25,12} - \frac{eV_D}{k_BT}\right]$ and $E_{F25,12}$ is the Fermi energy in the present case which can be determined from

$$\frac{C_{ins}}{e}(V_T - V_{GS}) = \frac{g_veB}{2\pi\hbar}\sum_{n=0}^{n_{\max}}\sum_{i=0}^{i_{\max}}[F_{-1}(\bar{\eta}_{25,12}) + F_{-1}(\bar{\theta}_{25,12})] \tag{16.36}$$

16.2.9 The $I_D - V_D$ equation for ballistic magneto HDQWFETs of bismuth telluride

From (11.141a), we can write

$$V_x = \left[\frac{2\sqrt{I_1}}{\hbar}\right]\left[\frac{\sqrt{\gamma_2(E,\eta_g)}}{\gamma'_2(E,\eta_g)}\right] \tag{16.37}$$

where

$$I_1 = \frac{\hbar^2}{2m_0}\bar{\bar{\alpha}}_{11}$$

Using (6.1)–(6.4), (15.76), (15.77) and (16.37), we get

$$\frac{J_D}{W} = \frac{2e^2Bg_v\sqrt{I_1}}{\pi^2\hbar^2}\sum_{n=0}^{n_{\max}}\sum_{i=0}^{i_{\max}}\left[\left[\frac{\sqrt{\gamma_2(e_{96},\eta_g)}}{\gamma'_2(e_{96},\eta_g)}\right][F_{-1}(\bar{\eta}_{25,13}) - F_{-1}(\bar{\theta}_{25,13})]\right] \tag{16.38}$$

where $\bar{\eta}_{25,13} = (k_BT)^{-1}[E_{F25,13} - e_{96}]$, $\bar{\theta}_{25,13} = \left[\bar{\eta}_{25,13} - \frac{eV_D}{k_BT}\right]$ and $E_{F25,13}$ is the Fermi energy in the present case which can be determined from

$$\frac{C_{ins}}{e}(V_T - V_{GS}) = \frac{g_veB}{2\pi\hbar}\sum_{n=0}^{n_{\max}}\sum_{i=0}^{i_{\max}}[F_{-1}(\bar{\eta}_{25,13}) + F_{-1}(\bar{\theta}_{25,13})] \tag{16.39}$$

16.2.10 The $I_D - V_D$ equation for ballistic magneto HDQWFETs of germanium

(a) Model of Cardona *et al.*

From (11.68), we can write

$$V_x = \left[\sqrt{\frac{2}{m_\perp^*}}\frac{\sqrt{\gamma_2(E,\eta_g)}}{\gamma'_2(E,\eta_g)}\right] \tag{16.40}$$

Using (6.1)–(6.4), (15.79), (15.80) and (16.40), we get

$$\frac{J_D}{W} = \frac{e^2 B g_v}{\pi^2 \hbar} \sqrt{\frac{2}{m_\perp^*}} \sum_{n=0}^{n_{\max}} \sum_{i=0}^{i_{\max}} \left[\left[\frac{\sqrt{\gamma_2(e_{97}, \eta_g)}}{\gamma_2'(e_{97}, \eta_g)} \right] [F_{-1}(\bar{\eta}_{25,14}) - F_{-1}(\bar{\theta}_{25,14})] \right] \tag{16.41}$$

where $\bar{\eta}_{25,14} = (k_B T)^{-1}[E_{F25,14} - e_{97}]$, $\bar{\theta}_{25,14} = \left[\bar{\eta}_{25,14} - \frac{eV_D}{k_B T}\right]$ and $E_{F25,14}$ is the Fermi energy in the present case which can be determined from

$$\frac{C_{ins}}{e}(V_T - V_{GS}) = \frac{g_v e B}{2\pi\hbar} \sum_{n=0}^{n_{\max}} \sum_{i=0}^{i_{\max}} [F_{-1}(\bar{\eta}_{25,14}) + F_{-1}(\bar{\theta}_{25,14})] \tag{16.42}$$

(b) Model of Wang and Ressler

From (11.73), we can write

$$V_x = \frac{\omega_{25}'(E, \eta_g)}{\sqrt{2m_\perp^*}} \tag{16.43}$$

where

$$\omega_{25}(E, \eta_g) = \left[\frac{1 - \sqrt{1 - 4c_1 \gamma_3(E, \eta_g)}}{2c_1} \right]^{1/2}$$

Using (6.1)–(6.4), (15.83), (15.84) and (16.43), we get

$$\frac{J_D}{W} = \frac{e^2 B g_v}{\pi^2 \hbar} \sqrt{\frac{1}{2m_\perp^*}} \sum_{n=0}^{n_{\max}} \sum_{i=0}^{i_{\max}} [[\omega_{25}'(e_{98}, \eta_g)][F_{-1}(\bar{\eta}_{25,15}) - F_{-1}(\bar{\theta}_{25,15})]] \tag{16.44}$$

where $\bar{\eta}_{25,15} = (k_B T)^{-1}[E_{F25,15} - e_{98}]$, $\bar{\theta}_{25,15} = \left[\bar{\eta}_{25,15} - \frac{eV_D}{k_B T}\right]$ and $E_{F25,15}$ is the Fermi energy in the present case. which can be determined from

$$\frac{C_{ins}}{e}(V_T - V_{GS}) = \frac{g_v e B}{2\pi\hbar} \sum_{n=0}^{n_{\max}} \sum_{i=0}^{i_{\max}} [F_{-1}(\bar{\eta}_{25,15}) + F_{-1}(\bar{\theta}_{25,15})] \tag{16.45}$$

16.2.11 The $I_D - V_D$ equation for ballistic magneto HDQWFETs of gallium antimonide

Equation (1.96) under the condition of heavy doping assume the form

$$\frac{\hbar^2 k^2}{2m_c} = I_{36}(E, \eta_g) \tag{16.46}$$

where

$$I_{36}(E,\eta_g)=\left[\gamma_3(E,\eta_g)+E'_g-\frac{m_c}{m_0}\cdot\frac{E'_g}{2}-\left[\left(\frac{E'_g}{2}\right)^2+\left[\frac{E'_g}{2}\left(1-\frac{m_c}{m_0}\right)\right]^2\right.\right.$$
$$\left.\left.+\left(\frac{E'_g}{2}\right)^2\left(1-\frac{m_c}{m_0}\right)+\gamma_3(E,\eta_g)E'_g\left(1-\frac{m_c}{m_0}\right)\right]^{1/2}\right]$$

From (16.46), we get

$$V_x=\sqrt{\frac{2}{m_c}}\left[\frac{\sqrt{I_{36}(E,\eta_g)}}{I'_{36}(E,\eta_g)}\right] \tag{16.47}$$

Using (6.1)–(6.4), (15.87), (15.88) and (16.47), we get

$$\frac{J_D}{W}=\frac{e^2Bg_v}{\pi^2\hbar}\sqrt{\frac{2}{m_\perp^*}}\sum_{n=0}^{n_{\max}}\sum_{i=0}^{i_{\max}}\left[\left[\frac{\sqrt{I_{36}(e_{99},\eta_g)}}{I'_{36}(e_{99},\eta_g)}\right]\right.$$
$$\left.\times\ [F_{-1}(\bar{\eta}_{25,15})-F_{-1}(\bar{\theta}_{25,15})]\right] \tag{16.48}$$

where $\bar{\eta}_{25,15}=(k_BT)^{-1}[E_{F25,15}-e_{99}]$, $\bar{\theta}_{25,15}=\left[\bar{\eta}_{25,15}-\frac{eV_D}{k_BT}\right]$ and $E_{F25,15}$ is the Fermi energy in the present case which can be determined from

$$\frac{C_{ins}}{e}(V_T-V_{GS})=\frac{g_veB}{2\pi\hbar}\sum_{n=0}^{n_{\max}}\sum_{i=0}^{i_{\max}}[F_{-1}(\bar{\eta}_{25,15})+F_{-1}(\bar{\theta}_{25,15})] \tag{16.49}$$

16.3 Summary and Conclusion

In Section 16.2.1, we have formulated $I_D - V_D$ for ballistic Magneto HDQWFETs made of nonlinear optical materials as given by (16.2) for the computation of which we need the expression of the 2D electron statistics which is given in (16.3). The equations for $I_D - V_D$ for ballistic Magneto HDQWFETs made of III–V, ternary and quaternary materials has been investigated in Section 16.2.2 in accordance with the models of three band model of Kane, two band model of Kane, Parabolic energy bands, Stillman *et al.* and Palik *et al.* are given by (16.5), (16.8), (16.11), (16.14) and (16.17) and the corresponding 2D electron statistics are expressed through (16.6), (16.9), (16.12), (16.15) and (16.18) respectively. In Section 16.2.3, we have studied the $I_D - V_D$ equation for Magneto HDQWFETs of II–VI materials which is given by (16.20) and the corresponding 2D electron statistics has been represented in (16.21). In Section 16.2.4, we have studied the $I_D - V_D$ equation in Magneto HDQWFETs of IV–VI materials which is given by (16.23) and the corresponding 2D electron statistics has been represented in (16.24). In Section 16.2.5, we have investigated the $I_D - V_D$ equation in Magneto HDQWFETs of stressed Kane type materials which is given by (16.26) and the corresponding 2D electron statistics has been represented in (16.27). In Section 16.2.6, we have studied the $I_D - V_D$ equation in Magneto HDQWFETs of Te which is given by (16.29) and the corresponding 2D electron statistics has been represented in (16.30). In Section 16.2.7, we have studied the $I_D - V_D$ equation in QWFETs of GaP which is given by (16.32) and the corresponding 2D electron statistics has been represented in (16.33). In Section 16.2.8, we have investigated the $I_D - V_D$ equation in Magneto HDQWFETs of $PtSb_2$ which is given by (16.35) and the corresponding 2D electron statistics has been represented in (16.36). In Section 16.2.9, we have studied the $I_D - V_D$ equation in Magneto HDQWFETs of Bismuth Telluride materials which is given by (16.38) and the corresponding 2D electron statistics has been represented in (16.39). In Section 16.2.10, we have studied the $I_D - V_D$ equation in Magneto HDQWFETs of Germanium materials in accordance with the models of Cardona *et al.* and Wang *et al.* are given by (16.41) and (16.44) and the corresponding 2D electron statistics are expressed through (16.42) and (16.45) respectively. In Section 16.2.11, we have studied the $I_D - V_D$ equation in Magneto HDQWFETs of Gallium Antimonide materials which is given by (16.48) and the corresponding 2D electron statistics has been represented in (16.49).

We have not given the graphs of the $I_D - V_D$ equations for ballistic Magneto HDQTFETs made of many materials as discussed here since we feel that our readers will perform the intricate numerical computation to obtain the oscillatory plots for better assessment and joy of understanding. It is the band structure which changes in a fundamental way and consequently all the physical properties of all the electronic materials changes radically leading to new physical concepts.

16.4 Open Research Problem

(R.16.1) Derive equations for ballistic HDQTs under alternating quantizing magnetic fields made of all the materials whose DRs are given in Chapter 1. Draw the corresponding graphs and explain the results.

Chapter 17

The MTP in HDQDSLs

The two most important places to keep clean and tidy are my mind and heart.

17.1 Introduction

The importance of quantum dot superlattices and its different applications are already well-known in the literature [1–32]. The influence of the Dispersion Relation (DR) is extremely important in understanding the electronic properties of the quantized structures. In this chapter, the MTP has been investigated in III–V, II–VI, IV–VI, HgTe/CdTe and strained layer quantum dot HD superlattices (QDSLs) with graded interfaces have been studied in Sections 17.2.1–17.2.5. From Sections 17.2.6 to 17.2.10, the MTP in III–V, II–VI, IV–VI, HgTe/CdTe and strained layer quantum dot effective mass HD superlattices respectively have been presented. Section 17.3 contains the summary and conclusion pertinent to this chapter. The last Section 17.4 presents one multi-dimensional open research problem.

17.2 Theoretical Background

17.2.1 The MTP in III–V HDQDSLs with graded interfaces

The energy spectrum of the conduction electrons in bulk specimens of the constituent materials of III–V SLs whose energy band structures are defined by three band model of Kane can be written as

$$(\hbar^2 k^2)/(2m_{ci}) = EG(E, E_{g_{0i}}, \Delta_i) \tag{17.1}$$

where $i = 1$ and 2

and

$$G(E, E_{g_{0i}}, \Delta_i) \equiv \frac{\left(E_{g_{0i}} + \frac{2}{3}\Delta_i\right)\left(E + E_{g_{0i}} + \Delta_i\right)\left(E + E_{g_{0i}}\right)}{\left[E_{g_{0i}}\left(E_{g_{0i}} + \Delta_i\right)\left(E + E_{g_{0i}} + \frac{2}{3}\Delta_i\right)\right]}$$

The electron dispersion law in bulk specimens of the heavily doped constituent materials of III–V SLs whose un-doped energy band structures are defined by three band model of Kane as written above can be expressed as

$$\frac{\hbar^2 k^2}{2m_{cj}} = T_{1j}(E, \Delta_j, E_{gj}, \eta_{gj}) + iT_{2j}(E, \Delta_j, E_{g_{0i}}, \eta_{gj}) \tag{17.2}$$

where

$$\begin{aligned} j &= 1, 2, T_{ij}(E, \Delta_j, E_{gj}, \eta_{gj}) \\ &= 2[1 + Erf(E/\eta_{gj})]^{-1}\Bigg[(\alpha_j b_j/c_j) \cdot \theta_0(E, \eta_{gj}) + [(\alpha_j c_j + b_j c_j - \alpha_j b_j)/c_j^2] \\ &\quad \times \gamma_0(E, \eta_{gj}) + \Bigg[(1/c_j)(1 - (\alpha_j/c_j))(1 - (b_j/c_j))\frac{1}{2}[1 + Erf(E/\eta_{gj})] \\ &\quad - (1/c_j)(1 - (\alpha_j/c_j))(1 - (b_j/c_j))(2/(c_j\eta_{gj}\sqrt{\pi}))\exp(-u_j^2) \\ &\quad \times \left[\sum_{p=1}^{\infty}(\exp(-p^2/4)/p)\sinh(pu_j)\right]\Bigg]\Bigg] \end{aligned}$$

$$b_j \equiv (E_{g_{0i}} + \Delta_j)^{-1}, \quad c_j \equiv \left(E_{g_{0i}} + \frac{2}{3}\Delta_j\right)^{-1}, \quad u_j \equiv \frac{1 + c_j E}{c_j \eta_{gj}},$$

$$\alpha_j = \frac{1}{E_{g_{0i}}}$$

and

$$T_{2j}(E, \Delta_j, E_{g_{0i}}, \eta_{gj}) \equiv 2[1 + Erf(E/\eta_{gj})]^{-1}\frac{1}{c_j}\left(1 - \frac{\alpha_j}{c_j}\right)\left(1 - \frac{b_j}{c_j}\right) \times \frac{\sqrt{\pi}}{c_j\eta_{gj}}\exp(-u_j^2)$$

Therefore, the dispersion law of the electrons of heavily doped III–V SLs with graded interfaces can be expressed as [24]

$$k_z^2 = G_8 + iH_8 \tag{17.3}$$

where

$$k_s^2 = k_x^2 + k_y^2$$

$$G_8 = \left[\frac{C_7^2 - D_7^2}{L_0^2} - k_s^2\right], \quad C_7 = \cos^{-1}(\bar{\omega}_7),$$

$$\bar{\omega}_7 = (2)^{\frac{-1}{2}}\left[(1 - G_7^2 - H_7^2) - \sqrt{(1 - G_7^2 - H_7^2)^2 + 4G_7^2}\right]^{\frac{1}{2}}$$

$$G_7 = [G_1 + (\rho_5 G_2/2) - (\rho_6 H_2/2) + (\Delta_0/2)\{\rho_7 G_3 - \rho_8 H_3 + \rho_9 G_4 - \rho_{10} H_4 + \rho_{11} G_5 - \rho_{12} H_5 + (1/12)(\rho_{13} G_6 - \rho_{14} H_6)\}],$$

$$G_1 = [(\cos(h_1))(\cosh(h_2))(\cosh(g_1))(\cos(g_2)) + (\sin(h_1))(\sinh(h_2)) \times (\sinh(g_1))(\sin(g_2))]$$

Δ_0 is the width of the interface $L_0(\equiv a_0 + b_0)$ is the length of the period a_0 and b_0 are the width of the barrier and the well respectively, V_0 is the potential barrier encountered by the electron

$$(V_0 = |E_{g_{02}} - E_{g_{01}}|), \quad h_1 = e_1(b_0 - \Delta_0), \quad e_1 = 2^{\frac{-1}{2}}\left(\sqrt{t_1^2 + t_1^2} + t_1\right)^{\frac{1}{2}},$$

$$t_1 = [(2m_{c1}/\hbar^2) \cdot T_{11}(E, E_{g_{01}}, \Delta_1, \eta_{g1}) - k_s^2],$$

$$t_2 = [(2m_{c1}/\hbar^2) \cdot T_{21}(E, E_{g_{01}}, \Delta_1, \eta_{g1})],$$

$$h_2 = e_2(b_0 - \Delta_0), \quad e_2 = 2^{\frac{-1}{2}}\left(\sqrt{t_1^2 + t_1^2} - t_1\right)^{\frac{1}{2}}, \quad g_1 = d_1(a_0 - \Delta_0),$$

$$d_1 = 2^{\frac{-1}{2}}\left(\sqrt{x_1^2 + y_1^2} + x_1\right)^{\frac{1}{2}},$$

$$x_1 = [-(2m_{c2}/\hbar^2) \cdot T_{12}(E - V_0, E_{g_{02}}, \Delta_2, \eta_{g2}) + k_s^2],$$

$$y_1 = [(2m_{c2}/\hbar^2) T_{22}(E - V_0, E_{g_{02}}, \Delta_2, \eta_{g2})]$$

$$g_2 = d_2(a_0 - \Delta_0), d_2 = 2^{\frac{-1}{2}}\left(\sqrt{x_1^2 + y_1^2} - x_1\right)^{\frac{1}{2}},$$

$$\rho_5 = (\rho_3^2 + \rho_4^2)^{-1}[\rho_1\rho_3 - \rho_2\rho_4], \quad \rho_1 = [d_1^2 + e_2^2 - d_2^2 - e_1^2],$$

$$\rho_3 = [d_1 e_1 + d_2 e_2], \quad \rho_2 = 2[d_1 d_2 + e_1 e_2], \quad \rho_4 = [d_1 e_2 - e_1 d_2],$$

$$G_2 = [(\sin(h_1))(\cosh(h_2))(\sinh(g_1))(\cos(g_2)) + (\cos(h_1))(\sinh(h_2)) \times (\cosh(g_1))(\sin(g_2))]$$

$$
\begin{aligned}
\rho_6 &= (\rho_3^2+\rho_4^2)^{-1}[\rho_1\rho_4+\rho_2\rho_3],\\
H_2 &= [(\sin(h_1))(\cosh(h_2))(\sin(g_2))(\cosh(g_1)) - (\cos(h_1))(\sinh(h_2))\\
&\quad \times (\sinh(g_1))(\cos(g_2))]\\
\rho_7 &= [(e_1^2+e_2^2)^{-1}[e_1(d_1^2-d_2^2)-2d_1d_2e_2]-3e_1],\\
G_3 &= [(\sin(h_1))(\cosh(h_2))(\cosh(g_1))(\cos(g_2)) + (\cos(h_1))(\sinh(h_2))\\
&\quad \times (\sinh(g_1))(\sin(g_2))]\\
\rho_8 &= [(e_1^2+e_2^2)^{-1}[e_2(d_1^2-d_2^2)+2d_1d_2e_1]+3e_2],\\
H_3 &= [(\sin(h_1))(\cosh(h_2))(\sin(g_2))(\sinh(g_1)) - (\cos(h_1))(\sinh(h_2))\\
&\quad \times (\cosh(g_1))(\cos(g_2))]\\
\rho_9 &= [(d_1^2+d_2^2)^{-1}[d_1(e_2^2-e_1^2)+2e_2d_1e_1]+3d_1],\\
G_4 &= [(\cos(h_1))(\cosh(h_2))(\cos(g_2))(\sinh(g_1)) - (\sin(h_1))(\sinh(h_2))\\
&\quad \times (\cosh(g_1))(\sin(g_2))]\\
\rho_{10} &= [-(d_1^2+d_2^2)^{-1}[d_2(-e_2^2+e_1^2)+2e_2d_1\rho_1]+3d_2],\\
H_4 &= [(\cos(h_1))(\cosh(h_2))(\cosh(g_1))(\sin(g_2)) + (\sin(h_1))(\sinh(h_2))\\
&\quad \times (\sinh(g_1))(\cos(g_2))]\\
\rho_{11} &= 2[d_1^2+e_2^2-d_2^2-e_1^2],\\
G_5 &= [(\cos(h_1))(\cosh(h_2))(\cos(g_2))(\cosh(g_1)) - (\sin(h_1))(\sinh(h_2))\\
&\quad \times (\sinh(g_1))(\sin(g_2))]\\
\rho_{12} &= 4[d_1d_2+e_1e_2],\\
H_5 &= [(\cos(h_1))(\cosh(h_2))(\sinh(g_1))(\sin(g_2)) + (\sin(h_1))(\sinh(h_2))\\
&\quad \times (\cosh(g_1))(\cos(g_2))]\\
\rho_{13} &= [\{5(d_1e_1^3-3e_1e_2^2d_1)+5d_2(e_1^3-3e_1^2e_2)\}(d_1^2+d_2^2)^{-1}+(e_1^2+e_2^2)^{-1}\\
&\quad \times \{5(e_1d_1^3-3d_2e_1^2d_1)+5(d_2^3e_2-3d_1^2d_2e_2)\}-34(d_1e_1+d_2e_2)],\\
G_6 &= [(\sin(h_1))(\cosh(h_2))(\sinh(g_1))(\cos(g_2)) + (\cos(h_1))(\sinh(h_2))\\
&\quad \times (\cosh(g_1))(\sin(g_2))],\\
\rho_{14} &= [\{5(d_1e_2^3-3e_2e_1^2d_1)+5d_2(-e_1^3+3e_2^2e_1)\}(d_1^2+d_2^2)^{-1}+(e_1^2+e_2^2)^{-1}\\
&\quad \times \{5(-e_1d_2^3+3d_1^2d_2e_1)+5(-d_1^3e_2+3d_2^2d_1e_2)\}+34(d_1e_2-d_2e_1)],
\end{aligned}
$$

$$H_6 = [(\sin(h_1))(\cosh(h_2))(\cosh(g_1))(\sin(g_2)) - (\cos(h_1))(\sinh(h_2)) \times (\sinh(g_1))(\cos(g_2))],$$

$$H_7 = [H_1 + (\rho_5 H_2/2) + (\rho_6 G_2/2) + (\Delta_0/2)\{\rho_8 G_3 + \rho_7 H_3 + \rho_{10} G_4 + \rho_9 H_4 + \rho_{12} G_5 + \rho_{11} H_5 + (1/12)(\rho_{14} G_6 + \rho_{13} H_6)\}],$$

$$H_1 = [(\sin(h_1))(\sinh(h_2))(\cosh(g_1))(\cos(g_2)) + (\cos(h_1))(\cosh(h_2)) \times (\sinh(g_1))(\sin(g_2))],$$

$$D_7 = \sinh^{-1}(\bar{\omega}_7), \quad H_8 = (2C_7 D_7/L_0^2)$$

The simplified dispersion relation of HDQD III–V superlattices with graded interfaces can be expressed as

$$\left(\frac{n_x \pi}{d_x}\right)^2 + \left(\frac{n_y \pi}{d_y}\right)^2 + \left(\frac{n_z \pi}{d_z}\right)^2 = (G_8 + iH_8)\Big|_{E=E_{17,1},\, k_x=\frac{n_x\pi}{d_x},\, k_y=\frac{n_y\pi}{d_y}} \tag{17.4}$$

where $E_{17,1}$ is the total quantized energy in this case.

The DOS function is given by

$$N(E) = \frac{2}{d_x d_y d_z} \sum_{n_x=1}^{n_{x\max}} \sum_{n_y=1}^{n_{y\max}} \sum_{n_z=1}^{n_{z\max}} \delta'(E - E_{17,1}) \tag{17.5}$$

The electron concentration assumes the form

$$n_0 = \frac{2}{d_x d_y d_z} \text{Real part of} \sum_{n_x=1}^{n_{x\max}} \sum_{n_y=1}^{n_{y\max}} \sum_{n_z=1}^{n_{z\max}} F_{-1}(\eta_{17,1}) \tag{17.6}$$

where $\eta_{17,1} = (k_B T)^{-1}(E_{F17,1} - E_{17,1})$ and $E_{F17,1}$ is the Fermi energy in this case.

Using (17.6) and (1.7), we can study the MTP in this case.

17.2.2 The MTP in II–VI HDQDSLs with graded interfaces

The energy spectrum of the conduction electrons of the constituent materials of II–VI SLs can be rewritten as

$$E = \frac{\hbar^2 k_s^2}{2m_{\perp,1}^*} + \frac{\hbar^2 k_z^2}{2m_{\|,1}^*} + \bar{\lambda}_0 k_s \tag{17.7}$$

and

$$\frac{\hbar^2 k^2}{2m_{c2}} = EG(E, E_{g2}, \Delta_2) \tag{17.8}$$

where $m^*_{\perp,1}$ and $m^*_{\parallel,1}$ are the transverse and longitudinal effective electron masses respectively at the edge of the conduction band of the first material.

The electron energy spectra of the heavily doped constituent materials of II–VI SLs are given by

$$\gamma_3(E, \eta_{g1}) = [(\hbar^2 k_s^2/2m^*_{\perp,1}) + (\hbar^2 k_z^2/2m^*_{\parallel,1})] \pm \bar{\lambda}_0 k_s \tag{17.9}$$

and

$$\frac{\hbar^2 k^2}{2m_{c2}} = T_{12}(E, \Delta_2, E_{g2}\eta_{g2}) + iT_{22}(E, \Delta_2, E_{g2}, \eta_{g2}) \tag{17.10}$$

The energy-wave vector dispersion relation of the conduction electrons in heavily doped II–VI SLs with graded interfaces can be expressed as

$$k_z^2 = G_{19} + iH_{19} \tag{17.11}$$

where

$$G_{19} = \left[\frac{C_{18}^2 - D_{18}^2}{L_0^2} - k_s^2\right]$$

$$C_{18} = \cos^{-1}(\omega_{18}),$$

$$\omega_{18} = (2)^{\frac{-1}{2}}\left[(1 - G_{18}^2 - H_{18}^2) - \sqrt{(1 - G_{18}^2 - H_{18}^2)^2 + 4G_{18}^2}\right]^{\frac{1}{2}}$$

$$G_{18} = \frac{1}{2}[G_{11} + G_{12} + \Delta_0(G_{13} + G_{14}) + \Delta_0(G_{15} + G_{16})],$$

$$G_{11} = 2(\cosh(g_1))(\cos(g_2))(\cos\gamma_{11}(E, k_s))$$

$$\gamma_{11}(E, k_s) = k_{21}(E, k_s)(b_0 - \Delta_0),$$

$$k_{21}(E, k_s) = \left\{\left[\gamma_3(E, \eta_{g1}) - \frac{\hbar^2 k_s^2}{2m^*_{\perp,1}} \pm \bar{\lambda}_0 k_s\right]\frac{2m^*_{\parallel,1}}{\hbar^2}\right\}^{\frac{1}{2}}$$

$$G_{12} = ([\Omega_1(E, k_s)(\sinh g_1)(\cos g_2) - \Omega_2(E, k_s)(\sin g_2)(\cosh g_1)] \times (\sin\gamma_{11}(E, k_s)))$$

$$\Omega_1(E,k_s) = \left[\frac{d_1}{k_{21}(E,k_s)} - \frac{k_{21}(E,k_s)d_1}{d_1^2+d_2^2}\right], \quad d_1 = \left[\frac{\sqrt{x_1^2+y_1^2}+x_1}{2}\right]^{1/2},$$

$$d_2 = \left[\frac{\sqrt{x_1^2+y_1^2}-x_1}{2}\right]^{1/2},$$

$$x_1 = \left[k_s^2 - \left(\frac{2m_{c2}}{\hbar^2}\right)T_{12}(E-V_0,\Delta_2,E_{g2},\eta_{g2})\right],$$

$$y_1 = \left[\left(\frac{2m_{c2}}{\hbar^2}\right)T_{22}(E-V_0,\Delta_2,E_{g2},\eta_{g2})\right],$$

$$\Omega_2(E,k_s) = \left[\frac{d_2}{k_{21}(E,k_s)} + \frac{k_{21}(E,k_s)d_2}{d_1^2+d_2^2}\right],$$

$$G_{13} = ([\Omega_3(E,k_s)(\cosh g_1)(\cos g_2) - \Omega_4(E,k_s)(\sinh g_1)(\sin g_2)] \times (\sin\gamma_{11}(E,k_s)))$$

$$\Omega_3(E,k_s) = \left[\frac{d_1^2-d_2^2}{k_{21}(E,k_s)} - 3k_{21}(E,k_s)\right], \quad \Omega_4(E,k_s) = \left[\frac{2d_1d_2}{k_{21}(E,k_s)}\right],$$

$$G_{14} = ([\Omega_5(E,k_s)(\sinh g_1)(\cos g_2) - \Omega_6(E,k_s)(\sin g_1)(\cosh g_2)] \times (\cos\gamma_{11}(E,k_s)))$$

$$\Omega_5(E,k_s) = \left[3d_1 - \frac{d_1}{d_1^2+d_2^2}k_{21}^2(E,k_s)\right],$$

$$\Omega_6(E,k_s) = \left[3d_2 + \frac{d_2}{d_1^2+d_2^2}k_{21}^2(E,k_s)\right],$$

$$G_{15} = ([\Omega_9(E,k_s)(\cosh g_1)(\cos g_2) - \Omega_{10}(E,k_s)(\sinh g_1)(\sin g_2)] \times (\cos\gamma_{11}(E,k_s)))$$

$$\Omega_9(E,k_s) = [2d_1^2 - 2d_2^2 - k_{21}^2(E,k_s)], \quad \Omega_{10}(E,k_s) = [2d_1d_2]$$

$$G_{16} = ([\Omega_7(E,k_s)(\sinh g_1)(\cos g_2) - \Omega_8(E,k_s)(\sin g_1)(\cosh g_2)] \times (\sin\gamma_{11}(E,k_s)(12)^{-1}))$$

$$\Omega_7(E,k_s) = \left[\frac{5d_1}{d_1^2+d_2^2}k_{21}^3(E,k_s) + \frac{5(d_1^3-3d_2^2d_1)}{k_{21}(E,k_s)} - 34k_{21}(E,k_s)d_2\right],$$

$$\Omega_8(E,k_s) = \left[\frac{5d_2}{d_1^2+d_2^2}k_{21}^3(E,k_s) + \frac{5(d_1^3-3d_1^2d_2)}{k_{21}(E,k_s)} + 34k_{21}(E,k_s)d_2\right],$$

$$
\begin{aligned}
H_{18} &= \frac{1}{2}[H_{11} + H_{12} + \Delta_0(H_{13} + H_{14}) + \Delta_0(H_{15} + H_{16})] \\
H_{11} &= 2(\sinh g_1 \sin g_2 \cos\gamma_{11}(E, k_s)), \\
H_{12} &= ([\Omega_2(E, k_s)(\sinh g_1)(\cos g_2) + \Omega_1(E, k_s)(\sin g_2)(\cosh g_1)] \\
&\quad \times (\sin\gamma_{11}(E, k_s))), \\
H_{13} &= ([\Omega_4(E, k_s)(\cosh g_1)(\cos g_2) + \Omega_3(E, k_s)(\sinh g_1)(\sin g_2)] \\
&\quad \times (\sin\gamma_{11}(E, k_s))), \\
H_{14} &= ([\Omega_6(E, k_s)(\sinh g_1)(\cos g_2) + \Omega_5(E, k_s)(\sin g_1)(\cosh g_2)] \\
&\quad \times (\cos\gamma_{11}(E, k_s))), \\
H_{15} &= ([\Omega_{10}(E, k_s)(\cosh g_1)(\cos g_2) + \Omega_9(E, k_s)(\sinh g_1)(\sin g_2)] \\
&\quad \times (\cos\gamma_{11}(E, k_s))), \\
H_{16} &= ([\Omega_8(E, k_s)(\sinh g_1)(\cos g_2) + \Omega_7(E, k_s)(\sin g_1)(\cosh g_2)] \\
&\quad \times (\sin\gamma_{11}(E, k_s)/12)), \\
H_{19} &= \left[\frac{2C_{18}D_{18}}{L_0^2}\right] \quad \text{and} \quad D_{18} = \sinh^{-1}(\omega_{18})
\end{aligned}
$$

The simplified dispersion relation in heavily doped quantum dot II–VI superlattices with graded interfaces can be expressed as

$$
\left(\frac{n_x\pi}{d_x}\right)^2 + \left(\frac{n_y\pi}{d_y}\right)^2 + \left(\frac{n_z\pi}{d_z}\right)^2 = G_{19} + iH_{19}\Big|_{E=E_{17,2},\, k_x=\frac{n_x\pi}{d_x},\, k_y=\frac{n_y\pi}{d_y}} \tag{17.12}
$$

where $E_{17,2}$ is the total quantized energy in this case.

The DOS function is given by

$$
N(E) = \frac{2}{d_x d_y d_z} \sum_{n_x=1}^{n_{x\max}} \sum_{n_y=1}^{n_{y\max}} \sum_{n_z=1}^{n_{z\max}} \delta'(E - E_{17,2}) \tag{17.13}
$$

The electron concentration assumes the form

$$
n_0 = \frac{2}{d_x d_y d_z} \text{Real part of} \sum_{n_x=1}^{n_{x\max}} \sum_{n_y=1}^{n_{y\max}} \sum_{n_z=1}^{n_{z\max}} F_{-1}(\eta_{17,2}) \tag{17.14}
$$

where $\eta_{17,2} = (k_B T)^{-1}(E_{F17,2} - E_{17,2})$ and $E_{F17,2}$ is the Fermi energy in this case.

Using (17.14) and (1.7), we can study the MTP in this case.

17.2.3 The MTP in IV–VI HDQDSLs with graded interfaces

The E–k dispersion relation of the conduction electrons of the constituent materials of the IV–VI SLs can be re-expressed as

$$E = a_i k_s^2 + b_i k_z^2 + \left[[c_i k_s^2 + d_i k_z^2] + \left(e_i k_s^2 + f_i k_z^2 + \frac{E_{g_i}}{2} \right)^2 \right]^{1/2} - \frac{E_{g_i}}{2} \tag{17.15}$$

where

$$a_i \equiv \left[\frac{\hbar^2}{2m_{\perp,i}^-} \right], \quad b_i \equiv \left(\frac{\hbar^2}{2m_{\|,i}^-} \right), \quad c_i \equiv P_{\perp,i}^2, \quad d_i \equiv P_{\perp,i}^2,$$

$$e_i \equiv \left[\frac{\hbar^2}{2m_{\perp,i}^+} \right] \quad \text{and} \quad f_i \equiv \left(\frac{\hbar^2}{2m_{\|,i}^+} \right)$$

The dispersion relation of the conduction electrons of the heavily doped constituent materials of the IV–VI SLs can be rewritten as

$$\begin{aligned} k_z^2 = {} & [2\bar{p}_{9,i}]^{-1}[+\bar{q}_{9,i}(E, k_s, \eta_{gi}) - [[\bar{q}_{9,i}(E, k_s, \eta_{gi})]^2 \\ & - 4\bar{p}_{9,i}\bar{R}_{9,i}(E, k_s, \eta_{gi})]^{\frac{1}{2}}] \end{aligned} \tag{17.16}$$

where

$$\begin{aligned} \bar{p}_{9,i} = (b_i^2 - f_i^2), \quad \bar{q}_{9,i}(E, k_s, \eta_{gi}) = {} & \left[2b_i \left\{ \gamma_3(E, \eta_{gi}) + \frac{E_{gi}}{2} \right\} \right. \\ & \left. -2a_i b_i k_s^2 + d_i - 2e_i f_i k_s^2 - f_i E_{gi} \right] \end{aligned}$$

and

$$\begin{aligned} \bar{R}_{9,i}(E, k_s, \eta_{gi}) = {} & \left[\left\{ \gamma_3(E, \eta_{gi}) + \frac{E_{gi}}{2} \right\}^2 + a_i^2 k_s^4 - c_i k_s^2 - e_i k_s^4 - \frac{E_{gi}^2}{4} \right. \\ & \left. - e_i k_s^2 E_{gi} - 2a_i \left\{ \gamma_3(E, \eta_{gi}) + \frac{E_{gi}}{2} \right\} k_s^2 \right] \end{aligned}$$

The electron dispersion law in heavily doped IV–VI SLs with graded interfaces can be expressed as

$$\cos(L_0 k) = \frac{1}{2}\Phi_2(E, k_s) \tag{17.17}$$

where

$$\begin{aligned}
\Phi_2(E,k_s) \equiv & \Bigg[2\cosh\{\beta_2(E,k_s)\}\cos\{\gamma_2(E,k_s)\} + \varepsilon_2(E,k_s)\sinh\{\beta_2(E,k_s)\} \\
& \times \sin\{\gamma_{22}(E,k_s)\} + \Delta_0\left[\left(\frac{\{K_{112}(E,k_s)\}^2}{K_{212}(E,k_s)} - 3K_{212}(E,k_s)\right)\right. \\
& \times \cosh\{\beta_2(E,k_s)\}\sin\{\gamma_{22}(E,k_s)\} + \left(3K_{112}(E,k_s) - \frac{\{K_{112}(E,k_s)\}^2}{K_{212}(E,k_s)}\right) \\
& \left.\times \sinh\{\beta_2(E,k_s)\}\cos\{\gamma_{22}(E,k_s)\}\right] + \Delta_0\Bigg[2(\{K_{112}(E,k_s)\}^2 \\
& - \{K_{212}(E,k_s)\}^2)\cosh\{\beta_2(E,k_s)\}\cos\{\gamma_{22}(E,k_s)\} \\
& + \frac{1}{12}\left[\frac{5\{K_{112}(E,k_s)\}^3}{K_{212}(E,k_s)} + \frac{5\{K_{212}(E,k_s)\}^3}{K_{112}(E,k_s)} - 34K_{212}(E,k_s)K_{112}(E,k_s)\right] \\
& \times \sinh\{\beta_2(E,k_s)\}\sin\{\gamma_{22}(E,k_s)\}\Bigg]\Bigg] \quad \beta_2(E,k_s) \equiv K_{112}(E,k_s)[a_0 - \Delta_0],
\end{aligned}$$

$$\begin{aligned}
k_{112}^2(E,k_s) = & [2\bar{p}_{9,2}]^{-1}[-\bar{q}_{9,2}(E - V_0, k_s, \eta_{g2}) + [[\bar{q}_{9,2}(E - V_0, k_s, \eta_{g2})]^2 \\
& - 4\bar{p}_{9,2}\bar{R}_{9,2}(E - V_0, k_s, \eta_{g2})]^{\frac{1}{2}}], \quad \gamma_{22}(E,k_s) \equiv K_{212}(E,k_s)[b_0 - \Delta_0], \\
k_{212}^2(E,k_s) = & [2\bar{p}_{9,1}]^{-1}[+\bar{q}_{9,1}(E, k_s, \eta_{g1}) - [[\bar{q}_{9,1}(E, k_s, \eta_{g1})]^2 \\
& - 4\bar{p}_{9,1}\bar{R}_{9,1}(E, k_s, \eta_{g1})]^{\frac{1}{2}}],
\end{aligned}$$

and

$$\varepsilon_2(E,k_s) \equiv \left[\frac{K_{112}(E,k_s)}{K_{212}(E,k_s)} - \frac{K_{212}(E,k_s)}{K_{112}(E,k_s)}\right]$$

The simplified dispersion relation in heavily doped quantum dot IV–VI superlattices with graded interfaces can be expressed as

$$\begin{aligned}
\left(\frac{n_z\pi}{d_z}\right)^2 = & \frac{1}{L_0^2}\left[\cos^{-1}\left\{\frac{1}{2}\Phi_2\left(E_{17,3}, \frac{n_x\pi}{d_x}, \frac{n_y\pi}{d_y}\right)\right\}\right]^2 \\
& - \left[\left(\frac{n_x\pi}{d_x}\right)^2 + \left(\frac{n_y\pi}{d_y}\right)^2\right]
\end{aligned} \tag{17.18}$$

where $E_{17,3}$ is the total quantized energy in this case.

The DOS function is given by

$$N(E) = \frac{2}{d_x d_y d_z} \sum_{n_x=1}^{n_{x\max}} \sum_{n_y=1}^{n_{y\max}} \sum_{n_z=1}^{n_{z\max}} \delta'(E - E_{17,3}) \tag{17.19}$$

The electron concentration assumes the form

$$n_0 = \frac{2}{d_x d_y d_z} \sum_{n_x=1}^{n_{x\max}} \sum_{n_y=1}^{n_{y\max}} \sum_{n_z=1}^{n_{z\max}} F_{-1}(\eta_{17,3}) \tag{17.20}$$

where $\eta_{17,3} = (k_B T)^{-1}(E_{F17,3} - E_{17,3})$ and $E_{F17,3}$ is the Fermi energy in this case.

Using (17.20) and (1.7), we can study the MTP in this case.

*17.2.4 The MTP in HgTe/CdTe HDQDSLs with graded interfaces

The electron energy spectra of the constituent materials of in HD HgTe/CdTe SLs are given by

$$k^2 = \left[\frac{B_{01}^2 + 2A_1\gamma_3(E,\eta_{g1}) - B_{01}\sqrt{B_{01}^2 + 4A_1\gamma_3(E,\eta_{g1})}}{2A_1^2}\right] \tag{17.21}$$

and

$$\frac{\hbar^2 k^2}{2m_{c2}} = T_{12}(E,\Delta_2,E_{g2}\eta_{g2}) + iT_{22}(E,\Delta_2,E_{g2},\eta_{g2}) \tag{17.22}$$

where $B_{01} = (3|e|^2/128\varepsilon_{sc1})$, $A_1 = (\hbar^2/2m_{c1}) \cdot \varepsilon_{sc1}$ is the semiconductor permittivity of the first material.

The energy-wave vector dispersion relation of the conduction electrons in heavily doped HgTe/CdTe SLs with graded interfaces can be expressed as

$$k_z^2 = G_{192} + iH_{192} \tag{17.23}$$

where

$$G_{192} = [((C_{182}^2 - D_{182}^2)/L_0^2) - k_s^2], \quad C_{182} = \cos^{-1}(\omega_{182}),$$

$$\omega_{182} = (2)^{\frac{-1}{2}}\left[(1 - G_{182}^2 - H_{182}^2) - \sqrt{(1 - G_{182}^2 - H_{182}^2)^2 + 4G_{182}^2}\right]^{\frac{1}{2}},$$

$$G_{182} = \frac{1}{2}[G_{112} + G_{122} + \Delta_0(G_{132} + G_{142}) + \Delta_0(G_{152} + G_{162})],$$

$$G_{112} = 2(\cos(g_{12}))(\cos(g_{22}))(\cos\gamma_8(E,k_s)),$$

$$\gamma_8(E,k_s) = k_8(E,k_s)(b_0 - \Delta_0),$$

$$
\begin{aligned}
k_8(E,k_s) &= \left[\frac{B_{01}^2 + 2A_1\gamma_3(E,\eta_{g1}) - B_{01}\sqrt{B_{01}^2 + 4A_1\gamma_3(E,\eta_{g1})}}{2A_1^2} - k_s^2\right]^{1/2} \\
G_{122} &= ([\Omega_{12}(E,k_s)(\sinh g_{12})(\cos g_{22}) - \Omega_{22}(E,k_s)(\sin g_{22})(\cosh g_{12})] \\
&\quad \times (\sin\gamma_8(E,k_s))) \\
\Omega_{12}(E,k_s) &= \left[\frac{d_{12}}{k_8(E,k_s)} - \frac{k_8(E,k_s)d_{12}}{d_{12}^2 + d_{22}^2}\right], \quad d_{12} = \left[\frac{\sqrt{x_{12}^2 + y_{12}^2} + x_{12}}{2}\right]^{1/2}, \\
d_{22} &= \left[\frac{\sqrt{x_{12}^2 + y_{12}^2} - x_{12}}{2}\right]^{1/2}, \\
x_{12} &= \left[k_s^2 - \left(\frac{2m_{c2}}{\hbar^2}\right) T_{12}(E - V_0, \Delta_2, E_{g2}, \eta_{g2})\right], \\
y_{12} &= \left[\left(\frac{2m_{c2}}{\hbar^2}\right) T_{22}(E - V_0, \Delta_2, E_{g2}, \eta_{g2})\right] \\
\Omega_{22}(E,k_s) &= \left[\frac{d_{22}}{k_8(E,k_s)} + \frac{k_8(E,k_s)d_{22}}{d_{12}^2 + d_{22}^2}\right], \\
G_{132} &= ([\Omega_{32}(E,k_s)(\cosh g_{12})(\cos g_{22}) - \Omega_{42}(E,k_s)(\sinh g_{12})(\sin g_{22})] \\
&\quad \times (\sin\gamma_8(E,k_s))), \\
\Omega_{32}(E,k_s) &= \left[\frac{d_{12}^2 - d_{22}^2}{k_8(E,k_s)} - 3k_8(E,k_s)\right], \quad \Omega_{42}(E,k_s) = \left[\frac{2d_{12}d_{22}}{k_8(E,k_s)}\right], \\
G_{142} &= ([\Omega_{52}(E,k_s)(\sinh g_{12})(\cos g_{22}) - \Omega_{62}(E,k_s)(\sin g_{12})(\cosh g_{22})] \\
&\quad \times (\cos\gamma_8(E,k_s))), \\
\Omega_{52}(E,k_s) &= \left[3d_{12} - \frac{d_{12}}{d_{12}^2 + d_{22}^2} k_8^2(E,k_s)\right], \\
\Omega_{62}(E,k_s) &= \left[3d_{22} + \frac{d_{22}}{d_{12}^2 + d_{22}^2} k_8^2(E,k_s)\right], \\
G_{152} &= ([\Omega_{92}(E,k_s)(\cosh g_{12})(\cos g_{22}) - \Omega_{102}(E,k_s)(\sinh g_{12})(\sin g_{22})] \\
&\quad \times (\cos\gamma_8(E,k_s))), \\
\Omega_{92}(E,k_s) &= [2d_{12}^2 - 2d_{22}^2 - k_8^2(E,k_s)], \quad \Omega_{102}(E,k_s) = [2d_{12}d_{22}], \\
G_{162} &= ([\Omega_{72}(E,k_s)(\sinh g_{12})(\cos g_{22}) - \Omega_{82}(E,k_s)(\sin g_{12})(\cosh g_{22})] \\
&\quad \times (\sin\gamma_8(E,k_s)/12))
\end{aligned}
$$

$$\Omega_{72}(E,k_s) = \left[\frac{5d_{12}}{d_{12}^2+d_{22}^2}k_8^3(E,k_s) + \frac{5(d_{12}^3-3d_{22}^2d_{12})}{k_8(E,k_s)} - 34k_8(E,k_s)d_{22}\right],$$

$$\Omega_{82}(E,k_s) = \left[\frac{5d_{22}}{d_{12}^2+d_{22}^2}k_8^3(E,k_s) + \frac{5(d_{12}^3-3d_{22}d_{12}^2)}{k_8(E,k_s)} + 34k_8(E,k_s)d_{22}\right],$$

$$H_{182} = \frac{1}{2}[H_{112}+H_{122}+\Delta_0(H_{132}+H_{142})+\Delta_0(H_{152}+H_{162})],$$

$$H_{112} = 2((\sinh g_{12})(\sin g_{22})(\cos\gamma_8(E,k_s))), \quad g_{12} = d_{12}(a_0-\Delta_0),$$

$$g_{22} = d_{22}(a_0-\Delta_0),$$

$$H_{122} = ([\Omega_{22}(E,k_s)(\sinh g_{12})(\cos g_{22}) + \Omega_{12}(E,k_s)(\sin g_{22})(\cosh g_{12})] \times (\sin\gamma_8(E,k_s))),$$

$$H_{132} = ([\Omega_{42}(E,k_s)(\cosh g_{12})(\cos g_{22}) + \Omega_{32}(E,k_s)(\sinh g_{12})(\sin g_{22})] \times (\sin\gamma_8(E,k_s))),$$

$$H_{142} = ([\Omega_{62}(E,k_s)(\sinh g_{12})(\cos g_{22}) + \Omega_{52}(E,k_s)(\sin g_{12})(\cosh g_{22})] \times (\cos\gamma_8(E,k_s))),$$

$$H_{152} = ([\Omega_{102}(E,k_s)(\cosh g_{12})(\cos g_{22}) + \Omega_{92}(E,k_s)(\sinh g_{12})(\sin g_{22})] \times (\cos\gamma_8(E,k_s))),$$

$$H_{162} = ([\Omega_{82}(E,k_s)(\sinh g_{12})(\cos g_{22}) + \Omega_{72}(E,k_s)(\sin g_{12})(\cosh g_{22})] \times (\sin\gamma_8(E,k_s)/12)),$$

$$H_{192} = [((2C_{182}D_{182})/L_0^2)] \quad \text{and} \quad D_{182} = \sinh^{-1}(\omega_{182})$$

The simplified dispersion relation of HDQD HgTe/CdTe superlattices with graded interfaces can be expressed as

$$\left(\frac{n_x\pi}{d_x}\right)^2 + \left(\frac{n_y\pi}{d_y}\right)^2 + \left(\frac{n_z\pi}{d_z}\right)^2 = (G_{192}+iH_{192})\Big|_{E=E_{17,4},\,k_x=\frac{n_x\pi}{d_x},\,k_y=\frac{n_y\pi}{d_y}} \tag{17.24}$$

where $E_{17,4}$ is the total quantized energy in this case.

The DOS function is given by

$$N(E) = \frac{2}{d_xd_yd_z}\sum_{n_x=1}^{n_{x\max}}\sum_{n_y=1}^{n_{y\max}}\sum_{n_z=1}^{n_{z\max}}\delta'(E-E_{17,4}) \tag{17.25}$$

The electron concentration assumes the form

$$n_0 = \frac{2}{d_x d_y d_z} \text{Real part of} \sum_{n_x=1}^{n_{x\max}} \sum_{n_y=1}^{n_{y\max}} \sum_{n_z=1}^{n_{z\max}} F_{-1}(\eta_{17,4}) \tag{17.26}$$

where $\eta_{17,4} = (k_B T)^{-1}(E_{F17,4} - E_{17,4})$ and $E_{F17,4}$ is the Fermi energy in this case.

Using (17.26) and (1.7), we can study the MTP in this case.

*17.2.5 The MTP in strained layer HDQDSLs with graded interfaces

The electron energy spectrum in HD stressed materials can be written from (11.116) in this case as

$$P_{11,i}(E,\eta_g)k_x^2 + Q_{11,i}(E,\eta_g)k_y^2 + S_{11,i}(E,\eta_g)k_z^2 = 1 \tag{17.27}$$

The energy-wave vector dispersion relation of the conduction electrons in heavily doped strained layer SLs with graded interfaces can be expressed as

$$\cos(L_0 k) = \frac{1}{2}\bar{\Phi}_6(E,k_s,\eta_{g1},\eta_{g2}) \tag{17.28}$$

where

$$\begin{aligned}
\bar{\Phi}_6(E,k_s,\eta_{g1},\eta_{g2}) &= [2\cosh[T_4(E,\eta_{g2})]\cos[T_5(E,\eta_{g1})]] \\
&+ [T_6(E,n,\eta_{g1},\eta_2)]\sinh[T_4(E,\eta_{g2})]\sin[T_5(E,\eta_{g1})] \\
&+ \Delta_0\left[\left(\frac{k_0^2(E,\eta_{g2})}{k_{70}(E,\eta_{g1})} - 3k_{70}(E,\eta_{g1})\right)\cosh[T_4(E,\eta_{g2})]\sin[T_5(E,\eta_{g1})]\right. \\
&+ \left.\left(3k_0(E,\eta_{g2}) - \frac{k_{70}^2(E,\eta_{g1})}{k_0(E,\eta_{g2})}\right)\sinh[T_4(E,\eta_{g2})]\cos[T_5(E,\eta_{g1})]\right] \\
&+ \Delta_0[2(k_0^2(E,\eta_{g2}) - k_{70}^2(E,\eta_{g1}))\cosh[T_4(E,\eta_{g2})]\sin[T_5(E,\eta_{g1})]] \\
&+ \frac{1}{12}\left(\frac{5k_0^3(E,\eta_{g2})}{k_{70}(E,\eta_{g1})} + \frac{5k_{70}^3(E,\eta_{g1})}{k_0(E,\eta_{g2})} - 34k_0(E,\eta_{g2})k_{70}(E,\eta_{g1})\right) \\
&\times \sinh[T_4(E,\eta_{g2})]\sin[T_5(E,\eta_{g1})][T_4(E,\eta_{g2})] = k_0(E,\eta_{g2})[a_0 - \Delta_0],
\end{aligned}$$

$$\begin{aligned}
k_0(E,\eta_{g2}) &= [\bar{S}_{11,2}(E-V_0,\eta_{g2})]^{-1/2}[\bar{P}_{11,2}(E-V_0,\eta_{g2})k_x^2 \\
&+ \bar{Q}_{11,2}(E-V_0,\eta_{g2})k_y^2 - 1]^{1/2},
\end{aligned}$$

$$T_5(E,\eta_{g1}) = k_{70}(E,\eta_{g1})[b_0 - \Delta_0],$$
$$k_{70}(E,\eta_{g1}) = [\bar{S}_{11,1}(E,\eta_{g1})]^{-1/2}[1 - \bar{P}_{11,1}(E,\eta_{g1})k_x^2 - \bar{Q}_{11,1}(E,\eta_{g1})k_y^2]^{1/2}$$

and

$$T_6(E,\eta_{g1},\eta_{g2}) = \left[\frac{k_0(E,\eta_{g2})}{k_{70}(E,\eta_{g1})} - \frac{k_{70}(E,\eta_{g1})}{k_0(E,\eta_{g2})}\right]$$

The simplified dispersion relation in heavily doped quantum dot strained layer superlattices with graded interfaces can be expressed as

$$\left(\frac{n_z\pi}{d_z}\right)^2 = \frac{1}{L_0^2}\left[\cos^{-1}\left\{\frac{1}{2}\bar{\Phi}_6\left(E_{17,5},\frac{n_x\pi}{d_x},\frac{n_y\pi}{d_y}\right)\right\}\right]^2 - \left[\left(\frac{n_x\pi}{d_x}\right)^2 + \left(\frac{n_y\pi}{d_y}\right)^2\right] \tag{17.29}$$

where $E_{17,5}$ is the total quantized energy in this case.

The DOS function is given by

$$N(E) = \frac{2}{d_x d_y d_z}\sum_{n_x=1}^{n_{x\max}}\sum_{n_y=1}^{n_{y\max}}\sum_{n_z=1}^{n_{z\max}}\delta'(E - E_{17,5}) \tag{17.30}$$

The electron concentration assumes the form

$$n_0 = \frac{2}{d_x d_y d_z}\sum_{n_x=1}^{n_{x\max}}\sum_{n_y=1}^{n_{y\max}}\sum_{n_z=1}^{n_{z\max}}F_{-1}(\eta_{17,5}) \tag{17.31}$$

where $\eta_{17,5} = (k_B T)^{-1}(E_{F17,5} - E_{17,5})$ and $E_{F17,5}$ is the Fermi energy in this case.

Using (17.31) and (1.7), we can study the MTP in this case.

17.2.6 The MTP in III–V HDQD effective mass superlattices

Following Sasaki [32], the electron dispersion law in III–V heavily doped effective mass superlattices (EMSLs) can be written as

$$k_x^2 = \left[\frac{1}{L_0^2}\{\cos^{-1}(f_{21}(E,k_y,k_z))\}^2 - k_\perp^2\right] \tag{17.32}$$

in which

$$k_\perp^2 = k_y^2 + k_z^2$$

$$f_{21}(E, k_y, k_z) = \bar{a}_1 \cos[a_0 C_{21}(E, k_\perp, \eta_{g1}) + b_0 D_{21}(E, k_\perp, \eta_{g2})] - \bar{a}_2 \cos[a_0 C_{21}(E, k_\perp, \eta_{g2}) - b_0 D_{21}(E, k_\perp, \eta_{g2})], \quad k_\perp^2 = k_y^2 + k_z^2,$$

$$\bar{a}_1 = \left[\sqrt{\frac{M_2(0,\eta_{g2})}{M_1(0,\eta_{g1})}} + 1\right]^2 \left[4\left(\frac{M_2(0,\eta_{g2})}{M_1(0,\eta_{g1})}\right)^{1/2}\right]^{-1}$$

$$\bar{a}_2 = \left[\sqrt{\frac{M_2(0,\eta_{g2})}{M_1(0,\eta_{g1})}} - 1\right]^2 \left[4\left(\frac{M_2(0,\eta_{g2})}{M_1(0,\eta_{g1})}\right)^{1/2}\right]^{-1}$$

$$M_i(0,\eta_{gi}) = m_{ci}\left[\frac{-2}{\sqrt{\pi}} T_i(0,\eta_{gi}) + 2\left[\frac{\alpha_i b_i}{c_i}\frac{\eta_{gi}}{\sqrt{\pi}} + \frac{1}{2}\left(\frac{\alpha_i c_i + c_i b_i - \alpha_i b_i}{c_i^2}\right) + \frac{1}{c_i\sqrt{\pi}}\left(1 - \frac{\alpha_i}{c_i}\right)\left(1 - \frac{b_i}{c_i}\right) - \frac{1}{c_i}\left(1 - \frac{\alpha_i}{c_i}\right)\left(1 - \frac{b_i}{c_i}\right)\frac{2}{c_i \eta_{gi}\sqrt{\pi}} \times \left\{\frac{-2}{c_i \eta_{gi}^2}\exp\left(\frac{-1}{c_i^2\eta_{gi}^2}\right)\left(\sum_{p=1}^{\alpha}\exp\left(\frac{-p^2}{4}\right)\frac{1}{p}\sinh\left(\frac{p}{c_i\eta_{gi}}\right)\right) + \exp\left(\frac{-1}{c_i^2\eta_{gi}^2}\right)\left(\sum_{p=1}^{\alpha}\exp\left(\frac{-p^2}{4}\right)\frac{1}{\eta_{gi}}\cosh\left(\frac{p}{c_i\eta_{gi}}\right)\right)\right\}\right]\right],$$

$$T_i(0,\eta_{gi}) = 2\left[\frac{\alpha_i b_i}{c_i}\frac{\eta_{gi}^2}{4} + \left(\frac{\alpha_i c_i + b_i c_i - \alpha_i b_i}{c_i^2}\right)\frac{\eta_{gi}}{2\sqrt{\pi}} + \frac{1}{2c_i}\left(1 - \frac{\alpha_i}{c_i}\right) \times \left(1 - \frac{b_i}{c_i}\right) - \frac{1}{c_i}\left(1 - \frac{\alpha_i}{c_i}\right)\left(1 - \frac{b_i}{c_i}\right)\frac{2}{c_i\eta_{gi}\sqrt{\pi}}\exp\left(\frac{1}{c_i^2\eta_{gi}^2}\right) \times \sum_{p=1}^{\alpha}\frac{\exp(-p^2/4)}{p}\sinh\left(\frac{p}{c_i\eta_{gi}}\right)\right],$$

$$C_{21}(E, k_\perp, \eta_{g1}) = e_1 + ie_2, \quad D_{21}(E, k_\perp, \eta_{g1}) = e_3 + ie_4,$$

$$e_1 = \left[\left(\left(\sqrt{t_1^2 + t_2^2} + t_1\right)/2\right)\right]^{\frac{1}{2}}, \quad e_2 = \left[\left(\left(\sqrt{t_1^2 + t_2^2} - t_1\right)/2\right)\right]^{\frac{1}{2}},$$

$$t_1 = \left[\frac{2m_{c1}}{\hbar^2} T_{11}(E, \Delta_1, \eta_{g1}, E_{g1}) - k_\perp^2\right], \quad t_2 = \frac{2m_{c1}}{\hbar^2} T_{21}(E, \Delta_1, \eta_{g1}, E_{g1}),$$

$$e_3 = \left[\frac{\sqrt{t_3^2 + t_4^2} + t_3}{2}\right]^{1/2}, \quad e_4 = \left[\frac{\sqrt{t_3^2 + t_4^2} - t_3}{2}\right]^{1/2}$$

$$t_3 = \left[\frac{2m_{c2}}{\hbar^2} T_{12}(E, \Delta_2, \eta_{g2}, E_{g2}) - k_\perp^2\right] \quad \text{and}$$

$$t_4 = \left[\frac{2m_{c2}}{\hbar^2} T_{22}(E, \Delta_2, \eta_{g2}, E_{g2})\right]$$

Therefore, (17.32) can be expressed as

$$k_x^2 = \delta_7 + i\delta_8 \tag{17.33}$$

where

$$\delta_7 = \left[\frac{1}{L_0^2}(\delta_5^2 - \delta_6^2) - k_\perp^2\right], \quad \delta_5 = \cos^{-1} p_5,$$

$$p_5 = \left[\frac{1 - \delta_3^2 - \delta_4^2 - \sqrt{(1 - \delta_3^2 - \delta_4^2)^2 + 4\delta_4^2}}{2}\right]^{1/2},$$

$$\delta_3 = (\bar{a}_1 \cos\Delta_1 \cosh\Delta_2 - \bar{a}_2 \cos\Delta_3 \cosh\Delta_4),$$

$$\delta_4 = (\bar{a}_1 \sin\Delta_1 \sinh\Delta_2 - \bar{a}_2 \sin\Delta_3 \sinh\Delta_4),$$

$$\Delta_1 = (a_0 e_1 + b_0 e_3), \quad \Delta_2 = (a_0 e_2 + b_0 e_4), \quad \Delta_3 = (a_0 e_1 - b_0 e_3),$$

$$\Delta_4 = (a_0 e_1 - b_0 e_4), \quad \delta_6 = \sinh^{-1} p_5 \quad \text{and} \quad \delta_8 = [2\delta_5\delta_6/L_0^2]$$

The dispersion relation in III–V heavily doped effective mass quantum dot superlattices can be written as

$$\left(\frac{n_x \pi}{d_x}\right)^2 = (\delta_7 + i\delta_8)\bigg|_{E=E_{17,6},\, k_y=\frac{n_y\pi}{d_y},\, k_z=\frac{n_z\pi}{d_z}} \tag{17.34}$$

where $E_{17,6}$ is the total quantized energy in this case.

The DOS function is given by

$$N(E) = \frac{2}{d_x d_y d_z} \sum_{n_x=1}^{n_{x\max}} \sum_{n_y=1}^{n_{y\max}} \sum_{n_z=1}^{n_{z\max}} \delta'(E - E_{17,6}) \tag{17.35}$$

The electron concentration assumes the form

$$n_0 = \frac{2}{d_x d_y d_z} \text{Real part of} \sum_{n_x=1}^{n_{x\max}} \sum_{n_y=1}^{n_{y\max}} \sum_{n_z=1}^{n_{z\max}} F_{-1}(\eta_{17,6}) \tag{17.36}$$

where $\eta_{17,6} = (k_BT)^{-1}(E_{F17,6} - E_{17,6})$ and $E_{F17,6}$ is the Fermi energy in this case.

Using (17.36) and (1.7), we can study the MTP in this case.

17.2.7 The MTP in II–VI HDQD effective mass superlattices

Following Sasaki [24], the electron dispersion law in heavily doped II–VI EMSLs can be written as

$$k_z^2 = \Delta_{13} + i\Delta_{14} \tag{17.37}$$

where

$$\Delta_{13} = \left[\frac{1}{L_0^2}(\Delta_{11}^2 - \Delta_{12}^2) - k_s^2\right], \quad \Delta_{11} = \cos^{-1} p_6,$$

$$p_6 = \left[\frac{1 - \Delta_9^2 - \Delta_{10}^2 - \sqrt{(1 - \Delta_9^2 - \Delta_{10}^2)^2 + 4\Delta_{10}^2}}{2}\right]^{1/2},$$

$$\Delta_9 = (\bar{a}_3 \cos \Delta_6 \cosh \Delta_7 - \bar{a}_4 \cos \Delta_8 \cosh \Delta_7),$$

$$\Delta_{10} = (\bar{a}_3 \sin \Delta_6 \sinh \Delta_7 + \bar{a}_4 \sin \Delta_8 \sinh \Delta_7),$$

$$\Delta_6 = [a_0 C_{22}(E, k_s, \eta_{g1}) + b_0 e_3], \quad \Delta_7 = b_0 e_4,$$

$$\Delta_8 = [a_0 C_{22}(E, k_s, \eta_{g1}) - b_0 e_3],$$

$$C_{22}(E, k_s \eta_{g1}) = \left[\frac{2m_{\parallel,1}^*}{\hbar^2}\left\{\gamma_3(E, \eta_{g1}) - \frac{\hbar^2 k_s^2}{2m_{\perp,1}^*} \mp \bar{\lambda}_0 k_s\right\}\right]^{1/2},$$

$$e_3 = \left[\frac{\sqrt{t_3^2 + t_4^2} + t_3}{2}\right]^{1/2}, \quad e_4 = \left[\frac{\sqrt{t_3^2 + t_4^2} - t_3}{2}\right]^{1/2},$$

$$t_3 = \left[\left(\frac{2m_{c2}}{\hbar^2}\right) T_{12}(E, \Delta_2, \eta_{g2}, E_{g2}) - k_s^2\right],$$

$$t_4 = \left[\left(\frac{2m_{c2}}{\hbar^2}\right) T_{22}(E, \Delta_2, \eta_{g2}, E_{g2}) - k_s^2\right],$$

$$\bar{a}_3 = \left[\sqrt{\frac{M_2(0, \eta_{g2})}{\bar{M}_1(0, \eta_{g1})}} + 1\right]^2 \left[4\left(\frac{M_2(0, \eta_{g2})}{\bar{M}_1(0, \eta_{g1})}\right)^{1/2}\right]^{-1}, \quad \bar{M}_2(0, \eta_{g2})$$

can be obtained by substituting $i = 2$, in the definition of $\bar{M}_i(0, \eta_{gi})$ which is given in the definitions of the symbols of (13.46),

$$\bar{M}_1(0, \eta_{g1}) = m^*_{\|,1}\gamma'_3(E, \eta_{g1})\,|_{E=0}$$

$$\bar{a}_4 = \left[\sqrt{\frac{M_2(0,\eta_{g2})}{\bar{M}_1(0,\eta_{g1})}} - 1\right]^2 \left[4\left(\frac{M_2(0,\eta_{g2})}{\bar{M}_1(0,\eta_{g1})}\right)^{1/2}\right]^{-1},$$

$$\Delta_{12} = \sinh^{-1} p_6 \quad \text{and} \quad \Delta_{14} = \frac{2\Delta_{11}\Delta_{12}}{L_0^2}$$

The dispersion relation in II–VI heavily doped effective mass quantum dot superlattices can be written as

$$\left(\frac{n_z\pi}{d_z}\right)^2 = (\delta_{13} + i\delta_{14})\Big|_{E=E_{17,7},\ k_x=\frac{n_x\pi}{d_x},\ k_y=\frac{n_y\pi}{d_y}} \tag{17.38}$$

where $E_{17,7}$ is the total quantized energy in this case.

The DOS function is given by

$$N(E) = \frac{2}{d_x d_y d_z}\sum_{n_x=1}^{n_{x\max}}\sum_{n_y=1}^{n_{y\max}}\sum_{n_z=1}^{n_{z\max}} \delta'(E - E_{17,7}) \tag{17.39}$$

The electron concentration assumes the form

$$n_0 = \frac{2}{d_x d_y d_z}\text{Real part of}\sum_{n_x=1}^{n_{x\max}}\sum_{n_y=1}^{n_{y\max}}\sum_{n_z=1}^{n_{z\max}} F_{-1}(\eta_{17,7}) \tag{17.40}$$

where $\eta_{17,7} = (k_BT)^{-1}(E_{F17,7} - E_{17,7})$ and $E_{F17,7}$ is the Fermi energy in this case.

Using (17.40) and (1.7), we can study the MTP in this case.

17.2.8 The MTP in IV–VI HDQD effective mass superlattices

Following Sasaki [32], the electron dispersion law in HD IV–VI, EMSLs can be written as

$$k_z^2 = \left[\frac{1}{L_0^2}\{\cos^{-1}(f_{23}(E, k_x, k_y))\}^2 - k_s^2\right] \tag{17.41}$$

where

$$f_{23}(E, k_x, k_y) = \bar{a}_{30}\cos[a_0 C_{23}(E, k_x, k_y\,\eta_{g1}) + b_0 D_{23}(E, k_x, k_y\,\eta_{g1})]$$
$$- \bar{a}_{40}\cos[a_0 C_{23}(E, k_x, k_y, \eta_{g2}) - b_0 D_{23}(E, k_x, k_y, \eta_{g2})],$$

$$\bar{a}_{30} = \left[\sqrt{\frac{M_3(0,\eta_{g2})}{M_3(0,\eta_{g1})}} + 1\right]^2 \left[4\left(\frac{M_3(0,\eta_{g2})}{M_3(0,\eta_{g1})}\right)^{1/2}\right]^{-1},$$

$$\bar{a}_{40} = \left[\sqrt{\frac{M_3(0,\eta_{g2})}{M_3(0,\eta_{g1})}} - 1\right]^2 \left[4\left(\frac{M_3(0,\eta_{g2})}{M_3(0,\eta_{g1})}\right)^{1/2}\right]^{-1}$$

$$M_3(0,\eta_{gi}) = \frac{\hbar^2}{4\bar{p}_{9,i}}\Bigg[\{\bar{q}_{9,i}(0,0,\eta_{gi})\}' - \frac{[\{\bar{q}_{9,i}(0,0,\eta_{gi})\}\{\bar{q}_{9,i}(0,0,\eta_{gi})\}' - 2\bar{p}_{9,i}\{\bar{R}_{9,i}(0,0,\eta_{gi})\}']}{\sqrt{\bar{q}_{9,i}[(0,0,\eta_{gi})]^2 - [4\bar{p}_{9,i}\bar{R}_{9,i}(0,0,\eta_{gi})]}}\Bigg],$$

$$D_{23}(E,k_s,\eta_{g2}) = [[2\bar{p}_{9,2}]^{-1}[+\bar{q}_{9,2}(E,k_s,\eta_{g2}) - [\{\bar{q}_{9,2}(E,k_s,\eta_{g2})\}^2 - (4\bar{p}_{9,2})\bar{R}_{9,2}(E,k_s,\eta_{g2})]^{1/2}]]^{1/2} \quad \text{and}$$

$$C_{23}(E,k_s,\eta_{g1}) = [[2\bar{p}_{9,1}]^{-1}[+\bar{q}_{9,2}(E,k_s,\eta_{g1}) - [\{\bar{q}_{9,2}(E,k_s,\eta_{g1})\}^2 - (4\bar{p}_{9,1})\bar{R}_{9,1}(E,k_s,\eta_{g1})]^{1/2}]]^{1/2}$$

The symbols in the definitions of $M_3(0,\eta_{gi})$, $\bar{q}_{9,i}(E,k_s,\eta_{gi})$ and $\bar{R}_{9,i}(E,k_s,\eta_{gi})$ are defined in Chapter 7.

Therefore, the dispersion relation in heavily doped IV–VI, quantum well EMSLs can be written as

$$\left(\frac{n_z\pi}{d_z}\right)^2 = \left[\frac{1}{L_0^2}\left\{\cos^{-1}\left(f_{23}\left(E_{17,8},\frac{n_x\pi}{d_x},\frac{n_y\pi}{d_y}\right)\right)\right\}^2 - \left[\left(\frac{n_x\pi}{d_x}\right)^2 + \left(\frac{n_y\pi}{d_y}\right)^2\right]\right] \tag{17.42}$$

The DOS function is given by

$$N(E) = \frac{2}{d_x d_y d_z}\sum_{n_x=1}^{n_{x\max}}\sum_{n_y=1}^{n_{y\max}}\sum_{n_z=1}^{n_{z\max}} \delta'(E - E_{17,8}) \tag{17.43}$$

The electron concentration assumes the form

$$n_0 = \frac{2}{d_x d_y d_z}\sum_{n_x=1}^{n_{x\max}}\sum_{n_y=1}^{n_{y\max}}\sum_{n_z=1}^{n_{z\max}} F_{-1}(\eta_{17,8}) \tag{17.44}$$

where $\eta_{17,8} = (k_B T)^{-1}(E_{F17,8} - E_{17,8})$ and $E_{F17,8}$ is the Fermi energy in this case.

Using (17.44) and (1.7), we can study the MTP in this case.

*17.2.9 The MTP in HgTe/CdTe HDQD effective mass superlattices

Following Sasaki [32], the dispersion relation in heavily doped HgTe/CdTe EMSLs assume the form

$$k_z^2 = \Delta_{13H} + i\Delta_{14H} \tag{17.45}$$

where

$$\Delta_{13H} = \left[\frac{1}{L_0^2}(\Delta_{11H}^2 - \Delta_{12H}^2) - k_s^2\right],$$

$$\Delta_{11H} = \cos^{-1} p_{6H},$$

$$p_{6H} = (2)^{-1/2}\left[1 - \Delta_{9H}^2 - \Delta_{10H}^2 - \sqrt{(1-\Delta_{9H}^2) - \Delta_{10H}^2{}^2 + 4\Delta_{10H}^2}\right]^{1/2},$$

$$\Delta_{9H} = (\bar{a}_{1H}\cos\Delta_{5H}\cosh\Delta_{7H} - \bar{a}_{2H}\cos\Delta_{8H}\cosh\Delta_{7H}),$$

$$\Delta_{10H} = (\bar{a}_{1H}\sin\Delta_{6H}\sinh\Delta_{7H} + \bar{a}_{2H}\sin\Delta_{8H}\sinh\Delta_{7H}),$$

$$\Delta_{6H} = [a_0 C_{22H}(E, k_s, \eta_{g1}) + b_0 e_3], \quad \Delta_{7H} = b_0 e_4,$$

$$\Delta_{8H} = [a_0 C_{22H}(E, k_s, \eta_{g1}) - b_0 e_3],$$

$$C_{22H}(E, k_s, \eta_{g1}) = \left[\frac{B_{01}^2 + 2A_1\gamma(E,\eta_{g1}) - B_{01}(B_{01}^2 + 4A_1\gamma(E,\eta_{g1}))}{2A_1^2} - k_\perp^2\right]^{1/2},$$

$$e_3 = \left[\frac{\sqrt{t_3^2 + t_4^2} + t_3}{2}\right]^{1/2}, \quad e_4 = \left[\frac{\sqrt{t_3^2 + t_4^2} - t_3}{2}\right]^{1/2},$$

$$t_3 = \left[\frac{2m_{c2}^*}{\hbar^2}T_{12}(E, \Delta_2, E_{g2}, \eta_{g2}) - k_s^2\right], \quad t_4 = \left[\frac{2m_{c2}^*}{\hbar^2}T_{22}(E, \Delta_2, E_{g2}, \eta_{g2})\right],$$

$$\bar{a}_{1H} = \left[\sqrt{\frac{M_2(0,\eta_{g2})}{M_9(0,\eta_{g1})}} + 1\right]^2\left[4\left(\frac{M_2(0,\eta_{g2})}{M_9(0,\eta_{g1})}\right)\right]^{-1},$$

$$\bar{a}_{2H} = \left[\sqrt{\frac{M_2(0,\eta_{g2})}{M_9(0,\eta_{g1})}} - 1\right]^2\left[4\left(\frac{M_2(0,\eta_{g2})}{M_9(0,\eta_{g1})}\right)\right]^{-1}$$

$M_2(0, \eta_{g2})$ can be obtained by putting $i = 2$ in the expression of $M_i(0, \eta_{gi})$ as given in (13.46),

$$M_9(0, \eta_{g1}) = m_{c1} \left[\gamma'(E, \eta_{g1}) \left[1 - \frac{B_0}{\sqrt{B_0^2 + 4A_1 \gamma(E, \eta_{g1})}} \right] \right] \Bigg|_{E \to 0},$$

$$\Delta_{12H} = \sinh^{-1} p_{6H} \quad \text{and} \quad \Delta_{14H} = \frac{2\Delta_{11H}\Delta_{12H}}{L_0^2}$$

The dispersion relation in heavily doped QD HgTe/CdTe EMSLs can be written as

$$\left(\frac{n_z \pi}{d_z} \right)^2 = (\Delta_{13H} + i\Delta_{14H}) \Big|_{E=E_{17,9},\, k_x = \frac{n_x \pi}{d_x},\, k_y = \frac{n_y \pi}{d_y}} \tag{17.46}$$

The DOS function is given by

$$N(E) = \frac{2}{d_x d_y d_z} \sum_{n_x=1}^{n_{x\max}} \sum_{n_y=1}^{n_{y\max}} \sum_{n_z=1}^{n_{z\max}} \delta'(E - E_{17,9}) \tag{17.47}$$

The electron concentration assumes the form

$$n_0 = \frac{2}{d_x d_y d_z} \text{Real part of} \sum_{n_x=1}^{n_{x\max}} \sum_{n_y=1}^{n_{y\max}} \sum_{n_z=1}^{n_{z\max}} F_{-1}(\eta_{17,9}) \tag{17.48}$$

where $\eta_{17,9} = (k_B T)^{-1}(E_{F17,9} - E_{17,9})$ and $E_{F17,9}$ is the Fermi energy in this case.

Using (17.48) and (1.7), we can study the MTP in this case.

*17.2.10 The MTP in strained layer HDQD effective mass superlattices

The electron energy spectrum in heavily doped strained layer effective mass superlattices can be written in this case following (17.44) as

$$k_z^2 = \left[\frac{1}{L_0^2} \{\cos^{-1}(f_{40}(E, k_s))\}^2 - k_s^2 \right] \tag{17.49}$$

where

$$f_{40}(E,k_s) = a_{20}\cos[a_0C_{40}(E,k_s,\eta_{g1}) + b_0D_{40}(E,k_s,\eta_{g1})] - a_{21}\cos[a_0C_{40}(E,k_s,\eta_{g2}) - b_0D_{40}(E,k_s,\eta_{g2})],$$

$$a_{20} = \left[\sqrt{\frac{M_{s2}(0,\eta_{g2})}{M_{s1}(0,\eta_{g1})}}+1\right]^2 4\left[\left(\frac{M_{s2}(0,\eta_{g2})}{M_{s1}(0,\eta_{g1})}\right)^{1/2}\right]^{-1},$$

$$a_{21} = \left[\sqrt{\frac{M_{s2}(0,\eta_{g2})}{M_{s1}(0,\eta_{g1})}}+1\right]^2 \left[4\left(\frac{M_{s2}(0,\eta_{g2})}{M_{s1}(0,\eta_{g1})}\right)^{1/2}\right]^{-1},$$

$$M_{si}(0,\eta_{gi}) = (-\hbar^2/2)[S_{11,i}(0,\eta_{gi})]'[S_{11,i}(0,\eta_{gi})]^{-2}$$

$$C_{40}(E,k_s,\eta_{g1}) = [1 - P_{11,1}(E,\eta_{g1})k_x^2 - \bar{Q}_{11,1}(E,\eta_{g1})k_y^2]^{1/2} \times [\bar{S}_{11,1}(E,\eta_{g1})]^{-1/2}$$

and

$$D_{40}(E,k_s,\eta_{g2}) = [1 - \bar{P}_{11,2}(E,\eta_{g2})k_x^2 - \bar{Q}_{11,2}(E,\eta_{g2})k_y^2]^{1/2} \times [\bar{S}_{11,2}(E,\eta_{g2})]^{-1/2}$$

Therefore, the dispersion relation in heavily doped strained layer effective mass quantum dot superlattices can be expressed as

$$\left(\frac{n_z\pi}{d_z}\right)^2 = \left[\frac{1}{L_0^2}\left\{\cos^{-1}\left(f_{40}\left(E_{17,10},\frac{n_x\pi}{d_x},\frac{n_y\pi}{d_y}\right)\right)\right\}^2 - \left[\left(\frac{n_x\pi}{d_x}\right)^2+\left(\frac{n_y\pi}{d_y}\right)^2\right]\right] \tag{17.50}$$

The DOS function is given by

$$N(E) = \frac{2}{d_xd_yd_z}\sum_{n_x=1}^{n_{x\max}}\sum_{n_y=1}^{n_{y\max}}\sum_{n_z=1}^{n_{z\max}}\delta'(E-E_{17,10}) \tag{17.51}$$

The electron concentration assumes the form

$$n_0 = \frac{2}{d_xd_yd_z}\sum_{n_x=1}^{n_{x\max}}\sum_{n_y=1}^{n_{y\max}}\sum_{n_z=1}^{n_{z\max}}F_{-1}(\eta_{17,10}) \tag{17.52}$$

where $\eta_{17,10} = (k_B T)^{-1}(E_{F17,10} - E_{17,10})$ and $E_{F17,10}$ is the Fermi energy in this case.

Using (17.52) and (1.7), we can study the MTP in this case.

17.3 Summary and Conclusion

This chapter explores the MTP in III–V, II–VI, IV–VI, HgTe/CdTe and strained layer QDHDSLs with graded interfaces has been studied in Sections 17.2.1–17.2.5. From Sections 17.2.6 to 17.2.10, the MTP in III–V, II–VI, IV–VI, HgTe/CdTe and strained layer quantum dot heavily doped effective mass superlattices respectively has been presented. The presence of essential poles in the constituent materials of HD III-V and HgTe/CdTe SLs makes the dispersion relations in the said SLs complex (for both SLs with graded interfaces and effective mass SLs). The $E-k_s^2$ plots for HD III–V and HgTe/CdTe QWSLs are quantized closed 2D surfaces and they exist in the complex energy plane. The dispersion relations for other cases are quantized closed 2D surfaces and they exist in the real energy plane. The dispersion relations in all the cases are concentration dependent for any value of the electron energy. The sub-band energies will be real or complex in accordance with said logic. The DR and the SE will be wave vector dependent for quantum well and quantum wire heavily doped superlattices and analytical expressions of the DOS and the DOS dependent electronic properties are not possible so that the numerical analysis should be used for the determinations of the all types of electronic quantities in the respective cases.

17.4 Open Research Problem

(R.17.1) Investigate the influence of arbitrarily oriented alternating quantizing magnetic field and strain on the ER, the ΔC_{44} the ΔC_{456}, the Einstein's photoemission, the field emission and the other ES dependent electronic properties for all types of HD superlattices whose carrier energy spectra are described in this book.

References

[1] S. Mukherjee, S. N. Mitra, P. K. Bose, A. R. Ghatak, A. Neoigi, J. P. Banerjee, A. Sinha, M. Pal, S. Bhattacharya, K. P. Ghatak, *J. Comput. Theor. Nanosci.*, 4, 550 (**2007**); N. G. Anderson, W. D. Laidig, R. M. Kolbas,

Y. C. Lo, *J. Appl. Phys.*, 60, 2361 (**1986**); N. Paitya, K. P. Ghatak, *J. Adv. Phys.*, 1, 161 (**2012**).

[2] N. Paitya, S. Bhattacharya, D. De, K. P. Ghatak, *Adv. Sci. Engg. Medi.*, 4, 96 (**2012**); S. Bhattacharya, D. De, S. M. Adhikari, K. P. Ghatak, *Superlatt. Microstruct.*, 51, 203 (**2012**); D. De, S. Bhattacharya, S. M. Adhikari, A. Kumar, P. K. Bose, K. P. Ghatak, *Beilstein J. Nanotech.*, 2, 339 (**2012**).

[3] D. De, A. Kumar, S. M. Adhikari, S. Pahari, N. Islam, P. Banerjee, S. K. Biswas, S. Bhattacharya, K. P. Ghatak, *Superlatt. Microstruct.*, 47, 377 (**2010**); S. Pahari, S. Bhattacharya, S. Roy, A. Saha, D. De, K. P. Ghatak, *Superlatt. Microstruct.*, 46, 760 (**2009**); S. Pahari, S. Bhattacharya, K. P. Ghatak, *J. Comput. Theor. Nanosci.*, 6, 2088 (**2009**).

[4] S. K. Biswas, A. R. Ghatak, A. Neogi, A. Sharma, S. Bhattacharya, K. P. Ghatak, *Phys. E: Low-dimen. Syst. Nanostruct.*, 36, 163 (**2007**); L. J. Singh, S. Choudhury, D. Baruah, S. K. Biswas, S. Pahari, K. P. Ghatak, *Phys. B: Conden. Matter*, 368, 188 (**2005**); S. Chowdhary, L. J. Singh, K. P. Ghatak, *Phys. B: Conden. Matter*, 365, 5 (**2005**).

[5] L. J. Singh, S. Choudhary, A. Mallik, K. P. Ghatak, *J. Comput. Theor. Nanosci.*, 2, 287 (**2005**); K. P. Ghatak, J. Mukhopadhyay, J. P. Banerjee, *SPIE Proc. Ser.*, 4746, 1292 (**2002**); K. P. Ghatak, S. Dutta, D. K. Basu, B. Nag, *Il Nuovo Cimento D*, 20, 227 (**1998**).

[6] K. P. Ghatak, D. K. Basu, B. Nag, *J. Phys. Chem. Solids*, 58, 133 (**1997**); K. P. Ghatak, B. De, *Mat. Res. Soc. Proc.*, 300, 513 (**1993**); K. P. Ghatak, B. Mitra, *Il Nuovo Cimento D*, 15, 97 (**1993**).

[7] K. P. Ghatak, *Inter. Soc. Opt. Photon. Proc. Soc. Photo Opt. Instru. Engg.*, 1626, 115 (**1992**); K. P. Ghatak, A. Ghoshal, *Phys. Stat. Sol. (b)*, 170, K27 (**1992**); K. P. Ghatak, S. Bhattacharya, S. N. Biswas, *Proc. Soc. Photo Opt. Instru. Engg.*, 836, 72 (**1988**); K. P. Ghatak, A. Ghoshal, S. N. Biswas, M. Mondal, *Proc. Soc. Photo Opt. Instru. Engg.*, 1308, 356 (**1990**).

[8] K. P. Ghatak, B. De, *Proc. Wide Bandgap Semi. Symp., Mat. Res. Soc.*, 377 (**1992**); K. P. Ghatak, B. De, *Defect Engg. Semi. Growth, Process. Device Tech. Symp., Mat. Res. Soc.*, 262, 911 (**1992**); S. N. Biswas, K. P. Ghatak, *Int. J. Electron. Theor. Exp.*, 70, 125 (**1991**).

[9] B. Mitra, K. P. Ghatak, *Phys. Lett. A*, 146, 357 (**1990**); B. Mitra, K. P. Ghatak, *Phys. Lett. A*, 142, 401 (**1989**); K. P. Ghatak, B. Mitra, A. Ghoshal, *Phys. Stat. Sol. (b)*, 154, K121 (**1989**).

[10] L. V. Keldysh, *Sov. Phys. Sol. Stat.*, 4, 1658 (**1962**); L. Esaki, R. Tsu, *IBM J. Res. Dev.*, 14, 61 (**1970**); G. Bastard, *Wave Mechanics Applied to Heterostructures*, (Editions de Physique, Les Ulis, France, **1990**); E. L. Ivchenko, G. Pikus, *Superlattices and Other Heterostructures*, (Springer-Berlin, **1995**); R. Tsu, *Superlattices to Nanoelectronics*, (Elsevier, The Netherlands, **2005**).

[11] P. Fürjes, Cs. Dücs, M. Ádám, J. Zettner, I. Bársony, *Superlatt. Microstruct.*, 35, 455 (**2004**); T. Borca-Tasciuc, D. Achimov, W. L. Liu, G. Chen, H. W. Ren, C. H. Lin, S. S. Pei, *Microscale Thermophys. Engg.*, 5, 225 (**2001**).

[12] B. S. Williams, *Nat. Photon.*, 1, 517 (**2007**); A. Kosterev, G. Wysocki, Y. Bakhirkin, S. So, R. Lewicki, F. Tittel, R. F. Curl, *Appl. Phys. B*, 90, 165 (**2008**); M. A. Belkin, F. Capasso, F. Xie, A. Belyanin, M. Fischer, A. Wittmann, J. Faist, *Appl. Phys. Lett.*, 92, 201101 (**2008**).
[13] G. J. Brown, F. Szmulowicz, R. Linville, A. Saxler, K. Mahalingam, C.-H. Lin, C. H. Kuo, W. Y. Hwang, *IEEE Photon. Technol. Lett.*, 12, 684 (**2000**); H. J. Haugan, G. J. Brown, L. Grazulis, K. Mahalingam, D. H. Tomich, *Phys. E: Low-dimen. Syst. Nanostruct.*, 20, 527 (**2004**).
[14] S. A. Nikishin, V. V. Kuryatkov, A. Chandolu, B. A. Borisov, G. D. Kipshidze, I. Ahmad, M. Holtz, H. Temkin, *Jpn. J. Appl. Phys.*, 42, L1362 (**2003**); Y. K. Su, H. C. Wang, C. L. Lin, W. B. Chen, S. M. Chen, *Jpn. J. Appl. Phys.*, 42, L751 (**2003**); C. H. Liu, Y. K. Su, L. W. Wu, S. J. Chang, R. W. Chuang, *Semicond. Sci. Technol.*, 18, 545 (**2003**); S. B. Che, I. Nomura, A. Kikuchi, K. Shimomura, K. Kishino, *Phys. Stat. Sol. (b)*, 229, 1001 (**2002**).
[15] C. P. EnEESes, F. Lewen, T. F. Giesen, S. SchlEEMer, D. G. Paveliev, Y. I. Koschurinov, V. M. Ustinov, A. E. Zhucov, *Rev. Sci. Instrum.*, 78, 043106 (**2007**).
[16] F. Klappenberger, K. F. Renk, P. Renk, B. Rieder, Y. I. Koshurinov, D. G. Pavelev, V. Ustinov, A. Zhukov, N. Maleev, A. Vasilyev, *Appl. Phys. Lett.*, 84, 3924 (**2004**).
[17] X. Jin, Y. Maeda, T. Saka, M. Tanioku, S. Fuchi, T. Ujihara, Y. Takeda, N. Yamamoto, Y. Nakagawa, A. Mano, S. Okumi, M. Yamamoto, T. Nakanishi, H. Horinaka, T. Kato, T. Yasue, T. Koshikawa, *J. Crys. Growth*, 310, 5039 (**2008**); X. Jin, N. Yamamoto, Y. Nakagawa, A. Mano, T. Kato, M. Tanioku, T. Ujihara, Y. Takeda, S. Okumi, M. Yamamoto, T. Nakanishi, T. Saka, H. Horinaka, T. Kato, T. Yasue, T. Koshikawa, *Appl. Phys. Express*, 1, 045002 (**2008**).
[18] B. H. Lee, K. H. Lee, S. Im, M. M. Sung, *Organic Electron.*, 9, 1146 (**2008**).
[19] P. H. Wu, Y. K. Su, I. L. Chen, C. H. Chiou, J. T. Hsu, W. R. Chen, *Jpn. J. Appl. Phys.*, 45, L647 (**2006**); A. C. Varonides, *Renew. Energy*, 33, 273 (**2008**).
[20] M. Walther, G. Weimann, *Phys. Stat. Sol. (b)*, 203, 3545 (**2006**); R. Rehm, M. Walther, J. Schmitz, J. Fleißner, F. Fuchs, J. Ziegler, W. Cabanski, *Opto-Electronics Rev.*, 14, 19 (**2006**); R. Rehm, M. Walther, J. Scmitz, J. Fleissner, J. Ziegler, W. Cabanski, R. Breiter, *Electron. Lett.*, 42, 577 (**2006**).
[21] G. J. Brown, F. Szmulowicz, H. Haugan, K. Mahalingam, S. Houston, *Microelectron. J.*, 36, 256 (**2005**).
[22] K. V. Vaidyanathan, R. A. Jullens, C. L. Anderson, H. L. Dunlap, *Sol. Stat. Electron.*, 26, 717 (**1983**).
[23] B. A. Wilson, *IEEE J. Quant. Electron.*, 24, 1763 (**1988**).
[24] M. Krichbaum, P. Kocevar, H. Pascher, G. Bauer, *IEEE J. Quant. Electron.*, 24, 717 (**1988**).
[25] J. N. Schulman, T. C. McGill, *Appl. Phys. Lett.*, 34, 663 (**1979**).
[26] H. Kinoshita, T. Sakashita, H. Fajiyasu, *J. Appl. Phys.*, 52, 2869 (**1981**).

[27] L. Ghenin, R. G. Mani, J. R. Anderson, J. T. Cheung, *Phys. Rev. B*, 39, 1419 (**1989**).
[28] C. A. Hoffman, J. R. Mayer, F. J. Bartoli, J. W. Han, J. W. Cook, J. F. Schetzina, J. M. Schubman, *Phys. Rev. B*, 39, 5208 (**1989**).
[29] V. A. Yakovlev, *Sov. Phys. Semicond.*, 13, 692 (**1979**).
[30] E. O. Kane, *J. Phys. Chem. Solids*, 1, 249 (**1957**).
[31] H. X. Jiang, J. Y. Lin, *J. Appl. Phys.*, 61, 624 (**1987**).
[32] H. Sasaki, *Phys. Rev. B*, 30, 7016 (**1984**).

Chapter 18

The EP in HDQWSLs Under Magnetic Quantization

Life is too short to waste time hating any one.

18.1 Introduction

The different physical properties of heavily doped quantum well superlattices (HDQWSLs) have recently being investigated in the literature [1–5]. In this chapter, the magneto ES in III–V, II–VI, IV–VI, HgTe/CdTe and strained layer HDQWSLs with graded interfaces has been studied in Sections 18.2.1–18.2.5. From Sections 18.2.6 to 18.2.10, the magneto ES in III–V, II–VI, IV–VI, HgTe/CdTe and strained layer quantum well heavily doped effective mass superlattices respectively has been presented. We have formulated the expressions for the Einstein's photo-emitted current density only for this chapter. Section 18.3 contains the summary and conclusion pertinent to this chapter. The last Section 18.4 presents four open research problems.

18.2 Theoretical Background

18.2.1 The EP in III–V HDQWSLs with graded interfaces under magnetic quantization

The dispersion relation in this case can be written from (17.3) as

$$\left(\frac{n_z \pi}{d_z}\right)^2 = G_{8E_{41,n}} + iH_{8E_{41,n}} \tag{18.1}$$

where $E_{41,n}$ is the total quantized energy in this case,

$$G_{8E,n} = \left[\frac{C_{7E,n}^2 - D_{7E,n}^2}{L_0^2} - \frac{2eB}{\hbar}\left(n+\frac{1}{2}\right)\right], \quad C_{7E,n} = \cos^{-1}(\bar{\omega}_{7E,n}),$$

$$\bar{\omega}_{7E,n} = (2)^{\frac{-1}{2}}\left[(1 - G_{7E,n}^2 - H_{7E,n}^2) - \sqrt{(1 - G_{7E,n}^2 - H_{7E,n}^2)^2 + 4G_{7E,n}^2}\,\right]^{\frac{1}{2}}$$

$$G_{7E,n} = [G_{1E,n} + (\rho_{5E,n}G_{2E,n}/2) - (\rho_{6E,n}H_{2E,n}/2) + (\Delta_0/2) \times \{\rho_{7E,n}G_{3E,n} - \rho_{8E,n}H_{3E,n} + \rho_{9E,n}G_{4E,n} - \rho_{10E,n}H_4 + \rho_{11E,n}G_{5E,n} - \rho_{12E,n}H_5 + (1/12)(\rho_{13E,n}G_{6E,n} - \rho_{14E,n}H_{6E,n})\}],$$

$$G_{1E,n} = [(\cos(h_{1E,n}))(\cosh(h_{2E,n}))(\cosh(g_{1E,n}))(\cos(g_{2E,n})) + (\sin(h_{1E,n}))(\sinh(h_{2E,n}))(\sinh(g_{1E,n}))(\sin(g_{2E,n}))],$$

$$(V_0 = |E_{g_2} - E_{g_1}|), \quad h_{1E,n} = e_{1E,n}(b_0 - \Delta_0),$$

$$e_{1E,n} = 2^{\frac{-1}{2}}\left(\sqrt{t_{1E,n}^2 + t_{1E,n}^2} + t_{1E,n}\right)^{\frac{1}{2}},$$

$$t_{1E,n} = \left[(2m_{c1}/\hbar^2)\cdot T_{11}(E, E_{g_{01}}, \Delta_1, \eta_{g1}) - \frac{2eB}{\hbar}\left(n+\frac{1}{2}\right)\right],$$

$$t_{2E,n} = [(2m_{c1}/\hbar^2)\cdot T_{21}(E, E_{g_{01}}, \Delta_1, \eta_{g1})],$$

$$h_{2E,n} = e_{2E,n}(b_0 - \Delta_0), \quad e_{2E,n} = 2^{\frac{-1}{2}}\left(\sqrt{t_{1E,n}^2 + t_{1E,n}^2} - t_{1E,n}\right)^{\frac{1}{2}},$$

$$g_{1E,n} = d_{1E,n}(a_0 - \Delta_0), \quad d_{1E,n} = 2^{\frac{-1}{2}}\left(\sqrt{x_{1E,n}^2 + y_{1E,n}^2} + x_{1E,n}\right)^{\frac{1}{2}},$$

$$x_{1E,n} = \left[-(2m_{c2}/\hbar^2)\cdot T_{12}(E - V_0, E_{g_{02}}, \Delta_2, \eta_{g2}) + \frac{2eB}{\hbar}\left(n+\frac{1}{2}\right)\right],$$

$$y_{1E,n} = [(2m_{c2}/\hbar^2)T_{22}(E - V_0, E_{g_{02}}, \Delta_2, \eta_{g2})]$$

$$g_{2E,n} = d_{2E,n}(a_0 - \Delta_0), \quad d_{2E,n} = 2^{\frac{-1}{2}}\left(\sqrt{x_{1E,n}^2 + y_{1E,n}^2} - x_{1E,n}\right)^{\frac{1}{2}},$$

$$\rho_{5E,n} = (\rho_{3E,n}^2 + \rho_{4E,n}^2)^{-1}[\rho_{1E,n}\rho_{3E,n} - \rho_{2E,n}\rho_{4E,n}],$$

$$\rho_{1E,n} = [d_{1E,n}^2 + e_{2E,n}^2 - d_{2E,n}^2 - e_{1E,n}^2], \quad \rho_{3E,n} = [d_{1E,n}e_1 + d_2e_2],$$

$$\rho_{2E,n} = 2[d_{1E,n}d_{2E,n} + e_{1E,n}e_{2E,n}], \quad \rho_{4E,n} = [d_{1E,n}e_{2E,n} - e_{1E,n}d_{2E,n}],$$

$$G_2 = [(\sin(h_1))(\cosh(h_2))(\sinh(g_1))(\cos(g_2)) + (\cos(h_1))(\sinh(h_2)) \times (\cosh(g_1))(\sin(g_2))]$$

$$\rho_{6E,n} = (\rho_{3E,n}^2 + \rho_{4E,n}^2)^{-1}[\rho_{1E,n}\rho_{4E,n} - \rho_{2E,n}\rho_{3E,n}],$$

$$H_{2E,n} = [(\sin(h_{1E,n}))(\cosh(h_{2E,n}))(\sin(g_{2E,n}))(\cosh(g_{1E,n})) - (\cos(h_{1E,n}))(\sinh(h_{2E,n}))(\sinh(g_{1E,n}))(\cos(g_{2E,n}))]$$

$$\rho_{7E,n} = [(e_{1E,n}^2 + e_{2E,n}^2)^{-1}[e_{1E,n}(d_{1E,n}^2 - d_{2E,n}^2) - 2d_{1E,n}d_{2E,n}e_{2E,n}] - 3e_{1E,n}],$$

$$G_{3E,n} = [(\sin(h_{1E,n}))(\cosh(h_{2E,n}))(\cosh(g_{1E,n}))(\cos(g_{2E,n})) + (\cos(h_{1E,n}))(\sinh(h_{2E,n}))(\sinh(g_{1E,n}))(\sin(g_{2E,n}))]$$

$$\rho_{8E,n} = [(e_{1E,n}^2 + e_{2E,n}^2)^{-1}[e_{2E,n}(d_{1E,n}^2 - d_{2E,n}^2) + 2d_{1E,n}d_{2E,n}e_{1E,n}] + 3e_{2E,n}],$$

$$H_{3E,n} = [(\sin(h_{1E,n}))(\cosh(h_{2E,n}))(\sin(g_{2E,n}))(\sinh(g_{1E,n})) - (\cos(h_{1E,n}))(\sinh(h_{2E,n}))(\cosh(g_{1E,n}))(\cos(g_{2E,n}))]$$

$$\rho_{9E,n} = [(d_{1E,n}^2 + d_{2E,n}^2)^{-1}[d_{1E,n}(e_{2E,n}^2 - e_{1E,n}^2) + 2e_{2E,n}d_{1E,n}e_{1E,n}] + 3d_{1E,n}],$$

$$G_{4E,n} = [(\cos(h_{1E,n}))(\cosh(h_{2E,n}))(\cos(g_{2E,n}))(\sinh(g_{1E,n})) - (\sin(h_{1E,n}))(\sinh(h_{2E,n}))(\cosh(g_{1E,n}))(\sin(g_{2E,n}))]$$

$$\rho_{10E,n} = [-(d_{1E,n}^2 + d_{2E,n}^2)^{-1}[d_{2E,n}(-e_{2E,n}^2 + e_{1E,n}^2) + 2e_{2E,n}d_{1E,n}\rho_{1E,n}] + 3d_{2E,n}],$$

$$H_{4E,n} = [(\cos(h_{1E,n}))(\cosh(h_{2E,n}))(\cosh(g_{1E,n}))(\sin(g_{2E,n})) + (\sin(h_{1E,n}))(\sinh(h_{2E,n}))(\sinh(g_{1E,n}))(\cos(g_{2E,n}))]$$

$$\rho_{11E,n} = 2[d_{1E,n}^2 + e_{2E,n}^2 - d_{2E,n}^2 - e_{1E,n}^2],$$

$$G_{5E,n} = [(\cos(h_{1E,n}))(\cosh(h_{2E,n}))(\cos(g_{2E,n}))(\cosh(g_{1E,n})) - (\sin(h_{1E,n}))(\sinh(h_{2E,n}))(\sinh(g_{1E,n}))(\sin(g_{2E,n}))]$$

$$\rho_{12E,n} = 4[d_{1E,n}d_{2E,n} + e_{1E,n}e_{2E,n}],$$

$$H_{5E,n} = [(\cos(h_{1E,n}))(\cosh(h_{2E,n}))(\sinh(g_{1E,n}))(\sin(g_{2E,n})) + (\sin(h_{1E,n}))(\sinh(h_{2E,n}))(\cosh(g_{1E,n}))(\cos(g_{2E,n}))]$$

$$\rho_{13E,n} = [\{5(d_{1E,n}e^3_{1E,n} - 3e_{1E,n}e^2_{2E,n}d_{1E,n}) + 5d_{2E,n} \times (e^3_{1E,n} - 3e^2_{1E,n}e_{2E,n})\}(d^2_{1E,n} + d^2_{2E,n})^{-1} + (e^2_{1E,n} + e^2_{2E,n})^{-1} \times \{5(e_{1E,n}d^3_{1E,n} - 3d_{2E,n}e^2_{1E,n}d_{1E,n}) + 5(d^3_{2E,n}e_{2E,n} - 3d^2_{1E,n}d_{2E,n}e_{2E,n})\} - 34(d_{1E,n}e_{1E,n} + d_{2E,n}e_{2E,n})],$$

$$G_{6E,n} = [(\sin(h_{1E,n}))(\cosh(h_{2E,n}))(\sinh(g_{1E,n}))(\cos(g_{2E,n})) + (\cos(h_{1E,n}))(\sinh(h_{2E,n}))(\cosh(g_{1E,n}))(\sin(g_{2E,n}))],$$

$$\rho_{14E,n} = [\{5(d_{1E,n}e^3_{2E,n} - 3e_{2E,n}e^2_{1E,n}d_{1E,n}) + 5d_{2E,n} \times (-e^3_{1E,n} + 3e^2_{2E,n}e_{1E,n})\}(d^2_{1E,n} + d^2_{2E,n})^{-1} + (e^2_{1E,n} + e^2_{2E,n})^{-1} \times \{5(-e_{1E,n}d^3_{2E,n} + 3d^2_{1E,n}d_{2E,n}e_{1E,n}) + 5(-d^3_{1E,n}e_{2E,n} + 3d^2_{2E,n}d_{1E,n}e_{2E,n})\} + 34(d_{1E,n}e_{2E,n} - d_{2E,n}e_{1E,n})],$$

$$H_{6E,n} = [(\sin(h_{1E,n}))(\cosh(h_{2E,n}))(\cosh(g_{1E,n}))(\sin(g_{2E,n})) - (\cos(h_{1E,n}))(\sinh(h_{2E,n}))(\sinh(g_{1E,n}))(\cos(g_{2E,n}))],$$

$$H_{7E,n} = [H_{1E,n} + (\rho_{5E,n}H_{2E,n}/2) + (\rho_{6E,n}G_{2E,n}/2) + (\Delta_0/2)\{\rho_{8E,n}G_{3E,n} + \rho_{7E,n}H_{3E,n} + \rho_{10E,n}G_{4E,n} + \rho_{9E,n}H_{4E,n} + \rho_{12E,n}G_{5E,n} + \rho_{11E,n}H_{5E,n} + (1/12)(\rho_{14E,n}G_{6E,n} + \rho_{13E,n}H_{6E,n})\}],$$

$$H_{1E,n} = [(\sin(h_{1E,n}))(\sinh(h_{2E,n}))(\cosh(g_{1E,n}))(\cos(g_{2E,n})) + (\cos(h_{1E,n}))(\cosh(h_{2E,n}))(\sinh(g_{1E,n}))(\sin(g_{2E,n}))],$$

$$D_{7E,n} = \sinh^{-1}(\bar{\omega}_{7E,n}), \quad H_{8E,n} = (2C_{7E,n}D_{7E,n}/L_0^2)$$

The DOS function in this case can be expressed as

$$N_{\text{MQWSL}}(E) = \frac{g_v eB}{\pi\hbar} \sum_{n_z=1}^{n_{z\max}} \sum_{n=0}^{n_{\max}} \delta'(E - E_{41,n}) \tag{18.2}$$

The electron concentration can be written as

$$n_0 = \frac{g_v eB}{\pi\hbar} \text{Real part of} \sum_{n_z=1}^{n_{z\max}} \sum_{n=0}^{n_{\max}} F_{-1}(\bar{\eta}_{41,n}) \tag{18.3}$$

where $\bar{\eta}_{41,n} = (k_B T)^{-1}[\bar{E}_{F41,n} - E_{41,n}]$ and $\bar{E}_{F41,n}$ is the Fermi energy in this case.

The Einstein's photo-emitted current density in this case is given by

$$J_{\text{PHOTO}} = \frac{\alpha_0 e^2 g_v B}{2\pi\hbar d_z} \text{Real part of} \sum_{n=0}^{n_{\max}} \sum_{n_{z\min}}^{n_{z\max}} F_{-1}(\bar{\eta}_{41,n}) V_1(\bar{E}_{15,11\text{SL}}) \tag{18.4}$$

where

$$V_1(\bar{E}_{15,11\text{SL}}) = (2\pi n_z/hd_z)[(G_8' + iH_8')\big|_{k_x=0,\ k_y=0,\ k_z=\frac{n_z\pi}{d_z}\text{ and } E=\bar{E}_{15,11\text{SL}}}]^{-1}$$

In which $\bar{E}_{15,11\text{SL}}$ is the quantized energy along z-direction and can be obtain by substituting $k_x = 0$, $k_y = 0$, $k_z = \frac{n_z\pi}{d_z}$ and $E = \bar{E}_{15,11\text{SL}}$ in (17.3).

18.2.2 The EP in II–VI HDQWSLs with graded interfaces under magnetic quantization

The dispersion relation in HDQW II–VI superlattices under magnetic quantization can be written from (17.11) as

$$\left(\frac{n_z\pi}{d_z}\right)^2 = G_{19E_{42,n}} + iH_{19E_{42,n}} \tag{18.5}$$

where $E_{42,n}$ is the total quantized energy in this case,

$$G_{19E,n} = \left[\frac{C_{18E,n}^2 - D_{18E,n}^2}{L_0^2} - \frac{2eB}{\hbar}\left(n + \frac{1}{2}\right)\right],$$

$$C_{18E,n} = \cos^{-1}(\omega_{18E,n}),$$

$$\omega_{18E,n} = (2)^{\frac{-1}{2}}\Big[(1 - G_{18E,n}^2 - H_{18E,n}^2) - \sqrt{(1 - G_{18E,n}^2 - H_{18E,n}^2)^2 + 4G_{18E,n}^2}\Big]^{\frac{1}{2}}$$

$$
\begin{aligned}
G_{18E,n} &= \frac{1}{2}[G_{11E,n} + G_{12E,n} + \Delta_0(G_{13E,n} + G_{14E,n}) \\
&\quad + \Delta_0(G_{15E,n} + G_{16E,n})], \\
G_{11E,n} &= 2(\cosh(g_{1E,n}))(\cos(g_{2E,n}))(\cosh\gamma_{11E,n}(E,n)), \\
\gamma_{11E,n}(E,n) &= k_{21E,n}(E,n)(b_0 - \Delta_0), \\
k_{21E,n}(E,n) &= \left\{\left[\gamma_3(E,\eta_{g1}) - \frac{\hbar^2}{2m^*_{\perp,1}}\left(\frac{2eB}{\hbar}\right)\left(n+\frac{1}{2}\right)\right.\right. \\
&\quad \left.\left. \pm\, \bar{\lambda}_0\sqrt{\left(\frac{2eB}{\hbar}\right)\left(n+\frac{1}{2}\right)}\right]\frac{2m^*_{\parallel,1}}{\hbar^2}\right\}^{\frac{1}{2}} \\
G_{12E,n} &= ([\Omega_1(E,n)(\sinh g_{1E,n})(\cos g_{2E,n}) - \Omega_{2E,n}(E,n)(\sin g_{2E,n}) \\
&\quad \times (\cosh g_{1E,n})](\sin\gamma_{11E,n}(E,n))) \\
\Omega_{1E,n}(E,n) &= \left[\frac{d_{1E,n}}{k_{21E,n}(E,n)} - \frac{k_{21E,n}(E,n)d_{1E,n}}{d^2_{1E,n}+d^2_{2E,n}}\right], \\
d_{1E,n} &= \left[\frac{\sqrt{x^2_{1E,n}+y^2_{1E,n}} + x_{1E,n}}{2}\right]^{1/2}, \\
d_{2E,n} &= \left[\frac{\sqrt{x^2_{1E,n}+y^2_{1E,n}} - x_{1E,n}}{2}\right]^{1/2}, \\
x_{1E,n} &= \left[\left(\frac{2eB}{\hbar}\right)\left(n+\frac{1}{2}\right) - \left(\frac{2m_{c2}}{\hbar^2}\right)T_{12}(E-V_0,\Delta_2,E_{g2},\eta_{g2})\right], \\
y_{1E,n} &= \left[\left(\frac{2m_{c2}}{\hbar^2}\right)T_{22}(E-V_0,\Delta_2,E_{g2},\eta_{g2})\right] \quad \text{and} \\
\Omega_{2E,n}(E,n) &= \left[\frac{d_{2E,n}}{k_{21E,n}(E,n)} + \frac{k_{21E,n}(E,n)d_{2E,n}}{d^2_{1E,n}+d^2_{2E,n}}\right] \\
G_{13E,n} &= ([\Omega_{3E,n}(E,n)(\cosh g_{1E,n})(\cos g_{2E,n}) \\
&\quad - \Omega_{4E,n}(E,n)(\sinh g_{1E,n})(\sin g_{2E,n})](\sin\gamma_{11E,n}(E,n))) \\
\Omega_{3E,n}(E,n) &= \left[\frac{d^2_{1E,n} - d^2_{2E,n}}{k_{21E,n}(E,n)} - 3k_{21E,n}(E,n)\right],
\end{aligned}
$$

$$\Omega_{4E,n}(E,n) = \left[\frac{2d_{1E,n}d_{2E,n}}{k_{21E,n}(E,n)}\right],$$

$$\begin{aligned}G_{14E,n} &= ([\Omega_{5E,n}(E,n)(\sinh g_{1E,n})(\cos g_{2E,n})\\ &\quad - \Omega_{6E,n}(E,n)(\sin g_{1E,n})(\cosh g_{2E,n})](\cos \gamma_{11E,n}(E,n)))\end{aligned}$$

$$\Omega_{5E,n}(E,n) = \left[3d_{1E,n} \times \frac{d_{1E,n}}{d_{1E,n}^2 + d_{2E,n}^2} - k_{21E,n}^2(E,n)\right],$$

$$\Omega_{6E,n}(E,n) = \left[3d_{2E,n} + \frac{d_{2E,n}}{d_{1E,n}^2 + d_{2E,n}^2} k_{21E,n}^2(E,n)\right],$$

$$\begin{aligned}G_{15E,n} &= ([\Omega_{9E,n}(E,n)(\cosh g_{1E,n})(\cos g_{2E,n})\\ &\quad - \Omega_{10E,n}(E,n)(\sinh g_{1E,n})(\sin g_{2E,n})](\cos \gamma_{11E,n}(E,n)))\end{aligned}$$

$$\Omega_{9E,n}(E,n) = [2d_{1E,n}^2 - 2d_{2E,n}^2 - k_{21E,n}^2(E,n)],$$

$$\Omega_{10E,n}(E,n) = [2d_{1E,n}d_{2E,n}]$$

$$\begin{aligned}G_{16E,n} &= ([\Omega_{7E,n}(E,n)(\sinh g_{1E,n})(\cos g_{2E,n})\\ &\quad - \Omega_{8E,n}(E,n)(\sin g_{1E,n})(\cosh g_{2E,n})](\sin \gamma_{11E,n}(E,n)))(12)^{-1}\end{aligned}$$

$$\begin{aligned}\Omega_{7E,n}(E,n) &= \left[\frac{5d_{1E,n}}{d_{1E,n}^2 + d_{2E,n}^2} k_{21E,n}^3(E,n) + \frac{5(d_{1E,n}^3 - 3d_{2E,n}^2 d_{1E,n})}{k_{21}(E,k_s)}\right.\\ &\quad \left. - 34k_{21E,n}(E,n)d_{2E,n}\right],\end{aligned}$$

$$\begin{aligned}\Omega_{8E,n}(E,n) &= \left[\frac{5d_{2E,n}}{d_{1E,n}^2 + d_{2E,n}^2} k_{21E,n}^3(E,n) + \frac{5(d_{1E,n}^3 - 3d_{1E,n}^2 d_{2E,n})}{k_{21}(E,n)}\right.\\ &\quad \left. + 34k_{21E,n}(E,n)d_{2E,n}\right],\end{aligned}$$

$$\begin{aligned}H_{18E,n} &= \frac{1}{2}[H_{11E,n} + H_{12E,n} + \Delta_0(H_{13E,n} + H_{14E,n})\\ &\quad + \Delta_0(H_{15E,n} + H_{16E,n})],\end{aligned}$$

$$H_{11E,n} = 2(\sinh g_{1E,n} \sin g_{2E,n} \cos \gamma_{11E,n}(E,n)),$$

$$\begin{aligned}H_{12E,n} &= ([\Omega_2(E,n)(\sinh g_{1E,n})(\cos g_{2E,n})\\ &\quad + \Omega_{1E,n}(E,n)(\sin g_{2E,n})(\cosh g_{1E,n})](\sin \gamma_{11E,n}(E,n))),\end{aligned}$$

$$
\begin{aligned}
H_{13E,n} &= ([\Omega_{4E,n}(E,n)(\cosh g_{1E,n})(\cos g_{2E,n}) \\
&\quad + \Omega_{3E,n}(E,n)(\sinh g_{1E,n})(\sin g_{2E,n})](\sin \gamma_{11E,n}(E,n))), \\
H_{14E,n} &= ([\Omega_{6E,n}(E,n)(\sinh g_{1E,n})(\cos g_{2E,n}) \\
&\quad + \Omega_{5E,n}(E,n)(\sin g_{1E,n})(\cosh g_{2E,n})](\cos \gamma_{11E,n}(E,n))), \\
H_{15E,n} &= ([\Omega_{10E,n}(E,n)(\cosh g_{1E,n})(\cos g_{2E,n}) \\
&\quad + \Omega_{9E,n}(E,n)(\sinh g_{1E,n})(\sin g_{2E,n})](\cos \gamma_{11E,n}(E,n))), \\
H_{16E,n} &= ([\Omega_{8E,n}(E,n)(\sinh g_{1E,n})(\cos g_{2E,n}) \\
&\quad + \Omega_{7E,n}(E,n)(\sin g_{1E,n})(\cosh g_{2E,n})](\sin \gamma_{11E,n}(E,n)/12)), \\
H_{19E,n} &= \left[\frac{2C_{18E,n}D_{18E,n}}{L_0^2}\right] \quad \text{and} \quad D_{18E,n} = \sinh^{-1}(\omega_{18E,n})
\end{aligned}
$$

The DOS function in this case can be expressed as

$$
N_{\mathrm{MQWSL}}(E) = \frac{g_v eB}{2\pi\hbar} \sum_{n_z=1}^{n_{z\max}} \sum_{n=0}^{n_{\max}} \delta'(E - E_{42,n}) \tag{18.6}
$$

The electron concentration can be written as

$$
n_0 = \frac{g_v eB}{2\pi\hbar} \text{Real part of} \sum_{n_z=1}^{n_{z\max}} \sum_{n=0}^{n_{\max}} F_{-1}(\bar{\eta}_{42,n}) \tag{18.7}
$$

where $\bar{\eta}_{42,n} = (k_B T)^{-1}[\bar{E}_{F42,n} - E_{42,n}]$.

The Einstein's photo-emitted current density in this case is given by

$$
J_{\mathrm{PHOTO}} = \frac{\alpha_0 e^2 g_v B}{4\pi\hbar d_z} \text{Real part of} \sum_{n=0}^{n_{\max}} \sum_{n_{z\min}}^{n_{z\max}} F_{-1}(\bar{\eta}_{42,n}) V_2(\bar{E}_{15,12\mathrm{SL}}) \tag{18.8}
$$

where

$$
\begin{aligned}
&V_2(\bar{E}_{15,12\mathrm{SL}}) \\
&\quad = (2\pi n_z/hd_z)[(G'_{19} + iH'_{19})|_{k_x=0,\ k_y=0,\ k_z=\frac{n_z\pi}{d_z} \text{ and } E=\bar{E}_{15,12\mathrm{SL}}}]^{-1}
\end{aligned}
$$

In which $\bar{E}_{15,12\mathrm{SL}}$ is the quantized energy along z-direction and can be obtain by substituting $k_x = 0$, $k_y = 0$, $k_z = \frac{n_z\pi}{d_z}$ and $E = \bar{E}_{15,12\mathrm{SL}}$ in (17.11).

18.2.3 The EP in IV–VI HDQWSLs with graded interfaces under magnetic quantization

The dispersion relation in HDQW IV–VI superlattices under magnetic quantization can be written as

$$\left(\frac{n_z\pi}{d_z}\right)^2 = \frac{1}{L_0^2}\left[\cos^{-1}\left\{\frac{1}{2}\Phi_2(E_{43,n},n)\right\}\right]^2 - \frac{2eB}{\hbar}\left(n+\frac{1}{2}\right) \quad (18.9)$$

where $E_{43,n}$ is the total quantized energy in this case,

$$\Phi_2(E,n) \equiv \Bigg[2\cosh\{\beta_2(E,n)\}\cos\{\gamma_2(E,n)\} + \varepsilon_2(E,n)\sinh\{\beta_2(E,n)\}$$

$$\times\sin\{\gamma_2(E,n)\} + \Delta_0\Bigg[((\{K_{112}(E,n)\}^2/K_{212}(E,n)) - 3K_{212}(E,n))$$

$$\times\cosh\{\beta_2(E,n)\}\sin\{\gamma_2(E,n)\} + \left(3K_{112}(E,n) - \frac{\{K_{112}(E,n)\}^2}{K_{212}(E,n)}\right)$$

$$\times\sinh\{\beta_2(E,n)\}\cos\{\gamma_2(E,n)\}\Bigg] + \Delta_0\Bigg[2(\{K_{112}(E,n)\}^2$$

$$-\{K_{212}(E,n)\}^2)\cosh\{\beta_2(E,n)\}\cos\{\gamma_2(E,n)\}$$

$$+\frac{1}{12}\left[\frac{5\{K_{112}(E,n)\}^3}{K_{212}(E,n)} + \frac{5\{K_{212}(E,n)\}^3}{K_{112}(E,n)} - 34K_{212}(E,n)K_{112}(E,n)\right]$$

$$\times\sinh\{\beta_2(E,n)\}\sin\{\gamma_2(E,n)\}\Bigg]\Bigg]$$

$$\beta_2(E,n) \equiv K_{112}(E,n)[a_0 - \Delta_0],$$

$$k_{112}^2(E,n) = [2\bar{p}_{9,2n}]^{-1}[-\bar{q}_{9,2n}(E-V_0,n,\eta_{g2}) + [[\bar{q}_{9,2n}(E-V_0,n,\eta_{g2})]^2$$

$$+4\bar{p}_{9,2n}\bar{R}_{9,2n}(E-V_0,n,\eta_{g2})]^{\frac{1}{2}}]$$

$$\bar{p}_{9,i} = (b_i^2 - f_i^2),$$

$$\bar{q}_{9,i}(E,n,\eta_{gi}) = \Bigg[2b_i\left\{\gamma_3(E,\eta_{gi}) + \frac{E_{gi}}{2}\right\} - 2a_ib_i\frac{2eB}{\hbar}\left(n+\frac{1}{2}\right) + d_i$$

$$-2e_if_i\frac{2eB}{\hbar}\left(n+\frac{1}{2}\right) - f_iE_{gi}\Bigg]$$

$$\bar{R}_{9,i}(E,n,\eta_{gi}) = \left[\left\{\gamma_3(E,\eta_{gi}) + \frac{E_{gi}}{2}\right\}^2 + a_i^2\left[\frac{2eB}{\hbar}\left(n+\frac{1}{2}\right)\right]^2\right.$$

$$- c_i\frac{2eB}{\hbar}\left(n+\frac{1}{2}\right) - e_i\left[\frac{2eB}{\hbar}\left(n+\frac{1}{2}\right)\right]^2 - \frac{E_{gi}^2}{4}$$

$$\left. - e_i\frac{2eB}{\hbar}\left(n+\frac{1}{2}\right)E_{gi} - 2a_i\left\{\gamma_3(E,\eta_{gi}) + \frac{E_{gi}}{2}\right\}\frac{2eB}{\hbar}\left(n+\frac{1}{2}\right)\right],$$

$$\gamma_{22}(E,n) = K_{212}(E,n)[b_0 - \Delta_0],$$

$$K_{212}^2(E,n) = [2\bar{p}_{9,1}]^{-1}[+\bar{q}_{9,1n}(E,n,\eta_{g1}) - [[\bar{q}_{9,1n}(E,n,\eta_{g1})]^2$$

$$+ 4\bar{p}_{9,1}\bar{R}_{9,1n}(E,n,\eta_{g1})]^{\frac{1}{2}}] \quad \text{and}$$

$$\varepsilon_2(E,n) \equiv \left[\frac{K_{112}(E,n)}{K_{212}(E,n)} - \frac{K_{212}(E,n)}{K_{112}(E,n)}\right]$$

The DOS function in this case can be expressed as

$$N_{\text{MQWSL}}(E) = \frac{g_v eB}{\pi\hbar}\sum_{n_z=1}^{n_{z\max}}\sum_{n=0}^{n_{\max}}\delta'(E - E_{43,n}) \tag{18.10}$$

The electron concentration can be written as

$$n_0 = \frac{g_v eB}{\pi\hbar}\sum_{n_z=1}^{n_{z\max}}\sum_{n=0}^{n_{\max}}F_{-1}(\bar{\eta}_{43,n}) \tag{18.11}$$

where $\bar{\eta}_{43,n} = (k_B T)^{-1}[\bar{E}_{F43,n} - E_{43,n}]$ and $\bar{E}_{F43,n}$ is the Fermi energy in this case.

The Einstein's photo-emitted current density in this case is given by

$$J_{\text{PHOTO}} = \frac{\alpha_0 e^2 g_v B}{2\pi\hbar d_z}\sum_{n=0}^{n_{\max}}\sum_{n_{z\min}}^{n_{z\max}}F_{-1}(\bar{\eta}_{43,n})V_3(\bar{E}_{15,13\text{SL}}) \tag{18.12}$$

where

$$V_3(\bar{E}_{15,13\text{SL}}) = \frac{2L_0\sin(L_0 n_z\pi/d_z)}{\hbar[\Phi_2(\bar{E}_{15,13\text{SL}},0)]'}$$

in which $\bar{E}_{15,13\text{SL}}$ can be obtained as

$$\cos(L_0 n_z\pi/d_z) = \frac{1}{2}\Phi_2(\bar{E}_{15,13\text{SL}},0) \tag{18.13}$$

18.2.4 The EP in HgTe/CdTe HDQWSLs with graded interfaces under magnetic quantization

The dispersion relation in HDQW HgTe/CdTe superlattices under magnetic quantization can be written from (17.25) as

$$\left(\frac{n_z\pi}{d_z}\right)^2 = G_{192E_{44,n}} + iH_{192E_{44,n}} \tag{18.14}$$

where $E_{44,n}$ is the total quantized energy in this case,

$$G_{192E,n} = \left[((C^2_{182E,n} - D^2_{182E,n})/L_0^2) - \left(\frac{2eB}{\hbar}\right)\left(n+\frac{1}{2}\right)\right],$$

$$C_{182E,n} = \cos^{-1}(\omega_{182E,n}),$$

$$\omega_{182E,n} = (2)^{\frac{-1}{2}}\left[(1 - G^2_{182E,n} - H^2_{182E,n}) - \sqrt{(1 - G^2_{182E,n} - H^2_{182E,n})^2 + 4G^2_{182E,n}}\right]^{\frac{1}{2}},$$

$$G_{182E,n} = \frac{1}{2}[G_{112E,n} + G_{122E,n} + \Delta_0(G_{132E,n} + G_{142E,n}) + \Delta_0(G_{152E,n} + G_{162E,n})],$$

$$G_{112E,n} = 2(\cos(g_{12E,n}))(\cos(g_{22E,n}))(\cos\gamma_8(E,n))$$

$$\gamma_8(E,n) = k_8(E,n)(b_0 - \Delta_0),$$

$$k_8(E,n) = \left[\frac{B_{01}^2 + 2A_1\gamma_3(E,\eta_{g1}) - B_{01}\sqrt{B_{01}^2 + 4A_1\gamma_3(E,\eta_{g1})}}{2A_1^2} - \left(\frac{2eB}{\hbar}\right)\left(n+\frac{1}{2}\right)\right]^{1/2}$$

$$G_{122E,n} = ([\Omega_{12E,n}(E,n)(\sinh g_{12E,n})(\cos g_{22E,n}) - \Omega_{22E,n}(E,n)(\sin g_{22E,n})(\cosh g_{12E,n})](\sin\gamma_8(E,n)))$$

$$\Omega_{12E,n}(E,n) = \left[\frac{d_{12E,n}}{k_{8E,n}(E,n)} - \frac{k_{8E,n}(E,n)d_{12E,n}}{d^2_{12E,n} + d^2_{22E,n}}\right],$$

$$d_{12E,n} = \left[\frac{\sqrt{x^2_{12E,n} + y^2_{12E,n}} + x_{12E,n}}{2}\right]^{1/2},$$

$$d_{22E,n} = \left[\frac{\sqrt{x_{12E,n}^2 + y_{12E,n}^2} - x_{12E,n}}{2}\right]^{1/2},$$

$$x_{12E,n} = \left[\left(\frac{2eB}{\hbar}\right)\left(n+\frac{1}{2}\right) - \left(\frac{2m_{c2}}{\hbar^2}\right)T_{12}(E-V_0,\Delta_2,E_{g2},\eta_{g2})\right],$$

$$y_{12E,n} = \left[\left(\frac{2m_{c2}}{\hbar^2}\right)T_{22}(E-V_0,\Delta_2,E_{g2},\eta_{g2})\right],$$

$$\Omega_{22E,n}(E,k_s) = \left[\frac{d_{22E,n}}{k_{8E,n}(E,n)} + \frac{k_{8E,n}(E,n)d_{22E,n}}{d_{12E,n}^2 + d_{22E,n}^2}\right],$$

$$G_{122E,n} = ([\Omega_{12E,n}(E,n)(\sinh g_{12E,n})(\cos g_{22E,n}) - \Omega_{22E,n}(E,n)(\sin g_{22E,n})(\cosh g_{12E,n})](\sin\gamma_8(E,n)))$$

$$\Omega_{12E,n}(E,n) = \left[\frac{d_{12E,n}}{k_{8E,n}(E,n)} - \frac{k_{8E,n}(E,n)d_{12E,n}}{d_{12E,n}^2 + d_{22E,n}^2}\right],$$

$$d_{12E,n} = \left[\frac{\sqrt{x_{12E,n}^2 + y_{12E,n}^2} + x_{12E,n}}{2}\right]^{1/2},$$

$$d_{22E,n} = \left[\frac{\sqrt{x_{12E,n}^2 + y_{12E,n}^2} - x_{12E,n}}{2}\right]^{1/2},$$

$$x_{12E,n} = \left[\left(\frac{2eB}{\hbar}\right)\left(n+\frac{1}{2}\right) - \left(\frac{2m_{c2}}{\hbar^2}\right)T_{12}(E-V_0,\Delta_2,E_{g2},\eta_{g2})\right],$$

$$y_{12E,n} = \left[\left(\frac{2m_{c2}^*}{\hbar^2}\right)T_{22}(E-V_0,\Delta_2,E_{g2},\eta_{g2})\right],$$

$$\Omega_{22E,n}(E,k_s) = \left[\frac{d_{22E,n}}{k_{8E,n}(E,n)} + \frac{k_{8E,n}(E,n)d_{22E,n}}{d_{12E,n}^2 + d_{22E,n}^2}\right],$$

$$G_{132E,n} = ([\Omega_{32E,n}(E,n)(\cosh g_{12E,n})(\cos g_{22E,n}) - \Omega_{42E,n}(E,n)(\sinh g_{12E,n})(\sin g_{22E,n})](\sin\gamma_8(E,n))),$$

$$\Omega_{32E,n}(E,n) = \left[\frac{d_{12E,n}^2 - d_{22E,n}^2}{k_8(E,n)} - 3k_8(E,n)\right],$$

$$\Omega_{42E,n}(E,n) = \left[\frac{2d_{12E,n}d_{22E,n}}{k_8(E,n)}\right],$$

$$G_{142E,n} = ([\Omega_{52E,n}(E,n)(\sinh g_{12E,n})(\cos g_{22E,n})$$

$$- \Omega_{62E,n}(E,n)(\sin g_{12E,n})(\cosh g_{22E,n})](\cos\gamma_8(E,n))),$$

$$\Omega_{52E,n}(E,n) = \left[3d_{12E,n} - \frac{d_{12E,n}}{d_{12E,n}^2 + d_{22E,n}^2}k_8^2(E,n)\right],$$

$$\Omega_{62E,n}(E,n) = \left[3d_{22E,n} + \frac{d_{22E,n}}{d_{12E,n}^2 + d_{22E,n}^2}k_8^2(E,n)\right],$$

$$G_{152E,n} = ([\Omega_{92}(E,n)(\cosh g_{12E,n})(\cos g_{22E,n})$$

$$- \Omega_{102E,n}(E,n)(\sinh g_{12E,n})(\sin g_{22E,n})](\cos\gamma_8(E,n))),$$

$$\Omega_{92E,n}(E,n) = [2d_{12E,n}^2 - 2d_{22E,n}^2 - k_8^2(E,n)],$$

$$\Omega_{102E,n}(E,n) = [2d_{12E,n}d_{22E,n}],$$

$$G_{162E,n} = ([\Omega_{72E,n}(E,n)(\sinh g_{12E,n})(\cos g_{22E,n})$$

$$- \Omega_{82E,n}(E,n)(\sin g_{12E,n})(\cosh g_{22E,n})](\sin\gamma_8(E,n)/12))$$

$$\Omega_{72E,n}(E,n) = \left[\frac{5d_{12E,n}}{d_{12E,n}^2 + d_{22E,n}^2}k_8^3(E,n) + \frac{5(d_{12E,n}^3 - 3d_{22E,n}^2 d_{12E,n})}{k_8(E,n)}\right.$$

$$\left. - 34k_8(E,n)d_{22E,n}\right],$$

$$\Omega_{82E,n}(E,n) = \left[\frac{5d_{22E,n}}{d_{12E,n}^2 + d_{22E,n}^2}k_8^3(E,n) + \frac{5(d_{12E,n}^3 - 3d_{22E,n}d_{12E,n}^2)}{k_8(E,n)}\right.$$

$$\left. + 34k_8(E,n)d_{22E,n}\right],$$

$$H_{182E,n} = \frac{1}{2}[H_{112E,n} + H_{122E,n} + \Delta_0(H_{132E,n} + H_{142E,n})$$

$$+ \Delta_0(H_{152E,n} + H_{162E,n})]$$

$$H_{112E,n} = 2(\sinh g_{12E,n}\sin g_{22E,n}\cos\gamma_8(E,n)),$$

$$
\begin{aligned}
g_{12E,n} &= d_{12E,n}(a_0 - \Delta_0), \quad g_{22E,n} = d_{22E,n}(a_0 - \Delta_0), \\
H_{122E,n} &= ([\Omega_{22E,n}(E,n)(\sinh g_{12E,n})(\cos g_{22E,n}) \\
&\quad + \Omega_{12E,n}(E,n)(\sin g_{22E,n})(\cosh g_{12E,n})](\sin \gamma_8(E,n))), \\
H_{132E,n} &= ([\Omega_{42E,n}(E,n)(\cosh g_{12E,n})(\cos g_{22E,n}) \\
&\quad + \Omega_{32E,n}(E,n)(\sinh g_{12E,n})(\sin g_{22E,n})](\sin \gamma_8(E,n))), \\
H_{142E,n} &= ([\Omega_{62E,n}(E,n)(\sinh g_{12E,n})(\cos g_{22E,n}) \\
&\quad + \Omega_{52E,n}(E,n)(\sin g_{12E,n})(\cosh g_{22E,n})](\cos \gamma_8(E,n))), \\
H_{152E,n} &= ([\Omega_{102E,n}(E,n)(\cosh g_{12E,n})(\cos g_{22E,n}) \\
&\quad + \Omega_{92E,n}(E,n)(\sinh g_{12E,n})(\sin g_{22E,n})](\cos \gamma_8(E,n))), \\
H_{162E,n} &= ([\Omega_{82E,n}(E,n)(\sinh g_{12E,n})(\cos g_{22E,n}) \\
&\quad + \Omega_{72E,n}(E,n)(\sin g_{12E,n})(\cosh g_{22E,n})](\sin \gamma_8(E,n)/12)), \\
H_{192E,n} &= [((2C_{182E,n}D_{182E,n})/L_0^2)] \quad \text{and} \quad D_{182E,n} = \sinh^{-1}(\omega_{182E,n})
\end{aligned}
$$

The DOS function in this case can be expressed as

$$
N_{\mathrm{MQWSL}}(E) = \frac{g_v eB}{\pi\hbar} \sum_{n_z=1}^{n_{z\max}} \sum_{n=0}^{n_{\max}} \delta'(E - E_{44,n}) \tag{18.15}
$$

The electron concentration can be written as

$$
n_0 = \frac{g_v eB}{\pi\hbar} \text{ Real part of } \sum_{n_z=1}^{n_{z\max}} \sum_{n=0}^{n_{\max}} F_{-1}(\bar{\eta}_{44,n}) \tag{18.16}
$$

where $\bar{\eta}_{44,n} = (k_B T)^{-1}[\bar{E}_{F44,n} - E_{44,n}]$.

The Einstein's photo-emitted current density in this case is given by

$$
J_{\mathrm{PHOTO}} = \frac{\alpha_0 e^2 g_v B}{2\pi\hbar d_z} \text{ Real part of } \sum_{n=0}^{n_{\max}} \sum_{n_{z_{\min}}}^{n_{z\max}} F_{-1}(\bar{\eta}_{44,n}) V_4(\bar{E}_{15,14\mathrm{SL}}) \tag{18.17}
$$

where

$$
\begin{aligned}
&V_4(\bar{E}_{15,14\mathrm{SL}}) \\
&= (2\pi n_z/hd_z)[(G'_{192} + iH'_{192})\big|_{k_x=0,\,k_y=0,\,k_z=\frac{n_z\pi}{d_z},\ \text{and}\ E=\bar{E}_{15,14\mathrm{SL}}}]^{-1}
\end{aligned}
$$

In which $\bar{E}_{15,14\mathrm{SL}}$ is the quantized energy along z-direction and can be obtain by substituting $k_x = 0$, $k_y = 0$, $k_z = \frac{n_z\pi}{d_z}$ and $E = \bar{E}_{15,14\mathrm{SL}}$ in (17.25).

*18.2.5 The EP in strained layer HDQWSLs with graded interfaces under magnetic quantization

The dispersion relation of the conduction electrons in HDQW strained layer SLs with graded interfaces can be expressed from (17.31) as

$$\left(\frac{n_z\pi}{d_z}\right)^2 = \frac{1}{L_0^2}\left[\cos^{-1}\left\{\frac{1}{2}\bar{\Phi}_6(E_{47,n}, n, \eta_{g1}, \eta_{g2})\right\}\right]^2 - \frac{2eB}{\hbar}\left(n+\frac{1}{2}\right) \tag{18.18}$$

where $E_{47,n}$ is the total quantized energy in this case,

$$\begin{aligned}
&\bar{\Phi}_6(E, n, \eta_{g1}, \eta_{g2}) = [2\cosh[T_{4n}(E, n, \eta_{g2})]\cos[T_{5n}(E, n, \eta_{g1})]] \\
&\quad + [T_{6n}(E, n, \eta_{g1}, \eta_{g2})]\sinh[T_{4n}(E, n, \eta_{g2})]\sin[T_{5n}(E, n, \eta_{g1})] \\
&\quad + \Delta_0\left[\left(\frac{k_{0n}^2(E, n, \eta_{g2})}{k_{70n}(E, n, \eta_{g1})} - 3k_{70n}(E, n, \eta_{g1})\right)\cosh[T_{4n}(E, n, \eta_{g2})]\right. \\
&\quad \times \sin[T_{5n}(E, n, \eta_{g1})] + \left(3k_{0n}(E, n, \eta_{g2}) - \frac{k_{70n}^2(E, n, \eta_{g1})}{k_{0n}(E, n, \eta_{g2})}\right) \\
&\quad \left.\times \sinh[T_{4n}(E, n, \eta_{g2})]\cos[T_{5n}(E, n, \eta_{g1})]\right] + \Delta_0[2(k_{0n}^2(E, n, \eta_{g2}) \\
&\quad - k_{70n}^2(E, n, \eta_{g1}))\cosh[T_{4n}(E, n, \eta_{g2})]\sin[T_{5n}(E, n, \eta_{g1})]] \\
&\quad + \frac{1}{12}\left(\frac{5k_{0n}^3(E, n, \eta_{g2})}{k_{70n}(E, n, \eta_{g1})} + \frac{5k_{70n}^3(E, n, \eta_{g1})}{k_{0n}(E, n, \eta_{g2})}\right. \\
&\quad \left. - 34k_{0n}(E, n, \eta_{g2})k_{70n}(E, n, \eta_{g1})\right)\sinh[T_{4n}(E, n, \eta_{g2})] \\
&\quad \times \sin[T_{5n}(E, n, \eta_{g1})][T_{4n}(E, n, \eta_{g2})] = k_{0n}(E, n, \eta_{g2})[a_0 - \Delta_0], \\
&k_{0n}(E, n, \eta_{g2}) = [\bar{S}_{11,2}(E - V_0, \eta_{g2})]^{-1/2} \\
&\quad \times \left[\left(n+\frac{1}{2}\right)\hbar\omega_{17}(E - V_0, \eta_{g2}) - 1\right]^{1/2}, \\
&\omega_{17}(E - V_0, \eta_{g2}) = \frac{eB}{\sqrt{m_{17}m_{18}}}, \quad m_{17} = \frac{\hbar^2}{2\bar{P}_{11,2}(E - V_0, \eta_{g2})},
\end{aligned}$$

$$m_{18} = \frac{\hbar^2}{2\bar{Q}_{11,2}(E - V_0, \eta_{g2})}, \quad T_{5n}(E, \eta_{g1}) = k_{70n}(E, \eta_{g1})[b_0 - \Delta_0],$$

$$k_{70n}(E, n, \eta_{g1}) = [\bar{S}_{11,1}(E, \eta_{g1})]^{-1/2}\left[1 - \left(n + \frac{1}{2}\right)\hbar\omega_{18}(E, \eta_{g1})\right]^{1/2},$$

$$\omega_{18}(E, \eta_{g1}) = \left(\frac{eB}{\sqrt{m_{19}^* m_{20}^*}}\right), \quad m_{19}^* = \frac{\hbar^2}{2\bar{P}_{11,1}(E, \eta_{g1})},$$

$$m_{20}^* = \frac{\hbar^2}{2\bar{Q}_{11,1}(E, \eta_{g1})} \quad \text{and}$$

$$T_{6n}(E, n, \eta_{g1}, \eta_{g2}) = \left[\frac{k_{0n}(E, n, \eta_{g2})}{k_{70n}(E, n, \eta_{g1})} - \frac{k_{70n}(E, n, \eta_{g1})}{k_{0n}(E, n, \eta_{g2})}\right]$$

The DOS function in this case can be expressed as

$$N_{\text{MQWSL}}(E) = \frac{g_v eB}{\pi\hbar}\sum_{n_z=1}^{n_{z\max}}\sum_{n=0}^{n_{\max}}\delta'(E - E_{47,n}) \tag{18.19}$$

The electron concentration can be written as

$$n_0 = \frac{g_v eB}{\pi\hbar}\sum_{n_z=1}^{n_{z\max}}\sum_{n=0}^{n_{\max}} F_{-1}(\bar{\eta}_{47,n}) \tag{18.20}$$

where $\bar{\eta}_{47,n} = (k_B T)^{-1}[\bar{E}_{F47,n} - E_{47,n}]$.

The Einstein's photo-emitted current density in this case is given by

$$J_{\text{PHOTO}} = \frac{\alpha_0 e^2 g_v B}{2\pi\hbar d_z}\sum_{n=0}^{n_{\max}}\sum_{n_{z\min}}^{n_{z\max}} F_{-1}(\bar{\eta}_{47,n})V_5(\bar{E}_{15,15\text{SL}}) \tag{18.21}$$

where

$$V_5(\bar{E}_{15,15\text{SL}}) = \frac{2L_0\sin(L_0 n_z\pi/d_z)}{\hbar[\bar{\Phi}_6(\bar{E}_{15,15\text{SL}}, 0, \eta_{g1}, \eta_{g2})]'}$$

which $\bar{E}_{15,15\text{SL}}$ can be obtained from (17.31) as

$$\cos(L_0 n_z\pi/d_z) = \frac{1}{2}\bar{\Phi}_6(\bar{E}_{15,15\text{SL}}, 0, \eta_{g1}, \eta_{g2}) \tag{18.22}$$

18.2.6 The EP in III–V HDQW effective mass superlattices under magnetic quantization

The dispersion relation in HDQW III–V effective mass superlattices under magnetic quantization can be expressed from (17.35) as form

$$\left(\frac{n_z\pi}{d_z}\right)^2 = (\delta_{7A1,n} + i\delta_{7A1,n})\Big|_{E=A1,n} \tag{18.23}$$

where $A1,n$ is the total quantized energy in this case,

$$\delta_{7E,n} = \left[\frac{1}{L_0^2}(\delta_{5E,n}^2 - \delta_{6E,n}^2) - \left(\frac{2eB}{\hbar}\right)\left(n+\frac{1}{2}\right)\right], \quad \delta_{5E,n} = \cos^{-1} p_{5E,n},$$

$$p_{5E,n} = \left[\frac{1-\delta_{3E,n}^2 - \delta_{4E,n}^2 - \sqrt{(1-\delta_{3E,n}^2-\delta_{4E,n}^2)^2 + 4\delta_{4E,n}^2}}{2}\right]^{\frac{1}{2}},$$

$$\delta_{3E,n} = (\bar{a}_1 \cos\Delta_{1E,n}\cosh\Delta_{2E,n} - \bar{a}_2\cos\Delta_{3E,n}\cosh\Delta_4),$$

$$\delta_{4E,n} = (\bar{a}_1 \sin\Delta_{1E,n}\sinh\Delta_{2E,n} - \bar{a}_2\sin\Delta_{3E,n}\sinh\Delta_{4E,n}),$$

$$\Delta_{1E,n} = (a_0 e_{1E,n} + b_0 e_{3E,n}), \quad \Delta_{2E,n} = (a_0 e_{2E,n} + b_0 e_{4E,n}),$$

$$\Delta_{3E,n} = (a_0 e_{1E,n} - b_0 e_{3E,n}), \quad \Delta_{4E,n} = (a_0 e_{1E,n} - b_0 e_{4E,n}),$$

$$\delta_{6E,n} = \sinh^{-1} p_{5E,n} \quad \text{and} \quad \delta_{8E,n} = [2\delta_{5E,n}\delta_{6E,n}/L_0^2],$$

$$e_{1E,n} = \left[\left(\left(\sqrt{t_{1E,n}^2 + t_{2E,n}^2} + t_{1E,n}\right)/2\right)\right]^{\frac{1}{2}},$$

$$e_{2E,n} = \left[\left(\left(\sqrt{t_{1E,n}^2 + t_{2E,n}^2} - t_{1E,n}\right)/2\right)\right]^{\frac{1}{2}},$$

$$t_{1E,n} = \left[\frac{2m_{c1}}{\hbar^2}T_{11}(E,\Delta_1,\eta_{g1},E_{g1}) - \left(\frac{2eB}{\hbar}\right)\left(n+\frac{1}{2}\right)\right],$$

$$t_{2E,n} = \frac{2m_{c1}}{\hbar^2}T_{21}(E,\Delta_1,\eta_{g1},E_{g1}), \quad e_{3E,n} = \left[\frac{\sqrt{t_{3E,n}^2 + t_{4E,n}^2} + t_{3E,n}}{2}\right]^{\frac{1}{2}},$$

$$e_{4E,n} = \left[\frac{\sqrt{t_{3E,n}^2 + t_{4E,n}^2} - t_{3E,n}}{2}\right]^{\frac{1}{2}},$$

$$t_{3E,n} = \left[\frac{2m_{c2}}{\hbar^2}T_{12}(E, \Delta_2, \eta_{g2}, E_{g2}) - \left(\frac{2eB}{\hbar}\right)\left(n + \frac{1}{2}\right)\right] \quad \text{and}$$

$$t_{4E,n} = \left[\frac{2m_{c2}}{\hbar^2}T_{22}(E, \Delta_2, \eta_{g2}, E_{g2})\right]$$

The DOS function in this case can be expressed as

$$N_{\text{MQWSL}}(E) = \frac{g_v eB}{\pi\hbar} \sum_{n_z=1}^{n_{z\max}} \sum_{n=0}^{n_{\max}} \delta'(E - A1, n) \tag{18.24}$$

The electron concentration can be written as

$$n_0 = \frac{g_v eB}{\pi\hbar} \text{Real part of} \sum_{n_z=1}^{n_{z\max}} \sum_{n=0}^{n_{\max}} F_{-1}(\bar{\eta}_{7A1,n}) \tag{18.25}$$

where $\bar{\eta}_{7A1,n} = (k_B T)^{-1}[\bar{E}_{F41,n} - A1, n]$ and $\bar{E}_{F41,n}$ is the Fermi energy in this case.

The Einstein's photo-emitted current density in this case is given by

$$J_{\text{PHOTO}} = \frac{\alpha_0 e^2 g_v B}{2\pi\hbar d_z} \text{Real part of} \sum_{n=0}^{n_{\max}} \sum_{n_{z\min}}^{n_{z\max}} F_{-1}(\bar{\eta}_{7A1,n}) V_6(\bar{E}_{15,16\text{SL}}) \tag{18.26}$$

where

$$V_6(\bar{E}_{15,16\text{SL}}) = (2\pi n_z/hd_z)[(\delta_7' + i\delta_8')|_{k_x=0,\ k_y=0,\ k_z=\frac{n_z\pi}{d_z} \text{ and } E=\bar{E}_{15,16\text{SL}}}]^{-1}$$

In which $\bar{E}_{15,16\text{SL}}$ is the quantized energy along z-direction and can be obtain by substituting $k_x = 0$, $k_y = 0$, $k_z = \frac{n_z\pi}{d_z}$ and $E = \bar{E}_{15,16\text{SL}}$ in (17.35).

18.2.7 The EP in II–VI HDQW effective mass superlattices under magnetic quantization

The dispersion relation in HDQW III–V effective mass superlattices under magnetic quantization can be expressed as

$$\left(\frac{n_z\pi}{d_z}\right)^2 = (\Delta_{13A2,n} + i\Delta_{14A2,n})\Big|_{E=A2,n} \tag{18.27}$$

where $A2, n$ is the total quantized energy in this case,

$$\Delta_{13E,n} = \left[\frac{1}{L_0^2}(\Delta_{11E,n}^2 - \Delta_{12E,n}^2) - \left(\frac{2eB}{\hbar}\right)\left(n + \frac{1}{2}\right)\right],$$

$$\Delta_{11E,n} = \cos^{-1} p_{6E,n},$$

$$p_{6E,n} = \left[\frac{1 - \Delta_{9E,n}^2 - \Delta_{10E,n}^2 - \sqrt{(1 - \Delta_{9E,n}^2 - \Delta_{10E,n}^2)^2 + 4\Delta_{10E,n}^2}}{2}\right]^{1/2},$$

$$\Delta_{9E,n} = (\bar{a}_3 \cos\Delta_{6E,n} \cosh\Delta_{7E,n} - \bar{a}_4 \cos\Delta_{8E,n} \cosh\Delta_{7E,n}),$$

$$\Delta_{10E,n} = (\bar{a}_3 \sin\Delta_{6E,n} \sinh\Delta_{7E,n} + \bar{a}_4 \sin\Delta_{8E,n} \sinh\Delta_{7E,n}),$$

$$\Delta_{6E,n} = [a_0 C_{22}(E, n, \eta_{g1}) + b_0 e_{3E,n}], \quad \Delta_{7E,n} = b_0 e_{4E,n},$$

$$\Delta_{8E,n} = [a_0 C_{22}(E, n, \eta_{g1}) - b_0 e_{3E,n}],$$

$$C_{22}(E, n, \eta_{g1}) = \left[\frac{2m_{\|,1}^*}{\hbar^2}\left\{\gamma_3(E, \eta_{g1}) - \frac{\hbar^2\left(\frac{2eB}{\hbar}\right)\left(n + \frac{1}{2}\right)}{2m_{\perp,1}^*}\right.\right.$$

$$\left.\left.\mp \bar{\lambda}_0\sqrt{\left(\frac{2eB}{\hbar}\right)\left(n + \frac{1}{2}\right)}\right\}\right]^{1/2},$$

$$e_{3E,n} = \left[\frac{\sqrt{t_{3E,n}^2 + t_{4E,n}^2} + t_{3E,n}}{2}\right]^{1/2},$$

$$e_{4E,n} = \left[\frac{\sqrt{t_{3E,n}^2 + t_{4E,n}^2} - t_{3E,n}}{2}\right]^{1/2},$$

$$t_{3E,n} = \left[\left(\frac{2m_{c2}}{\hbar^2}\right) T_{12}(E, \Delta_2, \eta_{g2}, E_{g2}) - \left(\frac{2eB}{\hbar}\right)\left(n + \frac{1}{2}\right)\right],$$

$$t_{4E,n} = \left[\left(\frac{2m_{c2}}{\hbar^2}\right) T_{22}(E, \Delta_2, \eta_{g2}, E_{g2}) - \left(\frac{2eB}{\hbar}\right)\left(n + \frac{1}{2}\right)\right],$$

$$\Delta_{12E,n} = \sinh^{-1} p_{6E,n} \quad \text{and} \quad \Delta_{14E,n} = \frac{2\Delta_{11E,n}\Delta_{12E,n}}{L_0^2}$$

The DOS function in this case can be expressed as

$$N_{\mathrm{MQWSL}}(E) = \frac{g_v eB}{2\pi\hbar} \sum_{n_z=1}^{n_{z\max}} \sum_{n=0}^{n_{\max}} \delta'(E - A2, n) \tag{18.28}$$

The electron concentration can be written as

$$n_0 = \frac{g_v eB}{2\pi\hbar} \text{ Real part of } \sum_{n_z=1}^{n_{z\max}} \sum_{n=0}^{n_{\max}} F_{-1}(\bar{\eta}_{13A2,n}) \tag{18.29}$$

where $\bar{\eta}_{13A2,n} = (k_B T)^{-1}[\bar{E}_{F41,n} - A2,n]$.

The Einstein's photo-emitted current density in this case is given by

$$J_{\text{PHOTO}} = \frac{\alpha_0 e^2 g_v B}{4\pi\hbar d_z} \text{Real part of} \sum_{n=0}^{n_{\max}} \sum_{n_{z\min}}^{n_{z\max}} F_{-1}(\bar{\eta}_{13A2,n}) V_7(\bar{E}_{15,17\text{SL}}) \tag{18.30}$$

where

$$\begin{aligned} &V_7(\bar{E}_{15,17\text{SL}}) \\ &= (2\pi n_z/hd_z)[(\Delta'_{13} + i\Delta'_{14})|_{k_x=0,\ k_y=0,\ k_z=\frac{n_z\pi}{d_z} \text{ and } E=\bar{E}_{15,17\text{SL}}}]^{-1} \end{aligned}$$

In which $\bar{E}_{15,17\text{SL}}$ is the quantized energy along z-direction and can be obtain by substituting $k_x = 0$, $k_y = 0$, $k_z = \frac{n_z\pi}{d_z}$ and $E = \bar{E}_{15,17\text{SL}}$ in (17.40).

18.2.8 The EP in IV–VI HDQW effective mass superlattices under magnetic quantization

The dispersion relation of the conduction electrons in HDQW strained layer effective mass SLs with graded interfaces can be expressed from (17.44) as

$$\left(\frac{n_z\pi}{d_z}\right)^2 = \left[\frac{1}{L_0^2}[\cos^{-1}\{f_{23}(A3,n)\}]^2 - \frac{2eB}{\hbar}\left(n + \frac{1}{2}\right)\right]\Bigg|_{E=A3,n} \tag{18.31}$$

where $A3, n$ is the total quantized energy in this case,

$$\begin{aligned} f_{23}(E,n) &= \bar{a}_{30}\cos[a_0 C_{23}(E,n,\eta_{g1}) + b_0 D_{23}(E,n,\eta_{g1})] \\ &\quad - \bar{a}_{40}\cos[a_0 C_{23}(E,n,\eta_{g2}) - b_0 D_{23}(E,n,\eta_{g2})], \end{aligned}$$

$$\bar{a}_{30} = \left[\sqrt{\frac{M_3(0,\eta_{g2})}{M_3(0,\eta_{g1})}} + 1\right]^2 \left[4\left(\frac{M_3(0,\eta_{g2})}{M_3(0,\eta_{g1})}\right)^{1/2}\right]^{-1},$$

$$\bar{a}_{40} = \left[\sqrt{\frac{M_3(0,\eta_{g2})}{M_3(0,\eta_{g1})}} - 1\right]^2 \left[4\left(\frac{M_3(0,\eta_{g2})}{M_3(0,\eta_{g1})}\right)^{1/2}\right]^{-1}$$

$$M_3(0,\eta_{gi}) = \frac{\hbar^2}{4\bar{p}_{9,i}}\left[\{\bar{q}_{9,i}(0,0,\eta_{gi})\}' - \frac{[\{\bar{q}_{9,i}(0,0,\eta_{gi})\}\{\bar{q}_{9,i}(0,0,\eta_{gi})\}' - 2\bar{p}_{9,i}\{\bar{R}_{9,i}(0,0,\eta_{gi})\}']}{\sqrt{\bar{q}_{9,i}[(0,0,\eta_{gi})]^2 - [4\bar{p}_{9,i}\bar{R}_{9,i}(0,0,\eta_{gi})]}}\right],$$

$$D_{23}(E,n,\eta_{g2}) = [[2\bar{p}_{9,2}]^{-1}[+\bar{q}_{9,2}(E,n,\eta_{g2}) - [\{\bar{q}_{9,2}(E,n,\eta_{g2})\}^2 - (4\bar{p}_{9,2})\bar{R}_{9,2}(E,n,\eta_{g2})]^{1/2}]]^{1/2},$$

$$C_{23}(E,n,\eta_{g1}) = [[2\bar{p}_{9,1}]^{-1}[+\bar{q}_{9,2}(E,n,\eta_{g1}) - [\{\bar{q}_{9,2}(E,n,\eta_{g1})\}^2 - (4\bar{p}_{9,1})\bar{R}_{9,1}(E,n,\eta_{g1})]^{1/2}]]^{1/2},$$

$$\bar{q}_{9,i}(E,n,\eta_{gi}) = \left[2b_i\left\{\gamma_3(E,\eta_{gi}) + \frac{E_{gi}}{2}\right\} - 2a_ib_i\left(\frac{2eB}{\hbar}\right)\left(n+\frac{1}{2}\right) + d_i - 2e_if_i\left(\frac{2eB}{\hbar}\right)\left(n+\frac{1}{2}\right) - f_iE_{gi}\right]$$

and

$$\bar{R}_{9,i}(E,n,\eta_{gi}) = \left[\left\{\gamma_3(E,\eta_{gi}) + \frac{E_{gi}}{2}\right\}^2 + a_i^2\left[\left(\frac{2eB}{\hbar}\right)\left(n+\frac{1}{2}\right)\right]^2 - c_i\left(\frac{2eB}{\hbar}\right)\left(n+\frac{1}{2}\right) - e_i\left[\left(\frac{2eB}{\hbar}\right)\left(n+\frac{1}{2}\right)\right]^2 - \frac{E_{gi}^2}{4} - e_i\left(\frac{2eB}{\hbar}\right)\left(n+\frac{1}{2}\right)E_{gi} - 2a_i\left\{\gamma_3(E,\eta_{gi}) + \frac{E_{gi}}{2}\right\}\left(\frac{2eB}{\hbar}\right)\left(n+\frac{1}{2}\right)\right]$$

The DOS function in this case can be expressed as

$$N_{\text{MQWSL}}(E) = \frac{g_veB}{\pi\hbar}\sum_{n_z=1}^{n_{z\max}}\sum_{n=0}^{n_{\max}}\delta'(E - A3,n) \tag{18.32}$$

The electron concentration can be written as

$$n_0 = \frac{g_veB}{\pi\hbar}\sum_{n_z=1}^{n_{z\max}}\sum_{n=0}^{n_{\max}}F_{-1}(\bar{\eta}_{1528,n}) \tag{18.33}$$

where $\bar{\eta}_{1528,n} = (k_BT)^{-1}[\bar{E}_{F41,n} - A3,n]$.

The Einstein's photo-emitted current density in this case is given by

$$J_{\text{PHOTO}} = \frac{\alpha_0 e^2 g_v B}{2\pi\hbar d_z} \sum_{n=0}^{n_{\max}} \sum_{n_{z_{\min}}}^{n_{z_{\max}}} F_{-1}(\bar{\eta}_{15,28,n}) V_8(\bar{E}_{15,18\text{SL}}) \tag{18.34}$$

where

$$V_8(\bar{E}_{15,18\text{SL}}) = \frac{2L_0 \sin(L_0 n_z \pi/d_z)}{\hbar[f_{23}(\bar{E}_{15,18\text{SL}}, 0)]'}$$

In which $\bar{E}_{15,18\text{SL}}$ can be obtained from the (17.44) as

$$\cos(L_0 n_z \pi/d_z) = f_{23}(\bar{E}_{15,18\text{SL}}, 0) \tag{18.35}$$

18.2.9 The EP in HgTe/CdTe HDQW effective mass superlattices under magnetic quantization

The dispersion relation in HDQW III–V effective mass superlattices under magnetic quantization can be expressed from (17.48) as

$$\left(\frac{n_z \pi}{d_z}\right)^2 = (\Delta_{13A2,n} + i\Delta_{14A2,n})\Big|_{E=A4,n} \tag{18.36}$$

where $A4, n$ is the total quantized energy in this case,

$$\Delta_{13\text{HE},n} = \left[\frac{1}{L_0^2}(\Delta_{11\text{HE},n}^2 - \Delta_{12\text{HE},n}^2) - \left(\frac{2eB}{\hbar}\right)\left(n + \frac{1}{2}\right)\right],$$

$$\Delta_{11\text{HE},n} = \cos^{-1} p_{6\text{HE,n}}, p_{6\text{HE},n} = (2)^{-1/2}\left[1 - \Delta_{9\text{HE},n}^2 - \Delta_{10\text{HE},n}^2 \right.$$

$$\left. - \sqrt{(1 - \Delta_{9\text{HE},n}^2) - \Delta_{10\text{HE},n}^2{}^2 + 4\Delta_{10\text{HE},n}^2}\right]^{1/2},$$

$$\Delta_{9\text{HE},n} = (\bar{a}_{1H} \cos \Delta_{5\text{HE},n} \cosh \Delta_{7HE,n}$$

$$- \bar{a}_{2H} \cos \Delta_{8\text{HE},n} \cosh \Delta_{7\text{HE},n}),$$

$$\Delta_{10\text{HE},n} = (\bar{a}_{1H} \sin \Delta_{6\text{HE},n} \sinh \Delta_{7\text{HE},n}$$

$$+ \bar{a}_{2\text{HE},n} \sin \Delta_{8\text{HE},n} \sinh \Delta_{7\text{HE},n}),$$

$$\Delta_{6\text{HE},n} = [a_0 C_{22H}(E, n, \eta_{g1}) + b_0 e_{3E,n}], \quad \Delta_{7\text{HE},n} = b_0 e_{4E,n},$$

$$\Delta_{8\text{HE},n} = [a_0 C_{22H}(E, n, \eta_{g1}) - b_0 e_{3E,n}],$$

$$C_{22H}(E,n,\eta_{g1}) = \left[\frac{B_{01}^2 + 2A_1\gamma(E,\eta_{g1}) - B_{01}(B_{01}^2 + 4A_1\gamma(E,\eta_{g1}))}{2A_1^2} - \left(\frac{2eB}{\hbar}\right)\left(n+\frac{1}{2}\right)\right]^{1/2},$$

$$e_{3E,n} = \left[\frac{\sqrt{t_{3E,n}^2 + t_{4E,n}^2} + t_{3E,n}}{2}\right]^{1/2},$$

$$e_{4E,n} = \left[\frac{\sqrt{t_{3E,n}^2 + t_{4E,n}^2} - t_{3E,n}}{2}\right]^{1/2},$$

$$t_{3E,n} = \left[\frac{2m_{c2}}{\hbar^2}T_{12}(E,\Delta_2,E_{g2},\eta_{g2}) - \left(\frac{2eB}{\hbar}\right)\left(n+\frac{1}{2}\right)\right],$$

$$t_{4E,n} = \left[\frac{2m_{c2}}{\hbar^2}T_{22}(E,\Delta_2,E_{g2},\eta_{g2})\right],$$

$$\bar{a}_{1H} = \left[\sqrt{\frac{M_2(0,\eta_{g2})}{M_9(0,\eta_{g1})}} + 1\right]^2\left[4\left(\frac{M_2(0,\eta_{g2})}{M_9(0,\eta_{g1})}\right)\right]^{-1},$$

$$\bar{a}_{2H} = \left[\sqrt{\frac{M_2(0,\eta_{g2})}{M_9(0,\eta_{g1})}} - 1\right]^2\left[4\left(\frac{M_2(0,\eta_{g2})}{M_9(0,\eta_{g1})}\right)\right]^{-1}$$

$M_2(0,\eta_{g2})$ can be obtained by putting $i=2$ in the expression of $M_i(0,\eta_{gi})$ as given in (13.46),

$$M_9(0,\eta_{g1}) = m_{c1}\left[\gamma'(E,\eta_{g1})\left[1 - \frac{B_0}{\sqrt{B_0^2 + 4A_1\gamma(E,\eta_{g1})}}\right]\right]\Bigg|_{E\to 0},$$

$$\Delta_{12\mathrm{HE},n} = \sinh^{-1} p_{6\mathrm{HE},n} \quad \text{and} \quad \Delta_{14\mathrm{HE},n} = \frac{2\Delta_{11\mathrm{HE},n}\Delta_{12\mathrm{HE},n}}{L_0^2}$$

The DOS function in this case can be expressed as

$$N_{\mathrm{MQWSL}}(E) = \frac{g_v eB}{\pi\hbar}\sum_{n_z=1}^{n_{z\max}}\sum_{n=0}^{n_{\max}}\delta'(E - A4, n) \tag{18.37}$$

The electron concentration can be written as

$$n_0 = \frac{g_v eB}{\pi\hbar} \text{Real part of} \sum_{n_z=1}^{n_{z\max}} \sum_{n=0}^{n_{\max}} F_{-1}(\bar{\eta}_{13,A4,n}) \tag{18.38}$$

where $\bar{\eta}_{13,A4,n} = (k_B T)^{-1}[\bar{E}_{F41,n} - A4, n]$.

The Einstein's photo-emitted current density in this case is given by

$$J_{\text{PHOTO}} = \frac{\alpha_0 e^2 g_v B}{2\pi\hbar d_z} \text{Real part of} \sum_{n=0}^{n_{\max}} \sum_{n_{z_{\min}}}^{n_{z\max}} F_{-1}(\bar{\eta}_{13A4,n}) V_9(\bar{E}_{15,19\text{SL}}) \tag{18.39}$$

where

$$V_9(\bar{E}_{15,19\text{SL}}) = (2\pi n_z/hd_z)[(\Delta'_{13H} + i\Delta'_{14H})\big|_{k_x=0,\, k_y=0,\, k_z=\frac{n_z\pi}{d_z} \text{ and } E=\bar{E}_{15,19\text{SL}}}$$

In which $\bar{E}_{15,19\text{SL}}$ is the quantized energy along z-direction and can be obtain by substituting $k_x = 0$, $k_y = 0$, $k_z = \frac{n_z\pi}{d_z}$ and $E = \bar{E}_{15,19\text{SL}}$ in (17.48).

*18.2.10 The EP in strained layer HDQW effective mass superlattices under magnetic quantization

The dispersion relation of the conduction electrons in HDQW strained layer effective mass SLs with graded interfaces can be expressed from (17.52) as

$$\left(\frac{n_z\pi}{d_z}\right)^2 = \left[\frac{1}{L_0^2}[\cos^{-1}\{f_{40}(A8,n)\}]^2 - \frac{2eB}{\hbar}\left(n + \frac{1}{2}\right)\right]\Bigg|_{E=A8,n} \tag{18.40}$$

where $A8, n$ is the total quantized energy in this case,

$$f_{40}(E,n) = a_{20}\cos[a_0 C_{40}(E,n,\eta_{g1}) + b_0 D_{40}(E,n,\eta_{g1})]$$
$$- a_{21}\cos[a_0 C_{40}(E,n,\eta_{g2}) - b_0 D_{40}(E,n,\eta_{g2})],$$

$$a_{20} = \left[\sqrt{\frac{M_{s2}(0,\eta_{g2})}{M_{s1}(0,\eta_{g1})}} + 1\right]^2 4\left[\left(\frac{M_{s2}(0,\eta_{g2})}{M_{s1}(0,\eta_{g1})}\right)^{1/2}\right]^{-1},$$

$$a_{21} = \left[\sqrt{\frac{M_{s2}(0,\eta_{g2})}{M_{s1}(0,\eta_{g1})}} + 1\right]^2 \left[4\left(\frac{M_{s2}(0,\eta_{g2})}{M_{s1}(0,\eta_{g1})}\right)^{1/2}\right]^{-1},$$

$$M_{si}(0,\eta_{gi}) = (-\hbar^2/2)[S_{11,i}(0,\eta_{gi})]'[S_{11,i}(0,\eta_{gi})]^{-2}$$

$$C_{40}(E,k_s,\eta_{g1}) = \left[1-\left(n+\frac{1}{2}\right)\hbar\omega_{18}(E,\eta_{g1})\right]^{1/2}[\bar{S}_{11,1}(E,\eta_{g1})]^{-1/2},$$

$$D_{40}(E,k_s,\eta_{g2}) = \left[1-\left(n+\frac{1}{2}\right)\hbar\omega_{19}(E,\eta_{g2})\right]^{1/2}[\bar{S}_{11,2}(E,\eta_{g2})]^{-1/2},$$

$$\omega_{19}(E,\eta_{g2}) = \left(\frac{eB}{\sqrt{m_{21}^*m_{22}^*}}\right),\quad m_{21}^* = \frac{\hbar^2}{2\bar{P}_{11,2}(E,\eta_{g1})}\quad \text{and}$$

$$m_{22}^* = \frac{\hbar^2}{2\bar{Q}_{11,2}(E,\eta_{g1})}$$

The DOS function in this case can be expressed as

$$N_{\text{MQWSL}}(E) = \frac{g_v eB}{\pi\hbar}\sum_{n_z=1}^{n_{z\max}}\sum_{n=0}^{n_{\max}}\delta'(E-A8,n) \tag{18.41}$$

The electron concentration can be written as

$$n_0 = \frac{g_v eB}{\pi\hbar}\sum_{n_z=1}^{n_{z\max}}\sum_{n=0}^{n_{\max}}F_{-1}(\bar{\eta}_{1529,n}) \tag{18.42}$$

where $\bar{\eta}_{1529,n} = (k_BT)^{-1}[\bar{E}_{F41,n} - A8,n]$.

The Einstein's photo-emitted current density in this case is given by

$$J_{\text{PHOTO}} = \frac{\alpha_0 e^2 g_v B}{2\pi\hbar d_z}\sum_{n=0}^{n_{\max}}\sum_{n_{z\min}}^{n_{z\max}}F_{-1}(\bar{\eta}_{1529,n})V_{10}(\bar{E}_{15,20\text{SL}}) \tag{18.43}$$

where

$$V_{10}(\bar{E}_{15,20\text{SL}}) = \frac{2L_0\sin(L_0 n_z\pi/d_z)}{\hbar[f_{40}(\bar{E}_{15,20\text{SL}},0)]'}$$

In which $\bar{E}_{15,20\text{SL}}$ can be obtained from (17.52) as

$$\cos(L_0 n_z\pi/d_z) = f_{40}(\bar{E}_{15,20\text{SL}},0) \tag{18.44}$$

18.3 Summary and Conclusion

In this chapter, the magneto EP in III–V, II–VI, IV–VI, HgTe/CdTe and strained layer HDQWSLs with graded interfaces has been studied in Sections 18.2.1–18.2.5. From Sections 18.2.6 to 18.2.10, the magneto EP in

III–V, II–VI, IV–VI, HgTe/CdTe and strained layer heavily doped quantum well effective mass superlattices respectively has been presented.

The magneto dispersion relations in HDQWSLs exhibit the fact that the total energy is quantized since the corresponding wave vector space is totally quantized by quantizing magnetic field and size quantization along z-direction.

The DOS functions for all the cases are series of non-uniformly distributed Dirac's Delta functions at specified quantized points in the respective energy axis. The spacing between the consecutive Delta functions are functions of energy band constants and quantization of the wave vector space of a particular material. The DOS function needs two summations namely one summation over the Landau quantum number and the other one is due to size quantization.

It may be noted that the energy levels in this case lead to the discrete energy levels, somewhat like atomic energy levels, which produce very large changes. This follows from the inherent nature of the quantum confinement of the carrier gas dealt with here. In HDQWSLs, under magnetic quantization there remain no free carrier states in between any two allowed sets of size-quantized levels unlike that found for QWs, NWs and quantum dot superlattices where the quantum confinements are 1D, 2D and 0D respectively. Consequently, the crossing of the Fermi level by the quantized levels in this case would have much greater impact on the redistribution of the carriers among the allowed levels, as compared to that found for QWs, NWs and quantum dot superlattices respectively.

It is the band structure which changes in a fundamental way and consequently all the physical properties of all the electronic materials changes radically leading to new physical concepts. We have not given any plot of EP in this case since we believe our able readers will plot the same and enjoy intricate physics, computational analysis, advanced mathematics, and will create new problems both theoretical and experimental.

18.4 Open Research Problems

(R.18.1) Investigate the magneto transport properties in the presence of an arbitrarily oriented non-quantizing magnetic field in III–V, II–VI, IV–VI, HgTe/CdTe and strained layer heavily doped quantum well superlattices with graded interfaces by including the electron spin.

(R.18.2) Investigate the magneto transport properties in III–V, II–VI, IV–VI, HgTe/CdTe and strained layer heavily doped effective mass quantum well superlattices in the presence of an arbitrarily oriented non-quantizing magnetic field by including the electron spin.

(R.18.3) Investigate the transport properties for both the problems, (R.18.1) and (R.18.2) in the presence of an additional arbitrarily oriented electric field.

(R.18.4) Investigate the transport properties for all the problems from (R.18.1) to (R.18.3) in the presence of arbitrarily oriented crossed electric and magnetic fields.

References

[1] H. Cao, Z. Yu, P. Lu, *Superlatt. Microstruct.*, 86, 501 **(2015)**.

[2] M. Hostu, M. Alyoruk, T. Tansel, A. Kilic, R. Turan, A. Aydinli, Y. Ergun, *Superlatt. Microstruct.*, 79, 116 **(2015)**.

[3] W. Huang *et al.*, *IEEE Photon. Technol. Lett.*, 31, 137 **(2019)**.

[4] Z.-L. Deng, X. Li, T. Fu and G. P. Wang, *IEEE Photon. J.*, 9, 1 **(2017)**.

[5] S. Y. A. Wang, H. Yin, N. Gu and L. Liu, *J. Display Technol.*, 12, 1112 **(2016)**.

Chapter 19

The Heavily Doped Non-Parabolic Materials Under Magneto-Size Quantization and the Righi–Leduc Coefficient

Failure is a tragedy when seen in close-up but a comedy in long-shot.

19.1 Introduction

In recent years, there has been considerable interest in studying the various physical properties of different heavily doped non-parabolic materials under magneto-size quantization [1–25]. In Section 19.2.1, of the theoretical background, the RLC (the importance of studying the RLC has been explained in the preface) have been investigated in HD nonlinear optical and tetragonal materials in the presence of magneto-size quantization. Section 19.2.2 contains the results for RLC of HD III–V, ternary and quaternary compounds in accordance with the three and the two band models of Kane. In the same section the RLC in accordance with the models of Stillman *et al.* and Palik *et al.* have also been studied for the purpose of relative comparison. Section 19.2.3 contains the study of the RLC in HD II–VI semiconductors under magneto-size quantization [26–30]. In Section 19.2.4, the RLC in HD IV–VI materials under magneto-size quantization have been discussed in accordance with the models of Cohen, Lax, Dimmock, Bangert and Kastner and Foley and Landenberg respectively. In Section 19.2.5, the magneto RLC under magneto-size quantization in stressed QWs of HD Kane type semiconductors have been investigated. In Section 19.2.6, the DOS and RLC in HD Te under magneto-size quantization have been studied. In Section 19.2.7, the RLC in HD n-GaP have been studied under magneto-size quantization. In Section 19.2.8, the RLC in HD $PtSb_2$ have been explored under magneto-size quantization. In Section 19.2.9, the RLC in HD Bi_2Te_3 have been studied under magneto-size quantization. In Section 19.2.10, the RLC in HD Ge have been studied under magneto-size quantization in accordance with the models of Cardona *et al.* and Wang and Ressler respectively. In Sections 19.2.11 and 19.2.12, the RLC in HD n-GaSb and II–V compounds have been studied under magneto-size quantization. In Sections 19.2.13 and 19.2.14, the magneto RLC in HD $Pb_{1-x}Ge_xTe$ and HD Zinc and Cadmium Diphosphides have been discussed under magneto-size quantization. Section 19.3 explores the summary and conclusion and the last Section 19.4 contains 19 open research problems for this chapter.

19.2 Theoretical Background

19.2.1 The HD nonlinear optical and tetragonal materials under magneto-size quantization and the RLC

Using (1.205) together with (15.5) and (15.6) and the allied definitions we can write the RLC in this case as

$$R_{81} = \left(\frac{cn_0}{eB^2\chi_{ph}}\right) \text{Real part of } \left[\frac{g_v eB(k_BT)^2}{2\pi\hbar n_0}[2F_1(\eta_{81}) + \eta_{81}^2 F_{-1}(\eta_{81}) - 2\eta_{81}F_1(\eta_{81})] - \left(\frac{g_v eBk_BT}{2\pi\hbar n_0}\right)^2 [F_0(\eta_{81}) - \eta_{81}F_{-1}(\eta_{81})]^2\right] \quad (19.1)$$

19.2.2 The HD III–V materials under magneto-size quantization and the RLC

(a) Three band model of Kane

Using (1.205) together with (15.12b) and (15.13) and the allied definitions we can write the RLC in this case as

$$R_{82} = \left(\frac{cn_0}{eB^2\chi_{ph}}\right) \text{Real part of } \left[\left(\frac{g_v eB(k_BT)^2}{2\pi\hbar n_0}\right)[2F_1(\eta_{82}) + \eta_{82}^2 F_{-1}(\eta_{82}) - 2\eta_{82}F_1(\eta_{82})] - \left(\frac{g_v eBk_BT}{2\pi\hbar n_0}\right)^2 [F_0(\eta_{82}) - \eta_{82}F_{-1}(\eta_{82})]^2\right] \quad (19.2)$$

(b) Two band model of Kane

Using (1.205) together with (15.15) and (15.16) and the allied definitions we can write the RLC in this case as

$$R_{84} = \left(\frac{cn_0}{eB^2\chi_{ph}}\right) \left[\left(\frac{g_v eB(k_BT)^2}{2\pi\hbar n_0}\right)[2F_1(\eta_{84}) + \eta_{84}^2 F_{-1}(\eta_{84}) - 2\eta_{84}F_1(\eta_{84})] - \left(\frac{g_v eBk_BT}{2\pi\hbar n_0}\right)^2 [F_0(\eta_{84}) - \eta_{84}F_{-1}(\eta_{84})]^2\right] \quad (19.3)$$

(c) The model of Stillman *et al.*

Using (1.205) together with (15.22) and (15.23) and the allied definitions we can write the RLC in this case as

$$R_{85} = \left(\frac{cn_0}{eB^2\chi_{ph}}\right)\left[\frac{g_v eB(k_BT)^2}{\pi\hbar n_0}[2F_1(\eta_{85}) + \eta_{85}^2F_{-1}(\eta_{85}) - 2\eta_{85}F_1(\eta_{85})] - \left(\frac{g_v eBk_BT}{\pi\hbar n_0}\right)^2[F_0(\eta_{85}) - \eta_{85}F_{-1}(\eta_{85})]^2\right] \tag{19.4}$$

(d) The model of Palik *et al.*

Using (1.205) together with (15.28) and (15.29a) and the allied definitions we can write the RLC in this case as

$$R_{86} = \left(\frac{cn_0}{eB^2\chi_{ph}}\right)\left[\frac{g_v eB(k_BT)^2}{2\pi\hbar n_0}[2F_1(\eta_{86}) + \eta_{86}^2F_{-1}(\eta_{86}) - 2\eta_{86}F_1(\eta_{86})] - \left(\frac{g_v eBk_BT}{2\pi\hbar n_0}\right)^2[F_0(\eta_{86}) - \eta_{86}F_{-1}(\eta_{86})]^2\right] \tag{19.5}$$

19.2.3 The HD II–VI materials under magneto-size quantization and the RLC

Using (1.205) together with (15.31) and (15.32) and the allied definitions we can write the RLC in this case as

$$R_{87} = \left(\frac{cn_0}{eB^2\chi_{ph}}\right)\left[\frac{g_v eB(k_BT)^2}{2\pi\hbar n_0}[2F_1(\eta_{87}) + \eta_{87}^2F_{-1}(\eta_{87}) - 2\eta_{87}F_1(\eta_{87})] - \left(\frac{g_v eBk_BT}{2\pi\hbar n_0}\right)^2[F_0(\eta_{87}) - \eta_{87}F_{-1}(\eta_{87})]^2\right] \tag{19.6}$$

19.2.4 The HD IV–VI materials under magneto-size quantization and the RLC

The electron energy spectrum in IV–VI semiconductors are defined by the models of Cohen, Lax, Dimmock, Bangert and Kastner and Foley and Landenberg respectively. The magneto DR in HD IV–VI semiconductors and RLC are discussed in accordance with the DR the said models for the purpose of relative comparisons:

(a) Cohen model

Using (1.205) together with (15.36) and (15.37) and the allied definitions we can write the RLC in this case as

$$R_{88} = \left(\frac{cn_0}{eB^2\chi_{ph}}\right)\left[\frac{g_v eB(k_BT)^2}{2\pi\hbar n_0}[2F_1(\eta_{88}) + \eta_{88}^2 F_{-1}(\eta_{88}) - 2\eta_{88}F_1(\eta_{88})]\right.$$
$$\left. - \left(\frac{g_v eBk_BT}{2\pi\hbar n_0}\right)^2 [F_0(\eta_{88}) - \eta_{88}F_{-1}(\eta_{88})]^2\right] \tag{19.7}$$

(b) Lax model

Using (1.205) together with (15.43) and (15.44) and the allied definitions we can write the RLC in this case as

$$R_{89} = \left(\frac{cn_0}{eB^2\chi_{ph}}\right)\left[\frac{g_v eB(k_BT)^2}{2\pi\hbar n_0}[2F_1(\eta_{89}) + \eta_{89}^2 F_{-1}(\eta_{89}) - 2\eta_{89}F_1(\eta_{89})]\right.$$
$$\left. - \left(\frac{g_v eBk_BT}{2\pi\hbar n_0}\right)^2 [F_0(\eta_{89}) - \eta_{89}F_{-1}(\eta_{89})]^2\right] \tag{19.8}$$

(c) Dimmock model

Using (1.205) together with (15.48a) and (15.48b) and the allied definitions we can write the RLC in this case as

$$R_{90} = \left(\frac{cn_0}{eB^2\chi_{ph}}\right)\left[\frac{g_v eB(k_BT)^2}{\pi\hbar n_0}[2F_1(\eta_{90}) + \eta_{90}^2 F_{-1}(\eta_{90}) - 2\eta_{90}F_1(\eta_{90})]\right.$$
$$\left. - \left(\frac{g_v eBk_BT}{\pi\hbar n_0}\right)^2 [F_0(\eta_{90}) - \eta_{90}F_{-1}(\eta_{90})]^2\right] \tag{19.9}$$

(d) Model of Bangert and Kastner

Using (1.205) together with (15.52) and (15.53) and the allied definitions we can write the RLC in this case as

$$R_{91} = \left(\frac{cn_0}{eB^2\chi_{ph}}\right) \text{Real part of} \left[\frac{g_v eB(k_BT)^2}{\pi\hbar n_0}[2F_1(\eta_{91}) + \eta_{91}^2 F_{-1}(\eta_{91})\right.$$
$$\left. - 2\eta_{91}F_1(\eta_{91})] - \left(\frac{g_v eBk_BT}{\pi\hbar n_0}\right)^2 [F_0(\eta_{91}) - \eta_{91}F_{-1}(\eta_{91})]^2\right] \tag{19.10}$$

(e) Model of Foley and Langenberg

Using (1.205) together with (15.58) and (15.59) and the allied definitions we can write the RLC in this case as

$$R_{92} = \left(\frac{cn_0}{eB^2\chi_{ph}}\right)\left[\frac{g_v eB(k_BT)^2}{\pi\hbar n_0}[2F_1(\eta_{92}) + \eta_{92}^2 F_{-1}(\eta_{92}) - 2\eta_{92}F_1(\eta_{92})]\right.$$
$$\left. - \left(\frac{g_v eBk_BT}{\pi\hbar n_0}\right)^2 [F_0(\eta_{92}) - \eta_{92}F_{-1}(\eta_{92})]^2\right] \quad (19.11)$$

*19.2.5 The HD stressed Kane type materials under magneto-size quantization and the RLC

Using (1.205) together with (15.62) and (15.63) and the allied definitions we can write the RLC in this case as

$$R_{93} = \left(\frac{cn_0}{eB^2\chi_{ph}}\right)\left[\frac{g_v eB(k_BT)^2}{\pi\hbar n_0}[2F_1(\eta_{93}) + \eta_{93}^2 F_{-1}(\eta_{93}) - 2\eta_{93}F_1(\eta_{93})]\right.$$
$$\left. - \left(\frac{g_v eBk_BT}{\pi\hbar n_0}\right)^2 [F_0(\eta_{93}) - \eta_{93}F_{-1}(\eta_{93})]^2]\right] \quad (19.12)$$

19.2.6 The HD Te under magneto-size quantization and the RLC

Using (1.205) together with (15.66) and (15.67) and the allied definitions we can write the RLC in this case as

$$R_{94} = \left(\frac{cn_0}{eB^2\chi_{ph}}\right)\left[\frac{g_v eB(k_BT)^2}{2\pi\hbar n_0}[2F_1(\eta_{94}) + \eta_{94}^2 F_{-1}(\eta_{94}) - 2\eta_{94}F_1(\eta_{94})]\right.$$
$$\left. - \left(\frac{g_v eBk_BT}{2\pi\hbar n_0}\right)^2 [F_0(\eta_{94}) - \eta_{94}F_{-1}(\eta_{94})]^2\right] \quad (19.13)$$

19.2.7 The HD gallium phosphide under magneto-size quantization and the RLC

Using (1.205) together with (15.70) and (15.69) and the allied definitions we can write the RLC in this case as

$$R_{95} = \left(\frac{cn_0}{eB^2\chi_{ph}}\right)\left[\frac{g_v eB(k_BT)^2}{\pi\hbar n_0}[2F_1(\eta_{95}) + \eta_{95}^2 F_{-1}(\eta_{95}) - 2\eta_{95}F_1(\eta_{95})]\right.$$
$$\left. - \left(\frac{g_v eBk_BT}{\pi\hbar n_0}\right)^2 [F_0(\eta_{95}) - \eta_{95}F_{-1}(\eta_{95})]^2\right] \quad (19.14)$$

19.2.8 The HD platinum antimonide under magneto-size quantization and the RLC

Using (1.205) together with (15.72) and (15.73) and the allied definitions we can write the RLC in this case as

$$R_{96} = \left(\frac{cn_0}{eB^2\chi_{ph}}\right)\left[\frac{g_v eB(k_BT)^2}{\pi\hbar n_0}[2F_1(\eta_{96}) + \eta_{96}^2 F_{-1}(\eta_{96}) - 2\eta_{96}F_1(\eta_{96})]\right.$$

$$\left. - \left(\frac{g_v eBk_BT}{\pi\hbar n_0}\right)^2 [F_0(\eta_{96}) - \eta_{96}F_{-1}(\eta_{96})]^2\right] \tag{19.15}$$

19.2.9 The HD bismuth telluride under magneto-size quantization and the RLC

Using (1.205) together with (15.77) and (15.78) and the allied definitions we can write the RLC in this case as

$$R_{97} = \left(\frac{cn_0}{eB^2\chi_{ph}}\right)\left[\frac{g_v eB(k_BT)^2}{\pi\hbar n_0}[2F_1(\eta_{97}) + \eta_{97}^2 F_{-1}(\eta_{97}) - 2\eta_{97}F_1(\eta_{97})]\right.$$

$$\left. - \left(\frac{g_v eBk_BT}{\pi\hbar n_0}\right)^2 [F_0(\eta_{97}) - \eta_{97}F_{-1}(\eta_{97})]^2\right] \tag{19.16}$$

*19.2.10 The HD germanium under magneto-size quantization and the RLC

(a) Model of Cardona *et al.*

Using (1.205) together with (15.80) and (15.81) and the allied definitions we can write the RLC in this case as

$$R_{98} = \left(\frac{cn_0}{eB^2\chi_{ph}}\right)\left[\frac{g_v eB(k_BT)^2}{\pi\hbar n_0}[2F_1(\eta_{98}) + \eta_{98}^2 F_{-1}(\eta_{98}) - 2\eta_{98}F_1(\eta_{98})]\right.$$

$$\left. - \left(\frac{g_v eBk_BT}{\pi\hbar n_0}\right)^2 [F_0(\eta_{98}) - \eta_{98}F_{-1}(\eta_{98})]^2\right] \tag{19.17}$$

(b) Model of Wang and Ressler

Using (1.205) together with (15.84) and (15.85) and the allied definitions we can write the RLC in this case as

$$R_{99} = \left(\frac{cn_0}{eB^2\chi_{ph}}\right)\left[\frac{g_v eB(k_BT)^2}{\pi\hbar n_0}[2F_1(\eta_{99}) + \eta_{99}^2 F_{-1}(\eta_{99}) - 2\eta_{99}F_1(\eta_{99})]\right.$$
$$\left. - \left(\frac{g_v eBk_BT}{\pi\hbar n_0}\right)^2 [F_0(\eta_{99}) - \eta_{99}F_{-1}(\eta_{99})]^2\right] \quad (19.18)$$

19.2.11 The HD gallium antimonide under magneto-size quantization and the RLC

Using (1.205) together with (15.88) and (15.89) and the allied definitions we can write the RLC in this case as

$$R_{100} = \left(\frac{cn_0}{eB^2\chi_{ph}}\right)\left[\frac{g_v eB(k_BT)^2}{\pi\hbar n_0}[2F_1(\eta_{100}) + \eta_{100}^2 F_{-1}(\eta_{100})\right.$$
$$\left. - 2\eta_{100}F_1(\eta_{100})] - \left(\frac{g_v eBk_BT}{\pi\hbar n_0}\right)^2 [F_0(\eta_{100}) - \eta_{100}F_{-1}(\eta_{100})]^2\right] \quad (19.19)$$

19.2.12 The HD II–V materials under magneto-size quantization and the RLC

Using (1.205) together with (15.95) and (15.96) and the allied definitions we can write the RLC in this case as

$$R_{101} = \left(\frac{cn_0}{eB^2\chi_{ph}}\right)\left[\left(\frac{g_v eB(k_BT)^2}{2\pi\hbar n_0}\right)[2F_1(\eta_{101}) + \eta_{101}^2 F_{-1}(\eta_{101})\right.$$
$$\left. - 2\eta_{101}F_1(\eta_{101})] - \left(\frac{g_v eBk_BT}{2\pi\hbar n_0}\right)^2 [F_0(\eta_{101}) - \eta_{101}F_{-1}(\eta_{101})]^2\right] \quad (19.20)$$

*19.2.13 The HD lead germanium telluride under magneto-size quantization and the RLC

Using (1.205) together with (15.100) and (15.101) and the allied definitions we can write the RLC in this case as

$$R_{102} = \left(\frac{cn_0}{eB^2\chi_{ph}}\right)\left[\frac{g_veB(k_BT)^2}{2\pi\hbar n_0}[2F_1(\eta_{102}) + \eta_{102}^2F_{-1}(\eta_{102}) - 2\eta_{102}F_1(\eta_{102})] - \left(\frac{g_veBk_BT}{2\pi\hbar n_0}\right)^2[F_0(\eta_{102}) - \eta_{102}F_{-1}(\eta_{102})]^2\right] \tag{19.21}$$

*19.2.14 The HD zinc and cadmium diphosphides under magneto-size quantization and the RLC

Using (1.205) and the allied definitions we can write the RLC in this case as

$$R_{103} = \left(\frac{cn_0}{eB^2\chi_{ph}}\right)\left[\frac{g_veB(k_BT)^2}{2\pi\hbar n_0}[2F_1(\eta_{103}) + \eta_{103}^2F_{-1}(\eta_{103}) - 2\eta_{103}F_1(\eta_{103})] - \left(\frac{g_veBk_BT}{2\pi\hbar n_0}\right)^2[F_0(\eta_{103}) - \eta_{103}F_{-1}(\eta_{103})]^2\right] \tag{19.22}$$

19.3 Summary and Conclusion

The DRs in magneto-size quantized of HD materials exhibit the fact that the total energy is quantized since the corresponding wave vector space is totally quantized by quantizing magnetic field and size quantization along z-direction.

- The DOS functions for all the materials in this case are series of non-uniformly distributed Dirac's Delta functions at specified quantized points in the respective energy axis. The spacing between the consecutive Delta functions are functions of energy band constants and quantization of the wave vector space of a particular material. The DOS function needs two summations namely one summation over the Landau quantum number and the other one is due to size quantization.
- It may be noted that the energy levels in magneto-size quantized HD materials lead to the discrete energy levels, somewhat like atomic energy

levels, which produce very large changes. This follows from the inherent nature of the quantum confinement of the carrier gas dealt with here. In QBs, there remain no free carrier states in between any two allowed sets of size-quantized levels unlike that found for QWs, NWs and QBs where the quantum confinements are 1D, 2D and 3D respectively. Consequently, the crossing of the Fermi level by the size-quantized levels in this case would have much greater impact on the redistribution of the carriers among the allowed levels, as compared to that found for QWs, NWs and QBs respectively for the electronic properties in general.

- By using the appropriate equations we can plot the RLC with respect to many variables in all the cases of this chapter. We have not given any plot of the RLC in this case since we believe our able readers will plot the same and enjoy intricate physics, computational analysis, advanced mathematics, and will create new problems both theoretical and experimental.
- It is the band structure which changes in a fundamental way and consequently all the physical properties of all the electronic materials changes radically leading to new physical concepts.

19.4 Open Research Problems

(R.19.1) Investigate the RLC in the presence of an arbitrarily oriented quantizing magnetic field for all the QW HD materials as given in problems in Chapter 11.

(R.19.2) Investigate the RLC in the presence of an arbitrarily oriented quantizing magnetic field in QW HD nonlinear optical semiconductors by including broadening and the electron spin. Study all the special cases for QW HD III–V, ternary and quaternary materials in this context.

(R.19.3) Investigate the RLCs for QW HD IV–VI, II–VI and stressed Kane type compounds in the presence of an arbitrarily oriented quantizing magnetic field by including broadening and electron spin.

(R.19.4) Investigate the RLC for all the QW HD materials as stated in Chapter 11 in the presence of an arbitrarily oriented quantizing magnetic field by including broadening and electron spin.

(R.19.5) Investigate the RLC in the presence of an arbitrarily oriented quantizing magnetic field and crossed electric fields in QW HD nonlinear optical and tetragonal semiconductors by including broadening and the electron spin. Study all the special cases for HD III–V, ternary and quaternary materials in this context.

(R.19.6) Investigate the RLCs for QW HD IV–VI, II–VI and stressed Kane type compounds in the presence of an arbitrarily oriented quantizing magnetic field and crossed electric field by including broadening and electron spin.

(R.19.7) Investigate the RLC for all the QW HD materials as considered in Chapter 11 in the presence of an arbitrarily oriented quantizing magnetic field and crossed electric field by including broadening and electron spin.

(R.19.8) Investigate the 2D RLC in QW HD nonlinear optical and tetragonal, III–V, II–VI, IV–VI and stressed Kane type semiconductors.

(R.19.9) Investigate the 2D RLC for QW HD for all the materials of Chapter 11.

(R.19.10) Investigate the 2D RLC in the presence of an arbitrarily oriented non-quantizing magnetic field for the QWs HD nonlinear optical semiconductors by including the electron spin. Study all the special cases for III–V, ternary and quaternary materials in this context.

(R.19.11) Investigate the RLCs in QWs HD IV–VI, II–VI and stressed Kane type compounds in the presence of an arbitrarily oriented non-quantizing magnetic field by including the electron spin.

(R.19.12) Investigate the 2D RLC for QW HD for all the materials as stated in Chapter 11 in the presence of an arbitrarily oriented magnetic field by including electron spin and broadening.

(R.19.13) Investigate the 2D RLC for all the problems from Chapter 11 in the presence of an additional arbitrarily oriented non-quantizing electric field.

(R.19.14) Investigate the 2D RLC for all the problems from Chapter 11 in the presence of arbitrarily oriented crossed electric and magnetic fields.

(R.19.15) Investigate all the problems from of Chapter 1 for Chapter 19.

(R.19.16) Investigate all the problems of Chapter 1 for Chapter 19 in the presence of exponential, Kane, Halperin and Lax and Bonch-Bruevich band tails [12].

(R.19.17) Investigate all the problems of this chapter by removing all the mathematical approximations and establishing the uniqueness conditions in each case.

(R.19.18) (a) Investigate the RLC in all the QW HD semiconductors as considered in this chapter in the presence of defects and magnetic quantization.

(b) Investigate the RLC for the all the materials of Chapter 1 in the presence of an arbitrarily oriented quantizing magnetic field including broadening and the electron spin (applicable under magnetic quantization) for all the QW HD semiconductors whose unperturbed carrier energy spectra are defined in Chapter 1.

(R.19.19) Investigate all the problems of this section by removing all the mathematical approximations and establishing the respective appropriate uniqueness conditions.

References

[1] N. Miura, *Physics of Semiconductors in High Magnetic Fields*, Series on Semiconductor Science and Technology, (Oxford University Press, USA, **2007**); K. H. J. Buschow, F. R. de Boer, *Physics of Magnetism and Magnetic Materials*, (Springer, New York, **2003**); D. Sellmyer, R. Skomski (Eds.), *Advanced Magnetic Nanostructures*, (Springer, New York, **2005**).

[2] J. A. C. Bland, B. Heinrich (Eds.), *Ultrathin Magnetic Structures III: Fundamentals of Nanomagnetism (Pt. 3)*, (Springer-Verlag, Germany, **2005**); B. K. Ridley, *Quantum Processes in Semiconductors*, Fourth Edition, (Oxford publications, Oxford, **1999**); J. H. Davies, *Physics of Low Dimensional Semiconductors*, (Cambridge University Press, UK, **1998**).

[3] S. Blundell, *Magnetism in Condensed Matter*, Oxford Master Series in Condensed Matter Physics, (Oxford University Press, USA, **2001**); C. Weisbuch, B. Vinter, *Quantum Semiconductor Structures: Fundamentals and Applications*, (Academic Publishers, USA, **1991**); D. Ferry, *Semiconductor Transport*, (CRC, USA, **2000**).

[4] M. Reed (Ed.), *Semiconductors and Semimetals: Nanostructured Systems*, (Academic Press, USA, **1992**); T. Dittrich, *Quantum Transport and Dissipation*, (Wiley-VCH Verlag GmbH, Germany, **1998**); A. Y. Shik, *Quantum Wells: Physics & Electronics of Two Dimensional Systems*, (World Scientific, USA, **1997**).

[5] K. P. Ghatak, M. Mondal, *Zietschrift fur Naturforschung A*, 41a, 881 **(1986)**; K. P. Ghatak, M. Mondal, *J. Appl. Phys.*, 62, 922 **(1987)**; K. P. Ghatak, S. N. Biswas, *Phys. Stat. Sol. (b)*, 140, K107 **(1987)**.

[6] K. P. Ghatak, M. Mondal, *J. Mag. and Mag. Mat.*, 74, 203 **(1988)**; K. P. Ghatak, M. Mondal, *Phys. Stat. Sol. (b)*, 139, 195 **(1987)**; K. P. Ghatak, M. Mondal, *Phys. Stat. Sol. (b)*, 148, 645 **(1988)**.

[7] K. P. Ghatak, B. Mitra, A. Ghoshal, *Phys. Stat. Sol. (b)*, 154, K121 **(1989)**; K. P. Ghatak, S. N. Biswas, *J. Low Temp. Phys.*, 78, 219 **(1990)**; K. P. Ghatak, M. Mondal, *Phys. Stat. Sol. (b)*, 160, 673 **(1990)**.

[8] K. P. Ghatak, B. Mitra, *Phys. Lett. A*, 156, 233 **(1991)**; K. P. Ghatak, A. Ghoshal, B. Mitra, *Nuovo Cimento D*, 13, 867 **(1991)**; K. P. Ghatak, M. Mondal, *Phys. Stat. Sol. (b)*, 148, 645 **(1989)**.

[9] K. P. Ghatak, B. Mitra, *Int. J. Electron.*, 70, 345 **(1991)**; K. P. Ghatak, S. N. Biswas, *J. Appl. Phys.*, 70, 299 **(1991)**; K. P. Ghatak, A. Ghoshal, *Phys. Stat. Sol. (b)*, 170, K27 **(1992)**.

[10] K. P. Ghatak, *Nuovo Cimento D*, 13, 1321 **(1992)**; K. P. Ghatak, B. Mitra, *Int. J. Electron.*, 72, 541 **(1992)**; K. P. Ghatak, S. N. Biswas, *Nonlinear Opt.*, 4, 347 **(1993)**.

[11] K. P. Ghatak, M. Mondal, *Phys. Stat. Sol. (b)*, 175, 113 **(1993)**; K. P. Ghatak, S. N. Biswas, *Nonlinear Opt.*, 4, 39 **(1993)**; K. P. Ghatak, B. Mitra, *Nuovo Cimento D*, 15, 97 **(1993)**.

[12] K. P. Ghatak, S. N. Biswas, *Nanostruct. Mater.*, 2, 91 **(1993)**; K. P. Ghatak, M. Mondal, *Phys. Stat. Sol. (b)*, 185, K5 **(1994)**; K. P. Ghatak, B. Goswami, M. Mitra, B. Nag, *Nonlinear Opt.*, 16, 9 **(1996)**; K. P. Ghatak, M. Mitra, B. Goswami, B. Nag, *Nonlinear Opt.*, 16, 167 **(1996)**; K. P. Ghatak, B. Nag, *Nanostruct. Mater.*, 10, 923 **(1998)**.

[13] D. Roy Choudhury, A. K. Choudhury, K. P. Ghatak, A. N. Chakravarti, *Phys. Stat. Sol. (b)*, 98, K141 **(1980)**; A. N. Chakravarti, K. P. Ghatak, A. Dhar, S. Ghosh, *Phys. Stat. Sol. (b)*, 105, K55 **(1981)**; A. N. Chakravarti, A. K. Choudhury, K. P. Ghatak, *Phys. Stat. Sol. (a)*, 63, K97 **(1981)**.

[14] A. N. Chakravarti, A. K. Choudhury, K. P. Ghatak, S. Ghosh, A. Dhar, *Appl. Phys.*, 25, 105 **(1981)**; A. N. Chakravarti, K. P. Ghatak, G. B. Rao, K. K. Ghosh, *Phys. Stat. Sol. (b)*, 112, 75 **(1982)**; A. N. Chakravarti, K. P. Ghatak, K. K. Ghosh, H. M. Mukherjee, *Phys. Stat. Sol. (b)*, 116, 17 **(1983)**; M. Mondal, K. P. Ghatak, *Phys. Stat. Sol. (b)*, 133, K143 **(1984)**.

[15] M. Mondal, K. P. Ghatak, *Phys. Stat. Sol. (b)*, 126, K47 **(1984)**; M. Mondal, K. P. Ghatak, *Phys. Stat. Sol. (b)*, 126, K41 **(1984)**; M. Mondal, K. P. Ghatak, *Phys. Stat. Sol. (b)*, 129, K745 **(1985)**.

[16] M. Mondal, K. P. Ghatak, *Phys. Scr.*, 31, 615 **(1985)**; M. Mondal, K. P. Ghatak, *Phys. Stat. Sol. (b)*, 135, 239 **(1986)**; M. Mondal, K. P. Ghatak, *Phys. Stat. Sol. (b)*, 93, 377 **(1986)**.

[17] M. Mondal, K. P. Ghatak, *Phys. Stat. Sol. (b)*, 135, K21 **(1986)**; M. Mondal, S. Bhattacharyya, K. P. Ghatak, *Appl. Phys. A*, 42, 331 **(1987)**; S. N. Biswas, N. Chattopadhyay, K. P. Ghatak, *Phys. Stat. Sol. (b)*, 141, K47 **(1987)**.

[18] B. Mitra, K. P. Ghatak, *Phys. Stat. Sol. (b)*, 149, K117 **(1988)**; B. Mitra, A. Ghoshal, K. P. Ghatak, *Phys. Stat. Sol. (b)*, 150, K67 **(1988)**; M. Mondal, K. P. Ghatak, *Phys. Stat. Sol. (b)*, 147, K179 **(1988)**.

[19] M. Mondal, K. P. Ghatak, *Phys. Stat. Sol. (b)*, 146, K97 **(1988)**; B. Mitra, A. Ghoshal, K. P. Ghatak, *Phys. Stat. Sol. (b)*, 153, K209 **(1989)**; B. Mitra, K. P. Ghatak, *Phys. Lett. A*, 142, 401 **(1989)**.

[20] B. Mitra, A. Ghoshal, K. P. Ghatak, *Phys. Stat. Sol. (b)*, 154, K147 **(1989)**; B. Mitra, K. P. Ghatak, *Sol. Stat. Electron.*, 32, 515 **(1989)**; B. Mitra, A. Ghoshal, K. P. Ghatak, *Phys. Stat. Sol. (b)*, 155, K23 **(1989)**.
[21] B. Mitra, K. P. Ghatak, *Phys. Lett. A*, 135, 397 **(1989)**; B. Mitra, K. P. Ghatak, *Phys. Lett. A*, 146, 357 **(1990)**; B. Mitra, K. P. Ghatak, *Phys. Stat. Sol. (b)*, 164, K13 **(1991)**; S. N. Biswas, K. P. Ghatak, *Int. J. Electron.*, 70, 125 **(1991)**.
[22] P. R. Wallace, *Phys. Stat. Sol. (b)*, 92, 49 **(1979)**.
[23] B. R. Nag, *Electron Transport in Compound Semiconductors*, Springer Series in Solid State Sciences, Vol. 11, (Springer-Verlag, Germany, **1980**).
[24] K. P. Ghatak, S. Bhattacharya, D. De, *Einstein Relation in Compound Semiconductors and Their Nanostructures*, Springer Series in Materials Science, Vol. 116, (Springer-Verlag, Germany, **2009**).
[25] C. C. Wu and C. J. Lin, *J. Low Temp. Phys.*, 57, 469 **(1984)**; M. H. Chen, C. C. Wu, C. J. Lin, *J. Low Temp. Phys.*, 55, 127 **(1984)**.
[26] E. Bangert, P. Kastner, *Phys. Stat. Sol. (b)*, 61, 503 **(1974)**.
[27] G. M. T. Foley, P. N. Langenberg, *Phys. Rev. B*, 15, 4850 **(1977)**.
[28] M. Singh, P. R. Wallace, S. D. Jog, E. Arushanov, *J. Phys. Chem. Solids*, 45, 409 **(1984)**.
[29] Y. Yamada, *Phys. Soc. Jap.*, 35, 1600 **(1973)**.
[30] L. A. Vassilev, *Phys. Stat. Sol. (b)*, 121, 203 **(1984)**.
[31] R. W. Cunningham, *Phys. Rev.*, 167, 761 **(1968)**.
[32] A. I. Yekimov, A. A. Onushchenko, A. G. Plyukhin, A. L. Efros, *J. Exp. Theor. Phys.*, 88, 1490 **(1985)**.
[33] B. J. Roman, A. W. Ewald, *Phys. Rev. B*, 5, 3914 **(1972)**.

Chapter 20

The Landau's Dia (χ_d) and Pauli's Para (χ_p) Magnetic Susceptibilities in Quantized 0D Structures

What other people think of me is none of my business.

20.1 Introduction

In recent years there has been considerable interest in studying the different physical properties of heavily doped superlattices (HDSLs) under different physical conditions [1–28]. We wish to note that the susceptibility due to the free motion of electrons in the presence of a magnetic field is called the diamagnetic susceptibility of Landau (χ_d) and the susceptibility due to spin of the electrons is called paramagnetic susceptibility of Pauli (χ_p), For 2D semiconductors having parabolic energy bands, the paramagnetic-to-diamagnetic ratio is 1/3. It has been shown that both the susceptibilities depend on the density-of-states function [26–28]. In this chapter, in Sections 20.2.1–20.2.4 we study the χ_d and χ_p in magneto-size quantized nonlinear optical and tetragonal, III–V, ternary, quaternary, II–VI and IV–VI materials respectively. We have studied the χ_d and χ_p in Magneto size quantized stressed Kane type, Bi_2Te_3, Ge, GaAs, GaSb, InSb and Te in Sections 20.2.5–20.2.11 respectively. The χ_d and χ_p in Magneto inversion layers of nonlinear optical and tetragonal, III–V, ternary, quaternary, II–VI, IV–VI, stressed Kane type, Bi_2Te_3, Ge, GaAs, GaSb, InSb and Te in Sections 20.2.12–20.2.22 respectively. The χ_d and χ_p in magneto III–V, II–VI, IV–VI, HgTe/CdTe and strained layer QW superlattices with graded interfaces have been studied in Sections 20.2.23–20.2.27 respectively. The χ_d and χ_p in magneto III–V, II–VI, IV–VI, HgTe/CdTe and strained layer QW effective mass superlattices have been studied in Sections 20.2.28–20.2.32 respectively. The χ_d and χ_p in doping superlattices of nonlinear optical and tetragonal, III–V, ternary, quaternary, II–VI, IV–VI and, stressed Kane type under magnetic quantization have been studied in Sections 20.2.33–20.2.37 respectively. Sections 20.3 contains summary and conclusion pertinent to this chapter. The last Section 20.4 has been reserved for 18 open research problems.

20.2 Theoretical Background

*20.2.1 The χ_d and χ_p in magneto-size quantized HD nonlinear optical and tetragonal materials

From (1.199), in this case, we can write that

$$F(B) = \frac{g_v eB}{2\pi\hbar} A_{20} \left[\sum_{n=0}^{n_{\max}} \sum_{n_z=1}^{n_{z\max}} [E_{F_{81}} F_{-1}(\eta_{81}) - k_B T[F_0(\eta_{81})]] \right] \tag{20.1}$$

where

$$A_{20} = \text{Real part of} \tag{20.2}$$

Using (1.198) and (20.1), we can study the χ_p and neglecting spin we can study the χ_d also in this case.

20.2.2 The χ_d and χ_p in magneto-size quantized HD III–V, ternary and quaternary materials

(a) Three band model of Kane

From (1.199), in this case, we can write that

$$F(B) = \frac{g_v eB}{2\pi\hbar} \text{ Real part of } \left[\sum_{n=0}^{n_{\max}} \sum_{n_z=1}^{n_{z\max}} [E_{F_{82}} F_{-1}(\eta_{82}) - k_B T[F_0(\eta_{82})]] \right] \tag{20.3}$$

Using (1.198) and (20.3), we can study the χ_p and neglecting spin we can study the χ_d also in this case.

(b) Two band model of Kane

From (1.199), in this case, we can write that

$$F(B) = \frac{g_v eB}{2\pi\hbar} \left[\sum_{n=0}^{n_{\max}} \sum_{n_z=1}^{n_{z\max}} [E_{F_{83}} F_{-1}(\eta_{83}) - k_B T[F_0(\eta_{83})]] \right] \tag{20.4}$$

Using (1.198) and (20.4), we can study the χ_p and neglecting spin we can study the χ_d also in this case.

(c) Parabolic energy bands

From (1.199), in this case, we can write that

$$F(B) = \frac{g_v eB}{2\pi\hbar}\left[\sum_{n=0}^{n_{\max}}\sum_{n_z=1}^{n_{z\max}}[E_{F_{84}}F_{-1}(\eta_{84}) - k_B T[F_0(\eta_{84})]]\right] \tag{20.5}$$

Using (1.198) and (20.5), we can study the χ_p and neglecting spin we can study the χ_d also in this case.

(d) The model of Stillman *et al.*

From (1.199), in this case, we can write that

$$F(B) = \frac{g_v eB}{\pi\hbar}\left[\sum_{n=0}^{n_{\max}}\sum_{n_z=1}^{n_{z\max}}[E_{F_{85}}F_{-1}(\eta_{85}) - k_B T[F_0(\eta_{85})]]\right] \tag{20.6}$$

Using (1.198) and (20.6), we can study the χ_d and including spin we can study the χ_p also in this case.

(e) The model of Palik *et al.*

From (1.199), in this case, we can write that

$$F(B) = \frac{g_v eB}{2\pi\hbar}\left[\sum_{n=0}^{n_{\max}}\sum_{n_z=1}^{n_{z\max}}[E_{F_{86}}F_{-1}(\eta_{86}) - k_B T[F_0(\eta_{86})]]\right] \tag{20.7}$$

Using (1.198) and (20.7), we can study the χ_p and neglecting spin we can study the χ_d also in this case.

20.2.3 The χ_d and χ_p in magneto-size quantized HD II–VI materials

From (1.199), in this case, we can write that

$$F(B) = \frac{g_v eB}{2\pi\hbar}\left[\sum_{n=0}^{n_{\max}}\sum_{n_z=1}^{n_{z\max}}[E_{F_{87}}F_{-1}(\eta_{87}) - k_B T[F_0(\eta_{87})]]\right] \tag{20.8}$$

Using (1.198) and (20.8), we can study the χ_d and by including spin we can study the χ_p also in this case.

*20.2.4 The χ_d and χ_p in magneto-size quantized HD IV–VI materials

(a) Cohen model

From (1.199), in this case, we can write that

$$F(B) = \frac{g_v eB}{2\pi\hbar}\left[\sum_{n=0}^{n_{\max}}\sum_{n_z=1}^{n_{z\max}}[E_{F_{88}}F_{-1}(\eta_{88}) - k_B T[F_0(\eta_{88})]]\right] \tag{20.9}$$

Using (1.198) and (20.9), we can study the χ_p and by neglecting spin we can study the χ_d also in this case.

(b) Lax model

From (1.199), in this case, we can write that

$$F(B) = \frac{g_v eB}{2\pi\hbar}\left[\sum_{n=0}^{n_{\max}}\sum_{n_z=1}^{n_{z\max}}[E_{F_{89}}F_{-1}(\eta_{89}) - k_B T[F_0(\eta_{89})]]\right] \tag{20.10}$$

Using (1.198) and (20.10), we can study the χ_p and by neglecting spin we can study the χ_d also in this case.

(c) Dimmock model

From (1.199), in this case, we can write that

$$F(B) = \frac{g_v eB}{\pi\hbar}\left[\sum_{n=0}^{n_{\max}}\sum_{n_z=1}^{n_{z\max}}[E_{F_{90}}F_{-1}(\eta_{90}) - k_B T[F_0(\eta_{90})]]\right] \tag{20.11}$$

Using (1.198) and (20.11), we can study the χ_d and by including spin we can study the χ_p also in this case.

(d) Model of Bangert and Kastner

From (1.199), in this case, we can write that

$$F(B) = \frac{g_v eB}{\pi\hbar}\left[\sum_{n=0}^{n_{\max}}\sum_{n_z=1}^{n_{z\max}}[E_{F_{91}}F_{-1}(\eta_{91}) - k_B T[F_0(\eta_{91})]]\right] \tag{20.12}$$

Using (1.198) and (20.12), we can study the χ_d and by including spin we can study the χ_p also in this case.

(e) Model of Foley and Langenberg

From (1.199), in this case, we can write that

$$F(B) = \frac{g_v eB}{\pi\hbar}\left[\sum_{n=0}^{n_{\max}}\sum_{n_z=1}^{n_{z\max}}[E_{F_{92}}F_{-1}(\eta_{92}) - k_B T[F_0(\eta_{92})]]\right] \quad (20.13)$$

Using (1.198) and (20.13), we can study the χ_d and by including spin we can study the χ_p also in this case.

20.2.5 The χ_d and χ_p in magneto-size quantized HD stressed Kane type materials

From (1.199), in this case, we can write that

$$F(B) = \frac{g_v eB}{\pi\hbar}\left[\sum_{n=0}^{n_{\max}}\sum_{n_z=1}^{n_{z\max}}[E_{F_{93}}F_{-1}(\eta_{93}) - k_B T[F_0(\eta_{93})]]\right] \quad (20.14)$$

Using (1.198) and (20.14), we can study the χ_d and by including spin we can study the χ_p also in this case.

20.2.6 The χ_d and χ_p in magneto-size quantized HD bismuth telluride

From (1.199), in this case, we can write that

$$F(B) = \frac{g_v eB}{\pi\hbar}\left[\sum_{n=0}^{n_{\max}}\sum_{n_z=1}^{n_{z\max}}[E_{F_{97}}F_{-1}(\eta_{97}) - k_B T[F_0(\eta_{97})]]\right] \quad (20.15)$$

Using (1.198) and (20.15), we can study the χ_d and by including spin we can study the χ_p also in this case.

20.2.7 The χ_d and χ_p in magneto-size quantized HD germanium

(a) Model of Cardona *et al.*

From (1.199), in this case, we can write that

$$F(B) = \frac{g_v eB}{\pi\hbar}\left[\sum_{n=0}^{n_{\max}}\sum_{n_z=1}^{n_{z\max}}[E_{F_{98}}F_{-1}(\eta_{98}) - k_B T[F_0(\eta_{98})]]\right] \quad (20.16)$$

Using (1.198) and (20.16), we can study the χ_d and by including spin we can study the χ_p also in this case.

(b) Model of Wang and Ressler

From (1.199), in this case, we can write that

$$F(B) = \frac{g_v eB}{\pi\hbar}\left[\sum_{n=0}^{n_{\max}}\sum_{n_z=1}^{n_{z\max}}[E_{F_{99}}F_{-1}(\eta_{99}) - k_B T[F_0(\eta_{99})]]\right] \tag{20.17}$$

Using (1.198) and (20.17), we can study the χ_d and by including spin we can study the χ_p also in this case.

20.2.8 The χ_d and χ_p in magneto-size quantized HD gallium arsenide

The dispersion relation in HD n-GaAs including band non-parabolacity, warping and spin-splitting can approximately be written following (5.81) as

$$\begin{aligned}\gamma_3(\bar{E},\eta_g) = {} & \frac{\hbar^2k^2}{2m_c} + \alpha_0 k^4 + \beta_0[k_x^2k_y^2 + k_y^2k_z^2 + k_z^2k_x^2] \\ & \pm\bar{\gamma}_0[k^2(k_x^2k_y^2 + k_y^2k_z^2 + k_z^2k_x^2) - 9k_x^2k_y^2k_z^2]^{1/2}\end{aligned} \tag{20.18}$$

Using the method of spherical averaging [12] we can write

$$k^2 = \bar{\gamma}_{4\pm}(\bar{E},\eta_g) \tag{20.19}$$

where

$$\bar{\gamma}_{4\pm}(\bar{E},\eta_g) = (2e_2)^{-1}\left[+e_1 - \sqrt{e_1^2 - 4e_2[\gamma_3(\bar{E},\eta_g) \pm \Gamma_4(\bar{E},\eta_g)]}\right],$$

$$\Gamma_4(\bar{E},\eta_g) = \left(e_3\left(\frac{1}{2e_2}\right)^{3/2}\right)\left[+e_1 - \sqrt{e_1^2 - 4\gamma_3(\bar{E},\eta_g)e_2}\right]^{3/2}$$

The DR under magneto-size quantization can be expressed in this case by using (20.19) as

$$\frac{2eB}{\hbar}\left(n+\frac{1}{2}\right) + \left(\frac{n_z\pi}{d_z}\right)^2 = \bar{\gamma}_{4\pm}(\bar{e}_{100},\eta_g) \tag{20.20}$$

where $\bar{e}_{100}$ is the total quantized energy in this case.

The DOS function is given by

$$N_{Bn_z} = \frac{g_v eB}{\pi\hbar} \sum_{n=0}^{n_{\max}} \sum_{n_z=1}^{n_{z\max}} \delta'(E - \bar{e}_{100}) \tag{20.21}$$

The electron concentration is given by

$$n_0 = \frac{g_v eB}{\pi\hbar} \sum_{n=0}^{n_{\max}} \sum_{n_z=1}^{n_{z\max}} F_{-1}(\bar{\eta}_{100}) \tag{20.22}$$

where $\bar{\eta}_{100} = (k_B T)^{-1}[\bar{E}_{f_{100}} - \bar{e}_{100}]$ and $\bar{E}_{f_{100}}$ is the Fermi energy in this case.

From (1.199), in this case, we can write that

$$F(B) = \frac{g_v eB}{\pi\hbar} \left[\sum_{n=0}^{n_{\max}} \sum_{n_z=1}^{n_{z\max}} [\bar{E}_{f_{100}} F_{-1}(\bar{\eta}_{100}) - k_B T[F_0(\bar{\eta}_{100})]] \right] \tag{20.23}$$

Using (1.198), (20.21) and (20.23), we can study the χ_d and by including spin we can study the χ_p also in this case.

20.2.9 The χ_d and χ_p in magneto-size quantized HD gallium antimonide

From (1.199), in this case, we can write that

$$F(B) = \frac{g_v eB}{\pi\hbar} \left[\sum_{n=0}^{n_{\max}} \sum_{n_z=1}^{n_{z\max}} [E_{F_{100}} F_{-1}(\eta_{100}) - k_B T[F_0(\eta_{100})]] \right] \tag{20.24}$$

Using (1.198), (19.117) and (20.24), we can study the χ_d and by including spin we can study the χ_p also in this case.

20.2.10 The χ_d and χ_p in magneto-size quantized HD indium antimonide

From (1.199), in this case, we can write that

$$F(B) = \frac{g_v eB}{\pi\hbar} \left[\sum_{n=0}^{n_{\max}} \sum_{n_z=1}^{n_{z\max}} [E_{F_{55}} F_{-1}(\eta_{55}) - k_B T[F_0(\eta_{55})]] \right] \tag{20.25}$$

Using (1.198), (5.262) and (20.25), we can study the χ_d and by including spin we can study the χ_p also in this case.

20.2.11 The χ_d and χ_p in magneto-size quantized HD tellurium

From (1.199), in this case, we can write that

$$F(B) = \frac{g_v eB}{2\pi\hbar}\left[\sum_{n=0}^{n_{\max}}\sum_{n_z=1}^{n_{z\max}}[E_{F_{94}}F_{-1}(\eta_{94}) - k_BT[F_0(\eta_{94})]]\right] \tag{20.26}$$

Using (1.198), (19.85) and (20.26), we can study the χ_d and by including spin we can study the χ_p also in this case.

*20.2.12 The χ_d and χ_p in magneto inversion layers of nonlinear optical and tetragonal materials

(a) Weak electric field limit

From (1.199), in this case, we can write that

$$F(B) = \frac{g_v eB}{\pi\hbar}\left[\sum_{n=0}^{n_{\max}}\sum_{n_z=1}^{n_{z\max}}[\bar{E}_{F_1}F_{-1}(\bar{\eta}_1) - k_BT[F_0(\bar{\eta}_1)]]\right] \tag{20.27}$$

where $\bar{\eta}_1 = (k_BT)^{-1}[\bar{E}_{F_1} - E_{41}]$, $\bar{E}_{F_1}$ is the Fermi energy in this case and E_{41} can be obtained from (5.9) of Chapter 5.

Using (1.198), (5.11) and (20.27), we can study the χ_d and by including spin we can study the χ_p also in this case.

(b) Strong electric field limit

From (1.199), in this case, we can write that

$$F(B) = \frac{g_v eB}{\pi\hbar}\left[\sum_{n=0}^{n_{\max}}\sum_{n_z=1}^{n_{z\max}}[\bar{E}_{F_1}F_{-1}(\bar{\eta}_2) - k_BT[F_0(\bar{\eta}_2)]]\right] \tag{20.28}$$

where $\bar{\eta}_2 = (k_BT)^{-1}[\bar{E}_{F_1} - E_{42}]$ and E_{42} can be obtained from (5.13) of Chapter 5.

Using (1.198), (5.15) and (20.28), we can study the χ_d and by including spin we can study the χ_p also in this case.

20.2.13 The χ_d and χ_p in magneto inversion layers of III–V, ternary and quaternary materials

(a) Weak electric field limit

(i) The three band Kane model

From (1.199), in this case, we can write that

$$F(B) = \frac{g_v eB}{\pi\hbar}\left[\sum_{n=0}^{n_{\max}}\sum_{n_z=1}^{n_{z\max}}[\bar{E}_{F_1}F_{-1}(\bar{\eta}_3) - k_BT[F_0(\bar{\eta}_3)]]\right] \tag{20.29}$$

where $\bar{\eta}_3 = (k_BT)^{-1}[\bar{E}_{F_1} - E_{43}]$ and E_{43} can be obtained from (5.17) of Chapter 5.

Using (1.198), (5.19) and (20.29), we can study the χ_d and by including spin we can study the χ_p also in this case.

(b) Strong electric field limit

(i) The three band Kane model

From (1.199), in this case, we can write that

$$F(B) = \frac{g_v eB}{\pi\hbar}\left[\sum_{n=0}^{n_{\max}}\sum_{n_z=1}^{n_{z\max}}[\bar{E}_{F_1}F_{-1}(\bar{\eta}_6) - k_BT[F_0(\bar{\eta}_6)]]\right] \tag{20.30}$$

where $\bar{\eta}_6 = (k_BT)^{-1}[\bar{E}_{F_1} - E_{47}]$ and E_{47} can be obtained from (5.27) of Chapter 5.

Using (1.198), (5.29) and (20.30), we can study the χ_d and by including spin we can study the χ_p also in this case.

(ii) Two band model of Kane

From (1.199), in this case, we can write that

$$F(B) = \frac{g_v eB}{\pi\hbar}\left[\sum_{n=0}^{n_{\max}}\sum_{n_z=1}^{n_{z\max}}[\bar{E}_{F_1}F_{-1}(\bar{\eta}_7) - k_BT[F_0(\bar{\eta}_7)]]\right] \tag{20.31}$$

where $\bar{\eta}_7 = (k_BT)^{-1}[\bar{E}_{F_1} - E_{48}]$ and E_{48} can be obtained from (5.31) of Chapter 5.

Using (1.198), (5.33) and (20.31), we can study the χ_d and by including spin we can study the χ_p also in this case.

20.2.14 The χ_d and χ_p in magneto inversion layers of II–VI materials

From (1.199), in this case, we can write that

$$F(B) = \frac{g_v eB}{2\pi\hbar}\left[\sum_{n=0}^{n_{\max}}\sum_{n_z=1}^{n_{z\max}}[\bar{E}_{F_1}F_{-1}(\bar{\eta}_8) - k_B T[F_0(\bar{\eta}_8)]]\right] \tag{20.32}$$

where $\bar{\eta}_8 = (k_B T)^{-1}[\bar{E}_{F_1} - E_{49}]$ and E_{49} can be obtained from (5.36) of Chapter 5.

Using (1.198), (5.38) and (20.32), we can study the χ_d and by including spin we can study the χ_p also in this case.

*20.2.15 The χ_d and χ_p in magneto inversion layers of IV–VI materials

(a) Weak electric field limit

From (1.199), in this case, we can write that

$$F(B) = \frac{g_v eB}{\pi\hbar}\left[\sum_{n=0}^{n_{\max}}\sum_{n_z=1}^{n_{z\max}}[\bar{E}_{F_1}F_{-1}(\bar{\eta}_9) - k_B T[F_0(\bar{\eta}_9)]]\right] \tag{20.33}$$

where $\bar{\eta}_9 = (k_B T)^{-1}[\bar{E}_{F_1} - E_{50}]$ and E_{50} can be obtained from (5.42) of Chapter 5.

Using (1.198), (5.44) and (20.33), we can study the χ_d and by including spin we can study the χ_p also in this case.

(b) Strong electric field limit

From (1.199), in this case, we can write that

$$F(B) = \frac{g_v eB}{\pi\hbar}\left[\sum_{n=0}^{n_{\max}}\sum_{n_z=1}^{n_{z\max}}[\bar{E}_{F_1}F_{-1}(\bar{\eta}_{10}) - k_B T[F_0(\bar{\eta}_{10})]]\right] \tag{20.34}$$

where $\bar{\eta}_{10} = (k_B T)^{-1}[\bar{E}_{F_1} - E_{51}]$ and E_{51} can be obtained from (5.47) of Chapter 5.

Using (1.198), (5.49) and (20.34), we can study the χ_d and by including spin we can study the χ_p also in this case.

20.2.16 The χ_d and χ_p in magneto inversion layers of stressed Kane type materials

From (1.199), in this case, we can write that

$$F(B) = \frac{g_v eB}{\pi\hbar}\left[\sum_{n=0}^{n_{\max}}\sum_{n_z=1}^{n_{z\max}}[\bar{E}_{F_1}F_{-1}(\bar{\eta}_{45}) - k_BT[F_0(\bar{\eta}_{45})]]\right] \tag{20.35}$$

where $\bar{\eta}_{45} = (k_BT)^{-1}[\bar{E}_{F_1} - E_{88}]$ and E_{88} can be obtained from (5.213) of Chapter 5.

Using (1.198), (5.215) and (20.35), we can study the χ_d and by including spin we can study the χ_p also in this case.

20.2.17 The χ_d and χ_p in magneto inversion layers of bismuth telluride

From (1.199), in this case, we can write that

$$F(B) = \frac{g_v eB}{\pi\hbar}\left[\sum_{n=0}^{n_{\max}}\sum_{n_z=1}^{n_{z\max}}[\bar{E}_{F_1}F_{-1}(\bar{\eta}_{46}) - k_BT[F_0(\bar{\eta}_{46})]]\right] \tag{20.36}$$

where $\bar{\eta}_{46} = (k_BT)^{-1}[\bar{E}_{F_1} - E_{89}]$ and E_{89} can be obtained from (5.218) of Chapter 5.

Using (1.198), (5.220) and (20.36), we can study the χ_d and by including spin we can study the χ_p also in this case.

20.2.18 The χ_d and χ_p in magneto inversion layers of germanium

(a) Model of Cardona *et al.*

From (1.199), in this case, we can write that

$$F(B) = \frac{g_v eB}{\pi\hbar}\left[\sum_{n=0}^{n_{\max}}\sum_{n_z=1}^{n_{z\max}}[\bar{E}_{F_1}F_{-1}(\bar{\eta}_{15}) - k_BT[F_0(\bar{\eta}_{15})]]\right] \tag{20.37}$$

where $\bar{\eta}_{15} = (k_BT)^{-1}[\bar{E}_{F_1} - E_{57}]$ and E_{57} can be obtained from (5.224) of Chapter 5.

Using (1.198), (5.226) and (20.37), we can study the χ_d and by including spin we can study the χ_p also in this case.

(b) The model of Wang *et al.*

From (1.199), in this case, we can write that

$$F(B) = \frac{g_v eB}{\pi\hbar}\left[\sum_{n=0}^{n_{\max}}\sum_{n_z=1}^{n_{z\max}}[\bar{E}_{F_1}F_{-1}(\bar{\eta}_{48}) - k_BT[F_0(\bar{\eta}_{48})]]\right] \tag{20.38}$$

where $\bar{\eta}_{48} = (k_BT)^{-1}[\bar{E}_{F_1} - E_{91}]$ and E_{91} can be obtained from (5.230) of Chapter 5.

Using (1.198), (5.232) and (20.38), we can study the χ_d and by including spin we can study the χ_p also in this case.

20.2.19 The χ_d and χ_p in magneto inversion layers of gallium aresenide

(a) Rossler model

From (1.199), in this case, we can write that

$$F(B) = \frac{g_v eB}{2\pi\hbar}\left[\sum_{n=0}^{n_{\max}}\sum_{n_z=1}^{n_{z\max}}[\bar{E}_{F_1}F_{-1}(\bar{\eta}_{49}) - k_BT[F_0(\bar{\eta}_{49})]]\right] \tag{20.39}$$

where $\bar{\eta}_{49} = (k_BT)^{-1}[\bar{E}_{F_1} - E_{92}]$ and E_{92} can be obtained from (5.236) of Chapter 5.

Using (1.198), (5.238) and (20.39), we can study the χ_d and by including spin we can study the χ_p also in this case.

(b) Model of Stillman *et al.*

From (1.199), in this case, we can write that

$$F(B) = \frac{g_v eB}{\pi\hbar}\left[\sum_{n=0}^{n_{\max}}\sum_{n_z=1}^{n_{z\max}}[\bar{E}_{F_1}F_{-1}(\bar{\eta}_{50}) - k_BT[F_0(\bar{\eta}_{50})]]\right] \tag{20.40}$$

where $\bar{\eta}_{50} = (k_BT)^{-1}[\bar{E}_{F_1} - E_{93}]$ and E_{93} can be obtained from (5.240) of Chapter 5.

Using (1.198), (5.242) and (20.40), we can study the χ_d and by including spin we can study the χ_p also in this case.

20.2.20 The χ_d and χ_p in magneto inversion layers of gallium antimonide

(a) Model of Seiler *et al.*

From (1.199), in this case, we can write that

$$F(B) = \frac{g_v eB}{2\pi\hbar}\left[\sum_{n=0}^{n_{\max}}\sum_{n_z=1}^{n_{z\max}}[\bar{E}_{F_1}F_{-1}(\bar{\eta}_{51}) - k_BT[F_0(\bar{\eta}_{51})]]\right] \tag{20.41}$$

where $\bar{\eta}_{51} = (k_BT)^{-1}[\bar{E}_{F_1} - E_{94}]$ and E_{94} can be obtained from (5.244) of Chapter 5.

Using (1.198), (5.246) and (20.41), we can study the χ_d and by including spin we can study the χ_p also in this case.

(b) Model of Mathur *et al.*

From (1.199), in this case, we can write that

$$F(B) = \frac{g_v eB}{\pi\hbar}\left[\sum_{n=0}^{n_{\max}}\sum_{n_z=1}^{n_{z\max}}[\bar{E}_{F_1}F_{-1}(\bar{\eta}_{52}) - k_BT[F_0(\bar{\eta}_{52})]]\right] \tag{20.42}$$

where $\bar{\eta}_{52} = (k_BT)^{-1}[\bar{E}_{F_1} - E_{95}]$ and E_{95} can be obtained from (5.248) of Chapter 5.

Using (1.198), (5.250) and (20.42), we can study the χ_d and by including spin we can study the χ_p also in this case.

(c) Model of Zhang

From (1.199), in this case, we can write that

$$F(B) = \frac{g_v eB}{\pi\hbar}\left[\sum_{n=0}^{n_{\max}}\sum_{n_z=1}^{n_{z\max}}[\bar{E}_{F_1}F_{-1}(\bar{\eta}_{53}) - k_BT[F_0(\bar{\eta}_{53})]]\right] \tag{20.43}$$

where $\bar{\eta}_{53} = (k_BT)^{-1}[\bar{E}_{F_1} - E_{96}]$ and E_{96} can be obtained from (5.252) of Chapter 5.

Using (1.198), (5.254) and (20.43), we can study the χ_d and by including spin we can study the χ_p also in this case.

20.2.21 The χ_d and χ_p in magneto inversion layers of indium antimonide

(a) Model of Palik *et al.*

From (1.199), in this case, we can write that

$$F(B) = \frac{g_v eB}{\pi\hbar}\left[\sum_{n=0}^{n_{\max}}\sum_{n_z=1}^{n_{z\max}}[\bar{E}_{F_1}F_{-1}(\bar{\eta}_{54}) - k_BT[F_0(\bar{\eta}_{54})]]\right] \quad (20.44)$$

where $\bar{\eta}_{54} = (k_BT)^{-1}[\bar{E}_{F_1} - E_{97}]$ and E_{97} can be obtained from (5.256) of Chapter 5.

Using (1.198), (5.258) and (20.44), we can study the χ_d and by including spin we can study the χ_p also in this case.

(b) Model of Johnson *et al.*

From (1.199), in this case, we can write that

$$F(B) = \frac{g_v eB}{\pi\hbar}\left[\sum_{n=0}^{n_{\max}}\sum_{n_z=1}^{n_{z\max}}[\bar{E}_{F_1}F_{-1}(\bar{\eta}_{55}) - k_BT[F_0(\bar{\eta}_{55})]]\right] \quad (20.45)$$

where $\bar{\eta}_{55} = (k_BT)^{-1}[\bar{E}_{F_1} - E_{98}]$ and E_{98} can be obtained from (5.260) of Chapter 5.

Using (1.198), (5.260) and (20.45), we can study the χ_d and by including spin we can study the χ_p also in this case.

(c) Model of Agafonov *et al.*

From (1.199), in this case, we can write that

$$F(B) = \frac{g_v eB}{\pi\hbar}\left[\sum_{n=0}^{n_{\max}}\sum_{n_z=1}^{n_{z\max}}[\bar{E}_{F_1}F_{-1}(\bar{\eta}_{56}) - k_BT[F_0(\bar{\eta}_{56})]]\right] \quad (20.46)$$

where $\bar{\eta}_{56} = (k_BT)^{-1}[\bar{E}_{F_1} - E_{99}]$ and E_{99} can be obtained from (5.264) of Chapter 5.

Using (1.198), (5.266) and (20.46), we can study the χ_d and by including spin we can study the χ_p also in this case.

20.2.22 The χ_d and χ_p in magneto inversion layers of tellurium

From (1.199), in this case, we can write that

$$F(B) = \frac{g_v eB}{2\pi\hbar}\left[\sum_{n=0}^{n_{\max}}\sum_{n_z=1}^{n_{z\max}}[\bar{E}_{F_1}F_{-1}(\bar{\eta}_{57}) - k_BT[F_0(\bar{\eta}_{57})]]\right] \tag{20.47}$$

where $\bar{\eta}_{57} = (k_BT)^{-1}[\bar{E}_{F_1} - E_{100}]$ and E_{100} can be obtained from (5.268) of Chapter 5.

Using (1.198), (5.270) and (20.47), we can study the χ_d and by including spin we can study the χ_p also in this case.

20.2.23 The χ_d and χ_p in magneto HD III–V QW effective mass superlattices with graded interfaces

From (1.199), in this case, we can write that

$$F(B) = \frac{g_v eB}{\pi\hbar}\ \text{Real part of}\left[\sum_{n=0}^{n_{\max}}\sum_{n_z=1}^{n_{z\max}}[\bar{E}_{F_{41,n}}F_{-1}(\bar{\eta}_{41,n}) - k_BT[F_0(\bar{\eta}_{41,n})]]\right] \tag{20.48}$$

where $\bar{\eta}_{41,n} = (k_BT)^{-1}[\bar{E}_{F_{41,n}} - E_{41,n}]$ and $E_{F41,n}$ can be obtained from (18.1) of Chapter 18.

Using (1.198), (18.3) and (20.48), we can study the χ_d and by including spin we can study the χ_p also in this case.

20.2.24 The χ_d and χ_p in magneto HD II–VI QW superlattices with graded interfaces

From (1.199), in this case, we can write that

$$F(B) = \frac{g_v eB}{2\pi\hbar}\ \text{Real part of}\left[\sum_{n=0}^{n_{\max}}\sum_{n_z=1}^{n_{z\max}}[\bar{E}_{F_{42,n}}F_{-1}(\bar{\eta}_{42n}) - k_BT[F_0(\bar{\eta}_{42n})]]\right] \tag{20.49}$$

where $\bar{\eta}_{42n} = (k_BT)^{-1}[\bar{E}_{F_{42,n}} - E_{42,n}]$ and $E_{42,n}$ can be obtained from (18.5) of Chapter 18.

Using (1.198), (18.7) and (20.49), we can study the χ_d and by including spin we can study the χ_p also in this case.

*20.2.25 The χ_d and χ_p in magneto HD IV–VI QW superlattices with graded interfaces

From (1.199), in this case, we can write that

$$F(B) = \frac{g_v eB}{\pi\hbar}\left[\sum_{n=0}^{n_{\max}}\sum_{n_z=1}^{n_{z\max}}[\bar{E}_{F_{43,n}}F_{-1}(\bar{\eta}_{1528,n}) - k_BT[F_0(\bar{\eta}_{1528,n})]]\right] \tag{20.50}$$

where $\bar{\eta}_{1528,n} = (k_BT)^{-1}[\bar{E}_{F_{43,n}} - E_{43,n}]$ and $E_{43,n}$ can be obtained from (18.9) of Chapter 18.

Using (1.198), (18.11) and (20.50), we can study the χ_d and by including spin we can study the χ_p also in this case.

20.2.26 The χ_d and χ_p in magneto HD HgTe/CdTe QW superlattices with graded interfaces

From (1.199), in this case, we can write that

$$F(B) = \frac{g_v eB}{\pi\hbar}\text{ Real part of }\left[\sum_{n=0}^{n_{\max}}\sum_{n_z=1}^{n_{z\max}}[\bar{E}_{F_{44,n}}F_{-1}(\bar{\eta}_{44,n}) - k_BT[F_0(\bar{\eta}_{44,n})]]\right] \tag{20.51}$$

where $\bar{\eta}_{44,n} = (k_BT)^{-1}[\bar{E}_{F_{44,n}} - E_{44,n}]$ and $E_{44,n}$ can be obtained from (18.14) of Chapter 18.

Using (1.198), (18.16) and (20.51), we can study the χ_d and by including spin we can study the χ_p also in this case.

20.2.27 The χ_d and χ_p in magneto HD strained layer QW superlattices with graded interfaces

From (1.199), in this case, we can write that

$$F(B) = \frac{g_v eB}{\pi\hbar}\left[\sum_{n=0}^{n_{\max}}\sum_{n_z=1}^{n_{z\max}}[\bar{E}_{F_{47,n}}F_{-1}(\bar{\eta}_{47,n}) - k_BT[F_0(\bar{\eta}_{47,n})]]\right] \tag{20.52}$$

where $\bar{\eta}_{47,n} = (k_BT)^{-1}[\bar{E}_{F_{47,n}} - E_{47,n}]$ and $E_{47,n}$ can be obtained from (18.18) of Chapter 18.

Using (1.198), (18.20) and (20.52), we can study the χ_d and by including spin we can study the χ_p also in this case.

20.2.28 The χ_d and χ_p in magneto HD III–V QW effective mass superlattices

From (1.199), in this case, we can write that

$$F(B) = \frac{g_v eB}{\pi\hbar} \text{ Real part of } \left[\sum_{n=0}^{n_{\max}} \sum_{n_z=1}^{n_{z\max}} [\bar{E}_{F_{41,n}} F_{-1}(\bar{\eta}_{7,A1,n}) - k_B T[F_0(\bar{\eta}_{7,A1,n})]]\right] \tag{20.53}$$

where $\bar{\eta}_{7,A1,n} = (k_B T)^{-1}[\bar{E}_{F_{41,n}} - A1, n]$ and $A1, n$ can be obtained from (18.23) of Chapter 18.

Using (1.198), (18.25) and (20.53), we can study the χ_d and by including spin we can study the χ_p also in this case.

20.2.29 The χ_d and χ_p in magneto HD II–VI QW effective mass superlattices

From (1.199), in this case, we can write that

$$F(B) = \frac{g_v eB}{2\pi\hbar} \text{ Real part of } \left[\sum_{n=0}^{n_{\max}} \sum_{n_z=1}^{n_{z\max}} [\bar{E}_{F_{41,n}} F_{-1}(\bar{\eta}_{13,A2,n}) - k_B T[F_0(\bar{\eta}_{13,A2,n})]]\right] \tag{20.54}$$

where $\bar{\eta}_{13,A2,n} = (k_B T)^{-1}[\bar{E}_{F_{41,n}} - A2, n]$ and $A2, n$ can be obtained from (18.27) of Chapter 18.

Using (1.198), (18.29) and (20.54), we can study the χ_d and by including spin we can study the χ_p also in this case.

20.2.30 The χ_d and χ_p in magneto HD IV–VI QW effective mass superlattices

From (1.199), in this case, we can write that

$$F(B) = \frac{g_v eB}{2\pi\hbar} \left[\sum_{n=0}^{n_{\max}} \sum_{n_z=1}^{n_{z\max}} [\bar{E}_{F_{41,n}} F_{-1}(\bar{\eta}_{1528,n}) - k_B T[F_0(\bar{\eta}_{1528,n})]]\right] \tag{20.55}$$

where $\bar{\eta}_{1528,n} = (k_B T)^{-1}[\bar{E}_{F_{41,n}} - A3, n]$ and $A3, n$ can be obtained from (18.31) of Chapter 18.

Using (1.198), (18.31) and (20.55), we can study the χ_d and by including spin we can study the χ_p also in this case.

20.2.31 The χ_d and χ_p in magneto HD HgTe/CdTe QW effective mass superlattices

From (1.199), in this case, we can write that

$$F(B) = \frac{g_v eB}{\pi\hbar} \text{ Real part of } \left[\sum_{n=0}^{n_{\max}} \sum_{n_z=1}^{n_{z\max}} [\bar{E}_{F_{41,n}} F_{-1}(\bar{\eta}_{13,A4,n}) - k_B T[F_0(\bar{\eta}_{13,A4,n})]] \right] \quad (20.56)$$

where $\bar{\eta}_{13,A4,n} = (k_B T)^{-1}[\bar{E}_{F_{41,n}} - A4, n]$ and $A4, n$ can be obtained from (18.36) of Chapter 18.

Using (1.198), (18.38) and (20.56), we can study the χ_d and by including spin we can study the χ_p also in this case.

20.2.32 The χ_d and χ_p in magneto HD strained layer QW effective mass superlattices

From (1.199), in this case, we can write that

$$F(B) = \frac{g_v eB}{\pi\hbar} \left[\sum_{n=0}^{n_{\max}} \sum_{n_z=1}^{n_{z\max}} [\bar{E}_{F_{41,n}} F_{-1}(\bar{\eta}_{1529,n}) - k_B T[F_0(\bar{\eta}_{1529,n})]] \right] \quad (20.57)$$

where $\bar{\eta}_{1529,n} = (k_B T)^{-1}[\bar{E}_{F_{41,n}} - A8, n]$ and $A8, n$ can be obtained from (18.40) of Chapter 18.

Using (1.198), (18.42) and (20.57), we can study the χ_d and by including spin we can study the χ_p also in this case.

20.2.33 The χ_d and χ_p in doping superlattices of HD nonlinear optical and tetragonal materials under magnetic quantization

From (1.199), in this case, we can write that

$$F(B) = \frac{g_v eB}{\pi\hbar}\left[\sum_{n=0}^{n_{\max}}\sum_{n_z=1}^{n_{z\max}}[\bar{E}_{F_{11,1}}F_{-1}(\eta_{11,1}) - k_BT[F_0(\eta_{11,1})]]\right] \quad (20.58)$$

where $\eta_{11,1} = (k_BT)^{-1}[\bar{E}_{F_{11,1}} - E_{11,1}]$ and $E_{11,1}$ can be obtained from (13.2) of Chapter 18.

Using (1.198), (13.4) and (20.58), we can study the χ_d and by including spin we can study the χ_p also in this case.

20.2.34 The χ_d and χ_p in doping superlattices of HD III–V, ternary and quaternary materials under magnetic quantization

(a) Three band model of Kane

From (1.199), in this case, we can write that

$$F(B) = \frac{g_v eB}{\pi\hbar}\left[\sum_{n=0}^{n_{\max}}\sum_{n_z=1}^{n_{z\max}}[\bar{E}_{F_{11,3}}F_{-1}(\eta_{11,3}) - k_BT[F_0(\eta_{11,3})]]\right] \quad (20.59)$$

where $\eta_{11,3} = (k_BT)^{-1}[\bar{E}_{F_{11,3}} - E_{11,3}]$ and $E_{11,3}$ can be obtained from (13.6) of Chapter 13.

Using (1.198), (13.8) and (20.59), we can study the χ_d and by including spin we can study the χ_p also in this case.

(b) Two band model of Kane

From (1.199), in this case, we can write that

$$F(B) = \frac{g_v eB}{\pi\hbar}\left[\sum_{n=0}^{n_{\max}}\sum_{n_z=1}^{n_{z\max}}[\bar{E}_{F_{11,5}}F_{-1}(\eta_{11,5}) - k_BT[F_0(\eta_{11,5})]]\right] \quad (20.60)$$

where $\eta_{11,5} = (k_BT)^{-1}[\bar{E}_{F_{11,5}} - E_{11,5}]$ and $E_{11,5}$ can be obtained from (13.10) of Chapter 13.

Using (1.198), (13.12) and (20.60), we can study the χ_d and by including spin we can study the χ_p also in this case.

(c) Parabolic band model of Kane

From (1.199), in this case, we can write that

$$F(B) = \frac{g_v eB}{\pi\hbar}\left[\sum_{n=0}^{n_{\max}}\sum_{n_z=1}^{n_{z\max}}[\bar{E}_{F_{11,7}}F_{-1}(\eta_{11,7}) - k_BT[F_0(\eta_{11,7})]]\right] \quad (20.61)$$

where $\eta_{11,7} = (k_BT)^{-1}[\bar{E}_{F_{11,7}} - E_{11,7}]$ and $E_{11,7}$ can be obtained from (13.14) of Chapter 13.

Using (1.198), (13.16) and (20.61), we can study the χ_d and by including spin we can study the χ_p also in this case.

20.2.35 The χ_d and χ_p in doping superlattices of HD II–VI materials under magnetic quantization

From (1.199), in this case, we can write that

$$F(B) = \frac{g_v eB}{\pi\hbar}\left[\sum_{n=0}^{n_{\max}}\sum_{n_z=1}^{n_{z\max}}[\bar{E}_{F_{11,9}}F_{-1}(\eta_{11,9}) - k_BT[F_0(\eta_{11,9})]]\right] \quad (20.62)$$

where $\eta_{11,9} = (k_BT)^{-1}[\bar{E}_{F_{11,9}} - E_{11,9}]$ and $E_{11,9}$ can be obtained from (13.19) of Chapter 13.

Using (1.198), (13.21) and (20.62), we can study the χ_d and by including spin we can study the χ_p also in this case.

20.2.36 The χ_d and χ_p in doping superlattices of HD IV–VI materials under magnetic quantization

From (1.199), in this case, we can write that

$$F(B) = \frac{g_v eB}{\pi\hbar}\left[\sum_{n=0}^{n_{\max}}\sum_{n_z=1}^{n_{z\max}}[\bar{E}_{F_{11,11}}F_{-1}(\eta_{11,11}) - k_BT[F_0(\eta_{11,11})]]\right] \quad (20.63)$$

where $\eta_{11,11} = (k_BT)^{-1}[\bar{E}_{F_{11,11}} - E_{11,11}]$ and $E_{11,11}$ can be obtained from (13.24) of Chapter 13.

Using (1.198), (13.26) and (20.63), we can study the χ_d and by including spin we can study the χ_p also in this case.

*20.2.37 The χ_d and χ_p in doping superlattices of HD stressed Kane type materials under magnetic quantization

From (1.199), in this case, we can write that

$$F(B) = \frac{g_v eB}{\pi\hbar}\left[\sum_{n=0}^{n_{\max}}\sum_{n_z=1}^{n_{z\max}}[\bar{E}_{F_{11,13}}F_{-1}(\eta_{11,13}) - k_BT[F_0(\eta_{11,13})]]\right] \quad (20.64)$$

where $\eta_{11,13} = (k_BT)^{-1}[\bar{E}_{F_{11,13}} - E_{11,13}]$ and $\bar{E}_{11,13}$ can be obtained from (13.29) of Chapter 13.

Using (1.198), (13.31) and (20.64), we can study the χ_d and by including spin we can study the χ_p also in this case.

20.3 Summary and Conclusion

The DRs in magneto-size quantized of HD superlattices (HDSLs) exhibit the fact that the total energy is quantized since the corresponding wave vector space is totally quantized by quantizing magnetic field and size quantization along z-direction.

- The DOS functions for all the HDSLs in this case are series of non-uniformly distributed Dirac's Delta functions at specified quantized points in the respective energy axis. The spacing between the consecutive Delta functions are functions of energy band constants and quantization of the wave vector space of a particular material. The DOS function needs two summations one summation over the Landau quantum number and the other one is due to size quantization.
- It may be noted that the energy levels in magneto-size quantized HDSLs lead to the discrete energy levels, somewhat like atomic energy levels, which produce very large changes. This follows from the inherent nature of the quantum confinement of the carrier gas dealt with here. In QBs, there remain no free carrier states in between any two allowed sets of size-quantized levels unlike that found for QWs, NWs and QBs where the quantum confinements are 1D, 2D and 3D respectively. Consequently, the crossing of the Fermi level by the size-quantized levels in this case would have much greater impact on the redistribution of the carriers among

the allowed levels, as compared to that found for QW HDSLs and NW HDSLs respectively for the electronic properties in general.

- By using the appropriate equations we can plot the χ_d and χ_p with respect to many variables in all the cases of this chapter. We have not given any plot of the χ_d and χ_p here since we believe our able readers will plot the same and enjoy intricate physics, computational analysis, advanced mathematics, and will create new problems both theoretical and experimental.
- It is the band structure which changes in a fundamental way and consequently all the physical properties of all the electronic materials changes radically leading to new physical concepts.

20.4 Open Research Problems

(R.20.1) Investigate the χ_d and χ_p in the presence of an arbitrarily oriented quantizing magnetic field for all the QWHDSLs as given in Chapter 17.

(R.20.2) Investigate the χ_d and χ_p in the presence of an arbitrarily oriented quantizing magnetic field in all QWHDSLs of Chapter 17 by including broadening and the electron spin. Study all the special cases for QWHDSLs of III–V, ternary and quaternary materials in this context.

(R.20.3) Investigate the χ_d and χ_p for QWHDSLs of IV–VI, II–VI and stressed Kane type compounds in the presence of an arbitrarily oriented quantizing magnetic field by including broadening and electron spin.

(R.20.4) Investigate the χ_d and χ_p for all the QWHDSLs in the presence of an arbitrarily oriented quantizing magnetic field by including broadening and electron spin.

(R.20.5) Investigate the χ_d and χ_p in the presence of an arbitrarily oriented quantizing magnetic field and crossed electric fields in QWHDSLs of nonlinear optical materials by including broadening and the electron spin. Study all the special cases for QWHDSLs of III–V, ternary and quaternary materials in this context.

(R.20.6) Investigate the χ_d and χ_p for QWHDSLs of IV–VI, II–VI and stressed Kane type compounds in the presence of an arbitrarily oriented quantizing magnetic field and crossed electric field by including broadening and electron spin.

(R.20.7) Investigate the χ_d and χ_p for all the QWHDSLs in the presence of an arbitrarily oriented quantizing magnetic field and crossed electric field by including broadening and electron spin.

(R.20.8) Investigate the χ_d and χ_p in QWHDSLs of nonlinear optical and tetragonal materials, III–V, II–VI, IV–VI and stressed Kane type semiconductors in the presence of arbitrarily oriented quantizing electric field.

(R.20.9) Investigate the χ_d and χ_p in the presence of an arbitrarily oriented non-quantizing magnetic field for the QWHDSLs of nonlinear optical and tetragonal materials by including the electron spin. Study all the special cases for III–V, ternary and quaternary materials in this context.

(R.20.10) Investigate the χ_d and χ_p in QWHDSLs of IV–VI, II–VI and stressed Kane type compounds in the presence of an arbitrarily oriented non-quantizing magnetic field by including the electron spin.

(R.20.11) Investigate the χ_d and χ_p for QWHDSLs for all the materials as stated in Chapter 17 in the presence of an arbitrarily oriented magnetic field by including electron spin and broadening.

(R.20.12) Investigate the χ_d and χ_p for all the HDSLs as considered in this chapter in the presence of an additional arbitrarily oriented non-quantizing electric field.

(R.20.13) Investigate the χ_d and χ_p for all the HDSLs as considered in this chapter in the presence of arbitrarily oriented crossed electric and magnetic fields.

(R.20.14) Investigate χ_d and χ_p for HD Fibonacci SLs, in the presence of arbitrarily oriented light waves and magnetic quantization.

(R.20.15) Investigate χ_d and χ_p for HD Fibonacci SLs in the presence of exponential, Kane, Halperin and Lax and Bonch-Bruevich band tails.

(R.20.17) Investigate all the problems of this chapter by removing all the mathematical approximations and establishing the uniqueness conditions in each case.

(R.20.18) Investigate the χ_d and χ_p for HD Fibonacci SLs in the presence of defects and magnetic quantization.

References

[1] S. Rakshit, S. N. Biswas, A. N. Chakravarti, *Phys. Lett.*, 37A, 64 **(1971)**; S. Rakshit, S. N. Biswas, A. N. Chakravarti, *Sol. Stat. Electron.*, 5, 702, **(1972)**; A. N. Chakravarti, D. P. Parui, *Phys. Lett.*, 40A, 113 **(1972)**.

[2] A. N. Chakravarti, D. P. Parui, *Phys. Stat. Sol. (a)*, 14, K23 **(1972)**; A. N. Chakravarti, D. P. Parui, *Phys. Stat. Sol. (a)*, 14, K139 **(1972)**; A. N. Chakravarti, D. P. Parui, *Phys. Stat. Sol. (a)*, 14, K55 **(1972)**.

[3] A. N. Chakravarti, D. P. Parui, *Czech. J. Phys. B*, 23, 548 **(1973)**; A. N. Chakravarti, D. P. Parui, *Phys. Lett.*, 43A, 60 **(1973)**.

[4] A. N. Chakravarti, D. P. Parui, *Int. J. Electron.*, 34, 709 **(1973)**; A. N. Chakravarti, D. P. Parui, *Czech. J. Phys. B*, 23, 923 **(1973)**; B. R. Nag and A. N. Chakravarti, *Phys. Stat. Sol. (b)*, 60, K49 **(1973)**.

[5] A. N. Chakravarti, D. Mukherjee, *Phys. Lett.*, 53A, 403 **(1975)**; A. N. Chakravarti and S. Swaminathan, *Phys. Stat. Sol. (a)*, 23, K191 **(1974)**.

[6] A. N. Chakravarti, K. P. Ghatak, K. K. Ghosh and A. Dhar, *Phys. Stat. Sol. (b)*, 103, K55 **(1981)**; A. N. Chakravarti, D. P. Parui, *Int. J. Electron.*, 36, 275 **(1974)**; A. N. Chakravarti, B. R. Nag, *Phys. Stat. Sol. (a)*, 22, K153 **(1974)**; A. N. Chakravarti, S. Swaminathan, *Phys. Stat. Sol. (a)*, 25, K47 **(1974)**.

[7] B. R. Nag, A. N. Chakravarti and P. K. Basu, *Phys. Stat. Sol. (a)*, 68, K75 **(1981)**; B. R. Nag and A. N. Chakravarti, *Phys. Stat. Sol. (a)*, 67, K113 **(1981)**; A. N. Chakravarti and D. P. Parui, *Phys. Lett.*, 40A, 113 **(1972)**; A. N. Chakravarti, D. Mukherjee, *Phys. Lett.*, 53A, 57 **(1975)**; B. R. Nag and A. N. Chakravarti, *Sol. Stat. Sol. B*, 18, 109 **(1975)**; A. N. Chakravarti, S. Swaminathan, *Czech. J. Phys. B*, 25, 712 **(1975)**.

[8] A. N. Chakravarti and D. P. Parui, *Canad. J. Phys.*, 51, 451 **(1973)**; D. Mukherjee, A. N. Chakravarti and B. R. Nag, *Phys. Stat. Sol. (a)*, 26, K27 **(1974)**; A. N. Chakravarti, *Czech. J. Phys. B*, 25, 778 **(1975)**.

[9] S. Ghosh and A. N. Chakravarti, *Phys. Stat. Sol. (b)*, 147, 355 **(1988)**; A. N. Chakravarti, K. P. Ghatak, K. K. Ghosh, H. M. Mukherjee, *Phys. Stat. Sol. (b)*, 118, 843 **(1983)**.

[10] A. N. Chakravarti, K. P. Ghatak, A. Dhar, K. K. Ghosh, *Czech. J. Phys.*, 33, 65 **(1983)**; B. R. Nag, A. N. Chakravarti, *Phys. Sol. Stat. Sol. (b)*, 1, K45 **(1975)**.

[11] A. N. Chakravarti, K. P. Ghatak, K. K. Ghosh, S. Ghosh, A. Dhar, *Zeitschrift für Physik B Condensed Matter*, 47, 149 **(1983)**; D. R. Choudhury, P. K. Basu, A. N. Chakravarti, *Phys. Stat. Sol. (a)*, 36, K65 **(1976)**.

[12] A. N. Chakravarti, K. P. Ghatak, K. K. Ghosh, G. B. Rao, *Phys. Stat. Sol. (b)*, 111, K61 **(1983)**; D. R. Choudhury, P. K. Basu, A. N. Chakravarti, *Phys. Stat. Sol. (a)*, 38, K35 **(1976)**.

[13] A. N. Chakravarti, K. P. Ghatak, S. Ghosh, A. K. Chowdhury, *Phys. Stat. Sol. (b)*, 109, 705 **(1982)**; A. K. Chowdhury, A. N. Chakravarti, *Czech. J. Phys. B*, 28, 228 **(1978)**; A. N. Chakravarti, A. K. Chowdhury, *Phys. Stat. Sol. (b)*, 1, K31 **(1977)**.

[14] A. N. Chakravarti, K. P. Ghatak, K. K. Ghosh, S. Ghosh, H. M. Mukherjee, *Phys. Stat. Sol. (b)*, 108, 609 **(1981)**; A. N. Chakravarti, A. K. Chowdhury, S. Ghosh, K. P. Ghatak, A. Dhar, *Czech. J. Phys. B*, 31, 1037 **(1981)**; A. N. Chakravarti, A. K. Chowdhury, K. P. Ghatak, A. Dhar, D. R. Choudhury, *Czech. J. Phys. B*, 31, 905 **(1981)**.

[15] A. N. Chakravarti, K. P. Ghatak, A. Dhar, K. K. Ghosh, S. Ghosh, *Acta Phys. Polonica A*, 60, 151 **(1981)**; A. N. Chakravarti, K. P. Ghatak, A. Dhar, S. Ghosh, *Phys. Stat. Sol. (b)*, 105, 691 **(1981)**; K. P. Ghatak, A. K. Chowdhury, S. Ghosh, A. N. Chakravarti, *Appl. Phys.*, 23, 241 **(1980)**.

[16] A. N. Chakravarti, A. K. Chowdhury, K. P. Ghatak, D. R. Choudhury, *Czech. J. Phys. B*, 30, 1161 **(1980)**; K. P. Ghatak, A. K. Chowdhury, S. Ghosh, A. N. Chakravarti, *Acta Phys. Polonica A*, 58, 389 **(1980)**; K. P. Ghatak, A. K. Chowdhury, S. Ghosh, A. N. Chakravarti, *Czech. J. Phys. B*, 30, 925 **(1980)**.

[17] A. N. Chakravarti, A. K. Chowdhury, K. P. Ghatak, D. R. Choudhury, *Phys. Stat. Sol. (a)*, 59, K211 **(1980)**; A. N. Chakravarti, A. K. Chowdhury, K. P. Ghatak, D. R. Choudhury, *Acta Phys. Polonica A*, 58, 251 **(1980)**.

[18] C. Majumdar, A. B. Maity, A. N. Chakravarti, *Phys. Stat. Sol. (b)*, 140, K7 **(1987)**; C. Majumdar, A. B. Maity, A. N. Chakravarti, *Phys. Stat. Sol. (b)*, 141, K35 **(1987)**; N. R. Das, K. K. Ghosh, D. Ghoshal, *Phys. Stat. Sol. (b)*, 197, 97 **(1996)**; A. N. Chakravarti, A. K. Chowdhury, *Czech. J. Phys.*, 27, 1161 **(1977)**; A. K. Chowdhury, A. N. Chakravarti, *Phys. Stat. Sol. (a)*, 44, K111 **(1977)**.

[19] C. Majumdar, A. B. Maity, A. N. Chakravarti, *Phys. Stat. Sol. (b)*, 144, K13 **(1987)**; N. R. Das, A. N. Chakravarti, *Phys. Stat. Sol. (b)*, 176, 335 **(1993)**; S. Sen, N. R. Das and A. N. Chakravarti, *J. Phys. Conden. Mat.*, 19, 186205 **(2007)**; K. P. Ghatak, A. K. Chowdhury, S. Ghosh, A. N. Chakravarti, *Phys. Stat. Sol. (b)*, 99, K55 **(1980)**.

[20] N. R. Das, S. Ghosh, A. N. Chakravarti, *Phys. Stat. Sol. (b)*, 174, 45 **(1992)**; A. B. Maity, C. Majumdar, A. N. Chakravarti, *Phys. Stat. Sol. (b)*, 144, K93 **(1987)**; A. B. Maity, C. Majumdar, A. N. Chakravarti, *Phys. Stat. Sol. (b)*, 149, 565 **(1988)**; K. P. Ghatak, A. K. Chowdhury, S. Ghosh, A. N. Chakravarti, *Appl. Phys.*, 23, 241 **(1980)**.

[21] S. Bhattacharya, D. De, S. M. Adhikari, K. P. Ghatak, *Superlatt. Microstruct.*, 51, 203 **(2012)**; S. Bhattacharya, D. De, S. M. Adhikari, S. Saha, K. M. Chatterjee, S. Choudhury, K. P. Ghatak, *Superlatt. Microstruct.*, 50, 609 **(2011)**; S. Bhattacharya, S. Choudhury, K. P. Ghatak, *Superlatt. Microstruct.*, 48, 257 **(2010)**.

[22] D. De, A. Kumar, S. M. Adhikari, S. Pahari, N. Islam, P. Banerjee, S. K. Biswas, S. Bhattacharya, K. P. Ghatak, *Superlatt. Microstruct.*, 47, 277 **(2010)**; S. Pahari, S. Bhattacharya, S. Roy, A. Saha, D. De, K. P. Ghatak, *Superlatt. Microstruct.*, 46, 760 **(2009)**; K. P. Ghatak, S. Bhattacharya, S. Pahari, D. De, R. Benedictus, *Superlatt. Microstruct.*, 46, 387 **(2009)**.

[23] S. Debbarma, K. P. Ghatak, *Nanosci. Nanotech.*, 16, 1095 **(2016)**; M. Chakraborty, K. P. Ghatak, *Quant. Matt.*, 4, 104 **(2015)**; S. M. Adhikari, A. Sakar, K. P. Ghatak, *Quant. Matt.*, 2, 455 **(2013)**.

[24] P. K. Bose, N. Paitya, S. Bhattacharya, D. De, S. Saha, K. M. Chatterjee, S. Pahari, K. P. Ghatak, *Quant. Matt.*, 1, 89 (**2012**); S. Pahari, S. Bhattacharya, K. P. Ghatak, *Comput. Theor. Nanosci.*, 6, 2088 (**2009**); S. K. Biswas, A. R. Ghatak, A. Neogi, A. Sharma, S. Bhattacharya, K. P. Ghatak, *Phys. E: Low-dimen. Syst. Nanostruct.*, 36, 263 (**2007**).

[25] L. J. Singh, S. Choudhury, D. Baruah, S. K. Biswas, S. Pahari, K. P. Ghatak, *Phys. B: Conden. Matter*, 368, 188 (**2005**); S. Chowdhary, L. J. Singh, K. P. Ghatak, *Phys. B: Conden. Matter*, 365, 5 (**2005**); L. J. Singh, S. Choudhary, A. Mallik, K. P. Ghatak, *Comput. Theor. Nanosci.*, 2, 287 (**2005**).

[26] N. R. Das, A. N. Chakravarti, *Phys. Stat. Sol. (b)*, 160, 603 (**1990**).

[27] K. P. Ghatak, M. Mondal, *Phys. Stat. Sol. (b)*, 160, 673 (**1990**).

[28] K. P. Ghatak, B. De, *Proc. Mat. Res. Soc.*, EA16, 173 (**1991**).

Chapter 21

The Magneto Elastic Constants in 0D Systems Under Intense Light Waves

Falling down is not my DEFEAT. Defeat is when I refuse to get back up.

21.1 Introduction

With the advent of nano-photonics, there has been a considerable interest in studying the optical processes in materials and their nanostructures [1]. It appears from the literature, that the investigations have been carried out on the assumption that the carrier energy spectra are invariant quantities in the presence of intense light waves, *which is not fundamentally true.* The physical properties of materials in the presence of light waves which change the basic dispersion relation have been relatively less investigated in the literature [2–9]. In this chapter, we shall study the MECs' in HD III–V, ternary and quaternary materials and their quantized counter parts on the basis of newly formulated electron dispersion law under external photo excitation.

In Section 21.2.1, of the theoretical background 21.2, we have formulated the DRs' of the conduction electrons of HD III–V, ternary and quaternary materials in the presence of light waves whose unperturbed electron energy spectrum is described by the three-band model of Kane in the absence of band tailing.

In Section 21.2.2, we have studied the MECs' in QDs of HD Kane type materials in the presence of light waves. In Section 21.2.3, we have studied the magneto MECs' in QWs of HD Kane type materials in the presence of light Waves. In Section 21.2.4, we have studied the magneto MECs' in accumulation and inversion layers of Kane type materials in the presence of light waves. In Section 21.2.5, we have investigated the magneto MECs' in doping superlattices of Kane type materials in the presence of light waves Section 21.2.6, we have studied the magneto MECs' in QWHD effective mass superlattices of Kane type materials in the presence of light waves. In Section 21.2.7, we have studied the MECs' in Quantum dot HD superlattices of Kane type materials with graded interfaces in the presence of light waves. In Section 21.2.8, we have studied the magneto MEC in QWHD superlattices of Kane type materials with graded interfaces in the presence of light waves. Section 21.3 contains the summary and conclusion pertinent to this chapter. The last Section 21.4 presents one open research problem.

21.2 Theoretical Background

21.2.1 The formulation of the electron dispersion law in the presence of light waves in HD III–V, ternary and quaternary materials

The Hamiltonian ($\hat{H}$) of an electron in the presence of light wave characterized by the vector potential $\vec{A}$ can be written following [3] as

$$\hat{H} = [|(\hat{p} + |e|\vec{A})|^2/2m] + V(\bar{r}) \tag{21.1}$$

in which $\hat{p}$ is the momentum operator, $V(\bar{r})$ is the crystal potential and m is the free electron mass. (21.1) can be expressed as

$$\hat{H} = \hat{H}_0' + \hat{H}' \tag{21.2}$$

where

$$\hat{H}_0' = \frac{\hat{p}^2}{2m} + V(\bar{r})$$

and

$$\hat{H}' = \frac{|e|}{2m}\bar{A} \cdot \hat{p} \tag{21.3}$$

The perturbed Hamiltonian $\hat{H}$ can be written as

$$\hat{H}' = \left(\frac{-i\hbar|e|}{2m}\right)(\bar{A} \cdot \nabla) \tag{21.4}$$

where

$$i = \sqrt{-1} \quad \text{and} \quad \hat{p} = -i\hbar\nabla$$

The vector potential $((\vec{A}))$ of the monochromatic light of plane wave can be expressed as

$$\vec{A} = A_0\vec{\varepsilon}_s \cos(\vec{s}_0 \cdot \vec{r} - \omega t) \tag{21.5}$$

where A_0 is the amplitude of the light wave, $\vec{\varepsilon}_s$ is the polarization vector, $\vec{s}_0$ is the momentum vector of the incident photon, $\vec{r}$ is the position vector, ω is the angular frequency of light wave and t is the time scale. The matrix element of $\hat{H}'_{nl}$ between initial state, $\psi_1(\vec{q}, \vec{r})$ and final state $\psi_n(\vec{k}, \vec{r})$ in different bands can be written as

$$\hat{H}'_{nl} = \frac{|e|}{2m}\langle n\vec{k}|\vec{A} \cdot \hat{p}|l\vec{q}\rangle \tag{21.6}$$

Using (21.4) and (21.5), we can re-write (21.6) as

$$\hat{H}'_{nl} = \left(\frac{-i\hbar|e|A_0}{4m}\right)\vec{\varepsilon}_s \cdot \left[\{\langle n\vec{k}|e^{(i\vec{s}_0\cdot\vec{r})}\nabla|l\vec{q}\rangle e^{-i\omega t}\} + \{\langle n\vec{k}|e^{(i\vec{s}_0\cdot\vec{r})}\nabla|l\vec{q}\rangle e^{i\omega t}\}\right] \tag{21.7}$$

The first matrix element of (21.7) can be written as

$$\langle n\vec{k}|e^{(i\vec{s}_0\cdot\vec{r})}\nabla|l\vec{q}\rangle = \int e^{(i[\vec{q}+\vec{s}_0-\vec{k}]\cdot\vec{r})} i\vec{q}u_n^*(\vec{k}\cdot\vec{r})u_l(\vec{q},\vec{r})d^3r + \int e^{(i[\vec{q}+\vec{s}_0-\vec{k}]\cdot\vec{r})}u_n^*(\vec{k},\vec{r})\nabla u_l(\vec{q},\vec{r})d^3r \tag{21.8}$$

The functions $u_n^* u_l$ and $u_n^*\nabla u_l$ are periodic. The integral over all space can be separated into a sum over unit cells times an integral over a single unit cell. It is assumed that the wave length of the electromagnetic wave is sufficiently large so that if $\vec{k}$ and $\vec{q}$ are within the Brillouin zone, $(\vec{q}+\vec{s}_0-\vec{k})$ is not a reciprocal lattice vector.

Therefore, we can write (21.8) as

$$\langle n\vec{k}|e^{(i\vec{s}_0-\vec{r})}\nabla|l\vec{q}\rangle = \left[\frac{(2\pi)^3}{\Omega}\right]\left\{i\vec{q}\delta(\vec{q}+\vec{s}_0-\vec{k})\delta_{nl} + \delta(\vec{q}+\vec{s}_0-\vec{k})u_n^* \times \int_{cell}(\vec{k},\vec{r})\nabla u_1(\vec{q},\vec{r})d^3r\right\} = \left[\frac{(2\pi)^3}{\Omega}\right]\left\{\delta(\vec{q}+\vec{s}_0-\vec{k}) \times \int_{cell} u_n^*(\vec{k},\vec{r})\nabla u_1(\vec{q},\vec{r})d^3r\right\} \tag{21.9}$$

where Ω is the volume of the unit cell and

$$\int u_n^*(\vec{k},\vec{r})u_l(\vec{q},\vec{r})d^3r = \delta(\vec{q}-\vec{k})\delta_{nl} = 0,$$

since $n \neq l$.

The delta function expresses the conservation of wave vector in the absorption of light wave and $\vec{s}_0$ is small compared to the dimension of a typical Brillouin zone and we set $\vec{q} = \vec{k}$.

From (21.8) and (21.9), we can write,

$$\hat{H}'_{nl} = \frac{|e|A_0}{2m}\vec{\varepsilon}_s \cdot \hat{p}_{nl}(\vec{k})\delta(\vec{q}-\vec{k})\cos(\omega t) \tag{21.10}$$

where

$$\hat{p}_{nl}(\vec{k}) = -i\hbar \int u_n^* \nabla u_1 d^3t = \int u_n^*(\vec{k},\vec{r})\hat{p}u_l(\vec{k},\vec{r})d^3r$$

Therefore, we can write

$$\hat{H}'_{nl}\frac{|e|A_0}{2m}\vec{\varepsilon}_s \cdot \hat{p}_{nl}(\vec{k}) \tag{21.11}$$

where

$$\vec{\varepsilon} = \vec{\varepsilon}_s \cos\omega t$$

When a photon interacts with a semiconductor, the carriers (i.e., electrons) are generated in the bands which are followed by the inter-band transitions. For example, when the carriers are generated in the valence band, they then make inter-band transition to the conduction band. The transition of the electrons within the same band i.e., $\hat{H}'_{nn} = \langle n\vec{k}|\hat{H}'|n\vec{k}\rangle$ is neglected. Because, in such a case, i.e., when the carriers are generated within the same bands by photons, are lost by recombination within the aforementioned band resulting in zero carriers.

Therefore,

$$\langle n\vec{k}|\hat{H}'|n\vec{k}\rangle = 0 \tag{21.12}$$

With $n = c$ stands for conduction band and $l = v$ stand for valance band, the energy equation for the conduction electron can approximately be written as

$$I_{11}(E) = \left(\frac{\hbar^2k^2}{2m_c}\right) + \frac{(\frac{|e|A_0}{2m})^2\langle|\vec{\varepsilon}\cdot\hat{p}_{cv}(\vec{k})|^2\rangle_{av}}{E_c(\vec{k}) - E_v(\vec{k})} \tag{21.13}$$

where

$$I_{11}(E) \equiv E(aE+1)(bE+1)/(cE+1), \quad a \equiv 1/E_{g0}, \quad a \equiv 1/E_{g0},$$

E_{g0} is the un-perturbed band-gap, $b \equiv 1/(E_{g0}+\Delta)$, $c \equiv 1/(E_{g0}+2\Delta/3)$, and $\langle|\vec{\varepsilon}\cdot\hat{p}_{cv}(\vec{k})|^2\rangle_{av}$ represents the average of the square of the optical matrix element (OME).

For the three-band model of Kane, we can write,

$$\xi_{1k} = E_c(\vec{k}) - E_v(\vec{k}) = (E_{g0}^2 + E_{g0}\hbar^2k^2/m_r)^{1/2} \tag{21.14}$$

where m_r is the reduced mass and is given by $m_r^{-1} = (m_c)^{-1} + m_v^{-1}$, and m_v is the effective mass of the heavy hole at the top of the valance band in the absence of any field.

The doubly degenerate wave functions $u_1(\vec{k},\vec{r})$ and $u_2(\vec{k},\vec{r})$ can be expressed as [16]

$$u_1(\vec{k},\vec{r}) = a_{k+}[(is)\downarrow'] + b_{k+}\left[ds\frac{X'-iY'}{\sqrt{2}}\uparrow'\right] + c_{k+}[Z'\downarrow'] \tag{21.15}$$

and

$$u_2(\vec{k},\vec{r}) = a_{k-}[(is)\uparrow'] - b_{k-}\left[\frac{X'+iY'}{\sqrt{2}}\downarrow'\right] + c_{k-}[Z'\uparrow'] \tag{21.16}$$

s is the s-type atomic orbital in both unprimed and primed coordinates, $\downarrow'$ indicates the spin down function in the primed coordinates,

$$a_{k\pm} \equiv \beta[E_{g_0} - (\gamma_{0k\pm})^2(E_{g_0} - \delta')]^{1/2}(E_{g_0} + \delta')^{-1/2}$$

$$\beta \equiv [(6(E_{g_0} + 2\Delta/3)(E_{g_0} + \Delta))/\chi]^{1/2}, \chi \equiv (6E_{g_0}^2 + E_{g_0}\Delta + 4\Delta^2),$$

$$\gamma_{0k\pm} \equiv \left[\frac{(\xi_{1k} \mp E_{g_0})}{2(\xi_{1k} + \delta')}\right]^{1/2}, \quad \xi_{1k} = E_c(\vec{k}) - E_v(\vec{k})$$

$$= E_{g0}\left[1 + 2\left(1 + \frac{m_c}{m_v}\right)\frac{I_{11}(E)}{E_{g_0}}\right]^{1/2}, \quad \delta' \equiv (E_{g0}^2\Delta)(\chi)^{-1}, \quad X', \quad Y',$$

and Z' are the p-type atomic orbital's in the primed coordinates, $\uparrow'$ indicates the spin-up function in the primed coordinates,

$$b_{k\pm} \equiv \rho\gamma_{0k\pm}, \quad \rho \equiv (4\Delta^2/3\chi)^{1/2}, \quad c_{k\pm} \equiv t\gamma_{0k\pm} \quad \text{and}$$

$$t \equiv [6(E_{g0} + 2\Delta/3)^2/\chi]^{1/2}$$

We can, therefore, write the expression for the optical matrix element (OME) as

$$OME = \hat{p}_{cv}(\vec{k}) = \langle u_1(\vec{k},\vec{r})|\hat{p}|u_2(\vec{k},r')\rangle \tag{21.17}$$

Since the photon vector has no interaction in the same band for the study of inter-band optical transition, we can therefore write

$$\langle S|\hat{p}|S\rangle = \langle X|\hat{p}|X\rangle = \langle Y|\hat{p}|Y\rangle = \langle Z|\hat{p}|Z\rangle = 0$$

and

$$\langle X|\hat{p}|Y\rangle = \langle Y|\hat{p}|Z\rangle = \langle Z|\hat{p}|X\rangle = 0$$

There are finite interactions between the conduction band (CB) and the valance band (VB) and we can obtain

$$\begin{aligned}\langle S|\hat{p}|X\rangle &= \hat{i}\cdot\hat{p} = \hat{i}\cdot\hat{p}_x\\ \langle S|\hat{p}|Y\rangle &= \hat{j}\cdot\hat{p} = \hat{j}\cdot\hat{p}_x\\ \langle S|\hat{p}|Z\rangle &= \hat{k}\cdot\hat{p} = \hat{k}\cdot\hat{p}_x\end{aligned}$$

where $\hat{i},\hat{j}$ and $\hat{k}$ are the unit vectors along x, y and z-axes respectively.

It is well known that

$$\begin{bmatrix}\uparrow'\\ \downarrow'\end{bmatrix} = \begin{bmatrix} e^{-i\phi/2}\cos(\theta/2) & e^{i\phi/2}\sin(\theta/2)\\ -e^{-i\phi/2}\sin(\theta/2) & e^{i\phi/2}\cos(\theta/2)\end{bmatrix}\begin{bmatrix}\uparrow\\ \downarrow\end{bmatrix}$$

and

$$\begin{bmatrix}X'\\ Y'\\ Z'\end{bmatrix} = \begin{bmatrix}\cos\theta\cos\phi & \cos\theta\sin\phi & -\sin\theta\\ -\sin\phi & \cos\phi & 0\\ \sin\theta\cos\phi & \sin\theta\sin\phi & \cos\theta\end{bmatrix}\begin{bmatrix}X\\ Y\\ Z\end{bmatrix}$$

Besides, the spin vector can be written as

$$\vec{S} = \frac{\hbar}{2}\vec{\sigma},\ \text{where } \sigma_x = \begin{bmatrix}0 & 1\\ 1 & 0\end{bmatrix},\quad \sigma_y = \begin{bmatrix}0 & -i\\ i & 0\end{bmatrix}\quad \text{and}\quad \sigma_z = \begin{bmatrix}1 & 0\\ 0 & -1\end{bmatrix}$$

From above, we can write

$$\begin{aligned}\hat{P}_{CV}(\vec{k}) &= \langle u_1(\vec{k},\vec{r})|\hat{p}|u_2(\vec{k},r')\rangle\\ &= \left\langle\left\{a_{k_+}[(iS)\downarrow'] + b_{k_+}\left[\left(\frac{X'-iY'}{\sqrt{2}}\right)\uparrow'\right] + c_{k_+}[Z'\downarrow']\right\}|\hat{P}|\left\{a_{k_-}(iS)\uparrow'\right.\right.\\ &\quad\left.\left. - b_{k_-}\left[\left(\frac{X'+iY'}{\sqrt{2}}\right)\downarrow' + c_{k_-}[Z'\uparrow']\right]\right\}\right\rangle\end{aligned}$$

Using above relations, we get

$$\begin{aligned}\hat{p}_{CV}(\vec{k}) &= \langle u_1(\vec{k},\vec{r})|\hat{P}|u_2(\vec{k},r')\rangle \\ &= \frac{b_{k_+}a_{k_-}}{\sqrt{2}}\{\langle(X'-iY')|\hat{P}|iS\rangle\langle\uparrow'|\uparrow'\rangle\} + c_{k_+}a_{k_-}\{\langle Z'|\hat{P}|iS\rangle\langle\downarrow'|\uparrow'\rangle\} \\ &\quad - \frac{a_{k_+}b_{k_-}}{\sqrt{2}}\{\langle iS|\hat{P}|(X'+iY')\rangle\langle\downarrow'|\downarrow'\rangle\} + a_{k_+}c_{k_-}\{\langle iS|\hat{P}|Z'\rangle\langle\downarrow'|\uparrow'\rangle\}\end{aligned} \tag{21.18}$$

We can also write

$$\begin{aligned}\langle(X'-iY')|\hat{P}|iS\rangle = \langle(X')|\hat{P}|iS\rangle - \langle(iY')|\hat{P}|iS\rangle = i\int u_{X+}^{*}\hat{P}S \\ - \int -iu_{Y+}^{*}\hat{P}iu_x = i\langle X'|\hat{P}|iS\rangle - \langle Y'|\hat{P}|iS\rangle\end{aligned}$$

From the above relations, for X', Y' and Z', we get

$$|X'\rangle = \cos\theta\cos\phi|X\rangle + \cos\theta\sin\phi|Y\rangle - \sin\theta|Z\rangle$$

Thus,

$$\langle X'|\hat{P}|S\rangle = \cos\theta\cos\phi\langle X'|\hat{P}|S\rangle + \cos\theta\sin\phi\langle Y|\hat{P}|S\rangle - \sin\theta\langle Z|\hat{P}|S\rangle = \hat{P}\hat{r}_1$$

where

$$\hat{r}_1 = \hat{i}\cos\theta\cos\phi + \hat{j}\cos\theta\sin\phi - \hat{k}\sin\theta$$

$$|Y'\rangle = -\sin\phi|X\rangle + \cos\phi|Y\rangle + 0|Z\rangle$$

Thus,

$$\langle Y'|\hat{P}|S\rangle = -\sin\phi\langle X|\hat{P}|S\rangle + \cos\phi\langle Y|\hat{P}|S\rangle + 0\langle Z|\hat{P}|S\rangle = \hat{P}\hat{r}_2$$

where

$$\hat{r}_2 = -\hat{i}\sin\phi + \hat{j}\cos\phi$$

so that

$$\langle(X'-iY')|\hat{P}|S\rangle = \hat{P}(i\hat{r}_1 - \hat{r}_2)$$

Thus,

$$\frac{a_{k_-}b_{k_+}}{\sqrt{2}}\langle(X'-iY')|\hat{P}|S\rangle\langle\uparrow'|\uparrow'\rangle = \frac{a_{k_-}b_{k_+}}{\sqrt{2}}(i\hat{r}_1 - \hat{r}_2)\langle\uparrow'|\uparrow'\rangle \tag{21.19}$$

Now since,

$$\langle iS|\hat{P}|(X'+iY')\rangle = i\langle S|\hat{P}|X'\rangle - \langle S|\hat{P}|Y'\rangle = \hat{P}(i\hat{r}_1 - \hat{r}_2)$$

We can write,

$$-\left[\frac{a_{k_+}b_{k_-}}{\sqrt{2}}\{\langle iS|\hat{P}|(X'+iY')\rangle\langle\downarrow'|\downarrow'\rangle\}\right] = \left[\frac{a_{k_+}b_{k_-}}{\sqrt{2}}\hat{P}(i\hat{r}_1 - \hat{r}_2)\langle\downarrow'|\downarrow'\rangle\right] \tag{21.20}$$

Similarly, we get

$$|Z'\rangle = \sin\theta\cos\phi|X\rangle + \sin\theta\sin\phi|Y\rangle + \cos\theta|Z\rangle$$

So that,

$$\langle Z'|\hat{P}|S\rangle = i\langle Z'|\hat{P}|S\rangle - i\hat{P}\{\sin\theta\cos\phi\hat{i} + \sin\theta\sin\phi\hat{j} + \cos\theta\hat{k}\} = i\hat{P}\hat{r}_3$$

where

$$\hat{r}_3 = \hat{\imath}\sin\theta\cos\phi + \hat{\jmath}\sin\theta\sin\phi + \hat{k}\cos\theta$$

Thus,

$$c_{k_+}a_{k_-}\langle Z'|\hat{P}|S\rangle\langle\downarrow'|\uparrow'\rangle = c_{k_+}a_{k_-}i\hat{P}\hat{r}_3\langle\downarrow'|\uparrow'\rangle \tag{21.21}$$

Similarly, we can write,

$$c_{k_+}a_{k_-}\langle iS|\hat{P}|Z'\rangle\langle\downarrow'|\uparrow'\rangle = c_{k_-}a_{k_+}i\hat{P}\hat{r}_3\langle\downarrow'|\uparrow'\rangle \tag{21.22}$$

Therefore, we obtain

$$\begin{aligned}&\frac{a_{k_-}b_{k_+}}{\sqrt{2}}\{\langle(X'+iY')|\hat{P}|S\rangle\langle\uparrow'|\uparrow'\rangle\} - \frac{a_{k_-}b_{k_+}}{\sqrt{2}}\{\langle iS|\hat{P}|(X'+iY')\rangle\langle\downarrow'|\downarrow'\rangle\}\\ &= \frac{\hat{P}}{\sqrt{2}}(-a_{k_+}b_{k_-}\langle\downarrow'|\downarrow'\rangle + a_{k_-}b_{k_+}\langle\uparrow'|\uparrow'\rangle)(i\hat{r}_2 - \hat{r}_2)\end{aligned} \tag{21.23}$$

Also, we can write,

$$\begin{aligned}&c_{k_+}a_{k_-}\langle Z'|\hat{P}|iS\rangle\langle\downarrow'|\uparrow'\rangle + c_{k_-}a_{k_+}\langle iS|\hat{P}|Z'\rangle\langle\downarrow'|\uparrow'\rangle\\ &= i\hat{P}(c_{k_+}a_{k_-} + c_{k_-}a_{k_+})\hat{r}_3\langle\downarrow'|\downarrow'\rangle\end{aligned} \tag{21.24}$$

Combining (21.23) and (21.24), we find

$$\hat{p}_{CV}(\vec{k}) = \frac{\hat{p}}{\sqrt{2}}(i\hat{r}_1 - \hat{r}_2)\{(b_{k_+}a_{k_-})\langle\uparrow'|\uparrow'\rangle - (b_{k_-}a_{k_+})\langle\downarrow'|\downarrow'\rangle\}$$
$$+ i\hat{P}\hat{r}_3(c_{k_+}a_{k_-} - c_{k_-}c_{k_+})\langle\downarrow'|\uparrow'\rangle \quad (21.25)$$

From the above relations, we obtain,

$$\begin{Bmatrix} \uparrow' = e^{-i\phi/2}\cos(\theta/2)\uparrow + e^{i\phi/2}\sin(\theta/2)\downarrow \\ \downarrow' = e^{-i\phi/2}\sin(\theta/2)\uparrow + e^{i\phi/2}\cos(\theta/2)\downarrow \end{Bmatrix} \quad (21.26)$$

Therefore,

$$\langle\downarrow'|\uparrow'\rangle_x = -\sin(\theta/2)\cos(\theta/2)\langle\uparrow|\uparrow\rangle_x + e^{-i\theta}\cos^2(\theta/2)\langle\downarrow|\uparrow\rangle_x$$
$$- e^{i\phi}\sin^2(\theta/2)\langle\uparrow|\downarrow\rangle_x + \sin(\theta/2)\cos(\theta/2)\langle\downarrow|\downarrow\rangle_x \quad (21.27)$$

But we know from above that

$$\langle\uparrow|\uparrow\rangle_x = 0, \quad \langle\downarrow|\uparrow\rangle = \frac{1}{2}, \quad \langle\downarrow|\uparrow\rangle_x = \frac{1}{2} \quad \text{and} \quad \langle\downarrow|\downarrow\rangle_x = 0$$

Thus, we get

$$\begin{aligned}\langle\downarrow'|\uparrow'\rangle_x &= \frac{1}{2}[e^{-i\phi}\cos^2(\theta/2) - e^{i\phi}\sin^2(\theta/2)] \\ &= \frac{1}{2}[(\cos\phi - i\sin\phi)\cos^2(\theta/2) - (\cos\phi + i\sin\phi)\sin^2(\theta/2)] \\ &= \frac{1}{2}[\cos\phi\cos\theta - i\sin\phi] \end{aligned} \quad (21.28)$$

Similarly, we obtain

$$\langle\downarrow'|\uparrow\rangle_y = \frac{1}{2}[i\cos\phi + \sin\phi\cos\theta] \quad \text{and} \quad \langle\downarrow'|\uparrow\rangle_z = \frac{1}{2}[-\sin\theta]$$
$$\langle\downarrow'|\uparrow'\rangle = \hat{\imath}\langle\downarrow'|\uparrow'\rangle_x + \hat{\jmath}\langle\downarrow'|\uparrow'\rangle_y + \hat{k}\langle\downarrow'|\uparrow'\rangle$$

Therefore,

$$\frac{1}{2}\{(\cos\theta\cos\phi - i\sin\phi)\hat{\imath} + (i\cos\phi + \sin\phi\cos\theta)\hat{\jmath} - \sin\theta\hat{k}\}$$
$$= \frac{1}{2}[\{(\cos\theta\cos\phi)\hat{\imath} + (\sin\phi\cos\theta)\hat{\jmath} - \sin\theta\hat{k}\} + i\{-\hat{\imath}\sin\phi + \hat{\jmath}\cos\phi\}]$$
$$= \frac{1}{2}[\hat{r}_1 + i\hat{r}_2] = -\frac{1}{2}i[i\hat{r}_1 - \hat{r}_2]$$

Similarly, we can write

$$\langle\uparrow'|\uparrow'\rangle = \frac{1}{2}[\hat{i}\sin\theta\cos\phi + \hat{j}\sin\theta\sin\phi + \hat{k}\cos\theta] = \frac{1}{2}\hat{r}_3 \quad \text{and} \quad \langle\downarrow'|\downarrow'\rangle = -\frac{1}{2}\hat{r}_3$$

Using the above results, we can write

$$\begin{aligned}\hat{p}_{CV}(\vec{k}) &= \frac{\hat{p}}{\sqrt{2}}(i\hat{r}_2 - \hat{r}_2)\{(a_{k_-}b_{k_+})\langle\uparrow'|\uparrow'\rangle - (b_{k_-}a_{k_+})\langle\downarrow'|\downarrow'\rangle\}\\ &\quad + i\hat{P}\hat{r}_3\{(c_{k_+}a_{k_-} - c_{k_-}a_{k_+})\langle\downarrow'|\uparrow'\rangle\}\\ &= \frac{\hat{P}}{2}\hat{r}_3(i\hat{r}_1 - \hat{r}_2)\left\{\left(\frac{a_{k_-}b_{k_+}}{\sqrt{2}} + \frac{b_{k_-}a_{k_+}}{\sqrt{2}}\right)\right\}\\ &\quad + \frac{\hat{P}}{2}\hat{r}_3(i\hat{r}_2 - \hat{r}_2)\left\{(c_{k_+}a_{k_-} + c_{k_-}c_{k_+})\right\}\end{aligned}$$

Thus,

$$\hat{p}_{CV}(\vec{k}) = \frac{\hat{p}}{2}\hat{r}_3(i\hat{r}_2 - \hat{r}_2)\left\{a_{k_+}\left(\frac{b_{k_-}}{\sqrt{2}} + c_{k_-}\right) + a_{k_-}\left(\frac{b_{k_+}}{\sqrt{2}} + c_{k_+}\right)\right\} \tag{21.29}$$

we can write that,

$|\hat{r}_1| = |\hat{r}_2| = |\hat{r}_2| = 1$, also, $\hat{P}\hat{r}_3 = \hat{P}_x \sin\theta\cos\phi\hat{i} + \hat{P}_y\sin\theta\sin\phi\hat{j} + \hat{P}_z\cos\theta\hat{k}$

where

$$\begin{aligned}\hat{P} &= \langle S|P|X\rangle = \langle S|P|Y\rangle = \langle S|P|Z\rangle,\\ \langle S|P'|X\rangle &= \int u_C^*(0,\vec{r})\hat{P}u_{VX}(0,\vec{r})d^3r = \hat{P}_{CVX}(0) \quad \text{and}\\ \langle S|\hat{P}|Z\rangle &= \hat{P}_{CVZ}(0)\end{aligned}$$

Thus,

$$\hat{P} = \hat{P}_{CVX}(0) = \hat{P}_{CVY}(0) = \hat{P}_{CVZ}(0) = \hat{P}_{CV}(0)$$

where

$$\hat{P}_{CV}(0) = \int u_C^*(0,\vec{r})\hat{P}_{u_V}(0,\vec{r})d^3r \equiv \hat{P}$$

For a plane polarized light wave, we have the polarization vector $\vec{\varepsilon}_s = \hat{k}$, when the light wave vector is traveling along the z-axis. Therefore, for a plane polarized light-wave, we have considered $\vec{\varepsilon}_s = \hat{k}$.

Then, from (21.29), we get

$$(\vec{\varepsilon}\cdot\hat{p}_{CV}(\vec{k})) = \vec{k}\cdot\frac{\hat{P}}{2}\hat{r}_3(i\hat{r}_1 - \hat{r}_2)[A(\vec{k}) + B(\vec{k})]\cos\omega t \qquad (21.30)$$

and

$$\left\{\begin{aligned} A(\vec{k}) &= a_{k_-}\left(\frac{b_{k_+}}{\sqrt{2}} + c_{k_+}\right) \\ B(\vec{k}) &= a_{k_+}\left(\frac{b_{k_-}}{\sqrt{2}} + c_{k_-}\right) \end{aligned}\right\} \qquad (21.31)$$

Thus,

$$\begin{aligned} |\vec{\varepsilon}\cdot\hat{p}_{CV}(\vec{k})|^2 &= \left|\vec{k}\cdot\frac{\hat{P}}{2}\hat{r}_3\right|^2 |i\hat{r}_1 - \hat{r}_2|^2[A(\vec{k}) + B(\vec{k})]^2\cos\omega t \\ &= \frac{1}{4}|\hat{P}_z\cos\theta|^2[A(\vec{k}) + B(\vec{k})]^2\cos^2\omega t \end{aligned} \qquad (21.32)$$

So, the average value of

$$|\vec{\varepsilon}\cdot\hat{p}_{CV}(\vec{k})|^2$$

for a plane polarized light-wave is given by

$$\begin{aligned} \langle|\vec{\varepsilon}\cdot\hat{p}_{CV}(\vec{k})|^2\rangle_{av} &= \frac{2}{4}|\hat{P}_z|^2[A(\vec{k}) + B(\vec{k})]^2\left(\int_0^{2\pi}d\phi\int_0^{\pi}\cos^2\theta\sin\theta d\theta\right)\left(\frac{1}{2}\right) \\ &= \frac{2\pi}{3}|\hat{P}_z|^2[A(\vec{k}) + B(\vec{k})]^2 \end{aligned} \qquad (21.33)$$

where

$$|\hat{P}_z|^2 = \left(\frac{1}{2}\right)|\vec{k}\cdot\hat{p}_{cv}(0)|^2$$

and

$$|\vec{k}\cdot\hat{p}_{cv}(0)|^2 = \frac{m^2}{4m_r}\frac{E_{g_0}(E_{g_0} + \Delta)}{(E_{g_0} + \frac{2}{3}\Delta)} \qquad (21.34)$$

We shall express $A(\vec{k})$ and $B(\vec{k})$ in terms of constants of the energy spectra in the following way:

Substituting

$$a_{k_\pm}, \quad b_{k_\pm}, \quad c_{k_\pm} \quad \text{and} \quad \gamma_{0k_\pm}$$

in $A(\vec{k})$ and $B(\vec{k})$ in (21.31), we get

$$A(\vec{k}) = \beta\left(t + \frac{\rho}{\sqrt{2}}\right)\left\{\left(\frac{E_{g_0}}{E_{g_0} + \delta'}\right)\gamma_{0k_+}^2 - \gamma_{0k_+}^2\gamma_{0k_-}^2\left(\frac{E_{g_0} - \delta'}{E_{g_0} + \delta'}\right)\right\}^{1/2} \tag{21.35}$$

$$B(\vec{k}) = \beta\left(t + \frac{\rho}{\sqrt{2}}\right)\left\{\left(\frac{E_{g_0}}{E_{g_0} + \delta'}\right)\gamma_{0k_-}^2 - \gamma_{0k_+}^2\gamma_{0k_-}^2\left(\frac{E_{g_0} - \delta'}{E_{g_0} + \delta'}\right)\right\}^{1/2} \tag{21.36}$$

in which

$$\gamma_{0k_+}^2 = \frac{\xi_{1k} - E_{g_0}}{2(\xi_{1k} + \delta')} \equiv \frac{1}{2}\left[1 - \left(\frac{E_{g_0} + \delta'}{\xi_{1k} + \delta'}\right)\right]$$

and

$$\gamma_{0k_-}^2 = \frac{\xi_{1k} + E_{g_0}}{2(\xi_{1k} + \delta')} \equiv \frac{1}{2}\left[1 - \left(\frac{E_{g_0} - \delta'}{\xi_{1k} + \delta'}\right)\right]$$

Substituting

$$x \equiv \xi_{1k} + \delta' \,\text{in}\, \gamma_{0k_\pm}^2,$$

we can write,

$$A(\vec{k}) = \beta\left(t + \frac{\rho}{\sqrt{2}}\right)\left\{\left(\frac{E_{g_0}}{E_{g_0} + \delta'}\right)\frac{1}{2}\left(1 - \frac{E_{g_0} + \delta'}{x}\right) - \frac{1}{4}\left(\frac{E_{g_0} - \delta'}{E_{g_0} + \delta'}\right)\left(1 - \frac{E_{g_0} + \delta'}{x}\right)\left(1 - \frac{E_{g_0} - \delta'}{x}\right)\right\}^{1/2}$$

Thus,

$$A(\vec{k}) = \frac{\beta}{2}\left(t + \frac{\rho}{\sqrt{2}}\right)\left\{1 - \frac{2a_0}{x} + \frac{a_1}{x^2}\right\}^{1/2}$$

where

$$a_0 \equiv (E_{g_0}^2 + \delta'^2)(E_{g_0} + \delta')^{-1} \quad \text{and} \quad a_1 \equiv (E_{g_0} - \delta')^2$$

After tedious algebra, one can show that

$$A(\vec{k}) = \frac{\beta}{2}\left(t + \frac{\rho}{\sqrt{2}}\right)(E_{g_0} - \delta')\left[\frac{1}{\xi_{1k} + \delta'} - \frac{1}{E_{g_0} + \delta'}\right]^{1/2} \times \left[\frac{1}{\xi_{1k} + \delta'}\frac{(E_{g_0} + \delta')}{(E_{g_0} - \delta')^2}\right]^{1/2} \tag{21.37}$$

Similarly, from (21.36), we can write,

$$B(\vec{k}) = \beta\left(t + \frac{\rho}{\sqrt{2}}\right)\left\{\left(\frac{E_{g_0}}{E_{g_0}+\delta'}\right)\frac{1}{2}\left(1+\frac{E_{g_0}-\delta'}{x}\right)\right.$$
$$\left. -\frac{1}{4}\left(\frac{E_{g_0}-\delta'}{E_{g_0}+\delta'}\right)\left(1-\frac{E_{g_0}+\delta'}{x}\right)\left(1+\frac{E_{g_0}-\delta'}{x}\right)\right\}^{1/2}$$

So that, finally we get,

$$B(\vec{k}) = \frac{\beta}{2}\left(t+\frac{\rho}{\sqrt{2}}\right)\left(1+\frac{E_{g_0}-\delta'}{\xi_{1k}+\delta'}\right) \tag{21.38}$$

Using the above equations, we can write

$$\left(\frac{|e|A_0}{2m}\right)^2\frac{\langle|\vec{\varepsilon}\cdot\hat{p}_{CV}(\vec{k})|^2\rangle_{av}}{E_c(\vec{k})-E_v(\vec{k})} = \left(\frac{|e|A_0}{2m}\right)^2\frac{2\pi}{3}|\vec{k}\cdot\hat{p}_{CV}(0)|^2\frac{\beta^2}{4}\left(t+\frac{\rho}{\sqrt{2}}\right)^2$$
$$\frac{1}{\xi_{1k}}\left\{\left(1+\frac{E_{g_0}-\delta'}{\xi_{1k}+\delta'}\right)+(E_{g_0}-\delta')\left[\frac{1}{\xi_{1k}+\delta'}-\frac{1}{E_{g_0}+\delta'}\right]^{1/2}\right.$$
$$\left.\times\left[\frac{1}{\xi_{1k}+\delta'}-\frac{E_{g_0}+\delta'}{(E_{g_0}-\delta')^2}\right]^{1/2}\right\}^2 \tag{21.39}$$

Following Nag [9], it can be shown that

$$A_0^2 = \frac{I\lambda^2}{2\pi^2c^3\sqrt{\varepsilon_{sc}\varepsilon_0}} \tag{21.40}$$

where I is the light intensity of wavelength λ, ε_0 is the permittivity of free space and c is the velocity of light. Thus, the simplified electron energy spectrum in III–V, ternary and quaternary materials in the presence of light waves can approximately be written as

$$\frac{\hbar^2k^2}{2m_c^*} = \beta_0(E,\lambda) \tag{21.41}$$

where

$$\beta_0(E,\lambda) \equiv [I_{11}(E)-\theta_0(E,\lambda)],$$
$$I_{11}(E) \equiv \frac{E(E+E_{g_0})(E+E_{g_0}+\Delta)\left(E_{g_0}+\frac{1}{3}\Delta\right)}{E_{g_0}(E_{g_0}+\Delta)\left(E+E_{g_0}+\frac{2}{3}\Delta\right)}$$

$$\theta_0(E,\lambda) \equiv \frac{|e|^2}{96 m_r \pi c^3} \frac{I\lambda^2}{\sqrt{\varepsilon_{sc}\varepsilon_0}} \frac{E_{g_0}(E_{g_0}+\Delta)}{(E_{g_0}+\frac{2}{3}\Delta)} \frac{\beta^2}{4}\left(t+\frac{\rho}{\sqrt{2}}\right)^2 \frac{1}{\phi_0(E)}$$
$$\times\left\{\left(1+\frac{E_{g_0}-\delta'}{\phi_0(E)+\delta'}\right)+(E_{g_0}-\delta')\left[\frac{1}{\phi_0(E)+\delta'}-\frac{1}{E_{g_0}+\delta'}\right]^{1/2}\right.$$
$$\left.\times\left[\frac{1}{\phi_0(E)+\delta'}-\frac{E_{g_0}+\delta'}{(E_{g_0}-\delta')^2}\right]^{1/2}\right\}^2 \text{ and}$$

$$\phi_0(E) \equiv E_{g_0}\left(1+2\left(1+\frac{m_c}{m_v}\right)\frac{I_{11}(E)}{E_{g_0}}\right)^{1/2}$$

Thus, under the limiting condition $\vec{k} \to 0$, from (21.41), we observe that $E \neq 0$ and is positive. Therefore, in the presence of external light waves, the energy of the electron does not tend to zero when $\vec{k} \to 0$, where as for the un-perturbed three band model of Kane, $I_{11}(E) = [\hbar^2 k^2/(2m_c)]$ in which $E \to 0$ for $\vec{k} \to 0$. As the conduction band is taken as the reference level of energy, therefore the lowest positive value of E for $k' \to 0$ provides the increased band gap (ΔE_g) of the semiconductor due to photon excitation. The values of the increased band gap can be obtained by computer iteration processes for various values of I and λ respectively.

Special Cases

For the two-band model of Kane, we have $\Delta \to 0$ under this condition, $I_{11}(E) \to E(1+aE) = \frac{\hbar^2 k^2}{2m_c}$. Since, $\beta \to 1, t \to 1, \rho \to 0, \delta' \to 0$ for $\Delta \to 0$ from (21.41), we can write the energy spectrum of III–V, ternary and quaternary materials in the presence of external photo-excitation whose unperturbed conduction electrons obey the two band model of Kane as

$$\frac{\hbar^2 k^2}{2m_c} = \tau_0(E,\lambda) \tag{21.42}$$

where

$$\tau_0(E,\lambda) \equiv E(1+aE) - B_0(E,\lambda),$$

$$B_0(E,\lambda) \equiv \frac{|e|^2}{384\pi c^3 m_r} \frac{I\lambda^2 E_{g_0}}{\sqrt{\varepsilon_{sc}\varepsilon_0}} \frac{1}{\phi_1(E)}\left\{\left(1+\frac{E_{g_0}}{\phi_0(E)}\right)\right.$$
$$\left.+E_{g_0}\left[\frac{1}{\phi_1(E)}-\frac{1}{E_{g_0}}\right]\right\}^2,$$

$$\phi_1(E) \equiv E_{g_0}\left\{1+\frac{2m_c}{m_r}aE(1+aE)\right\}^{1/2}$$

For relatively wide band gap materials, one can write,

$$a \to 0, \quad b \to 0, \quad c \to 0 \quad \text{and} \quad I_{11}(E) \to E$$

Thus, from (21.42), we get,

$$\frac{\hbar^2 k^2}{2m_c} = \rho_0(E, \lambda) \tag{21.43}$$

where

$$\rho_0(E, \lambda) \equiv E - \frac{|e|^2 I\lambda^2}{96\pi C^3 m_r \sqrt{\varepsilon_{sc}\varepsilon_0}} \left[1 + \left(\frac{2m_c}{m_r}\right) aE\right]^{-3/2}$$

Equations (21.41)–(21.43) can approximately be written as

$$\frac{\hbar^2 k^2}{2m_c} = U_\lambda I_{11}(E) - P_\lambda \tag{21.44}$$

$$\frac{\hbar^2 k^2}{2m_c} = t_{1\lambda} E + t_{2\lambda} E^2 - \delta_\lambda \tag{21.45}$$

and

$$\frac{\hbar^2 k^2}{2m_c} = t_{1\lambda} E - \delta_\lambda \tag{21.46a}$$

where

$$U_\lambda = (1 + \theta_\lambda), \quad \theta_\lambda = \frac{C_0}{A}(t_\lambda + \frac{BJ_\lambda}{A}),$$

$$C_0 = \left[\frac{|e|^2}{96\pi c^3 m_r} \frac{I\lambda^2 E_{g_0}(E_{g_0} + \Delta)}{\sqrt{\varepsilon_{sc}\varepsilon_0}\,(E_{g_0} + \frac{2}{3}\Delta)} \frac{\beta^2}{4}\left(1 + \frac{\rho}{\sqrt{2}}\right)^2\right],$$

$$A = E_{g_0}, \quad B = \left[1 + \frac{m_c}{m_V}\right], \quad G_\lambda = \left[\frac{2B}{(A + \delta')^3} - \frac{BC_\lambda}{(A + \delta')}\right],$$

$$C_\lambda = [(E_{g_0} + \delta')^{-1} + (E_{g_0} + \delta')(E_{g_0} - \delta')^{-2}](A + \delta')^{-1},$$

$$P_\lambda = \frac{C_0}{A} J_\lambda, \quad J_\lambda = (D_\lambda + 2(E_{g_0} - \delta')\sqrt{f_\lambda}), \quad D_\lambda = \left(1 + \frac{2(E_{g_0} - \delta')}{(A + \delta')}\right),$$

$$f_\lambda = \left[\frac{1}{(A + \delta')^2} + \frac{1}{(E_{g_0} - \delta')^2} - C_\lambda\right],$$

$$t_{1\lambda} = \left(1 + \frac{3m_c}{m_r}\alpha\delta_\lambda\right), \quad \alpha = \frac{1}{E_{g_0}}, \quad \delta_\lambda = \frac{|e|^2 I\lambda^2}{96 m_r \pi c^3 \sqrt{\varepsilon_{sc}\varepsilon_0}} \quad \text{and}$$

$$t_{2\lambda} = \alpha t_{1\lambda}$$

Under the condition of heavy doping, using the method of averaging the kinetic energy of the electron, the HD dispersion relations in this case in the presence of light waves can be written as

$$\frac{\hbar^2 k^2}{2m_c} = T_1(E, \eta_g, \lambda) \tag{21.46b}$$

$$\frac{\hbar^2 k^2}{2m_c} = T_2(E, \eta_g, \lambda) \tag{21.47}$$

and

$$\frac{\hbar^2 k^2}{2m_c} = T_3(E, \eta_g, \lambda) \tag{21.48}$$

where

$$T_1(E, \eta_g, \lambda) = [U_\lambda[T_{31}(E, \eta_g) + iT_{32}(E, \eta_g)] - P_\lambda],$$

$$\begin{aligned} T_{31}(E, \eta_g) \equiv& \left(\frac{2}{1 + Erf(E/\eta_g)}\right) \left[\frac{\alpha b}{c}\theta_0(E, \eta_g) + \left[\frac{\alpha c + bc - \alpha b}{c^2}\right] \gamma_0(E, \eta_g)\right. \\ &+ \frac{1}{c}\left(1 - \frac{\alpha}{c}\right)\left(1 - \frac{b}{c}\right)\frac{1}{2}\left[1 + Erf\left(\frac{E}{\eta_g}\right)\right] \\ &- \frac{1}{c}\left(1 - \frac{\alpha}{c}\right)\left(1 - \frac{b}{c}\right)\frac{2}{c\eta_g\sqrt{\pi}}\exp(-u_2^2) \\ &\times \left.\left[\sum_{p=1}^{\infty} \frac{\exp(-p^2/4)}{p}\sinh(pu_2)\right]\right] \end{aligned}$$

$$b \equiv \left(\frac{1}{E_g + \Delta}\right), \quad c \equiv \left(\frac{1}{E_g + \frac{2}{3}\Delta}\right), \quad u_2 \equiv \frac{1 + cE}{c\eta_g},$$

$$T_{32}(E, \eta_g) \equiv \left(\frac{2}{1 + Erf(E/\eta_g)}\right) \frac{1}{c}\left(1 - \frac{\alpha}{c}\right)\left(1 - \frac{b}{c}\right)\frac{\sqrt{\pi}}{c\eta_g}\exp(-u_2^2)$$

$$\theta_0(E, \eta_g) = \frac{\eta_g E}{2\sqrt{\pi}}\exp\left(\frac{-E^2}{\eta_g^2}\right) + \frac{1}{4}(\eta_g^2 + 2E^2)\left[1 + Erf\left(\frac{E}{\eta_g}\right)\right],$$

$$\gamma_0(E, \eta_g) = \eta_g \exp(-E^2/\eta_g^2)(2\sqrt{\pi})^{-1} + \frac{E}{2}(1 + Erf(E/\eta_g)),$$

$$T_2(E, \eta_g, \lambda) = [U_{1\lambda}\gamma_3(E, \eta_g) + (t_{2\lambda})2\theta_0(E, \eta_g)[1 + Erf(E/\eta_g)]^{-1} - \delta_\lambda]$$

$$\gamma_3(E,\eta_g) \equiv \left[\frac{2}{(1+Erf(E/\eta_g))}\right]\gamma_0(E,\eta_g) \quad \text{and}$$

$$T_3(E,\eta_g,\lambda) = [t_{1\lambda}\gamma_3(E,\eta_g) - \delta_\lambda]$$

21.2.2 The MEC in QDs of HD Kane type materials in the presence of light waves

(a) The DR in QDs of HD III–V, ternary and quaternary materials in the presence of external photo-excitation whose unperturbed electrons are defined by the three band model of Kane can be expressed following (21.46b)

$$\frac{\hbar^2(n_z\pi/d_z)^2}{2m_c} + \frac{\hbar^2(n_y\pi/d_y)^2}{2m_c} + \frac{\hbar^2(n_x\pi/d_x)^2}{2m_c} = T_1(E_{17,1},\eta_g,\lambda) \quad (21.49)$$

where $E_{17,1}$ is the total quantized energy in this case.

The DOS function is given by

$$N_{0\mathrm{DT}}(E) = \frac{2g_v}{d_x d_y d_z}\sum_{n_x=1}^{n_{x\max}}\sum_{n_y=1}^{n_{y\max}}\sum_{n_z=1}^{n_{z\max}}\delta'(E - E_{17,1}) \quad (21.50)$$

The electron concentration can be written as

$$n_0 = \text{Real part of } \bar{n}_0 \quad (21.51)$$

$$\bar{n}_0 = \frac{2g_v}{d_x d_y d_z}\sum_{n_x=1}^{n_{x\max}}\sum_{n_y=1}^{n_{y\max}}\sum_{n_z=1}^{n_{z\max}} F_{-1}(\eta_{126}) \quad (21.52)$$

where

$$\eta_{126} = \frac{E_{F126} - E_{17.1}}{k_B T} \quad \text{and} \quad E_{F126}$$

is the Fermi energy in this case.

The electronic contribution to the second- and third-order elastic constants for HD materials in this case can be written as

$$\Delta C_{44} = \frac{-G_0^2}{9}\text{ Real part of }\left[\frac{\partial n_0}{\partial(E_{F126} - E_{17.1})}\right] \quad (21.53)$$

and

$$\Delta C_{456} = \frac{G_0^3}{27}\text{ Real part of }\left[\frac{\partial^2 n_0}{\partial(E_{F126} - E_{17.1})^2}\right] \quad (21.54)$$

Using (21.51), (21.53) and (21.54), we can find ΔC_{44} and ΔC_{456} in this case.

(b) The DR in QDs of HD III–V, ternary and quaternary materials in the presence of external photo-excitation whose unperturbed electrons are defined by the two band model of Kane can be expressed from (21.47) as

$$\frac{\hbar^2(n_z\pi/d_z)^2}{2m_c}+\frac{\hbar^2(n_y\pi/d_y)^2}{2m_c}+\frac{\hbar^2(n_x\pi/d_x)^2}{2m_c}=T_2(E_{17,3},\eta_g,\lambda) \quad (21.55)$$

where $E_{17,3}$ is the total quantized energy in this case.

The DOS function is given by

$$N_{\mathrm{0DT}}(E)=\frac{2g_v}{d_xd_yd_z}\sum_{n_x=1}^{n_{x\max}}\sum_{n_y=1}^{n_{y\max}}\sum_{n_z=1}^{n_{z\max}}\delta'(E-E_{17,3}) \quad (21.56)$$

The electron concentration can be written as

$$n_0=\frac{2g_v}{d_xd_yd_z}\sum_{n_x=1}^{n_{x\max}}\sum_{n_y=1}^{n_{y\max}}\sum_{n_z=1}^{n_{z\max}}F_{-1}(\eta_{127}) \quad (21.57)$$

where

$$\eta_{127}=\frac{E_{F127}-E_{17.3}}{k_BT} \quad \text{and} \quad E_{F127}$$

is the Fermi energy in this case.

The electronic contribution to the second- and third-order elastic constants for HD materials in this case can be written as

$$\Delta C_{44}=\frac{-G_0^2}{9}\left[\frac{\partial n_0}{\partial(E_{F127}-E_{17.3})}\right] \quad (21.58)$$

and

$$\Delta C_{456}=\frac{G_0^3}{27}\left[\frac{\partial^2 n_0}{\partial(E_{F127}-E_{17.3})^2}\right] \quad (21.59)$$

Using (21.57)–(21.59), we can find ΔC_{44} and ΔC_{456} in this case.

(c) The DR in QDs of HD III–V, ternary and quaternary materials in the presence of external photo-excitation whose unperturbed electrons are defined by the parabolic energy bands can be expressed following (21.48) as

$$\frac{\hbar^2(n_z\pi/d_z)^2}{2m_c}+\frac{\hbar^2(n_y\pi/d_y)^2}{2m_c}+\frac{\hbar^2(n_x\pi/d_x)^2}{2m_c}=T_3(E_{17,5},\eta_g,\lambda) \quad (21.60)$$

where $E_{17,5}$ is the total quantized energy in this case

The DOS function is given by

$$N_{0\mathrm{DT}}(E)=\frac{2g_v}{d_xd_yd_z}\sum_{n_x=1}^{n_{x\max}}\sum_{n_y=1}^{n_{y\max}}\sum_{n_z=1}^{n_{z\max}}\delta'(E-E_{17,5}) \quad (21.61)$$

The electron concentration can be written as

$$n_0=\frac{2g_v}{d_xd_yd_z}\sum_{n_x=1}^{n_{x\max}}\sum_{n_y=1}^{n_{y\max}}\sum_{n_z=1}^{n_{z\max}}F_{-1}(\eta_{1275}) \quad (21.62)$$

where

$$\eta_{1275}=\frac{E_{F1275}-E_{17.5}}{k_BT} \quad \text{and} \quad E_{F1275}$$

is the Fermi energy in this case.

The electronic contribution to the second- and third-order elastic constants for HD materials in this case can be written as

$$\Delta C_{44}=\frac{-G_0^2}{9}\left[\frac{\partial n_0}{\partial(E_{F1275}-E_{17.5})}\right] \quad (21.63)$$

and

$$\Delta C_{456}=\frac{G_0^3}{27}\left[\frac{\partial^2 n_0}{\partial(E_{F1275}-E_{17.5})^2}\right] \quad (21.64)$$

Using (21.62)–(21.64), we can find ΔC_{44} and ΔC_{456} in this case.

21.2.3 The MEC in QWs of HD Kane type materials in the presence of light waves

(a) The magneto DR in QWs of HD III–V, ternary and quaternary materials, whose unperturbed band structure is defined by the three band model of Kane, in the presence of light waves can be expressed from (21.46b) as

$$\left(n+\frac{1}{2}\right)\hbar\omega_0+\frac{\hbar^2}{2m_c}\left(\frac{n_z\pi}{d_z}\right)^2=T_1(E_{17,8},\eta_g,\lambda) \tag{21.65}$$

where $E_{17,8}$ is the total quantized energy in this case.

The DOS function is given by

$$N_{0\mathrm{DT}}(E)=\frac{eBg_v}{\pi\hbar d_z}\sum_{n_z=1}^{n_{z\max}}\sum_{n=0}^{n_{\max}}\delta'(E-E_{17,8}) \tag{21.66}$$

The electron concentration can be written as

$$n_0 = \text{Real part of } \bar{n}_0 \tag{21.67}$$

$$n_0 = \frac{eBg_v}{\pi\hbar d_z}\sum_{n=0}^{n_{\max}}\sum_{n_z=1}^{n_{z\max}}F_{-1}(\eta_{128}) \tag{21.68}$$

where

$$\eta_{128}=\frac{E_{F128}-E_{17.8}}{k_BT} \quad \text{and} \quad E_{F128}$$

is the Fermi energy in this case.

The electronic contribution to the second- and third-order elastic constants for HD materials in this case can be written as

$$\Delta C_{44}=\frac{-G_0^2}{9}\text{ Real part of }\left[\frac{\partial n_0}{\partial(E_{F128}-E_{17,8})}\right] \tag{21.69}$$

and

$$\Delta C_{456}=\frac{G_0^3}{27}\text{ Real part of }\left[\frac{\partial^2 n_0}{\partial(E_{F128}-E_{17,8})^2}\right] \tag{21.70}$$

Using (21.67), (21.69) and (21.70), we can find ΔC_{44} and ΔC_{456} in this case.

(b) The magneto DR in QWs of HD III–V, ternary and quaternary materials, whose unperturbed band structure is defined by the two band model of Kane, in the presence of light waves can be expressed from (21.47) as

$$\left(n+\frac{1}{2}\right)\hbar\omega_0+\frac{\hbar^2}{2m_c}\left(\frac{n_z\pi}{d_z}\right)^2=T_2(E_{17,9},\eta_g,\lambda) \tag{21.71}$$

where $E_{17,9}$ is the total quantized energy in this case.

The DOS function is given by

$$N_{\mathrm{0DT}}(E)=\frac{eBg_v}{\pi\hbar d_z}\sum_{n_z=1}^{n_{z\max}}\sum_{n=0}^{n_{\max}}\delta'(E-E_{17,9}) \tag{21.72}$$

The electron concentration can be written as

$$n_0=\frac{eBg_v}{\pi\hbar d_z}\sum_{n=0}^{n_{\max}}\sum_{n_z=1}^{n_{z\max}}F_{-1}(\eta_{129}) \tag{21.73}$$

where

$$\eta_{129}=\frac{E_{F129}-E_{17.9}}{k_BT} \quad \text{and} \quad E_{F129}$$

is the Fermi energy in this case.

The electronic contribution to the second- and third-order elastic constants for HD materials in this case can be written as

$$\Delta C_{44}=\frac{-G_0^2}{9}\left[\frac{\partial n_0}{\partial(E_{F129}-E_{17.9})}\right] \tag{21.74}$$

and

$$\Delta C_{456}=\frac{G_0^3}{27}\left[\frac{\partial^2 n_0}{\partial(E_{F129}-E_{17.9})^2}\right] \tag{21.75}$$

Using (21.73)–(21.75), we can find ΔC_{44} and ΔC_{456} in this case.

(c) The magneto DR in QWs of HD III–V, ternary and quaternary materials, whose unperturbed band structure is defined by the parabolic energy bands in the presence of light waves, can be expressed from (21.48) as

$$\left(n+\frac{1}{2}\right)\omega_0+\frac{\hbar^2}{2m_c}\left(\frac{n_z\pi}{d_z}\right)^2=T_3(E_{17,10},\eta_g,\lambda) \tag{21.76}$$

where $E_{17,10}$ is the total quantized energy in this case.

The DOS function is given by

$$N_{\mathrm{0DT}}(E) = \frac{eBg_v}{\pi\hbar d_z} \sum_{n_z=1}^{n_{z\max}} \sum_{n=0}^{n_{\max}} \delta'(E - E_{17,10}) \tag{21.77}$$

The electron concentration can be written as

$$n_0 = \frac{eBg_v}{\pi\hbar d_z} \sum_{n=0}^{n_{\max}} \sum_{n_z=1}^{n_{z\max}} F_{-1}(\eta_{1210}) \tag{21.78}$$

where

$$\eta_{1210} = \frac{E_{F1210} - E_{17.10}}{k_B T} \quad \text{and} \quad E_{F1210}$$

is the Fermi energy in this case.

The electronic contribution to the second- and third-order elastic constants for HD materials in this case can be written as

$$\Delta C_{44} = \frac{-G_0^2}{9} \left[\frac{\partial n_0}{\partial(E_{F1210} - E_{17.10})} \right] \tag{21.79}$$

and

$$\Delta C_{456} = \frac{G_0^3}{27} \left[\frac{\partial^2 n_0}{\partial(E_{F1210} - E_{17.10})^2} \right] \tag{21.80}$$

Using (21.78)–(21.80), we can find ΔC_{44} and ΔC_{456} in this case.

*21.2.4 The magneto MEC in accumulation and inversion layers of Kane type materials in the presence of light waves

(a) The 2D DR in accumulation layers of HD III–V, ternary and quaternary materials, whose unperturbed band structure is defined by the three band model of Kane, in the presence of light waves under magnetic quantization, can be expressed as

$$T_1(E_{17,30}, \eta_g, \lambda) = \left(n + \frac{1}{2} =\right) \hbar\omega_0 + S_i \left[\frac{\hbar|e|F_s[T_1(E_{17,30}, \eta_g, \lambda)]'}{\sqrt{2m_c}} \right]^{\frac{2}{3}} \tag{21.81}$$

where $E_{17,30}$ is the total quantized energy in this case. Equation (21.81) represents the magneto DR of the 2D electrons in accumulation layers of HD III–V, ternary and quaternary materials under the weak electric field limit in the presence of light waves whose bulk electrons obey the HD three band model of Kane. Since the DR in accordance with the HD three-band model of Kane is complex in nature, Equation (21.81) will also be complex in the energy plane. The total energy is quantized since the wave vector space is totally quantized.

The DOS function is given by

$$N_{2\mathrm{DALB}}(E) = \frac{g_v eB}{\pi\hbar} \sum_{n=0}^{n_{\max}} \sum_{i=0}^{i_{\max}} \delta'(E - E_{17,30}) \tag{21.82}$$

The electron concentration can be written as

$$n_0 = \text{ Real part of } \bar{n}_0 \tag{21.83}$$

where

$$\bar{n}_0 = \frac{g_v eB}{\pi\hbar} \sum_{n=0}^{n_{\max}} \sum_{i=0}^{i_{\max}} F_{-1}(\eta_{17,30}) \tag{21.84}$$

in which

$$\eta_{17,30} = \frac{E_{F17,30} - E_{17,30}}{k_B T} \quad \text{and} \quad E_{F17,30}$$

is the Fermi energy in this case.

The electronic contribution to the second- and third-order elastic constants for HD materials in this case can be written as

$$\Delta C_{44} = \frac{-G_0^2}{9} \text{ Real part of } \left[\frac{\partial n_0}{\partial(E_{F17,30} - E_{17,30})}\right] \tag{21.85}$$

and

$$\Delta C_{456} = \frac{G_0^3}{27} \text{ Real part of } \left[\frac{\partial^2 n_0}{\partial(E_{F17,30} - E_{17,30})^2}\right] \tag{21.86}$$

Thus, using (21.83), (21.85) and (21.86), we can find ΔC_{44} and ΔC_{456} in this case where n_0 is replaced by $(\bar{n}_0/t_i)$.

(b) The 2D DR in accumulation layers of HD III–V, ternary and quaternary materials, whose unperturbed band structure is defined by the two band model of Kane, in the presence of light waves under magnetic quantization, can be expressed as

$$T_2(E_{17,32}, \eta_g, \lambda) = \left(n + \frac{1}{2}\right)\hbar\omega_0 + S_i\left[\frac{\hbar|e|F_s\,[T_2(E_{17,32}, \eta_g, \lambda)]'}{\sqrt{2m_c}}\right]^{\frac{2}{3}} \tag{21.87}$$

where $E_{17,32}$ is the total quantized energy in this case.

Equation (21.87) represents the magneto DR of the 2D electrons in accumulation layers of HD III–V, ternary and quaternary materials under the weak electric field limit in the presence of light waves whose bulk electrons obey the HD two band model of Kane. Since the DR in accordance with the HD two-band model of Kane is real in nature, Equation (21.87) will also be real in the energy plane. The total energy is quantized since the wave vector space is totally quantized.

The DOS function is given by

$$N_{2\mathrm{DALB}}(E) = \frac{g_v eB}{\pi\hbar}\sum_{n=0}^{n_{\max}}\sum_{i=0}^{i_{\max}}\delta'(E - E_{17,32}) \tag{21.88}$$

The electron concentration can be written as

$$n_0 = \frac{g_v eB}{\pi\hbar}\sum_{n=0}^{n_{\max}}\sum_{i=0}^{i_{\max}}F_{-1}(\eta_{17,32}) \tag{21.89}$$

where

$$\eta_{17,32} = \frac{E_{F17,32} - E_{17,32}}{k_B T} \quad \text{and} \quad E_{F17,32}$$

is the Fermi energy in this case.

The electronic contribution to the second- and third-order elastic constants for HD materials in this case can be written as

$$\Delta C_{44} = \frac{-G_0^2}{9}\left[\frac{\partial n_0}{\partial(E_{F17,32} - E_{17,32})}\right] \tag{21.90}$$

and

$$\Delta C_{456} = \frac{G_0^3}{27}\left[\frac{\partial^2 n_0}{\partial(E_{F17,32} - E_{17,32})^2}\right] \tag{21.91}$$

Thus, using (21.89)–(21.91), we can find ΔC_{44} and ΔC_{456} in this case where n_0 is replaced by (n_0/t_i).

(c) The 2D DR in accumulation layers of HD III–V, ternary and quaternary materials, whose unperturbed band structure is defined by the parabolic energy bands, in the presence of light waves under magnetic quantization, can be expressed as

$$T_3(E_{17,323}, \eta_g, \lambda) = \left(n + \frac{1}{2}\right)\hbar\omega_0 + S_i\left[\frac{\hbar|e|F_s[T_3(E_{17,323}, \eta_g, \lambda)]'}{\sqrt{2m_c}}\right]^{\frac{2}{3}} \tag{21.92}$$

where $E_{17,323}$ is the total quantized energy in this case.

The DOS function is given by

$$N_{2\text{DALB}}(E) = \frac{g_v eB}{\pi\hbar}\sum_{n=0}^{n_{\max}}\sum_{i=0}^{i_{\max}}\delta'(E - E_{17,323}) \tag{21.93}$$

The electron concentration can be written

$$n_0 = \frac{g_v eB}{\pi\hbar}\sum_{n=0}^{n_{\max}}\sum_{i=0}^{i_{\max}}F_{-1}(\eta_{17,323}) \tag{21.94}$$

where

$$\eta_{17,323} = \frac{E_{F17,323} - E_{17,323}}{k_B T} \quad \text{and} \quad E_{F17,323}$$

is the Fermi energy in this case.

The electronic contribution to the second- and third-order elastic constants for HD materials in this case can be written as

$$\Delta C_{44} = \frac{-G_0^2}{9}\left[\frac{\partial n_0}{\partial(E_{F17,323} - E_{17,323})}\right] \tag{21.95}$$

and

$$\Delta C_{456} = \frac{G_0^3}{27}\left[\frac{\partial^2 n_0}{\partial(E_{F17,323} - E_{17,323})^2}\right] \tag{21.96}$$

Thus, using (21.94)–(21.96), we can find ΔC_{44} and ΔC_{456} in this case where n_0 is replaced by (n_0/t_i).

*21.2.5 The magneto MEC in doping superlattices of HD Kane type materials in the presence of light waves

(a) The magneto DR in doping superlattices of HD III–V, ternary and quaternary materials in the presence of external photo-excitation whose unperturbed electrons are defined by the three band model of Kane can be expressed as

$$T_1(E_{17,40},\eta_g,\lambda)=\left(n_i+\frac{1}{2}\right)\hbar\omega_{91\text{HD}}(E_{17,40},\eta_g,\lambda)+\left(n+\frac{1}{2}\right)\hbar\omega_0 \tag{21.97}$$

where $E_{17,40}$ is the total energy in this case and

$$\omega_{91\text{HD}}(E_{17,40},\eta_g,\lambda)=\left[\frac{n_0e^2}{\varepsilon_{sc}T_1'(E_{17,40},\eta_g,\lambda)m_c}\right]^{1/2}$$

The DOS function is given by

$$N_{2\text{DALB}}(E)=\frac{g_v eB}{\pi\hbar d_0}\sum_{n=0}^{n_{\max}}\sum_{n_i=0}^{n_{i\max}}\delta'(E-E_{17,40}) \tag{21.98}$$

The electron concentration can be expressed as

$$n_0=\text{Real part of }\bar{n}_0 \tag{21.99}$$

where

$$\bar{n}_0=\frac{g_v eB}{\pi\hbar d_0}\sum_{n=0}^{n_{\max}}\sum_{n_i=0}^{n_{i\max}}F_{-1}(\eta_{17,40}) \tag{21.100}$$

in which

$$\eta_{17,40}=\frac{E_{F17,40}-E_{17,40}}{k_BT}\quad\text{and}\quad E_{F17,40}$$

is the Fermi energy in this case.

The electronic contribution to the second- and third-order elastic constants for HD materials in this case can be written as

$$\Delta C_{44}=\frac{-G_0^2}{9}\text{ Real part of }\left[\frac{\partial n_0}{\partial(E_{F17,40}-E_{17,40})}\right] \tag{21.101}$$

and

$$\Delta C_{456}=\frac{G_0^3}{27}\text{ Real part of }\left[\frac{\partial^2 n_0}{\partial(E_{F17,40}-E_{17,40})^2}\right] \tag{21.102}$$

Thus, using (21.100)–(21.102), we can find ΔC_{44} and ΔC_{456} in this case where n_0 is replaced by $\bar{n}_0$

(b) The magneto DR in doping superlattices of HD III–V, ternary and quaternary materials in the presence of external photo-excitation whose unperturbed electrons are defined by the two band model of Kane can be expressed as

$$T_2(E_{17,41}, \eta_g, \lambda) = \left(n_i + \frac{1}{2}\right) \hbar\omega_{92\mathrm{HD}}(E_{17,41}, \eta_g, \lambda) + \left(n + \frac{1}{2}\right) \hbar\omega_0 \tag{21.103}$$

where $E_{17,41}$ is the total energy in this case and

$$\omega_{92\mathrm{HD}}(E_{17,41}, \eta_g, \lambda) = \left[\frac{n_0|e^2|}{\varepsilon_{sc} T_2'(E_{17,41}, \eta_g, \lambda) m_c}\right]^{1/2}$$

The DOS function is given by

$$N_{2\mathrm{DALB}}(E) = \frac{g_v eB}{\pi\hbar d_0} \sum_{n=0}^{n_{\max}} \sum_{i=0}^{i_{\max}} \delta'(E - E_{17,41}) \tag{21.104}$$

The electron concentration can be expressed as

$$n_0 = \frac{g_v eB}{\pi\hbar d_0} \sum_{n=0}^{n_{\max}} \sum_{n_i=0}^{n_{i\max}} F_{-1}(\eta_{17,41}) \tag{21.105}$$

where

$$\eta_{17,41} = \frac{E_{F17,41} - E_{17,41}}{k_B T} \quad \text{and} \quad E_{F17,41}$$

is the Fermi energy in this case.

The electronic contribution to the second- and third-order elastic constants for HD materials in this case can be written as

$$\Delta C_{44} = \frac{-G_0^2}{9} \left[\frac{\partial n_0}{\partial(E_{F17,41} - E_{17,41})}\right] \tag{21.106}$$

and

$$\Delta C_{456} = \frac{G_0^3}{27} \left[\frac{\partial^2 n_0}{\partial(E_{F17,41} - E_{17,41})^2}\right] \tag{21.107}$$

Thus, using (21.105)–(21.107), we can find ΔC_{44} and ΔC_{456} in this case.

(c) The magneto DR in doping superlattices of HD III–V, ternary and quaternary materials in the presence of external photo-excitation whose unperturbed electrons are defined by the parabolic energy bands can be expressed as

$$T_3(E_{17,42},\eta_g,\lambda) = \left(n_i + \frac{1}{2}\right)\hbar\omega_{93\text{HD}}(E_{17,42},\eta_g,\lambda) + \left(n + \frac{1}{2}\right)\hbar\omega_0 \tag{21.108}$$

where $E_{17,42}$ is the total energy in this case and

$$\omega_{93\text{HD}}(E_{17,42},\eta_g,\lambda) = \left[\frac{n_0|e^2|}{\varepsilon_{sc}T_3'(E_{17,42},\eta_g,\lambda)m_c}\right]^{1/2}$$

The DOS function is given by

$$N_{2\text{DALB}}(E) = \frac{g_v eB}{\pi\hbar d_0}\sum_{n=0}^{n_{\max}}\sum_{i=0}^{i_{\max}}\delta'(E - E_{17,42}) \tag{21.109}$$

The electron concentration can be expressed as

$$n_0 = \frac{g_v eB}{\pi\hbar d_0}\sum_{n=0}^{n_{\max}}\sum_{i=0}^{i_{\max}}F_{-1}(\eta_{17,42}) \tag{21.110}$$

where

$$\eta_{17,42} = \frac{E_{F17,42} - E_{17,42}}{k_B T} \quad \text{and} \quad E_{F17,42}$$

is the Fermi energy in this case.

The electronic contribution to the second- and third-order elastic constants for HD materials in this case can be written as

$$\Delta C_{44} = \frac{-G_0^2}{9}\left[\frac{\partial n_0}{\partial(E_{F17,42} - E_{17,42})}\right] \tag{21.111}$$

and

$$\Delta C_{456} = \frac{G_0^3}{27}\left[\frac{\partial^2 n_0}{\partial(E_{F17,42} - E_{17,42})^2}\right] \tag{21.112}$$

Thus, using (21.110)–(21.112), we can find ΔC_{44} and ΔC_{456} in this case.

*21.2.6 The magneto MEC in QWHD effective mass superlattices of Kane type materials in the presence of light waves

(a) In the presence of an external magnetic field along x-direction, the simplified magneto dispersion law in QWHD III–V effective mass superlattices (EMSLs) in the presence of light waves, the dispersion relations of whose constituent materials in the absence of any perturbation are defined by the three band model of Kane can be written as

$$\left(\frac{n_x\pi}{d_x}\right)^2 = [\rho_{4\mathrm{HGD}1}(n,E,\lambda)]\,|_{E=E_{17,50}} \tag{21.113}$$

where

$$\rho_{4\mathrm{HD}1}(n,E,\lambda) = \left[\frac{1}{L_0^2}[\cos^{-1}(f_{\mathrm{HD}1}(n,E,\lambda))]^2 - \left\{\frac{2|e|B}{\hbar}\left(n+\frac{1}{2}\right)\right\}\right],$$

$$f_{\mathrm{HD}1}(n,E,\lambda) = a_{1\mathrm{HD}1}\cos[a_0C_{1\mathrm{HD}1}(E,n,\lambda) + b_0D_{1\mathrm{HD}1}(E,n,\lambda)]$$
$$- a_{2\mathrm{HD}1}\cos[a_0C_{1\mathrm{HD}1}(E,n,\lambda) + b_0D_{1\mathrm{HD}1}(E,n,\lambda)],$$

$$C_{1\mathrm{HD}1}(E,n,\lambda) \equiv \left[\left(\frac{2m_{c1}}{\hbar^2}\right)T_1(E,\eta_{g1},\lambda) - \left\{\frac{2|e|B}{\hbar}\left(n+\frac{1}{2}\right)\right\}\right]^{1/2},$$

$$D_{1\mathrm{HD}1}(E,n,\lambda) \equiv \left[\left(\frac{2m_{c2}}{\hbar^2}\right)T_1(E,\eta_{g2},\lambda) - \left\{\frac{2|e|B}{\hbar}\left(n+\frac{1}{2}\right)\right\}\right]^{1/2},$$

$$a_{1\mathrm{HD}_1} = \left[\sqrt{\frac{m_{c2}\text{ Real part of }[T_1(0,\eta_{g2},\lambda)]}{m_{c1}\text{ Real part of }[T_1(0,\eta_{g1},\lambda)]}}+1\right]$$
$$\times\left[4\left(\sqrt{\frac{m_{c2}\text{ Real part of }[T_1(0,\eta_{g2},\lambda)]}{m_{c1}\text{ Real part of }[T_1(0,\eta_{g1},\lambda)]}}\right)^{1/2}\right]^{-1},$$

$$a_{2\mathrm{HD}_1} = \left[-1+\sqrt{\frac{m_{c2}\text{ Real part of }[T_1(0,\eta_{g2},\lambda)]}{m_{c1}\text{ Real part of }[T_1(0,\eta_{g1},\lambda)]}}\right]$$
$$\times\left[4\left(\sqrt{\frac{m_{c2}\text{ Real part of }[T_1(0,\eta_{g2},\lambda)]}{m_{c1}\text{ Real part of }[T_1(0,\eta_{g1},\lambda)]}}\right)^{1/2}\right]^{-1}$$

and $E_{17.50}$ is the total quantized energy in this case.

The DOS function is given by

$$N_{\mathrm{BQWHDEMSL}}(E) = \frac{g_v eB}{\pi\hbar d_x} \sum_{n=0}^{n_{\max}} \sum_{n_x=0}^{n_{x\max}} \delta'(E - E_{17.50}) \tag{21.114}$$

The electron concentration can be written as

$$n_0 = \text{ Real part of } \bar{n}_0 \tag{21.115}$$

$$\bar{n}_0 = \frac{g_v eB}{\pi\hbar d_x} \sum_{n=0}^{n_{\max}} \sum_{n_x=0}^{n_{x\max}} F_{-1}(\eta_{17.50}) \tag{21.116}$$

in which

$$\eta_{17.50} = \frac{E_{F17.50} - E_{17.50}}{k_B \mathrm{T}} \quad \text{and} \quad E_{F17.50}$$

is the Fermi energy in this case.

The electronic contribution to the second- and third-order elastic constants for HD materials in this case can be written as

$$\Delta C_{44} = \frac{-G_0^2}{9} \text{ Real part of } \left[\frac{\partial n_0}{\partial(E_{F17.50} - E_{17.50})}\right] \tag{21.117}$$

and

$$\Delta C_{456} = \frac{G_0^3}{27} \text{ Real part of } \left[\frac{\partial^2 n_0}{\partial(E_{F17.50} - E_{17.50})^2}\right] \tag{21.118}$$

Thus, using (21.116)–(21.118), we can find ΔC_{44} and ΔC_{456} in this case where n_0 is replaced by $\bar{n}_0$.

(b) In the presence of an external magnetic field along x-direction, the simplified magneto dispersion law in QWHD III–V effective mass superlattices (EMSLs) in the presence of light waves, the dispersion relations of whose constituent materials in the absence of any perturbation are defined by the two band model of Kane can be written as

$$\left(\frac{n_x \pi}{d_x}\right)^2 = [\rho_{4\mathrm{HGD2}}(n, E, \lambda)]|_{E=E_{17,51}} \tag{21.119}$$

where

$$\rho_{4\mathrm{HD}2}(n, E, \lambda) = \left[\frac{1}{L_0^2}[\cos^{-1}(f_{\mathrm{HD}2}(n, E, \lambda))]^2 - \left\{\frac{2|e|B}{\hbar}\left(n + \frac{1}{2}\right)\right\}\right],$$

$$f_{\mathrm{HD}2}(n, E, \lambda) = a_{1\mathrm{HD}2}\cos[a_0 C_{1\mathrm{HD}1}(E, n, \lambda) + b_0 D_{1\mathrm{HD}2}(E, n, \lambda)] - a_{2\mathrm{HD}2}\cos[a_0 C_{1\mathrm{HD}2}(E, n, \lambda) + b_0 D_{1\mathrm{HD}2}(E, n, \lambda)],$$

$$C_{1\mathrm{HD}2}(E, n, \lambda) \equiv \left[\left(\frac{2m_{c1}}{\hbar^2}\right) T_2(E, \eta_{g2}, \lambda) - \left\{\frac{2|e|B}{\hbar}\left(n + \frac{1}{2}\right)\right\}\right]^{1/2},$$

$$D_{1\mathrm{HD}2}(E, n, \lambda) \equiv \left[\left(\frac{2m_{c2}}{\hbar^2}\right) T_2(E, \eta_{g2}, \lambda) - \left\{\frac{2|e|B}{\hbar}\left(n + \frac{1}{2}\right)\right\}\right]^{1/2},$$

$$a_{1\mathrm{HD}_2} = \left[\sqrt{\frac{m_{c2}[T_2(0, \eta_{g2}, \lambda)]}{m_{c1}[T_2(0, \eta_{g1}, \lambda)]}} + 1\right]\left[4\left(\sqrt{\frac{m_{c2}[T_2(0, \eta_{g2}, \lambda)]}{m_{c1}[T_2(0, \eta_{g1}, \lambda)]}}\right)^{1/2}\right]^{-1},$$

$$a_{2\mathrm{HD}_2} = \left[-1 + \sqrt{\frac{m_{c2}[T_2(0, \eta_{g2}, \lambda)]}{m_{c1}[T_2(0, \eta_{g1}, \lambda)]}}\right]\left[4(\sqrt{\frac{m_{c2}[T_2(0, \eta_{g2}, \lambda)]}{m_{c1}[T_2(0, \eta_{g1}, \lambda)]}})^{1/2}\right]^{-1}$$

and $E_{17.51}$ is the total quantized energy in this case.

The DOS function is given by

$$N_{\mathrm{BQWHDEMSL}}(E) = \frac{g_v eB}{\pi\hbar d_x}\sum_{n=0}^{n_{\max}}\sum_{n_x=0}^{n_{x\max}} \delta'(E - E_{17.51}) \tag{21.120}$$

The electron concentration can be written as

$$n_0 = \frac{g_v eB}{\pi\hbar d_x}\sum_{n=0}^{n_{\max}}\sum_{n_x=0}^{n_{x\max}} F_{-1}(\eta_{17.51}) \tag{21.121}$$

where

$$\eta_{17.51} = \frac{E_{F17.51} - E_{17.51}}{k_B T} \quad \text{and} \quad E_{F17.51}$$

is the Fermi energy in this case.

The electronic contribution to the second- and third-order elastic constants for HD materials in this case can be written as

$$\Delta C_{44} = \frac{-G_0^2}{9}\frac{\partial n_0}{\partial(E_{F17.51} - E_{17.51})}\Bigg] \tag{21.122}$$

and

$$\Delta C_{456} = \frac{G_0^3}{27}\left[\frac{\partial^2 n_0}{\partial(E_{F17.51} - E_{17.51})^2}\right] \tag{21.123}$$

Thus, using (21.121)–(21.123), we can find ΔC_{44} and ΔC_{456} in this case.

(c) In the presence of an external magnetic field along x-direction, the simplified magneto dispersion law in QWHD III–V effective mass superlattices (EMSLs) in the presence of light waves, the dispersion relations of whose constituent materials in the absence of any perturbation are defined by the isotropic parabolic energy bands can be written as

$$\left(\frac{n_x \pi}{d_x}\right)^2 = [\rho_{4\mathrm{HGD3}}(n, E, \lambda)]\,|_{E=E_{17,52}} \tag{21.124}$$

$$\rho_{4\mathrm{HD3}}(n, E, \lambda) = \left[\frac{1}{L_0^2}[\cos^{-1}(f_{\mathrm{HD3}}(n, E, \lambda))]^2 - \left\{\frac{2|e|B}{\hbar}\left(n + \frac{1}{2}\right)\right\}\right],$$

$$f_{\mathrm{HD3}}(n, E, \lambda) = a_{1\mathrm{HD3}} \cos[a_0 C_{1\mathrm{HD1}}(E, n, \lambda) + b_0 D_{1\mathrm{HD3}}(E, n, \lambda)] - a_{2\mathrm{HD3}} \cos[a_0 C_{1\mathrm{HD3}}(E, n, \lambda) + b_0 D_{1\mathrm{HD3}}(E, n, \lambda)],$$

$$C_{1\mathrm{HD3}}(E, n, \lambda) \equiv \left[\left(\frac{2m_{c1}}{\hbar^2}\right) T_3(E, \eta_{g1}, \lambda) - \left\{\frac{2|e|B}{\hbar}\left(n + \frac{1}{2}\right)\right\}\right]^{1/2},$$

$$D_{1\mathrm{HD3}}(E, n, \lambda) \equiv \left[\left(\frac{2m_{c2}}{\hbar^2}\right) T_3(E, \eta_{g2}, \lambda) - \left\{\frac{2|e|B}{\hbar}\left(n + \frac{1}{2}\right)\right\}\right]^{1/2},$$

$$a_{1\mathrm{HD}_3} = \left[\sqrt{\frac{m_{c2}[T_3(0, \eta_{g2}, \lambda)]}{m_{c1}[T_3(0, \eta_{g1}, \lambda)]}} + 1\right]\left[4\left(\sqrt{\frac{m_{c2}[T_3(0, \eta_{g2}, \lambda)]}{m_{c1}[T_3(0, \eta_{g1}, \lambda)]}}\right)^{1/2}\right]^{-1},$$

$$a_{2\mathrm{HD}_3} = \left[-1 + \sqrt{\frac{m_{c2}[T_3(0, \eta_{g2}, \lambda)]}{m_{c1}[T_3(0, \eta_{g1}, \lambda)]}}\right]\left[4\left(\sqrt{\frac{m_{c2}[T_3(0, \eta_{g2}, \lambda)]}{m_{c1}[T_3(0, \eta_{g1}, \lambda)]}}\right)^{1/2}\right]^{-1}$$

and $E_{17.52}$ is the total quantized energy in this case.

The DOS function is given by

$$N_{\mathrm{BQWHDEMSL}}(E) = \frac{g_v eB}{\pi \hbar d_x} \sum_{n=0}^{n_{\max}} \sum_{n_x=1}^{n_{x\max}} \delta'(E - E_{17,52}) \tag{21.125}$$

The electron concentration can be written as

$$n_0 = \frac{g_v eB}{\pi \hbar d_x} \sum_{n=0}^{n_{\max}} \sum_{n_x=0}^{n_{x\max}} F_{-1}(\eta_{17.52}) \tag{21.126}$$

where

$$\eta_{17.52} = \frac{E_{F17.52} - E_{17.52}}{k_B T} \quad \text{and} \quad E_{F17.52}$$

is the Fermi energy in this case.

The electronic contribution to the second- and third-order elastic constants for HD materials in this case can be written as

$$\Delta C_{44} = \frac{-G_0^2}{9}\left[\frac{\partial n_0}{\partial(E_{F17.52} - E_{17.52})}\right] \tag{21.127}$$

and

$$\Delta C_{456} = \frac{G_0^3}{27}\left[\frac{\partial^2 n_0}{\partial(E_{F17.52} - E_{17.52})^2}\right] \tag{21.128}$$

Thus, using (21.126)–(21.128), we can find ΔC_{44} and ΔC_{456} in this case.

*21.2.7 The magneto MEC in quantum dot HD superlattices of Kane type materials with graded interfaces in the presence of light waves

The electron dispersion law in bulk specimens of the heavily doped constituent materials of III–V SLs in the presence of light waves can be expressed as

$$\frac{\hbar^2 k^2}{2m_{cj}} = V_{3j}(E, \eta_{gj}, \lambda, \Delta_j, E_{g0j}) + iV_{3j}(E, \eta_{gj}, \lambda, \Delta_j, E_{g0j}) \tag{21.129}$$

where

$$j = 1, 2, V_{3j}(E, \eta_{gj}, \lambda, \Delta_j, E_{g0j}) = [U_{\lambda j} T_{3j}(E, \Delta_j, E_{g0j}, \eta_{gj}) - P_{\lambda j}]$$

$$U_{\lambda j} = (1 + \theta_{\lambda j}), \quad \theta_{\lambda j} = \frac{C_{0j}}{A_j}\left(t_{\lambda j} + \frac{B_j J_{\lambda j}}{A_j}\right),$$

$$C_{0j} = \left[\frac{|e|^2 I \lambda^2 E_{g0j}(E_{g0j} + \Delta_j)\beta_j^2\left(1 + \frac{\rho_j}{\sqrt{2}}\right)^2}{384 m_{rj}\pi c^3\sqrt{\varepsilon_{scj}\varepsilon_0}(E_{g0j} + \frac{2}{3}\Delta_j)}\right],$$

$$\beta_j = \left[6\left(E_{g0j} + \frac{2}{3}\Delta_j\right)\frac{(E_{g0j} + \Delta_j)}{\chi_j}\right]^{1/2},$$

$$\chi_j = (6E_{g0}^2 + 9E_{g0j}\Delta_j + 4\Delta_j^2), \quad \rho_j = \left[\frac{4\Delta_j^2}{3\chi_j}\right]^{1/2},$$

$$A_j = E_{g0j}, \quad t_{\lambda j} = \left[E_{\lambda j} - \frac{G_{\lambda j}(E_{g0j} - \delta_j')}{\sqrt{t_{\lambda j}}}\right],$$

$$E_{\lambda j} = \frac{2B_j(E_{g0j} - \delta'_j)}{(A_j + \delta'_j)^2}, \quad B_j = \left[1 + \frac{m_{cj}}{m_{vj}}\right],$$

$$G_{\lambda j} = \left[\frac{2B_j}{(A_j + \delta'_j)^3} - \frac{B_j C_{\lambda j}}{(A_j + \delta'_j)}\right],$$

$$C_{\lambda j} = [(E_{g0j} + \delta'_j)^{-1} + (E_{g0j} + \delta'_j)(E_{g0j} - \delta'_j)^{-2}]\ (A_j + \delta'_j)^{-1},$$

$$P_{\lambda j} = \frac{C_{0j}}{A_j} J_{\lambda j}, \quad J_{\lambda j} = (D_{\lambda j} + 2(E_{g0j} - \delta'_j)\sqrt{f_{\lambda j}}),$$

$$D_{\lambda j} = \left(1 + \frac{2(E_{g0j} - \delta'_j)}{(A_j + \delta'_j)}\right),$$

$$f_{\lambda j} = \left[\frac{1}{(A_j + \delta'_j)^2} + \frac{1}{(E_{g0j} - \delta'_j)^2} - C_{\lambda j}\right],$$

$$T_{3j}(E, \Delta_j, E_{g0j}, \eta_{gj}) = \Big(2/(1 + Erf(E/\eta_{gj}))\Big[(\alpha_j b_j/c_j)\Big]\Big)$$
$$\times\, \theta_0(E, \eta_{gj}) + [\alpha_j c_j + b_j c_j - \alpha_j b_j/c_j^2].$$

$$[\gamma_0(E, \eta_{gj})] + \Bigg[(1/c_j)(1 - (\alpha_j/c_j))(1 - b_j/c_j)\frac{1}{2}[1 + Erf(E/\eta_{gj})]$$
$$- (1/c_j)(1 - (\alpha_j/c_j))(1 - b_j/c_j)(2/c_j\eta_{gj}\sqrt{\pi})\exp(-u_j^2)$$
$$\times \left[\sum_{p=1}^{\infty}(\exp(-p^2/4)/p)\sinh(pu_j)\right]\Bigg],$$

$$b_j \equiv (E_{g0j} + \Delta_j)^{-1}, \quad c_j \equiv (E_{g0j} + \frac{2}{3}\Delta_j)^{-1}, \quad u_j \equiv (1 + c_j E)(c_j \eta_{gj})^{-1},$$

$$V_{3j}(E, \eta_{gj}, \lambda, \Delta_j, E_{g0j}) = [U_{\lambda j} T_{3j}(E, \Delta_j, E_{g0j}, \eta_{gj})] \quad \text{and}$$

$$T_{3j}(E, \Delta_j, E_{g0j}, \eta_{gj}) \equiv \left(\frac{2}{(1 + Erf(E/\eta_{gj})}\right)\frac{1}{c_j}\left(1 - \frac{\alpha_j}{c_j}\right)$$
$$\times \left(1 - \frac{b_j}{c_j}\right)\frac{\sqrt{\pi}}{c_j \eta_{gj}}\exp(-u_j^2)$$

The DR in QDHD superlattices of Kane type materials with graded interfaces in the presence of light waves can be expressed as

$$\left(\frac{n_z \pi}{d_z}\right)^2 = [G_{8,17,50} + iH_{8,17,50}]|_{E=E_{17,53}} \tag{21.130}$$

where $E_{17,53}$ is the total quantized energy in this case,

$$G_{8,17,50} = \left[\frac{C_{7,50}^2 - D_{7,50}^2}{L_0^2} - \left[\left(\frac{n_x\pi}{d_x}\right)^2 + \left(\frac{n_y\pi}{d_y}\right)^2\right]\right],$$

$$C_{7,50} = \cos^{-1}(\bar{\omega}_{7,50}),$$

$$\bar{\omega}_{7,50} = (2)^{\frac{-1}{2}}[(1 - G_{7,50}^2 - H_{7,50}^2) - \sqrt{(1 - G_{7,50}^2 - H_{7,50}^2)^2 + 4G_{7,50}^2}]^{\frac{1}{2}},$$

$$G_{7,50} = [G_{1,50} + (\rho_{5,,50}G_{2,,50}/2) - (\rho_{6,50}H_{2,50}/2) + (\Delta_0/2)$$
$$\{\rho_{7,50}G_{3,50} - \rho_{8,50}H_{3,50} + \rho_{9,50}G_{4,50} - \rho_{10,50}H_{4,50}$$
$$+ \rho_{11,50}H_{5,50} - \rho_{12,50}H_{5,50} + (1/12)(\rho_{13,50}G_{6,50} - \rho_{14,50}H_{6,50})\}],$$

$$G_{1,50} = [(\cos(h_{1,50}))(\cosh(h_{2,50}))(\cosh(g_{1,50}))(\cos(g_{2,50}))$$
$$+ (\sin(h_{1,50}))(\sinh(h_{2,50}))(\sinh(g_{1,50}))(\sin(g_{2,50}))],$$

$$h_{1,50} = e_{1,50}(b_0 - \Delta_0), \quad e_{1,50} = 2^{\frac{-1}{2}}\left(\sqrt{t_{1,50}^2 + t_{2,50}^2} + t_{1,50}\right)^{\frac{1}{2}},$$

$$t_{1,50} = \left[(2m_{c1}/\hbar^2)\cdot[U_{\lambda_1}T_{31}(E, E_{g1}, \Delta_1, \eta_{g1}) - P_\lambda]\right.$$
$$\left.- \left[\left(\frac{n_x\pi}{d_x}\right)^2 + \left(\frac{n_y\pi}{d_y}\right)^2\right]\right],$$

$$t_{2,50} = [(2m_{c1}/\hbar^2)[U_{\lambda_1}T_{32}(E, E_{g1}, \Delta_1, \eta_{g1})]],$$

$$h_{2,50} = e_{2,50}(b_0 - \Delta_0), \quad e_{2,50} = 2^{\frac{-1}{2}}\left(\sqrt{t_{1,50}^2 + t_{2,50}^2} - t_{1,50}\right)^{\frac{1}{2}},$$

$$g_{1,50} = d_{1,50}(a_0 - \Delta_0), \quad d_{1,50} = 2^{\frac{-1}{2}}\left(\sqrt{x_{1,50}^2 + y_{1,50}^2} + x_{1,50}\right)^{\frac{1}{2}},$$

$$x_{1,50} = \left[(-2m_{c2}/\hbar^2)\cdot[U_{\lambda_2}T_{31}(E - V_0, E_{g2}, \Delta_2, \eta_{g2}) - P_{\lambda 2}]\right.$$
$$\left.+ \left[\left(\frac{n_x\pi}{d_x}\right)^2 + \left(\frac{n_y\pi}{d_y}\right)^2\right]\right],$$

$$y_{1,50} = [(2m_{c2}/\hbar^2)\cdot[U_{\lambda_2}T_{32}(E - V_0, E_{g2}, \Delta_2, \eta_{g2})]],$$

$$g_{2,50} = d_{2,50}(a_0 - \Delta_0), \quad d_{2,50} = 2^{\frac{-1}{2}}\left(\sqrt{x_{1,50}^2 + y_{1,50}^2} - x_{1,50}\right)^{\frac{1}{2}},$$

$$\rho_{5,50} = (\rho_{3,50}^2 + \rho_{4,50}^2)^{-1}[\rho_{1,50}\rho_{3,50} - \rho_{2,50}\rho_{4,50}],$$

$$\rho_{1,50} = [d_{1,50}^2 + e_{2,50}^2 - d_{2,50}^2 - e_{1,50}^2], \quad \rho_{3,50} = [d_{1,50}e_{1,50} + d_{2,50}e_{2,50}],$$

$$
\begin{aligned}
\rho_{2,50} &= 2[d_{1,50}d_{2,50} + e_{1,50}e_{2,50}],\\
\rho_{4,50} &= [d_{1,50}e_{2,50} - e_{1,50}d_{2,50}],\\
G_{2,50} &= [(\sin(h_{1,50}))(\cosh(h_{2,50}))(\sinh(g_{1,50}))(\cos(g_{2,50}))\\
&\quad + (\cos(h_{1,50}))(\sinh(h_{2,50}))(\cosh(g_{1,50}))(\sin(g_{2,50}))],\\
\rho_{6,50} &= (\rho_{3,50}^2 + \rho_{4,50}^2)^{-1}[\rho_{1,50}\rho_{4,50} + \rho_{2,50}\rho_{3,50}],\\
H_{2,50} &= [(\sin(h_{1,50}))(\cosh(h_{2,50}))(\sinh(g_{2,50}))(\cos(g_{1,50}))\\
&\quad - (\cos(h_{1,50}))(\sinh(h_{2,50}))(\sinh(g_{1,50}))(\cos(g_{2,50}))],\\
\rho_{7,50} &= [(e_{1,50}^2 + e_{2,50}^2)^{-1}[e_{1,50}(d_{1,50}^2 - d_{2,50}^2) - 2d_{1,50}d_{2,50}e_{2,50}] - 3e_{1,50}],\\
G_{3,50} &= [(\sin(h_{1,50}))(\cosh(h_{2,50}))(\cosh(g_{1,50}))(\cos(g_{2,50}))\\
&\quad + (\cos(h_{1,50}))(\sinh(h_{2,50}))(\sinh(g_{1,50}))(\sin(g_{2,50}))],\\
\rho_{8,50} &= [(e_{1,50}^2 + e_{2,50}^2)^{-1}[e_{2,50}(d_{1,50}^2 - d_{2,50}^2) + 2d_{1,50}d_{2,50}e_{1,50}] + 3e_{2,50}],\\
H_{3,50} &= [(\sin(h_{1,50}))(\cosh(h_{2,50}))(\sin(g_{2,50}))(\sinh(g_{1,50}))\\
&\quad - (\cos(h_{1,50}))(\sinh(h_{2,50}))(\cosh(g_{1,50}))(\cos(g_{2,50}))],\\
\rho_{9,50} &= [(d_{1,50}^2 + d_{2,50}^2)^{-1}[d_{1,50}(e_{2,50}^2 - e_{1,50}^2) + 2e_{2,50}d_{1,50}e_{1,50}] + 3d_{1,50}],\\
G_{4,50} &= [(\cos(h_{1,50}))(\cosh(h_{2,50}))(\cos(g_{2,50}))(\sinh(g_{1,50}))\\
&\quad - (\sin(h_{1,50}))(\sinh(h_{2,50}))(\cosh(g_{1,50}))(\sin(g_{2,50}))],\\
\rho_{10,50} &= [-(d_{1,50}^2 + d_{2,50}^2)^{-1}[d_{2,50}(-e_{2,50}^2 + e_{1,50}^2) + 2e_{2,50}d_{2,50}\rho_{1,50}]\\
&\quad + 3d_{2,50}],\\
H_{4,50} &= [(\cos(h_{1,50}))(\cosh(h_{2,50}))(\cosh(g_{1,50}))(\sin(g_{2,50}))\\
&\quad + (\sin(h_{1,50}))(\sinh(h_{2,50}))(\sinh(g_{1,50}))(\cos(g_{2,50}))],\\
\rho_{11,50} &= 2[d_{1,50}^2 + e_{2,50}^2 - d_{2,50}^2 - e_{1,50}^2],\\
G_{5,50} &= [(\cos(h_{1,50}))(\cosh(h_{2,50}))(\cos(g_{2,50}))(\cosh(g_{1,50}))\\
&\quad - (\sin(h_{1,50}))(\sinh(h_{2,50}))(\sinh(g_{1,50}))(\sin(g_{2,50}))],\\
\rho_{12,50} &= 4[d_{1,50}d_{2,50} + e_{1,50}e_{2,50}],\\
H_{5,50} &= [(\cos(h_{1,50}))(\cosh(h_{2,50}))(\sinh(g_{1,50}))(\sin(g_{2,50}))\\
&\quad + (\sin(h_{1,50}))(\sinh(h_{2,50}))(\cosh(g_{1,50}))(\cos(g_{2,50}))],\\
\rho_{13,50} &= [\{5(d_{1,50}e_{1,50}^3 - 3e_{1,50}e_{2,50}^2d_{1,50}) + 5d_{2,50}(e_{1,50}^3 - 3e_{1,50}^2e_{2,50})\}\\
&\quad \times (d_{1,50}^2 + d_{2,50}^2)^{-1} + (e_{1,50}^2 + e_{2,50}^2)^{-1}\{5(e_{1,50}d_{1,50}^3
\end{aligned}
$$

$$
\begin{aligned}
&- 3d_{2,50}e_{1,50}^2 d_{1,50}) + 5(d_{2,50}^3 e_{2,50} - 3d_{1,50}^2 d_{2,50}e_{2,50})\} \\
&-34(d_{1,50}e_{1,50} + d_{2,50}e_{2,50})],
\end{aligned}
$$

$$
\begin{aligned}
G_{6,50} &= [(\sin(h_{1,50}))(\cosh(h_{2,50}))(\sinh(g_{1,50}))(\cos(g_{2,50})) \\
&\quad + (\cos(h_{1,50}))(\sinh(h_{2,50}))(\cosh(g_{1,50}))(\sin(g_{2,50}))],
\end{aligned}
$$

$$
\begin{aligned}
\rho_{14,50} &= [\{5(d_{1,50}e_{2,50}^3 - 3e_{2,50}e_{1,50}^2 d_{1,50}) + 5d_{2,50}(-e_{1,50}^3 + 3e_{2,50}^2 e_{1,50})\} \\
&\quad \times (d_{1,50}^2 + d_{2,50}^2)^{-1} + (e_{1,50}^2 + e_{2,50}^2)^{-1} \\
&\quad \times \{5(-e_{1,50}d_{2,50}^3 + 3d_{1,50}^2 d_{2,50}e_{1,50}) \\
&\quad + 5(-d_{1,50}^3 e_{2,50} + 3d_{2,50}^2 d_{1,50}e_{2,50})\} + 34(d_{1,50}e_{2,50} - d_{2,50}e_{1,50})],
\end{aligned}
$$

$$
\begin{aligned}
H_{6,50} &= [(\sin(h_{1,50}))(\cosh(h_{2,50}))(\cosh(g_{1,50}))(\sin(g_{2,50})) \\
&\quad - (\cos(h_{1,50}))(\sinh(h_{2,50}))(\sinh(g_{1,50}))(\cos(g_{2,50}))],
\end{aligned}
$$

$$
\begin{aligned}
H_{7,50} &= [H_{1,50} + (\rho_{5,50}H_{2,50}/2) + (\rho_{6,50}G_{2,50}/2) \\
&\quad + (\Delta_0/2)\{\rho_{8,50}G_{3,50} + \rho_{7,50}H_{3,50} + \rho_{10,50}G_{4,50} + \rho_{9,50}H_{4,50} \\
&\quad + \rho_{12,50}G_{5,50} + \rho_{11,50}H_{5,50} + (1/12)(\rho_{14,50}G_{6,50} + \rho_{13,50}H_{6,50})\}],
\end{aligned}
$$

$$
\begin{aligned}
H_{1,50} &= [(\sin(h_{1,50}))(\sinh(h_{2,50}))(\cosh(g_{1,50}))(\cos(g_{2,50})) \\
&\quad + (\cos(h_{1,50}))(\cosh(h_{2,50}))(\sinh(g_{1,50}))(\sin(g_{2,50}))],
\end{aligned}
$$

$$D_{7,50} = \sinh^{-1}(\bar{\omega}_{7,50}), \quad H_{8,50} = (2C_{7,50}D_{7,50}/L_0^2)$$

The DOS function is given by

$$N_{\text{QDHDSL}}(E, \eta_g, \lambda) = \frac{2g_v}{d_x d_y d_z} \sum_{n_x=1}^{n_{x\max}} \sum_{n_y=1}^{n_{y\max}} \sum_{n_z=1}^{n_{z\max}} \delta'(E - E_{17,53}) \quad (21.131)$$

The electron concentration can be expressed as

$$n_0 = \text{Real part of } \bar{n}_0 \quad (21.132)$$

where

$$\bar{n}_0 = \frac{2g_v}{d_x d_y d_z} \sum_{n_x=1}^{n_{x\max}} \sum_{n_y=1}^{n_{y\max}} \sum_{n_z=1}^{n_{z\max}} F_{-1}(\eta_{17,53}) \quad (21.133)$$

in which

$$\eta_{17,53} = \frac{E_{F17.53} - E_{17.53}}{k_B T} \quad \text{and} \quad E_{F17.53}$$

is the Fermi energy in this case.

The electronic contribution to the second- and third-order elastic constants for HD materials in this case can be written as

$$\Delta C_{44} = \frac{-G_0^2}{9} \text{ Real part of } \left[\frac{\partial n_0}{\partial(E_{F17.52} - E_{17.52})}\right] \tag{21.134}$$

and

$$\Delta C_{456} = \frac{G_0^3}{27} \text{ Real part of } \left[\frac{\partial^2 n_0}{\partial(E_{F17.52} - E_{17.52})^2}\right] \tag{21.135}$$

Thus, using (21.132), (21.134) and (21.135), we can find ΔC_{44} and ΔC_{456} in this case.

*21.2.8 The magneto MEC in QWHD superlattices of Kane type materials with graded interfaces in the presence of light waves

The magneto DR in QWHD superlattices of Kane type materials with graded interfaces in the presence of light waves can be expressed as

$$\left(\frac{n_z\pi}{d_z}\right)^2 = [G_{8,17,54} + iH_{8,17,54}]|_{E=E_{17,55}} \tag{21.136}$$

where $E_{17,55}$ is the total quantized energy in this case,

$$G_{8,17,54} = \left[\frac{C_{7,54}^2 - D_{7,54}^2}{L_0^2} - \left[\frac{2eB}{\hbar}\left(n+\frac{1}{2}\right)\right]\right], \quad C_{7,54} = \cos^{-1}(\bar{\omega}_{7,54}),$$

$$\bar{\omega}_{7,54} = (2)^{\frac{-1}{2}}[(1 - G_{7,54}^2 - H_{7,54}^2) - \sqrt{(1 - G_{7,54}^2 - H_{7,54}^2)^2 + 4G_{7,54}^2}]^{\frac{1}{2}},$$

$$\begin{aligned} G_{7,54} = [&G_{1,54} + (\rho_{5,,54}G_{2,54}/2) - (\rho_{6,54}H_{2,54}/2) + (\Delta_0/2)\{\rho_{7,54}G_{3,54} \\ &- \rho_{8,54}H_{3,54} + \rho_{9,54}G_{4,54} - \rho_{10,54}H_{4,54} + \rho_{11,54}G_{5,54} \\ &- \rho_{12,54}G_{5,54} + (1/12)(\rho_{13,54}G_{6,54} - \rho_{14,54}H_{6,54})\}], \end{aligned}$$

$$\begin{aligned} G_{1,54} = [&(\cos(h_{1,54}))(\cosh(h_{2,54}))(\cosh(g_{1,54}))(\cos(g_{2,54})) \\ &+ (\sin(h_{1,54}))(\sinh(h_{2,54}))(\sinh(g_{1,54}))(\sin(g_{2,54}))], \end{aligned}$$

$$h_{1,54} = e_{1,54}(b_0 - \Delta_0), e_{1,54} = 2^{\frac{-1}{2}}\left(\sqrt{t_{1,54}^2 + t_{2,54}^2} + t_{1,54}\right)^{\frac{1}{2}},$$

$$\begin{aligned} t_{1,54} = \Bigg[&(2m_{c1}/\hbar^2)\cdot[U_{\lambda 1}T_{31}(E, E_{g1}, \quad \Delta_1, \eta_{g1}) - P_{\lambda 1}] \\ &- \frac{2eB}{\hbar}\left(n+\frac{1}{2}\right)\Bigg], \quad t_{2,54} = [(2m_{c1}/\hbar^2)[U_{\lambda 1}T_{32}(E, E_{g1}, \Delta_1, \eta_{g1})]], \end{aligned}$$

$$h_{2,54} = e_{2,54}(b_0 - \Delta_0), \quad e_{2,54} = 2^{\frac{-1}{2}}\left(\sqrt{t_{1,54}^2 + t_{2,54}^2} - t_{1,54}\right)^{\frac{1}{2}},$$

$$g_{1,54} = d_{1,54}(a_0 - \Delta_0), \quad d_{1,54} = 2^{\frac{-1}{2}}\left(\sqrt{x_{1,54}^2 + y_{1,54}^2} + x_{1,54}\right)^{\frac{1}{2}},$$

$$x_{1,54} = \left[(-2m_{c2}/\hbar^2)\cdot[U_{\lambda 2}T_{31}(E - V_0, E_{g2}, \Delta_2, \eta_{g2}) - P_{\lambda 2}] + \frac{2eB}{\hbar}\left(n + \frac{1}{2}\right)\right],$$

$$y_{1,54} = [(2m_{c2}/\hbar^2)\cdot[U_{\lambda 2}T_{32}(E - V_0, E_{g2}, \Delta_2, \eta_{g2})]],$$

$$g_{2,54} = d_{2,54}(a_0 - \Delta_0), \quad d_{2,54} = 2^{\frac{-1}{2}}\left(\sqrt{x_{1,54}^2 + y_{1,54}^2} - x_{1,54}\right)^{\frac{1}{2}},$$

$$\rho_{5,54} = (\rho_{3,54}^2 + \rho_{4,54}^2)^{-1}[\rho_{1,54}\rho_{1,54} - \rho_{2,54}\rho_{4,54}],$$

$$G_{2,54} = [(\sin(h_{1,54}))(\cosh(h_{2,54}))(\sinh(g_{1,54}))(\cos(g_{2,54})) + (\cos(h_{1,54}))(\sinh(h_{2,54}))(\cosh(g_{1,54}))(\sin(g_{2,54}))],$$

$$\rho_{6,54} = (\rho_{3,54}^2 + \rho_{4,54}^2)^{-1}[\rho_{1,54}\rho_{4,54} + \rho_{2,54}\rho_{3,54}],$$

$$H_{2,54} = [(\sin(h_{1,54}))(\cosh(h_{2,54}))(\sinh(g_{2,54}))(\cos(g_{1,54})) - (\cos(h_{1,54}))(\sinh(h_{2,54}))(\sinh(g_{1,54}))(\cos(g_{2,54}))],$$

$$\rho_{7,54} = [(e_{1,54}^2 + e_{2,54}^2)^{-1}[e_{1,54}(d_{1,54}^2 - d_{2,54}^2) - 2d_{1,54}d_{2,54}e_{2,54}] - 3e_{1,54}],$$

$$G_{3,54} = [(\sin(h_{1,54}))(\cosh(h_{2,54}))(\cosh(g_{1,54}))(\cos(g_{2,54})) + (\cos(h_{1,54}))(\sinh(h_{2,54}))(\sinh(g_{1,54}))(\sin(g_{2,54}))],$$

$$\rho_{8,54} = [(e_{1,54}^2 + e_{2,54}^2)^{-1}[e_{2,54}(d_{1,54}^2 - d_{2,54}^2) + 2d_{1,54}d_{2,54}e_{1,54}] + 3e_{2,54}],$$

$$H_{3,54} = [(\sin(h_{1,54}))(\cosh(h_{2,54}))(\sin(g_{2,54}))(\sinh(g_{1,54})) - (\cos(h_{1,54}))(\sinh(h_{2,54}))(\cosh(g_{1,54}))(\cos(g_{2,54}))],$$

$$\rho_{9,54} = [(d_{1,54}^2 + d_{2,54}^2)^{-1}[d_{1,54}(e_{2,54}^2 - e_{1,54}^2) + 2e_{2,54}d_{2,54}e_{1,54}] + 3d_{1,54}],$$

$$G_{4,54} = [(\cos(h_{1,54}))(\cosh(h_{2,54}))(\cos(g_{2,54}))(\sinh(g_{1,54})) - (\sin(h_{1,54}))(\sinh(h_{2,54}))(\cosh(g_{1,54}))(\sin(g_{2,54}))],$$

$$\rho_{10,54} = \left[-(d_{1,54}^2 + d_{2,54}^2)^{-1}\left[d_{2,54}(-e_{2,54}^2 + e_{1,54}^2) + 2e_{2,54}d_{1,54}\rho_{1,54}\right] + 3d_{2,54}\right],$$

$$H_{4,54} = [(\cos(h_{1,54}))(\cosh(h_{2,54}))(\cosh(g_{1,54}))(\sin(g_{2,54})) + (\sin(h_{1,54}))(\sinh(h_{2,54}))(\sinh(g_{1,54}))(\cos(g_{2,54}))],$$

$$
\begin{aligned}
\rho_{11,54} &= 2[d_{1,54}^2 + e_{2,54}^2 - d_{2,54}^2 - e_{1,54}^2],\\
G_{5,54} &= [(\cos(h_{1,54}))(\cosh(h_{2,54}))(\cos(g_{2,54}))(\cosh(g_{1,54}))\\
&\quad - (\sin(h_{1,54}))(\sinh(h_{2,54}))(\sinh(g_{1,54}))(\sin(g_{2,54}))],\\
\rho_{12,54} &= 4[d_{1,54}d_{2,54} + e_{1,54}e_{2,54}],\\
H_{5,54} &= [(\cos(h_{1,54}))(\cosh(h_{2,54}))(\sinh(g_{1,54}))(\sin(g_{2,54}))\\
&\quad + (\sin(h_{1,54}))(\sinh(h_{2,54}))(\cosh(g_{1,54}))(\cos(g_{2,54}))],\\
\rho_{13,54} &= [\{5(d_{1,54}e_{1,54}^3 - 3e_{1,54}e_{2,54}^2 d_{1,54}) + 5d_{2,54}(e_{1,54}^3 - 3e_{1,54}^2 e_{2,54})\}\\
&\quad \times (d_{1,54}^2 + d_{2,54}^2)^{-1} + (e_{1,54}^2 + e_{2,54}^2)^{-1}\{5(e_{1,54}d_{1,54}^3\\
&\quad - 3d_{2,54}e_{1,54}^2 d_{1,54}) + 5(d_{2,54}^3 e_{2,54} - 3d_{1,54}^2 d_{2,54}e_{2,54})\}\\
&\quad - 34(d_{1,54}e_{1,54} + d_{2,54}e_{2,54})],\\
G_{6,54} &= [(\sin(h_{1,54}))(\cosh(h_{2,54}))(\sinh(g_{1,54}))(\cos(g_{2,54}))\\
&\quad + (\cos(h_{1,54}))(\sinh(h_{2,54}))(\cosh(g_{1,54}))(\sin(g_{2,54}))],\\
\rho_{14,54} &= \left[\left\{5(d_{1,54}e_{2,54}^3 - 3e_{2,54}e_{1,54}^2 d_{1,54}) + 5d_{2,54}\left(-e_{1,54}^3\right.\right.\right.\\
&\quad \left.\left. + \; 3e_{2,54}^2 e_{1,54})\right\}(d_{1,54}^2 + d_{2,54}^2)^{-1}\\
&\quad + (e_{1,54}^2 + e_{2,54}^2)^{-1}\left\{5(-e_{1,54}d_{2,54}^3 + 3d_{1,54}^2 d_{2,54}e_{1,54})\right.\\
&\quad \left.\left. + 5\;(-d_{1,54}^3 e_{2,54} + 3d_{2,54}^2 d_{1,54}e_{2,54})\right\} + 34(d_{1,54}e_{2,54} - d_{2,54}e_{1,54})\right],\\
H_{6,54} &= [(\sin(h_{1,54}))(\cosh(h_{2,54}))(\cosh(g_{1,54}))(\sin(g_{2,54}))\\
&\quad - (\cos(h_{1,54}))(\sinh(h_{2,54}))(\sinh(g_{1,54}))(\cos(g_{2,54}))],\\
H_{7,54} &= [H_{1,54} + (\rho_{5,54}H_{2,54}/2) + (\rho_{6,54}G_{2,54}/2) + (\Delta_0/2)\{\rho_{8,54}G_{3,54}\\
&\quad + \rho_{7,54}H_{3,54} + \rho_{10,54}G_{4,54} + \rho_{9,54}H_{4,54}\\
&\quad + \; \rho_{12,54}G_{5,54} + \rho_{11,54}H_{5,54} + (1/12)(\rho_{14,54}G_{6,54} + \rho_{13,54}H_{6,54})\}],\\
H_{1,54} &= [(\sin(h_{1,54}))(\sinh(h_{2,54}))(\cosh(g_{1,54}))(\cos(g_{2,54}))\\
&\quad + (\cos(h_{1,54}))(\cosh(h_{2,54}))(\sinh(g_{1,54}))(\sin(g_{2,54}))],\\
D_{7,54} &= \sinh^{-1}(\bar{\omega}_{7,54}), \quad H_{8,17,54} = (2C_{7,54}D_{7,54}/L_0^2)
\end{aligned}
$$

The DOS function is given by

$$
N_{\text{QWHDSLB}}(E, \eta_g, \lambda) = \frac{g_v eB}{\pi \hbar d_z} \sum_{n_z=1}^{n_{z\max}} \sum_{n=0}^{n_{\max}} \delta'(E - E_{17,55}) \tag{21.137}
$$

The electron concentration can be written as

$$n_0 = \text{Real part of } \bar{n}_0 \tag{21.138}$$

$$\bar{n}_0 = \frac{g_v eB}{\pi\hbar d_z} \sum_{n_z=1}^{n_{z\max}} \sum_{n=0}^{n_{\max}} F_{-1}(\eta_{17,55}) \tag{21.139}$$

in which

$$\eta_{17,55} = \frac{E_{F17,55} - E_{17,55}}{k_B T} \quad \text{and} \quad E_{F1755}$$

is the Fermi energy in the present case.

The electronic contribution to the second- and third-order elastic constants for HD materials in this case can be written as

$$\Delta C_{44} = \frac{-G_0^2}{9} \text{ Real part of } \left[\frac{\partial n_0}{\partial(E_{F17.55} - E_{17.55})}\right] \tag{21.140}$$

and

$$\Delta C_{456} = \frac{G_0^3}{27} \text{ Real part of } \left[\frac{\partial^2 n_0}{\partial(E_{F17.55} - E_{17.55})^2}\right] \tag{21.141}$$

Thus, using (21.138), (21.140) and (21.141), we can find ΔC_{44} and ΔC_{456} in this case.

21.3 Summary and Conclusion

For the purpose of condensation we have not plotted any case with the hope that the readers will perform all the computer programming for all the quantized materials under different physical conditions for the purpose of creating new physics and other important transport quantities which are totally band structure dependent. The theoretical results as given here would be useful in analyzing various other experimental data related to the transport phenomenon. We must note that the study of transport phenomena and the formulation of the electronic properties of low dimensional field aided HD compounds are based on the dispersion relations in such materials.

It is imperative to state that the present investigation excludes the many-body, hot electron, broadening and the allied effects in the simplified theoretical formalism due to the absence of proper analytical techniques for including them for generalized systems as considered here. Our simplified approach will be appropriate for the purpose of comparisons when the methods of tackling the formidable problems after inclusion of the said effects for the generalized systems emerge. The results of this simplified approach get transformed to the well-known formulation of the MECs' for wide gap materials having parabolic energy bands. This indirect test not only exhibits the mathematical compatibility of the formulation but also shows the fact that this simple analysis is more generalized one, since one can obtain the corresponding results for materials having parabolic energy bands under certain limiting conditions from the present derivation. For the purpose of computer simulations for obtaining the plots of MECs' versus various external variables, we have taken very low temperatures since the quantization effects are basically low temperature phenomena together with the fact that the temperature dependence of all the energy band constants of all the materials and their nanostructures as considered in this chapter are not available in the literature. The inclusion of the said effects would certainly increase the accuracy of the results although the qualitative features of MECs' would not change in the presence of the aforementioned relations of the various quantized structures as discussed above. It may finally be noted that the basic aim of this chapter is not solely to demonstrate the influence of quantum confinement on the MECs for a wide class of quantized materials but also to formulate the appropriate carrier statistics in the most generalized form, since the transport and other phenomena in modern nano-structured devices having different band structures and the derivation of the expressions of many important carrier properties are based on the carrier statistics in such systems in the presence of intense light waves.

21.4 Open Research Problem

(R.21.1) Investigate the MTP, HC, RLC and QC in this case in the presence of arbitrarily oriented quantizing magnetic field and crossed electric field for all types of HD materials of Chapter 11.

References

[1] P. K. Basu, *Theory of Optical Processes in Materials, Bulk and Microstructures*, (Oxford University Press, Oxford, **1997**).

[2] K. P. Ghatak, S. Bhattacharya, D. De, *Einstein Relation in Compound Semiconductors and Their Nanostructures*, Springer Series in Materials Science, Vol. 116, (Springer-Verlag, Germany, **2008**).

[3] K. P. Ghatak, J. Pal, *Quantum Capacitance in Quantized Transistors*, Series on the Foundations of Natural Science and Technology, Vol. 17, (World Scientific Publishing Co. Pte. Ltd., **2024**)
[4] K. P. Ghatak and S. Bhattacharya, *J. Appl. Phys.*, 102, 073704 **(2007)**.
[5] K. P. Ghatak, S. Bhattacharya, S. K. Biswas, A. De and A. K. Dasgupta, *Phys. Scr.*, 75, 820 **(2007)**.
[6] K. Seeger, *Semiconductor Physics*, 6th ed. (Springer Verlag, **1997**).
[7] B. R. Nag, *Electron Transport in Compound Semiconductors*, (Springer Verlag, **1980**).
[8] B. R. Nag, *Physics of Quantum Well Devices*, (Kluwer Academic Publisher, The Netherlands, **2000**).
[9] H. Sasaki, *Phys. Rev. B*, 30, 7016 **(1984)**.

Chapter 22

The Acoustic Mode Scattering Under Intense Electric Field in HD Kane Type Materials

The illiterate of the present century will not be those who cannot read and write but those who cannot learn, unlearn and relearn.

22.1 Introduction

With the advent of modern nano devices, there has been considerable interest in studying the electric field induced processes in semiconductors having different band structures. It appears from the literature that the studies have been made on the assumption that the carrier dispersion laws are invariant quantities in the presence of intense electric field, which is not fundamentally true. In this chapter, we shall present the simplified formulation of the relaxation time due to acoustic mode scattering (RAS) in quantum confined optoelectronic semiconductors under strong electric field. In Section 22.2.1, we have formulated the DR in the presence of intense electric field in HD III–V, ternary and quaternary semiconductors. Section 22.2.2, the RAS in QDs of HD Kane type semiconductors in the presence of intense electric field has been studied. In Section 22.2.3, we have studied the magneto RAS in QWs of HD Kane Type materials in the presence of intense electric field. In Section 22.2.4, the magneto RAS in accumulation and inversion layers of Kane type semiconductors in the presence of intense electric field has been studied. In Section 22.2.5, the magneto RAS in doping superlattices of HD Kane type semiconductors in the presence of intense electric field has been investigated. In Section 22.2.6, the RAS in QDHD effective mass superlattices of Kane Type materials in the presence of intense electric field has been studied. In Section 22.2.7, the magneto RAS in QWHD effective mass superlattices of Kane type semiconductors in the presence of intense electric field has been studied.

In Section 22.2.8, the RAS in QDHD superlattices of Kane type semiconductors with graded interfaces in the presence of intense electric field has been studied. In Section 22.2.9, the magneto RAS in QWHD superlattices of Kane type materials with graded interfaces in the presence of intense electric field has been investigated. Section 22.3 contains a brief overview of the summary and conclusion as pertinent to this chapter. Section 22.4 presents six open research problems which challenges the first order creativity from the readers of diverse fields.

22.2 Theoretical Background

22.2.1 The formulation of the electron dispersion law in the presence of intense electric field in HD III–V, ternary and quaternary materials

The expression of the inter-band transition matrix element (X_{12}) in this case can be written as

$$X_{12} = i \int u^*_{\bar{k}1}(\bar{r}) \cdot \frac{\partial}{\partial k_x} u_{\bar{k}2}(\bar{r}) d^3 r \tag{22.1}$$

where $u_{\bar{k}1}(\bar{r}) \equiv u_1(\bar{k}, \bar{r})$ and $u_{\bar{k}2}(\bar{r}) \equiv u_2(\bar{k}, \bar{r})$ in which $u_1(\bar{k}, \bar{r})$ and $u_2(\bar{k}, \bar{r})$ are given by (21.15) and (21.16) respectively.

In the case of the presence of an external electric field, F_s along x-axis, using the appropriate equations, we can write

$$X_{12} = i \int d^3 r \left\{ a_{k_+} [(iS) \downarrow'] + b_{k_+} \left[\left(\frac{X' - iY'}{\sqrt{2}} \right) \uparrow' \right] + c_{k_+} [Z' \downarrow'] \right\}^* \frac{\partial}{\partial k_x},$$
$$\left\{ a_{k_-} [(iS) \uparrow'] - b_{k_-} \left[\left(\frac{X' + iY'}{\sqrt{2}} \downarrow' \right) \right] + c_{k_-} [Z' \uparrow'] \right\}$$
$$= i \int d^3 r \left[\left\{ \left(a_{k_+} \frac{\partial}{\partial k_x} a_{k_-} \right) \cdot [(iS) \downarrow']^* \cdot [(iS) \uparrow'] \right.\right.$$
$$+ \left(\frac{b_{k_+}}{\sqrt{2}} \frac{\partial}{\partial k_x} a_{k_-} \right) [(X' - iY') \uparrow']^* \cdot [(iS) \uparrow']$$
$$\left.\left. + \left(c_{k_+} \frac{\partial}{\partial k_x} a_{k_-} \right) [(Z' \downarrow')]^* \cdot [(iS) \uparrow'] \right\} \right]$$
$$+ i \int d^3 r \left[\left\{ \left(-\frac{a_{k_+}}{\sqrt{2}} \frac{\partial}{\partial k_x} b_{k_-} \right) \cdot [(iS) \downarrow']^* \cdot [(X' + iY') \downarrow'] \right.\right.$$

$$
\begin{aligned}
&- \left(\frac{b_{k_+} \frac{\partial}{\partial k_x} b_{k_-}}{2} \right) [(X' - iY') \uparrow']^* \cdot [(X' + iY') \downarrow'] \\
&- \left(\frac{c_{k_+} \frac{\partial}{\partial k_x} b_{k_-}}{\sqrt{2}} \right) [(Z' \downarrow')]^* \cdot [(X' + iY') \downarrow'] \Bigg\} \Bigg] \\
&= i \int d^3 r \left[\left\{ \left(a_{k_+} \frac{\partial}{\partial k_x} c_{k_-} \right) \cdot ([(iS) \downarrow']^* \cdot [Z' \uparrow']) \right. \right. \\
&+ \left(\frac{b_{k_+}}{\sqrt{2}} \frac{\partial}{\partial k_x} c_{k_-} \right) [(X' - iY') \uparrow']^* \cdot [Z' \uparrow'] \\
&+ \left. \left. \left(c_{k_+} \frac{\partial}{\partial k_x} c_{k_-} \right) [(Z' \downarrow')]^* \cdot [Z' \uparrow'] \right\} \right]
\end{aligned} \tag{22.2}
$$

Therefore,

$$
\begin{aligned}
\frac{X_{12}}{i} = &\left[\left\{ \left[\left(a_{k_+} \frac{\partial}{\partial k_x} a_{k_-} \right) \cdot \langle iS \mid iS \rangle \langle \downarrow' \mid \uparrow' \rangle \right] \right. \right. \\
&+ \left[\left(\frac{b_{k_+}}{\sqrt{2}} \frac{\partial}{\partial k_x} a_{k_-} \right) \langle (X' - iY') \mid iS \rangle \langle \uparrow' \mid \downarrow' \rangle \right] \\
&+ \left. \left. \left[\left(c_{k_+} \frac{\partial}{\partial k_x} a_{k_-} \right) \langle Z' \mid iS \rangle \langle \downarrow' \mid \uparrow' \rangle \right] \right\} \right] \\
&- \left[\left\{ \left[\left(\frac{a_{k_+}}{\sqrt{2}} \frac{\partial}{\partial k_x} b_{k_-} \right) \cdot \langle iS \mid (X' + iY') \rangle \langle \downarrow' \mid \downarrow' \rangle \right] \right. \right. \\
&+ \left[\left(\frac{b_{k_+}}{2} \frac{\partial}{\partial k_x} b_{k_+} \right) \langle (X' - iY') \mid (X' + iY') \rangle \langle \uparrow' \mid \downarrow' \rangle \right] \\
&+ \left. \left. \left[\left(\frac{c_k}{\sqrt{2}} \frac{\partial}{\partial k_x} b_{k_-} \right) \langle Z' \mid (X' + iY') \rangle \langle \downarrow' \mid \downarrow' \rangle \right] \right\} \right] \\
&+ \left\{ \left[\left(a_k \frac{\partial}{\partial k_x} c_{k_-} \right) \cdot \langle iS \mid Z' \rangle \langle \downarrow' \mid \uparrow' \rangle \right] \right. \\
&+ \left[\left(\frac{b_{k_+}}{\sqrt{2}} \frac{\partial}{\partial k_x} c_{k_-} \right) \cdot \langle (X' - iY') \mid Z' \rangle \langle \uparrow' \mid \uparrow' \rangle \right] \\
&+ \left. \left[\left(c_k \frac{\partial}{\partial k_x} c_{k_-} \right) \cdot \langle Z' \mid Z' \rangle \langle \downarrow' \mid \uparrow' \rangle \right] \right\}
\end{aligned} \tag{22.3}
$$

Therefore, we can write,

$$
X_{12} = i \left\{ - \left(a_{k_+} \frac{\partial}{\partial k_x} a_{k_-} \right) \langle \downarrow' \mid \uparrow' \rangle + \left(c_{k_+} \frac{\partial}{\partial k_x} c_{k_-} \right) \langle \downarrow' \mid \uparrow' \rangle \right\} \tag{22.4}
$$

We can prove that

$$\langle \downarrow' \mid \uparrow' \rangle = \frac{1}{2}\left(\hat{r}_1 + i\hat{r}_2\right) \tag{22.5}$$

Therefore, (22.4) and (22.5), we get

$$\begin{aligned} X_{12} &= i\left\{-\left(a_{k+}\frac{\partial}{\partial k_x}a_{k-}\right) + \left(c_{k+}\frac{\partial}{\partial k_x}c_{k-}\right)\right\}\langle \downarrow' \mid \cdot \uparrow' \rangle \\ &= -i\left\{\left(a_{k_+}\frac{\partial}{\partial k_x}a_{k-}\right) - \left(c_{k+}\frac{\partial}{\partial k_x}c_{k-}\right)\right\} \cdot \frac{1}{2}(\hat{r}_1 + i\hat{r}_2) \\ &= \frac{-iA(k)}{2}(\hat{r}_1 + i\hat{r}_2) \end{aligned} \tag{22.6}$$

where

$$A(k) = \left(a_{k+}\frac{\partial}{\partial k_x}a_{k-}\right) - \left(c_{k+}\frac{\partial}{dk_x}c_{k-}\right) \tag{22.7}$$

From (22.6), we find,

$$|X_{12}|^2 = \frac{1}{4}A^2(k)(1+1) \quad [\text{since, } |\hat{r}_1| = |\hat{r}_2| = 1] \tag{22.8}$$

Considering spin-up and spin-down, we have to multiply (22.8) by 2

$$|X_{12}|^2 = 2 \times \frac{1}{2}A^2(k) = A^2(k) \tag{22.9}$$

We can evaluate X_{11} and X_{21} in the following way:

$$\begin{aligned} X_{11} &= i\int u^*_{\bar{k}1}(\bar{r}) \cdot \frac{\partial}{\partial k_x}u_{\bar{k}1}(\bar{r}) \cdot d^3r \\ &= i\int d^3r\left\{\left(a_{k_+}\frac{\partial}{\partial k_x}a_{k_+}\right) + \left(b_{k_+}\frac{\partial}{\partial k_x}b_{k_+}\right) + \left(c_{k_+}\frac{\partial}{\partial k_x}c_{k_+}\right)\right\} \\ &= \frac{1}{2}i\int d^3r\left\{\frac{\partial}{\partial k_x}(a_k^2 + b_k^2 + c_k^2)\right\} \\ &= \frac{1}{2}i\int d^3r\left\{\frac{\partial}{\partial k_x}(1)\right\} = 0, \quad \text{since } a_{k+}^2 + b_{k+}^2 + c_{k+}^2 = 1 \end{aligned} \tag{22.10}$$

Therefore, $X_{11} = 0$, and similarly, we can prove $X_{22} = 0$. Thus, we conclude that intraband momentum matrix element due to external electric field (X_{CC}) is zero. From the expression of $a_{k\pm}$, we can write

$$a_{k_+}^2 = r_0^2 \left[\frac{E_{go} - \gamma_{k_+}^2 (E_{go} - \delta')}{E_{go} + \delta'}\right]^2 \quad \text{and} \quad a_{k-}^2 = r_0^2 \left[\frac{E_{go} - \gamma_{k-}^2 (E_{go} - \delta')}{E_{go} + \delta'}\right]^2$$

Therefore,

$$2a_{k-} \frac{\partial}{\partial k_x} a_{k-} = r_0^2 \left[-\left(\frac{E_{go} - \delta'}{E_{go} + \delta'} r\right)\right] \frac{\partial \gamma_{k-}^2}{\partial k_x} \quad \text{and}$$

$$\frac{\partial a_{k-}}{\partial k_x} = -\frac{r_0^2}{2}\left(\frac{E_{go} - \delta'}{E_{go} + \delta'}\right) \frac{1}{a_{k-}} \frac{\partial \gamma_{k-}^2}{\partial k_x}$$

Combining, we can write

$$a_{k+} \frac{\partial a_{k-}}{\partial k_x} = -\frac{r_0^2}{2}\left(\frac{E_{go} - \delta'}{E_{go} + \delta'}\right) \frac{a_{k+}}{a_{k-}} \frac{\partial \gamma_{k-}^2}{\partial k_x}$$

Similarly, $c_{k_+} = t\gamma_{k_+}$ and $c_{k_-} = t\gamma_{k_-}$.

Therefore,

$$c_{k+} \frac{\partial}{\partial k_x} c_{k-} = \frac{t^2}{2} \frac{c_{k+}}{c_{k-}} \frac{\partial \gamma_{k-}^2}{\partial k_x},$$

$$A(k) = \left\{-\frac{r_0^2}{2}\left[\frac{E_{go} - \delta'}{E_{go} + \delta'}\right] \frac{a_{k+}}{a_{k-}} - \frac{t^2}{2} \frac{c_{k+}}{c_{k-}}\right\} \frac{\partial \gamma_{k-}^2}{\partial k_x}$$

Now,

$$\left(\frac{a_{k+}}{a_{k-}}\right)^2 = \frac{E_{go} - \gamma_{k+}^2 (E_{go} - \delta')}{E_{go} - \gamma_{k-}^2 (E_{go} - \delta')} = \frac{E_{go} - \frac{\eta - E_{go}}{2(\eta + \delta')}(E_{go} - \delta')}{E_{go} - \frac{\eta + E_{go}}{2(\eta + \delta')}(E_{go} - \delta')}$$

$$= \frac{2E_{go} - (\eta + \delta') - (\eta - E_{go})(E_{go} - \delta')}{2E_{go}(\eta + \delta') - (\eta + E_{go})(E_{go} - \delta')}$$

$$= \frac{\eta(E_{go} + \delta') + E_{go}(E_{go} + \delta')}{\eta(E_{go} + \delta') - E_{go}(E_{go} - 3\delta')}$$

Therefore,

$$\left(\frac{a_{k+}}{a_{k-}}\right)^2 = \frac{\eta + E_{go}}{\eta - E_{go}\left(\frac{E_{go} - 3\delta'}{E_{go} + \delta'}\right)} = \frac{\eta + E_{go}}{\eta - E_{go}} \quad \text{where } E'_{go} = \frac{E_{go}(E_{go} - 3\delta')}{E_{go} + \delta'}$$

Thus,

$$\frac{a_{k+}}{a_{k-}} = \sqrt{\frac{\eta + E_{g_0}}{\eta - E'_{g_0}}}$$

Similarly,

$$\frac{c_{k+}}{c_{k-}} = \frac{\gamma_{k+}}{\gamma_{k-}} = \sqrt{\frac{\eta - E_{g_0}}{\eta + E'_{g_0}}}$$

and thus,

$$A(k) = -\left\{P\left(\frac{\eta + E_{g_0}}{\eta - E'_{g_0}}\right)^{1/2} + Q\left(\frac{\eta - E_{g_0}}{\eta + E_{g_0}}\right)^{1/2}\right\}\frac{\partial \gamma_{k-}^2}{\partial k_x} \tag{22.11}$$

where

$$P = \frac{r_0^2}{2}\left(\frac{E_{g_0} - \delta'}{E_{g_0} + \delta'}\right) \quad \text{and} \quad Q = t^2/2$$

Now,

$$\gamma_{k-}^2 = \frac{\eta + E_{g_0}}{2(\eta + \delta')} \quad \text{so that } \frac{\partial \gamma_{k-}^2}{\partial k_x} = \frac{1}{2}\left[\frac{\partial \eta / \partial k_x}{(\eta + \delta')} - \frac{\eta + E_{g_0}}{(\eta + \delta')^2}\frac{\partial \eta}{\partial k_x}\right]$$

Thus,

$$\frac{\partial \gamma_{k-}^2}{\partial k_x} = \frac{1}{2}\left[\frac{\eta + \delta' - \eta - E_{g_0}}{(\eta + \delta')}\right]\frac{\partial \eta}{\partial k_x} = -\frac{1}{2}\frac{(E_{g_0} - \delta')}{(\eta + \delta')^2}\cdot\frac{\partial \eta}{\partial k_x} \tag{22.12}$$

From (22.11) and (22.12), we get

$$A(k) = \frac{1}{2}\frac{(E_{g_0} - \delta')}{(\eta + \delta')^2}\cdot\frac{\partial \eta}{\partial k_u}\cdot\left\{P\left(\frac{\eta + E_{g_0}}{\eta - E'_{g_0}}\right)^{1/2} + Q\left(\frac{\eta - E_{g_0}}{\eta + E_{g_0}}\right)^{1/2}\right\} \tag{22.13}$$

This implies

$$\frac{\partial \eta}{\partial k_x} = \frac{E_{g_0}\hbar^2}{m_r}\cdot\frac{k_x}{\eta} \tag{22.14}$$

From (22.13) and (22.14), one can write

$$A(k) = \frac{E_{g_0}\hbar^2}{2m_r}\cdot\frac{k_u}{\eta}\cdot\frac{(E_{g_0} - \delta')}{(\eta + \delta')^2}\cdot\left\{P\left(\frac{\eta + E_{g_0}}{\eta - E'_{g_0}}\right)^{1/2} + Q\left(\frac{\eta - E_{g_0}}{\eta + E_{g_0}}\right)^{1/2}\right\} \tag{22.15}$$

Thus,

$$|A(k)|^2 = \frac{E_{g_0}^2(E_{g_0} - \delta')\hbar^2}{4m_r}\frac{\hbar^2 k_x^2}{m_r}\frac{1}{\eta^2}\frac{1}{(\eta+\delta')^4}$$
$$\times \left\{P\left(\frac{\eta + E_{g_0}}{\eta - E'_{g_0}}\right)^{1/2} + Q\left(\frac{\eta - E_{g_0}}{\eta + E_{g_0}}\right)^{1/2}\right\}^2 \quad (22.16)$$

and

$$|X_{12}|^2 = |A('k)|^2 \quad (22.17)$$

From (22.16) and (22.17), we can write the square of the magnitude of the inter-band transition matrix element due to external electric field ($|X_{CV}|^2$).

It is well-known that the energy Eigen value, $E_n^{(2)}(\bar{k})$ in the presence of a perturbed Hamiltonian, H', is given by [1]

$$E_n^{(2)}(\bar{k}) = E_n(\bar{k}) + \langle n\bar{k}|H'|n\bar{k}\rangle + \{|\langle n\bar{k}|H'|n\bar{k}\rangle|^2[E_n(\bar{k}) - E_m(\bar{k})]\} \quad (22.18)$$

where

$$H\psi_n(\bar{k},\bar{r}) = E\psi(\bar{k},\bar{r}) \quad (22.19)$$

$$H = H_0 + H' \quad (22.20)$$

in which H is the total Hamiltonian, $\psi(\bar{k},\bar{r})$ is the wave function, where $u_n(\bar{k},\bar{r})$ is the periodic function of it, H_0 is the unperturbed Hamiltonian, n is the band index, and $E_n(\bar{k})$ is the energy of an electron in the periodic lattice.

For an external electric field (F_S) applied along the x-axis, the perturbed Hamiltonian (H') can be written as

$$H' = -F \cdot x \quad (22.21)$$

where

$$F(= eF_s)$$

Therefore, we get

$$E_n^{(2)}(\bar{k}) = E_n(\bar{k}) - F\langle n\bar{k}|H'|n\bar{k}\rangle + F^2\{|\langle n\bar{k}|H'|n\bar{k}\rangle|^2/[E_n(\bar{k}) - E_m(\bar{k})]\} \quad (22.22)$$

In (22.22), the second and the third terms are due to the perturbation factor.

For

$$X_{nm}(\bar{k}) = \langle n\bar{k}|x|m\bar{k}\rangle \tag{22.23}$$

we find

$$X_{nm}(\bar{k}) = i\int u_n^*(\bar{k},\bar{r})(\partial/\partial u)[u_m(\bar{k},\bar{r})]d^3r \tag{22.24}$$

where k_x is the x component of the $\bar{k}$ and the integration in (22.24) extends over the unit cell. From (22.22), (22.23) and (22.24), with the n corresponding to the conduction band (C) and m corresponding to the valance band (V), we get

$$E_c^{(2)}(\bar{k}) = E_c(\bar{k}) - FX_{cc} + \{F^2|X_{\text{CV}}|^2/E_C(\bar{k}) - E_V(\bar{k})]\} \tag{22.25}$$

Thus, combining the appropriate equations, the dispersion relation of the conduction electrons in the presence of electric field along x-axis can be written as

$$\begin{aligned} I_{11}(E) &= \frac{\hbar^2k_x^2}{2m_c} + \frac{\hbar^2k_y^2}{2m_c} + \frac{\hbar^2k_z^2}{2m_c} + \frac{F^2|X_{12}|^2}{\eta} = \left[\frac{\hbar^2k_x^2}{2m_c} + \frac{\hbar^2k_y^2}{2m_c} + \frac{\hbar^2k_z^2}{2m_c}\right. \\ &\quad + \left\{\frac{\hbar^2k_x^2}{2m_c}\cdot\frac{2m_c}{m_r}\cdot\frac{F^2\hbar^2E_{g_0}^2(E_{g_0}-\delta')^2}{4m_r}\frac{1}{\eta^3}\frac{1}{(\eta+\delta')^4}\right. \\ &\quad \left.\left.\times\left[P\left(\frac{\eta+E_{g_0}}{\eta-E'_{g_0}}\right)^{1/2} + Q\left(\frac{\eta-E_{g_0}}{\eta+E_{g_0}}\right)^{1/2}\right]^2\right\}\right] \end{aligned} \tag{22.26}$$

When $F \to 0$, we have from (22.26),

$$k^2 \to \frac{2m_c}{\hbar^2}I_{11}(E) \quad \text{and} \quad \eta_1^2 = \left[E_g^2 + E_g\frac{2m_c}{m_r}I_{11}(E)\right]$$

Using the method of successive approximation one can write

$$1 = \frac{\hbar^2k_x^2}{2m_cI_{11}(E)} + \frac{\hbar^2k_y^2}{2m_cI_{11}(E)} + \frac{\hbar^2k_z^2}{2m_cI_{11}(E)} + \frac{\hbar^2k_x^2}{2m_cI_{11}(E)}\cdot\Phi(E,F) \tag{22.27}$$

where

$$\begin{aligned} \Phi(E,F) &= \frac{2m_c}{m_r}\frac{F^2\hbar^2E_{g_0}^2(E_{g_0}-\delta')^2}{4m_r}\frac{1}{\eta_1^3}\frac{1}{(\eta_1+\delta')^4}\left[P\left(\frac{\eta_1+E_{g_0}}{\eta_1-E'_{g_0}}\right)^{1/2}\right. \\ &\quad \left.+Q\left(\frac{\eta_1-E_{g_0}}{\eta_1+E_{g_0}}\right)^{1/2}\right]^2 \end{aligned}$$

Therefore, the E–k dispersion relation in the presence of an external electric field for III–V, ternary and quaternary materials whose unperturbed energy band structures are defined by the three band model of Kane can be expressed as

$$\frac{k_x^2}{\frac{2m_c}{\hbar^2}\left[\frac{I_{11}(E)}{1+\Phi(E,F)}\right]} + \frac{k_y^2}{\frac{2m_c}{\hbar^2}I_{11}(E)} + \frac{k_z^2}{\frac{2m_c}{\hbar^2}I_{11}(E)} = 1 \tag{22.28}$$

In (22.28), the coefficients of k_x, k_y and k_z are not same and for this reason, this basic equation is "anisotropic" in nature together with the fact that the anisotropic dispersion relation is the ellipsoid of revolution in the k-space.

From (22.28), the expressions of the effective electron masses along x-, y-, and z-directions can, respectively, be written as

$$m_x^*(E,F) = \hbar^2 k_x \left.\frac{\partial k_x}{\partial E}\right|_{k_y=0,k_z=0} = m_c[1+\Phi(E,F)]^{-2}[[1+\Phi(E,F)]I_{11}'(E) - I_{11}(E)(\Phi'(E,F))] \tag{22.29}$$

$$m_y^*(E,F) = \hbar^2 k_y \left.\frac{\partial k_y}{\partial E}\right|_{k_x=0,k_z=0} = m_0 I_{11}'(E) \tag{22.30}$$

$$m_z^*(E,F) = \hbar^2 k_z \left.\frac{\partial k_z}{\partial E}\right|_{k_x=0,k_y=0} = m_0 I_{11}'(E) \tag{22.31}$$

where

$$I_{11}'(E) = \frac{\partial}{\partial E}(I_{11}(E)) \quad \text{and} \quad \Phi'(E,F) = \frac{\partial}{\partial E}[\Phi(E,F)]$$

It may be noted from (22.29) that the effective mass along x-direction is a function of both electron energy and electric field respectively whereas from (22.30) and (22.31), we can infer the expressions of the effective masses along y- and z-directions are same and they depend on the electron energy only. Thus, in the presence of an electric field, the mass anisotropy for Kane type semiconductors depends both on electron energy and electric field respectively.

The use of the usual approximation [2]

$$k_x^2 \approx \frac{1}{3}k^2 \tag{22.32}$$

in (22.28), leads to the simplified expression of the electron energy spectrum in the present case as

$$I_{11}(E) = \frac{\hbar^2 k^2}{2m_c} + \frac{F^2\hbar^2 E_g^2(E_g-\delta')^2}{12m_r}\frac{2m_c}{m_r}I_{11}(E)\frac{1}{\eta_1}\frac{1}{(\eta_1+\delta')^4} \times \left\{P\left(\frac{\eta_1+E_{g_0}}{\eta_1-E'_{g_0}}\right)^{1/2}+Q\left(\frac{\eta_1-E_{g_0}}{\eta_1+E_{g_0}}\right)^{1/2}\right\}^2 \tag{22.33}$$

Equation (22.33) can approximately be written as

$$\frac{\hbar^2 k^2}{2m_c} = \left[e_1E^4+e_2E^3+e_3E^2+e_4E+e_5-\frac{e_6}{1+CE}+e_7(1+CE)^{-2}\right] \tag{22.34}$$

where

$$e_1 = Q_f\cdot\omega_1,\quad Q_f=\frac{m_c}{m_r}E_{g_0}^{-4}[5e_fE_{g_0}^{-2}-6G_f+7h_fE_{g_0}^{-4}],$$

$$e_f = A_fP_f,\quad A_f=[F\hbar E_{g_0}(E_{g_0}-\delta')]^2m_c(6m_r^2(\delta')^4)^{-1},$$

$$F=eF_s,\quad G_f=e_f(4\delta'+C_f),$$

$$C_f=(2E_{g_0}Q^2+PQ(E_{g_0}-E'_{g_0})-2P^2E_{g_0}),$$

$$E'_{g_0}=\frac{E_{g_0}(E_{g_0}-3\delta')}{E_{g_0}+\delta'},\quad P=\frac{r_0^2}{2}\left(\frac{E_{g_0}-\delta'}{E_{g_0}+\delta'}\right),$$

$$r_0=\left[\frac{6}{\chi}(E_{g_0}+\Delta)\left(E_{g_0}+\frac{2}{3}\Delta\right)\right]^{1/2},\quad Q=\frac{t^2}{2},$$

$$t=\left[\frac{6}{\chi}\left(E_{g_0}+\frac{2}{3}\Delta\right)\right]^{1/2},\quad h_f=(4\delta'e_fC_f)(B_f)^{-1},\quad B_f=(P+Q)^2$$

$$P_f=E_{g_0}^{-3}(e_fE_{g_0}^{-2}-G_f+h_fE_g^{-4}),\quad \omega_1=a_1^2,\quad a_1=\frac{ab}{c},\quad a=\frac{1}{E_{g_0}},$$

$$b=\frac{1}{E_{g_0}+\Delta},\quad c=\left(E_{g_0}+\frac{2}{3}\Delta\right)^{-1},$$

$$e_2=Q_f\cdot\omega_2,\quad \omega_2=2a_1b_1,\quad b_1=(c)^{-2}(ac+bc-ab),$$

$$e_3=(1-P_f)a_1+Q_f\cdot\omega_3,\quad \omega_3=(b_1^2+2a_1c_1),$$

$$c_1 = \left[\frac{1}{c}\left(1-\frac{a}{c}\right)\left(1-\frac{b}{c}\right)\right], \quad e_4 = [(1-P_f)b_1 + Q_f\omega_4], \quad \omega_4 = 2b_1c_1,$$

$$e_5 = [(1-P_f)c_1 + \omega_5 Q_f]$$

$$\omega_5 = (c_1^2 - 2c_1b_1), \quad e_7 = Q_f\omega_7, \quad \omega_7 = c_1^2, \quad e_6 = [(1-P_f)c_1 - Q_f\omega_6]$$

and $\omega_6 = \left[\frac{2c_1b_1}{c}\left(1-\frac{cc_1}{b_1}\right)\right]$

Using (2.4) and (22.34), we get

$$\begin{aligned}
&\frac{\hbar^2k^2}{2m_c}\int_{-\infty}^{E} F(V)dV \\
&= e_1\int_{-\infty}^{E}(E-V)^4F(V)dV + e_2\int_{-\infty}^{E}(E-V)^3F(V)dV \\
&\quad + e_3\int_{-\infty}^{E}(E-V)^2F(V)dV + e_4\int_{-\infty}^{E}(E-V)F(V)dV \\
&\quad + e_5\int_{-\infty}^{E}F(V)dV - e_6\int_{-\infty}^{E}\frac{F(V)dV}{[1+c(E-V)]} \\
&\quad + e_7\int_{-\infty}^{E}\frac{F(V)dV}{[1+c(E-V)]^2}
\end{aligned} \tag{22.35}$$

where

$$c = [E_g]^{-1}$$

We can prove that

$$\begin{aligned}
&\int_{-\infty}^{E}(E-V)^4F(V)dV \\
&= \frac{E^4}{2}[1+Erf(E/\eta_g)] + \frac{3\eta_g^4}{8\pi}\left[1+Erf(E/\eta_g) - \frac{2\eta_g}{3E}\exp\left(\frac{-E^2}{\eta_g^2}\right)\right] \\
&\quad + \frac{3}{2}(E\eta_g)^2[1+Erf(E/\eta_g)] + \frac{E^3\eta_g}{\sqrt{\pi}}\exp\left(\frac{-E^2}{\eta_g^2}\right) \\
&\quad + \frac{2}{\sqrt{\pi}}E\eta_g^3\exp\left(\frac{-E^2}{\eta_g^2}\right)\left[1+\frac{E^2}{\eta_g^2}\right] = \psi_0(E,\eta_g)
\end{aligned} \tag{22.36}$$

We can also prove that

$$\int_{-\infty}^{E}\frac{F(V)dV}{[1+c(E-V)]} = c_1(c,E,\eta_g) - ic_2(c,E,\eta_g) \tag{22.37}$$

where

$$c_1(c,E,\eta_g)=\left[\frac{2}{c\eta_g\sqrt{\pi}}\exp(-u^2)\left[\sum_{p=1}^{\infty}\exp\left(\frac{-p^2}{4}\right)(p)^{-1}\mathrm{Sinh}(pu)\right]\right],$$

$$c_2(c,E,\eta_g)=\frac{\sqrt{\pi}}{c\eta_g}\exp(-u^2)$$

and

$$u=\frac{1+cE}{c\eta_g}$$

We know that

$$\frac{\partial}{\partial x}\int_{A(x)}^{B(x)}F(x,\zeta)d\zeta=\int_{A(x)}^{B(x)}\frac{\partial}{\partial x}[F(x,\zeta)]d\zeta+F(x,B(x))\frac{\partial B(x)}{\partial x}$$
$$-F(x,A(x))\frac{\partial A(x)}{\partial x} \tag{22.38}$$

Using (22.37) and (22.38), we get

$$\int_{-\infty}^{E}\frac{F(V)dV}{[1+c(E-V)]^2}=c_3(c,E,\eta_g)-iD_3(c,E,\eta_g) \tag{22.39}$$

where

$$c_3(c,E,\eta_g)=\left[\frac{-4u\exp(-u^2)}{c^2\eta_g^2\sqrt{\pi}}\left[\sum_{p=1}^{\infty}\exp\left(\frac{-p^2}{4}\right)(p)^{-1}\mathrm{Sinh}(pu)\right]\right.$$
$$\left.+\frac{1}{\pi c\eta_g}\exp\left(\frac{-E^2}{\eta_g^2}\right)-\frac{2\exp(-u^2)}{c^2\eta_g^2\sqrt{\pi}}\sum_{p=1}^{\infty}\exp\left(\frac{-p^2}{4}\right)\mathrm{Cosh}(pu)\right]$$

$$\text{and}\quad D_3(c,E,\eta_g)=\frac{2u}{c^2\eta_g^2}\exp(-u^2)$$

Therefore, the DR in HD Kane type semiconductors can be written using (22.35)–(22.37) and (22.39) as

$$\frac{\hbar^2k^2}{2m_c}=J_1(E,c,\eta_g,F)+iJ_2(E,c,\eta_g,F) \tag{22.40}$$

where

$$J_1(E,c,\eta_g,F)$$
$$=2[1+Erf(E/\eta_g)]^{-1}\left[e_1\psi_0(E,\eta_g)+e_2\psi_1(E,\eta_g)+e_3\theta_0(E,\eta_g)\right.$$
$$+e_4\gamma_0(E,\eta_g)+e_5\frac{1}{2}[1+Erf(E/\eta_g)]-e_6c_1(E,c,\eta_g)$$

$$+ e_7c_3(E,c,\eta_g)\Big], \quad \psi_1(E,\eta_g)$$

$$= \left[\frac{E}{2}[1+Erf(E/\eta_g)]\left[E^2+\frac{3}{2}\eta_g^2\right]+\frac{\eta_g}{2\sqrt{\pi}}\exp\left(\frac{-E^2}{\eta_g^2}\right)(4E^2+\eta_g^2)\right]$$

and

$$J_2(E,c,\eta_g,F) = 2[1+Erf(E/\eta_g)]^{-1}[e_6c_2(E,c,\eta_g)+e_7D_3(E,c,\eta_g)]$$

22.2.2 The RAS in QDs in HD Kane type materials in the presence of intense electric field

The DR in this case is given by

$$\frac{\hbar^2}{2m_c}\left(\frac{n_z}{d_z}\right)^2+\frac{\hbar^2}{2m_c}\left(\frac{n_y}{d_y}\right)^2+\frac{\hbar^2}{2m_c}\left(\frac{n_x}{d_x}\right)^2$$
$$= [J_1(E_{n_z18,3HD},c,\eta_g,F)+iJ_1(E_{n_z18,3HD},c,\eta_g,F) \tag{22.41}$$

where $E_{n_z18,3HD}$ is total quantized energy in this case.

The DOS function in this case is given by

$$N_{QBHDL}(E,\eta_g,F) = \frac{2g_v}{d_xd_yd_z}\sum_{n_x=1}^{n_{x\max}}\sum_{n_y=1}^{n_{y\max}}\sum_{n_z=1}^{n_{z\max}}\delta'(E-E_{n_z18,3HD}) \tag{22.42}$$

Following (22.42) and (1.210), the RAS in this case is given by

$$\tau_{RAS}(E,\eta_g,F) = \tau_0\left[\sum_{n_x=1}^{n_{x\max}}\sum_{n_y=1}^{n_{y\max}}\sum_{n_z=1}^{n_{z\max}}\delta'(E-E_{n_z18,3HD})\right]^{-1} \tag{22.43}$$

22.2.3 The magneto RAS in QWs of HD Kane type materials in the presence of intense electric field

The DR in this case can be written (22.43) as

$$\frac{\hbar^2}{2m_c}\left(\frac{n_x\pi}{d_x}\right)^2+\left(n+\frac{1}{2}\right)\hbar\omega_0$$
$$= [J_1(E_{n_z18,4HD},c,\eta_g,F)+iJ_2(E_{n_z18,4HD},c,\eta_g,F)] \tag{22.44}$$

where $E_{n_z18,4HD}$ is the total quantized energy in this case.

The DOS function in this case is given by

$$N_{\mathrm{QWBHD}}(E,\eta_g,F)=\frac{g_v eB}{\pi\hbar}\sum_{n_x=1}^{n_{x\max}}\sum_{n=0}^{n_{\max}}\delta'(E-E_{n_z 18,4\mathrm{HD}}) \tag{22.45}$$

Following (22.45) and (1.139), the RAS in this case is given by

$$\tau_{\mathrm{RAS}}(E,\eta_g,F)=\tau_0\left[\sum_{n_x=1}^{n_{x\max}}\sum_{n=0}^{n_{\max}}\delta'(E-E_{n_z 18,4\mathrm{HD}})\right]^{-1} \tag{22.46}$$

*22.2.4 The magneto RAS in accumulation and inversion layers of Kane type materials in the presence of intense electric field

The 2D DR in accumulation layers of HD III–V, ternary and quaternary materials, in this case can be expressed as

$$[J_1(E,c,\eta_g,F)+iJ_2(E,c,\eta_g,F)]$$
$$=\frac{\hbar^2k_s^2}{2m_c}+\frac{\hbar|e|F_s}{\sqrt{2m_c}}\left(\frac{2\sqrt{2}S_i^{3/2}}{3}\right)[J_1''(E,c,\eta_g,F)+iJ_2''(E,c,\eta_g,F)]^{1/2} \tag{22.47}$$

Since the DR in accordance with the HD three-band model of Kane is complex in nature, Equation (22.47) will also be complex. Both the complexities occur due to the presence of poles in the finite complex plane of the dispersion relation of the materials in the absence of band tails.

The magneto DR in accumulation layers of HD III–V, ternary and quaternary materials, in this case can be expressed as

$$[J_1(E_{450},c,\eta_g,F)+iJ_2(E_{450},c,\eta_g,F)]=\left(n+\frac{1}{2}\right)\hbar\omega_o$$
$$+\frac{\hbar|e|F_s}{\sqrt{2m_c}}\left(\frac{2\sqrt{2}S_i^{3/2}}{3}\right)[J_1''(E_{450},c,\eta_g,F)+iJ_2''(E_{450},c,\eta_g,F)]^{1/2} \tag{22.48}$$

where E_{450}, is the total energy in this case.

The DOS function is given by

$$N_{\mathrm{HDA}}(E,\eta_g)=\frac{g_v eB}{\pi\hbar}\sum_{i=0}^{i_{\max}}\sum_{n=0}^{n_{\max}}\delta'(E-E_{450}) \tag{22.49}$$

$$\tau_{\mathrm{RAS}}(E,\eta_g)=\tau_0\left[\sum_{i=0}^{i_{\max}}\sum_{n=0}^{n_{\max}}\delta'(E-E_{450})\right]^{-1} \tag{22.50}$$

*22.2.5 The magneto RAS in doping superlattices of HD Kane type materials in the presence of intense electric field

The DR in doping superlattices of HD III–V, ternary and quaternary materials in the presence of intense electric field whose unperturbed electrons are defined by the three band model of Kane can be expressed as

$$[J_1(E,c,\eta_g,F)+iJ_2(E,c,\eta_g,F)]$$
$$=\left(n_i+\frac{1}{2}\right)\hbar\omega_{91\mathrm{HD}1}(E,c,\eta_g,F)+\frac{\hbar^2k_s^2}{2m_c} \tag{22.51}$$

where

$$\omega_{91\mathrm{HD}1}(E,c,\eta_g,F)=\left(\frac{n_s|e|^2}{d_0\varepsilon_{sc}[J_1(E,c,\eta_g,F)+iJ_2(E,c,\eta_g,F)]'m_c}\right)^{\frac{1}{2}}$$

The magneto DR in doping superlattices of HD III–V, ternary and quaternary materials in the presence of intense electric field whose unperturbed electrons are defined by the three band model of Kane can be expressed following (22.51) as

$$[J_1(E_{555},c,\eta_g,F)+iJ_2(E_{555},c,\eta_g,F)]$$
$$=\left(n_i+\frac{1}{2}\right)\hbar\omega_{91\mathrm{HD}1}(E_{555},c,\eta_g,F)+\left(n+\frac{1}{2}\right)\hbar\omega_0 \tag{22.52}$$

where E_{555} is the totally quantized energy in this case.

The DOS function is given by

$$N_{\mathrm{HDA}}(E,\eta_g)=\frac{g_v eB}{\pi\hbar}\sum_{i=0}^{i_{\max}}\sum_{n=0}^{n_{\max}}\delta'(E-E_{555}) \tag{22.53}$$

Following (22.56) and (1.139), the RAS in this case is given by

$$\tau_{\text{RAS}}(E,\eta_g) = \tau_0 \left[\sum_{i=0}^{i_{\max}} \sum_{n=0}^{n_{\max}} \delta'(E - E_{555}) \right]^{-1} \tag{22.54}$$

22.2.6 The RAS in quantum dot HD effective mass superlattices of Kane type materials in the presence of intense electric field

Following Sasaki [3], the DR in QDHD III–V effective mass superlattices (EMSLs) in the presence of intense electric field, the dispersion relations of whose constituent materials in the absence of any perturbation are defined by the three band model of Kane can be written as

$$\left(\frac{n_x \pi}{d_x}\right)^2 = \left[\frac{1}{L_0^2}\{\cos^{-1}(f_{18\text{HD}4}(E_{600}, n_y, n_z, F))\}^2 - G_2(n_y, n_z)\right] \tag{22.55}$$

In which

$$f_{18\text{HD}4}(E_{600}, n_y, n_z, F) = a_{1\text{HD}1,18} \cos[a_0 C_{1\text{HD}4,18}(E_{600}, n_y, n_z, F)$$
$$+ b_0 D_{1\text{HD}4,18}(E_{600}, n_y, n_z, F)]$$
$$- a_{2\text{HD}1,18} \cos[a_0 C_{1\text{HD}4,18}(E_{600}, n_y, n_z, F) + b_0 D_{1\text{HD}4,18}(E_{600}, n_y, n_z, F)]$$

E_{600} is the totally quantized energy in this case,

$$a_{1\text{HD}1,18} = \left[\sqrt{\frac{m_{c2} \text{ Real part of } \{[J_3(E, c, \eta_{g2}, F)]'\}|_{E\to 0}}{m_{c1} \text{ Real part of } \{[J_3(E, c, \eta_{g1}, F)]'\}|_{E\to 0}}} + 1\right]^2$$
$$\left[4\left(\frac{m_{c2} \text{ Real part of } \{[J_3(E, c, \eta_{g2}, F)]'\}|_{E\to 0}}{m_{c1} \text{ Real part of } \{[J_3(E, c, \eta_{g1}, F)]'\}|_{E\to 0}}\right)^{\frac{1}{2}}\right]^{-1},$$

$$a_{2\text{HD}1,18} = \left[\sqrt{\frac{m_{c2} \text{ Real part of } \{[J_3(E, c, \eta_{g2}, F)]'\}|_{E\to 0}}{m_{c1} \text{ Real part of } \{[J_3(E, c, \eta_{g1}, F)]'\}|_{E\to 0}}} - 1\right]^2$$
$$\left[4\left(\frac{m_{c2} \text{ Real part of } \{[J_3(E, c, \eta_{g2}, F)]'\}|_{E\to 0}}{m_{c1} \text{ Real part of } \{[J_3(E, c, \eta_{g1}, F)]'\}|_{E\to 0}}\right)^{\frac{1}{2}}\right]^{-1}$$

$$J_1(E, c, \eta_g, F) + iJ_2(E, c, \eta_g, F) = J_3(E, c, \eta_g, F),$$

$$C_{1\mathrm{HD}4,18}(E_{600},n_y,n_z,F) \equiv \left[\left(\frac{2m_{c1}}{\hbar^2}\right)J_3(E_{600},n_y,n_z,F) - G_2(n_y,n_z)\right]^{\frac{1}{2}},$$

$$D_{1\mathrm{HD}4,18}(E_{600},n_y,n_z,F) \equiv \left[\left(\frac{2m_{c2}}{\hbar^2}\right)J_3(E_{600},n_y,n_z,F) - G_2(n_y,n_z)\right]^{\frac{1}{2}}$$

and

$$G_2(n_y,n_z) = \left(\frac{n_y\pi}{d_y}\right)^2 + \left(\frac{n_z\pi}{d_z}\right)^2$$

The DOS function in this case can be written as

$$N_{600}(E,c,F) = \frac{2g_v}{d_x d_y d_z}\sum_{n_x=1}^{n_{x\max}}\sum_{n_y=1}^{n_{y\max}}\sum_{n_z=1}^{n_{z\max}}\delta'(E - E_{600}) \qquad (22.56)$$

Following (22.59) and (1.139), the RAS in this case is given by

$$\tau_{\mathrm{RAS}}(E,c,F) = \tau_0\left[\sum_{n_x=1}^{n_{x\max}}\sum_{n_y=1}^{n_{y\max}}\sum_{n_z=1}^{n_{z\max}}\delta'(E - E_{600})\right]^{-1} \qquad (22.57)$$

*22.2.7 The magneto RAS in QWHD effective mass superlattices of Kane type materials in the presence of intense electric field

Following Sasaki [3], the electron dispersion law in HD III–V effective mass superlattices (EMSLs) in the presence of light waves, the dispersion relations of whose constituent materials in the absence of any perturbation are defined by the three band model of Kane can be written as

$$\left(\frac{n_x\pi}{d_x}\right)^2 = \left[\frac{1}{L_0^2}\{\cos^{-1}(f_{18\mathrm{HD}2}(E_{601},n,F))\}^2 - \frac{2eB}{\hbar}\left(n+\frac{1}{2}\right)\right] \qquad (22.58)$$

where E_{601} is the totally quantized energy in this case,

$$f_{18\mathrm{HD}2}(E,n,F) = a_{1\mathrm{HD}1,18}\cos[a_0 C_{1\mathrm{HD}2,18}(E,n,F) + b_0 D_{1\mathrm{HD}2,18}(E,n,F)]$$
$$- a_{2\mathrm{HD}1,18}\cos[a_0 C_{1\mathrm{HD}2,18}(E,n,F) + b_0 D_{1\mathrm{HD}2,18}(E,n,F)],$$

$$a_{1\mathrm{HD}1,18} = \left[\sqrt{\frac{m_{c2}\ \text{Real part of}\ [J_3(0,c,\eta_{g2},F)]}{m_{c1}\ \text{Real part of}\ [J_3(0,c,\eta_{g1},F)]}} + 1\right]$$

$$\times \left[4\left(\sqrt{\frac{m_{c2} \text{ Real part of } [J_3(0,c,\eta_{g2},F)]}{m_{c1} \text{ Real part of } [J_3(0,c,\eta_{g1},F)]}}\right)^{1/2}\right]^{-1},$$

$$a_{2\text{HD}1} = \left[-1+\sqrt{\frac{m_{c2} \text{ Real part of } [J_3(0,c,\eta_{g2},F)]}{m_{c1} \text{ Real part of } [J_3(0,c,\eta_{g1},F)]}}\right]$$

$$\times \left[4\left(\sqrt{\frac{m_{c2} \text{ Real part of } [J_3(0,c,\eta_{g2},F)]}{m_{c1} \text{ Real part of } [J_3(0,c,\eta_{g1},F)]}}\right)^{1/2}\right]^{-1}$$

$$J_1(E,c,\eta_g,F) + iJ_2(E,c,\eta_g,F) = J_3(E,c,\eta_g,F),$$

$$C_{1\text{HD}2,18}(E,n,F) \equiv \left[\left(\frac{2m_c}{\hbar^2}\right) \times J_3(E,c,\eta_g,F) - \frac{2eB}{\hbar}\left(n+\frac{1}{2}\right)\right]$$

and

$$D_{1\text{HD}2,18}(E,n,F) \equiv \left[\left(\frac{2m_{c1}}{\hbar^2}\right) \times J_3(E,c,\eta_{g2},F) - \frac{2eB}{\hbar}\left(n+\frac{1}{2}\right)\right]$$

The DOS function is given by

$$N_{601}(E,F,\eta_g) = \frac{g_v eB}{\pi\hbar} \sum_{n_x=1}^{n_{x\max}} \sum_{n=0}^{n_{\max}} \delta'(E - E_{601}) \tag{22.59}$$

Following (22.59) and (1.139), the RAS in this case is given by

$$\tau_{\text{RAS}}(E,F,\eta_g) = \tau_0 \left[\sum_{n_x=1}^{n_{x\max}} \sum_{n=0}^{n_{\max}} \delta'(E - E_{601})\right]^{-1} \tag{22.60}$$

22.2.8 The RAS in quantum dot HD superlattices of with graded interfaces Kane type materials in the presence of intense electric field

The DR in bulk specimens of the heavily doped constituent materials of III–V SLs in the presence of intense electric field can be expressed as

$$\frac{\hbar^2 k^2}{2m_{cj}} = J_{1j}(E,\Delta_j,E_{g0j},\eta_j,F) + iJ_{2j}(E,\Delta_j,E_{g0j},\eta_j,F) \tag{22.61}$$

where $j = 1, 2$,

$$J_{1j}(E, \Delta_j, E_{g_{0j}}, \eta_j, F) = 2[1 + Erf(E/\eta_{g_j})]^{-1}[e_{1j}\psi_{0j}(E, \eta_{g_j}) + e_{2j}\psi_{1j}(E, \eta_{g_j}) + e_{3j}\theta_{0j}(E, \eta_{g_j}) + e_{4j}\gamma_0(E, \eta_{g_j}) + e_{5j}\frac{1}{2}[1 + Erf(E/\eta_{g_j})] - e_{6j}c_{1j}(E, c, \eta_{g_j}) + e_{7j}c_{3j}(E, c, \eta_{g_j})],$$

$$\psi_{1j}(E, \eta_{g_j}) = \left[\frac{E}{2}[1 + Erf(E/\eta_{g_j})]\left[E^2 + \frac{3}{2}\eta_{g_j}^2\right] + \frac{\eta_{g_j}}{2\sqrt{\pi}}\exp\left(\frac{-E^2}{\eta_{g_j}^2}\right)(4E^2 + \eta_{g_j}^2)\right],$$

$$J_{2j}(E, \Delta_j, E_{g_{0j}}, \eta_j, F) = 2[1 + Erf(E/\eta_{g_j})]^{-1}[e_{6j}c_{2j}(E, c, \eta_{g_j}) + e_{7j}D_{3j}(E, c, \eta_{g_j})],$$

$$e_{1j} = Q_{fj}\cdot\omega_{1j}, \quad Q_{fj} = \frac{m_{cj}}{m_{rj}}E_{g_j}^{-4}[5e_{fj}E_{g_j}^{-2} - 6G_{fj} + 7h_{fj}E_{g_j}^{-4}],$$

$$e_{fj} = A_{fj}P_{fj},$$

$$a_j = \frac{1}{E_{g_{0j}}}, \quad b_j = \frac{1}{E_{g_{0j}} + \Delta_j}, \quad c_j = \left(E_{g_{0j}} + \frac{2}{3}\Delta_j\right)^{-1},$$

$$e_{2j} = Q_{fj}\cdot\omega_{2j}, \quad \omega_{2j} = 2a_{1j}b_{1j}, \quad b_{1j} = (c_j)^{-2}(a_jc_j + b_jc_j - a_jb_j),$$

$$e_{3j} = (1 - P_{fj})a_{1j} + Q_{fj}\cdot\omega_{3j},$$

$$\omega_{3j} = (b_{1j}^2 + 2a_{1j}c_{1j}), \quad c_{1j} = \left[\frac{1}{c_j}\left(1 - \frac{a_j}{c_j}\right)\left(1 - \frac{b_j}{c_j}\right)\right],$$

$$e_{4j} = [(1 - P_{fj})b_{1j} + Q_{fj}\omega_{4j}],$$

$$\omega_{4j} = 2b_{1j}c_{1j}, \quad e_{5j} = [(1 - P_{fj})c_{1j} + \omega_{5j}Q_{fj}]$$

$$\omega_{5j} = (c_{1j}^2 - 2c_{1j}b_{1j}), \quad e_{7j} = Q_{fj}\omega_{7j}, \quad \omega_{7j} = c_{1j}^2,$$

$$e_{6j} = [(1 - P_{fj})c_{1j} - Q_{fj}\omega_{6j}] \quad \text{and} \quad \omega_{6j} = \left[\frac{2c_{1j}b_{1j}}{c_j}\left(1 - \frac{c_jc_{1j}}{b_{1j}}\right)\right]$$

$$A_{fj} = [F\hbar E_{g_j}(E_{g_j} - \delta'_j)]^2 m_{cj}(6m_{rj}^2(\delta'_j)^4)^{-1},$$

$$F = eF_s, \quad G_{fj} = e_{fj}(4\delta'_j + C_{fj}),$$

$$C_{fj} = (2E_{g_j}Q_j^2 + P_jQ_j(E_{g_j} - E'_{g_j}) - 2P_j^2E_{g_j}),$$

$$E'_{g_{0j}} = \frac{E_{g_{0j}}(E_{g_{0j}} - 3\delta'_j)}{E_{g_{0j}} + \delta'_j}, \quad P_j = \frac{r_{0j}^2}{2}\left(\frac{E_{g_{0j}} - \delta'_j}{E_{g_{0j}} + \delta'_j}\right),$$

$$r_{0j} = \left[\frac{6}{\chi_j}(E_{g_{0j}} + \Delta_j)(E_{g_j} + \frac{2}{3}\Delta_j)\right]^{1/2}, \quad Q_j = \frac{t_j^2}{2},$$

$$t_j = \left[\frac{6}{\chi_j}\left(E_{g_{0j}} + \frac{2}{3}\Delta_j\right)\right]^{1/2}, \quad h_{fj} = (4\delta'_j e_{fj} C_{fj})(B_{fj})^{-1},$$

$$B_{fj} = (P_j + Q_j)^2$$

$$P_{fj} = E_{g_{0_j}}^{-3}(e_{fj}E_{g_{0j}}^{-2} - G_{fj} + h_{fj}E_{g_{0j}}^{-4}), \quad \omega_{1j} = a_{1j}^2, \quad a_{1j} = \frac{a_j b_j}{c_j},$$

The DR in QDHD superlattices of Kane type semiconductors with graded interfaces in the presence of intense electric field can be expressed as

$$\left(\frac{n_z\pi}{d_z}\right)^2 = [G_{8,20} + iH_{8,20}]_{E=E_{620}} \tag{22.62}$$

where E_{620} is the totally quantized energy in this case.

$$G_{8,20} = \left[\frac{C_{7,20}^2 - D_{7,20}^2}{L_0^2} - \left[\left(\frac{n_x\pi}{d_x}\right)^2 + \left(\frac{n_y\pi}{d_y}\right)^2\right]\right], \quad C_{7,20} = \cos^{-1}(\bar{\omega}_{7,20}),$$

$$\bar{\omega}_{7,20} = (2)^{-1/2}\left[(1 - G_{7,20}^2 - H_{7,20}^2) - \sqrt{(1 - G_{7,20}^2 - H_{7,20}^2)^2 + 4G_{7,20}^2}\right]^{1/2}$$

$$\begin{aligned} G_{7,20} = {} & [G_{1,20} + (\rho_{5,20}G_{2,20}/2) - (\rho_{6,20}H_{2,20}/2) + (\Delta_0/2) \\ & \{\rho_{6,20}G_{3,20} - \rho_{8,20}H_{3,20} + \rho_{9,20}G_{4,20} - \rho_{10,20}H_{4,20} + \rho_{11,20}H_{5,20} \\ & - \rho_{12,20}H_{5,20} + (1/12)(\rho_{12,20}G_{6,20} - \rho_{14,20}H_{6,20})\}] \end{aligned}$$

$$\begin{aligned} G_{1,20} = {} & [(\cos(h_{1,20}))(\cosh(h_{2,20}))(\cosh(g_{1,20}))(\cos(g_{2,20})) \\ & + (\sin(h_{1,20}))(\sinh(h_{2,20}))(\sinh(g_{1,20}))(\sin(g_{2,20}))], \end{aligned}$$

$$h_{1,20} = e_{1,20}(b_0 - \Delta_0), \quad e_{1,20} = 2^{-1/2}(\sqrt{t_{1,20}^2 + t_{2,20}^2} + t_{1,20})^{1/2},$$

$$t_{1,20} = \left[\left(\frac{2m_{c1}}{\hbar^2}\right)J_{11}(E, \Delta_1, E_{g_{01}}, F) - \left[\left(\frac{n_x\pi}{d_x}\right)^2 + \left(\frac{n_y\pi}{d_y}\right)^2\right]\right]$$

$$t_{2,20} = \left[\left(\frac{2m_{c1}}{\hbar^2}\right) J_{21}(E, \Delta_1, E_{g_{01}}, F)\right], \quad h_{2,20} = e_{2,20}(b_0 - \Delta_0),$$

$$e_{2,20} = 2^{-1/2}\left(\sqrt{t_{1,20}^2 + t_{2,20}^2} - t_{1,20}\right)^{1/2},$$

$$g_{1,20} = d_{1,20}(a_0 - \Delta_0), \quad d_{1,20} = 2^{-1/2}\left(\sqrt{x_{1,20}^2 + y_{1,20}^2} + x_{1,20}\right)^{1/2},$$

$$x_{1,20} = [-(2m_{c2}/\hbar^2) \cdot J_{12}(E - V_0, E_{g2}, \Delta_2, \eta_{g2}, F) + G_2(n_y, n_z)],$$

$$y_{1,20} = [(2m_{c2}/\hbar^2) \cdot J_{21}(E - V_0, E_{g2}, \Delta_2, \eta_{g2}, F)],$$

$$g_{2,20} = d_{2,20}(a_0 - \Delta_0), \quad d_{2,20} = 2^{-1/2}\left(\sqrt{x_{1,20}^2 + y_{1,20}^2} - x_{1,20}\right)^{1/2},$$

$$\rho_{5,20} = (\rho_{3,20}^2 + \rho_{4,20}^2)^{-1}[\rho_{1,20}\rho_{3,20} - \rho_{2,20}\rho_{4,20}],$$

$$\rho_{1,20} = [d_{1,20}^2 + e_{2,20}^2 - d_{2,20}^2 - e_{1,20}^2], \quad \rho_{3,20} = [d_{1,20}e_{1,20} + d_{2,20}e_{2,20}],$$

$$\rho_{2,20} = 2[d_{1,20}d_{2,20} + e_{1,20}e_{2,20}], \quad \rho_{4,20} = [d_{1,20}e_{2,20} - e_{1,20}d_{2,20}],$$

$$\begin{aligned} G_{2,20} = {} & [(\sin(h_{1,20}))(\cosh(h_{2,20}))(\sinh(g_{1,20}))(\cos(g_{2,20})) \\ & + (\cos(h_{1,20}))(\sinh(h_{2,20}))(\cosh(g_{1,20}))(\sin(g_{2,20}))], \end{aligned}$$

$$\rho_{6,20} = (\rho_{3,20}^2 + \rho_{4,20}^2)^{-1}[\rho_{1,20}\rho_{4,20} + \rho_{2,20}\rho_{3,20}],$$

$$\begin{aligned} H_{2,20} = {} & [(\sin(h_{1,20}))(\cosh(h_{2,20}))(\sin(g_{1,20}))(\cosh(g_{1,20})) \\ & - (\cos(h_{1,20}))(\sinh(h_{2,20}))(\sinh(g_{1,20}))(\cos(g_{2,20}))], \end{aligned}$$

$$\rho_{7,20} = [(e_{1,20}^2 + e_{2,20}^2)^{-1}[e_{1,20}(d_{1,20}^2 - d_{2,20}^2) - 2d_{1,20}d_{2,20}e_{2,20}] - 3e_{1,20}],$$

$$\begin{aligned} G_{3,20} = {} & [(\sin(h_{1,20}))(\cosh(h_{2,20}))(\cosh(g_{1,20}))(\cos(g_{2,20})) \\ & + (\cos(h_{1,20}))(\sinh(h_{2,20}))(\sinh(g_{1,20}))(\sin(g_{2,20}))], \end{aligned}$$

$$\rho_{8,20} = [(e_{1,20}^2 + e_{2,20}^2)^{-1}[e_{2,20}(d_{1,20}^2 - d_{2,20}^2) + 2d_{1,20}d_{2,20}e_{1,20}] + 3e_{2,20}],$$

$$\begin{aligned} H_{3,20} = {} & [(\sin(h_{1,20}))(\cosh(h_{2,20}))(\sin(g_{2,20}))(\sinh(g_{1,20})) \\ & - (\cos(h_{1,20}))(\sinh(h_{2,20}))(\cosh(g_{1,20}))(\cos(g_{2,20}))], \end{aligned}$$

$$\rho_{9,20} = [(d_{1,20}^2 + d_{2,20}^2)^{-1}[d_{1,20}(e_{2,20}^2 - e_{1,20}^2) + 2e_{2,20}d_{2,20}e_{1,20}] + 3d_{1,20}],$$

$$\begin{aligned} G_{4,20} = {} & [(\cos(h_{1,20}))(\cosh(h_{2,20}))(\cos(g_{2,20}))(\sinh(g_{1,20})) \\ & - (\sin(h_{1,20}))(\sinh(h_{2,20}))(\cosh(g_{1,20}))(\sin(g_{2,20}))], \end{aligned}$$

$$\begin{aligned} \rho_{10,20} = {} & [-(d_{1,20}^2 + d_{2,20}^2)^{-1}[d_{2,20}(-e_{2,20}^2 + e_{1,20}^2) \\ & + 2e_{2,20}d_{2,20}\rho_{1,20}] + 3d_{2,20}], \end{aligned}$$

$$
\begin{aligned}
H_{4,20} &= [(\cos(h_{1,20}))(\cosh(h_{2,20}))(\cosh(g_{1,20}))(\sin(g_{2,20}))\\
&\quad + (\sin(h_{1,20}))(\sinh(h_{2,20}))(\sinh(g_{1,20}))(\cos(g_{2,20}))],\\
\rho_{11,20} &= 2[d_{1,20}^2 + e_{2,20}^2 - d_{2,20}^2 - e_{1,20}^2],\\
G_{5,20} &= [(\cos(h_{1,20}))(\cosh(h_{2,20}))(\cos(g_{2,20}))(\cosh(g_{1,20}))\\
&\quad - (\sin(h_{1,20}))(\sinh(h_{2,20}))(\sinh(g_{1,20}))(\sin(g_{2,20}))],\\
\rho_{12,20} &= 4[d_{1,20}d_{2,20} + e_{1,20}e_{2,20}],\\
H_{5,20} &= [(\cos(h_{1,20}))(\cosh(h_{2,20}))(\sinh(g_{1,20}))(\sin(g_{2,20}))\\
&\quad + (\sin(h_{1,20}))(\sinh(h_{2,20}))(\cosh(g_{1,20}))(\cos(g_{2,20}))],\\
\rho_{13,20} &= [\{5(d_{1,20}e_{1,20}^3 - 3e_{1,20}e_{2,20}^2 d_{1,20})\\
&\quad + 5d_{2,20}(e_{1,20}^3 - 3e_{1,20}^2 e_{2,20})\}(d_{1,20}^2 + d_{2,20}^2)^{-1}\\
&\quad + (e_{1,20}^2 + e_{2,20}^2)^{-1}\{5(e_{1,20}d_{1,20}^3 - 3d_{2,20}e_{1,20}^2 d_{1,20})\\
&\quad + 5(d_{2,20}^3 e_{2,20} - 3d_{1,20}^2 d_{2,20}e_{2,20})\} - 34(d_{1,20}e_{1,20} + d_{2,20}e_{2,20})],\\
G_{6,20} &= [(\sin(h_{1,20}))(\cosh(h_{2,20}))(\sinh(g_{1,20}))(\cos(g_{2,20}))\\
&\quad + (\cos(h_{1,20}))(\sinh(h_{2,20}))(\cosh(g_{1,20}))(\sin(g_{2,20}))],\\
\rho_{14,20} &= [\{5(d_{1,20}e_{2,20}^3 - 3e_{2,20}e_{1,20}^2 d_{1,20})\\
&\quad + 5d_{2,20}(-e_{1,20}^3 + 3e_{2,20}^2 e_{1,20})\}(d_{1,20}^2 + d_{2,20}^2)^{-1}\\
&\quad + (e_{1,20}^2 + e_{2,20}^2)^{-1}\{5(-e_{1,20}d_{2,20}^3 + 3d_{1,20}^2 d_{2,20}e_{1,20})\\
&\quad + 5(-d_{1,20}^3 e_{2,20} + 3d_{2,20}^2 d_{1,20}e_{2,20})\} + 34(d_{1,20}e_{2,20} - d_{2,20}e_{1,20})],\\
H_{6,20} &= [(\sin(h_{1,20}))(\cosh(h_{2,20}))(\cosh(g_{1,20}))(\sin(g_{2,20}))\\
&\quad - (\cos(h_{1,20}))(\sinh(h_{2,20}))(\sinh(g_{1,20}))(\cos(g_{2,20}))],\\
H_{7,20} &= [H_{1,20} + (\rho_{5,20}H_{2,20}/2) + (\rho_{6,20}G_{2,20}/2)\\
&\quad + (\Delta_0/2)\{\rho_{8,20}G_{3,20} + \rho_{7,20}H_{3,20}\\
&\quad + \rho_{10,20,}G_{4,20} + \rho_{9,20}H_{4,20} + \rho_{12,20}G_{5,20} + \rho_{11,20}H_{5,20}\\
&\quad + (1/12)(\rho_{14,20}G_{6,20} + \rho_{14,20}H_{6,20})\}],\\
H_{1,20} &= [(\sin(h_{1,20}))(\sinh(h_{2,20}))(\cosh(g_{1,20}))(\cos(g_{2,20}))\\
&\quad + (\cos(h_{1,20}))(\cosh(h_{2,20}))(\sinh(g_{1,20}))(\sin(g_{2,20}))],\\
D_{7,20} &= \sinh^{-1}(\bar{\omega}_{7,20}), \quad H_{8,20} = (2C_{7,20}D_{7,20}/L_0^2)
\end{aligned}
$$

The DOS function is given by

$$N_{\text{QDHDSL}}(E, \eta_g, F) = \frac{2g_v}{d_x d_y d_z} \sum_{n_x=1}^{n_{x\max}} \sum_{n_y=1}^{n_{y\max}} \sum_{n_z=1}^{n_{z\max}} \delta'(E - E_{620}) \quad (22.63)$$

Following (22.63) and (1.139), the RAS in this case is given by

$$\tau_{\text{RAS}}(E, \eta_g, F) = \tau_0 \left[\sum_{n_x=1}^{n_{x\max}} \sum_{n_y=1}^{n_{y\max}} \sum_{n_z=1}^{n_{z\max}} \delta'(E - E_{620}) \right]^{-1} \quad (22.64)$$

22.2.9 The magneto RAS in QWHD superlattices of Kane type materials with graded interfaces in the presence of intense electric field

The magneto DR in QWHD superlattices of Kane type semiconductors with graded interfaces in the presence of intense electric field can be expressed as

$$\left(\frac{n_z \pi}{d_z}\right)^2 = [G_{8,19n} + iH_{8,19n}]_{E=E_{650}} \quad (22.65)$$

where E_{650} is the totally quantized energy in this case,

$$G_{8,19n} = \left[\frac{C_{7,19n}^2 - D_{7,19n}^2}{L_0^2} - \frac{2eB}{\hbar}\left(n + \frac{1}{2}\right)\right], \quad C_{7,19n} = \cos^{-1}(\bar{\omega}_{7,19n}),$$

$$\bar{\omega}_{7,19n} = (2)^{\frac{-1}{2}} \left[(1 - G_{7,19n}^2 - H_{7,19n}^2) - \sqrt{(1 - G_{7,19n}^2 - H_{7,19n}^2)^2 + 4G_{7,19n}^2}\right]^{\frac{1}{2}},$$

$$G_{7,19n} = \Big[G_{1,19n} + (\rho_{5,19n} G_{2,19n}/2) - (\rho_{6,19n} H_{2,19n}/2) + (\Delta_0/2)\Big\{\rho_{6,19n} G_{3,19n} - \rho_{8,19n} H_{3,19n} + \rho_{9,19n} H_{4,19n} - \rho_{10,19n} H_{4,19n} + \rho_{11,19n} H_{5,19n} - \rho_{12,19n} H_{5,19n} + \frac{1}{12}(\rho_{11,19n} G_{6,19n} - \rho_{14,19n} H_{6,19n})\Big\}\Big]$$

$$h_{1,19n} = e_{1,19n}(b_0 - \Delta_0), \quad e_{1,19n} = 2^{\frac{-1}{2}} \left(\sqrt{t_{1,19n}^2 + t_{2,19n}^2} + t_{1,19n}\right)^{1/2}$$

$$t_{1,19n} = \left[(2m_{c1}/\hbar^2)J_{11}(E,\Delta_1,E_{g01},F) - \frac{2eB}{\hbar}\left(n+\frac{1}{2}\right)\right],$$

$$t_{2,19n} = [(2m_{c1}/\hbar^2)J_{21}(E,\Delta_1,E_{g01},F)],$$

$$h_{2,19n} = e_{2,19n}(b_0-\Delta_0), \quad e_{2,19n} = 2^{\frac{-1}{2}}\left(\sqrt{t_{1,19n}^2+t_{2,19n}^2} - t_{1,19n}\right)^{1/2},$$

$$g_{1,19n} = d_{1,19n}(a_0-\Delta_0),$$

$$d_{1,19n} = 2^{\frac{-1}{2}}\left(\sqrt{x_{1,19n}^2+y_{2,19n}^2} + x_{1,19n}\right)^{1/2},$$

$$x_{1,19n} = \left[-(2m_{c2}/\hbar^2)\cdot J_{11}(E-V_0,E_{g2}\Delta_2,\eta_{g2},F) + \frac{2eB}{\hbar}\left(n+\frac{1}{2}\right)\right],$$

$$y_{1,19n} = [(2m_{c2}/\hbar^2)J_{21}(E-V_0,E_{g2}\Delta_2,\eta_{g2},F)],$$

$$g_{2,19n} = d_{2,19n}(a_0-\Delta_0), \quad d_{2,19n} = 2^{\frac{-1}{2}}\left(\sqrt{x_{1,19n}^2+y_{2,19n}^2} - x_{1,19n}\right)^{1/2},$$

$$\rho_{5,19n} = (\rho_{3,19n}^2+\rho_{4,19n}^2)^{-1}[\rho_{1,19n}\rho_{3,19n} - \rho_{2,19n}\rho_{4,19n}],$$

$$\rho_{1,19n} = [d_{1,19n}^2 + e_{2,19n}^2 - d_{2,19n}^2 - e_{1,19n}^2],$$

$$\rho_{3,19n} = [d_{1,19n}e_{1,19n} + d_{2,19n}e_{2,19n}],$$

$$\rho_{2,19n} = 2[d_{1,19n}d_{2,19n} + e_{1,19n}e_{2,19n}],$$

$$\rho_{4,19n} = [d_{1,19n}e_{2,19n} - e_{1,19n}d_{2,19n}],$$

$$\begin{aligned} G_{2,19n} &= [(\sin(h_{1,19n}))(\cosh(h_{2,19n}))(\sinh(g_{1,19n}))(\cos(g_{2,19n})) \\ &\quad + [(\cos(h_{1,19n}))(\sinh(h_{2,19n}))(\cosh(g_{1,19n}))(\sin(g_{2,19n}))], \end{aligned}$$

$$\rho_{6,19n} = (\rho_{3,19n}^2+\rho_{4,19n}^2)^{-1}[\rho_{1,19n}\rho_{4,19n} + \rho_{2,19n}\rho_{3,19n}]$$

$$\begin{aligned} H_{2,19n} &= [(\sin(h_{1,19n}))(\cos(h_{2,19n}))(\sin(g_{2,19n}))(\cosh(g_{1,19n})) \\ &\quad - (\cos(h_{1,19n}))((\sinh(h_{2,19n}))((\sinh(g_{2,19n}))(\cos(g_{2,19n}))], \end{aligned}$$

$$\begin{aligned} \rho_{7,19n} &= [(e_{1,19n}^2+e_{2,19n}^2)^{-1}[e_{1,19n}(d_{1,19n}^2 - d_{2,19n}^2) \\ &\quad - 2d_{1,19n}d_{2,19n}e_{2,19n}] - 3e_{1,19n}], \end{aligned}$$

$$\begin{aligned} G_{3,19n} &= [(\sin(h_{1,19n}))(\cosh(h_{2,19n}))(\cosh(g_{1,19n}))(\cos(g_{2,19n})) \\ &\quad + (\cos(h_{1,19n}))((\sinh(h_{2,19n}))((\sinh(g_{2,19n}))(\cos(g_{2,19n}))], \end{aligned}$$

$$\begin{aligned} \rho_{8,19n} &= [(e_{1,19n}^2+e_{2,19n}^2)^{-1}[e_{2,19n}(d_{1,19n}^2 - d_{2,19n}^2) \\ &\quad + 2d_{1,19n}d_{2,19n}e_{1,19n}] + 3e_{2,19n}], \end{aligned}$$

$$
\begin{aligned}
H_{3,19n} &= [(\sin(h_{1,19n}))(\cosh(h_{2,19n}))(\sin(g_{2,19n}))(\sinh(g_{1,19n}))\\
&\quad - (\cos(h_{1,19n}))((\sinh(h_{2,19n}))(\cosh(g_{1,19n}))(\cos(g_{2,19n}))],\\
\rho_{9,19n} &= [(d_{1,19n}^2 + d_{2,19n}^2)^{-1}[d_{1,19n}(e_{2,19n}^2 - e_{1,19n}^2)\\
&\quad + 2e_{2,19n}d_{2,19n}e_{1,19n}] + 3d_{1,19n}],\\
G_{4,19n} &= [(\cos(h_{1,19n}))(\cosh(h_{2,19n}))(\cos(g_{2,19n}))(\sinh(g_{1,19n}))\\
&\quad - (\sin(h_{1,19n}))((\sinh(h_{2,19n}))(\cosh(g_{1,19n}))(\sin(g_{2,19n}))],\\
\rho_{10,19n} &= [-(d_{1,19n}^2 + d_{2,19n}^2)^{-1}[d_{2,19n}(-e_{2,19n}^2 + e_{1,19n}^2)\\
&\quad + 2e_{2,19n}d_{2,19n}\rho_{1,19n}] + 3d_{2,19n}],\\
H_{4,19n} &= [(\cos(h_{1,19n}))(\cosh(h_{2,19n}))(\cosh(g_{1,19n}))(\sin(g_{2,19n}))\\
&\quad + (\sin(h_{1,19n}))((\sinh(h_{2,19n}))((\sin(g_{1,19n}))(\cos(g_{2,19n}))],\\
\rho_{11,19n} &= 2[d_{1,19n}^2 + e_{2,19n}^2 - d_{2,19n}^2 - e_{1,19n}^2],\\
G_{5,19n} &= [(\cos(h_{1,19n}))(\cosh(h_{2,19n}))(\cos(g_{2,19n}))(\cosh(g_{1,19n}))\\
&\quad - (\sin(h_{1,19n}))((\sinh(h_{2,19n}))(\sinh(g_{1,19n}))(\sin(g_{2,19n}))],\\
\rho_{12,19n} &= 4[d_{1,19n}d_{2,19n} + e_{1,19n}e_{2,19n}],\\
H_{5,19n} &= [(\cos(h_{1,19n}))(\cosh(h_{2,19n}))(\sinh(g_{1,19n}))(\sin(g_{2,19n}))\\
&\quad + (\sin(h_{1,19n}))((\sinh(h_{2,19n}))(\cosh(g_{1,19n}))(\cos(g_{2,19n}))],\\
\rho_{13,19n} &= [\{5(d_{1,19n}e_{1,19n}^3 - 3e_{1,19n}e_{2,19n}^2 d_{1,19n})\\
&\quad + 5d_{2,19n}(e_{1,19n}^3 - 3e_{1,19n}^2 e_{2,19n})\}(d_{1,19n}^2 + d_{2,19n}^2)\\
&\quad + (e_{1,19n}^2 + e_{2,19n}^2)^{-1}\{5(e_{1,19n}d_{1,19n}^3 - 3d_{2,19n}e_{1,19n}^2 d_{1,19n})\\
&\quad + 5(d_{2,19n}^3 e_{2,19n} - 3d_{1,19n}^2 d_{2,19n}e_{2,19n})\}\\
&\quad - 34(d_{1,19n}e_{1,19n} + d_{2,19n}e_{2,19n})]\\
G_{6,19n} &= [(\sin(h_{1,19n}))(\cosh(h_{2,19n}))(\sinh(g_{1,19n}))(\cos(g_{2,19n}))\\
&\quad + (\cos(h_{1,19n}))((\sinh(h_{2,19n}))(\cosh(g_{1,19n}))(\sin(g_{2,19n}))],\\
\rho_{14,19n} &= [\{5(d_{1,19n}e_{1,19n}^3 - 3e_{2,19n}e_{1,19n}^2 d_{1,19n})\\
&\quad + 5d_{2,19n}(-e_{1,19n}^3 + 3e_{2,19n}^2 e_{1,19n})\}(d_{1,19n}^2 + d_{2,19n}^2)\\
&\quad + (e_{1,19n}^2 + e_{2,19n}^2)^{-1}\{5(-e_{1,19n}d_{2,19n}^3 + 3d_{1,19n}^2 d_{2,19n}e_{1,19n})\\
&\quad + 5(-d_{1,19n}^3 e_{2,19n} + 3d_{2,19n}^2 d_{1,19n}e_{2,19n})\}\\
&\quad - 34(d_{1,19n}e_{2,19n} - d_{2,19n}e_{1,19n})],
\end{aligned}
$$

$$H_{6,19n} = [(\sin(h_{1,19n}))(\cosh(h_{2,19n}))(\cosh(g_{1,19n}))(\sin(g_{2,19n}))$$
$$- (\cos(h_{1,19n}))((\sinh(h_{2,19n}))(\sinh(g_{1,19n}))(\cos(g_{2,19n}))],$$
$$H_{7,19n} = [H_{1,19n} + (\rho_{5,19}H_{2,19n}/2) + (\rho_{6,19n}G_{2,19n}/2) + (\Delta_0/2)$$
$$\times \{\rho_{8,19n}G_{3,19n} + \rho_{7,19n}H_{3,19n} + \rho_{10,19n}G_{4,19n} + \rho_{9,19n}H_{4,19n}$$
$$\rho_{12,19n}G_{5,19n} + \rho_{11,19n}H_{5,19n} + (1/12)(\rho_{14,19n}G_{6,19n}$$
$$+ \rho_{14,19n}H_{6,19n})\}]$$
$$H_{1,19n} = [(\sin(h_{1,19n}))(\sinh(h_{2,19n}))(\cosh(g_{1,19n}))(\cos(g_{2,19n}))$$
$$+ (\cos(h_{1,19n}))((\cosh(h_{2,19n}))(\sinh(g_{1,19n}))(\sin(g_{2,19n}))],$$
$$D_{7,19n} = \sinh^{-1}(\bar{\omega}_{7,19n}), H_{8,19n} = (2C_{7,19n}D_{7,19n}/L_0^2)$$

The DOS function is given by

$$N_{\text{QDHDSL}}(E, \eta_g, F) = \frac{g_v eB}{\pi\hbar} \sum_{n_z=1}^{n_{z\max}} \sum_{n=0}^{n_{\max}} \delta'(E - E_{650}) \tag{22.66}$$

Following (22.66) and (1.139), the RAS in this case is given by

$$\tau_{\text{RAS}}(E, \eta_g, F) = \tau_0 \left[\sum_{n_z=1}^{n_{z\max}} \sum_{n=0}^{n_{\max}} \delta'(E - E_{650}) \right]^{-1} \tag{22.67}$$

22.3 Summary and Conclusion

In this chapter, we have studied the relaxation time for the *acoustic mode scattering* in the appropriate cases under different physical conditions of various heavily doped quantized structures in the presence of intense electric field. For the purpose of condensation we have not given the plots for RAS any case with the hope that the readers will perform all the computer programming for all the quantized materials under different physical conditions for the purpose of creating new physics and other important transport quantities which are totally band structure dependent. The theoretical results as given here would be useful in analyzing various other experimental data related to the transport phenomenon. We must note that the study of transport phenomena and the formulation of the electronic properties of low dimensional field aided HD compounds are based on the dispersion relations in such materials.

22.4 Open Research Problems

(R.22.1) Investigate the relaxation time due to acoustic mode scattering (RAS) in the presence of intense external non-uniform electric field for all the HD superlattices whose respective dispersion relations of the carriers are given in this chapter.

(R.22.2) Investigate the RAS for the heavily doped semiconductors in SLs the presences of Gaussian, exponential, Kane, Halperian, Lax and Bonch-Burevich types of band tails for all SL systems as discussed in this chapter in the presence of external oscillatory and non-uniform electric field.

(R.22.3) Investigate the RAS in the presence of external non-uniform electric field for short period, strained layer, random and Fibonacci HD superlattices in the presence of an arbitrarily oriented alternating electric field.

(R.22.4) Investigate all the appropriate problems of this chapter for a Dirac electron.

(R.22.5) Investigate all the appropriate problems of this chapter by including the many body, broadening and hot carrier effects respectively.

(R.22.6) Investigate all the appropriate problems of this chapter by removing all the mathematical approximations and establishing the respective appropriate uniqueness conditions.

References

[1] L. Schiff, *Quantum Mechanics*, (McGraw-Hill, USA, **1968**).
[2] B. R. Nag, *Electron Transport in Compound Semiconductors*, (Springer, Heidelberg, **1980**).
[3] H. Sasaki, *Phys. Rev. B*, 30, 7016 **(1984)**.
[4] H. X. Jiang, J. Y. Lin, *J. Appl. Phys.*, 61, 624 **(1987)**.

Chapter 23

The Generalized Distribution Function in Heavily Doped Materials

The only mandatory condition to get the permanent visa to live in the wonderful world called SOLITUDE is that I must die before my body dies.

23.1 Introduction

The concept of distribution functions of the carriers in different materials occupies a singular position in the arena of materials science in general for not only the characterization of different systems but also for the study of the transport features in different technologically important materials [1, 2]. Fermi and Dirac (both independently) formulated the Fermi–Dirac (FD) statistics for electrons [3,4] whereas the Bose–Einstein (BE) statistics appeared in the literature in 1924 for bosons [5] together with the fact that both of them get simplified into well-known Maxwell–Boltzmann statistics under certain limiting conditions. The FD statistics is generally applicable for studying the transport features of degenerate materials, because in this case, the band structure prevails [6] and the Fermi energy (E_F) is above the edge of the conduction band (CB). Also, we know by Pauli exclusion principle, the occupational probability of an electron at an energy state is (1/2) at E_F [7]. The simple FD function cannot explain the existence of band tailing phenomena observed in the case of heavily doped systems at low temperatures where the quantum effects become prominent. Therefore, some modifications must be expected in the mathematical form of the well-known FD distribution function. It might be noted that in the case of disordered materials, the conduction band penetrates within the Forbidden Band (FB) and the carriers do exist in the tail region. We know that for a semiconductor with heavily doped conditions, the doping's levels must satisfy the inequality $(a_D \cdot Ni^{1/3}) > 1$, where a_D is the Bohr radius and Ni is the doping concentration of the carriers, and that for "non-degenerately doped case", the doping condition should satisfy the inequality [9–11] $\{(0.02 \leq [a_D \cdot Ni^{1/3}] < 1.0)\}$ [8–12]. As we know that the normal FD distribution function cannot explain the occupational probability of the carriers in the Tail-Zone (TZ), because it is valid only for the Band gap zone (BZ). Also, in general, the FB zone is known as imaginary band (IB). Therefore, we might conclude that for the TZ region, the probability of the distribution of the carriers must be different from normal FD distribution function. In this chapter, an attempt is made in the theoretical background to derive the generalize FD and BE in a very simplified manner and we also note various important conclusion which emerge from it. This new probability of distribution in the TZ region is termed as "Extended and Modified distribution functions". Sections 23.2–23.4 contain the theoretical background, the result and discussions and the single open research problem respectively in this context.

23.2 Theoretical Background

Let us assume that the probability of occupation of the energy level $\bar{E}$ by an electron is $F(\bar{E})$. Let us also assume that there are two initial energy levels namely $\bar{E}_1$ and $\bar{E}_2$ together with two final energy levels $\bar{E}_3$ and $\bar{E}_4$ respectively as shown in Fig. 23.1.

The probability of forward transmission of an electron in the initial energy level $\bar{E}_1$ to the final energy level $\bar{E}_3$ and an electron in the initial energy level $\bar{E}_2$ to the final energy level $\bar{E}_4$ is proportional to

$$F(\bar{E}_1)[1 - F(\bar{E}_3)]F(\bar{E}_2)[1 - F(\bar{E}_4)]$$

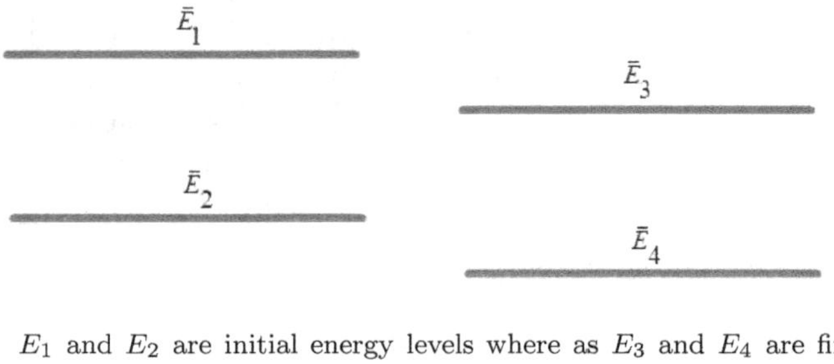

Fig. 23.1. E_1 and E_2 are initial energy levels where as E_3 and E_4 are final energy levels.

Besides, the probability of reverse transmission is similarly proportional to

$$F(\bar{E}_3)[1-F(\bar{E}_1)]F(\bar{E}_4)[1-F(\bar{E}_2)]$$

The application of the principle of detailed balance of Einstein leads to the result

$$F(\bar{E}_1)[1-F(\bar{E}_3)]F(\bar{E}_2)[1-F(\bar{E}_4)] \\ = F(\bar{E}_3)[1-F(\bar{E}_1)]F(\bar{E}_4)[1-F(\bar{E}_2)] \tag{23.1}$$

Dividing both sides by $[F(\bar{E}_1)F(\bar{E}_2)F(\bar{E}_3)F(\bar{E}_4)]$ and assuming $[F(\bar{E}_1)F(\bar{E}_2)F(\bar{E}_3)F(\bar{E}_4)] \neq 0$, we get

$$\left[\frac{1}{F(\bar{E}_1)}-1\right]\left[\frac{1}{F(\bar{E}_2)}-1\right]=\left[\frac{1}{F(\bar{E}_3)}-1\right]\left[\frac{1}{F(\bar{E}_4)}-1\right] \tag{23.2}$$

Energy conservation rule gives us

$$\bar{E}_1+\bar{E}_2=\bar{E}_3+\bar{E}_4 \tag{23.3}$$

Let us assume

$$\frac{1}{F(\bar{E})}-1=A+\exp\left(\frac{\bar{E}-E_F}{k_BT}\right) \tag{23.4}$$

where A is a constant. Using (23.2)–(23.4), we get

$$\left[A+\exp\left(\frac{\bar{E}_1-E_F}{k_BT}\right)\right]\left[A+\exp\left(\frac{\bar{E}_2-E_F}{k_BT}\right)\right] \\ =\left[A+\exp\left(\frac{\bar{E}_3-E_F}{k_BT}\right)\right]\left[A+\exp\left(\frac{\bar{E}_4-E_F}{k_BT}\right)\right] \tag{23.5}$$

From (23.5), we can write

$$\left[\exp\left(\frac{\bar{E}_1}{k_BT}\right)+\exp\left(\frac{\bar{E}_2}{k_BT}\right)\right]^{-1}=\left[\exp\left(\frac{\bar{E}_3}{k_BT}\right)+\exp\left(\frac{\bar{E}_4}{k_BT}\right)\right]^{-1} \tag{23.6}$$

If (23.6) is satisfied by (23.4) then the $F(\bar{E})$ as given by (23.4) is a solution. Now, for heavily doped nano-materials which are being discussed in this chapter is a low temperature phenomena i.e., $T \to 0$ and under this condition, (23.6) is being satisfied by (23.4).

Thus, the modified Fermi–Dirac function can be written from (23.4) as

$$F(\bar{E}) = [1 + A + \exp(y)]^{-1} \tag{23.7}$$

where

$$y = ((\bar{E} - E_F)/(k_B T))$$

The conclusions from (23.7) are written as follows:

1. When $A = 0$, we get the ordinary Fermi–Dirac function from (23.7) as
$$F(\bar{E}) = [1 + \exp((\bar{E} - E_F)/(k_B T))]^{-1} \tag{23.8}$$
2. When $A = -1$, from (23.7), we can write that the $F(\bar{E}) = C\exp(-\bar{E}/k_B T)$, where $C = \exp(E_F/k_B T)$ which the well-known Maxwell–Boltzmann distribution.
3. The substitutions $A = 0$ and $\bar{E} = E_F$ together with $A = -4$ and $\bar{E} = E_F$ in (23.7) generates the well-known Pauli's exclusion principle as $(\pm 1/2)$.
4. For $A = i - 1 (i = \sqrt{-1})$ and $\bar{E} = E_F$ together with $A = -3 + i$ and $\bar{E} = E_F$ in (23.7) leads to complex value of the Pauli's spin in the tail zone as $1/2(\pm 1 \mp i)$ respectively.
5. The substitution $A = 1 + B_0 i$ (where B_0 is a real constant) leads to the modified and extended complex Fermi–Dirac function as
$$F_c(\bar{E}) = [1 + B_0 i + \exp(y)]^{-1} \tag{23.9}$$
From (23.9), we can write
$$\mathrm{Re}[F(\bar{E})] = [1 + \exp(y)][B_0^2 + (1 + \exp y)^2]^{-2} \tag{23.10}$$
and
$$\mathrm{Im}[F(\bar{E})] = B_0[B_0^2 + (1 + \exp y)^2]^{-2} \tag{23.11}$$
Few important points can also be noted in this context.
6. We know the concept of "spin g-factor" is very important for semiconductors; while studying the electron energy $(\bar{E}_5)$ in presence of a magnetic field B, applied to a semiconductor along z-axis which in this case assumes the form
7. $$\bar{E}_5 = \pm(g/2)[\hbar\omega_B] \tag{23.12}$$

where $\hbar\omega_B$ is the cyclotron resonance frequency of electron in this case.

Equation (23.12), exhibits the fact that it has only magnitude but no phase value.

The electron energy in the presence of the complex spin value $\pm(1/2)(1-i)$ in the tail zone can be written as

$$\bar{E}_6 = \pm(g/2)[\hbar\omega_B] \tag{23.13}$$

from (23.13), we observe that the electron energy ($\bar{E}_6$) in the presence of a magnetic field, has both magnitude and the phase value,

Thus, the magnitude is given by

$$\bar{E}_6 = \frac{g}{2}\hbar\omega_B(\sqrt{2}) = 0.707\left[\frac{g}{2}\hbar\omega_c\right] \approx 71\% \text{ of } (g\hbar\omega_B) \tag{23.14}$$

and the phase value can be written as

$$\phi_{\text{EX}} = \tan^{-1}\left\{-\frac{\left[\pm\frac{g}{2}\hbar\omega_B\right]}{\left[\pm\frac{g}{2}\hbar\omega_B\right]}\right\} = \pm\frac{\pi}{4} \tag{23.15}$$

In the conventional case, Pauli's value is 50% and the phase value $\phi_c = 0^0$. When an electron rotates in its orbit around the nucleus, the average electron energy is about 71% of its maximum value. The phase value ϕ_{EX}, in this case is $\pm\ \pi/4$.

For Boson, Pauli's exclusion principle is not obeyed. Replacing (-1) in (23.2) and (23.44) by $(+1)$

$$\left[\frac{1}{F(\bar{E}_1)}+1\right]\left[\frac{1}{F(\bar{E}_2)}+1\right] = \left[\frac{1}{F(\bar{E}_3)}+1\right]\left[\frac{1}{F(\bar{E}_4)}+1\right] \tag{23.16}$$

$$\frac{1}{F(\bar{E})}+1 = A + \exp\left(\frac{\bar{E}-E_F}{k_BT}\right)$$

Following the method as given above, we note that (23.17) is a solution of equation (23.16). Substituting $A = 0$, we note that

$$F(\bar{E}) = \left[\exp\left(\frac{\bar{E}-E_F}{k_BT}\right) - 1\right]^{-1} \tag{23.17}$$

which is the well-known BE statistics.

23.3 Result and Discussions

Figures 23.2–23.4 exhibit the ordinary Fermi–Dirac function, real and imaginary parts of the extended Fermi–Dirac functions as given by (23.8), (23.10) and (23.11), respectively, where for Figs. 23.3 and 23.4, we have

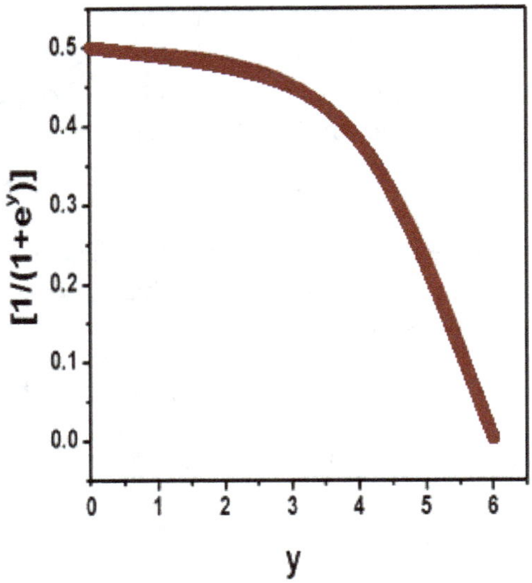

Fig. 23.2. Plot of the normal Fermi–Dirac probability distribution factor (23.8) as a function of y.

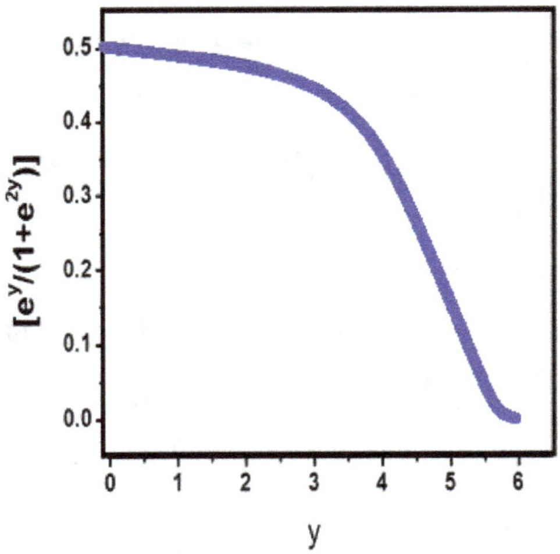

Fig. 23.3. Plot of the real part of the extended and modified Fermi–Dirac probability distribution factor as given by (23.10) as a function of y.

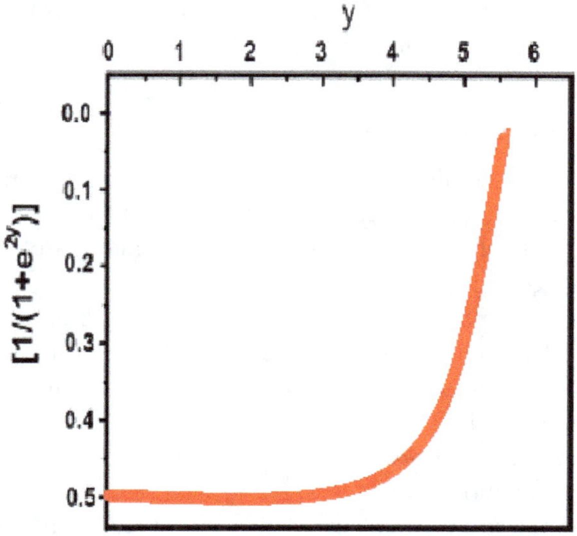

Fig. 23.4. Plot of the imaginary part of the extended and modified Fermi–Dirac probability distribution factor as given by (23.11) as a function of y.

taken $B_0 = 1$ for the purpose of simplifying numerical computations. The graphs (23.3) and (23.4) are new and should be used in dealing with all types of carrier properties in the tail zone of HD quantum material.

In this chapter, we have derived the modified and extended form of (FD) distribution function for the carriers in the Tail-zone (TZ) region, where the carriers penetrate with is the (FB zone) in the form of tails.

These carriers are free carriers as well as the carriers have finite effective masses. The transport properties of these carriers are not same as those of the carriers in the normal semiconductor bands [2]. In this chapter, we have derived the modified FD distribution function as given by (23.7) which is valid for carriers in the tail-zone (TZ).

The validity of the generalized equation (23.7) for which we have given a very simplified derivation can easily be assessed from six special cases as given above where the first three lead to the well-known results and the last three generate new concepts.

In this chapter, we have proposed the complex distribution function which may be used for studying the transport properties of quantized structures to generate new physics, new graphs and inferences from them which all together is a sea in itself.

23.4 Open Research Problem

(R.23.1) Derive generalized distribution functions in degenerate materials in the presence of strong light waves and intense electric field respectively and also derive the corresponding Boltzmann transport equation.

(R.23.2) Formulate the ionized impurity scattering limited mobility for highly degenerate III–V materials by formulating the Boltzmann transport equation where the generalized distribution function is given in this chapter.

(R.23.3) Formulate the problem as given in (R.23.2) under strong magnetic quantization.

(R.23.4) Formulate the problem as given in (R.23.2) under cross fields configurations.

(R.23.5) Formulate the acoustic scattering limited mobility for highly degenerate III–V materials by formulating the Boltzmann transport equation where the generalized distribution function is given in this chapter.

(R.23.6) Formulate the problem as given in (R.23.5) under strong magnetic quantization.

(R.23.7) Formulate the problem as given in (R.23.5) under cross fields configurations.

(R.23.8) Formulate the thermoelectric power for the problem (R.23.2).

(R.23.9) Formulate the ionized impurity scattering mobility for nonlinear optical and tetragonal materials in the context of the problem (R.23.2).

(R.23.10) Calculate the Piezoelectric scattering mobility for heavily doped II–V materials.

(R.23.11) Calculate the scattering cross-section for ionized impurity scattering for heavily doped IV–VI materials.

(R.23.12) Calculate the ionized impurity scattering limited mobility for quantum dots of heavily doped II–VI materials.

(R.23.13) Derive generalized distribution functions in degenerate materials in the presence of strong light waves and intense electric field respectively and also derive the corresponding Boltzmann transport equation.

References

[1] O. Madelung, *Introduction to Solid-State Theory*, Springer Series in Solid State Sciences, Vol. 2, p. 25 (**1978**).
[2] B. R. Nag, *Electron Transport in Compound Semiconductors*, Springer Series in Solid State Sciences, Vol. 11, p. 40 (**1980**).
[3] E. Fermi, *Z. Physik*, 36, 902 **(1926)**.
[4] P. A. M. Dirac, *Proc. Roy. Soc.*, 112, 661 **(1926)**.
[5] S. N. Bose, *Z. Physik*, 26, 178 **(1924)**.
[6] B. I. Shaklovskii and A. L. Efros, *Electronic Properties of Doped Semiconductors*, Springer-Verlag, Berlin, New York, Vol. 45, p. 253 (**1984**).
[7] W. Pauli, *Z. Physik*, 31, 765 **(1925)**.
[8] B. K. Choudhuri, B. N. Mondal and P. K. Chakraborty, *Pramana J. Phys.*, 90, 18 **(2019)**.
[9] P. K. Chakraborty, B. N. Mondal and B. K. Choudhury, *Open Access J. Math. Theor. Phys.*, 2, 22 **(2019)**.
[10] P. K. Chakraborty and J. C. Biswas, *J. Appl. Phys.*, 82, 7 **(1997)**.
[11] P. K. Chakraborty and B. N. Narayanan, *Q. Phys. Rev.*, 3, 1 **(2017)**.
[12] C. Kittle, *Elementary Statistical Mechanics*, (Dover Publications, Inc, New York, **1980**), pp. 86–88.

Chapter 24

The χ_d and χ_p in HD 0D Structures in the Presence of Magnetic Field

There is one mental light year distance between a "HUMAN BEING" and "BEING HUMAN".

In this chapter, we study χ_d and χ_p in cylindrical QD of HD III–V semiconductors in the presence of crossed electric and magnetic fields. In the presence of an external magnetic field, B, and an external electric field (E') along the z-direction, the Schrodinger equation becomes

$$\frac{(p-eA)^2}{2m_c}\psi - \alpha\left(\frac{(p-eA)^2}{2m_c}\right)^2\Psi - eE'z = E\psi \tag{24.1}$$

For a cylindrical quantum dot, the above equation has to be solved for the following boundary conditions:

$$\Psi = 0 \text{ at } \rho = a \text{ and at } z = 0 \text{ and } d$$

where a and d are the radius and the height of the cylindrical QD respectively.

Assuming that the magnetic field is perpendicular to the z-direction, one may choose the vector potential components as $A_\rho = A_\varphi = 0$, and $A_z = B_\rho \sin\varphi$, and consequently the (24.1) may further be expressed as

$$\begin{aligned}
&-\frac{\hbar^2}{2m_c}\left\{\frac{1}{\rho}\frac{\partial}{\partial\rho}\left(\rho\frac{\partial\Psi}{\partial\rho}\right)+\frac{1}{\rho^2}\frac{\partial^2\Psi}{\partial\varphi^2}+\frac{\partial^2\Psi}{\partial z^2}\right\}\\
&\quad+\frac{ie\hbar B}{m_c}\rho\sin\varphi\frac{\partial\Psi}{\partial z}+\frac{e^2B^2}{2m^*}\rho^2\sin^2\varphi\Psi - eE'z\Psi\\
&-\left[-\frac{\hbar^2}{2m_c}\left\{\frac{1}{\rho}\frac{\partial}{\partial\rho}\left(\rho\frac{\partial\Psi}{\partial\rho}\right)+\frac{1}{\rho^2}\frac{\partial^2}{\partial\varphi^2}+\frac{\partial^2}{\partial z^2}\right\}\right.\\
&\quad\left.+\frac{ie\hbar B}{m_c}\rho\sin\varphi\frac{\partial}{\partial z}+\frac{e^2B^2}{2m^*}\rho^2\sin^2\varphi\right]^2\Psi = E\Psi
\end{aligned} \tag{24.2}$$

Equation (24.2) can be written as

$$(\hat{H}_a - eE'z)\Psi + \{\hat{H}_b - \alpha(\hat{H}_a + \hat{H}_b)^2\}\Psi = E\Psi \tag{24.3}$$

where

$$\hat{H}_a = -\frac{\hbar^2}{2m_c}\left\{\frac{1}{\rho}\frac{\partial}{\partial\rho}\left(\rho\frac{\partial}{\partial\rho}\right) + \frac{1}{\rho^2}\frac{\partial^2}{\partial\varphi^2} + \frac{\partial^2}{\partial z^2}\right\} \text{ and}$$

$$\hat{H}_b \equiv \frac{ie\hbar B}{m_c}\rho\sin\varphi\frac{\partial}{\partial z} + \frac{e^2B^2}{2m_c}\rho^2\sin^2\varphi$$

Assuming that the effect of the magnetic field and of the band non-parabolicity is to perturb the system slightly in the presence of the electric field, we can break up the effective Hamiltonian into two parts:

The unperturbed Hamiltonian $(\hat{H}_a - eE'\text{z})$

The perturbing Hamiltonian $[\hat{H}_b - \alpha(\hat{H}_a + \hat{H}_b)^2]$

The unperturbed wave function (Ψ_0) can be obtained from the Eigen value equation as

$$(\hat{H}_a - eE'\text{z})\Psi_0 = E_0\Psi_0 \tag{24.4}$$

where E_0 is the unperturbed Eigen value. Substituting the values of $\hat{H}_a$ into (24.4) and using $\Psi_0 = R(\rho)\phi(\varphi)Z(z)$ for applying the method of separation of variables, we get

$$-\frac{\hbar^2}{2m_c}\left\{\frac{1}{\rho R}\frac{\partial}{\partial\rho}\left(\rho\frac{\partial R}{\partial\rho}\right) + \frac{1}{\rho^2\phi}\frac{\partial^2\phi}{\partial\varphi^2} + \frac{1}{Z}\frac{\partial^2 Z}{\partial z^2}\right\} - eE'z = E_0 \tag{24.5}$$

As the terms depending on z are not functions of ρ and φ, the z-dependent terms can be equated to a constant, say, E_{zn}. So, the z-dependent equation becomes

$$\frac{\partial^2 Z}{\partial z^2} + \frac{2m_c eE'}{\hbar^2}\left(z + \frac{E_{zn}}{eE'}\right)Z = 0 \tag{24.6}$$

With the boundary conditions $Z = 0$ at $z = 0$ and d. Making the substitution

$$\xi = \left(\frac{2m_c eE'}{\hbar^2}\right)^{1/3}\left(z + \frac{E_{zn}}{eE'}\right)$$

in the above equation, we can write

$$\frac{\partial^2 Z}{\partial \xi^2} + \xi Z = 0 \tag{24.7}$$

with the boundary conditions $Z = 0$ at for $\xi_1 \le \xi \le \xi_2$, where

$$\xi_1 = \left(\frac{2m_c eE'}{\hbar^2}\right)^{1/3} \frac{E_{zn}}{eE'} \quad \text{and} \quad \xi_2 = \left(\frac{2m_c eE'}{\hbar^2}\right)^{1/3} \left(d + \frac{E_{zn}}{eE'}\right) \tag{24.8}$$

The general solution of (24.7) is given by

$$Z = CAi(-\xi) + DBi(-\xi) \tag{24.9}$$

where C and D are constants and $Ai(-\xi)$ and $Bi(-\xi)$ are Airy functions. The condition to be satisfied for the non-trivial solutions of Z can be shown to be

$$Ai\,(-\xi_2)\,Bi\,(-\xi_1) - Ai\,(-\xi_1)\,Bi\,(-\xi_2) = 0 \tag{24.10}$$

For appropriate combinations of ξ_1 and ξ_2, (24.10) can be solved numerically. Then using the relations given by (24.9), one can get E_{zn}, which is the energy Eigen value of an electron inside the sample in the presence of the electric field, considering the motion along the direction of the electric field only. As E_{zn} takes only some discrete values, a suffix n has to be used to denote the quantum number on which the energy E_{zn} depends. Using the appropriate expressions for $Ai(-\eta)$ and $Bi(-\eta)$ for large arguments, the condition can be reduced to the form

$$\xi_2^{3/2} - \xi_1^{3/2} = \frac{3}{2} n\pi$$

where $n = 1, 2, 3, \ldots$. This equation can be solved for ξ_1 and ξ_2 to get E_{zn}. It can be shown that the φ-dependent equation is given by

$$-\frac{1}{\phi}\frac{\partial^2 \phi}{\partial \varphi^2} = m^2 \tag{24.11}$$

which gives

$$\phi = C_\phi \exp(\pm im\varphi)$$

where m is a positive integer including zero, C_φ, being a constant determined by the normalization condition. Proceeding in the same manner, we can write the radial part of (24.5) as

$$\rho^2 \frac{\partial^2 R}{\partial \rho^2} + \rho \frac{\partial R}{\partial \rho} + \left(\lambda^2 \rho^2 - m^2\right) R = 0 \tag{24.12}$$

where

$$\lambda^2 = \frac{2m_c\left(E_0 - E_{zn}\right)}{\hbar^2} \tag{24.13}$$

This leads to the solution for R as

$$R = C_\rho J_m(\lambda\rho)$$

where $J_m(\lambda\rho)$ is the Bessel function of the first kind and order m, C_ρ being a constant obtained by the normalization of the radial wave function R. The boundary condition to be imposed on R is $R = 0$ at $\rho = a$. This gives $\lambda = \alpha_{ml}/a$, where α_{ml} is the lth zero of the mth-order Bessel function. Hence, we can write

$$R = C_\rho J_m\left(\frac{\alpha_{ml}}{a}\rho\right) \tag{24.14}$$

Putting $\lambda = \alpha_{ml}/a$ into (24.13), the unperturbed energy Eigen value may be shown to be given by

$$E_0 = \frac{\hbar^2}{2m_c}\left(\frac{\alpha_{ml}}{a}\right)^2 + E_{zn} \tag{24.15}$$

We can denote E_0 by E_{lmn}, as E_0 depends on the quantum number l, m and n.

To calculate the influence of the perturbing Hamiltonian, we first split it into two parts:

$$\hat{H}_a \quad \text{and} \; -\alpha(\hat{H}_a + \hat{H}_b)^2 = \alpha\hat{H}^2_{\text{ab}}, \quad \text{where} \quad \hat{H}_{\text{ab}} = (\hat{H}_a + \hat{H}_b)$$

We shall use only corrections arising from the first-order perturbation. The first-order correction due to $\hat{H}_b$ is H_{bss}, where the suffix ss means that both ψ_0^* and ψ_0 correspond to the same s-state of the electron. It can be shown that

$$H_{\text{bss}} = \frac{1}{2}\frac{e^2B^2}{2m_c}\langle\rho^2\rangle \tag{24.16}$$

where $\langle\rho^2\rangle$ is the expectation value of ρ^2, given by

$$\langle\rho^2\rangle = \frac{\int_0^a \rho^3 R^2 d\rho}{\int_0^a \rho R^2 d\rho} \tag{24.17}$$

Similarly, the first-order correction due to $-\alpha\hat{H}^2_{\text{ab}}$ is H_{abss}, where

$$H_{\text{abss}} = -\alpha\frac{\int_V \psi^*_{0s}\hat{H}^2_{\text{ab}}\psi_{\text{os}}dV}{\int_V \psi^*_{0s}\psi_{\text{os}}dV}$$

Using the values of $\hat{H}_{\text{ab}}$ and (24.3), where necessary, we can write

$$H_{\text{abss}} = I_1 + I_2 + I_3 + I_4 + I_5 + I_6 \tag{24.18}$$

where $I_1, \ldots, I_6$ represent separate integral terms as given below,

$$I_1 = -\alpha E_0 \frac{\int_V \psi_{0s}^* \hat{H}_a \psi_{\text{os}} dV}{\int_V \psi_{0s}^* \psi_{\text{os}} dV}, \quad I_2 = -\alpha \frac{\int_V \psi_{0s}^* \hat{H}_a eE' z \psi_{\text{os}} dV}{\int_V \psi_{0s}^* \psi_{\text{os}} dV},$$

$$I_3 = -\alpha \frac{\int_V \psi_{0s}^* \hat{H}_a \hat{H}_b \psi_{\text{os}} dV}{\int_V \psi_{0s}^* \psi_{\text{os}} dV}, \quad I_4 = -\alpha \frac{\int_V E_0 \psi_{0s}^* \hat{H}_b \psi_{\text{os}} dV}{\int_V \psi_{0s}^* \psi_{\text{os}} dV},$$

$$I_5 = -\alpha \frac{\int_V \psi_{0s}^* \hat{H}_b eE' z \psi_{\text{os}} dV}{\int_V \psi_{0s}^* \psi_{\text{os}} dV} \quad \text{and} \quad I_6 = -\alpha \frac{\int_V \psi_{0s}^* \hat{H}_b \hat{H}_b \psi_{\text{os}} dV}{\int_V \psi_{0s}^* \psi_{\text{os}} dV}$$

Using the expression for the wave function $\psi_{\text{os}}, I_1, \ldots, I_6$ can be evaluated to be given by

$$\begin{aligned}
I_1 &= \left[-2i\hbar e^2 B_y^2 z u_u - 2i\hbar^2 k_x e B_y u_u\right] \qquad (24.19)\\
I_2 &= -\alpha \left[eE'\langle z\rangle E_0 + e^2 E'^2 \left\langle z^2\right\rangle\right]\\
I_3 &= -\alpha \left[\frac{\hbar^2}{2m_c}\left(\frac{\alpha_{\text{ml}}}{a}\right)^2 \frac{1}{2}\frac{e^2B^2}{2m_c}\left\langle \rho^2\right\rangle + \frac{1}{2}\frac{e^2B^2}{2m_c}\left\langle \rho^2\right\rangle (E_{zn} + eE'\langle z\rangle)\right]\\
I_4 &= -\alpha \frac{1}{2}\frac{e^2B^2}{2m^*}\left\langle \rho^2\right\rangle E_0\\
I_5 &= -\alpha \frac{1}{2}\frac{e^2B^2}{2m_c}\left\langle \rho^2\right\rangle eE'\langle z\rangle\\
I_6 &= -\alpha \left[4\frac{1}{2}\frac{e^2B^2}{2m_c}\left\langle \rho^2\right\rangle (E_{zn} + eE'\langle z\rangle) + \frac{3}{2}\left(\frac{1}{2}\frac{e^2B^2}{2m_c}\right)^2 \left\langle \rho^4\right\rangle\right]
\end{aligned}$$

$\langle z\rangle, \left\langle z^2\right\rangle$ and $\left\langle \rho^4\right\rangle$ being the expectation values of z, z^2 and ρ^4, respectively, and may easily be evaluated using the expression for ψ_{os}. For example, $\left\langle \rho^4\right\rangle$ may be obtained in a similar manner as shown in (24.19). For the z-dependent function $f(z)$, one may write

$$\langle f(z)\rangle = \frac{\int_0^d Z^2 f(z) dz}{\int_0^d Z^2 dz} \tag{24.20}$$

Now, inserting the values of $I_1, \ldots, I_6$, we can write, after some tedious algebraic manipulations,

$$H_{\text{abss}} = -\alpha \left[\left(E_0 + \frac{1}{2}\frac{e^2B^2}{2m_c}\left\langle \rho^2 \right\rangle + eE'\langle z\rangle \right)^2 + \left(\frac{1}{2}\frac{e^2B^2}{2m_c} \right)^2 \right.$$
$$\cdot \left(\frac{3}{2}\left\langle \rho^4 \right\rangle - \left\langle \rho^2 \right\rangle^2 \right) + e^2E'^2(\langle z^2\rangle - \langle z\rangle^2)$$
$$\left. + 4\frac{1}{2}\frac{e^2B^2}{2m_c}\left\langle \rho^2 \right\rangle (E_{zn} + eE'\langle z\rangle) \right] \tag{24.21}$$

The energy (E'_{lmn}) after the first-order correction is, therefore, given by

$$E'_{\text{lmn}} = E_{\text{lmn}} + \frac{1}{2}\frac{e^2B^2}{2m_c}\left\langle \rho^2 \right\rangle - \left[\left(E_{\text{lmn}} + \frac{1}{2}\frac{e^2B^2}{2m_c}\left\langle \rho^2 \right\rangle + eE'\langle z\rangle \right)^2 \right.$$
$$+ \left(\frac{1}{2}\frac{e^2B^2}{2m_c} \right)^2 \cdot \left(\frac{3}{2}\left\langle \rho^4 \right\rangle - \left\langle \rho^2 \right\rangle^2 \right) + e^2E'^2\left(\langle z^2\rangle - \langle z\rangle^2\right)$$
$$\left. + 4\frac{1}{2}\frac{e^2B^2}{2m_c}\left\langle \rho^2 \right\rangle (E_{zn} + eE'\langle z\rangle) \right] \tag{24.22}$$

The DR in cylindrical QD of HD III–V materials in the presence of crossed electric and magnetic fields can be written as

$$\gamma_3\left(E, \eta_g\right) = E_{\text{lmn}} + \frac{1}{2}\frac{e^2B^2}{2m_c}\left\langle \rho^2 \right\rangle - \left[\left(E_{\text{lmn}} + \frac{1}{2}\frac{e^2B^2}{2m_c}\left\langle \rho^2 \right\rangle + eE'\langle z\rangle \right)^2 \right.$$
$$+ \left(\frac{1}{2}\frac{e^2B^2}{2m_c} \right)^2 \cdot \left(\frac{3}{2}\left\langle \rho^4 \right\rangle - \left\langle \rho^2 \right\rangle^2 \right) + e^2E'^2\left(\langle z^2\rangle - \langle z\rangle^2\right)$$
$$\left. + 4\frac{1}{2}\frac{e^2B^2}{2m_c}\left\langle \rho^2 \right\rangle (E_{zn} + eE'\langle z\rangle) \right] \tag{24.23}$$

The DOS function in this case is given by

$$N(E) = \frac{2g_v}{\pi a^2 d} \sum_{n=0}^{n_{\max}} \sum_{i=0}^{i_{\max}} \sum_{r=0}^{r_{\max}} \delta'\left(E - E_{19,15}\right) \tag{24.24}$$

where $E_{19,15}$ is the total quantized energy in this case and can be expressed as

$$\left[\gamma_3(E,\eta_g) - \left[E_{\text{lmn}} + \frac{1}{2}\frac{e^2B^2}{2m_c}\langle\rho^2\rangle - \left[\left(E_{\text{lmn}} + \frac{1}{2}\frac{e^2B^2}{2m_c}\langle\rho^2\rangle + eE'\langle z\rangle\right)^2 + \left(\frac{1}{2}\frac{e^2B^2}{2m_c}\right)^2\left(\frac{3}{2}\langle\rho^4\rangle - \langle\rho^2\rangle\right) + e^2E'^2\left(\langle z^2\rangle - \langle z\rangle^2\right) + 4\frac{1}{2}\frac{e^2B^2}{2m_c}\langle\rho^2\rangle\left(E_{zn} + eE'\langle z\rangle\right)\right]\right]\right]\Bigg|_{E=E_{19,15}} = 0 \quad (24.25)$$

Therefore, the electron concentration is given by

$$N(E) = \frac{2g_v}{\pi a^2 d}\sum_{n=0}^{n_{\max}}\sum_{i=0}^{i_{\max}}\sum_{r=0}^{r_{\max}} F_{-1}(\eta_{24}) \quad (24.26)$$

where $\eta_{24} = (k_BT)^{-1}[E_{F24} - E_{19,15}]$ and E_{F24} is the Fermi energy in the present case.

From (24.26), in this case, we can write that

$$F(B) = \frac{2g_v}{\pi a^2 d}\left[\sum_{n=0}^{n_{\max}}\sum_{i=0}^{i_{\max}}\sum_{r=0}^{r_{\max}}\left[E_{F_{24}}F_{-1}(\eta_{24}) - k_BT\left[F_0(\eta_{24})\right]\right]\right] \quad (24.27)$$

Using (1.198), (24.26) and (24.27), we can study the χ_d and including spin we can study the χ_p also in this case.

For the purpose of condensation, we have not plotted any case with the hope that the readers will perform all the computer programming for all the quantized materials under different physical conditions for the purpose of creating new physics and other important transport quantities which are totally band structure dependent. The theoretical results as given here would be useful in analyzing various other experimental data related to the transport phenomenon. We must note that the study of transport phenomena and the formulation of the electronic properties HD compounds in the presence of crossed electric and magnetic fields are based on the dispersion relations in such materials.

Chapter 25

Conclusion and Scope for Future Research

The fortunate are those who can give without remembering but take without forgetting.

This book deals with the different quantum signatures of various 0D systems in various types of materials and their quantized counter parts under varying physical conditions. We have obtained the exact E–k dispersion relations of the various 3D quantized materials and formulated in detail the various electronic properties in HD nonlinear optical and tetragonal, III–V, II–VI, Gallium Phosphide, Germanium, Platinum Antimonide, stressed III–V materials, IV–VI, Lead Germanium Telluride, Tellurium, II–V, Zinc and Cadmium diphosphides, Bismuth Telluride, III–V, II–VI, IV–VI, HgTe/CdTe, strained layers quantum well HD superlattices with graded interfaces under magnetic quantization, III–V, II–VI, IV–VI, HgTe/CdTe, strained layers HD effective mass superlattices under magnetic quantization, quantum confined effective mass superlattices and HD superlattices with graded interfaces and Fibonacci superlattices under different physical conditions respectively. From the electron energy spectrum, one can obtain the DOS but the DOS technique, as used in the literature cannot generate the Dispersion Relations. Therefore, our study is more fundamental than those in the existing literature, because the Boltzmann transport equation and the carrier statistics which control the study of the charge transport properties of the quantum effect devices can be solved if and only if the Dispersion Relations are known.

The external photo excitation, quantization and strong electric field alter profoundly the basic band structures, which, in turn, generate pinpointed knowledge regarding different electronic properties in different types of materials and the devices made from them. In Chapters 5 and 6, we have formulated the Quantum Capacitance and the $I_D - V_D$ equations for various types of MOSFETs and QWFETs. Chapters 15 and 16 discuss the same for their heavily doped counter parts. In Chapter 20, we have formulated the χ_d and χ_p in Quantized HD0D structures and the same has been studied in Chapter 24 for cylindrical QDs of III–V materials in the presence of crossed electric and magnetic fields. We have also formulated the generalized distribution function in heavily doped materials which contain three new concepts and it may be used for studying the transport properties of quantized structures to generate new physics, new graphs and inferences from them which all together is a sea in itself. We hope that we have given enough materials to excite the multi-dimensional thinking of our readers together and we further hope that they can only convert this cozy cottage of quantum signatures in 0D structures into a sky scraper by solving all the problems of this book and extending the same in many other new directions both experimental and theoretical since we feel that their creativity is far more superior than that of ours. The in-depth experimental investigations covering the whole spectrum of 0D structures and allied sciences in general, are extremely important to uncover the underlying physics and the related mathematics in this particular aspect. Our investigations are based on the simplified $\vec{\mathbf{k}} \cdot \vec{\mathbf{p}}$ formalism of solid-state science without incorporating the advanced field theoretic techniques. In spite of such constraints, the role of band structure, which generates, in turn, new concepts are truly amazing and discussed throughout the text.

We present the last bouquet of open research problems in this pin pointed topic of 3D quantized structures to enhance further in depth investigations in this context.

(R.25.1) Investigate the χ_p and χ_d in the presence of a quantizing magnetic field for different 0D HD materials under exponential, Kane, Halperin, Lax and Bonch-Bruevich band tails [1] for all the problems of this book whose unperturbed carrier energy spectra are defined in Chapter 1 by including broadening effects.

(R.25.2) Investigate the influence of defect traps and surface states separately on the EP from the 3D quantized materials for all the appropriate problems of all the chapters after proper modifications.

(R.25.3) Investigate the MTP for the 3D quantized materials under the condition of non-equilibrium of the carrier states for all the appropriate problems of this book.

(R.25.4) Investigate the RLC for all the appropriate problems of this book for the corresponding HD Fibonacci superlattices.

(R.25.5) Investigate the field emissions for all the appropriate problems of this book for all types of quantized materials under arbitrarily oriented electric field in the presence of strain.

(R.25.6) Investigate the quantum capacitance for all the appropriate problems of this book for all types of HD quantized FETs in the presence of hot electron effects.

(R.25.7) Investigate the different electronic properties for all the appropriate problems of this book for all types of quantized materials for nonlinear charge transport.

(R.25.8) Investigate the ER for all the appropriate problems of this book for all types of HD quantized materials in the presence of strain in an arbitrary direction.

(R.25.9) Investigate all the appropriate problems of this book for strongly correlated electronic HD quantized materials in the presence of strain.

(R.25.10) Investigate all the appropriate problems of this chapter in the presence of arbitrarily oriented photon field and strain.

(R.25.11) Investigate all the appropriate problems of this book for HD Bi_2Te_3–Sb_2Te_3 and Graphene in the presence of strain.

(R.25.12) Investigate the influence of the localization of carriers on the different electronic properties in HD quantized materials for all the appropriate problems of this book in the presence of crossed fields.

(R.25.13) Investigate the different electronic properties for Si/SiGe and SiGe/SiGe superlattices under different appropriate physical conditions as discussed in this book in the presence of strain and crossed fields.

(R.25.14) Investigate the different electronic properties for all types of quantized materials under different appropriate physical conditions as discussed in this book in the presence of strain and arbitrarily oriented crossed fields.

(R.25.15) Investigate all the appropriate problems of this book in the presence of crossed electric and alternating quantizing magnetic fields.

(R.25.16) Investigate all the appropriate problems of this book in the presence of crossed alternating electric and quantizing magnetic fields.

(R.25.17) Investigate all the appropriate problems of this book in the presence of crossed alternating non-uniform electric and alternating quantizing magnetic fields.

(R.25.18) Investigate all the appropriate problems of this book in the presence of alternating crossed electric and alternating quantizing magnetic fields.
(R.25.19) Investigate all the appropriate problems of this book in the presence of arbitrarily oriented pulsed electric and quantizing magnetic fields.
(R.25.20) Investigate all the appropriate problems of this book in the presence of arbitrarily oriented alternating electric and quantizing magnetic fields.
(R.25.21) Investigate all the appropriate problems of this book in the presence of crossed non-homogeneous electric and alternating quantizing magnetic fields.
(R.25.22) Investigate all the appropriate problems of this book in the presence of arbitrarily oriented electric and alternating quantizing magnetic fields under strain.
(R.25.23) Investigate all the appropriate problems of this book in the presence of arbitrarily oriented electric and alternating quantizing magnetic fields under ultra-strong light waves.
(R.25.24) Investigate the different electronic properties for all types of quantized materials this book in the presence of many body effects, strain and arbitrarily oriented alternating ultra-strong light waves respectively.
(R.25.25) Investigate all the problems of this book by removing all the physical and mathematical approximations and establishing the respective appropriate uniqueness conditions.

With this note, we end our journey, for the time being, with you, our creative readers on this very pin-pointed topic of research and the new journey with you will start soon. Incidentally, the temple of knowledge is located beyond infinity and we believe that the end is where we start from. This book is born out of the 50 years of concentrated effort by the senior author on this very research arena of quantum materials in general. Dear readers, we personally deeply believe that it is of no use being merely as creative great physicist. One must be a good human being also, concerned especially about the less fortunate and profoundly conscious of all that one owes as well as all that one receives. It has been really nice to be with you all along a journey which started in 2009 with our first book on the Einstein's Relation which incidentally being the first book on this very specific topic [2]. In the last 15 years of this guided tour, with us as your joint guides, you have traveled to enjoy the other nineteen beautiful cities of the world of the quantum materials [3–23]. We further hope that you will enjoy throughout the rest of your academic life, a good number of other new cities of the said world in general.

The formulation of all the transport properties for all types of materials and their quantum confined counter parts considering the influence of all the bands created due to all types of quantization's after removing all the assumptions and establishing the respective appropriate uniqueness conditions is, in general, an extremely difficult problem. The circa 250 open research problems have been presented in this book and we hope that our readers will not only solve them but also will generate new concepts, both theoretical and experimental in the process. Incidentally, we can easily infer how little is presented and how much more is yet to be investigated in this exciting topic which is the signature of coexistence of new physics, advanced mathematics combined with the inner fire for performing creative researches in this context from the young scientists since like Kikoin [22], we firmly believe that "A young scientist is no good if his teacher learns nothing NEW from him and gives his teacher nothing to be proud of". In the meantime our research interest has been shifted and we are leaving this particular deep and beautiful topic with the hope that (R.25.25) alone is sufficient to draw the attention of the researchers from diverse fields and our readers are surely in tune with the fact that "Exposition, criticism, appreciation is the work for second-rate minds" [23].

Please realize that success is never final, failure is never fatal, it is the courage that counts and courage is the ladder on which all the other virtues mounts.

Enjoy your creative journey to see the research sea in front you in this context.

Good luck!!

References

[1] B. R. Nag, *Electron Transport in Compound Semiconductors*, Springer Series in Solid State Sciences, Vol. 11, (Springer-Verlag, Germany, **1980**).

[2] K. P. Ghatak, S. Bhattacharya, D. De, *Einstein Relation in Compound Semiconductors and Their Heterostructures*, Springer Series in Materials Science, Vol. 116, (Springer, Germany, **2009**).

[3] K. P. Ghatak, D. De, S. Bhattacharya, *Photoemission from Optoelectronic Materials and Their Nanostructures*, Springer Series in Nanostructure Science and Technology, (Springer, Germany, **2009**).

[4] K. P. Ghatak , S. Bhattacharya, *Thermoelectric Power in Nanostructured Materials Strong Magnetic Fields*, Springer Series in Materials Science, 137, (Springer, Germany, **2010**).

[5] S. Bhattacharya and K. P. Ghatak, *Fowler–Nordheim Field Emission: Effects in Semiconductor Nanostructures*, Springer Series in Solid State Sciences, Vol. 170, (Springer, Germany, **2012**).

[6] K. P. Ghatak, S. Bhattacharya, *Bismuth Characteristics, Production and Applications*, Materials Science and Technologies, (NOVA, New York, **2012**).

[7] S. Bhattacharya, K. P. Ghatak, *Effective Electron Mass in Low-Dimensional Semiconductors*, Springer Series in Materials Science, 167, (Springer-Verlag, Berlin Heidelberg, New York, **2013**).

[8] D. De, S. Bhattacharya, K. P. Ghatak, *Quantum Dots and Quantum Cellular Automata: Recent Trends and Applications*, Physics Research and Technology, (NOVA, New York, **2013**).

[9] K. P. Ghatak, S. Bhattacharya, *Debye Screening Length: Effects of Nanostructured Materials*, Springer Tracts in Modern Physics, Vol. 255, (Springer, Germany, **2014**).

[10] K. P. Ghatak, S. Bhattacharya, *Heavily Doped 2D Quantized Structures and the Einstein Relation*, Springer Tracts in Modern Physics, Vol. 260, (Springer, Germany, **2015**).

[11] K. P. Ghatak, *Einstein's Photo-emission: Emissions from Heavily Doped Nanostructures*, Springer Tracts in Modern Physics, Vol. 262, (Springer, Germany, **2015**).

[12] K. P. Ghatak, *Dispersion Relations in Heavily Doped Nanostructures*, Springer Tracts in Modern Physics, 265, (Springer, Germany, **2016**).

[13] K. P. Ghatak, *Magneto Thermoelectric Power in Heavily Doped Quantized Structures*, Series on the Foundations of Natural Science and Technology, 7, (World Scientific Publishing Co. Pte. Ltd., **2016**).

[14] K. P. Ghatak, *Quantum Effects, Heavy Doping, and the Effective Mass*, Series on the Foundations of Natural Science and Technology, 8, (World Scientific Publishing Co. Pte. Ltd., **2017**).

[15] K. P. Ghatak, M. Mitra, *Electronic Properties*, Series in Nanomaterials, 1, (De Guyter, Germany, **2020**).

[16] K. P. Ghatak, M. Mitra, *Quantization and Entropy*, Series in Nanomaterials, 2, (De Guyter, Germany, **2020**).

[17] K. P. Ghatak, *Quantum Wires an Overview*, Series on the NanoTechnology Science and Technology, (NOVA, New York, **2020**).

[18] K. P. Ghatak and M. Mitra, *Elastic Constants in Heavily Doped Low Dimensional Materials*, Series on the Foundations of Natural Science and Technology, 14, (World Scientific Publishing Co. Pte. Ltd., **2021**).

[19] K. P. Ghatak, M. Mitra, A. Biswas, *Heisenberg's Uncertainty Principle and the Electron Statistics in Quantized Structures*, (Springer Nature, Singapore, **2022**).

[20] K. P. Ghatak, A. Biswas, *Electron Statistics in Quantum Confined Superlattices*, Series on the Foundations of Natural Science and Technology, 16, (World Scientific Publishing Co. Pte. Ltd., **2023**).
[21] K. P. Ghatak, J. Pal, *Quantum Capacitance in Quantized Transistors*, Series on the Foundations of Natural Science and Technology, (World Scientific Publishing Co. Pte. Ltd., **2024**).
[22] I. K. Kikoin, *Science for Everyone: Encounters with Physicists and Physics*, (Mir Publishers, Russia, **1989**), p. 154.
[23] G. H. Hardy, *A Mathematician's Apology*, (Cambridge University Press, **1990**), p. 61.

Series on the Foundations of Natural Science and Technology

(Continued from page ii)

Vol. 12 *The Origin of Natural Order: An Axiomatic Theory of Biology*
by Q. Zhao

Vol. 9 *The Timeless Approach: Frontier Perspectives in 21st Century Physics*
by D. Fiscaletti

Vol. 8 *Quantum Effects, Heavy Doping, and the Effective Mass*
by K. P. Ghatak

Vol. 7 *Magneto Thermoelectric Power in Heavily Doped Quantized Structures*
by K. P. Ghatak

Vol. 6 *Nano-Engineering in Science and Technology: An Introduction to the World of Nano-Design*
by M. Rieth

Vol. 5 *Grasping Reality: An Interpretation-Realistic Epistemology*
by H. Lenk

Vol. 4 *What is Life? Scientific Approaches and Philosophical Positions*
by H.-P. Dürr, F.-A. Popp and W. Schommers

Vol. 3 *The Visible and the Invisible: Matter and Mind in Physics*
by W. Schommers

Vol. 2 *Symbols, Pictures and Quantum Reality: On the Theoretical Foundations of the Physical Universe*
by W. Schommers

Vol. 1 *Space and Time, Matter and Mind: The Relationship between Reality and Space-Time*
by W. Schommers

Forthcoming:

Vol. 10 *Space-Time Reality: Mach's Principle, Relativity and Novel Aspects*
by W. Schommers

Vol. 11 *The Origin of Mind: Interplay between Matter and Mind*
by W. Schommers

www.ingramcontent.com/pod-product-compliance
Lightning Source LLC
LaVergne TN
LVHW010321220926
848484LV00004BB/821